Structural Proteomics

METHODS IN MOLECULAR BIOLOGY™

John M. Walker, SERIES EDITOR

Structural Proteomics

High-Throughput Methods

Edited by

Bostjan Kobe
School of Molecular and Microbial Sciences, The University of Queensland, Brisbane, Queensland, Australia

Mitchell Guss
School of Molecular and Microbial Biosciences, University of Sydney, Sydney, Australia

Thomas Huber
School of Molecular and Microbial Sciences and Australian Institute for Bioengineering and Nanotechnology, The University of Queensland, Brisbane, Australia

 Humana Press

Bostjan Kobe
School of Molecular and Microbial Sciences
The University of Queensland
Brisbane, Queensland
Australia

Mitchell Guss
School of Molecular and Microbial Biosciences
University of Sydney, Sydney
Australia

Thomas Huber
School of Molecular and Microbial Sciences
    and Australian Institute for Bioengineering
    and Nanotechnology
The University of Queensland, Brisbane
Australia

ISBN: 978-1-58829-809-6   e-ISBN: 978-1-60327-058-8

Library of Congress Control Number: 2007937390

Printed on acid-free paper

9 8 7 6 5 4 3 2 1

springer.com

Preface

Structural genomics is a newly emerging field that has arisen following the successful footsteps of the major sequencing efforts generally bundled under the heading genomics. Practical considerations and the diversity of funding mechanisms in different countries have led to different interpretations of what structural genomics actually is. By a strict analogy to sequencing, one might envisage structural genomics to be the determination of the three-dimensional structures of all the proteins coded by a genome. This is an impractical goal because many proteins are not amenable to purification in a form suitable for structure determination, or in fact may be inherently unstructured. Furthermore, the numbers of distinct polypeptides produced in a eukaryotic organism—when one takes into account the possibility of splice variants and posttranslational modification—is so large that determining the structures of all forms would be an impossible task at the present time.

The National Institute of Health (NIH) in the United States defined the task of their structural genomics effort as the determination of a representative of every possible protein fold. Estimates of the total number of different folds vary from a few as 1,000 to nearly 10,000; therefore, the structures in these projects would not come from a single genome. Having at least one representative of each fold would enable related structures in other organisms to be modeled by homology and therefore provide the basis for modeling the structure for every protein in an organism even if the structures could not be determined experimentally. Substantial progress has been made toward this goal. The success of this strategy for determining new folds is evidenced by new submissions to the Protein Data Bank, in which more than 50% of structures submitted in 2006 with folds unrelated to those already deposited in the Protein Data Bank were determined by structural genomics consortia.

Structural genomics in Europe, Japan, and elsewhere have generally had a slightly different focus than the NIH-funded initiatives. These have targeted pathogenic organisms or potential targets for drug intervention. Other smaller-scale structural genomics projects might more properly be termed "structural proteomics," in which proteins are chosen based on their involvement in a particular disease state, biochemical pathway, or environmental niche.

Whatever their overall aim, structural genomics consortia throughout the world have made a major contribution to the development of new techniques and methods for all aspects of structural biology, ranging from the identification of target genes to the refinement of structures using either nuclear magnetic resonance or x-ray crystallographic techniques. Given that structural genomics of whatever flavor deals with large numbers of structures, the new and optimized methods enable high throughput processing of large numbers of samples. It may fairly be said that high throughput is really the defining characteristic of structural genomics.

The objective of this volume of Methods in Molecular Biology, *Structural Proteomics: High-Throughput Methods,* is to provide readers with a current view of all aspects of the "pipeline" that takes protein targets to structures and how these have been optimized. Given the wide variety of approaches taken in different laboratories, some individual methods will no doubt have been omitted despite a genuine attempt on the part of the editors and authors to cover their topic areas as widely as possible. This volume includes chapters describing the individual steps in the structural genomics pipeline in depth, as well as less detailed overviews of individual structural genomics initiatives. The overviews give some insight into the diversity of approaches adopted by different laboratories. The chapters are grouped in sections ordered in progression along the structural genomics pipeline: "Protein Target Selection, Bioinformatic Approaches and Data Management," "Protein Production," "Biophysical and Functional Characterization of Proteins," "Structural Characterization of Proteins," and "Structural Proteomics Initiatives Overviews." Readers are encouraged to access further details on the methodologies in online resources such as PepcDB (protein expression purification and crystallization database: pepcdb.pdb.org) and the cited literature. It should be emphasized that most methods are as amenable to small laboratories as large consortia, and do not require major investments in facilities. It is hoped that this volume will help smaller laboratories establish high-throughput techniques.

Bostjan Kobe, BSc, PhD
Mitchell Guss, BSc, PhD
Thomas Huber, PhD

Contents

Contributors

Aniruddha Achari
Raytheon IS, Huntsville, Alabama

Paul D. Adams
Berkeley Structural Genomics Center, Lawrence Berkeley National
Laboratory, and Department of Chemistry, University of California,
Berkeley, California

Pavel V. Afonine
Lawrence Berkeley National Laboratory, Berkeley, California

Margit A. Apponyi
Research School of Chemistry, Australian National University, Canberra,
Australia

Tracy Arakaki
Department of Biochemistry, University of Washington, Seattle, Washington

Cheryl H. Arrowsmith
Banting and Best Department of Medical Research, University of Toronto,
Toronto, Ontario, Canada

Anelia Atanassova
Department of Pharmacy, University of Toronto, Toronto, Ontario, Canada

Shane Atwell
SGX Pharmaceuticals, Inc., San Diego, California

Kevin Bain
SGX Pharmaceuticals, Inc., San Diego, California

David Baker
Department of Biochemistry, University of Washington, Seattle, Washington

Helen M. Berman
Research Collaboratory for Structural Bioinformatics Protein Data Bank
(RCSB PDB), Department of Chemistry and Chemical Biology,
Rutgers, The State University of New Jersey, Piscataway,
New Jersey

Georgina Berridge
The Structural Genomics Consortium, Botnar Research Centre,
University of Oxford, Oxford, United Kingdom

Alexey Bochkarev
Structural Genomics Consortium, University of Toronto,
Toronto, Ontario, Canada

Jürgen Bosch
Department of Biochemistry, University of Washington, Seattle,
Washington

Yan Boucher
Department of Chemistry and Biomolecular Sciences,
Macquarie University, Australia

Greg Brown
Banting and Best Department of Medical Research,
University of Toronto, Toronto, Ontario, Canada

James Brown
Division of Structural Biology, University of Oxford, United Kingdom

Fred Buckner
Department of Medicine, University of Washington, Seattle, Washington

Nicola A. Burgess-Brown
The Structural Genomics Consortium, Botnar Research Centre,
University of Oxford, Oxford, United Kingdom

Kyle Burkhardt
Research Collaboratory for Structural Bioinformatics Protein Data Bank
(RCSB PDB), Department of Chemistry and Chemical Biology, Rutgers,
The State University of New Jersey, Piscataway, New Jersey

Stephen K. Burley
SGX Pharmaceuticals, Inc., San Diego, California

Didier Busso
Structural Biology and Genomics Platform, IGBMC,
CNRS/INSERM/Université Louis Pasteur, Illkirch, France

Konrad Büssow
Max Planck Institute for Molecular Genetics, Department of Vertebrate
Genomics, Protein Structure Factory, Berlin, Germany

Brian Carroll
Department of Biochemistry, University of Washington, Seattle, Washington

Jonathan Caruthers
Department of Biochemistry, University of Washington, Seattle, Washington

John-Marc Chandonia
Berkeley Structural Genomics Center, Lawrence Berkeley National
Laboratory, and Department of Chemistry, University of California,
Berkeley, California

Yunjia Chen
Department of Microbiology, University of Alabama at Birmingham,
Birmingham, Alabama

Steve M. Colebrook
The Structural Genomics Consortium, Botnar Research Centre,
University of Oxford, Oxford, United Kingdom

Nathan P. Cowieson
Institute for Molecular Bioscience, University of Queensland,
Brisbane, Australia

Paul M.G. Curmi
School of Physics, University of New South Wales, Sydney, Australia

Miroslaw Cygler
Biotechnology Research Institute, National Research Council Canada,
Montreal, Canada

Larry DeSoto
Department of Biochemistry, University of Washington, Seattle, Washington

George DeTitta
Hauptman-Woodward Medical Research Institute, Buffalo, New York

Nicholas E. Dixon
Research School of Chemistry, Australian National University, Canberra,
Australia

Zsuzsanna Dosztányi
Institute of Enzymology, Biological Research Center, Hungarian Academy of
Sciences, Budapest, Hungary

Mark Dumont
University of Rochester, Rochester, New York

Shuchismita Dutta
Research Collaboratory for Structural Bioinformatics Protein Data Bank
(RCSB PDB), Department of Chemistry and Chemical Biology, Rutgers,
The State University of New Jersey, Piscataway, New Jersey

Orly Dym
Israel Structural Proteomics Center (ISPC), Department of Structural
Biology, Weizmann Institute of Science, Rehovot, Israel

Thomas Earnest
Berkeley Center for Structural Biology, Physical Biosciences Division,
Lawrence Berkeley National Laboratory, Berkeley, California

Aled M. Edwards
Banting and Best Department of Medical Research,
University of Toronto, Canada

Spencer Emtage
SGX Pharmaceuticals, Inc., San Diego, California

David Eramian
Departments of Biopharmaceutical Sciences and Pharmaceutical
Chemistry and California Institute for Quantitative Biomedical Research,
University of California at San Francisco, San Francisco, California

Asgar Ergin
Max Planck Institute for Molecular Genetics, Department of Vertebrate
Genomics, Protein Structure Factory, Berlin, Germany

Narayanan Eswar
Departments of Biopharmaceutical Sciences and Pharmaceutical Chemistry
and California Institute for Quantitative Biomedical Research,
University of California at San Francisco, San Francisco, California

Erkang Fan
Department of Biochemistry, University of Washington, Seattle, Washington

Stanley Fields
Department of Genome Sciences, University of Washington, Seattle,
Washington

Elizabeth Forsythe
Nektar Therapeutics, Huntsville, Alabama

Jade K. Forwood
School of Molecular and Microbial Sciences, University of Queensland,
Brisbane, Australia

Arie Geerlof
EMBL Hamburg Outstation, Hamburg, Germany

Michael H. Gelb
Department of Chemistry, University of Washington, Seattle, Washington

Tarun Gheyi
SGX Pharmaceuticals, Inc., San Diego, California

Opher Gileadi
The Structural Genomics Consortium, Botnar Research Centre,
University of Oxford, Oxford, United Kingdom

Claudio F. Gonzalez
Banting and Best Department of Medical Research,
University of Toronto, Toronto, Ontario, Canada

Elizabeth Grayhack
University of Rochester, Rochester, New York

Ralf W. Grosse-Kunstleve
Lawrence Berkeley National Laboratory, Berkeley, California

Matthew R. Groves
EMBL Hamburg Outstation, Hamburg, Germany

Amy P. Guilfoyle
Department of Chemistry and Biomolecular Sciences,
Macquarie University, Australia

Stacey Gulde
Hauptman-Woodward Medical Research Institute, Buffalo, New York

Gregor Guncar
School of Molecular and Microbial Sciences,
University of Queensland, Brisbane, Australia

Mitchell Guss
School of Molecular and Microbial Biosciences, University of Sydney,
Sydney, Australia

Sabrina Haquin
Yeast Structural Genomics, IBBMC, Université Paris-Sud, Orsay, France

Mark Harris
Department of Cell and Molecular Biology, University of Uppsala,
Uppsala, Sweden

Stephen J. Harrop
School of Physics, University of New South Wales, Sydney,
New South Wales, Australia

Kim Henrick
The Macromolecular Structure Database at the European Bioinformatics
Institute MSD-EBI, EMBL Outstation-Hinxton, Cambridge,
United Kingdom

Justine M. Hill
School of Molecular and Microbial Sciences, University of Queensland,
Brisbane, Australia

Martin Högbom
Stockholm Center for Biomembrane Research, Department Biochemistry
and Biophysics, Stockholm University, Stockholm, Sweden

Wim G. J. Hol
Department of Biochemistry, University of Washington, Seattle,
Washington

Mark Holl
Department of Electrical Engineering, University of Washington, Seattle,
Washington

Margaret Holmes
Department of Biochemistry, University of Washington, Seattle, Washington

Russell E. Hopson
Department of Chemistry, Brown University, Providence, Rhode Island

Michael Howell
Department of Chemistry and Biomolecular Sciences,
Macquarie University, Australia

Thomas Huber
School of Molecular and Microbial Sciences, and Australian Institute
for Bioengineering and Nanotechnology, University of Queensland,
Brisbane, Australia

David A. Hume
Institute for Molecular Bioscience, University of Queensland,
Brisbane, Australia

Li-Wei Hung
Biophysics Group, Los Alamos National Laboratory, Los Alamos,
New Mexico

Ming-ni Hung
Department of Biochemistry, McGill University, Montreal, Quebec, Canada

Helgi Ingolfsson
Department of Physiology and Biophysics, Weill Medical
College of Cornell University, Ithaca, New York

Thomas R. Ioerger
Department of Computer Science, Texas A&M University,
College Hill, Texas

Catrine Johansson
The Structural Genomics Consortium, Botnar Research Centre,
University of Oxford, Oxford, United Kingdom

Alwyn T. Jones
Department of Cell and Molecular Biology, University of Uppsala,
Uppsala, Sweden

Stuart Kellie
School of Molecular and Microbial Sciences, University of Queensland,
Brisbane, Australia

David Kim
Department of Biochemistry, University of Washington, Seattle,
Washington

Kyoung Hoon Kim
Department of Chemistry, College of Natural Sciences, Seoul National
University, Seoul, Korea

Rosalind Kim
Berkeley Structural Genomics Center, Lawrence Berkeley National
Laboratory, and Department of Chemistry, University of California,
Berkeley, California

Sung-Hou Kim
Berkeley Structural Genomics Center, Lawrence Berkeley National
Laboratory, and Department of Chemistry, University of California,
Berkeley, California

Gordon King
Institute for Molecular Bioscience, University of Queensland, Brisbane,
Australia

Bostjan Kobe
School of Molecular and Microbial Sciences, University of Queensland,
Brisbane, Australia

Takashi Kosada
Protein Data Bank Japan (PDBj), Institute for Protein Research, Osaka University, Osaka, Japan

Christopher M. Koth
Vertex Pharmaceuticals Inc., Cambridge, Massachusetts

Peter Kuhn
Department of Cellular Biology, The Scripps Research Institute, La Jolla, California

Ekaterina Kuznetsova
Banting and Best Department of Medical Research, University of Toronto, Toronto, Ontario, Canada

Doug LaCount
Department of Genome Sciences, University of Washington, Seattle, Washington

Isolde Le Trong
Department of Biochemistry, University of Washington, Seattle, Washington

Jizhen Li
Department of Biochemistry, University of Washington, Seattle, Washington

Pawel Listwan
School of Molecular and Microbial Sciences, University of Queensland, Brisbane, Australia

Peter Loppnau
The Structural Genomics Consortium, University of Toronto, Toronto, Ontario, Canada

Chi-Hao Luan
Center for Biophysical Sciences and Engineering, University of Alabama at Birmingham, Birmingham, Alabama

Joseph Luft
Hauptman-Woodward Medical Research Institute, Buffalo, New York

Ming Luo
Department of Microbiology, University of Alabama at Birmingham, Birmingham, Alabama

Bridget C. Mabbutt
Department of Chemistry and Biomolecular Sciences, Macquarie University, Australia

John L. Markley
Department of Biochemistry, University of Wisconsin-Madison, Madison, Wisconsin

Russell L. Marsden
Biochemistry and Molecular Biology Department, University College London, London, United Kingdom

Jennifer L. Martin
Institute for Molecular Bioscience, University of Queensland,
Brisbane, Australia

Archna Massey
Department of Biochemistry, University of Washington, Seattle,
Washington

Allan Matte
Biotechnology Research Institute, National Research Council Canada,
Montreal, Canada

Andrew P. May
Fluidigm Corp., San Francisco, California

Airlie J. McCoy
Department of Haematology, University of Cambridge,
Cambridge Institute for Medical Research, Cambridge, United Kingdom

Erik McKee
Department of Computer Science, Texas A&M University,
College Station, Texas

Ran Meged
Israel Structural Proteomics Center (ISPC), Department of Structural
Biology, Weizmann Institute of Science, Rehovot, Israel

Chris Mehlin
Department of Biochemistry, University of Washington, Seattle,
Washington

Deirdre R. Meldrum
Department of Electrical Engineering, University of Washington, Seattle,
Washington

Weining Meng
School of Molecular and Microbial Sciences, University of Queensland,
Brisbane, Australia

Ethan Merritt
Department of Biochemistry, University of Washington, Seattle,
Washington

Somnath Mondal
Department of Biochemistry, University of Washington, Seattle,
Washington

Dino Moras
Plate-forme de Biologie et de Génomique Structurales, IGBMC, CNRS/
INSERM/Université Louis Pasteur, Strasbourg, France

Nigel W. Moriarty
Lawrence Berkeley National Laboratory, Berkeley, California

Dmitri Mouradov
School of Molecular and Microbial Sciences, University of Queensland,
Brisbane, Australia

Peter J. Myler
Seattle Biomedical Research Institute, Seattle, Washington

Haruki Nakamura
Protein Data Bank Japan (PDBj), Institute for Protein Research,
Osaka University, Osaka, Japan

Joanne E. Nettleship
Oxford Protein Production Facility, Oxford, United Kingdom

Janet Newman
Commonwealth Scientific and Industrial Research Organization,
Bio21, Melbourne, Australia

Joseph D. Ng
Laboratory for Structural Biology, Department of Biological Sciences,
University of Alabama in Huntsville, Huntsville, Alabama

Eric Oeuillet
Yeast Structural Genomics, IBBMC, Université Paris-Sud, Orsay, France

Christine A. Orengo
Biochemistry and Molecular Biology Department, University College
London, London, United Kingdom

Gottfried Otting
Research School of Chemistry, Australian National University,
Canberra, Australia

Kiyoshi Ozawa
Research School of Chemistry, Australian National University,
Canberra, Australia

Rebecca Page
Department of Molecular Biology, Cell Biology and Biochemistry,
Brown University, Providence, Rhode Island

Adriana Pajon
EMBL Outstation, European Bioinformatics Institute,
Wellcome Trust Genome Campus, Hinxton, Cambridge, United Kingdom

Nadia H. Pantic
The Structural Genomics Consortium, Botnar Research Centre,
University of Oxford, Oxford, United Kingdom

Yoav Peleg
Israel Structural Proteomics Center (ISPC), Department of Structural
Biology, Weizmann Institute of Science, Rehovot, Israel

Anastassis Perrakis
NKI, Department of Molecular Carcinogenesis, Amsterdam,
The Netherlands

Wolfgang Peti
Department of Molecular Pharmacology, Physiology and Biotechnology,
Brown University, Providence, Rhode Island

Eric Phizicky
University of Rochester, Rochester, New York

Anne Poupon
Yeast Structural Genomics, IBBMC, Université Paris-Sud, Orsay, France

Michael Proudfoot
Banting and Best Department of Medical Research, University of Toronto,
Toronto, Ontario, Canada

Munish Puri
Institute for Molecular Bioscience, University of Queensland,
Brisbane, Australia

Marc Pusey
MI Research Inc., Huntsville, Alabama

Shihong Qiu
Department of Microbiology, University of Alabama at Birmingham,
Birmingham, Alabama

James Raftery
School of Chemistry, The University of Manchester, Manchester,
United Kingdom

Randy J. Read
Department of Haematology, University of Cambridge,
Cambridge Institute for Medical Research, Cambridge, United Kingdom

Mark Robien
Department of Biochemistry, University of Washington, Seattle, Washington

Gautier Robin
Institute for Molecular Bioscience, University of Queensland,
Brisbane, Australia

Andrew Robinson
Department of Chemistry and Biomolecular Sciences,
Macquarie University, Australia

Jodie Robinson
Institute for Molecular Bioscience, University of Queensland,
Brisbane, Australia

Christophe Romier
IGBMC, Département de Biologie et Génomique Structurales,
Illkirch, France

Isabelle Rooney
SGX Pharmaceuticals, Inc., San Diego, California

Ian L. Ross
Institute for Molecular Bioscience, University of Queensland,
Brisbane, Australia

Marc E. Rutter
SGX Pharmaceuticals, Inc., San Diego, California

James C. Sacchettini
Department of Biochemistry and Biophysics, Texas A&M University, College Station, Texas

Eidarus Salah
The Structural Genomics Consortium, Botnar Research Centre, University of Oxford, Oxford, United Kingdom

Andrej Sali
Departments of Biopharmaceutical Sciences and Pharmaceutical Chemistry and California Institute for Quantitative Biomedical Research, University of California at San Francisco, San Francisco, California

Stephen A. Sanders
Banting and Best Department of Medical Research, University of Toronto, Toronto, Ontario, Canada

Michael J. Sauder
SGX Pharmaceuticals, Inc., San Diego, California

Nicholas K. Sauter
Lawrence Berkeley National Laboratory, Berkeley, California

Pavel Savitsky
The Structural Genomics Consortium, Botnar Research Centre, University of Oxford, Oxford, United Kingdom

Lori Schoenfeld
Department of Biochemistry, University of Washington, Seattle, Washington

Brent W. Segelke
Chemistry, Materials, and Life Sciences, Lawrence Livermore National Laboratory, Livermore, CA

Min-Yi Shen
Departments of Biopharmaceutical Sciences and Pharmaceutical Chemistry and California Institute for Quantitative Biomedical Research, University of California at San Francisco, San Francisco, California

Dong-Hae Shin
Berkeley Structural Genomics Center, Lawrence Berkeley National Laboratory, and Department of Chemistry, University of California, Berkeley, California

Volker Sievert
Max Planck Institute for Molecular Genetics, Department of Vertebrate Genomics, Protein Structure Factory, Berlin, Germany

Andrea Sinz
Institute of Analytical Chemistry, Biotechnological-Biomedical Center, University of Leipzig, Leipzig, Germany

Carol E.A. Smee
The Structural Genomics Consortium, Botnar Research Centre, University of Oxford, Oxford, United Kingdom

Michael Soltis
SSRL, Stanford University, Stanford, California

Raymond C. Stevens
Department of Molecular Biology, The Scripps Research Institute,
La Jolla, California

Matthieu Stierlé
Plate-forme de Biologie et de Génomique Structurales, IGBMC,
CNRS/INSERM/Université Louis Pasteur, Strasbourg, France

Hatch W. Stokes
Department of Chemistry and Biomolecular Sciences,
Macquarie University, Australia

Laurent C. Storoni
Department of Haematology, University of Cambridge,
Cambridge Institute for Medical Research, Cambridge, United Kingdom

Se Won Suh
Department of Chemistry, College of Natural Sciences,
Seoul National University, Seoul, Korea

Mark Sullivan
University of Rochester, Rochester, New York

Visaahini Sureshan
Department of Chemistry and Biomolecular Sciences,
Macquarie University, Australia

Joel L. Sussman
Israel Structural Proteomics Center (ISPC), Department of Structural
Biology, Weizmann Institute of Science, Rehovot, Israel

Ganesh J. Swaminathan
The Macromolecular Structure Database at the European Bioinformatics
Institute MSD-EBI, EMBL Outstation-Hinxton, Cambridge,
United Kingdom

Wolfram Tempel
Structural Genomics Consortium, University of Toronto, Toronto,
Ontario, Canada

Thomas C. Terwilliger
Bioscience Division, Los Alamos National Laboratory, Los Alamos,
New Mexico

Anil S. Thakur
School of Molecular and Microbial Sciences, University of Queensland,
Brisbane, Australia

Jean-Claude Thierry
Plate-forme de Biologie et de Génomique Structurales, IGBMC,
CNRS/INSERM/Université Louis Pasteur, Strasbourg, France

Devon A. Thompson
SGX Pharmaceuticals, Inc., San Diego, California

Peter Tompa
Institute of Enzymology, Biological Research Center, Hungarian
Academy of Sciences, Budapest, Hungary

Tamar Unger
Israel Structural Proteomics Center (ISPC), Department of Structural
Biology, Weizmann Institute of Science, Rehovot, Israel

H. van Tilbeurgh
Yeast Structural Genomics, IBBMC, Université Paris-Sud, Orsay, France

Wes Van Voorhis
Department of Medicine, University of Washington, Seattle,
Washington

Christophe Verlinde
Department of Biochemistry, University of Washington, Seattle,
Washington

Marissa Vignali
Department of Genome Sciences, University of Washington, Seattle,
Washington

John Wagner
Biotechnology Research Institute, National Research Council Canada,
Montreal, Canada

Geoffrey S. Waldo
Bioscience Division, Los Alamos National Laboratory, Los Alamos,
New Mexico

Ben Webb
Departments of Biopharmaceutical Sciences and Pharmaceutical
Chemistry and California Institute for Quantitative Biomedical
Research, University of California at San Francisco, San Francisco,
California

Beth Wensley
University of Queensland, Brisbane, Australia

Melissa Swope Willis
American Type Culture Collection, Manassas, Virginia

Liz Worthey
Seattle Biomedical Research Institute, Seattle, Washington

Alexander F. Yakunin
Banting and Best Department of Medical Research,
University of Toronto, Toronto, Ontario, Canada

Jin Kuk Yang
Department of Chemistry, College of Natural Sciences,
Soongsil University, Seoul, Korea

Golan Yona
Department of Computer Science, Technion-Israel Institute of Technology,
Haifa, Israel

Deborah B. Zamble
Department of Chemistry, University of Toronto, Toronto, Ontario, Canada

Zsolt Zolnai
Department of Biochemistry, University of Wisconsin-Madison, Madison, Wisconsin

Frank Zucker
Department of Biochemistry, University of Washington, Seattle, Washington

Peter H. Zwart
Lawrence Berkeley National Laboratory, Berkeley, California

Section I

Protein Target Selection, Bioinformatic Approaches, and Data Management

Chapter 1

Target Selection for Structural Genomics: An Overview

Russell L. Marsden and Christine A. Orengo

The success of the whole genome sequencing projects brought considerable credence to the belief that high-throughput approaches, rather than traditional hypothesis-driven research, would be essential to structurally and functionally annotate the rapid growth in available sequence data within a reasonable time frame. Such observations supported the emerging field of structural genomics, which is now faced with the task of providing a library of protein structures that represent the biological diversity of the protein universe. To run efficiently, structural genomics projects aim to define a set of targets that maximize the potential of each structure discovery whether it represents a novel structure, novel function, or missing evolutionary link. However, not all protein sequences make suitable structural genomics targets: It takes considerably more effort to determine the structure of a protein than the sequence of its gene because of the increased complexity of the methods involved and also because the behavior of targeted proteins can be extremely variable at the different stages in the structural genomics "pipeline." Therefore, structural genomics target selection must identify and prioritize the most suitable candidate proteins for structure determination, avoiding "problematic" proteins while also ensuring the ultimate goals of the project are followed.

1. Introduction

The field of structural genomics has rapidly emerged as an international effort to systematically determine the three-dimensional shapes of all important biological macromolecules, with a significant emphasis placed on the elucidation of protein structure. Although traditional structural biology deposits of thousands of invaluable protein structures into the Protein Data Bank (PDB) (1) each year, many are identical or extremely similar to previously deposited structures, providing insights into the effect of sequence mutations or structures in an alternative ligand-bound state. To circumvent this redundancy and accelerate knowledge of protein structure space, structural genomics projects were proposed as a complementary approach, undertaking the high-throughput structure determination of carefully chosen structurally uncharacterized target sequences.

From: *Methods in Molecular Biology, Vol. 426: Structural Proteomics: High-throughput Methods*
Edited by: B. Kobe, M. Guss and T. Huber © Humana Press, Totowa, NJ

The success of whole genome sequencing, culminating in the release of the human genome, brought much credence to the premise that production-line efficiencies could be successfully implemented in the experimental laboratory. This belief, coupled with significant advances in x-ray crystallography, nuclear magnetic resonance (NMR), and gene cloning and expression, enabled the goal of high-throughput protein structure determination to become a feasible proposition. In 1999, an international meeting was held to outline the guiding principles that were to define the newly emerging field *(2)*. It was proposed that thousands of structures could be solved in less time and at a lower cost per structure than "business as usual" structural biology, bringing new perspectives into the relationship between protein structure and function. The capability to magnify the value of each structure determination across genome sequences by means of computational structure modeling would also considerably enhance the impact of structural genomics, further reducing the number of structures required to understand the structural repertoire encoded by genomes *(3)*.

Structural biology has traditionally supported a paradigm in which biochemical evidence of function is confirmed through the elucidation of structure. Studies on individual proteins are applied to the problems of enzyme–substrate or protein–ligand specificity, protein stability, allosteric control, and the elucidation of enzyme mechanisms. Structural genomics groups have now begun to challenge this paradigm because in their motivation to increase the coverage of known fold space, the structures of proteins with unknown biological roles are often solved. In such cases it is hoped that structure can inform function because the three-dimensional structure of a protein is intrinsically related to its function. The discovery of ligands fortuitously bound on crystallization, or active site residues can illuminate previously hypothesized or completely novel functional mechanisms. Furthermore, protein structure represents a powerful means of discovering distant evolutionary relationships that are invisible at the sequence level, because structure remains better conserved over evolutionary time than sequence. These structural matches can also suggest functional properties of proteins with as yet unknown function.

1.1. Target Selection

To run efficiently and achieve maximum value from each solved structure, structural genomics groups employ a process that identifies and prioritizes the most suitable candidate proteins for structure determination *(4,5)*. This process is termed target selection and, as with all of the stages in the high-throughput structural genomics pipeline, must be capable of routinely dealing with large quantities of data, reducing the set of potential candidate proteins to a prioritized target list.

Although often quoted, the analogy between structural genomics and high-throughput genome sequencing can be somewhat misleading, not least because the behavior of targeted proteins can vary considerably at the different stages in the structural genomics pipeline. Such behavior can be partly attributed to the comparably larger number and increased complexity of the experimental methods used within a structure determination pipeline, and also because some classes of protein, such as membrane proteins, are almost entirely inaccessible to most high-throughput structure determination procedures. Accordingly, it requires considerably more effort to determine the structure of a protein than the sequence of its gene. Therefore, it is essential to obtain a list of targets that maximize the potential of each structure discovery, whether it represents a novel structure, novel function, or missing evolutionary link. Avoiding problematic proteins while ensuring the goals of the project are adhered to requires a systematic target selection approach that should ultimately underpin the scientific validity of the structural genomics projects.

1.2. Chapter Overview

The goals of target selection, in their simplest terms, can be formulated as follows: Which structures from which organisms should be solved and in what order? An overview of some of the major steps in the target selection pipeline is illustrated in Fig. 1.1, and this chapter expands on these terms in more detail:

• Addressing scientific context of a project: Initial consideration is given to the identification of target proteins that fit within the scientific context of a structure genomics project. Specific emphasis is given to those projects

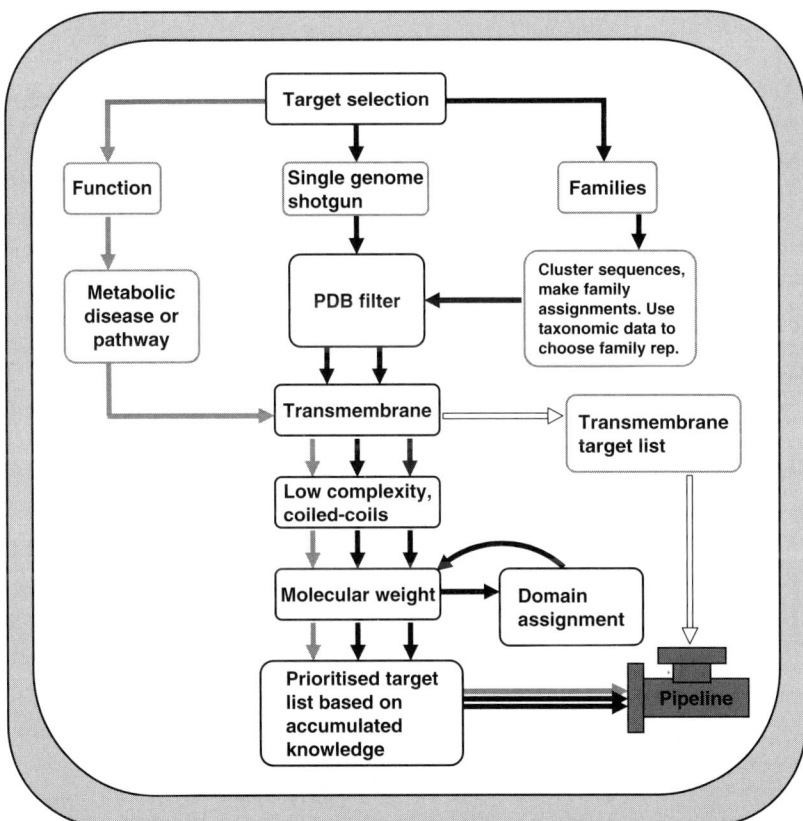

Fig. 1.1 Flowchart illustrating the major decision-making stages required to formulate a structural genomics target list. Depending on the project rationale, targets may initially be restricted to specific protein families, single organisms, or functional groupings.

aiming to advance coverage of fold space through computational structure modeling.

- Addressing limitations of high-throughput structure determination: Once the initial list of targets has been identified, the discussion is continued by describing the approaches and methods commonly used to exclude targets that are considered to be unsuitable for high-throughput structure determination.
- Future approaches: Finally some of the likely developments target selection must address in the future are considered.

2. Target Selection: Addressing Scientific Context of a Project

Structural genomics has been adopted as a broad research goal by a loosely structured coalition of researchers around the globe. Although the common feature of all these projects is the development and application of high-throughput methods for structure determination, the scientific rationale can vary considerably among projects:

- Augmenting fold space. The most frequently described approach is to achieve significant advance in structural coverage of sequence space by solving enough structures to enable the majority of proteins to be structurally modeled *(2)*. Frequently targets are selected on the basis of their likely structural novelty, and increasingly the first experimental information known about a protein is its three-dimensional structure *(6)*.
- Single genome approach. Also known as a "shotgun" approach, in which all accessible targets in an entire proteome, outside the reach of comparative modeling approaches, are passed through a structure determination pipeline *(7)*.
- Function. Targets are selected on the basis of known function, in which several members of a diverse protein family may be solved to obtain an in-depth view of how structure mediates function. Biomedical targets may be chosen because they appear unique to a given organism (ORFans) or are known to be implicated in a given disease state, whereas targets may also be selected to structurally characterize all members of a biochemical pathway *(8)*.
- Protein–protein interactions. Sometimes a target can only be solved in the presence of the correct protein partner, requiring the identification of known (or likely) partners involved in the protein complex. Candidate targets are often drawn from the literature, and may well have already been structurally characterized, albeit solved in a different context *(9)*.

2.1. Augmenting Fold Space

2.1.1. Overview of Fold Space

For many structural genomics projects, a considerable amount of effort and resources are directed toward the structural characterization of proteins that will embody a previously uncharacterized region of structure space. In so doing, solved targets are generally chosen to represent the first solved structure of a protein family in which, even if the fold is not novel, it is likely that the protein function is novel. At best, solved targets will be found to exhibit a protein fold that has not yet been characterized (a novel fold). The benefits of such an approach are twofold: First, solving the first representative structure of a protein fold is of considerable value to the structural biology community as a whole, because subsequent structural analysis of novel folds can often reveal valuable new insights into the relationship between sequence and structure—as outlined by the central dogma of structural biology. Second, the expansion of the current library of folds should permit a far larger expansion in fold annotation leverage across sequence space through the use of computational protein modeling *(3)*.

The power of combining known protein structures with computational modeling techniques can be appreciated if one considers the relative breadth of protein sequence space against the seemingly narrower universe of protein fold space: It has been hypothesized that structure space is populated by a limited number of protein folds. At present we have experimentally characterized structures of approximately 900 hundred distinct protein folds, as classified by the SCOP *(10)* and CATH *(11)* domain structure databases (Table 1.1). A series of studies have extrapolated this data estimating that altogether there could be between 1,000 and 10,000 naturally occurring globular (non-transmembrane) protein folds in nature (for review see *12*). Such estimates offer a thought-provoking glimpse toward the future because solving several thousand new protein structures is deemed to be within the capability of structure genomics projects. From this position, structural genomics' target selection is tasked with ensuring that a significant number of these new structures are indeed novel, bringing the biological diversity of the protein universe within our grasp.

It is also fair to say that there can also be a temptation to oversimplify the apparent organization of fold space. Evidently the estimation of the number of folds is open to considerable variation, largely due to the fact that the clustering of known fold space is dependent on the algorithms and similarity measurements used, and so far, have been mainly based on those proteins that are soluble and are relatively easy to crystallize. Furthermore, it is still somewhat open to question if protein structure space is really discrete, with a finite number of folds or continuous where each protein structure overlaps considerably with the next, sharing large structural motifs *(13)*.

In reality the relationship between structure and function is complex: It cannot be concluded that all proteins sharing a common fold have evolved from a common ancestor. Only in cases in which similarities in sequence or function are identified experimentally can such associations be accurately supported, enabling related proteins to be assigned to distant evolutionary groupings, commonly called superfamilies. Comparable folds that have not diverged from a common ancestor—in which it appears that different evolutionary paths have converged upon a similar structural solution—are described as fold analogues.

In many cases in which a protein structure has been solved it can reveal unsuspected evolutionary links to previously characterized structures forging connections between protein families that would otherwise be considered unrelated. In some cases this enables the function(s) of one family to be inherited by another. However, it does not always follow that structure automatically informs function and that structural similarity always implies functional similarity. Fold space is not evenly distributed, with some protein folds (most commonly described as superfolds) adopted by a disproportionately large number of sequences across a wide variety of superfamilies with a large number of different functions *(14,15)*.

2.1.2. Family Space

Proteins that have evolved from a common ancestor are often found to share a related structure, function, and sequence. The comparison of protein structures has the capability to identify very distant relationships between protein sequences. However, the three-dimensional structure of every sequence is not known. Instead the relationships hidden within sequence space usually must be teased out through the use of computational sequence comparison methods. Measuring sequence similarity to infer the evolutionary distance between proteins is a fundamental tenet of structural biology and has been drawn on to organize sequence space into clusters of proteins that have diverged from a common ancestor and therefore share a common protein fold. to detect as many related sequences as possible, powerful sequence comparison methods have been developed, such as PSI-BLAST *(16)* and hidden Markov models (HMMs) *(15)*, which use sequence profiles built up of groups of related sequences, to identify remote protein homologues. (See Section 2.1.5. for a more detailed explanation on how these techniques can be used to identify structural homologues.)

The application of these methods has led to the compilation of a number of comprehensive protein family classification databases. Such libraries can often be browsed via the internet or used to annotate sequences over the internet using comparison servers. Some classifications provide libraries of HMMs and comparison software to enable the user to generate automated up-to-date genome annotations on their local computer systems. Among the most popular databases are Pfam *(18)*, SMART *(19)* and TigrFam *(20)*, each of which attempt to classify protein sequences at the domain level (refer to Table 1.1).

To work toward an efficient structural coverage of sequence space, the selection of target sequences is often closely linked to protein families' classifications (Fig. 1.2A). Such a view of sequence space can facilitate the derivation of a "minimal set of targets" and enables the structural biologist to select the most suitable families and family representatives for structural analysis, as illustrated in Figs. 1.2B and 1.2C. Ideally, target selection should focus on sequences with the highest "leverage" value, in which details of a solved structured can be transferred across a large number of related sequences. Accordingly, members of large protein families are often preferred, although in practice, smaller families are often also targeted because of their biological or medical importance. The ability to leverage structure determination by means of computational structure modeling is one of the founding principles upon which many structural genomics projects are based *(3)*. If at least one three-dimensional structure of a family member is known, at the minimum, coarse-grained fold-level assignments can be assigned to the remaining members, with the possibility of more detailed structure models being provided for closer sequence relatives. A family member with a large number of "model-able" relatives is a preferred choice for structure determination, as this should optimize the computational structure modeling coverage of the family. Within this context it can be seen that the task faced by structural genomics is related directly to the sensitivity of methods used to define protein families (i.e., the diversity of sequence space permitted in each family) and the quality of computational structure modeling required. Understanding the interplay between computational structure modeling and how it relates to the

Table 1.1 Structures of protein folds.

Method	Database or program name	URL
Protein domains and families database	Pfam *(18)*	http://www.sanger.ac.uk/Software/Pfam/
Modular domain architecture search	SMART *(19)*	http://www.smart.embl-heidelberg.de
Protein families based on HMMs	TigrFam *(20)*	http://www.tigr.org/TIGRFAMs/
Structural classification of protein	SCOP *(10)*	http://scop.mrc-lmb.cam.ac.uk/scop/
Class architecture topology homology	CATH *(11)*	http://www.cathdb.info/latest/index.html
Protein domain database and search tool	InterPro *(54)*	http://www.ebi.ac.uk/interpro/scan.html
Secondary structure prediction using neural networks	PsiPred *(42)*	http://bioinf.cs.ucl.ac.uk/psipred/
Transmembrane helix prediction using neural networks	MEMSAT *(42)*	http://www.cs.ucl.ac.uk/staff/D.Jones/ memsat.html
Transmembrane helix prediction using HMMs	TMHMM *(43)*	http://www.cbs.dtu.dk/services/ TMHMM-2.0/
A prediction service for bacterial transmembrane beta barrels	PROFtmb *(44)*	http://cubic.bioc.columbia.edu/services/ proftmb/
Coiled-coils prediction	COILS *(40)*	http://toolkit.tuebingen.mpg.de/pcoils
Prediction of dimeric and trimeric coiled coils	MultiCoil *(41)*	http://multicoil.lcs.mit.edu/cgi-bin/ multicoil
Signal peptide prediction	SignalP *(45)*	http://www.cbs.dtu.dk/services/SignalP/
Prediction of low complexity regions	SEG *(46)*	http://helix.nih.gov/docs/gcg/seg.html
Disorder prediction server	DisoPred *(42)*	http://bioinf.cs.ucl.ac.uk/disopred/
Intrinsic protein disorder prediction	DisEMBL *(48)*	http://dis.embl.de/
Taxonomy database for most protein sequences	NCBI Entrez taxonomy *(24)*	http://www.ncbi.nlm.nih.gov/entrez/ query.fcgi?db=Taxonomy
Automatic analysis of genomic sequences by a large variety of bioinformatics tools	PEDANT *(32)*	http://pedant.gsf.de/
Structural and functional annotation of protein families	Gene3D *(33)*	http://cathwww.biochem.ucl.ac.uk:8080/ Gene3D/
Various bioinformatic methods to determine the biological function and/or structure of genes and the proteins they code for	EMBOSS *(50)*	http://www.ebi.ac.uk/Tools/sequence. html
Expert protein analysis system dedicated to the analysis of protein sequences	ExPASy *(51)*	http://www.expasy.org/
Sequence analysis, structure and function prediction	PredictProtein *(52)*	http://www.predictprotein.org/
Status and tracking information on the progress of the production and solutions of protein structures	TargetDB *(55)*	http://targetdb.pdb.org/
RCSB Protein Data Bank	PDB *(1)*	http://www.rcsb.org/

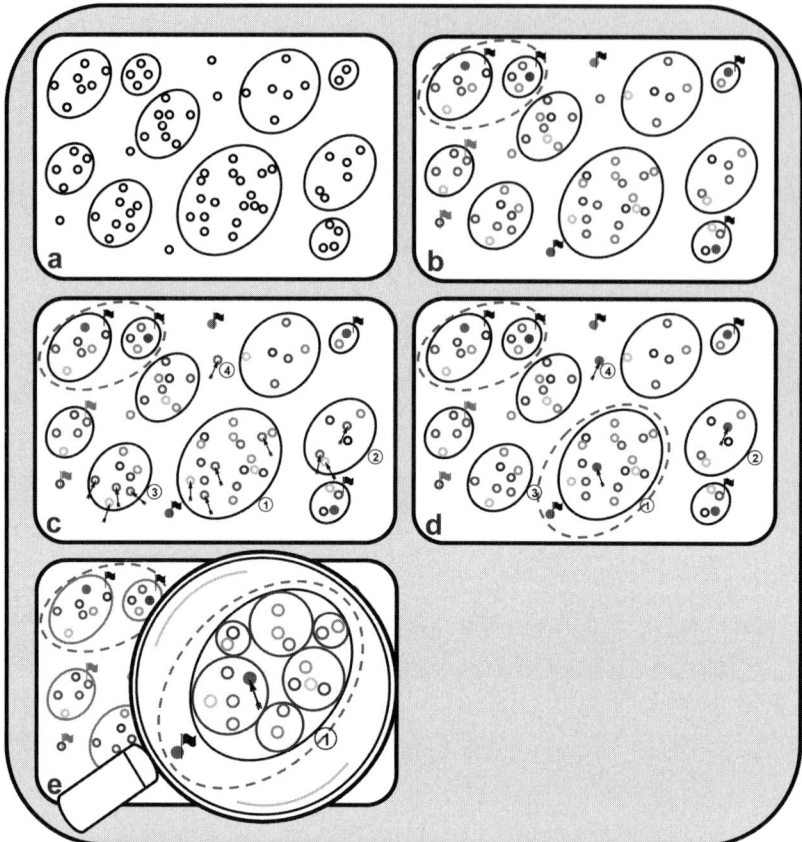

Fig. 1.2 Organizing sequence space for target selection. **A**. Sequence space, in which sequences are represented by small circles, is organized into families of distantly related proteins *(large ovals)* by sensitive sequence comparison methods such as Hidden Markov models or PSI-BLAST. **B**. Phylogenetic distributions are calculated, and problematic (gray flags) or structurally characterized *(filled circles, black flags)* sequences and families are identified and excluded from the target list. Structurally characterized families found to be distantly related by structure comparison (a relationship invisible to sequence comparison) are grouped into superfamilies *(ovals with a dashed line)*. **C**. Families are prioritized and targets selected *(arrows)*. Family 2 contains a human sequence; however, several alternative prokaryotic orthologues are selected instead to increase the chance of success. Target number 4 represents an ORFan sequence, for which no sequence relatives could be identified. **D**. Solved structures are now represented by filled circles: The characterization of a representative structure for Family 1 has revealed a hitherto unknown evolutionary link to a structure lacking any known sequence relatives *(new dashed oval)*. **E**. A coarse-grained approach to target selection suggests one structure per family is sufficient. Here, additional structures are required for Family 1 to provide a more comprehensive comparative modeling coverage. Subclusters have therefore been computed corresponding to a 30% pairwise sequence identity threshold.

structural coverage of sequence space is therefore of considerable value in defining the goals of target selection *(21)*.

Such observations naturally led to the question of how many protein families must be structurally characterized to obtain significant levels of genome annotation. The analysis of protein family classifications enables us to quantify the extent to which sequence space is already covered by structure, and how much remains out of reach of the current library of solved structures. At a "coarse-grained" level of structure annotation, one structure would be required for each protein family, in which each family may contain a broad spectrum of sequence relationships (closely to distantly related), compiled using methods such as PSI-BLAST or HMMs. Initial estimates suggested that a little over 16,000 structures would provide coarse-grained coverage for 90% of genome sequences *(22)*. More recent analyses have shown that this was a substantial underestimation, and as more genomes are sequenced, it seems more likely that over three times as many structures may be required to provide at least a fold annotation for the majority of protein families *(23)*.

A coarse-grained approach to structural genomics suggests that once a structure is solved for a protein family, investigators should move onto other structurally uncharacterized families. In the future, target selection will have to address a fine-grained approach to bring more protein sequences within the reach of a solved structure and accurate computational modeling (Fig. 1.2E). However, aiming for higher-quality structure annotations within large, diverse families would require an even greater number of solved structures *(23)*. Targets will have to be selected that add new, or complement existing structural information for a given protein family. In some cases it will be necessary to intelligently sample multiple experimental structures from each protein superfamily, providing valuable insights into protein evolution, particularly those cases in which function may have changed across a superfamily and may correlate with significant deviations in three-dimensional structure.

2.1.3. Comparative Modeling

The transfer of structural annotations by comparative modeling plays a significant role in target selection because it can be used to reduce the number of structures that are required to understand the structural repertoire encoded by genomes. A model-centric view of target selection requires that targets be selected on the basis that most of the remaining protein coding regions can be modeled with useful accuracy by comparative modeling. The number of structural determinations required to provide accurate comparative models across a significant proportion of genomes is therefore dependent on how model accuracy is defined. The higher the sequence identity required between the solved structure and modeled sequence, the greater the number of structures that are required.

Comparative modeling, also known as homology modeling, allows the construction of a three-dimensional model of a protein of unknown structure using a sequence of known structure as a template. Using today's methodology, accurate comparative modeling is currently restricted to protein sequences (targets) that share 30% or more sequence identity (over a significant overlap) to an experimentally solved protein structure template *(3)*. Below this sequence identity the reliability of the sequence alignment between target and template falls rapidly, leading to significant modeling errors; even if some details of the protein fold might still be assigned, low accuracy models should still be treated with caution. Medium accuracy models, obtained with a template-target sequence identity of 30–50%, tend to have approximately 85% of their C-alpha atoms within 3.5 Å of the correct position. These models are usually suitable for a variety of applications, including the testing of ligand binding states by designing site directed mutants with altered binding capacity, and computational screening of databases listing small molecules for potential lead compounds or inhibitors. Higher accuracy models, based on sequence identities above 50%, often have structures comparable to 3 Å resolution X-ray structures and may be used for more reliable calculations as required for ligand docking and directing medicinal chemistry in the optimization of lead compound candidates in drug design, although sequence identities above 90% may still be required to facilitate a meaningful biophysical description of the active site. It is important to remember that even very similar proteins can exhibit significant differences in their structures, and some of these differences may well be functionally important.

From a comparative modeling viewpoint, the effective contribution of each structure solved by the structure genomics initiatives can be measured as the number of structurally uncharacterized proteins, previously out of reach of comparative modeling (as assessed by a given threshold of accuracy), that can now be modeled using the newly solved protein as a structural template. The determination of at least one experimental structure for every protein sequence family defined at 30% sequence identity or above would bring all protein sequences within the scope of current methods, although the improvement of modeling methods would permit a coarser sampling of structure space in the future.

2.1.4. Phylogenetic Distribution

The species distribution of a protein family is an important consideration for target selection and prioritization, the knowledge of which can be exploited in a number of applications. A comprehensive taxonomy of protein sequences can be retrieved from the NCBI Taxonomy Database (24).

Families whose members span phylogenetically distant lineages in all three major kingdoms of life (Archaea, Bacteria, Eukaryota) are of interest since they may have an unknown but fundamental housekeeping function that a solved structure may help to define. Some projects target proteins found in a single organism lacking any homology to previously known proteins (ORFans). The structural characterization of such proteins could help clarify the existent ambiguity concerning their origin and purpose. Do ORFans correspond to expressed protein at all, and are they rapidly evolving or very recently evolved proteins that define a given species? It is possible that on structure characterization, distant membership of known superfamilies will be revealed, or is it possible that we must wait for future genome sequencing projects to pronounce the existence of sequence homologues?

Structural genomics expends a significant amount of effort attempting to improve protein solubility; the approaches include switching expression systems, replacing the expression host, and protein modification through mutagenesis (8). Many individual proteins are expected to be intractable without additional specialized effort. Therefore, parallel studies on related proteins from different species (orthologues) are often used to increase the chance of solving a structure for a family of proteins (Fig. 1.2C). In fact, several homologous targets may well be studied in parallel, with the higher probability of success outweighing the perceived redundancy in efforts (25).

Much emphasis is put on throughput in structural genomics. To expedite the structural characterization of protein families, most groups have focused on solving family members from prokaryote species (Fig. 1.2C). Cloning strategies for lower organisms are less complicated than those for higher eukaryotes, producing a greater frequency of small soluble domains that are easily expressed and purified in a bacterial host system. Structural data from prokaryotic protein can be used to annotate eukaryotic members (e.g., human proteins) through comparative modeling, circumventing the inherent difficulty in determining eukaryotic structures in a high-throughput approach.

2.1.5. *Assessing Similarity to Proteins of Known Structure*

Critical to the success of any target selection protocol is the use of comparative sequence analysis to limit the number of redundant structure determinations. Proteins that do not have a relative in the PDB are prioritized, so ideally all new structures will have a novel fold or represent a new superfamily of a previously observed fold. The exclusion of these PDB relatives is therefore dependent on how accurately structural characteristics can be inherited among sequences, which in turn is dependent on the evolutionary distance between those relatives.

Frequently a sequence identity cutoff of 30% is applied, corresponding to the lower threshold of comparative modeling accuracy, to exclude sequences within modeling distance of a solved protein. Pairwise sequence comparison methods such as Needleman and Wunsch *(26)* or Smith-Waterman *(27)* can be used to reliably identify close sequence homologues, corresponding to whole or partial sequence regions that share greater than 30% sequence identity, over a significant overlap. A note of warning, however: Short protein matches often have a high sequence identity that may not always be statistically significant, particularly for matches below 80 residues. To avoid this problem, a length-dependent sequence identity measure can be applied to the results of a pairwise sequence alignment, as depicted by the curve in Fig. 1.3A describing the length-dependent behavior between sequence identity and alignment length *(28)*.

Targets with more remote (<30% sequence identity) matches to the PDB can also be excluded to generate a list of high-priority targets in which there is an increased likelihood that they may assume a novel fold. However, standard pairwise sequence comparison methods cannot accurately detect distant evolutionary relationships between protein sequences that share 20–30% sequence identity. In this "twilight zone" *(29)* of sequence similarity, one can no longer determine whether two aligned protein sequences are related or not, based on the percentage sequence identity. The twilight zone curve (alignment length versus sequence identity) can be defined such that most protein pairs that appear above the curve can be reliably assigned as homologues. Around the curve, unrelated protein pairs start to appear and their numbers increase rapidly as one descends below 20% sequence identity (Fig. 1.3B) into the so-called midnight zone *(30)*.

Within the twilight zone, multiple alignment methods or consensus sequence profiles must be used to reliably infer homology. The expansion of the sequence databases over the past decade has increased the populations of protein families, and this has improved recognition of distant homologues in the twilight zone through the derivation of family-specific multiple sequence alignments. Such multiple alignments are an authoritative analytical tool as they create a consensus summary of the family, encapsulating the constraints of evolutionary change, and can reveal highly conserved residues that may correspond to function,

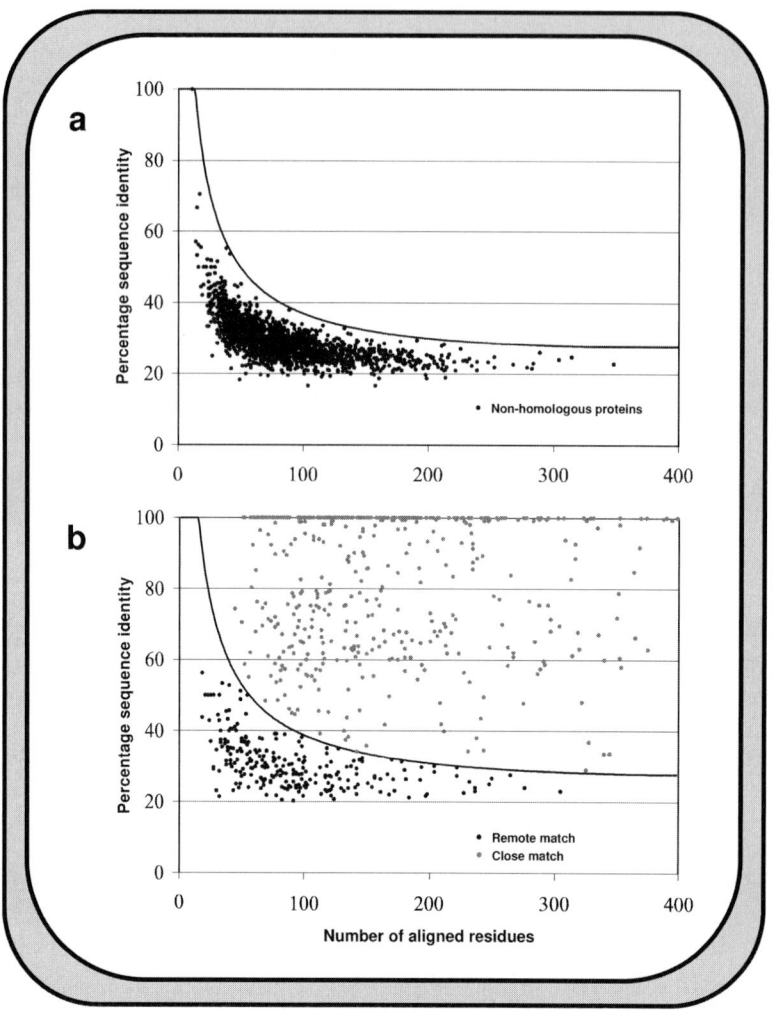

Fig. 1.3 Pairwise sequence identities measured between (**A**) evolutionarily unrelated sequences, and (**B**) evolutionary related sequences. The black line shows the sequence identity threshold that can be defined to assign true homologous relationships between sequence pairs for a given alignment length.

such as an active site. The unique features that define each protein family can then be captured in the form of patterns, or as profiles represented by position specific scoring matrices (as used by PSI-BLAST) or HMMs.

2.2. Single-Genome Approach

Whereas solving all proteins in nature seems an unfeasible task, obtaining a complete, or near-complete structural complement of a single gnome can seem more realistic. A single-genome approach to structural genomics transforms target selection from the rational sampling of all protein sequences into the comprehensive analysis of all genes and protein sequences in a particular genome and prioritizes them as targets for structural studies. In such an approach, target selection may be as simple as excluding all protein regions within a defined modeling distance of solved structures. Small genomes, such as *T. maritima* and *M. genitalium* are among those "model" organisms that have been targeted due to their small genome sizes, which are thought to encode the minimal set of genes and folds required for life. Other single-genome projects such as *M. tuberculosis* have been initiated with particular emphasis toward solving structures of novel potential drug targets and vaccine candidates in light of emerging strains with antibiotic resistance *(31)*. From this context a number of genome annotation resources prove particularly useful, containing exhaustive annotations of gene and protein sequences for completed genomes, including: the Protein Extraction, Description and Analysis Tool (PEDANT) *(32)*, and Gene3D *(33)*. It is also fair to say that ultimately most family-based approaches to target selection focus their efforts on a few organisms, being restricted to specific genomes for which DNA is available in the laboratory, rather than having the freedom to sample freely from the whole protein universe.

2.3. Function

In general, traditional structural biology (i.e., non-structural genomics laboratories) tends to structurally characterize proteins of known function. Structural evidence can then be used in verifying function, providing molecular-level detail of substrate- or ligand-binding mechanisms, with those structures representing the first structural characterization of a given protein function of considerable interest. In contrast to traditional biology, many (but not all) structural genomics projects solve structures in which little or no functional information is available. Only through discovery of distant homology through structural comparison, or the fortuitous scavenging of a ligand during structural characterization, can the first steps toward functional characterization be made.

However, some structural genomics initiatives also include functional characteristics when selecting targets, albeit on a larger scale. Gene function assignments using resources such as GO *(34)*, Kegg *(35)*, and EC *(36)* enable widespread function-based annotations across completed genomes, and have been used to build function-based target lists, such as structurally uncharacterized proteins thought to be involved in human disease *(37)*. The multiple sampling of specific regions of sequence space will also be extremely important for biomedical research as it allows an in-depth understanding of the structural diversity that can maintain similar functions. For example, the discovery of broad-spectrum antibacterial drugs requires multiple structures per family, in which the intricate distinctions that underlie the mechanisms of bacterial resistance can be compared across different pathogenic organisms. The development of human drug targets also benefits from the structural characterization of closely related proteins, allowing the design of a nontoxic drug that is selective toward only one of the many closely related structures *(38)*.

3. Target Selection: Addressing Limitations of High-Throughput Structure Determination

The previous section described some of the underlying principles and methods that can be used to build a structural genomics target list to reflect the research interests of a given project. However, there is no guarantee that all listed targets will be amenable to structure determination. It fact it is common for as little as 5% of the selected protein targets to successfully traverse the structure genomics pipeline and achieve a high-resolution protein structure. Frustratingly, each step of the structural genomics pipeline presents a bottleneck that can hamper the experimental success of structure determination. Significant obstacles include the cloning and expression of target sequences, protein purification, and achieving sufficient levels of protein solubility. It is not uncommon for many target proteins to be insoluble when expressed in a heterologous system, or when purified and concentrated at levels required for x-ray crystallographic or NMR structure determination. The choice of structure determination method can also present other complications, including the requirement of well-diffracting crystallized protein (x-ray crystallography) and the limitation of protein size (NMR).

As yet, the underlying factors determining expression, solubility, and crystallization are poorly understood. Nonetheless, certain sets of protein characteristics that are known to correlate with the various problem stages in structure determination can be predicted from protein sequence using a range of bioinformatics prediction tools, a variety of which are listed in Table 1.1. Many of these methods have been developed to provide predictions on a genomic scale; therefore, they are ideally suited for use in large-scale structural genomics target selection. By applying a range of methods, one can build a consensus picture describing how well-suited or tractable a target protein or family is likely to be for high-throughput structure determination.

It is also worth highlighting that although the high-throughput methodology and overall efficiency of structural genomics is hampered by a considerable target attrition rate, structure genomics groups are also in a unique position to expedite research in this key area. Most structure genomics efforts have recorded considerable quantities of information documenting the successes and failures of targets at different stages in the pipeline. Bioinformatics analysis of these data is beginning to untangle the complex links between the physical and chemical properties of target proteins and their progress (or lack of progress) through the pipeline *(39)*.

In the following section, the importance of identifying and excluding targets based on a number of protein characteristics, including coiled coils, transmembrane helices and protein disorder, through the use of available computational prediction methods is considered in more detail.

3.1. Coiled Coils

The coiled-coils structure is a ubiquitous protein motif, found in up to 5% of putative protein coding regions in sequenced genomes. Its structure is formed by the oligomerization of two or more alpha helices twisting around one another to generate a superhelical twist. One of the earliest characterized protein structures, a variety of coiled coil structures has now been structurally characterized; therefore, they are considered as extremely low-priority targets for structural genomics. In fact many groups attempt to exclude them from the target list entirely: *In vivo*, the coiled coil structure has been found to be used extensively to mediate protein oligomerization, and as a result cannot be easily studied without the appropriate binding partner, the provision of which would be a considerable hindrance to high-throughput methods. Furthermore, single crystal formation protocols suffer when applied to fibrous domains, which include the long extended structures frequently formed in coiled-coil proteins.

Sequence segments corresponding to coiled coils are characterized by a regular series of heptad repeats, making it possible to predict their location. Among the first programs to reliably identify coiled coils was the PCOILS algorithm *(40)*, which can be used to predict the likelihood and location of left-handed two-stranded coiled coils in protein sequences. More recently, the MultiCoil coil program *(41)* has been made available that predicts both two and three-stranded coiled coils.

3.2. Transmembrane Proteins

It has been estimated that 25% or more proteins in genomes belong to membrane protein families *(23)*, which act as a critical interface between the intracellular and extracellular environment. The extensive use of membrane proteins as therapeutic drug targets can be explained by the extensive variety of cellular functions they undertake, including transport, motility, regulation, interaction, and metabolism. Nevertheless, despite their biological importance, membrane structures are hugely underrepresented in the PDB compared with globular proteins. Such scarcity is born from the fact that membrane proteins are a notoriously difficult and inefficient class of molecule to target using traditional structural determination methods due to problems encountered with low solubility, aggregation, and precipitation in solution. The transmembrane topology requires the use of detergents during membrane protein purification, which can adversely affect protein yield and stability. Additionally, crystallization in the presence of detergent in combination with the dynamic nature of membrane structures can reduce crystal contacts, complicating structure determination. Accordingly, it is very difficult to purify sufficient quantity of active protein for crystallization, especially in high-throughput approaches. As such, transmembrane structures are routinely avoided in structural genomics target selection, with only specialized centers focusing on these proteins.

The prediction of the location of transmembrane helices has received much attention, and prediction methods now routinely achieve accuracy above 75%. Indeed, numerous prediction methods have been published that are often used in concert to provide more reliable consensus predictions. Among the most popular published methods are the neural network based MEMSAT *(42)* and hidden Markov model–based TMHMM *(43)*. Often prediction algorithms assign a membrane protein fold topology, describing the in–out orientations of each helix passing across the membrane. Membrane spanning beta-strands generally form beta-barrel structures across the outer membranes of gram-negative bacteria, mitochondria, and chloroplasts, constituting approximately 3% of the gram-negative proteome. The prediction of beta-strand membrane proteins has received less attention, although prediction methods do exist (e.g., PROFtmb) *(44)*.

It is also worthwhile to remember that the presence of a signal peptide in a target sequence can complicate transmembrane helix prediction, leading to false-positive assignments or incorrect topology prediction. This is because the hydrophobic core of a signal peptide is often mistaken for a transmembrane helix. Although some prediction methods circumvent this difficulty by masking out such regions, or ignoring membrane helices within a cutoff distance from the N-terminus, it is still worthwhile to check prediction results against signal peptide prediction results (see the next section).

3.3. Signal Peptides

Signal peptides play a key role in protein targeting in both prokaryotes and eukaryotes. These short peptide sequences direct the posttranslational transport of a protein through the crowded cytoplasm for secretion or for transfer to specific organelles. At destination, the transient signal peptide region is cleaved off to release the active protein. Interestingly, not all proteins contain signal peptides, suggesting that other mechanisms for protein targeting exist. Signal peptides tend to be characterized by short stretches of hydrophobic residues located close to the N-terminus of the sequence, making it possible to predict their location and cleavage site from sequence information alone. Targets containing N-terminal signal peptides are usually highlighted rather than excluded from structural genomics target lists. This is especially important for proteins that are not expressed in their host organism, in which it is safer to exclude likely signal peptides when designing expression constructs. The SignalP program *(45)* is widely used to predict the presence and cleavage position of signal peptide in gram-positive and -negative prokaryotes as well as eukaryotes. Careful analysis of the prediction scores enables the user to identify occasional cases in which membrane anchors are mis-predicted as signal peptides.

3.4. Low Complexity and Disordered Regions

Low-complexity and disordered proteins contain unstructured or only partially structured regions exhibiting extensive conformational flexibility that can present a host of complications for structural determination studies. Experimentalists often encounter difficulties with protein purification due to the vulnerability of proteolytic attack, crystallization failures, missing electron density during x-ray crystallographic structure determination, excessive broadening of side chain NMR peaks, and lack of chemical shift dispersion of NMR backbone data.

3.4.1. Low Complexity

Low complexity regions of sequences have comparatively little variation in residue, commonly composed of only two different amino acid types, and are traditionally associated with unstructured regions. These low-complexity regions frequently correlate with mobile regions in protein structure, and targets with a preponderance of these regions should be removed from the target list. Low-complexity detection software such as SEG *(46)* and CAST *(47)* can be used to analyze the information content of query sequences to distinguish subsequences with high or low complexity.

3.4.2. Disorder

Some proteins can be entirely or partially disordered in their native state. Although little is understood about the cellular and structural importance of disordered regions, it has been suggested that these regions become ordered upon interaction with specific binding partners or in response to changes in the biochemical environment, although their structural determination is difficult even with prior knowledge of required cofactors. Computational methods to help discern ordered globular domains from disordered regions are therefore key to target selection efforts—avoiding potentially disordered segments in protein expression constructs can increase the expression, foldability, and stability of the expressed protein. In the past few years, a wide variety of protein disorder prediction methods have been developed, including DisoPred *(42)* and DisEMBL *(48)*, and their use has revealed that these regions occur in nature with an unexpectedly high frequency. Although computational predictions of disorder (as with other prediction methods) are a useful front line in avoiding these problematic proteins, experimental evidence can often provide greater detail. For example, the technique of deuterium exchange combined with mass spectroscopy (DXMS) to identify unstructured regions has been used successfully to improve crystallization success *(49)*.

3.5. Additional Sequence Properties

Not all protein characteristics can be directly related to bottlenecks in the experimental pipeline; rather, they may prove more useful for guiding specific experimental procedures. A number of these properties are described in more detail in the following. Publicly available software packages and web servers are available for computing many of these characteristics, including the EMBOSS software package *(50)*, the ExPASy server *(51)* and PredictProtein *(52)*.

- Codon usage: Nucleotide sequence–based calculations are used to determine the encoding gene's GC content and its compatibility with that of the host expression system, a measure also known as the Codon Adaptation Index.
- Isoelectric point: The isoelectric point (pI) of a protein can be used as a useful predictor for the optimal choice of range and distribution of the pH sampling in crystallization trials.
- Hydrophobicity: Protein hydropathy can be established using the grand average of hydropathy (GRAVY) index using the Kyte and Doolittle hydrophobicity scale. The GRAVY index corresponds to the sum of hydrophobicity values for each residue in a protein normalized according to protein length. Positive GRAVY scores are a strong indicator of proteins that are likely to be difficult to crystallize.
- Methionine frequency: The use of multiple wavelength anomalous dispersion (MAD) phasing for structure determination requires a suitable number of methionines in the target gene sequence that can be replaced with selenomethionine on protein expression. The selenomethionine atom subset is used to produce small changes in the x-ray diffraction pattern that are used to increase the efficiency of obtaining phase information for structure determination.
- Molecular weight: Large proteins are traditionally considered as less favorable because they are more difficult to crystallize or their NMR spectra are difficult to resolve. Sequences may also be filtered by size to exclude proteins considered too large to be expressed and purified in the host expression organism. For example, studies have shown that proteins successfully expressed in *E. coli* tend to mirror the size of the organism's natural proteins, in the region of 30–60 kDa *(53)*.
- Thermostability: Thermostable proteins make excellent targets for high-throughput structure determination, with their compact size and density of salt bridges contributing to their thermostability and crystallizability.

Because of this, structural genomics efforts focusing on model organism have used thermophilic species such as *T. thermophilus* and *T. maritima*.

3.6. Domain Boundaries

Structural genomics targets are typically divided into their constituent domains in view of the fact that single domains are often more suited to high-throughput studies than their multidomain counterparts. For example, multidomain proteins are often more difficult to express in bacterial host systems, such as *E. coli*, and to crystallize due to their larger size and the presence of mobile polypeptide regions linking the different domains.

The most reliable form of domain assignments comes from analysis of solved proteins structures. However, in the absence of structure data, the accurate identification of domain boundaries in large multidomain proteins is fraught with difficulties and remains a major bottleneck for successful structure determination. The reliability of domain boundary prediction methods for target selection can be roughly divided into two levels; higher-accuracy predictions based on comparative sequence analysis, and lower-accuracy predictions provided by *ab initio* methods. Comparative domain boundary prediction methods use comprehensive sequence searches against known structural or sequence domains held in domain classification databases such as CATH and Pfam, or InterPro *(54)* (Fig. 1.4). Through sequence comparison using PSI-BLAST

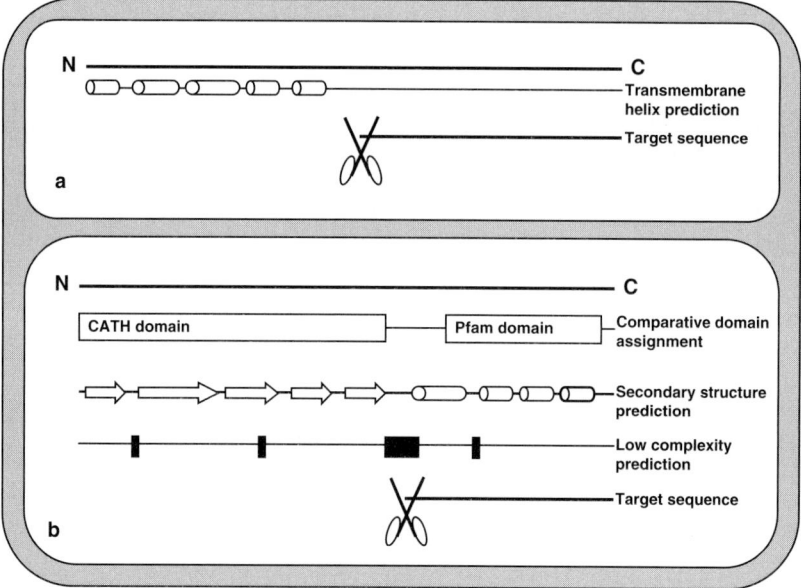

Fig. 1.4 Assignment of domain boundaries. **A.** A predicted N-terminal helical transmembrane domain is excised, providing a natural domain boundary for the target C-terminal globular domain. **B.** The domain boundary between a structurally uncharacterized target Pfam domain family and a CATH domain assignment is resolved using predicted secondary structures *(36)* and low complexity regions. Note; the N-terminus of the Pfam assignment falls within a predicted helix, the excised target sequence is therefore cut at the N-terminus of the helix, near a low complexity region known to correlate with domain linker regions.

or HMM based representations of domain families, the domain content and location of domain boundaries can be predicted across target sequences in cases in which a homologous domain is detectable (e.g., the Gene3D database provides comprehensive domain assignments of all sequences in completed genomes, Table 1.1).

The prediction of domain boundaries from sequence information alone (*ab initio* prediction) is more difficult and cannot always be relied on to provide accurate domain assignments. A series of methods are available (see Table 1.1 for some of the most popular) based on the physical properties of known domains, such as domain size, amino acid composition, predicted secondary structure, and *ab initio* structure prediction. Often a range of predictions is provided, which can be compared against other methods or other sequence predictions. For instance, the likely location of domain-linking regions can correlate to low-complexity or disordered regions or coil regions between predicted secondary structure. A typical domain parsing procedure occurs in which proteins predicted to contain helical transmembrane domains are split up into domains that surround the transmembrane region (Fig. 1.4). In these cases the transmembrane domain provides a natural domain boundary and all that remains is to determine how far from the transmembrane region to assign a domain boundary—endeavoring to pinpoint the exact interdomain cutpoint is critical to the success of obtaining a soluble expression product. Often a number of varied expression constructs, based around the predicted domain boundary, are used to increase the chance of expressing protein with the natural domain termini.

3.7. TargetDB

Structural genomics has transformed many of the traditional approaches to structure determination, a development that is exemplified by the TargetDB database. It was clear from the offset that the worldwide structural genomics effort would require coordination across the centers to maximize the efficiency of target selection. Such measures were initiated through the TargetDB database *(55)*, in which details of each structure genomics target list are made available to the public with progress reports, covering each initiative, updated on a weekly basis. The dissemination of target information to the scientific community enables centers to monitor the overlap between target lists, and also enables scientists from outside the field to access information on proteins that may coincide with their own research projects. The system works on the premise that after new targets are registered, the database is able to monitor progress of each target as it progresses along the structure determination pipeline. The data are organized according to the International Task Force in Target Tracking *(56)*. Therefore, unprecedented insights are given on the successes and failures of structure determination attempts in different structural biology laboratories, and much has been made of the potential of mining this information to develop prediction methods for protein solubility and likelihood of crystallization.

4. Target Selection: Future Approaches

Structural genomics hopes to make huge strides toward structural understanding of molecular biology, promising to provide a comprehensive library of experimentally and computationally derived three-dimensional protein structures.

Recent analyses of structures released by the initiatives have highlighted the significant contribution they are now making in both the scope and depth of structural and biological knowledge of protein families, especially when compared to the relative contribution of nonstructural genomics structures. The worldwide structural genomics initiatives now contribute approximately half of new structurally characterized families and over five times as many novel folds as mainstream structural biology, despite accounting for only approximately 20% of the new structures *(6,57)*.

Notwithstanding these achievements, future structural genomics target selection strategies must address some of the current challenges that will prevent researchers from gaining an accurate and comprehensive coverage of all protein structures in the cell. In some cases, even a number of "simple" proteins will not be tractable to high-throughput structure determination methods and will only be addressed by specific research programs focused on those particular systems. Nonetheless, considerable emphasis must be directed toward the structural characterization of integral membrane proteins, unstructured proteins, and the multiple sampling of large and diverse families to enable accurate comparative modeling for all sequence members.

The prediction of protein characteristics that correspond to the bottlenecks in the experimental structure determination pipeline will be essential in reducing the significant attrition rate of target proteins. Future analyses should put significant effort into the use of the vast quantity of "negative data" being generated by structural genomics efforts. Instead of attempting to ascertain why a target worked (from a bioinformatics perspective), it might be more profitable to identify why (many more) targets have failed given the depth of information available. One could argue that currently such information is not readily available, and that in the future databases will have to be developed that contain more pertinent information regarding the reasons for target failure.

Structural genomics initiatives will also have to address the fact that only a limited view of the protein world is gained when solving single domains. Although the complexity of an organism can, in part be measured according to its gene content, a more meaningful calculation accounts for the number and complexity of the protein–protein interaction networks that its gene products undertake. *In vivo* most proteins do not function as monomers; rather they interact with other proteins permanently or transiently, forming functional complexes. Therefore, to begin to understand cellular function requires detailed knowledge of these protein interactions *(19)*. At the experimental level, approaches such as yeast-two-hybrid and coaffinity purification methods have been developed to identify protein interactions *in vivo* and *in vitro*. Computational methods have also been developed to infer linkages among proteins on a genome-wide scale. Techniques include Rosetta Stone, phylogenetic profile, and conserved gene neighbor methods (for review see *58*), in which protein linkages revealed by these methods can be used to identify and target proteins that participate in protein complexes. The difficulty is identifying how to combine these methods for the inference of protein complexes with coexpression and cocrystallization of interaction partners— even in cases in which target selection can predict the components of a protein interaction complex; an automated approach for the structural characterization of protein complexes is limited. Current high-throughput structure determination pipelines are focused on individual domains or proteins, and are therefore not solved as a functional complex variant. Current restrictions include the availability of sufficient quantities of soluble protein complexes, the difficulty in reconstituting the complex from individual protein components, *in vivo*, and the half-life of the protein complex—highly stable systems are necessary for successful crystallographic studies. Although still beyond the scope of current structural genomics, future research must be directed toward defining the multitude of protein interactions to understand the cell in its entirety.

References

1. Bourne, P. E., Westbrook, J., and Berman, H. M. (2004) The Protein Data Bank and lessons in data management. *Brief. Bioinform.* **5**, 23–30.
2. Airlie Agreement (2001) http://www.nigms.nih.gov/news/meetings/airlie.html
3. Baker D, and Sali A. (2001) Protein structure prediction and structural genomics. *Science* **294**, 93–96.
4. Brenner, S. E., and Levitt, M. (2000) Expectations from structural genomics. *Protein Sci.* **9**, 197–200.
5. Chandonia, J. M., Earnest, T. N., and Brenner, S. E. (2004) Structural genomics and structural biology: compare and contrast. *Genome Biol.* **5**, 343.
6. Todd, A. E., Marsden, R. L., Thornton, J. M., and Orengo, C. A. (2005) Progress of structural genomics initiatives: an analysis of solved target structures. *J. Mol. Biol.* **348**, 1235–1260.
7. Bray, J. E., Marsden, R. L., Rison, S. C., Savchenko, A., Edwards, A. M., Thornton, J. M., and Orengo, C. A. (2004) A practical and robust sequence search strategy for structural genomics target selection. *Bioinformatics* **20**, 2288–2295.
8. Marsden, B. D., Sundstrom, M., and Knapp, S. (2006) High-throughput structural characterization of therapeutic protein targets. *Expert Opin. Drug Disc.* **1**, 123–136.
9. Bravo, J., and Aloy, P. (2006) Target selection for complex structural genomics. *Curr. Opin. Struct. Biol.* **16**, 385–392.
10. Murzin, A. G., Brenner, S. E., Hubbard, T., and Chothia, C. (2000) SCOP: a structural classification of proteins for the investigation of sequences and structures. *J. Mol. Biol.* **247**, 536–540.
11. Orengo, C. A., Mitchie, A. D., Jones, S., Jones, D. T., Swindells, M. B., and Thornton, J. M. (1997) CATH—a hierarchical classification of protein domain structures. *Structure* **5**, 1093–1108.
12. Grant, A., Lee, D., and Orengo, C. (2004) Progress towards mapping the universe of protein folds. *Genome Biol.* **5**, 107.
13. Harrison, A., Pearl, F., Mott, R., Thornton, J., and Orengo, C. (2002) Quantifying the similarities within fold space. *J. Mol. Biol.* **323**, 909–926.
14. Orengo, C. A., Jones, D. T., and Thornton, J. M. (1994) Protein superfamilies and domain superfolds. *Nature* **372**, 631–634.
15. Todd, A. E., Orengo, C. A., and Thornton, J. M. (2002) Sequence and structural differences between enzyme and nonenzyme homologs. *Structure* **10**, 1435–1451.
16. Altschul, S. F., Madden, T. L., Schaffer, A. A., Zhang, J., Zhang, Z., Miller, W., and Lipman, D. J. (1997) Gapped BLAST and PSI-BLAST: a new generation of protein database search programs. *Nucleic Acids Res.* **25**, 3389–3402.
17. Eddy, S. R. (1996) Hidden Markov models. *Curr. Opin. Struct. Biol.* **6**, 361–365.
18. Finn, R. D., Mistry, J., Schuster-Bockler, B., Griffiths-Jones, S., Hollich, V., Lassmann, T., Moxon, S., Marshall, M., Khanna, A., Durbin, R., Eddy, S. R., Sonnhammer, E. L., and Bateman, A. (2006) Pfam: clans, web tools and services. *Nucleic Acids Res.* **34**, D247–251.
19. Letunic, I., Copley, R. R., Pils, B., Pinkert, S., Schultz, J., and Bork, P. (2006) SMART 5: domains in the context of genomes and networks. *Nucleic Acids Res.* **34**, D257–260.

20. tigr fam protein families: http://www.tigr.org/TIGRFAMs
21. Friedberg, I., Jaroszewski, L., Ye, Y., and Godzik, A. (2004) The interplay of fold recognition and experimental structure determination in structural genomics. *Curr. Opin. Struct. Biol.* **14**, 307–312.
22. Vitkup, D., Melamud, E., Moult, J., and Sander, C. (2001) Completeness in structural genomics. *Nat. Struct. Biol.* **8**, 559–566.
23. Marsden, R. L., Lee, D., Maibaum, M., Yeats, C., and Orengo, C. A. (2006) Comprehensive genome analysis of 203 genomes provides structural genomics with new insights into protein family space. *Nucleic Acids Res.* **34**, 1066–1080.
24. Benson, D. A., Karsch-Mizrachi, I., Lipman, D. J., Ostell, J., and Wheeler, D. L. (2006) GenBank. *Nucleic Acids Res.* **34**, D16–20.
25. Savchenko, A., Yee, A., Khachatryan, A., Skarina, T., Evdokimova, E., Pavlova, M., Semesi, A., Northey, J., Beasley, S., Lan, N., Das, R., Gerstein, M., Arrowmith, C. H., and Edwards, A. M. (2003) Strategies for structural proteomics of prokaryotes: quantifying the advantages of studying orthologous proteins and of using both NMR and X-ray crystallography approaches. *Proteins* **50**, 392–329.
26. Needleman, S., and Wunsch, C. (1970) A general method applicable to the search for similarities in the amino acid sequence of two proteins. *J. Mol. Biol.* **48**, 443–453.
27. Smith, T., and Waterman, M. (1981) Identification of common molecular subsequences. *J. Mol. Biol.* **147**, 195–197.
28. Sander, C., and Schneider, R. (1991) Database of homology-derived protein structures and the structural meaning of sequence alignment. *Proteins* **9**, 56–68.
29. Doolittle, R. F. (1986) *Of URFs and ORFs: a primer on how to analyze derived amino acid sequences.* University Science Books, Mill Valley, California.
30. Rost, B. (1997). Protein structures sustain evolutionary drift. *Folding and Design* **2**, S19–S24.
31. Smith, C. V., and Sacchettini, J. C. (2003) Mycobacterium tuberculosis: a model system for structural genomics. *Curr. Opin. Struct. Biol.* **13**, 658–664.
32. Riley, M. L., Schmidt, T., Wagner, C., Mewes, H. W., and Frishman, D. (2005) The PEDANT genome database in 2005. *Nucleic Acids Res.* **33**, D308–310.
33. Yeats, C., Maibaum, M., Marsden, R., Dibley, M., Lee, D., Addou, S., and Orengo, C. A. (2006) Gene3D: modeling protein structure, function and evolution. *Nucleic Acids Res.* **34**, D281–284.
34. The Gene Ontology Consortium. (2000) Gene ontology: tool for the unification of biology. *Nature Genet.* **25**, 25–29.
35. Kanehisa, M., and Goto, S. (2000) KEGG: Kyoto Encyclopedia of Genes and Genomes. *Nucleic Acids Res.* **28**, 27–30.
36. Bairoch, A. (2000) The ENZYME database in 2000. *Nucleic Acids Res.* **28**, 304–305.
37. Xie, L., and Bourne P. E. (2005) Functional coverage of the human genome by existing structures, structural genomics targets, and homology models. *PLoS Comput. Biol.* **1**, e31.
38. Russell, R. B., and Eggleston, D. S. (2000) New roles for structure in biology and drug discovery. *Nat. Struct. Biol.* **7**, 928–930.
39. Goh, C. S., Lan, N., Douglas, S. M., Wu, B., Echols, N., Smith, A., Milburn, D., Montelione, G. T., Zhao, H., and Gerstein, M. (2004) Mining the structural genomics pipeline: identification of protein properties that affect high-throughput experimental analysis. *J. Mol. Biol.* **336**, 115–130.
40. Gruber, M., Soding, J., and Lupas, A. N. (2006) Comparative analysis of coiled-coil prediction methods. *J. Struct. Biol.* **155**, 140–145.
41. Wolf, E., Kim, P. S., and Berger, B. (1997) MultiCoil: a program for predicting two- and three-stranded coiled coils. *Protein Sci.* **6**, 1179–1189.
42. Bryson, K., McGuffin, L. J., Marsden, R. L., Ward, J. J., Sodhi, J. S., and Jones, D. T. (2005) Protein structure prediction servers at University College London. *Nucleic Acids Res.* **33**, W36–38.

43. Krogh, A., Larsson, B., von Heijne, G., and Sonnhammer, E. L. (2001) Predicting transmembrane protein topology with a hidden Markov model: application to complete genomes. *J. Mol. Biol.* **305**, 567–580.

44. Bigelow, H., and Rost, B. (2006) PROFtmb: a web server for predicting bacterial transmembrane beta barrel proteins. *Nucleic Acids Res.* **34**, W186–188.

45. Bendtsen, J. D., Nielsen, H., von Heijne, G., and Brunak, S. (2004) Improved prediction of signal peptides: SignalP 3.0. *J. Mol. Biol.* **340**, 783–795.

46. Wootton, J. C., and Federhen, S. (1996) Analysis of compositionally biased regions in sequence databases. *Methods Enzymol.* **266**, 554–571.

47. Promponas, V. J., Enright, A. J., Tsoka, S., Kreil, D. P., Leroy, C., Hamodrakas, S., Sander, C., and Ouzounis, C. A. (2000) CAST: an iterative algorithm for the complexity analysis of sequence tracts. Complexity analysis of sequence tracts. *Bioinformatics* **16**, 915–922.

48. Linding, R., Jensen, L. J., Diella, F., Bork, P., Gibson, T. J., and Russell, R. B. (2003) Protein disorder prediction: implications for structural proteomics. *Structure* **11**, 1453–1459.

49. Pantazatos, D., Kim, J. S., Klock, H. E., Stevens, R. C., Wilson, I. A., Lesely, S. A., and Woods, V. L. (2004) On the use of DXMS to produce more crystallizable proteins: structures of the *T. maritima* proteins TM0160 and TM1171. *Proc. Natl. Acad. Sci. USA* **101**, 751–756.

50. Sarachu, M., and Colet, M. (2005) wEMBOSS: a web interface for EMBOSS. *Bioinformatics* **21**, 540–541.

51. Gasteiger, E., Gattiker, A., Hoogland, C., Ivanyi, I., Appel, R. D., and Bairoch A. (2003) ExPASy: the proteomics server for in-depth protein knowledge and analysis. *Nucleic Acids Res.* **31**, 3784–3788.

52. Rost, B., Yachdav, G., and Liu, J. (2003) The PredictProtein Server. *Nucleic Acids Res.* **32**, W321–W326.

53. Canaves, J. M., Page, R., Wilson, I. A., and Stevens, R. C. (2004) Protein biophysical properties that correlate with crystallization success in Thermotoga maritima: maximum clustering strategy for structural genomics. *J. Mol. Biol.* **344**, 977–991.

54. Zdobnov, E. M., and Apweiler, R. (2001) InterProScan—an integration platform for the signature-recognition methods in InterPro. *Bioinformatics* **17**, 847–848.

55. Chen, L., Oughtred, R., Berman, H. M., and Westbrook, J. (2004) TargetDB: a target registration database for structural genomics projects. *Bioinformatics* **20**, 2860–2862.

56. Task Force on Target Tracking (2001) http://www.nigms.nih.gov/news/reports/airlie_tasks.html

57. Chandonia, J. M., and Brenner, S. E. (2006) The impact of structural genomics: expectations and outcomes. *Science* **311**, 347–351.

58. Pellegrini, M., Haynor, D., and Johnson, J. M. (2004) Protein interaction networks. *Expert Rev. Proteomics* **1**, 239–249.

Chapter 2

A General Target Selection Method for Crystallographic Proteomics

Gautier Robin, Nathan P. Cowieson, Gregor Guncar, Jade K. Forwood, Pawel Listwan, David A. Hume, Bostjan Kobe, Jennifer L. Martin, and Thomas Huber

Increasing the success in obtaining structures and maximizing the value of the structures determined are the two major goals of target selection in structural proteomics. This chapter presents an efficient and flexible target selection procedure supplemented with a Web-based resource that is suitable for small- to large-scale structural genomics projects that use crystallography as the major means of structure determination. Based on three criteria, biological significance, structural novelty, and "crystallizability," the approach first removes (filters) targets that do not meet minimal criteria and then ranks the remaining targets based on their "crystallizability" estimates. This novel procedure was designed to maximize selection efficiency, and its prevailing criteria categories make it suitable for a broad range of structural proteomics projects.

1. Introduction

The task that initiates all structural proteomics/genomics projects is choosing the targets—proteins, or regions of proteins—that will enter the structure determination process. This target selection step is particularly important because it embodies the goals of each specific project and can substantially influence the overall success rate of the process. Although each specific structural proteomics program may have a different biological focus, methodology, and throughput (*1–6*); nonetheless, common criteria can be used to select targets.

Target selection criteria can be divided into three categories: (a) biological significance/impact, (b) structural novelty, and (c) likelihood to crystallize. Clearly, the classification of biological significance/impact varies from one project to another, but structural novelty and likelihood to crystallize are common criteria for most target selection procedures. Furthermore, selection criteria can be categorized into two types: filters and sorters. Filters remove those targets that possess undesirable features (e.g., those having a known structure), whereas sorters allow the ranking of remaining targets so that the desired number of targets may be selected from the top-scoring targets.

The goal of the method described here is to provide a general approach for selecting targets in a small- to large-scale protein crystallography context.

From: *Methods in Molecular Biology, Vol. 426: Structural Proteomics: High-throughput Methods*
Edited by: B. Kobe, M. Guss and T. Huber © Humana Press, Totowa, NJ

2. Methods

2.1. Target Selection Frame Description

The process can be broken down into three sequential steps: automated filtering, ranking/sorting, and manual filtering (Fig. 2.1). The order of this sequence was designed to maximize efficiency of target selection while allowing the flexibility to cater for a specific focus of a structural proteomics project. This section overviews the strategy and methods used in the target selection procedure. The following final section presents the application of the procedure.

2.1.1. Automated Filtering (Step 1)

Automated filtering represents the major selection step. It includes criteria from all three of the selection categories mentioned in the preceding (biological significance/impact, structural novelty, likelihood to crystallize). It includes all criteria that are amenable to computer automation and are not components of the ranking parameters (listed in Section 2.1.2.). Examples of criteria that can be applied at this step are presented in Fig. 1.1. The outcome of this first step is a list of targets that have passed all the selected criteria (see Note 1).

2.1.2. Ranking of Targets (Step 2)

The automated filtering often results in more targets than can be handled in a single iteration of high-throughput protein production and structure determination. Therefore, it is important to be able to prioritize targets selected in step 1. In this procedure, the targets selected by the automatic filtering procedure are subsequently ranked according to their likelihood to crystallize. Recently, several groups have shown clear correlations between crystallization success and protein properties predicted from sequence only *(7–10)*.

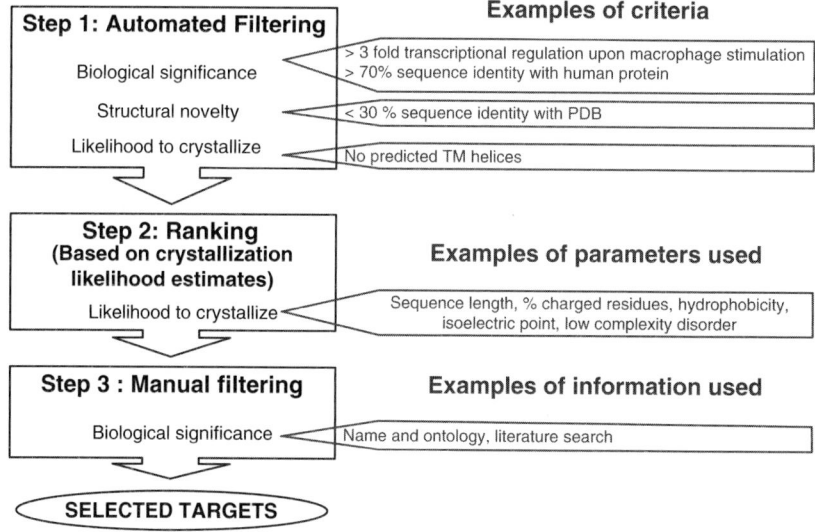

Fig. 2.1 A schematic diagram illustrating the target selection procedure.

Target selection methods take advantage of the data generated by structural genomics projects to identify correlations between protein attributes (determined by sequence analysis) and its success or failure through the expression–purification–crystallization process. This is then used to extract rules for target selection to optimize the output (Notes 7 and 8). One approach to detect such crystallization predictors consisted of generating distributions of various potentially relevant properties from a set of proteins (whole *Thermatoga maritima* proteome) and the subset of those that crystallized, to analyze trends for crystallization success *(7)*. The outcome was a list of crystallization predictors and target filtering strategies. Because the authors are interested at this stage in ranking rather than filtering, they used a similar approach to regenerate the distributions to quantify the likelihood of target crystallization, which is described in the following. They chose a larger and more representative initial set of proteins to represent the "whole universe" of proteins by using more than 3 million protein sequences from the nonredundant sequence database, and a nonredundant sample from the Protein Data Bank (PDB) as the subset representing the universe of successfully crystallized proteins. The normalized distributions are similar to those obtained from the *Thermatoga maritima* genome, thus validating the pertinence of both data sets. The predictive power of a given sequence characteristic is inversely proportional to the area of overlap between the global distribution and crystallized distribution. The protein properties used for estimating crystallization likelihood are: sequence length, predicted isoelectric point, percentage of charged residues, hydropathy, and a measure of low complexity disorder (Notes 2–4). These have already been reported as parameters influencing crystallization success *(7–10)*. The authors are currently in the process of systematically testing other parameters to further improve efficiency. Finally, the likelihood estimate *p* is calculated according to:

$$p = \prod_i \left(\frac{Y_i^C}{Y_i^U} \right) \tag{1}$$

where Y_i^C represents the frequency of proteins from the crystallized subset that match parameter *i* value (e.g., with a sequence length of 250–260 residues) of the protein evaluated (e.g., 253 residues); and Y_i^U represents the corresponding value from the whole universe set of proteins. This formula can be related to a probability calculation in which the ratio represents the number of successful events (crystallized proteins subset) over the whole set. The properties (e.g., length, pI, percentage of charged residues) are considered independent, which is reflected by the use of the product in Equation 1. The application of the ranking step is achieved through the Web-based UQSG Target Ranker (see Section 2.2.2.).

2.1.3. Manual Filtering (Step 3)

The final step is the manual evaluation of each of the ranked targets. This optional step is used because some criteria may be too difficult to program and the programmed selection procedures may have weaknesses. Typically the protein description, ontology, literature searches, and personal knowledge are employed to identify any specific problems in the remaining selected targets.

Evaluating each target manually is time consuming; thus, this step is performed last. In truly high-throughput programs manual intervention in the target selection is generally omitted; however, in small- and medium-throughput programs, additional quality control can help to focus on targets with higher biological significance and scientific impact.

2.2. Applying the Target Selection

2.2.1. Full-Length Protein or Protein Constructs?

Construct design may be considered for proteins of higher value. The constructs may be designed to pass given selection criteria (e.g., removal of transmembrane domains, deletion of low-complexity regions) and divide the protein into

Table 2.1. Useful links for protein construct design.

Software	URL
• **Splice variant database**	
H-InvDB *(14)*	http://hinvdb.ddbj.nig.ac.jp/ahg-db/index.jsp
LOCATE *(15)*	http://locate.imb.uq.edu.au/
Macrophages.com	http://www.macrophages.com/bioinfoweb/
Genome Browser *(16)*	http://genome.ucsc.edu/cgi-bin/hgGateway
• **Domain homology search**	
RPSBLAST *(17)*	http://www.ncbi.nlm.nih.gov/Structure/cdd/wrpsb.cgi
CDART *(18)*	http://www.ncbi.nlm.nih.gov/Structure/lexington/lexington.cgi?
BLAST *(19)*	http://www.ncbi.nlm.nih.gov/BLAST/Blast.cgi?
• **"Problematic" region prediction**	
	Unstructured regions
IUPred *(20,21)*	http://iupred.enzim.hu/index.html
DisEMBL *(22)*	http://dis.embl.de/
GlobPlot *(23)*	http://globplot.embl.de/
(for a more complete list see the chapter by Dosztáni and Tompa in this book)	
	Transmembrane regions
TMHMM *(24)*	http://www.cbs.dtu.dk/services/TMHMM-2.0/
	Signal sequence
SignalP *(25)*	http://www.cbs.dtu.dk/services/SignalP/
• **Secondary structure prediction**	
JPREP *(26,27)*	http://www.compbio.dundee.ac.uk/~www-jpred/
PredictProtein *(28)*	http://www.predictprotein.org/
PSIPRED *(29)*	http://bioinf.cs.ucl.ac.uk/psipred/
• **Tertiary structure prediction**	
Phyre *(30)*	http://www.sbg.bio.ic.ac.uk/~phyre
FUGUE *(31)*	http://www-cryst.bioc.cam.ac.uk/fugue/
WURST *(32)*	http://www.zbh.uni-hamburg.de/wurst/

predicted functional domains. In either case, the aim is to increase the chance of obtaining a crystal structure at the cost of limiting the structural information to a part of the protein, and lowering the protein flow rate through the pipeline (N constructs of a single protein instead of N different proteins).

Useful information to consider in designing constructs includes alternative splice variations (Note 5), domain homology searches in combination with secondary and tertiary structure prediction (to help defining domain boundaries), and "problematic" region prediction such as transmembrane domains (to omit in construct). Table 2.1 provides suggested links for accessing this information.

In any case, the selection procedure that follows is unchanged, as it does not distinguish between full-length or designed protein constructs; the target definition is any input sequence.

2.2.2. Selecting the Appropriate Criteria

Translating the project aims into criteria is a critical and subtle process. The vast amount of information that is readily available on protein targets through bioinformatics can retrieve an enormous amount of "interesting" information that is tempting to include in the selection process. It is recommended to restrict the criteria at this stage only to those that are able to remove targets (Note 1).

Figure 2.1 lists example of criteria. They are divided into the three categories mentioned, namely biological significance/impact, structural novelty, and likelihood to crystallize. Brief comments and tips are also given in the following.

The biological significance/impact criteria category segregate targets based on organism origin, amino acid sequence characteristics (e.g., protein family, presence of a particular domain, sequence similarity to human orthologue), and the results of a specific type of assay (e.g., biophysical, biochemical, clinical test). It is recommended to use available functional data if possible, as this can be highly valuable for biological significance and impact. For example, the authors use microarray data to select targets that are likely to have a role in inflammation (Note 6). Also, if the target proteins are not from *Homo sapiens*, assessing target relevance in human biology may be assessed based on sequence similarity. For example, the authors select those proteins with a minimum of 70% sequence identity with human proteins (probable human orthologues).

The structural novelty criterion is based on sequence alignment with proteins from the PDB. The 30% identity threshold is based on the assumption that a higher sequence identity enables homology modeling, thus reducing the value of structure determination. Although the percentage of sequence identity with known structures is indicative of structural novelty, more powerful tools such as threading (Table 2.1) based on tertiary structure prediction alignment can be used when obtaining new folds is of particular concern.

The "likelihood to crystallize" section is composed of one filter (no transmembrane helix) followed by a series of parameters (sorters) used to estimate the crystallization likelihood (method detailed in Section 2.1.2.). Although transmembrane sequences have a large detrimental effect on solubility and crystallization *(7)*, this criterion is optional in the Web-based tool to allow for the specific study of membrane proteins.

Finally, changing the order of application of these criteria does not modify the output; however, it does affect computer-processing time. Therefore, it is recommended to apply the less time-consuming criteria first.

2.2.3. Using the Web-Based Target Ranking Tool

The Web-based resource (UQSG Target Ranker, available at http://foo.maths. uq.edu.au/~huber/UQSG/ranker.pl), is a tool for automatically selecting and sorting targets based on predicted likelihood of crystallization. The inputs are target sequences and selected criteria; the output is a list of selected targets that are ranked according to their predicted crystallization likelihood estimates. A variety of sequence-based criteria have been proposed to predict the likelihood for a protein to crystallize *(7–10)* and the five most predictive ones have been implemented in the authors' Web server (see Section 2.1.2. for details). This tool gives the user the option to rank target sequences supplied in FASTA format according to all or individually selected parameters.

3. Notes

1. Why filter rather than rank? The use of a ranked grading system (e.g., 1: very bad; 2: bad; 3: OK; 4: good; 5: very good) leads to target comparisons based on arbitrary grades and grade combinations, resulting in inconsistent selection of targets. Instead, the authors use filter-type criteria (0: discard; 1: keep output per criterion), in which each criterion selects independently of the others, allowing unambiguous and precise control over the selection of target proteins. The ranking/sorting is achieved in the following step, on crystallizing likelihood estimates.

2. The length of the protein sequence has a strong influence on protein crystallization. When the size of a protein is too small, generally its thermodynamic stability is marginal and intrinsic thermal motion can inhibit crystallization. Very large proteins are also more difficult to crystallize because they are likely to exhibit higher flexibility and a reduced translational and rotational motion in solution, leading to kinetically inhibited nucleation.

3. The pI of a protein may influence crystallization success. At conditions with pH of the solution equal to the pI of the protein, the net charge on the protein is zero; as a result no overall electrostatic repulsion between protein molecules is present. Standard protein crystallization screens contain conditions optimized to crystallize a "typical" protein with only weakly repulsive (effective) interactions in stock solution. Given a standard (nonoptimized) protein buffer (typically pH 7.0–7.5), choosing proteins within the appropriate pI range, and thus appropriate effective interactions, can be beneficial.

4. Another criterion based on similar rationale as the pI is the percentage of charged amino acids in the sequence. Other physical properties of a protein that are known to influence crystallization, and thus can be beneficial when taken into account to rank targets, are the number of residues in regions of low complexity (associated with disordered regions) and the overall hydropathy.

5. It is useful to consider splice isoforms; nature may have already designed the construct, although one must keep in mind that splice forms may have different functions or may not be functional at the protein level *(11–13)*.

6. Functional data may add great value in terms of biological significance and impact. Microarray technology is particularly suitable for obtaining functional data in high throughput. The greater than three-fold transcriptional regulation upon stimulation criterion in Fig. 2.1 is an example of a criterion used for such experiment, although the threshold is specific to the data (for more details see elsewhere in this volume).

7. A note of caution. Although target selection can help improving the apparent success rate of structure determination, it simultaneously introduces a strong bias to narrow the diversity of proteins for which the structures are being determined. In an uncontrolled extreme, this can lead to a circular process, in which empirical target selection is based on previously successful structure determination, but future structure databases are confined to the large proportion of proteins that have been selected with this same bias.

8. Interestingly, despite many improvements in approaches to select new targets for structural genomics pipelines over the last decade, the success of these methods in practice is not well established. This is partly due to the difficulty of delineating effects as a result of changes in experimental procedures from improvements as a result of targeted selection itself. With experimental procedures becoming more established over time, it will be interesting to monitor the future development of target selection methods.

Acknowledgments

The authors thank Tim Ravasi, Munish Puri, Ian Ross, Tom Alber, and all the members of the macrophage protein group for their feedback and advice. This work was supported by an Australian Research Council (ARC) grant to JLM and BK. BK is an ARC Federation Fellow and a National Health and Medical Research Council Honorary Research Fellow, and NC an Australian Synchrotron Research Program Fellow.

References

1. Berry, I. M., Dym, O., Esnouf, R. M., Harlos, K., Meged, R., Perrakis, A., Sussman, J. L., Walter, T. S., Wilson, J., and Messerschmidt, A. (2006) SPINE high-throughput crystallization, crystal imaging and recognition techniques: current state, performance analysis, new technologies and future aspects. *Acta Crystallogr. D Biol. Crystallogr.* **62**, 1137–1149.

2. Bonanno, J. B., Almo, S. C., Bresnick, A., Chance, M. R., Fiser, A., Swaminathan, S., Jiang, J., Studier, F. W., Shapiro, L., Lima, C. D., Gaasterland, T. M., Sali, A., Bain, K., Feil, I., Gao, X., Lorimer, D., Ramos, A., Sauder, J. M., Wasserman, S. R., Emtage, S., D'Amico, K. L., and Burley, S. K. (2005) New York-Structural GenomiX Research Consortium (NYSGXRC): a large scale center for the protein structure initiative. *J. Struct. Funct. Genomics* **6**, 225–232.

3. Busso, D., Poussin-Courmontagne, P., Rose, D., Ripp, R., Litt, A., Thierry, J. C., and Moras, D. (2005) Structural genomics of eukaryotic targets at a laboratory scale. *J. Struct. Funct. Genom.* **6**, 81–88.

4. Lundstrom, K., Wagner, R., Reinhart, C., Desmyter, A., Cherouati, N., Magnin, T., Zeder-Lutz, G., Courtot, M., Prual, C., Andre, N., Hassaine, G., Michel, H., Cambillau, C., and Pattus, F. (2006) Structural genomics on membrane proteins: comparison of more than 100 GPCRs in 3 expression systems. *J. Struct. Funct. Genom.* Vol 7, Numb 2, pp. 77–91(15).

5. Moreland, N., Ashton, R., Baker, H. M., Ivanovic, I., Patterson, S., Arcus, V. L., Baker, E. N., and Lott, J. S. (2005) A flexible and economical medium-throughput strategy for protein production and crystallization. *Acta. Crystallogr. D Biol. Crystallogr.* **61**, 1378–1385.

6. Su, X. D., Liang, Y., Li, L., Nan, J., Brostromer, E., Liu, P., Dong, Y., and Xian, D. (2006) A large-scale, high-efficiency and low-cost platform for structural genomics studies. *Acta Crystallogr. D Biol. Crystallogr.* **62**, 843–851.

7. Canaves, J. M., Page, R., Wilson, I. A., and Stevens, R. C. (2004) Protein biophysical properties that correlate with crystallization success in Thermotoga maritima: maximum clustering strategy for structural genomics. *J. Mol. Biol.* **344**, 977–991.

8. Goh, C. S., Lan, N., Douglas, S. M., Wu, B., Echols, N., Smith, A., Milburn, D., Montelione, G. T., Zhao, H., and Gerstein, M. (2004) Mining the structural genomics pipeline: identification of protein properties that affect high-throughput experimental analysis. *J. Mol. Biol.* **336**, 115–130.

9. Rupp, B., and Wang, J. (2004) Predictive models for protein crystallization. *Methods* **34**, 390–407.

10. Smialowski, P., Schmidt, T., Cox, J., Kirschner, A., and Frishman, D. (2006) Will my protein crystallize? A sequence-based predictor. *Proteins* **62**, 343–355.

11. Homma, K., Kikuno, R. F., Nagase, T., Ohara, O., and Nishikawa, K. (2004) Alternative splice variants encoding unstable protein domains exist in the human brain. *J. Mol. Biol.* **343**, 1207–1220.

12. Stamm, S., Ben-Ari, S., Rafalska, I., Tang, Y., Zhang, Z., Toiber, D., Thanaraj, T. A., and Soreq, H. (2005) Function of alternative splicing. *Gene* **344**, 1–20.

13. Takeda, J., Suzuki, Y., Nakao, M., Barrero, R. A., Koyanagi, K. O., Jin, L., Motono, C., Hata, H., Isogai, T., Nagai, K., Otsuki, T., Kuryshev, V., Shionyu, M., Yura, K., Go, M., Thierry-Mieg, J., Thierry-Mieg, D., Wiemann, S., Nomura, N., Sugano, S., Gojobori, T., and Imanishi, T. (2006) Large-scale identification and characterization of alternative splicing variants of human gene transcripts using 56,419 completely sequenced and manually annotated full-length cDNAs. *Nucleic Acids Res.* **34**, 3917–3928.

14. Imanishi, T., Itoh, T., Suzuki, Y., O'Donovan, C., Fukuchi, S., Koyanagi, K. O., Barrero, R. A., Tamura, T., Yamaguchi-Kabata, Y., Tanino, M., Yura, K., Miyazaki, S., Ikeo, K., Homma, K., Kasprzyk, A., Nishikawa, T., Hirakawa, M., Thierry-Mieg, J., Thierry-Mieg, D., Ashurst, J., Jia, L., Nakao, M., Thomas, M. A., Mulder, N., Karavidopoulou, Y., Jin, L., Kim, S., Yasuda, T., Lenhard, B., Eveno, E., Suzuki, Y., Yamasaki, C., Takeda, J., Gough, C., Hilton, P., Fujii, Y., Sakai, H., Tanaka, S., Amid, C., Bellgard, M., Bonaldo Mde, F., Bono, H., Bromberg, S. K., Brookes, A. J., Bruford, E., Carninci, P., Chelala, C., Couillault, C., de Souza, S. J., Debily, M. A., Devignes, M. D., Dubchak, I., Endo, T., Estreicher, A., Eyras, E., Fukami-Kobayashi, K., Gopinath, G. R., Graudens, E., Hahn, Y., Han, M., Han, Z. G., Hanada, K., Hanaoka, H., Harada, E., Hashimoto, K., Hinz, U., Hirai, M., Hishiki, T., Hopkinson, I., Imbeaud, S., Inoko, H., Kanapin, A., Kaneko, Y., Kasukawa, T., Kelso, J., Kersey, P., Kikuno, R., Kimura, K., Korn, B., Kuryshev, V., Makalowska, I., Makino, T., Mano, S., Mariage-Samson, R., Mashima, J., Matsuda, H., Mewes, H. W., Minoshima, S., Nagai, K., Nagasaki, H., Nagata, N., Nigam, R., Ogasawara, O., Ohara, O., Ohtsubo, M., Okada, N., Okido, T., Oota, S., Ota, M., Ota, T., Otsuki, T., Piatier-Tonneau, D., Poustka, A., Ren, S. X., Saitou, N., Sakai, K., Sakamoto, S., Sakate, R., Schupp, I., Servant, F., Sherry, S., Shiba, R., Shimizu, N., Shimoyama, M., Simpson, A. J., Soares, B., Steward, C., Suwa, M., Suzuki, M., Takahashi, A., Tamiya, G., Tanaka, H., Taylor, T., Terwilliger, J. D., Unneberg, P., Veeramachaneni, V., Watanabe, S., Wilming, L., Yasuda, N., Yoo, H. S., Stodolsky, M., Makalowski, W., Go, M., Nakai, K., Takagi, T., Kanehisa, M., Sakaki, Y., Quackenbush, J., Okazaki, Y., Hayashizaki, Y., Hide, W., Chakraborty, R., Nishikawa, K., Sugawara, H., Tateno, Y., Chen, Z., Oishi, M., Tonellato, P., Apweiler, R., Okubo, K., Wagner, L., Wiemann, S., Strausberg, R. L., Isogai, T., Auffray, C., Nomura, N., Gojobori, T., and Sugano, S. (2004) Integrative annotation of 21,037 human genes validated by full-length cDNA clones. *PLoS Biol.* **2**, e162.

15. Fink, J. L., Aturaliya, R. N., Davis, M. J., Zhang, F., Hanson, K., Teasdale, M. S., Kai, C., Kawai, J., Carninci, P., Hayashizaki, Y., and Teasdale, R. D. (2006) LOCATE:

a mouse protein subcellular localization database. *Nucleic Acids Res.* **34**, D213–217.

16. Karolchik, D., Baertsch, R., Diekhans, M., Furey, T. S., Hinrichs, A., Lu, Y. T., Roskin, K. M., Schwartz, M., Sugnet, C. W., Thomas, D. J., Weber, R. J., Haussler, D., and Kent, W. J. (2003) The UCSC Genome Browser Database *Nucleic Acids Res.* **31**, 51–54.

17. Marchler-Bauer, A., Anderson, J. B., Cherukuri, P. F., DeWeese-Scott, C., Geer, L. Y., Gwadz, M., He, S., Hurwitz, D. I., Jackson, J. D., Ke, Z., Lanczycki, C. J., Liebert, C. A., Liu, C., Lu, F., Marchler, G. H., Mullokandov, M., Shoemaker, B. A., Simonyan, V., Song, J. S., Thiessen, P. A., Yamashita, R. A., Yin, J. J., Zhang, D., and Bryant, S. H. (2005) CDD: a Conserved Domain Database for protein classification. *Nucleic Acids Res.* **33**, D192–196.

18. Geer, L. Y., Domrachev, M., Lipman, D. J., and Bryant, S. H. (2002) CDART: protein homology by domain architecture. *Genome Res.* **12**, 1619–1623.

19. Altschul, S. F., Madden, T. L., Schaffer, A. A., Zhang, J., Zhang, Z., Miller, W., and Lipman, D. J. (1997) Gapped BLAST and PSI-BLAST: a new generation of protein database search programs. *Nucleic Acids Res.* **25**, 3389–3402.

20. Dosztanyi, Z., Csizmok, V., Tompa, P., and Simon, I. (2005) The pairwise energy content estimated from amino acid composition discriminates between folded and intrinsically unstructured proteins. *J Mol. Biol.* **347**, 827–839.

21. Dosztanyi, Z., Csizmok, V., Tompa, P., and Simon, I. (2005) IUPred: Web server for the prediction of intrinsically unstructured regions of proteins based on estimated energy content. *Bioinformatics* **21**, 3433–3434.

22. Linding, R., Jensen, L. J., Diella, F., Bork, P., Gibson, T. J., and Russell, R. B. (2003) Protein disorder prediction: implications for structural proteomics. *Structure* **11**, 1453–1459.

23. Linding, R., Russell, R. B., Neduva, V., and Gibson, T. J. (2003) GlobPlot: exploring protein sequences for globularity and disorder *Nucleic Acids Res.* **31**, 3701–3708.

24. Krogh, A., Larsson, B., von Heijne, G., and Sonnhammer, E. L. (2001) Predicting transmembrane protein topology with a hidden Markov model: application to complete genomes. *J. Mol. Biol.* **305**, 567–580.

25. Bendtsen, J. D., Nielsen, H., von Heijne, G., and Brunak, S. (2004) Improved prediction of signal peptides: SignalP 3.0. *J. Mol. Biol.* **340**, 783–795.

26. Cuff, J. A., and Barton, G. J. (2000) Application of multiple sequence alignment profiles to improve protein secondary structure prediction. *Proteins* **40**, 502–511.

27. Cuff, J. A., Clamp, M. E., Siddiqui, A. S., Finlay, M., and Barton, G. J. (1998) JPred: a consensus secondary structure prediction server. *Bioinformatics* **14**, 892–893.

28. Rost, B., and Liu, J. (2003) The PredictProtein server. *Nucleic Acids Res.* **31**, 3300–3304.

29. Jones, D. T. (1999) Protein secondary structure prediction based on position-specific scoring matrices. *J. Mol. Biol.* **292**, 195–202.

30. Kelley, L. A., MacCallum, R. M., and Sternberg, M. J. (2000) Enhanced genome annotation using structural profiles in the program 3D-PSSM. *J. Mol. Biol.* **299**, 499–520.

31. Shi, J., Blundell, T. L., and Mizuguchi, K. (2001) FUGUE: sequence-structure homology recognition using environment-specific substitution tables and structure-dependent gap penalties. *J. Mol. Biol.* **310**, 243–257.

32. Torda, A. E., Procter, J. B., and Huber, T. (2004) Wurst: a protein threading server with a structural scoring function, sequence profiles and optimized substitution matrices. *Nucleic Acids Res.* **32**, W532–535.

Chapter 3

Target Selection: Triage in the Structural Genomics Battlefield

James Raftery

A synopsis of some of the approaches to protein target selection is given, with an emphasis on using web resources to converge on well-ordered, readily crystallizable proteins that are maximally different from known structures. This is illustrated with the genomes of the pathogens causing tuberculosis and sleeping sickness.

1. Introduction

The original aim of structural genomics was for that for every gene there should be a structure. Experience from the structural genomics groups suggest that less than 30% of proteins expressed in soluble form in *E. coli* form crystals, of which only a portion yield x-ray quality specimens diffracting to 2.0 Å resolution or better *(1)*. The aim becomes a matter of trying to ensure your structure(s) are in that group (or solved by NMR!).

The governing principles of the process are:

• Inspired by Fermat's principle, the path of least work
• The expressed gene should result in a structure
• New information is maximized

These very general principles should provide a useful framework even when violated. This perspective will be from that of a small group without the resources to devote to extensive local bioinformatics facilities.

1.1. Biological Systems

The biological systems that are examined in detail are tuberculosis (TB) and, to a lesser extent, sleeping sickness.

1.1.1. Tuberculosis
• Family: Mycobacteriaceae
• Order: Actinomycetales

Tuberculosis, caused by *Mycobacterium tuberculosis*, and killing 2 to 3 million people per year, ranks among the world's most deadly infectious diseases.

From: *Methods in Molecular Biology, Vol. 426: Structural Proteomics: High-throughput Methods*
Edited by: B. Kobe, M. Guss and T. Huber © Humana Press, Totowa, NJ

One third of the world's population may carry a latent dormant TB infection, and in roughly 10% the disease will become active. Its synergy with HIV has increased its interest.

Relevant genomes are:

- *Mycobacterium tuberculosis (2)*
- *Mycobacterium leprae (3)*

1.1.2. Sleeping Sickness

- Family: Trypanosomatida
- Order: Kinetoplastida

African trypanosomiasis, caused by an extracellular eukaryotic flagellate protozoa of genus *Trypanosoma* (mainly *brucei gambiense/rhodesiense*), is a parasitic disease in human and other animals. Transmitted by the tsetse fly, the disease is endemic in regions of sub-Saharan Africa, covering about 36 countries and 60 million people. It is estimated that 300,000–500,000 people are infected, and about 40,000 die every year.

Some of the genomes to consider are:

- The pathogen(s): *Trypanosoma brucei gambiense/rhodesiens*
- Related Trypanosomatid parasitic protozoa: *Trypanosoma brucei brucei*, very similar to *Trypanosoma brucei rhodesiense* but nonpathogenic (in humans), *Trypanosoma cruzi*, and *Leishmania major*
- The vector *Glossina morsitans morsitans* (due 2007)
- The vector's obligate symbiote(s) *Wigglesworthia glossinidia, Sodalis glossinidius, Wolbachia pipientis* (ordered by degree of necessity)
- Host: *Homo sapiens*

Direct attacks on trypanosomiasis with approaches such as vaccination have failed against the extensive antigenic variation trypanosomes display in their mammalian host, whereas the pharmaceutical approach to disease management can prove expensive and with both host and pathogen being eukaryotic, problematic. The recent *(4)* analysis of the common genes of the trypanosomatids confirms that they are incapable of *de novo* purine synthesis as 9 of the 10 genes required to make inosine monophosphate (IMP), the "root" of the biosynthesis, from phosphoribosyl pyrophosphate (PRPP) are absent; no doubt other "weaknesses" will emerge but it is likely that a combination of methods will prove necessary *(5)*.

1.2. Comparative Genomics

Comparisons among genomes allow the identification of the insertions, deletions, mutations and horizontal gene transfers (HGTs) that result in the differing phenotypes of strains. More interestingly, as the distance between the genomes increases, the high level of evolutionary conservation of proteins becomes apparent, especially in prokaryotes and Archaea, with a large proportion of genes having homologues in distant genomes. The functions of many of these genes in newly sequenced genomes may be predicted by transfer of functional information from those of better-studied organisms, but with the caveat in mind of possible evolutionary divergence (i.e., genes being paralogues, not orthologues).

Dissection of protein functions may also be possible for multifunctional proteins. Vertebrate diversification probably featured two or more genome duplications during early chordate evolution. Fish may have undergone further genome duplication after the divergence of fish and tetrapods *(6)*, making the recent *Fugu rubripes* genome sequencing very useful, especially because it is a very condensed genome. The footprints of these duplications are seen in the tendency for genes found as single copies in invertebrates to be multiple copies in vertebrates, and for genes with several functions in mammals to have a fish gene devoted to each function *(7)*.

Phylogenetic patterns, examination of gene fusions, analysis of conserved gene strings, and inferences such as reconstruction of metabolic pathways and subcellular localization *(8,9)* are all enabled using comparative genomics.

Although the emphasis has been on proteins, a recent paper identifying the human accelerated region 1 *(10)*—a rapidly changing part of the genome in humans compared with chimpanzee, where it is very similar to mouse and chicken, as part of an RNA gene—is a timely reminder that conceptually at least, much of the genome is *terra incognita*.

1.2.1. Glossary

- COGs: Clusters of orthologous groups of proteins. Establish by doing all-against-all genome sequence comparisons; keep hits if more similar than any intragenomic hit (i.e., candidate orthologue); if hits from more than two phylogenetic lineages, it is a COG *(11)*.
- Orthologues: Direct evolutionary counterparts derived by vertical descent from a gene of the last common ancestor of the compared species.
- Paralogues: Products of gene duplications and freer to assume new roles.
- Synteny: Closely related genomes tend to have the same genes in the same order. Examination of run sequences can point to missing genes, HGTs, and generally disruptive genetic events. A mouse–human comparison can be seen at http://www.softberry.com/berry.phtml?topic=human-mouse&prg=none

1.3. Filters

1.3.1. Connectivity

With proteins as nodes, and interactions as lines the proteome can be represented as a graph *(12)*. It is found that the probability, *p*, of a connection number k is given by:

$$P(k) \cong 1/k^n : n = 2 - 3$$

that is, the "scale-free network" form in which a small number of proteins (hubs) interact with a lot of other proteins. This confers resistance to random mutation while having a few very vulnerable genes. Consequently genes with fewer links can mutate without unduly damaging the viability of an organism, "seeking" the mutations that will lead to a competitive advantage over rival organisms. Knockout studies on *S. cerevisiae* confirm this analysis with a correlation between lethality and connectivity of 0.75. Proteins with fewer than six connections constitute 93% of the total and 21% of the knockouts are lethal, whereas proteins with more than 15 connections constitute 0.7% and 62% are lethal *(13)*.

There are (at least) three sets of genes to be extracted from this analysis: the hubs, the genes that interact with a hub or hubs and authorities, and genes that interact with a high proportion of hubs.

To construct the network is a task of much labor and most readily available data are related to *S. cerevisiae*. Approaches such as the Rosetta Stone strategy *(14)* and allied approaches *(15,16)*, analysis of conserved binding modes (CBMs) in domains *(17)* and network alignment *(18)* combined with technical advances, especially in microarrays *(19)*, should lighten that labor as the (genomic) sequence and structural databases become larger. These are steps in the emergence of comparative interactomics *(20)* and a framework for thinking about setting upper limits to the number of types of interactions *(21)*.

Obvious elaborations consider the direction of interaction and the spatiotemporal distribution of connections (i.e., have regard to the cell not being a bag with all proteins expressed everywhere all the time). This leads naturally to systems biology, with all the advantages and otherwise that an *in silico* approach *(22)* (www.sys-bio.org) may bring.

As well as general interaction databases such as the Database of Interacting Proteins *(23)*, host–pathogen interaction databases are available (e.g., http://www.ncbi.nlm.nih.gov/RefSeq/HIVInteractions/index.html).

1.3.2. Crystallizability

In the "natural" state, it is usually undesirable for proteins to crystallize; therefore, it has been suggested that it should be selected against. The more concentrated the protein and the presence of structure features such as CBMs, the more severe the selective pressure. Because a necessary condition for crystallization is interaction—specifically surface interaction—surface amino acid residues with floppy side chains, to confer high entropy, are favored. Therefore, mutating K or E $\rightarrow$ A should increase the crystallizability; indeed, experimental evidence suggests that this is effective *(24)*. The Surface Entropy Reduction Prediction Server using these principles can be found at http://nih-server.mbi.ucla.edu/SER/.

Another common and older strategy is to move to a related naturally occurring protein *(25)*.

1.3.3. Disorder

At 33% prevalence in eukaryotes, admittedly much more than in the other domains, an association reflecting both compartmentalization (permitting protection from the degradation usual for disordered proteins) and their utility in nonhousekeeping roles more important in eukaryotes, this is not a negligible factor *(26)*. Its presence may prompt examination of the protein's connectivity to find another protein to form a complex with and possibly obviate the problem.

A straightforward method to predict whether a given protein sequence assumes a defined fold or is intrinsically unfolded has been proposed *(27)*, implemented, and made available at http://bioportal.weizmann.ac.il/fldbin/findex. It is based on $<H>$, the sum of the residue hydrophobicities, using the Kyte/Doolittle scale, rescaled to a range of 0–1, divided by the total number of residues, and on $|<R>|$, the absolute value of the difference between the numbers of positively and negatively charged residues at pH 7.0, divided by the total residue number. Using these two parameters as axes, proteins determined experimentally to be folded or intrinsically unfolded are separated by a straight boundary line.

The equation of the boundary line separating "folded" from "disordered" proteins is transformed into an index, shown below, that discriminates between folded and intrinsically unfolded proteins.

$$FoldIndex = 2.785\langle H\rangle - |\langle R\rangle| - 1.151$$

All positive values represent proteins (or domains) likely to be folded, and negative values represent those likely to be intrinsically unfolded *(28)*.

Another approach that seeks to abstract (dis)order tendencies from aligned PDB sequences using a neural net can be found in the Regional Order Neural Network *(29)* at http://www.strubi.ox.ac.uk/RONN.

1.3.4. Model Building

Being able to build a model of a protein using homology represents a slight conflict with novelty, but is increasingly attractive as fold space gets filled in. New domain combinations may square the circle.

1.3.5. Novelty

At the most general level this can be established by running the sequence, probably using BLAST, against a sequence database such as SwissProt. A more focused definition runs against the sequences of known (i.e., deposited with the PDB) structures, only keeping those that are sufficiently different. Difference can be defined by E, a parameter introduced by Altschul *(30)* to indicate how likely an alignment could be achieved by chance given the database being searched. A more sophisticated variation on this seeks to relate to known structural superfamilies by estimating the distance from them *(31)*. This is available at http://www.protarget.cs.huji.ac.il/. The underlying philosophy is similar to the Pfam5000 strategy of sampling weighted fold space to facilitate structure solution *(32)*.

A different way of regarding novelty is by using a probe such as the Conserved Domain Architecture Retrieval Tool at http://www.ncbi.nlm.nih. gov/Structure/lexington/lexington.cgi. Reverse position specific BLAST uses a protein sequence query to search a precalculated position-specific score matrices (as generated by PSI-BLAST) database of known domains, to deduce its domain structure. This domain architecture is then used to query the CDART database to find proteins with similar domain architecture, if any *(33)*.

The previous methods have attempted to match sequence against sequence, but another approach is to match sequence against structure. This is the tactic used in Directional Atomic Solvation EnergY *(34)*, which threads the sequence through known folds. It can be found at http://www.doe-mbi.ucla.edu/ Services/FOLD/ on the end of a pipeline of sequence-sequence methods such as Sequence Derived Properties *(35)*. Because it is rather time consuming, it is probably best used after some preselection.

1.3.6. Gene Overlap

The role of gene overlap is not clear, but it almost certainly is more than the original explanation of efficient use of nucleic acid, based on the initial discovery in viruses. It has been suggested that the roots lie deep in evolutionary history, and the existence of two classes of aminoacyl tRNA synthetases reflects this *(36)*. Whatever the reason(s) in viruses, it turns out to be both very common in microbes, with about one third of genes being over-lapped and not correlated with genome compactness *(37)*. Naturally, being overlapped produces restraints on change as at least two open reading frames (ORFs) have to be satisfied; if an orthologue overlaps it is probable that it does so with another orthologue.

1.3.7. Transmembrane Protein Potential

Membrane proteins are challenging to crystallize; therefore, they are generally best avoided. Various servers look for characteristic features in the sequence such as membrane spanning alpha helices. An example using the hidden Markov model approach *(38)* is to be found at http://www.cbs.dtu.dk/services/ TMHMM/ and has the advantage of allowing multiple sequences to be submitted in one run and output of one-line analysis summaries for each sequence.

1.4. Computational Considerations

Servers are accessed in a serial manner and the output analyzed. Enthusiasm wanes when this has to be done hundreds of times. Many sites recognize the problem and are making things easier for their users. For example, the disorder server at the Weizmann Institute allows XML output of the required data, considerably easing the subsequent analysis. A common approach is to automate the process and "screen scrape." This often involves accessing the server, not with a browser but with a script, usually Perl. An example available from the Weizmann Institute is shown in the following:

```perl
#!/usr/bin/perl
$| = 1;
my $VERSION = "1.02"; my $AUTHOR = "Prilusky"; my $YEAR = "2003";
use LWP::Simple;
1. # simple example to retrieve only the foldindex value
my @lines = <>;
my $sequence = qq{@lines};
my ($numres, $findex) = getFoldIndex($sequence);
print "Res = $numres\n";
print "FoldInd = $findex\n";
sub getFoldIndex {
my($aa) = @_;
my $content = get("http://bioportal.weizmann.ac.il/fldbin/findex?m=xml
    &sq=$aa");
my ($findex) = $content = ~ /<findex>([\-\.\d]+)<\/findex>/;
my ($numresidues) = $content = ~ /<numresidues>([\-\.\d]+)<\/numresi-
    dues>/;
return ($numresidues, $findex);
}
```

Indeed, for those comfortable with Perl, a very good first step is to download the BioPerl Toolkit *(39)*. It is a good idea to work in a UNIX environment; Cygwin is an excellent UNIX emulator for MS Windows users.

1.5. Data Management

Beyond a certain number of sequences, data management begins creeping up the priority list. The options are to buy a more or less general laboratory information management system (LIMS), obtain an open source version, or create your own. Collaborative Computational Project, Number 4 *(40)* is supporting the Protein Information Management System (PIMS) project, which aims to be a LIMS to meet the needs of the structural biology community. The web site http://www.pims-lims.org gives an idea of how complex things can become.

2. Materials

1. BLAST and associated files were downloaded from ftp://ftp.ncbi.nih.gov (*see* Notes 1 and 2).
2. Sequences and associated data were downloaded from The Institute for Genomic Research (TIGR) web site http://www.tigr.org/ as well as NCBI.
3. COGs analyses, when done externally, used the NCBI server at http://www.ncbi.nlm.nih.gov/COG/.
4. HTMLess was downloaded from http://www.oz.net/~sorth/.
5. Transmembrane tendencies were checked using the TMHMM server at http://www.cbs.dtu.dk/services/TMHMM/.
6. Disorder propensities were checked with the FoldIndex server at http://bioportal.weizmann.ac.il/fldbin/findex.
7. http://iubio.bio.indiana.edu/treeapp/treeprint-form.html produced the phenogram from data generated at http://www.igs.cnrs-mrs.fr/Tcoffee/tcoffee_cgi/index.cgi?stage1=1&daction=TCOFFEE::Regular.

3. Methods

3.1. Tuberculosis

Without access to gene knockout results to analyze the *M. tuberculosis* genome into nonessential and pathogenicity/viability genes, it was decided to exploit a natural experiment, the *M. leprae* genome; to a first approximation this is the *M. tuberculosis* genome after massive gene knockout (from 3,927 proteins to 1,605). It is almost as if *M. leprae* were an escaped obligate endosymbiont. This approach also neatly bypasses one of the problems of laboratory knockouts, namely that many knockouts have no distinctive phenotype (i.e., knockouts can appear to produce no effect), but it is possible that the right environment had not been chosen to show its necessity. As *M. leprae* has survived in the wild as a pathogenic organism, there is a built-in reality check.

1. Keeping common COGs:

$$A = M. \textit{tuberculosis} .AND. M. \textit{leprae} (915 \text{ COGs})$$

2. Originally *(41)*, to look at genes required by *M. tuberculosis* and organisms following similar strategies and not just those that every organism must have, the group with the smallest genome *(42)*, the Mycoplasmas, was used to remove the minimal set of genes necessary and sufficient for sustaining a functional cell. This is very much a broad brush approach, as for most essential cellular functions, two or more unrelated or distantly related proteins have evolved with only about 60 proteins, primarily those involved in translation, common to all cellular life *(43)*. To focus more tightly on

the core genome, the COGs present in all phyla that are also present in *M. genitalium* was utilized to arrive at 89 COGs.

$$B = A . AND. .NOT. (All\_phyla. AND. M. genitalium) (826 COGs)$$

3. Next, since part of the survival strategy of *M. tuberculosis* is the ability to survive in a vacuole as an intracellular inhabitant of macrophage and this is shared with Chlamydia:

$$C = B .AND. Chlamydia \ trahomatis \ (299 \ COGs)$$

4. To preempt any possible problems of a pharmaceutical nature, the COGs were expanded to 540 proteins and any with homologues in *H. sapiens* were removed using an $E > 10^{-20}$ criterion.

$$D = C. AND. .NOT. Hsapiens \ (373 \ proteins)$$

5. Those with related entries in the PDB($E < 0.001$) were removed, leaving 86.
6. Removing those with transmembrane tendencies left 50 proteins.
7. Checking these entries for a propensity to disorder revealed none with such a propensity.
8. Rather than just remove any closely related sequences by running them against each other, a phenogram was produced from Newick format data generated from the sequences. This allows more or less similar sequences to be selected as required.

3.2. Sleeping Sickness

The examination of this system is very much in the preliminary stages, but it does exemplify the opportunities to exploit the tide of genome sequences now being generated. At the moment the focus is on *Wigglesworthia glossinidia*, which appears to be the result of reductive evolution from an ancestor shared with *E. coli*. When the tsetse genome is completed it will be of interest to see how many originally *Wigglesworthia* genes have migrated to the tsetse nucleus. Almost certainly the "missing" gene coding for the DNA replication initiation protein, DnaA, will be there, providing a useful control on replication. *Wigglesworthia* has retained the ability to synthesize assorted vitamins and metabolites, including the vitamin B complex, lipoic acid, and protoheme, which are known to be low in the diet of tsetse flies, vertebrate blood, supporting indications from dietary supplementation experiments that providing vitamins plays a central role in this mutualistic relationship *(44,45)*.

1. Using the curation, biosynthesis of cofactors, prosthetic groups, and carrier gene sequences were downloaded from TIGR site and duplicates removed. In retrospect it would have been easier to download the whole genome and do the selection locally. At this stage the proteins number 90.
2. Comparing the sequences using a local copy of BLAST with PDB sequences left 14/90 proteins (see Note 3).
3. Checking for disorder propensity left 84/90 proteins.
4. Checking for transmembrane tendencies left 85/90 proteins; 2/90 additional hits were probably signal sequences.
5. The total effect was to leave 10/90 proteins. Using BLAST to run the 10 sequences against each other did not reveal any close relationships.

At a later stage the rest of the genome will be looked at in the light of the completed host genome and the adaptions made by the host organism, such as the specialized cells to accommodate the bacteria (bacteriocytes), which form a dedicated organ, the bacteriome, to fill in detail on the spectrum from pathogenic to mutualistic endosymbiosis.

3.3. Controlling Chaos

After sequences of interest have been garnered, an information management system becomes very useful. For the impecunious the natural approach is the Linux Apache MySQL P/HP/erl/ython approach. At this early stage only a few tables have been created for MySQL, but as work progresses they will be extended and incorporated into an open source LIMS such as PIMS.

4. Notes

1. Sometimes downloading from servers goes less than perfectly and you end up with an html file. In that case, a utility such as HTMLESS, which converts html to text using html formatting, is very useful.
2. Many nets are protected by a firewall that does not allow FTP. It would be helpful if more servers allowed http, certainly for the smaller downloads. Otherwise, install a proxy. A good ftp web server is http://www.net2ftp. com; however, it has a daily limit on the size of downloads. Less easy to circumvent are web servers using nonstandard ports. It is best to send a productive email requesting the standard port; otherwise, request access to the firewall rules.
3. If XML is chosen as output from BLAST(−m7) for convenience and ease of analysis, note that there is a default E,10, and sequences > E are ignored in the output. Define the threshold so that hits are only the (un)wanted sequences.

Acknowledgments

The author thanks John R. Helliwell for detailed discussions. This chapter is a contribution to the UK North West Structural Genomics Consortium (NWSGC) coordinated by S. S. Hasnain, CCLRC Daresbury Laboratory, UK.

References

1. Dale, G. E., Oefner, C., and D'Arcy, A. (2003) *J. Struct. Biol.* **142**, 88–97.
2. Cole, S. T., Brosch, R., Parkhill, J., Garnier, T., Churcher, C., Harris, D., Gordon, S. V., Eiglmeier, K., Gas, S., Barry, C. E., III, Tekaia, F., Badcock, K., Basham, D., Brown, D., Chillingworth, T., Connor, R., Davies, R., Devlin, K., Feltwell, T., Gentles, S., Hamlin, N., Holroyd, S., Hornsby, T., Jagels, K., Krogh, A., McLean, J., Moule, S., Murphy, L., Oliver, K., Osborne, J., Quail, M. A., Rajandream, M. A., Rogers, J., Rutter, S., Seeger, K., Skelton, J., Squares, R., Squares, S., Sulston, J. E., Taylor, K., Whitehead, S., and Barrell, B. G. (1998) *Nature* **393**, 537–544.
3. Cole, S. T., Eiglmeier, K., Parkhill, J., James, K. D., Thomson, N. R., Wheeler, P. R., Honore, N., Garnier, T., Churcher, C., Harris, D., Mungall, K., Basham, D.,

Brown, D., Chillingworth, T., Connor, R., Davies, R. M., Devlin, K., Duthoy, S., Feltwell, T., Fraser, A., Hamlin, N., Holroyd, S., Hornsby, T., Jagels, K., Lacroix, C., Maclean, J., Moule, S., Murphy, L., Oliver, K., Quail, M. A., Rajandream, M. A., Rutherford, K. M., Rutter, S., Seeger, K., Simon, S., Simmonds, M., Skelton, J., Squares, R., Squares, S., Stevens, K., Taylor, K., Whitehead, S., Woodward, J. R., and Barrell, B. G. (2001) *Nature* **409**, 1007–11.

4. El-Sayed, N. M., Myler, P. J., Blandin, G., Berriman, M., Crabtree, J., Aggarwal, G., Caler, E., Renauld, H., Worthey, E. A., Hertz-Fowler, C., Ghedin, E., Peacock, C., Bartholomeu, D. C., Haas, B. J., Tran, A.-N., Wortman, J. R., Alsmark, U. C. M., Angiuoli, S., Anupama, A., Badger, J., Bringaud, F., Cadag, E., Carlton, J. M., Cerqueira, G. C., Creasy, T., Delcher, A. L., Djikeng, A., Embley, T. M., Hauser, C., Ivens, A. C., Kummerfeld, S. K., Pereira-Leal, J. B., Nilsson, D., Peterson, J., Salzberg, S. L., Shallom, J., Silva, J. C., Sundaram, J., Westenberger, S., White, O., Melville, S. E., Donelson, J. E., Andersson, B., Stuart, K. D., and Hall, N. (2005) *Science* **309**, 404–409.

5. Aksoy, S., O'Neill, S. L., Maudlin, I., Dale, C., and Robinson, A. S. (2001) *Trends Parasitol.* **17**, 29–35.

6. Meyer, A., and Peer, Y. V. d. (2005) *BioEssays* **27**, 937–945.

7. Cottage, A. J., Edwards, Y. J. K., and Elgar, G. (2003) *Mammal. Gen.* **14**, 514–525.

8. Marcotte, E. M., Pellegrini, M., Ng, H.-L., Rice, D. W., Yeates, T. O., and Eisenberg, D. (1999) *Science* **285**, 751–753.

9. Marcotte, E. M., Xenarios, I., van der Bliek, A. M., and Eisenberg, D. (2000) *PNAS* **97**, 12115–12120.

10. Pollard, K. S., Salama, S. R., Lambert, N., Lambot, M.-A., Coppens, S., Pedersen, J. S., Katzman, S., King, B., Onodera, C., Siepel, A., Kern, A. D., Dehay, C., Igel, H., Ares, M., Vanderhaeghen, P., and Haussler, D. (2006) *Nature* **443**, 167–172.

11. Tatusov, R. L., Koonin, E. V., and Lipman, D. J. (1997) *Science* **278**, 631–637.

12. Albert, R., and Barabási, A.-L. (2002) *Rev. Mod. Phys.* **74**, 47.

13. Jeong, H., Mason, S. P., Barabasi, A. L., and Oltvai, Z. N. (2001) *Nature* **411**, 41–42.

14. Eisenberg, D., Marcotte, E. M., Xenarios, I., and Yeates, T. O. (2000) *Nature* **405**, 823–826.

15. Lee, I., Date, S. V., Adai, A. T., and Marcotte, E. M. (2004) *Science* **306**, 1555–1558.

16. Sprinzak, E., Altuvia, Y., and Margalit, H. (2006) *PNAS* **103**, 14718–14723.

17. Shoemaker, B. A., Panchenko, A. R., and Bryant, S. H. (2006) *Protein Sci.* **15**, 352–361.

18. Flannick, J., Novak, A., Srinivasan, B. S., McAdams, H. H., and Batzoglou, S. (2006) *Genome Res.* **16**, 1169–1181.

19. Quackenbush, J. (2003) *Science* **302**, 240–241.

20. Uetz, P., and Pankratz, M. J. (2004) *Nat. Biotech.* **22**, 43–44.

21. Aloy, P., and Russell, R. B. (2004) *Nat. Biotech.* **22**, 1317–1321.

22. Kitano, H. (2002) *Science* **295**, 1662.

23. Salwinski, L., Miller, C. S., Smith, A. J., Pettit, F. K., Bowie, J. U., and Eisenberg, D. (2004) *Nucl. Acids Res.* **32**, D449–451.

24. Derewenda, Z. S. (2004) *Structure* **12**, 529–535.

25. Kendrew, J. C., Parrish, R. G., Marrack, J. R., and Orlans, E. S. (1954) *Nature* **174**, 946–949.

26. Fink, A. L. (2005) *Curr. Opin. Struct. Biol.* **15**, 35–41.

27. Uversky, V. N., Gillespie, J. R., and Fink, A. L. (2000) *Prot. Struct. Funct. Genet.* **41**, 415–427.

28. Prilusky, J., Felder, C. E., Zeev-Ben-Mordehai, T., Rydberg, E. H., Man, O., Beckmann, J. S., Silman, I., and Sussman, J. L. (2005) *Bioinformatics* **21**, 3435–3438.

29. Yang, Z. R., Thomson, R., McNeil, P., and Esnouf, R. M. (2005) *Bioinformatics* **21**, 3369–3376.
30. Altschul, S. F., Gish, W., Miller, W., Meyers, E. W., and Lipman, D. J. (1990) *J. Mol. Biol.* **215**, 403–410.
31. Kifer, I., Sasson, O., and Linial, M. (2005) *Bioinformatics* **21**, 1020–1027.
32. Chandonia, J.-M., and Brenner, S. E. (2005) *Prot. Struct. Funct. Bioinformat.* **58**, 166–179.
33. Geer, L. Y., Domrachev, M., Lipman, D. J., and Bryant, S. H. (2002) *Genome Res.* **12**, 1619–1623.
34. Mallick, P., Weiss, R., and Eisenberg, D. (2002) *PNAS* **99**, 16041–16046.
35. Fischer, D., and Eisenberg, D. (1996) *Prot. Sci.* **5**, 947–955.
36. Carter, J. C. W., and Duax, W. L. (2002) *Mol. Cell* **10**, 705–708.
37. Johnson, Z. I., and Chisholm, S. W. (2004) *Genome Res.* **14**, 2268–2272.
38. Krogh, A., Larsson, B., von Heijne, G., and Sonnhammer, E. L. L. (2001) *J. Mol. Biol.* **305**, 567–580.
39. Stajich, J. E., Block, D., Boulez, K., Brenner, S. E., Chervitz, S. A., Dagdigian, C., Fuellen, G., Gilbert, J. G. R., Korf, I., Lapp, H., Lehvaslaiho, H., Matsalla, C., Mungall, C. J., Osborne, B. I., Pocock, M. R., Schattner, P., Senger, M., Stein, L. D., Stupka, E., Wilkinson, M. D., and Birney, E. (2002) *Genome Res.* **12**, 1611–1618.
40. Collaborative (1994) *Acta Crystallogr. Sect. D* **50**, 760–763.
41. Raftery, J., and Helliwell, J. R. (2002) *Acta Crystallogr. Sect. D* 58, 875–877.
42. Fraser, C. M., Gocayne, J. D., White, O., Adams, M. D., Clayton, R. A., Fleischmann, R. D., Bult, C. J., Kerlavage, A. R., Sutton, G., Kelley, J. M., Fritchman, J. L., Weidman, J. F., Small, K. V., Sandusky, M., Fuhrmann, J., Nguyen, D., Utterback, T. R., Saudek, D. M., Phillips, C. A., Merrick, J. M., Tomb, J.-F., Dougherty, B. A., Bott, K. F., Hu, P.-C., and Lucier, T. S. (1995) *Science* **270**, 397–404.
43. Koonin, E. V. (2003) *Nat. Rev. Microbiol.* **1**, 127–136.
44. Rio, R. V. M., Hu, Y., and Aksoy, S. (2004) *Trends Microbiol.* **12**, 325–336.
45. Zientz, E., Dandekar, T., and Gross, R. (2004) *Microbiol. Mol. Biol. Rev.* **68**, 745–770.

Chapter 4

Data Management in Structural Genomics: An Overview

Sabrina Haquin, Eric Oeuillet, Anne Pajon, Mark Harris, Alwyn T. Jones, Herman van Tilbeurgh, John L. Markley, Zolt Zolnai, and Anne Poupon

Data management has been identified as a crucial issue in all large-scale experimental projects. In this type of project, many different persons manipulate multiple objects in different locations; thus, unless complete and accurate records are maintained, it is extremely difficult to understand exactly what has been done, when it was done, who did it, and what exact protocol was used. All of this information is essential for use in publications, reusing successful protocols, determining why a target has failed, and validating and optimizing protocols. Although data management solutions have been in place for certain focused activities (e.g., genome sequencing and microarray experiments), they are just emerging for more widespread projects, such as structural genomics, metabolomics, and systems biology as a whole. The complexity of experimental procedures, and the diversity and high rate of development of protocols used in a single center, or across various centers, have important consequences for the design of information management systems. Because procedures are carried out by both machines and hand, the system must be capable of handling data entry both from robotic systems and by means of a user-friendly interface. The information management system needs to be flexible so it can handle changes in existing protocols or newly added protocols. Because no commercial information management systems have had the needed features, most structural genomics groups have developed their own solutions. This chapter discusses the advantages of using a LIMS (laboratory information management system), for day-to-day management of structural genomics projects, and also for data mining. This chapter reviews different solutions currently in place or under development with emphasis on three systems developed by the authors: Xtrack, Sesame (developed at the Center for Eukaryotic Structural Genomics under the US Protein Structural Genomics Initiative), and HalX (developed at the Yeast Structural Genomics Laboratory, in collaboration with the European SPINE project).

1. Introduction

The genome sequencing centers constituted the first large-scale experimental biological projects, and they were among the first centers to attack the problem of experimental data management. These projects quickly became automated,

From: *Methods in Molecular Biology, Vol. 426: Structural Proteomics: High-throughput Methods*
Edited by: B. Kobe, M. Guss and T. Huber © Humana Press, Totowa, NJ

and robots were soon used for all experimental steps. These projects required little in the way of standardization because it has not been necessary to compare experimental protocols. Currently, the most difficult tasks confronting genome sequencing centers concern the annotation of the genomes; much of this is automated, but hand curation is still mandatory.

A second large-scale biological activity is in the field of chip microarray experiments. Because the results of a microarray experiment cannot be considered without taking into account the experimental protocol, a list of the minimal data to be registered for these experiments has been developed *(1–3)*. MIAME (Minimum Information About Microarray Experiment) has been largely accepted as a standard, and nearly all LIMSs developed for microarray experiments are now "MIAME-compliant" *(4–6)*.

Metabolomics represents a third and newer large-scale activity and another in which experimental protocols are critical. The Metabolomics Standards Initiative recently recommended that metabolomics studies should report the details of study design, metadata, experimental, analytical, data processing, and statistical techniques used *(7)*. Capturing these details is imperative, because they can play a major role in data interpretation *(8,9)*. As a result, informatics resources need to be built on a data model that can capture all of the relevant information while maintaining sufficient flexibility for future development and integration into other resources *(10)*.

In the domain of data management, structural genomics projects share some of the same problems as microarray and metabolomics experiments. In all three cases, many targets are manipulated, and the results must be considered within the context of the experimental conditions. In all such large-scale efforts, data exchange standardization is a necessity *(11,12)*. These issues were discussed in the early meetings that led to the development of the US Protein Structure Initiative, and the Protein Data Bank and BioMagResBank (now partners in the World Wide Protein Data Bank) jointly developed data structures for capturing experimental details of protein production and structure determination by x-ray crystallography and NMR spectroscopy. These PDB standards were made publicly available and were picked up in part by European SPINE project and used in designing its data model *(13)*. The data model has been modified since its publication in 2005 and the latest version is available within the PiMS project (http://www.pims-lims.org). The data model the authors have designed allows the registration of experimental data from any protocol already in use in the numerous laboratories that have been consulted. It has also been designed to easily accommodate new protocols. Similar standards were incorporated into the TargetDB (Target Registration Database, http://targetdb.pdb.org/), which used by many of the worldwide structural genomics centers, and PEPCdb (Protein Expression, Purification and Crystallization Database, http://pepcdb.pdb.org/), which although publicly available is currently collecting information only from the US Protein Structure Initiative Network. All of these systems are based on data dictionaries that are freely available from the Web. With these standards for data exchange in place, it is possible for different laboratories and centers to exchange data even though their individual LIMS use site-specific data representations. The pairing of experimental data with protocols are critical features of the LIMS systems used in structural genomics. In recognition of this, the PEPCdb ties all information collected from centers in the US Protein Structure Initiative Network to specified protocols.

Projects aiming at developing a LIMS have to keep in mind that laboratory workers will not gladly use a LIMS just because it serves the scientific community; it is important that the LIMS simplifies their day-to-day work and that it becomes essential to their tasks. Because traditional laboratory notebooks cannot be used in a multiuser and multitask environment, a LIMS is an attractive alternative. A LIMS can bridge data flow involving individuals and robots and can provide very useful information to the experimenter carrying out a particular step in a complex pipeline. This chapter presents the rationale for laboratory information systems in structural genomics, and discusses the desirable features of such systems. It presents detailed information about three systems developed in the authors' laboratories: Xtrack, developed at the University of Uppsala, Sweden; Sesame, developed by the Center for Eukaryotic Structural Genomics in the context of the US Protein Structure Initiative *(14)*; and HalX *(15)*, developed in the Yeast Structural Genomics Project *(16,17)* and as part of the SPINE project *(13)*.

2. Rationale and Design Features of a LIMS for Structural Genomics

As stated, there are two main reasons for using a LIMS in structural genomics. The first is to facilitate the day-to-day work of the experimenters. The second is to allow the experimental data to be searched in different ways in the context of information available from other databases via the Web; this facilitates the management of projects and data mining used to evaluate the efficiency of protocols and develop hypotheses to be tested that may lead to improved protocols.

2.1. Limitations of Laboratory Notebooks and Spreadsheets

The laboratory notebook that almost all experimenters use for recording their day-to-day work is not well adapted to structural genomics projects. The main reasons are that persons laboring at each stage of a pipeline need to have access to complete information about the prior history of each target, need to add their information to this history, and need to pass this information to the next stage in the pipeline. Each person deals with a large number of targets over in a short period of time and can benefit from the organization of the task at hand that a LIMS system can offer. Typically, a cloning "campaign" in the Yeast Structural Genomics laboratory, or "workgroup" (Sesame nomenclature) consists of 96 different protein constructs (corresponding to a 96-titer well plate). These may be investigated as multiple constructs (coding for different N- and C-termini), subcloned into multiple vectors, or expressed in different hosts under multiple conditions. Although these tasks are routine from the structural genomics point of view, they generate a large amount of information specific to the exact protocol followed. It is highly unwieldy to capture experiments of this kind and their results in a laboratory notebook. Information can be entered into spreadsheets such as Excel, but because separate spreadsheets are needed for different stages of the pipeline at different locations, the data become fragmented. With multiple spreadsheets, it becomes difficult to trace all of the manipulations carried out with a particular target. Moreover, people working on a single part of the project are overwhelmed with information that is not relevant to their task.

In a typical structural genomics project, many persons work on a given target over its history. These persons often are not at the same site. Multiple targets are operated on by multiple persons and multiple robots at multiple sites. A given instrument can even be programmed to carry out different tasks. For example, a particular pipetting robot might be used for cloning, expression, and small-scale purification. This situation is not compatible with conventional laboratory notebooks or Excel spreadsheets. Even file sharing is problematic, because different persons cannot work with a file at the same time. Moreover, with file sharing, it is difficult to ensure the existence of a single true record of activity. From the preceding discussion, it is clear that structural genomics projects require information management quite different from that used in traditional structural biology projects. As discussed in the following, a well-structured LIMS can provide the solution.

2.2. Design Features of a LIMS for Structural Genomics

The ideal LIMS for structural genomics should have the following properties:

- The system should reside on a commercial-quality relational database management system to take advantage of the security, data organization, robustness, and other features these provide.
- It should have a user-friendly Web-based interface so data can be entered and accessed from multiple sites.
- The interface should provide flexible views of the database configured to individual stages in the pipeline so that persons entering data see only the parts relevant to their task and are not overwhelmed by unnecessary detail.
- The system should manage and facilitate data entry; it should associate entered data with a particular protocol, time stamp the entry and provide the name of the person entering the data or launching the run of a robot or other laboratory instrument. As indicated by the protocol, the outcome of a given step, leads to a decision or "action" that determines the next step to be followed. The system needs to be capable of recording that action and of reorganizing targets that follow different pathways.
- The system must make it possible to create variations on a given ORF (e.g., different domains, different constructs, chemically modified proteins) and to trace the relationships. The system must be capable of dealing with structural targets that consist of two or more different proteins (coexpressed or combined after isolation) or proteins and ligands.
- The system should be capable of flexible configuration by a laboratory manager (rather than software developers) so that the system can be tailored to fit the dynamic needs of a particular laboratory.
- It should be capable of flexible data input from external databases, a variety of laboratory robots, and laboratory personnel.
- The system should allow the user to search through and organize data from multiple files.
- The system should carry out routine calculations and organize information in useful ways; it should be capable of launching large-scale computations carried out on computer clusters or computational grids.
- The system should be capable of launching experiments carried out on complex instrumentation, such as crystallization robots and NMR spectrometers and of importing results achieved.

- The system should manage information about project personnel, laboratory resources (refrigerators and freezers), reagents used by the project, and products of the project (e.g., cloned genes, plasmids, expression hosts, proteins).
- It should create and read bar codes.
- It should have a flexible query interface that makes it possible to mine the data in useful ways.
- The system should be capable of creating reports of the type needed by laboratory personnel and project managers.
- Users should be able to attach a wide variety of files (text, images, data) at relevant positions and to view and manipulate these (e.g., carry out simple image processing) from the user interface.
- The LIMS should be capable of creating valid depositions for relevant databases, such as the TargetDB, the PEPCdb, and the World Wide Protein Data Bank.

2.3. The LIMS as a Project Management Tool

A well-organized LIMS automatically provides information needed by persons carrying out individual steps in a pipeline. For example, at the target selection level, it should be capable of displaying information about potential targets and should be able to sort these according to target selection protocols. At the small-scale expression level, it should carry forward information about the ORFs to be tested, the plasmids and expression hosts to be used as specified by the protocol. At some levels, such as cell growth and protein purification, tasks can be carried out by technicians who have minimal training. Because all operations are logged into the LIMS, it is possible for supervisors to create reports that trace the activities of individual technicians and compare them with results from the project as a whole. This helps to identify problems with training and understanding. The use of a LIMS also ensures that no target has been forgotten at a given stage of the experimental workflow. For example, a project manager might ask the list of proteins that were successfully tested for solubility but did not undergo large-scale production, and what amount of time these have been waiting. A LIMS can help to tackle bottlenecks in the experimental pipeline. If for example proteins spend a long time in the freezer after large-scale production and before purification, this might mean that additional purification hardware is needed.

2.4. The LIMS as a Data Mining Tool

It has often been hypothesized that some sequence features, or different sequence engineering techniques could improve protein expression, solubility, and stability or could lead to better crystals. However, it was never before possible to statistically validate these hypotheses. Indeed, such validation requires having a sufficient number of proteins prepared under the same conditions: using the same vectors, same constructions (tags in particular), same expression systems, and same purification protocols. Structural genomics renders such studies possible. A number of comparative studies emerging from structural genomics centers illustrate how information about the behavior of multiple targets can lead to improved protocols and better methods for predicting whether a given sequence is likely to succeed as a structural target.

Canaves et al. *(18)* presented a study on 539 proteins of *Thermotoga maritima* that had been successfully produced and purified in the Joint Center for Structural Genomics. For each protein, 480 crystallization conditions were tested, and 464 proteins were crystallized, leading to 86 crystal structures. From these results, the authors deduced that a subset of 67 crystallization conditions led to crystals for 86% of the 464 *(19)*. A second paper *(18)* presented correlations between crystallization results and different parameters, such as isoelectric point, sequence length, average hydropathy, or low complexity regions. The results showed that most proteins successfully crystallized could be distinguished on the basis of combinations of these parameters. Thus, these criteria can be used to select for sequences that are more likely to yield crystals and consequently save time and effort.

Another study carried out at the New York Structural Genomics Center *(20)* was based on access to database information about numerous proteins produced and crystallized under same experimental conditions. In this study, the authors used tree-based analyses to discover correlations between properties of the target and its behavior in the different stages of the experimental workflow. They showed that the parameters most predictive of success were existence of homologues in other organisms, the frequencies of charged residues, the grouping of hydrophobic residues, the number of interaction partners, and the length of the protein. Using these parameters to drive target choice should improve the success rate of the whole platform.

The Center for Eukaryotic Structural Genomics has used data collected in its LIMS to carry out three studies of this kind. In the first study, DNA arrays were used to analyze a cDNA library for *Arabidopsis thaliana*, and it was found that detection of the presence of an ORF was a good predictor of the success of cloning attempts *(21)*. CESG now uses available gene chip information in choosing targets. In the second study, a number of software packages were used to predict the degree of disorder on proteins corresponding to Arabidopsis ORFs. The results showed that PONDR predictions *(22)* could be used in target selection to improve the likelihood success *(23)*. In the third study, 96 targets were taken in parallel through CESG's *E. coli* cell–based and wheat germ cell–free pipelines, and the results for expression and solubility were compared. Whenever possible, each target was labeled with nitrogen-15 for analysis by ^{15}N HSQC NMR spectroscopy. The results showed that, in comparison with the *E. coli* cell–based approach, the wheat germ cell–free approach yielded a significantly larger number of soluble protein samples suitable for structural analysis by NMR spectroscopy *(24)*.

These studies did have an impact on the concerned projects, and have contributed to increase the efficiency of the experimental workflow. However, many of the results are, or could be, species specific, and their full demonstration requires statistical analysis on broader data sets. The generalization of LIMS usage, in structural genomics, but also to a wider audience in classical structural biology laboratories is the only possibility that all data can be used.

3. Examples of LIMS Developed for Structural Genomics and Structural Biology

3.1. A Crystallography LIMS: Xtrack

3.1.1. Introduction and Overview

Determining the three-dimensional structure of a protein by x-ray crystallography often involves the collection of a large number of data sets, at different wavelengths, and from multiple crystals. These crystals can be different from one another: native protein, selenomethionine-labeled protein, and crystals soaked in different heavy metals. Structure determination can be a lengthy task and often spreads over a few months. Finally, because the synchrotron where data are collected is frequently far from the laboratory, the people who collect the data are not necessarily those who determine the structure, and more than one person may work on the same structure. The multiuser and multitask problem was present in crystallography before it arose in molecular biology. This is why the Xtrack LIMS focused first on crystallography.

Xtrack *(25)* was originally designed to manage data from crystal to structure. More recent versions can manage crystallization data and contain the chemical compositions of the major crystallization screens. Xtrack can also store data concerning the expression and purification of the samples used in crystallization. Thus, after completion of the structure, all data needed for deposition in the PBD are present in the database. The use of Xtrack is very simple and intuitive. It contains scripts able to read log files from crystallography programs, such as CNS *(26)*, SCALEPACK *(27)*, or REFMAC *(28)*.

3.1.2. Parts and Pages

3.1.2.1. Protein Production and Target Description: The protein production part of Xtrack is separated into three different pages: Cloning, Expression, and Purification. Each page is very simple and allows only basic data entry of the kind required for structure deposition. The data that can be entered are:

- For cloning: the cloning vector, the tag if any, the storage location, the protocol as free text, and notes
- For expression: the organism, the expression system, the vector, the inducer, the temperature, the duration, the OD prior to induction, the storage location, and the protocol as free text and notes
- For purification: the purity, the storage location, the picture of the gel, the protocol as free text, and notes

A Chemistry page provides for entry of various data concerning the protein: mutations, assembly, ligands, molecular weight, isoelectric point (pI), number of cysteines, peptide sequence, and nucleic acid sequence. The protein production pages of Xtrack are available only for structure deposition needs and cannot be used as a LIMS for protein production.

3.1.2.2. Crystallization: Xtrack's crystallization interface handles the setup of plates, the examination of plates, and searching through plates. The plate setup interface offers all needed options for setting up a plate, either for a crystallization screen or screen optimization. The user first chooses a plate type, the number of drops in each well and names the plate. The interface (Fig. 4.1) allows for specification of the crystallization screens or gradients produced from stock solutions. The list of stock solutions can be edited. For convenience, most used crystallization screens are already available for use. The authors have been entering the new screens on demand for the past years. Entry of new screens is an involved manual process, because of the lack of an interface for this and as the result of manufacturers' inconsistencies in the naming of chemicals. The interface permits the user to fully customize the content of the drops by changing the proportions of the protein and reservoir solutions.

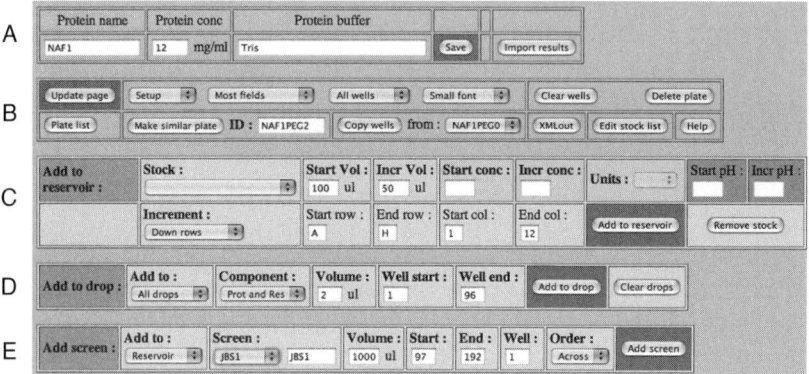

Fig. 4.1 Xtrack plate setup interface. While entering a plate, user defines which protein is used (**A**), and gives the concentration of the solution. The reservoir solutions can be defined either by using a crystallization screen (**E**), or designing gradients (**C**). The volumes of reservoir and protein solutions in the drop(s) are defined using the fields and drop-down menus in (**D**). Finally, the user can choose which data are displayed on the page, and which actions on is these data are editable using the drop-down menus in (**B**).

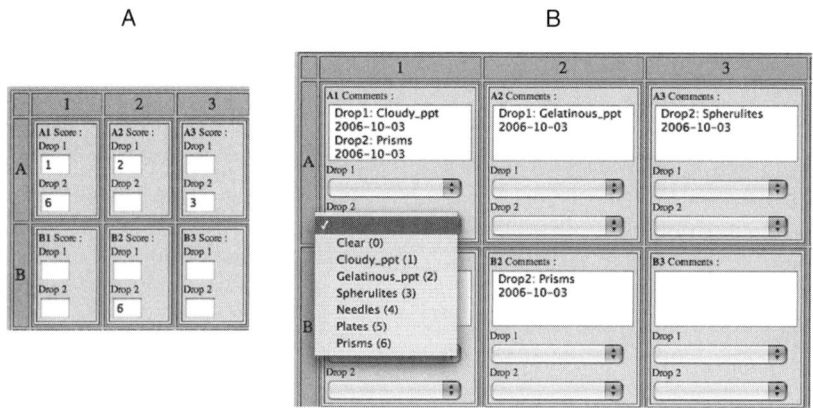

Fig. 4.2 Xtrack observation entry interface. The observations made on a given crystallization plates can be manually entered in two different ways: either using a number (**A**), or entering the corresponding observations with a drop-down menu (**B**). Both interfaces are displayed on the same page, and are synchronized. For example, entering "2" for drop 1, well A2, in the simple interface (**A**), the text "Drop1: Gelatinous ppt" (Gelatinous precipitate) appears in well A2 on the complete interface (**B**).

The plates can then be examined. The user can score the observation directly through the interface, either by providing a numerical code (0: clear; 1: cloudy precipitate; 2: gelatinous precipitate; 3: spherulites; 4: needles; 5: plates; 6: prisms) or by choosing from drop-down menus; the results of these two choices are synchronized (Fig. 4.2). For each drop, a photo can be uploaded. Observations and photos can also be entered as uploaded files. The required file format is very simple, and the output from most observation robots can easily be converted to it.

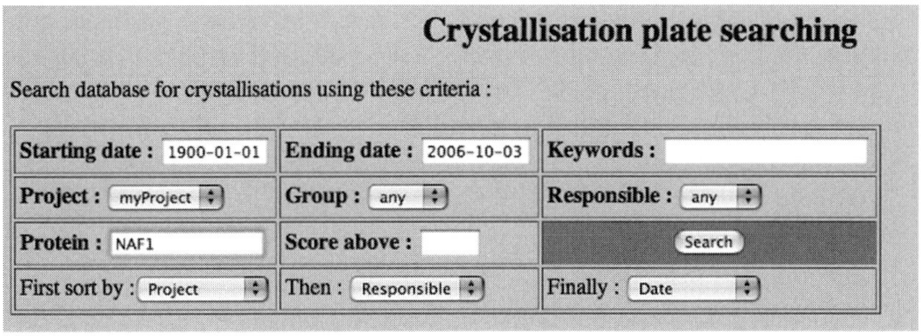

Fig. 4.3 Xtrack plate searching interface. Different criteria can be used to search plates in Xtrack's database: the date (plates done between starting and ending dates), keywords, project (all plates in a given project), group, responsible, protein, or "Score above." For the later, if entering "4" for example, only plates containing needles, plates or prisms will be returned.

A search interface (Fig. 4.3) allows for the listing of crystallization plates according to various filters, including the project, the protein, the date, or the person responsible. The list of plates can then be sorted on the basis of the best score in each plate.

3.1.2.3. Data Collection and Structure Determination: The data collection and structure resolution tracking system is divided into seven different pages: Crystallization, Collection, Data Reduction, Structure Solution, Refinement, Analysis, and Deposition.

The Crystallization page allows the user to define which crystal is going to be used for data collection. All data concerning the crystal can be extracted from the crystallization plate if it is part of the database. If not, the pH, temperature of crystallization, other conditions, and crystal type and size can be entered manually. In either case, additional data can be entered: heavy atom or ligand introduction and cryo conditions.

The Collection page concerns the data collection itself. The data relative to collection are numerous; however, many of these can be extracted from image files. Moreover, many fields are linked to one another. For example, upon choosing one of the listed beam lines, the "collection site" and "source" fill in automatically.

Similarly, the data to be entered on the Data Reduction page can be extracted from software, for example, from Denzo/Scalepack files.

The Structure Solution page registers the method used. A drop-down menu offers the choice of rigid body refinement, molecular replacement, MIR, or MAD. If another method is used, SAD for example, this can be entered in a text box. Similarly, a drop-down menu proposes a list of major structure solution programs; if a nonlisted program is used, this can be entered in a text box. The obtained R value can be registered.

For refinement, a list of programs is proposed, but again additional ones can be entered. This page also allows for tracking the number of refinement cycles done, the highest and lowest resolution, the R value, and the R free. At this stage one can define how the noncrystallographic symmetry was taken into account. This interface is also used to keep track of the latest coordinates file and the most recent backup file. "Analysis" concerns the final structure.

Various details concerning the crystal structure can be entered: the number of atoms of the protein, the error estimates, the RMSDs for bond length or angles. All of these parameters can be extracted from the last coordinates file.

The final page is filled in after deposition in the wwPDB (http://www.wwpdb.org/).

3.1.3. Implementation

Xtrack's user interface is programmed in (a server-side HTML embedded scripting language) and JavaScript. The data are stored in a PostgresSQL relational database. The model contains 10 main tables, corresponding to the different pages the user has access to: Structure, Chemistry, Expression, Crystallization, Collection, Data Reduction, Structure Solution, Refinement, Analysis, and Deposition. The first rows of each table contain the necessary data for the interface (e.g., field names, field types, field size, default values, help text). Because most of the data are in the database and not hard coded, the PHP code can be much more generic. Thus, a field name or a field size can be changed very easily in the interface.

Xtrack is distributed freely (http://xray.bmc.uu.se/xtrack/). It runs on most platforms and is easy to install. A demo version and a help file can also be found on the web site.

3.2. A Generalized LIMS: Sesame

3.2.1. Introduction and Overview

Sesame *(14)* is a set of software tools for information and hardware management. It was originally designed for the management of NMR experiments at the National Magnetic Resonance Facility in Madison, WI, and has been in use there since 1999. With the launch of structural genomics projects, Sesame was extended to include protein production and structure determination by x-ray crystallography and NMR spectroscopy. Sesame has evolved to be capable of handling and tracking back to the levels of gene(s) and compound(s): full-length genes, domains with varying termini, protein: protein complexes, and protein:ligand complexes. These developments have been supported since 2001 as part of the Center for Eukaryotic Structural Genomics (CESG). More recently the development of a Sesame module for metabolomics was developed with support from an NIH Roadmap Initiative Grant. Information on Sesame is available from its web page (http://www.sesame.wisc.edu).

Although, much of Sesame's development has been in the context of a structural genomics project, the software also is suitable for use by individual investigators pursuing research in structural biology. Six stand-alone instances of Sesame have been deployed around the world; however, several projects make use of the Sesame instance in Madison. Sesame is freely available to academic users and will be open source under a GPL license.

3.2.1.1. Overall Goals: The overall aims of the Sesame project have been to develop a LIMS that: (1) provides a flexible resource for storing, recovering and disseminating data that fits well in a research environment (multiple views of data and parameters); (2) allows the worldwide community of scientists to utilize the system; (3) allows remote collaborations at all stages; (4) provides full user data access security and data storage security; (5) permits data mining exercises and analysis of laboratory operations to identify best practices; and (6) simplifies the dissemination, installation, maintenance, and interoperability of software.

3.2.1.2. Sesame Operation Sesame was designed to support multiple, overlapping projects (multiple projects involving individual scientists or many collaborators located at the same or separate sites who may be working on more than one project). Individuals working in a Sesame environment may participate in a variety of associations (as ordinary users, members of labs, or members of collaborative groups).

3.2.1.3. Users Each user registers during his or her first access to the system, choosing a user name and password and supplying full name, e-mail, and organization. Every user is by default assigned to the "World" collaborative group and "World" lab, the parts of Sesame that are visible to all registered users.

3.2.1.4. Labs A virtual lab may represent a center, facility, research laboratory, or research project, and may consist of one or many members. Any user can create multiple labs or be a member of labs created by others. In a typical scenario, a principal investigator (PI) initiates a Sesame lab by designating a member of his or her laboratory to be the "Lab Master," who then creates the virtual lab by giving it a name and asking members of the PI's physical lab to register as Sesame users. The Lab Master then invites these persons (who can be located anywhere in the world) to join the Sesame lab. Individual users may accept or reject the invitation. Sesame users may leave a virtual lab or be removed by the Lab Master at any time. Thus, a Sesame lab can reflect the personnel in a real physical lab or project. A given PI may have multiple virtual labs with separate or overlapping membership. The Lab Master handles the configuration of the Sesame lab, that is, defines and maintains the membership and lab resources.

In the case of a lab that accepts user service requests (a facility), PIs can register and request access or be invited to use the facility. The PI maintains a list of the group members that are permitted to use the facility's services. If a Sesame instance supports multiple facilities, a PI may be associated with more than one facility and can differentially control the access of members to individual services.

3.2.1.5. Collaborative Groups These are a set of associated users who intend to share their data. Any user can initiate a group by defining a name for the virtual group. The initiating user becomes the "Group Master" and invites other users to join. Members of a collaborative group control visibility and/or access to their data records by defining *privilege levels*. Three levels are supported: user, group, and world. A user sees only his or her own personal records, if a member of a collaborative group, those flagged to be visible by members of the group, and those with a world flag.

3.2.1.6. Lab Resources The lab resources are: the status (e.g., active, inactive, request), which can be attached to individual items (e.g., workgroup, protein, sample, plasmid); lab protocols that correspond to different items and include the protocol name, type, contact, the whole protocol in text form, actions associated with it, and a protocol tree that contains tag-value nodes; record type (e.g., pipeline, research and development, medically relevant, outside user request, inactive); plates described by name, number of rows and columns, row and column labeling type and makers number; source organism (genus, species, strain, NCBI taxonomy id); barcode and label definitions for different items; barcode printers; locations (room, freezer, rack, shelf); 5′ and 3″ primer prefixes; plasmid components (plasmid source, host restriction, tag and cleavage); lock solvents (NMR); internal reference compounds (NMR); mass spectrum type; mass spectrometers; stock solutions for crystallization (precipitant, salt, buffer, additive, pH adjuster, pre-made solution, buffered precipitant, cryoprotectant) and associated data (reagent name, pH, concentration, concentration unit, type, molecular weight, form, volume, flush mode, dispense speed, dispense delay, pK_a, number of mixing cycles, and batch number); robot rack type; robot racks; crystal scores. The lab resources are only modifiable by the Lab Master; however, by using the "Can Do" resource, the Lab Master can delegate the maintenance of individual lab resources to specific lab members. This imparts extraordinary flexibility to the system and enables the virtual lab to adapt to changes in the operation of the associated real-world lab. Each Lab Master can export or import lab resources; this enables the dissemination of protocols and simplifies the setup of a new lab.

3.2.1.7. Objects and Records Objects in a Sesame lab can be thought of as those available to all lab members (e.g., reagents on shelves, instruments on benches, shared laboratory protocols). Objects attributed to a given lab are visible to all lab members, but only to them. Lab data are not visible to the world or to other labs. Each record (e.g., sample description) belongs to the person who created it. Other lab members can copy it and then modify the newly created record for their own use. Only the owner can update key fields or delete a record. Records usually contain fields that can be modified by other lab members: these include "actions" (reports on steps carried out), locations and quantities of intermediates (e.g., plasmids, cell pastes, purified protein, crystals, and solutions), and images (gel scans, spectra). Files can be attached to records by any lab member. Lab Masters are able to control the accessibility of records within the lab.

3.2.1.8. Data and System Access Security Passwords are encoded in the client using the MD5 message-digesting algorithm. The unencoded form of the password never leaves, nor is stored in the client, and only the encoded password is stored in the user's profile in the database. Security is maintained over all tiers throughout Sesame. Twice weekly, backups of the UW-Madison Sesame server are performed to a Network Attached Storage unit (RAID) located in a separate building.

3.2.1.9. Sesame Usage In addition to the UW-Madison Sesame instance described here, six external (stand-alone) Sesame instances have been installed. The UW-Madison Sesame instance currently has over 1,000 registered users who are members of more than 60 labs. The current content of the relational database is around 400 megabytes, and the file server contains around 7 gigabytes. Monitoring of the second and third Tier servers indicates that the existing Madison installation has never approached full capacity.

3.2.1.10. Sesame Reliability The Madison Sesame installation has proved exceptionally robust and stable. The Sesame system has been fully operational except during software upgrades and other routine maintenance. Over the past 4 years, the system has logged fewer than 7 lost days of operation (99.5%

reliability); most of these were the result of network related problems. Thus far, the second Tier of Sesame has never crashed or locked up.

3.2.2. Sesame Modules and Views

To simplify and customize the system, users access Sesame through "Modules" that are tailored to specific kinds of experimentation. A Module consists of a collection of "Views" that handle data specific to specific stages of experimentation. The Modules and Views provide simplified graphical user interfaces for data entry and retrieval (suitable for use with notebook or handheld computers); however, all data are stored in a single relational database. Current Sesame modules include: *Sheherazade* (structural genomics), *Jar* (outside user requests to a structural genomics facility), *Well* (crystallization trials), *Camel* (NMR spectroscopy), *Lamp* (metabolomics), *Rukh* (yeast two-hybrid molecular interactions), and *Sundial* (shared resource scheduling and management). These modules are described briefly in the following. All Sesame modules are publicly available from the Sesame web site http://www.sesame.wisc.edu ready for immediate use by all registered users. The features of Sesame are documented with Web-based help pages.

Modules share many common features. Users can attach any number of files and images (e.g., gel scans, images of spectra, text files) to most record types. Images can be attached to every individual data record and viewed, manipulated, and printed. For most Modules, the Lab Master maintains a set of lab resources that are used to construct Sesame records. Records in all Modules can be searched and retrieved on the basis of a variety of criteria (e.g., owner, type, status, components, location, different IDs, date created, date last modified).

Sesame supports data input from various devices, including barcode readers, plate readers and NMR spectrometers. Data can be exported in various forms, including barcoded labels, robot work lists, and text or XML reports of individual records.

Almost all Sesame Views contain "lab protocol" and "actions" cells to allow lab specific fields to be defined, and capture information about work carried out. To document lab activities, the Lab Master is able to define controlled vocabulary terms called *actions*. Actions are fully definable to meet the objectives of the project and can describe different steps for different protocols, such as a PCR, protein expression, protein purification, transferring a sample to a collaborator, or an instrument is down and a repair is in progress. The record contains the name of each person adding an action and a time stamp indicating when it was entered. This information can be used to analyze productivity, determine bottlenecks in procedures, and identify who should be coauthors on publications, or see what was ever done with a given ORF, protein, sample, or other Sesame item.

Every Sesame View contains fields that show when and who created the particular record, when and who modified it last, the status and type, external ID, user label, user- and time-stamped info entries, all linked items, and all attached files and images. This flexible combination of bioinformatics information, actions, images, files, and links to other records creates a comprehensive computerized record of all phases of a project accessible to all members of the lab.

Sesame users can retrieve data from various Views as CSV, FASTA, XML, or formatted text files and at the Module level as XML files. These files are readable by many spreadsheet software packages and are amenable for most general purpose computational exercises.

Sesame can generate specialized reports for depositing data into public databases (e.g., PEPCdb and TargetDB) and tracking progress on a lab/center level. Sesame currently is capable of generating the XML file required for depositing information into the first incarnation of PEPCdb. This file includes the protocols used to produce proteins for structure analysis, a dated status history for each protein target and a failure remark when appropriate, information on each trial used to produce a protein, references to target-related database accession codes, and information on the DNA and protein sequence. To be compliant with future PEPCdb requirements, an enhanced version is planned that will output quantitative information on protein cloning, purification, and crystallization trials from data captured in existing Sesame Views.

Sesame has the capability of printing and facilities for querying based on scanned barcodes. Printed barcodes can be attached to physical items, such as mass spectrometry samples, vials containing plasmids, NMR samples, and multiwell plates, and connect them to data records held inside the Sesame database. The Lab Master can customize the barcode and label definition, adding the contents of selected record fields (e.g., user name, date, linked record database id) to the unique database id.

Sheherazade is the client Module, that contains Views relevant to structural and functional genomics/biology. Sheherazade supports progress tracking and report generation. Views under Sheherazade include: "ORF," "ORF Details," "Workgroup," "Primer," "Plasmid," "Protein," "Sample," "Compound," "Mass Sample," "Screen," "Crystal," "NMR Experiment," "NMR Assignment," "Structure," "NMR Deposition," "X-Ray Deposition," "Software," "Hardware," "Vendor," "Recipe," "Amino Acid Calculator," "Citation," "DBNO List," "Job Request," and "Target Request." The different Views provide partially overlapping information. This makes it possible to traverse all parts of a project, both horizontally (following all the links and branches between different steps) and vertically (viewing and editing all the available records in a particular step of the project).

The "ORF" and "ORF Details" Views handle ORF records that contain available sequence information and annotation captured from genomics and bioinformatics web sites, for example, the target score, source organism, gene locus and gene locator, different IDs (e.g., main, lab, swissprot), name, category, predicted structure class, structure known, nucleotide code and length, amino acid code and length, number of introns, molecular weight, calculated absorbance coefficient (ε_{280}), pI, amino acid composition/count, signal peptide length, closest homologue, and actions applied in a lab. All these fields can be updated as the coverage and annotation improves. The ORF records can be searched, sorted, and grouped by various criteria, including name, number of introns, and sequence length.

The "Workgroup" View handles a set of ORFs that have been selected for experimentation. (The authors plan to generalize the concept of workgroups to handle other sets of Sesame items of the same type, e.g., protein, crystal.) The number of targets in the set is variable, but typically is 96 (or a multiple thereof) for compatibility with plates used by robots and high-throughput instrumentation. Once a workgroup of ORFs is formed, the Workgroup View can be used to construct primers for PCR, to create orders for primers, calculate digestion patterns, and attach images (e.g., gel scans). ORFs within a workgroup can also be linked directly to records from other Views, allowing data derived from experiments and calculations to be captured (see the following).

"Primer" View is designed to handle primer records. The primer-specific fields are: primer name, purpose, restriction enzyme, location (room, freezer, tower box, and position in box), reading direction, genetic code, length, meting temperature, stock concentration and PCR concentration, and date of synthesis.

"Plasmid" View handles plasmid records. Some of the plasmid specific fields are: protein name, clone type, clone utility, clone tissue source, insert source, stock type, variant name, host, host requirement, selectable marker, detailed construct description, amino acid sequence and data, and description of N-terminus modification during cloning.

"Protein" View manages information about the expression and purification of proteins. Protein records contain data about protein name, cleaved, uncleaved, and quality-controlled amino acid sequences (including automated calculation of a variety of physical parameters for each sequence), isotope labeling, storage location, and external ids.

"Compound" View is designed to track compounds. The compound-specific fields are: name, formula, molecular weight, isotope labeling, details, URL, storage temperature, storage location, synonyms, external ids, and amount tracking.

"Sample" View is designed to handle physical laboratory samples; describe where they are kept; specify the components (constituent type, name, concentration, and isotope labeling), pH, and ionic strength; and track changes in the location and quantities of a stored sample.

"Mass Sample" View manages information about samples used for mass spectrometry (amino acid sequence if the sample is a protein; expected molecular mass, concentrations of protein, buffer, and salt; list of impurities) and the results of the analysis (instrument; spectrum type, experimental mass[es], images of the obtained spectra, and raw data files).

"Screen" View handles information about x-ray crystallization and NMR sample solubility screens. It allows users to design screens and keeps a detailed description of the conditions of the trial (e.g., sample[s] used, volume, pH, date, temperature, droplet additives, screen type, all the screen component details and volumes, and location of the bar-coded plate), along with information on the observed results and progress toward obtaining diffraction-quality crystals or NMR samples suitable for solution structure determinations. The Screen View enables the user to store multiple images of the droplets, examine them using the included image processing tools, and finally score them. The Screen View can generate worklists for the Tecan Genesis and Gilson/ Cyberlab C-200G robots. In addition to the screening parameters, the Screen View and the associated robot-related Lab Resources maintain all the parameters required to operate the robot (racks and the solutions in them, viscosity of the solution, which is used to determine the pipetting speed).

"Crystal" View is for keeping track of crystals obtained from crystallization screens. The crystal-specific fields are: screen-, well- and droplet-number, pin and vial barcode, dimensions, morphology, effective and best resolution, quality, space group, unit cell, and parameter reliability.

"NMR Experiment" View is for managing NMR experiments. This tool provides a solution to the problem of storing, recovering, and disseminating parameters used in NMR data acquisitions over multiple spectrometers and spectrometer types. The NMR Experiment View beside the fields for spectrometer (manufacturer, model, field, and name), probe (manufacturer and model, experiment type, temperature, and pressure), provides representations of the parameters utilized by the two major NMR spectrometer manufacturers (Bruker and Varian) to launch and carry out a data acquisition protocol. The NMR Experiment View has the ability to upload the parameter and control files and download them to same or similar spectrometer. The user can access the database, query it, and restore files and/or directories needed to rerun the selected experiment(s). The downloaded environment can also serve as a starting point to set up a new experiment. An NMR experiment can be linked to a particular NMR sample (described by Sample View). In the future, NMR Experiment View will be capable of processing the acquired NMR data by supporting interfacing to script-derived software packages (e.g., NMRPipe or Felix) by using scripts associated with each experiment type.

"Assignment" View facilitates the capture of data related to NMR assignments: peak list files, related data, files, scripts. In future, it will allow launching of assignment programs, such as PISTACHIO and GARANT, using the files and scripts attached to the record and files attached to linked items (e.g., peak lists attached to NMR experiments). The output files from these programs will be attached to the particular record.

"Structure" View facilitates the capture of information related to structure determinations, such as related data, files, and scripts. In future, it will allow launching of structure determination programs such as CYANA and X-PLOR using the files and scripts attached to the record and files attached to linked items (e.g., NOESY peak intensities attached to NMR Experiments and assignment files attached to Assignment records). The output files from these programs will be attached to the particular record.

"NMR Deposition" View handles NMR deposition related data and ties together all the Sesame Items that are related to deposition. Future version will be able to generate and validate mmCIF (X-Ray) or NMR-STAR (NMR) depositions for direct submission to the wwPDB.

"X-Ray Deposition" View handles x-ray deposition related data and ties together all the Sesame Items that are related to deposition. Future version will be able to generate an mmCIF deposition that will be possible to directly submit to the wwPDB.

"Software" View keeps track of software packages used in different processing steps, e.g., assignment or structure determination. Attachments to software records can include: manuals, instructions, recommended parameter sets, or even the software installation package.

"Hardware" View handles different hardware (e.g., instruments, robots, beam lines) used in different stages of the project. Attachments to hardware records can include instrument specifications, instructions, manuals, and parameter sets.

"Vendor" View organized records of purchases; it handles standard fields such as category, name, address, phone, fax, URL, and e-mail. Catalogs and product specifications can be attached to these records.

"Citation" View is designed to handle reference information (book, journal, thesis, abstract, personal communication, PowerPoint presentation, Internet URL). The fields are supersets of the fields required for wwPDB deposition. If the PubMed id exists, it can be used to fetch the PubMed record and populate as many fields as possible. Within a lab, this View can be used to build up repositories of presentations, abstracts, and published articles, and to track and share versions of drafts in preparation.

"DBNO List" View allows users to create named lists of data base identifiers (DBNOs) for Sesame items of particular types. These named lists can be used as argument in queries and report generators.

"Recipe" View handles records that describe commonly used solutions or procedures, for example, how to create a "master mix for first PCR" or how to install a particular software package. Images and files can be attached to these records, and as with other records in Sesame, these can be linked to any Item.

"Job Request" View allows users to create and track job requests within a lab. The Job Request fields are: title, status, type, priority, severity, responsibility, date needed, and description. Job requests handle paths that differ from the normal pipeline, for example a request to re-grow a cell paste or remake a labeled protein.

"Target Request" View within Sheherazade allows lab members to submit target requests and to view and process external target requests submitted through the Jar Module (see the following).

"Lab Resources" View handles data for lab resources that are used by other Sesame Views, such as status, type, lab protocol, file type, plate, source organism, labeling pattern, barcode definitions, barcode printers, restriction, tag, host, lock solvent, robot rack type, robot rack, crystal score, crystal space group, salt, buffer, additive, cryoprotectant, primer prefix. The "CanDo" lab resource allows the Lab Master to delegate the maintenance of a lab resource.

"System Resources" View handles data that are available to all Sesame users, such as ORF type, NMR experiment type, NIH action, PEPCdb protocol, XML tag.

"Amino Acid Calculator" View handles basic nucleotide and amino acid sequence related calculations. For a given nucleotide sequence, it calculates the length, G+C content, amino acid sequence, and primers. It translates a DNA sequence to a protein sequence and for that sequence calculates its length (number of amino acid residues), amino acid composition (number of each residue type), molecular weight (depending on the selected labeling pattern), pI, and extinction coefficient (ε_{280}).

Jar is the Sesame Module accessible to external scientists who wish to nominate targets to be investigated. This module captures information about the requester along with the data that CESG requires for evaluation of each potential target (gene sequence, function if known, availability of a clone, availability of purified protein). Jar provides feedback to the requestors concerning decisions taken and progress along the structure determination pipeline.

Lamp is the Sesame Module for Metabolomics *(29)*. Lamp consists of Views used in metabolomics: "Standard Compounds," "Sample," "Mass Sample," "NMR Experiment," "Software," "Hardware," "Citation," "Vendor," "Recipe," "DBNO List." "Job Request," and "Experimental Sample." A "Small Molecule" View is under development that will support information imported from small molecule databases and structure-data files (SDFs) and will handle experimental data from mass spectrometry and NMR spectroscopy.

Sundial is the application Module used for requesting (Time Request View) and scheduling (Calendar View) data acquisition time on NMR spectrometers.

Sundial has been in use by all users of the National Magnetic Resonance Facility at Madison (NMRFAM) since October 1999. Since then, around 5,000 separate accepted time requests have been processed. Sundial provides multiple levels of access and functionality to users, principal investigators, and site administrators, while providing complete data security throughout. The incorporation of a relational database server allows for sophisticated querying and the ability to generate summative reports at a keystroke, not only for NMRFAM staff members, but also for facility users, principal investigators, and site administrators.

The "Calendar" View of Sundial Module is used by the Site Administrator of an NMR facility to schedule time on the facility's spectrometers. The Calendar View can access the records created by Time Request View for a given spectrometer over a given time period. The administrator then maps those requests onto a calendar. The finished schedule is then released to the public. Individual users are able to view their allotted time, and sign up for additional time or relinquish time slots that prove to be unnecessary. When time is released by one user, an e-mail notification is sent to all registered facility users; this helps minimize unused time resulting from cancellations. Sundial ensures more efficient utilization of a scarce and expensive resource.

The "Site Resources" View in Sundial handles site profile data, instrument descriptions and details, open time periods, and billing periods. The CanDo site resource allows the site administrator to delegate the maintenance of the site resource to one of more site staff members.

The "PI Resources" View handles the PI's profile, list of funding sources, and allows the PI to manage permits for use of the facility.

Rukh is the Module designed to manage yeast two-hybrid screens ("Bait View," "Y2H Screen View" and "Y2H Plate Read View") and handle user requests for screening (Y2H Request View). It also contains the Software, Hardware, Vendor, Recipe, DBNO List, Job Request, and Citation Views. Rukh was developed in collaboration with the Molecular Interaction Facility in the UW-Madison Biotechnology Center and has been in use for 3 years. Rukh is used to track and manage the screens, handle screening requests, and access the obtained results. The screen requests contain data about the bait, libraries, and user information (e.g., PI, billing info). Rukh is then used to manage the whole screening process from setting up the initial screening runs through the validation steps. The initial screens are set up by facility members on the basis of user requested baits and libraries. Screens can contain any number of plates. Rukh is then used to score positives on the basis of absorbance data obtained from a plate reader, selection from gel images, or other criteria, and generate a command file (worklist) for a Tecan Genesis robot to reformat the plates during the screening and validation process and generate the plate barcode. At the end, the obtained results are linked to the screening requests so they can be communicated to the users.

3.2.3. Implementation

The *software component model* has been used to simplify Sesame code development and maximize its reusability. The use of containers and components supports hierarchical software development and allows for maximal reuse of existing software components when creating new software components or improving existing ones.

At a higher level, the Sesame architecture is organized into a *Framework* and *Views*. Both the Framework and Views are built from Java objects. The Framework provides the basic three-tier infrastructure, the building blocks for Views, and system tables. Individual Views facilitate data capture, editing, processing, analysis, retrieval, or report generation for Sesame data items that reside in Tier three, or one or more tables in database or file system. Sesame *Modules* consist of sets of Sesame Views. A given Module simply provides the environment in which Sesame objects are executed. Views operate on particular subsets of the data, but all data reside in a common database. The schema is a partially normalized relational database model tuned for maximum performance and is under continuous development spurred by the scientific discovery process.

3.2.3.1. Sesame Framework   Sesame makes use of the Object Web paradigm, under which the object components of the client and server sides of an application can reside anywhere on the Web. They are connected by the Object Request Broker (ORB), which manages the communication between them. The Internet Inter-ORB Protocol (IIOP) is used for communications between different ORBs. Objects describe the concepts that underlie the data to be organized or processed. These objects are independent of any single application, and they can be used in unpredictable combinations.

The Sesame Framework has been implemented as a three-tier Java/CORBA client/server application *(30)*. The *first Tier* represents the visual aspects of the component, and these aspects usually reside on the client. The *second Tier* comprises the server objects that represent the persistent data and business and processing functions. The *third Tier* comprises the Sesame database and file system, and legacy databases. The Sesame database also contains administrative tables, which connect the component properties to users and Sesame records and govern the privileges. The server objects interact with their clients and the databases; they implement the logic of the Sesame object. Clients never interact directly with the third Tier. To keep as small a footprint as possible on the client side, all the "heavy" objects are loaded and instantiated on demand. Objects may reside in one or more servers. Under the CORBA architecture, all objects have IDL-defined interfaces and connect to one another on the same machine or different computers via ORBs (this provides transparent scalability).

Each Sesame Module has a user interface written in the platform-independent Java (currently Java 5) language. Each Module is accessible from the Sesame homepage. The client runs on any computer with an up-to-date Java Runtime installed by using Java Web Start. The client and server objects connect to one another across the network via Object Request Brokers (ORBs). The application server maintains the connection to the database, assembles and executes the SQL statements, and works with the resulting sets. Database programming uses the Java JDBC API, which decouples the database management system (DBMS) from the client and makes changes in the DBMS invisible to the user. The second and third Tier were designed so that the Sesame system can be adapted easily to different RDBMS engines. Sesame currently supports as the third Tier RDBMS either Oracle 8.1.7+, PostgreSQL 8+, or Microsoft SQL Server 2005.

One can carry out multiple queries against any type of Sesame item; in doing so, the "QueryPanel" presents the results in the client's main window in tree structure. In addition to standard menus, the interface presents selectable button bars and context-sensitive right-click pop-up menus (Fig. 4.4). "Board" is a class the authors have developed to display an arbitrary number of Sesame objects. These objects can be images, two-dimensional graphs, contour plots, text files, or views of Sesame records. The Board allows drag and drop of Sesame objects through the SesameTransferHandler class; alternatively, a Sesame object can be sent to the Board by firing a TransporterEvent that carries the object. In addition to its own menus, the Board also presents the menu for the selected inner object by grabbing it from a View. For example, when an image is active, the menu reflects the entire set of image processing options. For example, the Board class allows the user to compare data records for different proteins along with their attached gel scans. The objects in Board

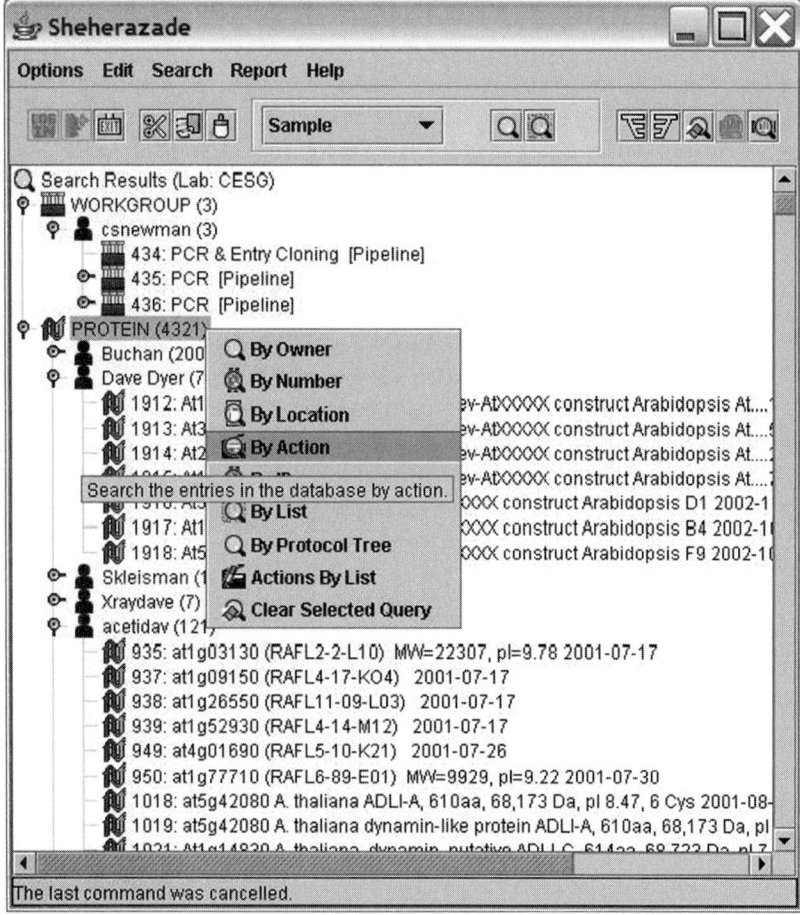

Fig. 4.4 An example of how the results from multiple queries can be viewed in the Sheherazade Module of Sesame. To display an item, one drags and drops it to the "Board" or simply double-clicks on it. Different records can be linked by dragging and dropping an item from QueryPanel onto an item in Board. From the Search menu, every query for items handled by the Module is accessible. The query buttons on button bar are selectable. As shown, they target samples. A right click on the PROTEIN node brought up the available queries for proteins.

can be cascaded, tiled (all or only selected items), minimized, and maximized. The SesameTransferHandler class supports standard window functionalities, such as "drag and drop," "cut," "copy," and "paste" on Sesame objects.

The GUI design (Fig. 4.5) provides considerable flexibility, since it allows the developers to easily add new fields when needed (e.g., if research identifies the need to track an additional parameter, or a new piece of equipment is added). This has shortened the development times for Views to handle new (complex) data types, especially those that require no new editors or renderers to be developed.

3.3. A LIMS for Small- to Large-Scale Laboratories: HalX

The Yeast Structural Genomics project (Université Paris-Sud, Orsay, France) was launched in 2001 *(16,17)*. From the start the authors decided to develop their own LIMS (named HalX). HalX *(15)* initially was developed as a simple web interface (written in PHP and Javascript) and gradually evolved into a complex database.

After a few years of experience, it became evident that the user interface required more interactive features. These features could not be implemented using the PHP and Javascript interface tools. Moreover, standardization efforts on the structure of the database within the SPINE project imposed drastic changes to HalX.

Therefore, it was decided to reimplement completely the software using the new data model, a three-tier architecture, and a java applet user interface. This effort has been on going for 2 years and the resulting HalX v2.0 was released in November 2006.

3.3.1. The Protein Production Data Model

Originally, structural genomics was aimed at determining the three-dimensional structure of one member of each structural family. This objective has been enlarged, and new objectives are now pursued, such as the determination of macromolecular complexes in the European projects 3D-Repertoire (http://www.3drepertoire.org/) or SPINE2 (http://www.spine2.eu/). Even for determining the three-dimensional structure of an isolated globular protein, the experimental process is quite complicated and poorly standardized. If cloning and expression steps are made using a small number of different protocols, purification often requires adjustments depending on the protein.

Structural genomics projects often group different laboratories responsible for different parts of the process. The consequence is that a huge amount of data has to be exchanged, for example, between the laboratory expressing the protein and the laboratory responsible for crystallization. In traditional structural biology laboratories, the same group of people was following the process form beginning to end; consequently, most of the information was only partially noted in lab books. Distributing the work among different people renders this system obsolete. In this new organization all experimental details have to be noted, and in a way that they will be understood by all.

These facts have put very strong constraints on the design of a data model. The main requirements were:

• The data model has to be able to represent any protocol that is in use in laboratories.

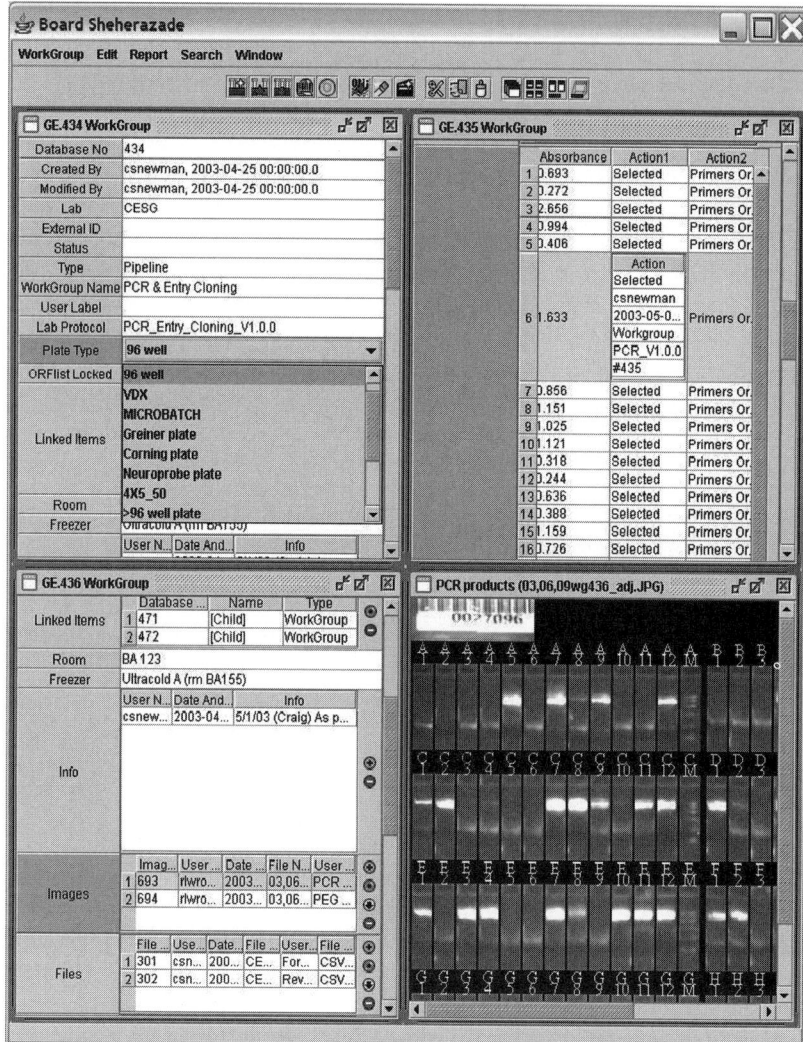

Fig. 4.5 Illustration of the versatility of the Sesame GUI. As shown, several tiled database records can be displayed simultaneously on the Sesame Board *(top left)*. Display of a workgroup record. The pull-down menu shows the available choices of "Plate Type" within the lab *(top right)*. Display of a second workgroup record illustrating how multiple workgroups can be is displayed on the Board for easy comparison of records. The figure shows how all the details associated with an action are displayed by clicking on the action *(lower left)*. In this display of a third workgroup, the rows containing "linked items," "info," "attached images," and "files" are shown. By clicking on the highlighted row, an image was selected and displayed *(lower right)*.

- It has to be flexible enough to be possible to add new protocols very fast.
- The samples at any stage of the experimental workflow can be traced all the way back to the gene in every detail.

Within the SPINE and eHTPX projects, and following the recommendations of the Task Force on the Deposition, Archiving, and Curation of the Primary Information, constituted at the Second International Structural Genomics Meeting, such a data model was designed *(13)*. This effort was led by the

Macromolecular Structure Database Group in EBI (European Bioinformatics Institute). Numerous structural genomics and structural biology groups then reviewed the model.

The model is divided in packages addressing different categories of data. A simplified view of the model is given Fig. 4.6. The UML diagrams of the complete model can be found on CCPN (Collaborative Computing Project for NMR) website. The main packages are "Target," "BluePrint" (complete name ExpBluePrint), "Experiment," "People," "Sample," and "MolComponent" (complete name RefSampleComponent). The package "Target" allows registering all relevant data concerning the targets: name, gene, biological and molecular function, project in which this target is included, and status (e.g., in progress, cloned). The package "BluePrint" registers the object that will be studied. This object can be one single and full-length target. It can also be more complex objects: a domain of a target, a complex between two or more targets, a complex between one target, and a small molecule. The package "Experiment" is used to store experimental details, mainly protocols and instruments. One experiment is linked to a single BluePrint. The package "People" registers details concerning the users, but also concerning other people, for example, a person who has provided a plasmid. Each experiment done manually is linked to the experimenter who has done it. The package "Sample" registers all the samples: input and output samples for the experiments, but also stock solutions and primers. The last package, "MolComponent," links "Sample" and "Target" by describing the molecule contained in the sample. In the authors' conception, a molecule is one given physical state of a given target, resulting from a given experiment. As an example, a PCR product is linked to the output sample of a PCR experiment, and the target that is studied. The complete model is of course considerably more detailed, and contains more than 100 classes.

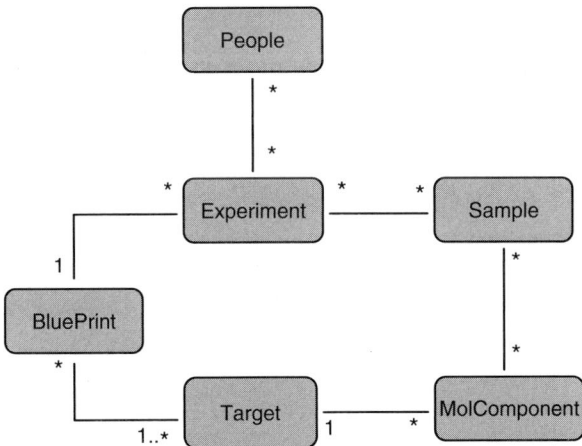

Fig. 4.6 Simplified view of the model used in the database underlying HalX. Each rounded box represents a different package. The links between packages give the cardinality of the relation. As an example the "*" on the BluePrint side of the BluePrint–Target relation means that one target can belong to any number of BluePrints, including none. The "1.*" on the Target side means that a BluePrint contains at least one target, and possibly more.

As stated, the original model was developed by a consortium of groups belonging to the SPINE and eHTPX projects. At that time, only the part of the data model needed for protein production had been designed. Within the PIMS and BIOXHIT, the data model is being extended to allow registration of crystallization experiments.

3.3.2. Code Generation from Data Model

We are using a Model Driven Architecture approach by using the CCPN framework that automatically generates code from the Unified Modeling Language (UML) model. One of the core ideas of the Model Driven Architecture (MDA) is to ease the change of the run-time platform by raising the level of abstraction in which just the business aspects are modeled, and separating business aspects from implementation details. This technology helps to build systems with good maintainability. In this architecture, the model is created and modified using a graphical interface, and the code needed to create the database and ensure communication among the database; the user interface is then automatically generated. This allows users to make changes in the data model and make them available in the user interface in a very short time.

CCPN aims at performing a service for NMR spectroscopists analogous to what CCP4 is for the x-ray community. The original observation was that NMR spectroscopists have difficulty moving data from one software package to the other. To develop conversion software among these different software packages, they have developed a data model that could sustain any of these applications. They have also developed a framework that automatically generated an Application Program Interface (API) from this model. In this operational model, a software uses the API to store the data in an ordered manner (either in XML files or a relational database), following the organization given by the data model. When the model needs to be changed, a new API is generated, in which only the modified parts of the model are different from the previous API. The consequence is that in most cases, nothing has to be changed in the legacy code of existing software. Another very important consequence is that the generation of the new API is very fast; thus, new entities such as those required for crystallization can be added in the data model, and the corresponding code are immediately available for user interface development.

3.3.3. HalX User Interface

The major reason for rewriting HalX was to get a more flexible and more user-friendly user interface. Indeed, PHP/Javascript interfaces require a lot of "field-filling," and submitting numerous pages one after the other in stages. The user feedback the authors received with HalX v1.x clearly indicated that users wanted to have more interactive interfaces that responded according to previously entered data, and the "checkbox/submit" system required for selecting items were replaced by drag-and-drop capabilities.

For HalX v2.0, the authors' concern was always to be as close as possible to the user's desires and never the complexity of the code needed to reach it. Its features mainly concentrate on experimental workflow management and experimental data entry.

3.3.3.1. Experimental Workflow Management Experimental protocol (which in the authors' understanding describes the details of a given experiment) entry is one of the weakest points in most existing LIMSs. In most systems, very few details are registered, or alternatively present through an attached file. However, this renders them useless for data mining purposes since there is no way to enforce strict file formats for such entries. Moreover, these protocols cannot be reused in new experiments.

HalX v2.0 allows users to register their experimental workflow, which is the description of a pipeline of experiments and the protocol for each (Fig. 4.7). For this purpose, the user selects individual experiments from the left panel. These individual experiments then appear in the working panel. Colour-coded boxes above and below each experiment indicate the nature of input and output samples. For example (see Fig. 4.7) a "Digest" experiment will have two inputs: digestion enzyme(s) and DNA, and one output: DNA. The experiments in the working panel can then be linked to one another through these

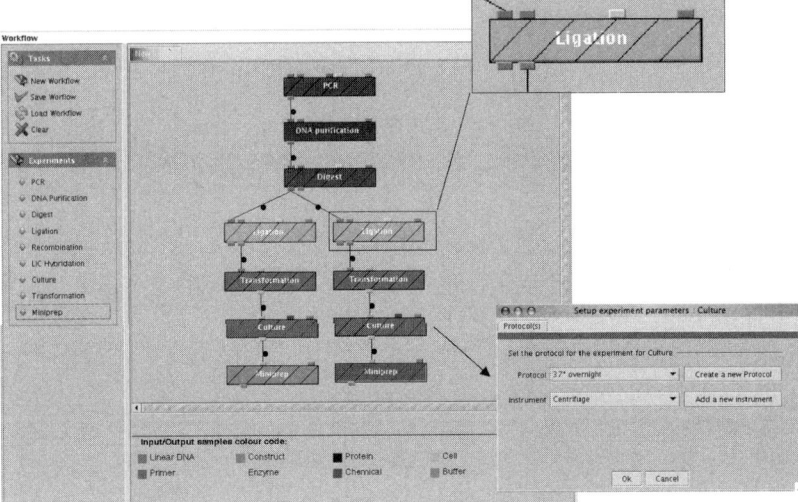

Fig. 4.7 HalX workflow construction interface. HalX's user interface is divided in three panels. The left panel gives access to different tasks. The top box allows the user to start a new workflow, save a finished workflow, load a workflow that can be subsequently modified and saved under a different name, and clear the working area. The central panel is the working area. Clicking on a given experiment type in the left panel makes the corresponding box appear in the working area. For example, clicking on "PCR" in the left panel makes a PCR box appear in the working area, like the one on top of the shown workflow. Each type of bow has smaller boxes above and below that give the nature of the input and output samples. These are color-coded, and the legend is given in the lower panel. An example is given for a ligation experiment. There are four inputs for this experiment: linear DNA or construct (circular DNA). Both are not necessarily used, for example, when ligating two linear DNA fragments, the "construct" input is not used. There is also an enzyme input and buffer input. There are two possible outputs, either a linear DNA or construct. Two consecutive experiments can be linked to each other by dragging with the mouse the output of the parent experiment on the input of the child experiment. This link is accepted only if these output and input are of the same type.

Clicking on one of the experiments opens a dialogue box *(bottom right)*, allowing the user to choose a predefined protocol for this experiment, and an instrument to realize it if needed. Alternatively the "new protocol" button gives access to the protocol creation interface.

input/output boxes using the mouse. Linking the DNA output of a "Digest" experiment to the DNA input of a "Ligation" experiment means that the DNA resulting from the first digest is ligated in a vector in the Ligation experiment. Experiments can be linked that way only if the linked boxes are of the same nature.

Any graph of experiment can be represented. This is a very important point since most systems only allow tree workflows: One parent experiment can be followed by multiple child experiments. However, none of these allow the many-to-one relationship: Two or more parent experiments are followed by a single child experiment. Such relationships are needed, for example, when producing polycistronic vectors, in which multiple ORFs coding sequences are introduced in the same vector.

For each individual experiment, the precise protocol can be defined. In the data model concept, a protocol is built as a succession of steps. Many different types of steps are available:

- "Temperature Step": These are, for example, the temperature variations needed for PCR cycles (Fig. 4.8).
- "Add Step": A new sample is added, for example, IPTG is added to a cell culture for induction.
- "Flow Step": This is used in purification protocols to describe the solution or solutions going through the chromatography columns.

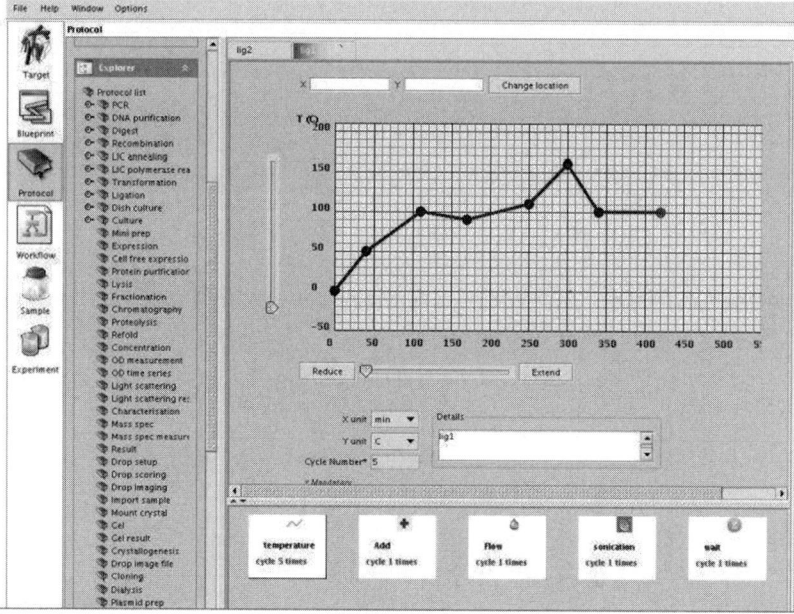

Fig. 4.8 Protocol entry interface. A protocol is build as a list of successive steps. These steps can be of different natures: incubation, adding solutions, chromatography column, etc. The example given here is the "Temperature" step, which is used when varying the temperature of the experiment. The user can place points on the grid in the center using the mouse. The position of these points can be adjusted using the boxes "x" and "y" above the grid. The scales can be reduced or extended and the units can be changed. Using the "Cycle number" field, which default value is 1, the user can cycle through this step.

The used interface proposes other types of steps, such as sonication or centrifugation. For each type of step, a specific user interface is proposed. Figure 4.8 presents the interface for a "Temperature Step." The user clicks on the grid to indicate the different temperatures. The positions of the points can be adjusted using the fields above the graph, the scales and units can be changed, and notes can be added in free-form text. When a step is validated, it appears in the bottom panel. It is possible to cycle through a step or group of steps. As long as the protocol has not been saved in the database, any step can be edited and eventually deleted.

Individual steps constituting a given protocol can be accessed through the user interface and reused to build a new protocol. Similarly, protocols constituting a given workflow can be reused as building blocks in a new workflow. However, saved workflows, protocols, and steps cannot be modified. The reason for that is they might have already been used in some experiments, and changing them would mean losing the exact experimental conditions for these saved experiments. Thus, already saved workflows, protocols, and steps can be edited and modified, but will then be saved under different names, thus consisting a new entry. Moreover, it should be noted that the American Food and Drug Administration imposes as a requirement that any data entered in a LIMS database cannot be modified.

3.3.3.2. Experiments Entry When entering experiments, the user will usually choose one workflow that has already been entered. Indeed, the protocols and experiments used in a given laboratory usually do not vary much from one protein to the other. When the workflow is loaded, the experiments to be done appear in the working panel with the layout chosen when entering the workflow.

By selecting one experiment, the user can access the protocol details as they have been defined. These details can be changed for this one experiment, which will not affect the protocols of the template workflow. This means, for example, that one can have a protocol with a cell culture at 37°C, but can choose, for a given protein, to make the cell culture at 25°C. The workflow will still indicate 37°C, but this one experiment will be registered as having been done at 25°C. Similarly, the user can decide to add experiments that are not present in the workflow. For example, the template workflow might indicate only one cell culture at 37°C, but the user can decide, for a given protein expression, to try two cultures at two different temperatures. Conversely, the user can decide not to do all the experiments present in the template workflow.

Thus, the workflow should really be seen as a guide, and is in no way as a constraint. However, the closer the experiments really done are to the workflow, the less data will have to be entered while registering the experiments. Indeed, if following the workflow, the only mandatory field is the date of the experiment (the default being the current day). The result of the experiment can be entered (successful, unsuccessful, unknown). Images can be attached to an experiment, for example gel or crystal images. Non-image documents can also be attached, for example chromatograms or sequencing files.

In version 2.0 it is not possible to register all experiments done in a 96-well plate in one click. However, this is possible in version 2.1 released in April 2007.

3.3.3.3. Searching Through Data and Generating Reports Once experiments have been entered, users are able to search through the data and generate reports.

- The user can search an experiment and access the details of this experiment.
- For a given sample, the list of experiments leading to this particular sample can be obtained, and a report with all corresponding experimental details can be generated.
- For a given experimental blueprint (for example "domain 1 of target 12"), the list of experiments done can be accessed, and the corresponding report generated.
- For a given target, the list of experiments done for all experimental blueprints attached to this particular target can be obtained, and the report generated.
- The user can also search for all blueprints that reached a certain stage, and did not go beyond. For example, all the blueprints that successfully passed the expression and solubility tests, but were not produced in large scale.

Access to the different searches depends on user privileges. The last point (searching for blueprints that have successfully passed a given step but did not go into the next), for example, is only accessible to a project manager. Similarly, while searching through experiments, a regular user only sees his or her experiments, whereas a project manager is able to see all the experiments done by people working in the managed project.

Previously described features are those present in HalX v2.0. Other features are most certainly needed, and the authors are counting on user feedback to prioritize these new features included in version 2.1.

3.3.3.4. Implementation   As for previous versions, HalX is distributed freely under GNU lesser general public license (GNU-lGPL, see http://www.gnu.org/licenses/lgpl.html). The software, demonstration version, and documentation are accessible on the website http://halx.genomics.eu.org.

4. Conclusion

Although all structural genomics projects have faced the problem of data management, very few use a real LIMS, and many still rely at best on Excel spreadsheets. Nevertheless, there are many reasons to use a LIMS. It facilitates the day-to-day work of experimenters by registering very detailed information on each experiment. LIMS organizes this information and allows its restoration in a convivial manner. It also simplifies the scientific management by centralizing the information and identifying bottlenecks from global data analysis. Finally, it allows data mining studies that could in turn help to choose better targets, thus improving the outcome of the project.

In addition to the three systems described here, the Northeast Structural Genomics Consortium has developed a LIMS as part of their SPINE-2 software *(31)*. The various systems developed to date are beginning to approach the ideal criteria presented in the preceding. One of the refreshing features of structural genomics projects developed under guidelines developed by the International Structural Genomics Organization (ISGO) (http://www.isgo.org/) has been the policy concerning the sharing information about targets selected, their progress, protocols being used, and prompt deposition of structures and related data. Laboratory information management systems and standards for data exchange are key factors in ensuring the success of this new research philosophy.

Acknowledgments

Development of Xtrack was supported by the European project TEMPLOR (grant QLRT-2001-00015), and the VIZIER project, funded under the 6th Framework Programme of the European Commission (LSHG-CT-2004-511960). Development of Sesame was supported by US National Institute of Health grants P41 RR02310, P50 GM64598, 1U54 GM074901, and R21 DK070297. Development of HalX was supported by the BIOXHIT project funded under the 6th Framework Programme of the European Commission (LSHG-CT-2003-503420), the VIZIER project and the European Science Foundation.

References

1. Ball, C., Brazma, A., Causton, H., Chervitz, S., Edgar, R., Hingamp, P., Matese, J. C., Parkinson, H., Quackenbush, J., Ringwald, M., Sansone, S. A., Sherlock, G., Spellman, P., Stoeckert, C., Tateno, Y., Taylor, R., White, J., and Winegarden, N. (2004) Standards for microarray data: an open letter. *Environ. Health Perspect.* **112**, A666–667.
2. Ball, C. A., Brazma, A., Causton, H., Chervitz, S., Edgar, R., Hingamp, P., Matese, J. C., Parkinson, H., Quackenbush, J., Ringwald, M., Sansone, S. A., Sherlock, G., Spellman, P., Stoeckert, C., Tateno, Y., Taylor, R., White, J., and Winegarden, N. (2004) Submission of microarray data to public repositories. *PLoS Biol* **2**, E317.
3. Brazma, A., Hingamp, P., Quackenbush, J., Sherlock, G., Spellman, P., Stoeckert, C., Aach, J., Ansorge, W., Ball, C. A., Causton, H. C., Gaasterland, T., Glenisson, P., Holstege, F. C., Kim, I. F., Markowitz, V., Matese, J. C., Parkinson, H., Robinson, A., Sarkans, U., Schulze-Kremer, S., Stewart, J., Taylor, R., Vilo, J., and Vingron, M. (2001) Minimum information about a microarray experiment (MIAME)-toward standards for microarray data. *Nat. Genet.* **29**, 365–371.
4. Ball, C. A., Awad, I. A., Demeter, J., Gollub, J., Hebert, J. M., Hernandez-Boussard, T., Jin, H., Matese, J. C., Nitzberg, M., Wymore, F., Zachariah, Z. K., Brown, P. O., and Sherlock, G. (2005) The Stanford Microarray Database accommodates additional microarray platforms and data formats. *Nucleic Acids Res.* **33**, D580–582.
5. Saal, L. H., Troein, C., Vallon-Christersson, J., Gruvberger, S., Borg, A., and Peterson, C. (2002) BioArray Software Environment (BASE): a platform for comprehensive management and analysis of microarray data. *Genome Biol.* **3**, SOFTWARE0003.
6. Webb, S. C., Attwood, A., Brooks, T., Freeman, T., Gardner, P., Pritchard, C., Williams, D., Underhill, P., Strivens, M. A., Greenfield, A., and Pilicheva, E. (2004) LIMaS: the JAVA-based application and database for microarray experiment tracking. *Mamm. Genome* **15**, 740–747.
7. Lindon, J. C., Nicholson, J. K., Holmes, E., Keun, H. C., Craig, A., Pearce, J. T., Bruce, S. J., Hardy, N., Sansone, S. A., Antti, H., Jonsson, P., Daykin, C., Navarange, M., Beger, R. D., Verheij, E. R., Amberg, A., Baunsgaard, D., Cantor, G. H., Lehman-McKeeman, L., Earll, M., Wold, S., Johansson, E., Haselden, J. N., Kramer, K., Thomas, C., Lindberg, J., Schuppe-Koistinen, I., Wilson, I. D., Reily, M. D., Robertson, D. G., Senn, H., Krotzky, A., Kochhar, S., Powell, J., van der Ouderaa, F., Plumb, R., Schaefer, H., and Spraul, M. (2005) Summary recommendations for standardization and reporting of metabolic analyses. *Nat. Biotechnol.* **23**, 833–838.

8. Castle, A. L., Fiehn, O., Kaddurah-Daouk, R., and Lindon, J. C. (2006) Metabolomics Standards Workshop and the development of international standards for reporting metabolomics experimental results. *Brief Bioinform.* **7**, 159–165.

9. Jenkins, H., Hardy, N., Beckmann, M., Draper, J., Smith, A. R., Taylor, J., Fiehn, O., Goodacre, R., Bino, R. J., Hall, R., Kopka, J., Lane, G. A., Lange, B. M., Liu, J. R., Mendes, P., Nikolau, B. J., Oliver, S. G., Paton, N. W., Rhee, S., Roessner-Tunali, U., Saito, K., Smedsgaard, J., Sumner, L. W., Wang, T., Walsh, S., Wurtele, E. S., and Kell, D. B. (2004) A proposed framework for the description of plant metabolomics experiments and their results. *Nat. Biotechnol.* **22**, 1601–1606.

10. Markley, J., Anderson, M., Cui, Q., Eghbalnia, H., Lewis, I., Hergerman, A., Li, J., Schulte, C., Sussman, M., Westler, W., Ulrich, E., and Zolnai, Z. (2007) New Bioinformatics Resources for Metabolomics. *Pac. Symp. Biocomput.* **12**, 157–168.

11. Achard, F., Vaysseix, G., and Barillot, E. (2001) XML, bioinformatics and data integration. *Bioinformatics* **17**, 115–125.

12. Brazma, A. (2001) On the importance of standardisation in life sciences. *Bioinformatics* **17**, 113–114.

13. Pajon, A., Ionides, J., Diprose, J., Fillon, J., Fogh, R., Ashton, A. W., Berman, H., Boucher, W., Cygler, M., Deleury, E., Esnouf, R., Janin, J., Kim, R., Krimm, I., Lawson, C. L., Oeuillet, E., Poupon, A., Raymond, S., Stevens, T., van Tilbeurgh, H., Westbrook, J., Wood, P., Ulrich, E., Vranken, W., Xueli, L., Laue, E., Stuart, D. I., and Henrick, K. (2005) Design of a data model for developing laboratory information management and analysis systems for protein production. *Proteins* **58**, 278–284.

14. Zolnai, Z., Lee, P. T., Li, J., Chapman, M. R., Newman, C. S., Phillips, G. N., Jr., Rayment, I., Ulrich, E. L., Volkman, B. F., and Markley, J. L. (2003) Project management system for structural and functional proteomics: Sesame. *J. Struct. Funct. Genom.* **4**, 11–23.

15. Prilusky, J., Oueillet, E., Ulryck, N., Pajon, A., Bernauer, J., Krimm, I., Quevillon-Cheruel, S., Leulliot, N., Graille, M., Liger, D., Tresaugues, L., Sussman, J. L., Janin, J., van Tilbeurgh, H., and Poupon, A. (2005) HalX: an open-source LIMS (Laboratory Information Management System) for small- to large-scale laboratories. *Acta Crystallogr. D Biol. Crystallogr.* **61**, 671–678.

16. Quevillon-Cheruel, S., Collinet, B., Zhou, C. Z., Minard, P., Blondeau, K., Henkes, G., Aufrere, R., Coutant, J., Guittet, E., Lewit-Bentley, A., Leulliot, N., Ascone, I., Sorel, I., Savarin, P., de La Sierra Gallay, I. L., de la Torre, F., Poupon, A., Fourme, R., Janin, J., and van Tilbeurgh, H. (2003) A structural genomics initiative on yeast proteins. *J. Synchrotron. Radiat.* **10**, 4–8.

17. Quevillon-Cheruel, S., Liger, D., Leulliot, N., Graille, M., Poupon, A., de La Sierra-Gallay, I. L., Zhou, C. Z., Collinet, B., Janin, J., and Van Tilbeurgh, H. (2004) The Paris-Sud yeast structural genomics pilot-project: from structure to function. *Biochimie* **86**, 617–623.

18. Canaves, J. M., Page, R., Wilson, I. A., and Stevens, R. C. (2004) Protein biophysical properties that correlate with crystallization success in Thermotoga maritima: maximum clustering strategy for structural genomics. *J. Mol. Biol.* **344**, 977–991.

19. Page, R., Grzechnik, S. K., Canaves, J. M., Spraggon, G., Kreusch, A., Kuhn, P., Stevens, R. C., and Lesley, S. A. (2003) Shotgun crystallization strategy for structural genomics: an optimized two-tiered crystallization screen against the Thermotoga maritima proteome. *Acta Crystallogr. D Biol. Crystallogr.* **59**, 1028–1037.

20. Goh, C. S., Lan, N., Douglas, S. M., Wu, B., Echols, N., Smith, A., Milburn, D., Montelione, G. T., Zhao, H., and Gerstein, M. (2004) Mining the structural genomics pipeline: identification of protein properties that affect high-throughput experimental analysis. *J. Mol. Biol.* **336**, 115–130.

21. Stolc, V., Samanta, M. P., Tongprasit, W., Sethi, H., Liang, S., Nelson, D. C., Hegeman, A., Nelson, C., Rancour, D., Bednarek, S., Ulrich, E. L., Zhao, Q.,

Wrobel, R. L., Newman, C. S., Fox, B. G., Phillips, G. N., Jr., Markley, J. L., and Sussman, M. R. (2005) Identification of transcribed sequences in Arabidopsis thaliana by using high-resolution genome tiling arrays. *Proc. Natl. Acad. Sci. USA* **102**, 4453–4458.

22. Vucetic, S., Brown, C. J., Dunker, A. K., and Obradovic, Z. (2003) Flavors of protein disorder. *Proteins* **52**, 573–584.

23. Oldfield, C. J., Ulrich, E. L., Cheng, Y., Dunker, A. K., and Markley, J. L. (2005) Addressing the intrinsic disorder bottleneck in structural proteomics. *Proteins* **59**, 444–453.

24. Tyler, R. C., Aceti, D. J., Bingman, C. A., Cornilescu, C. C., Fox, B. G., Frederick, R. O., Jeon, W. B., Lee, M. S., Newman, C. S., Peterson, F. C., Phillips, G. N., Jr., Shahan, M. N., Singh, S., Song, J., Sreenath, H. K., Tyler, E. M., Ulrich, E. L., Vinarov, D. A., Vojtik, F. C., Volkman, B. F., Wrobel, R. L., Zhao, Q., and Markley, J. L. (2005) Comparison of cell-based and cell-free protocols for producing target proteins from the *Arabidopsis thaliana* genome for structural studies. *Proteins* **59**, 633–643.

25. Harris, M., and Jones, T. A. (2002) Xtrack—a web-based crystallographic notebook. *Acta Crystallogr. D Biol. Crystallogr.* **58**, 1889–1891.

26. Brunger, A. T., Adams, P. D., Clore, G. M., DeLano, W. L., Gros, P., Grosse-Kunstleve, R. W., Jiang, J. S., Kuszewski, J., Nilges, M., Pannu, N. S., Read, R. J., Rice, L. M., Simonson, T., and Warren, G. L. (1998) Crystallography & NMR system: a new software suite for macromolecular structure determination. *Acta Crystallogr. D Biol. Crystallogr.* **54**, 905–921.

27. Otwinowski, Z., and Minor, W. (1997) Processing of X-ray diffraction data collected in oscillation mode. *Meth. Enz.* **276**, 307–326.

28. Murshudov, G. N., Vagin, A. A., and Dodson, E. J. (1997) Refinement of macromolecular structures by the maximum-likelihood method. *Acta Crystallogr. D Biol. Crystallogr.* **53**, 240–255.

29. Markley, J. L., Anderson, M. E., Cui, Q., Eghbalnia, H. R., Lewis, I. A., Hegeman, A. D., Li, J., Schulte, C. R., Sussman, M. R., Westler, W. M., Ulrich, E. L., and Zolnai, Z. (2007) New Bioinformatics Resources for Metabolomics. *Pac. Symp. Biocomput.* **12**, 157–168.

30. Orfali, R., and Harkey, D. (1998) *Client/Server Programming with JAVA and CORBA*, John Wiley and Sons, New York.

31. Goh, C. S., Lan, N., Echols, N., Douglas, S. M., Milburn, D., Bertone, P., Xiao, R., Ma, L. C., Zheng, D., Wunderlich, Z., Acton, T., Montelione, G. T., and Gerstein, M. (2003) SPINE 2: a system for collaborative structural proteomics within a federated database framework. *Nucleic Acids Res.* **31**, 2833–2838.

Chapter 5

Data Deposition and Annotation at the Worldwide Protein Data Bank

Shuchismita Dutta, Kyle Burkhardt, Ganesh J. Swaminathan, Takashi Kosada, Kim Henrick, Haruki Nakamura, and Helen M. Berman

The Protein Data Bank (PDB) is the repository for the three-dimensional structures of biological macromolecules, determined by experimental methods. The data in the archive are free and easily available via the Internet from any of the worldwide centers managing this global archive. These data are used by scientists, researchers, bioinformatics specialists, educators, students, and lay audiences to understand biological phenomena at a molecular level. Analysis of these structural data also inspires and facilitates new discoveries in science. This chapter describes the tools and methods currently used for deposition, processing, and release of data in the PDB. References to future enhancements are also included.

1. Introduction

The Protein Data Bank is the international repository for three-dimensional structures of macromolecular complexes of proteins, nucleic acids, and other biological molecules. Data in the PDB range from small protein fragments to large macromolecular assemblies such as viruses and ribosomes, whose structures have been determined by such experimental methods as x-ray crystallography (x-ray), nuclear magnetic resonance (NMR) spectroscopy, or electron microscopy (EM). These data are publicly accessible, and used by scientists, researchers, bioinformaticians, educators, students, and lay audiences. By annotating and archiving the data in an efficient and consistent way, the PDB supports the understanding of biological phenomena at a structural level and facilitates new discoveries in science.

   The PDB was founded in 1971 and was housed at the Brookhaven National Laboratories (1). The Research Collaboratory for Structural Bioinformatics (RCSB, http://home.rcsb.org) assumed management of the PDB in 1998 (2) and in the following years the Protein Data Bank Japan (PDBj, http://www.pdbj.org) and the Macromolecular Structure Database at the European Bioinformatics Institute (MSD-EBI, http://www.ebi.ac.uk/msd) also became deposition and processing centers for the PDB. In 2003, the Worldwide PDB (wwPDB, http://www.wwpdb.org/) (3,4) was established by these three sites. This formalized the agreement to ensure the continuance of PDB as the single

From: *Methods in Molecular Biology, Vol. 426: Structural Proteomics: High-throughput Methods*
Edited by: B. Kobe, M. Guss and T. Huber © Humana Press, Totowa, NJ

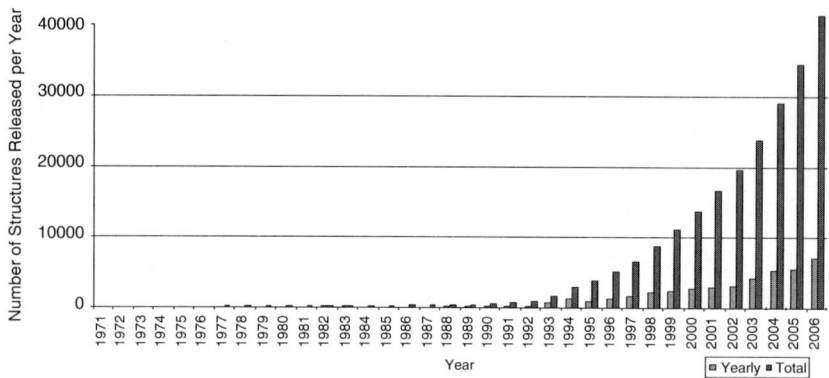

Fig. 5.1 The PDB growth curve.

archive of macromolecular structural data from where the data are freely and publicly available to the global community. The Biological Magnetic Resonance Data Bank (BMRB, http://www.bmrb.wisc.edu) *(5)* joined the wwPDB in 2006. In addition to providing access to the PDB archive, each wwPDB site provides databases and websites that provide different views and analyses of the structural data contained within the PDB archive *(6–12)*.

New methods for structure determination and advances in the existing ones have led to an exponential growth of the PDB, as seen in Fig. 5.1. From the seven structures at its inception, the PDB archive has grown to include more than 46,000 structures at the time of this writing. In 2006, the wwPDB received 6541 depositions. These depositions represent structures of important biological molecules deposited by scientists and projects around the world. One other factor contributing to this growth is the initiation of structural genomics (SG) projects around the world *(13)*. This international effort started in the late 1990s and aims to determine protein structures on a scale similar to the genome sequencing projects. Since there are many groups simultaneously working on a large number of target sequences, it is critical to track the progress of these projects in order to avoid duplication of efforts and share new techniques and protocols developed in this process.

This chapter provides an overview of the data representation used by the PDB and describes the tools and process of deposition and annotation of structural data at the PDB. Some details about tracking the progress of the various SG projects around the world are also included.

2. Materials

2.1. Data Representation

The structures archived in the PDB are determined by scientists around the world using experimental methods such as x-ray crystallography, nuclear magnetic resonance spectroscopy, and electron microscopy. When PDB users download coordinate and experimental data files from the PDB, they use various applications to visualize and/or analyze them. To facilitate the use of these data, it is critical that the files deposited to and distributed by the wwPDB are in a standard format. To this end, the PDB has created standards for representing and formatting these data. These are described in the following sections.

2.1.1. PDB File Format

The structure of the PDB file format *(14)* was based on the use of the 80-column punch card and the need to archive the data of structures determined by x-ray crystallography. Since this format archives structural data in a format easily read by humans, it has evolved only slightly over the past 35 years. The PDB format still endures as a commonly used format for data visualization and analysis.

Each PDB format file is composed of two main sections: header and coordinates. The header section carries details regarding the name(s) of molecule(s) in the structure, author, citation, sequence, chemical components, secondary structure, and information about the data collection and structure solution. The coordinate section includes atomic coordinates, residue and atom names and numbers, polymeric chain identifiers, alternate position identifiers, occupancy, and temperature factors. Through advances in structural biology, methods such as NMR and EM are also being used to determine structures. The PDB format has been modified to accommodate method-specific details in the header section.

Although archiving structural data from different experiments in a common format is convenient for some users, vestiges of the original design for x-ray data in the modified formats (for NMR, EM, and other methods) can be redundant or confusing for others. Additionally, archiving large and complex macromolecular assemblies presents many challenges to the PDB format *(15)*. The PDB format has limits on the maximum number of atoms in a structure, residues in a chain, and chains in the file. These limits have been exceeded in several structures deposited to the PDB. Large structures like ribosomes often have to be split into multiple PDB entries. For example, PDB entries 1GIX and 1GIY *(16)* together form the ribosome structure that was determined. Complex ligands or inhibitors, non-linear polymers like branched chain polysaccharides, and circular peptide or nucleic acid chains cannot be easily represented in this format, as seen in the entry 1ZA8 *(17)*. X-ray techniques in certain circumstances allow for a structure solution in which a large part of the protein structure needs to be represented in two or more conformations due statistical or dynamic disorder. These cases (e.g., in PDB ID 1ZY8) *(18)* are also not clearly represented in the PDB format.

2.1.2. mmCIF and the PDB Exchange Dictionary

The limitations of the PDB format in representing large macromolecular complexes have created the need for a more versatile representation of structural data. The data processing system at the PDB is based on the Macromolecular Crystallographic Information File format (mmCIF) *(19,20)*, which provides rigorous syntax and semantics for representing macromolecular structure and each experimental step in the structure determination. The system is driven by a software-accessible data dictionary that encodes definitions and examples for each item of data along with additional information that assists software in managing and validating data. The features of the dictionary include data types, boundary conditions, expression units, controlled vocabularies, and parent–child relationships. The mmCIF data dictionary was originally

```
_cell.entry_id          1QQQ
_cell.length_a     123.234
_cell.length_b      60.500
_cell.length_c      56.670
_cell.angle_alpha     90.00
_cell.angle_beta     119.05
_cell.angle_gamma     90.00
_cell.Z_PDB             4
_cell.ndb_unique_axis    ?
#
loop_
_entity_poly_seq.entity_id
_entity_poly_seq.num
_entity_poly_seq.mon_id
1    1      ASP
1    2      ILE
1    3      GLU
1    4      LEU
1    5      THR
1    6      GLY
1    7      SER
...truncated...
```

Fig. 5.2 Examples of mmCIF syntax for the categories cell and entity_poly.

developed to specifically document the macromolecular x-ray experiment. The PDB has used the mmCIF dictionary-based technology to develop an expanded data dictionary that covers the full scope of content collected and distributed by the resource. This data dictionary, called the PDB Exchange Dictionary *(21)*, has the same syntax as mmCIF but broader content. It also includes definitions for protein production, NMR, and EM. In addition to documenting the content of the PDB, the Exchange Dictionary provides the basis for standardizing annotation practices and accurately exchanging data among the wwPDB partners. A searchable version of the PDB Exchange Dictionary, related extension dictionaries, and the underlying Dictionary Definition Language (DDL) *(22)* dictionary are maintained at http://mmcif.rcsb.org. The dictionary-based architecture of the PDB has also facilitated supporting multiple electronic formats such as XML, as described in the following section.

Examples of major syntactical constructs used by the mmCIF format are illustrated in Fig. 5.2. Each data item or group of data items is preceded by an identifying keyword. Groups of related data items are organized in data categories and it is possible to accurately represent the hierarchy of macromolecular structure through the details of parent–child relationships in the data dictionary (Fig. 5.3).

2.1.3. PDBML File Format

A third data representation format, PDBML *(23)*, provides a representation of PDB data in XML format and was created by directly translating the files from mmCIF format. The syntax of these file formats is different, but they

Fig. 5.3 Examples of the relationships between data categories describing polymer molecular entities, their sequence, and related coordinate data are depicted here.

contain the same logical organization of data that allows easy translation between these formats. The XML format is equally flexible in accommodating large and complex structural data and provides users with the further advantage of using many off-the-shelf XML tools. Three different PDBML files are produced for each PDB entry—a fully marked-up file, a file without the three-dimensional coordinates, and a file with space-efficient encoding of three-dimensional coordinates of the atoms. Software tools for generating and reading the files in the various file formats are available from the RCSB (http://sw-tools.rcsb.org) and PDBj (http://www.pdbj.org/xpsss).

2.2. Tools for Data Deposition and Annotation

The data in the PDB are deposited by the scientists who determine the structures. Prior to release, these data are processed, annotated, and validated by annotators working at the wwPDB processing centers, who are themselves experienced structural scientists. The tools used for data deposition and annotation are described here.

Data deposition constitutes the communication of three-dimensional coordinates of structures, experimental data, and other information by the depositors (scientists) to the PDB archive. The wwPDB sites use different tools to collect and process structural data, but the types of information collected by each of the centers for any structure deposition is identical. A list of these deposition tools and resources is given in Table 5.1. Data annotation is conducted by various internal tools, many of which are extensions of the data deposition tools.

At the minimum, a PDB deposition consists of the three-dimensional coordinates of the structure and information about the composition of the structure. Other data are also required, such as information about the experiment performed and details of the structure determination steps. At the time of data deposition, it is helpful for depositors to have output and log files from all structure solution software on hand. These files provide information about the structure solution steps. When available, data harvest files (see Section 3.1.1.) generated by refinement programs should be used to automatically populate the required items for the deposition. Although the deposition of experimental

Table 5.1 wwPDB software tools for preparation and deposition of structural data.

wwPDB center	Software tool	Function	URL
	pdb_extract* (28)	Data harvesting (for x-ray and NMR structures)	pdb-extract.rcsb.org
	Validation Server	Data validation	deposit.rcsb.org/validate
RCSB	ADIT*	Data assembly and validation	deposit.rcsb.org/adit
	Ligand Depot (29)	Identification and checking of all molecules and chemical groups	ligand-depot.rcsb.org
	Validation Server	Data validation	pdbdep.protein.osaka-u.ac.jp/validate
PDBj	ADIT	Data assembly and validation	pdbdep.protein.osaka-u.ac.jp/adit
	Biotech Validation Server	Data validation	http://biotech.ebi.ac.uk:8400
	Autodep 4.0* (8)	Data harvesting, validation and assembly	www.ebi.ac.uk/msd-srv/autodep4/index.jsp
MSD-EBI	MSDchem (30)	Identification and checking of all molecules and chemical groups	www.ebi.ac.uk/msd-srv/chempdb/cgi-bin/cgi.pl
	MSDpisa (39)	Analysis of protein interfaces surfaces and assembly	www.ebi.ac.uk/msd-srv/prot_int/pistart.html
BMRB	ADIT-NMR	Assembly and validation of NMR structures	batfish.bmrb.wisc.edu/bmrb-adit

*Also available in a downloadable desktop version.

Table 5.2 Items depositors should have available during data deposition to the wwPDB.

All depositions	X-ray structures	NMR structures	EM structures
Coordinate file	Structure factor file (refinement and phasing data sets, as appropriate)	Constraint data	Volume data (EMDB)
Author contact information	Crystallization information	Chemical shifts (BMRB)	Sample preparation details
Release instructions	Crystal data	Ensemble description	Data acquisition
Title	Data collection details	Equipment used	Refinement details including source of fitted coordinates
Citation information (when available)	Data integration and phasing details	Sample solution and experiment details	
Molecule name	Refinement details		
Ligand information			
Sequence information and details			
Source information Software used			

data (like structure factor files for x-ray structures and constraint files for NMR structures) is optional, it is encouraged by the PDB and may be required to publish in some journals. A list of items, files, and information required for depositing structures solved by different methods is included in Table 5.2.

2.3. Tools for Tracking Progress of Structural Genomics Projects

The RCSB PDB has created an online portal to track the progress of structural genomics efforts (http://sg.pdb.org/) *(7)*. A target registration database, TargetDB (http://targetdb.pdb.org/) *(24)*, was founded in 2001 to track the sequence, status, and other information about the targets being studied at the SG centers. An extension of this resource, PepcDB (http://pepcdb.pdb.org/), was founded in 2004 and archives the cloning, expression, purification, and structure solution details in all trials for each target. Both TargetDB and PepcDB are housed at the RCSB PDB. MSDtarget (http://www.ebi.ac.uk/msd-srv/msdtarget/SGT_index.html), a tool developed by the MSD-EBI as part of the SPINE project *(25)* is another tool for searching and managing SG targets.

Aside from tracking the status of SG targets, the management of information generated by these projects has also presented a challenge. To address this issue, data models for laboratory information management and analysis have been proposed *(26)*. Based on this data model, a protein information management system (PIMS http://www.pims-lims.org/) is now also available.

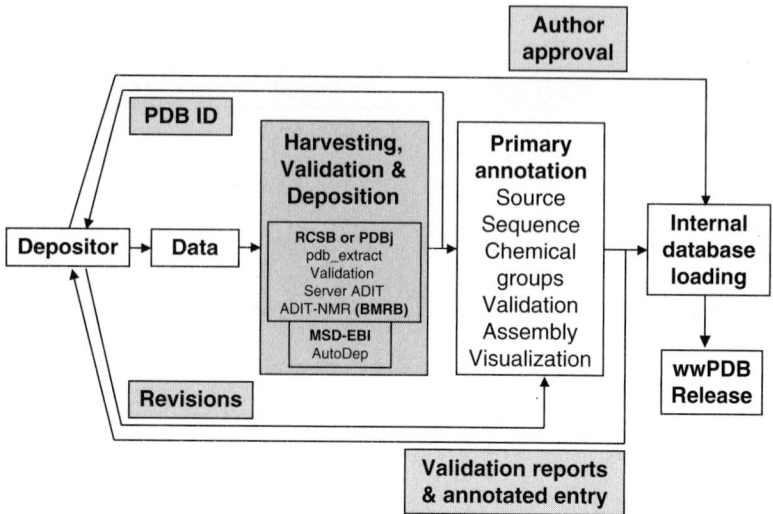

Fig. 5.4 The data deposition and annotation pipeline: steps and interactions between depositors and annotators in finalizing a PDB entry are shown.

3. Methods

3.1. Data Deposition and Annotation

Deposition of structural data to the PDB and its annotation is an interplay between scientists who determine the structure, structure determination software, deposition and annotation tools, and the annotation staff. Key aspects of this interplay are summarized in Fig. 5.4. The interaction between depositors and annotators is central to the presentation of the structure in an entry in its best possible format.

The deposition of three-dimensional coordinates, in PDB or mmCIF file format, is the entry point in the structure deposition and annotation pipeline. This process can be divided into three basic steps: data harvesting, entry building, and validation. Data harvesting involves the automatic or semiautomatic gathering of relevant data from the output, log, or harvest files of structure determination steps. The entry building step combines the harvested data with the coordinates and other information to form the structure deposition. The depositor then can validate the entry to check for format problems, internal consistency, and the geometry of the structure.

After the entry is deposited, the data are matched against and referenced to standard databases, checked for consistency, and validated by the wwPDB. Note that the validation during annotation is more detailed to ensure that all aspects of the deposition have been checked and any remaining inconsistencies or errors are brought to the depositor's attention in the validation report. The wwPDB accepts structures deposited from many different experimental methods, which vary in content, quality, and information. A major objective of all wwPDB annotators is to ensure that each entry is self-consistent, which in turn ensures that the PDB archive maintains its uniformity. Annotators at the wwPDB processing centers use common guidelines and policies for annotation. There is a high level of automation in the procedures that deal with the routine aspects of annotation of PDB entries. However, some aspects of this process are difficult to automate and need to be addressed on a case-by-case basis. The ultimate goal of this whole process is to present and preserve the structural data in a manner that is correct and annotated to a high level of understanding.

The principal steps in data deposition and annotation are described in the following sections.

3.1.1. Data Harvesting

Data harvesting is often the bottleneck in the deposition process. Structures may have been determined over a long period of time or by various people. Manually gathering specific information and details about each stage of structure solution can be time consuming and error prone. Several attempts have been made to automate the process of gathering information for data deposition to facilitate a richer, more complete and correct deposition without making higher demands on the depositor. Many refinement programs now generate a file that captures relevant information for structure deposition from that step of structure solution (27). These harvest files can be read in by deposition software programs such as ADIT (http://deposit.rcsb.org/adit/ and http://pdbdep.protein.osaka-u.ac.jp/adit/) and AutoDep 4.0 (http://www.ebi.ac.uk/msd-srv/autodep4/) (8) to automatically or semiautomatically communicate the appropriate values to the deposition. Data from harvest files can be combined with non-electronically generated information such as the molecule name, sequence, citation, and author information from a previous deposition using the AutoDep4.0 tool or a combination of the programs pdb_extract and the ADIT tool to complete the structure deposition.

The program pdb_extract (28) was designed to capture relevant details for x-ray structure depositions. This program can read in the output, log, or harvest files produced by different software programs used for experimental data reduction, phasing, density modification, and refinement of the structure and extract relevant information for structure deposition. It can also combine this electronically generated data with non-electronically generated information such as protein name, sequence and other information to create mmCIF format files for one or more related depositions (Fig. 5.5). Presently pdb_extract can

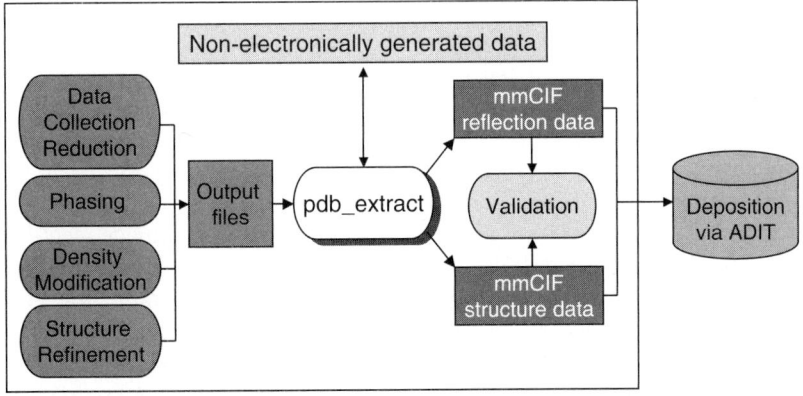

Fig. 5.5 A schematic representation of the use of pdb_extract for deposition of an x-ray structure.

help in preparing both x-ray and NMR depositions. These files can be deposited to the PDB using either ADIT or ADIT-NMR. Even though the specific functionalities of these tools appear to be different, their use at the different wwPDB centers leads to fast, easy, and accurate data deposition.

In addition to being available from the respective wwPDB web sites, most of the data deposition tools have desktop versions available for download. These versions can be used for preparing the data for deposition offline or creating an electronic notebook that records relevant information from all stages of structure solution.

3.1.2. Entry Building

The ADIT and AutoDep 4.0 tools can assemble all harvested and non-electronically generated data for validation and deposition. These tools also provide an interface in which the depositor can review, confirm, add, or edit the information that he or she has included in the deposition. Depending on the tool being used, format check and geometry validation of the deposition is either recommended or required during this process.

The non-electronically generated information associated with each structure deposition helps in classifying, organizing, and cross-referencing the deposited structure in the PDB. These need to be carefully checked and correctly communicated in the deposition to help in the annotation process and later use of the file. Some of the procedures to check these pieces of information are described in the following sections.

3.1.3. Validation: Checking Various Aspects of the Structure Deposition

Two types of data are required for a structural data deposition in the PDB: (1) coordinates and details about the structure determination; and (2) information about the source, sequence, chemical composition, and molecular assembly. Data related to the first aspect can be captured using data harvesting procedures and checked for accuracy, consistency, and completeness. Information about the composition and assembly of the structure cannot be automatically captured. They have to be checked against a number of external databases and resources. Some of this checking is part of the structure deposition process, whereas other checks are conducted during annotation. Use of appropriate tools and resources to check the data at the time of deposition ensures that the information provided is complete and correct. This in turn makes the annotation process efficient and minimizes the number of revisions before the entry is finalized for release.

3.1.3.1. Checking the Identity of Chemical Components in the Deposition: The various chemical components such as amino acids, nucleic acids, solvent molecules, and a large variety of ions, ligands, inhibitors, and other designed compounds associated with a structure need to be identified correctly in the structure. The wwPDB uses and maintains a common chemical component dictionary for this purpose. Each chemical component has a unique identifier of up to three alphanumeric characters that specifies its chemical name, formula, connectivity, bond order, and stereochemical information. For example, every instance of alpha-D-mannose in the PDB is identified with the unique identifier MAN, whereas the molecule beta-D-mannose, which differs in chirality at one carbon, is assigned a separate code BMA. All atoms and residue names for the standard amino acids and nucleotides have a specific nomenclature in this dictionary.

Through the process of structure deposition and annotation each chemical component in the structure must be correctly identified. Several free, Web-based tools help in matching the chemical component to the dictionary and identifying its unique identifier. Ligand Depot *(29)*, developed by the RCSB PDB, and MSDChem *(30)*, developed by the MSD-EBI, are data warehouses that integrate databases, services, tools, and methods related to all chemical components in the PDB. Both of these services are built on the same chemical component dictionary and provide chemical and structural information about all standard and nonstandard residues, ligands, inhibitors, or other chemical components associated with the structure. Depositors should use these tools to search for existing ligands in the PDB, either by a name, formula, substructure, or chemical fingerprint, and utilize the corresponding code in their deposition. For new ligands, authors should include a chemical drawing of the ligand (including bond types and hydrogens) during the deposition process, by uploading a drawing into ADIT or AutoDep 4.0.

During annotation, the nomenclature of the residue and order of the atoms in it are matched to that in the chemical component dictionary. The wwPDB centers have specialized tools that implement both the recognition of chemical identity and matching of atom names against the dictionary. These tools are also extremely important in the identification of nonstandard residues or bound ligands, also known as heterogens or hetgroups, associated with more than half the entries deposited with the PDB. Once the correct match is obtained, the ligand description in the PDB entry is changed to mirror that in the chemical component dictionary. If a new ligand is encountered during annotation, a new definition is created and added to the chemical component dictionary. One or more specialized software programs, such as OpenEye (http://www.eyesopen.com), ChemDraw (http://www.cambridgesoft.com/software/ChemDraw), ACDLabs (http://www.acdlabs.com), and CACTVS *(31,32)* are used to predict the stereochemistry, bond orders, chemical name, formula, and charge for the ligand. This is then verified by the annotator. When available, the predicted ligand definition is compared with a chemical drawing of the ligand provided by the author for verification. To prevent duplication and ensure a consistent approach toward handling of chemical components, ligand definitions are exchanged among the wwPDB sites on a daily basis.

3.1.3.2. Checking the Sequence, Source, and Sequence Database References Properties of structures deposited to the PDB depend on their composition. The structure of a small domain of a large protein or that of a mutant protein may be completely different from the native protein structure. Thus, the sequence of the polymers in a structure should be clearly defined. Authors should provide the sequence for all polymers (protein and nucleic acids) in the structure that were present in the experiment. This should include mutations, insertions, variants, uncleaved expression tags, and any residues missing from the coordinates due to disorder. Whenever possible, depositors should use standard sequence comparison tools and their knowledge about the sample to provide appropriate sequence database reference(s) for the sequence(s) used. The sequence(s) are deposited according to the standard one letter code. Nonstandard residues are listed using the code defined in the PDB chemical component dictionary. For example, a selenomethionine residue would be listed in the sequence as (MSE).

Identification of the biological source for the macromolecule(s) is the first step for identifying the correct sequence database reference. During annotation, the scientific name of the source organism is compared with standard databases such as the NCBI taxonomy database *(33)* or NEWT (http://www.ebi.ac.uk/newt) *(34)*. When one particular species has many synonyms, the official name is used. However, if a species does not exist in these taxonomy databases, information regarding the species is sent to them for further action.

Annotators identify the sequence database references by semiautomatically aligning the deposited sequence against the whole sequence database. When an appropriate sequence database reference is identified, information about the macromolecule's name, synonyms, and source are extracted from it and included in the PDB entry. In consultation with the depositor, the exact sequence that was used in the experiment is annotated to mark up the portions of the sequence that match the sequence database reference. Explanations for deviations from the reference are also included in the PDB entry. Comparison of the author's residue numbering (in the coordinates) to that in the sequence database reference is included in the file so that a user can map a specific residue in the coordinates to the one it corresponds to in the sequence database reference.

For proteins, a UniProt *(35)* reference is preferred. If a suitable sequence match is not found in UniProt, and the protein is not an antibody or a *de novo* protein, authors can submit their sequences directly to UniProt by using the SPIN (http://www.ebi.ac.uk/swissprot/Submissions/spin/) deposition system in order to obtain a UniProt accession code. Naturally occurring nucleic acids in structures should be referenced to sequences in GenBank *(36)*, DDBJ *(37)*, and EMBL *(38)*. After accounting for mutations, cloning artifacts, and expression tags, if any sequence discrepancies remain, the sequence and/or structure should be corrected, as necessary.

3.1.3.3. Checking the Molecular Assembly: An annotated PDB entry gives information on the oligomeric state or assembly of the macromolecules contained in the entry. One of the more challenging aspects of annotation of structures deposited with the PDB is the identification of this assembly (currently also referred to as biological unit, biological assembly, or quaternary structure). The molecular assembly seen in the structure has bearing on the understanding of the molecule's biological function. Although the oligomeric assembly observed in a crystal structure may not directly represent its known biologically relevant form, the possibility of forming the assembly may inspire additional experiments to test this. The deposited coordinates may contain one or more complete macromolecule(s), some parts of which may require application of crystallographic symmetry operations to generate the complete molecular assembly (Fig. 5.6).

Algorithms have been developed to determine the most likely oligomeric state, taking into account the symmetry-related chains, and overall surface area buried between different components in the structure. These algorithms form the basis of MSD-PISA *(39)* and the Protein Quaternary Structure server *(9)*. Other issues that impact on the accurate prediction of the molecular assembly include surface similarity and energies of association and dissociation.

(A)

(B)

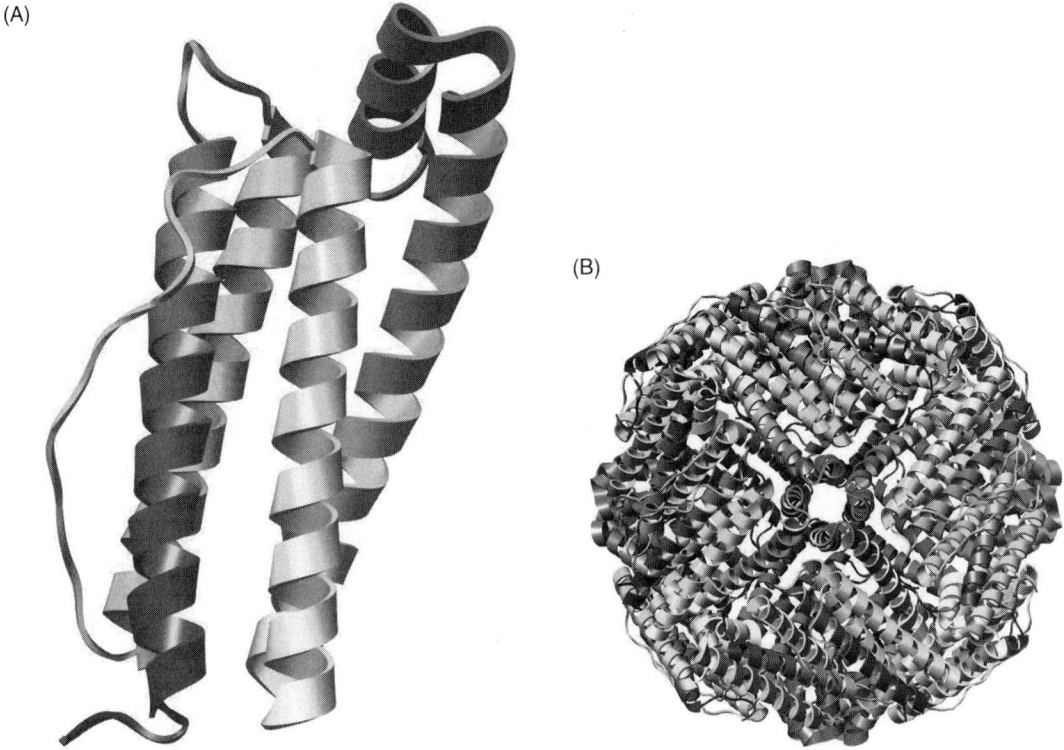

Fig. 5.6 PDB entry 1AEW *(52)*. **A**. The asymmetric unit is shown alongside **(B)** the complete molecular assembly that can be generated by applying 23 symmetry operations on the deposited coordinates.

However, since crystal structures often make many contacts with other protein molecules from neighboring unit cells, the prediction from such tools can never be completely accurate. Therefore, it is necessary to incorporate information from other sources, including biochemical data, visual inspection of surface complementarities at the oligomeric interface, and depositor input to determine the molecular assembly. It is important to note that the molecular assembly of a protein fragment can be substantially different from the molecular and biological assembly of the full-length macromolecule.

MSD-PISA is a publicly available resource that depositors should use to test molecular assembly in their deposition. Authors can compare the results from this tool to any biochemical evidence that they may have for the oligomerization state of the biologically relevant form of the molecule and point out any differences. In the majority of cases the PISA calculation of molecular assembly and the author's determination of biologically relevant assembly are the same. It is noted in the deposition when MSD-PISA predictions do not match the author-provided oligomeric assembly. For future depositions, the wwPDB has agreed to include both the author determined biological unit as well as the calculated molecular assembly.

3.1.4. Validation

Validation of structural data prior to deposition ensures that the files and information uploaded are complete and correct. Depositors are encouraged to validate their structure immediately before deposition as a final check. Reports generated at this stage help in identifying possible errors in the structure, and provide the depositors a chance to correct these issues prior to deposition. This in turn reduces the number of problems and corrections during annotation and review.

Although all the tools that help in building an entry (ADIT, ADIT-NMR and AutoDep 4.0) allow validation of data before deposition, separate validation servers are also available (see Table 5.1). Validation programs check the formats of coordinate and structure factor files, and create a variety of reports about the structure. The sequence of polymers in the structure are matched against the deposited sequences and all chemical modifications, ligands, or inhibitors associated with the structure are checked against the chemical component dictionary based on the identifier used in the deposition. After validation, the data can be deposited to the PDB and a unique PDB ID is assigned to the entry. Upon deposition, the PDB, mmCIF, and/or XML formatted files for the entry are generated for annotation.

In an attempt to make the data archived in PDB useful to a diverse community of users, annotators validate all deposited structures against standards set for evaluating these data *(40)*. The validation is conducted using internal programs based on methods and standards developed by others, including PROCHECK *(41)*, MolProbity *(42)*, and WHATCHECK *(40)*. Reports generated by the internal programs are similar to the ones generated prior to deposition where various quality indicators of the overall geometry are computed. The checks reported by the internal validation programs *(43)* include the identification of close contacts between atoms in the entry, unusual deviations in bond lengths and bond angles in covalently linked atoms, as well as residues whose backbone torsion angles lie outside predicted Ramachandran plot regions. For crystal structures, close contacts between atoms of symmetry-related molecules are also identified. All of these issues are brought to the attention of the depositor when the validation reports are communicated to them along with comments or questions regarding any remaining errors or discrepancies in the sequence, composition, geometry or file formats. After reviewing this, the depositor may provide a set of revised coordinates with corrected or improved geometry, which then replaces the initially deposited structure. The new coordinates are incorporated in the deposition and the entry is revalidated by annotators and reviewed again by the authors before the entry is finalized. In some cases this cycle may be repeated more than once. Through this validation process, annotators ensure that the data are as complete and correct as possible, consistent and accurate to within limits of the experiment and have indicators of quality.

The annotators use a variety of visualization tools, such as RasMol *(44)*, Chimera *(45)*, and AstexViewer *(46)*, to verify structures and identify potential issues with the deposition that may require corrective action by the depositor. Some common issues include identification of solvent or small molecules modeled at positions far removed from the macromolecule, bad bond lengths, and incorrect connectivities. Visualization tools are also used to assess the feasibility of predicted molecular assemblies. Visualization of the structure in conjunction with the structural data analysis provides an extra cognitive tool in the annotation toolbox.

Structure factor (SF) files, used for the refinement of an x-ray crystallographic structure are the experimental data that accompany an x-ray structure. These files may be deposited in a wide variety of formats, depending on the programs used for structure refinement. Additional SF data corresponding to data used for phasing the structure may also be deposited. As part of the annotation process, these files are converted to a single mmCIF format file. Programs and scripts like SFCHECK *(47)* and EDS *(48)* are used to calculate the correlation between the deposited coordinates and the SF data. Low correlation values may indicate errors in the uploaded file and are brought to the attention of the depositor.

Although NMR constraint data are accepted with the coordinates at all wwPDB centers, these files and other NMR-specific experimental files are currently checked and validated only at the BMRB *(49)*. Conversely, three-dimensional coordinate data deposited at BMRB through ADIT-NMR is forwarded to annotators at the RCSB for annotation.

The EM maps or volumes are the experimental data associated with electron microscopy depositions. The coordinate data for EM structures are accepted at most of the wwPDB centers but the EM maps can presently only be deposited with the EMDB *(50)* at EBI. As the guidelines for the validation for electron microscopy experiments are still under discussion and development, the majority of information in the EMDB provided by depositors is archived as is.

3.1.5. Interaction Between Depositors and Annotators

Communication between depositors and the annotation staff occurs daily, mostly through e-mail and sometimes via the telephone. While processing an entry, annotators may communicate with depositors to clarify the sequence of polymers or ligands and hetgroups in the file, or resolve any major errors or inconsistencies in the data. However, the majority of interactions are related to the communication of validation reports and processed files generated during the annotation of the structure. Through this interaction depositors get a chance to review their data after they have been processed according to the wwPDB annotation guidelines and check the accuracy of components in the deposition. The wwPDB plans to streamline the system by which authors receive notification about their entries, access their depositions, and respond to any questions, including providing any coordinate or other corrections or revisions.

Currently the wwPDB staff receives upward of 200 e-mail messages each day from depositors. Examples of typical author e-mails include questions on how to deposit and validate, notification of structure publication, and corrections to processed entries, including new coordinate sets. For some entries, the author may send revised coordinates multiple times, which leads to repeated processing and validation of the entry by the annotators. In order to facilitate rapid processing for all entries, the wwPDB strongly recommends that authors make use of validation tools before depositing their structures and use the check lists for deposition. Every communication between the author and annotator is electronically archived and easily accessible to the annotation staff.

Aside from exchanging e-mails regarding specific depositions, each member of the annotation staff is involved in community outreach. Even through the daily e-mail exchanges annotators educate depositors on better ways to deposit their data. User outreach also occurs through talks, posters, and discussions at various meetings, workshops, and demonstrations, and through informal exchanges.

3.2. wwPDB Annotation Policies

To maintain a consistent approach to various issues regarding data deposition and annotation, the wwPDB centers have agreed on some common policies that govern the deposition, processing, and release of entries at the wwPDB. Such policies are subject to constant review via e-mail exchanges, telephone and video conferences, and exchange meetings between members of the constituent sites and advisory committees. Some of these policies are discussed here.

3.2.1. Release of Entries

Depositors choose the release status for their entry at the time of deposition. Entries can either be released immediately (REL), upon publication of the primary citation (HPUB), or on a fixed date up to a year past deposition (HOLD). Entries deposited as HPUB that have not been published within the year after deposition must be either released or withdrawn. The citation information for any entry may be added or updated even after its release.

Any unreleased entry may be withdrawn upon author request. Entries that have already been released can not be withdrawn but are obsoleted. Most obsolete entries have corresponding superseding entries, deposited by the original author or their group. Users who search for an obsolete entry are automatically redirected to the superseding entry. The obsoleted entry is still available in the obsolete archive.

3.2.2. Citation Information

A few journals submit publication dates and citation information directly to the wwPDB. For all journals, however, the wwPDB scans the literature for publication dates. The wwPDB annotators also appreciate the citation information that is sent by depositors and the community.

3.2.3. Theoretical Models

The PDB is primarily an archive for structural data determined by experimental methods. In the past some theoretical and homology models have also been deposited with the PDB. As of October 15, 2006, the wwPDB follows the recommendations of a workshop held at the RCSB at Rutgers University in November, 2005. At this meeting, a working group of experts in a variety of related fields decided that PDB depositions should be restricted to atomic coordinates that are substantially determined by experimental measurements on specimens containing biological macromolecules. This includes, for example, structures determined by x-ray crystallography, NMR, and cryo-electron microscopy. It does not include models determined purely *in silico* using, for example, homology or *ab initio* methods. The outcome of the workshop and its recommendations have been published *(51)*.

4. Future of Data Deposition and Annotation

The PDB is growing at an exponential rate, and with structural genomics initiatives from around the world producing large amounts of data, it is now more important than ever to have systems in place that can handle the large number of expected depositions in an efficient and consistent manner. Members of the wwPDB have jointly developed common guidelines for the annotation of deposited structures to ensure that the level and quality of curation is consistent across all centers. With all the tools designed and implemented by the wwPDB partners, the stage is set for high-throughput curation of PDB entries.

A future goal is to develop a common deposition system for all wwPDB deposition sites that brings together the combined expertise and experience of the member sites in collecting and processing data. This new system would aim to provide automation, flexibility, and ease of use for the depositor, while at the same time ensuring that the data passed from the deposition to the annotation stream is of higher quality than at present. Another issue being addressed is the development of a better correspondence system between annotators and depositors that could allow the depositor to correspond and provide extra annotation or corrections from within the deposition system itself in a secure manner. To make the deposition process more rewarding for the depositor, there are plans to provide detailed validation statistics and additional information regarding the fold, motif, and site information using the tools already available to the annotators. It is hoped that all of these will go a long way toward contributing to a richer structural database for biological macromolecules.

Acknowledgments

The authors acknowledge the staff of all wwPDB sites, and our advisory committees.

At the RCSB PDB, we acknowledge the programming staff consisting of Li Chen, Zukang Feng, Vladimir Guranovic, Andrei Kouranov, John Westbrook, Huanwang Yang; and the annotation staff consisting of Jaroslaw Blaszczyk, Guanghua Gao, Irina Persikova, Massy Rajabzadeh, Bohdan Schneider, Monica Sekharan, Monica Sundd, Jasmine Young, and Muhammed Yousufuddin.

At MSD-EBI, we acknowledge annotators Adamandia Kapopoulou, Richard Newman, Gaurav Sahni, Glen van Ginkel, Sanchayita Sen, and Sameer Velankar.

At PDBj, we acknowledge annotators Reiko Igarashi, Yumiko Kengaku, Kanna Matsuura and Yasuyo Morita.

The RCSB PDB is operated by Rutgers, The State University of New Jersey and the San Diego Supercomputer Center and the Skaggs School of Pharmacy and Pharmaceutical Sciences at the University of California, San Diego. It is supported by funds from the National Science Foundation, the National Institute of General Medical Sciences, the Office of Science, Department of Energy, the National Library of Medicine, the National Cancer Institute, the National Center for Research Resources, the National Institute of Biomedical Imaging and Bioengineering, the National Institute of Neurological Disorders and Stroke, and the National Institute of Diabetes and Digestive and Kidney Diseases.

EBI-MSD is supported by funds from the Wellcome Trust (GR062025MA), the European Union (TEMBLOR, NMRQUAL, SPINE, AUTOSTRUCT and IIMS awards), CCP4, the Biotechnology and Biological Sciences Research Council (UK), the Medical Research Council (UK) and European Molecular Biology Laboratory. PDBj is supported by grant-in-aid from the Institute for Bioinformatics Research and Development, Japan Science and Technology Agency (BIRD-JST), and the Ministry of Education, Culture, Sports, Science and Technology (MEXT).

References

1. Bernstein, F. C., Koetzle, T. F., Williams, G. J. B., Meyer Jr., E. F., Brice, M. D., Rodgers, J. R., Kennard, O., Shimanouchi, T., and Tasumi, M. (1977) Protein Data Bank: a computer-based archival file for macromolecular structures. J. Mol. Biol. 112, 535–542.

2. Berman, H. M., Westbrook, J., Feng, Z., Gilliland, G., Bhat, T. N., Weissig, H., Shindyalov, I. N., and Bourne, P. E. (2000) The Protein Data Bank. Nucl. Acids Res. 28, 235–242.

3. Berman, H. M., Henrick, K., and Nakamura, H. (2003) Announcing the worldwide Protein Data Bank. Nat. Struct. Biol. 10, 980.

4. Berman, H., Henrick, K., Nakamura, H., and Markley, J. L. (2006) The worldwide Protein Data Bank (wwPDB): ensuring a single, uniform archive of PDB data. Nucl. Acids Res. doi: 10.1093/nar/gkl971.

5. Ulrich, E. L., Markley, J. L., and Kyogoku, Y. (1989) Creation of a Nuclear Magnetic Resonance Data Repository and Literature Database. Protein Seq. Data Anal. 2, 23–37.

6. Deshpande, N., Addess, K. J., Bluhm, W. F., Merino-Ott, J. C., Townsend-Merino, W., Zhang, Q., Knezevich, C., Xie, L., Chen, L., Feng, Z., Kramer Green, R., Flippen-Anderson, J. L., Westbrook, J., Berman, H. M., and Bourne, P. E. (2005) The RCSB Protein Data Bank: a redesigned query system and relational database based on the mmCIF schema. Nucl. Acids Res. 33, D233–D37.

7. Kouranov, A., Xie, L., de la Cruz, J., Chen, L., Westbrook, J., Bourne, P. E., and Berman, H. M. (2006) The RCSB PDB information portal for structural genomics Nucl. Acids Res. 34, D302–D305.

8. Tagari, M., Tate, J., Swaminathan, G. J., Newman, R., Naim, A., Vranken, W., Kapopoulou, A., Hussain, A., Fillon, J., Henrick, K., and Velankar, S. (2006) E-MSD: improving data deposition and structure quality. Nucl. Acids Res. 34, D287–290.

9. Henrick, K., and Thornton, J. M. (1998) PQS: A Protein Quarternary File Server. Trends Biochem. Sci. 23, 358–361.

10. Kinoshita, K., and Nakamura, H. (2004) eF-site and PDBjViewer: database and viewer for protein functional sites. Bioinformatics 20, 1329–1330.

11. Standley, D. M., Toh, H., and Nakamura, H. (2005) GASH: an improved algorithm for maximizing the number of equivalent residues between two protein structures. BMC Bioinformatics 6, 221.

12. Wako, H., Kato, M., and Endo, S. (2004) ProMode: a database of normal mode analyses on protein molecules with a full-atom model. Bioinformatics 20, 2035–2043.

13. Stevens, R. C., Yokoyama, S., and Wilson, I. A. (2001) Global efforts in structural genomics. Science 294, 89–92.

14. Callaway, J., Cummings, M., Deroski, B., Esposito, P., Forman, A., Langdon, P., Libeson, M., McCarthy, J., Sikora, J., Xue, D., Abola, E., Bernstein, F., Manning, N., Shea, R., Stampf, D., and Sussman, J. (1996) Protein Data Bank Contents Guide:

Atomic coordinate entry format description. Brookhaven National Laboratory, http://www.wwpdb.org/docs.html

15. Dutta, S., and Berman, H. M. (2005) Large macromolecular complexes in the Protein Data Bank: a status report. Structure 13, 381–3-88.

16. Yusupov, M. M., Yusupova, G. Z., Baucom, A., Lieberman, K., Earnest, T. N., Cate, J. H. D., and Noller, H. F. (2001) Crystal structure of the ribosome at 5.5 Å resolution. Science 282, 883–896.

17. Chen, B., Colgrave, M. L., Daly, N. L., Rosengren, K. J., Gustafson, K. R., Craik, D. J. (2005) Isolation and characterization of novel cyclotides from Viola hederaceae: solution structure and anti-HIV activity of vhl-1, a leaf-specific expressed cyclotide. J. Biol. Chem. 280, 22395–22405.

18. Ciszak, E. M., Makal, A., Hong, Y. S., Vettaikkorumakankauv, A. K., Korotchkina, L. G., Patel, M. S. (2006) How dihydrolipoamide dehydrogenase-binding protein binds dihydrolipoamide dehydrogenase in the human pyruvate dehydrogenase complex. J. Biol. Chem. 281, 648–655.

19. Bourne, P. E., Berman, H. M., Watenpaugh, K., Westbrook, J. D., and Fitzgerald, P. M. D. (1997) The macromolecular Crystallographic Information File (mmCIF). Meth. Enzymol. 277, 571–590.

20. Fitzgerald, P. M. D., Westbrook, J. D., Bourne, P. E., McMahon, B., Watenpaugh, K. D., and Berman, H. M. (2005) Macromolecular dictionary (mmCIF), in (Hall, S. R., and McMahon, B., eds.), International Tables for Crystallography Vol. G. Definition and exchange of crystallographic data, pp. 295–443, Springer, Dordrecht, The Netherlands.

21. Westbrook, J., Henrick, K., Ulrich, E. L., and Berman, H. M. (2005) The Protein Data Bank exchange data dictionary, in (Hall, S. R., and McMahon, B., eds.), International Tables for Crystallography Vol. G. Definition and exchange of crystallographic data, pp. 195–198, Springer, Dordrecht, The Netherlands.

22. Westbrook, J. D., Berman, H. M., and Hall, S. R. (2005) Specification of a relational Dictionary Definition Language (DDL2), in (Hall, S. R., and McMahon, B., eds.), International Tables for Crystallography Vol. G. Definition and exchange of crystallographic data, pp. 61–72, Springer, Dordrecht, The Netherlands.

23. Westbrook, J., Ito, N., Nakamura, H., Henrick, K., and Berman, H. M. (2005) PDBML: The representation of archival macromolecular structure data in XML. Bioinformatics 21, 988–992.

24. Chen, L., Oughtred, R., Berman, H. M., and Westbrook, J. (2004) TargetDB: a target registration database for structural genomics projects. Bioinformatics 20, 2860–2862.

25. Albeck, S., Alzari, P., Andreini, C., Banci, L., Berry, I. M., Bertini, I., Cambillau, C., Canard, B., Carter, L., Cohen, S. X., Diprose, J. M., Dym, O., Esnouf, R. M., Felder, C., Ferron, F., Guillemot, F., Hamer, R., Jelloul, M. B., Laskowski, R. A., Laurent, T., Longhi, S., Lopez, R., Luchinat, C., Malet, H., Mochel, T., Morris, R. J., Moulinier, L., Oinn, T., Pajon, A., Peleg, Y., Perrakis, A., Poch, O., Prilusky, J., Rachedi, A., Ripp, R., Rosato, A., Silman, I., Stuart, D. I., Sussman, J. L., Thierry, J.-C., Thompson, J. D., Thornton, J. M., Unger, T., Vaughan, B., Vranken, W., Watson, J. D., Whamond, G., and Henrick, K. (2006) SPINE bioinformatics and data-management aspects of high-throughput structural biology. Acta Cryst. D62, 1184–1195.

26. Pajon, A., Ionides, J., Diprose, J., Fillon, J., Fogh, R., Ashton, A. W., Berman, H., Boucher, W., Cygler, M., Deleury, E., Esnouf, R., Janin, J., Kim, R., Krimm, I., Lawson, C. L., Oeuillet, E., Poupon, A., Raymond, S., Stevens, T., van Tilbeurgh, H., Westbrook, J., Wood, P., Ulrich, E., Vranken, W., Xueli, L., Laue, E., Stuart, D. I., and Henrick, K. (2005) Design of a data model for developing laboratory information management and analysis systems for protein production. Proteins 58, 278–284.

27. Winn, M. D., Ashton, A.W., Briggs, P.J., Ballarda C.C. and Patel, P. (2002) Ongoing developments in CCP4 for high-throughput structure determination. Acta Crystallogr. D Biol. Crystallogr. 58, 1929–1936.

28. Yang, H., Guranovic, V., Dutta, S., Feng, Z., Berman, H. M., and Westbrook, J. (2004) Automated and accurate deposition of structures solved by X-ray diffraction to the Protein Data Bank. Acta Crystallogr. D Biol. Crystallogr. 60, 1833–1839.

29. Feng, Z., Chen, L., Maddula, H., Akcan, O., Oughtred, R., Berman, H. M., and Westbrook, J. (2004) Ligand Depot: a data warehouse for ligands bound to macromolecules. Bioinformatics 20, 2153–2155.

30. Golovin, A., Oldfield, T. J., Tate, J. G., Velankar, S., Barton, G. J., Boutselakis, H., Dimitropoulos, D., Fillon, J., Hussain, A., Ionides, J. M., John, M., Keller, P. A., Krissinel, E., McNeil, P., Naim, A., Newman, R., Pajon, A., Pineda, J., Rachedi, A., Copeland, J., Sitnov, A., Sobhany, S., Suarez-Uruena, A., Swaminathan, G. J., Tagari, M., Tromm, S., Vranken, W., and Henrick, K. (2004) E-MSD: an integrated data resource for bioinformatics. Nucl. Acids Res. 32, D211–216.

31. Ihlenfeldt, W.-D., Voigt, J. H., Bienfait, B., Oellien, F., and Nicklaus, M. C. (2002) Enhanced CACTVS Browser of the Open NCI Database. J. Chem. Inf. Comput. Sci. 42, 46–57.

32. Ihlenfeldt, W. D., Takahashi, Y., Abe, H., and Sasaki, S. (1994) Computation and management of chemical properties in CACTVS: an extensible networked approach toward modularity and flexibility. J. Chem. Inf. Comp. Sci. 34, 109–116.

33. Wheeler, D. L., Chappey, C., Lash, A. E., Leipe, D. D., Madden, T. L., Schuler, G. D., Tatusova, T. A., and Rapp, B. A. (2000) Database resources of the National Center for Biotechnology Information. Nucl. Acids Res. 28, 10–14.

34. Phan, I. Q., Pilbout, S. F., Fleischmann, W., and Bairoch, A. (2003) NEWT, a new taxonomy portal. Nucl. Acids Res. 31, 3822–3823.

35. Bairoch, A., Apweiler, R., Wu, C. H., Barker, W. C., Boeckmann, B., Ferro, S., Gasteiger, E., Huang, H., Lopez, R., Magrane, M., Martin, M. J., Natale, D. A., O'Donovan, C., Redaschi, N., and Yeh, L. S. (2005) The Universal Protein Resource (UniProt). Nucl. Acids Res. 33, D154–159.

36. Benson, D. A., Karsch-Mizrachi, I., Lipman, D. J., Ostell, J., and Wheeler, D. L. (2005) GenBank. Nucl. Acids Res. 33, D34–38.

37. Okubo, K., Sugawara, H., Gojobori, T., and Tateno, Y. (2006) DDBJ in preparation for overview of research activities behind data submissions. Nucl. Acids Res. 34, D6–9.

38. Kanz, C., Aldebert, P., Althorpe, N., Baker, W., Baldwin, A., Bates, K., Browne, P., Broek, A. v. d., Castro, M., Cochrane, G., Duggan, K., Eberhardt, R., Faruque, N., Gamble, J., Diez, F. G., Harte, N., Kulikova, T., Lin, Q., Lombard, V., Lopez, R., Mancuso, R., McHale, M., Nardone, F., Silventoinen, V., Sobhany, S., Stoehr, P., Tuli, M. A., Tzouvara, K., Vaughan, R., Wu, D., Zhu, W., and Apweiler, R. (2005) The EMBL Nucleotide Sequence Database. Nucl. Acids Res. 33, D29–33.

39. Krissinel, E., and Henrick, K. (2005) Detection of Protein Assemblies in Crystals, in (Berthold, M.R., Glen, R. Diederichs., K. Kohlbacher., O. Fischer., I. (eds.)), CompLife 2005, pp. 163–174, Springer-Verlag, Berlin, Heidelberg.

40. Hooft, R. W., Vriend, G., Sander, C., and Abola, E. E. (1996) Errors in protein structures. Nature 381, 272.

41. Laskowski, R. A., McArthur, M. W., Moss, D. S., and Thornton, J. M. (1993) PROCHECK: a program to check the stereochemical quality of protein structures. J. Appl. Cryst. 26, 283–291.

42. Lovell, S. C., Davis, I. W., Arendall, W. B., 3rd, de Bakker, P. I., Word, J. M., Prisant, M. G., Richardson, J. S., and Richardson, D. C. (2003) Structure validation by Calpha geometry: phi,psi and Cbeta deviation. Proteins 50, 437–450.

43. Westbrook, J., Feng, Z., Burkhardt, K., and Berman, H. M. (2003) Validation of protein structures for the Protein Data Bank. Meth. Enzymol. 374, 370–385.

44. Sayle, R., and Milner-White, E. J. (1995) RasMol: biomolecular graphics for all. Trends Biochem. Sci. 20, 374.

45. Pettersen, E. F., Goddard, T. D., Huang, C. C., Couch, G. S., Greenblatt, D. M., Meng, E. C., and Ferrin, T. E. (2004) UCSF Chimera—a visualization system for exploratory research and analysis. J. Comput. Chem. 25, 1605–1612.

46. Hartshorn, M. J. (2002) AstexViewer: a visualisation aid for structure-based drug design. J. Comput. Aided Mol. Des. 16, 871–881.

47. Vaguine, A. A., Richelle, J., and Wodak, S. J. (1999) SFCHECK: a unified set of procedures for evaluating the quality of macromolecular structure-factor data and their agreement with the atomic model. Acta Crystallogr. D Biol. Crystallogr. 55, 191–205.

48. Kleywegt, G. J., Harris, M. R., Zou, J., Taylor, T. C., Wählby, A., and Jones, T. A. (2004) The Uppsala Electron-Density Server. The Uppsala Electron-Density Server D60, 2240–2249.

49. Doreleijers, J. F., Nederveen, A. J., Vranken, W., Lin, J., Bonvin, A. M., Kaptein, R., Markley, J. L., and Ulrich, E. L. (2005) BioMagResBank databases DOCR and FRED containing converted and filtered sets of experimental NMR restraints and coordinates from over 500 protein PDB structures. J. Biomol. NMR 32, 1–12.

50. Henrick, K., Newman, R., Tagari, M., and Chagoyen, M. (2003) EMDep: a web-based system for the deposition and validation of high-resolution electron microscopy macromolecular structural information. J. Struct. Biol. 144, 228–237.

51. Berman, H. M., Burley, S. K., Chiu, W., Sali, A., Adzhubei, A., Bourne, P. E., Bryant, S. H., Roland L. Dunbrack, J., Fidelis, K., Frank, J., Godzik, A., Henrick, K., Joachimiak, A., Heymann, B., Jones, D., Markley, J. L., Moult, J., Montelione, G. T., Orengo, C., Rossmann, M. G., Rost, B., Saibil, H., Schwede, T., Standley, D. M., and Westbrook, J. D. (2006) Outcome of a workshop on archiving structural models of biological macromolecules. Structure 14, 1211–1217.

52. Hempstead, P. D., Yewdall, S. J., Fernie, A. R., Lawson, D. M., Artymiuk, P. J., Rice, D. W., Ford, G. C., and Harrison, P. M. (1997) Comparison of the three-dimensional structures of recombinant human H and horse L ferritins at high resolution. J. Mol. Biol. 268, 424–448.

Chapter 6

Prediction of Protein Disorder

Zsuzsanna Dosztányi and Peter Tompa

The recent advance in our understanding of the relation of protein structure and function cautions that many proteins, or regions of proteins, exist and function without a well-defined three-dimensional structure. These intrinsically disordered/unstructured proteins (IDP/IUP) are frequent in proteomes and carry out essential functions, but their lack of stable structures hampers efforts of solving structures at high resolution by x-ray crystallography and/or NMR. Thus, filtering such proteins/regions out of high-throughput structural genomics pipelines would be of significant benefit in terms of cost and success rate. This chapter outlines the theoretical background of structural disorder, and provides practical advice on the application of advanced bioinformatic predictors to this end, that is to recognize fully/mostly disordered proteins or regions, which are incompatible with structure determination. An emphasis is also given to a somewhat different approach, in which ordered/disordered regions are explicitly delineated to the end of making constructs amenable for structure determination even when disordered regions are present.

1. Introduction

Intrinsically disordered/unstructured proteins exist in a highly flexible conformational state, yet they carry out essential functions often related to signal transduction or regulation of gene transcription (1–4). Due to the inherent functional benefits, the frequency of structural disorder increases in evolution, and it reaches high proportions in eukaryotes, estimated to be about 5–15% for fully disordered proteins and 40–60% for proteins, which have at least one long (>30 consecutive residues) disordered segment (5–7). The functional importance and prevalence of these proteins/regions heightened interest in IUPs, which sprung into life a novel branch of protein science. The other side of the coin, however, is that the existence of IUPs is highly detrimental to traditional structure-solving efforts, which by definition demand a stable and uniform structure of protein molecules in the sample. IUPs or disordered regions exist as a rapidly interconverting ensemble of alternative conformations, which may hamper expression and purification, inhibit crystallization, and give rise to inadequate NMR signals for solving the structure in solution (8–10). Detection and filtering out such

From: *Methods in Molecular Biology, Vol. 426: Structural Proteomics: High-throughput Methods*
Edited by: B. Kobe, M. Guss and T. Huber © Humana Press, Totowa, NJ

proteins/regions of high-throughput structural genomics pipelines may save a lot of energy, and improve cost-effectiveness, as already demonstrated *(9)*. Although a wide range of biophysical techniques have been developed/adopted for studying IUPs *(11)*, and recent efforts have also aimed at developing proteomic techniques for their characterization *(12,13)*, only bioinformatic techniques can tackle this issue in general at the present. This chapter defines the rationales for predicting disorder, surveys the principles various predictors rely on, and provides practical descriptions of how to use them for optimal performance and greatest benefits.

1.1. Theoretical Considerations

Since there is no generally accepted definition of disorder, and in fact there exist no single—but rather several—kind(s) of disorder, no universal solution for disorder prediction can be given. Due to this limitation, it is important to familiarize ourselves with different aspects of the concept of disorder to select the solution most appropriate for a given situation. By definition, protein disorder denotes a structural state of proteins, which corresponds to an ensemble of alternative conformations *(1–4)*. Usually this state is characterized by a high flexibility corresponding to a low level, or no, repetitive secondary structure, and the lack of stable tertiary interactions. However, this definition encompasses several categories of disorder, ranging from molten globules through partially unstructured proteins to random coils, characterized by increasing levels of mobility and decreasing residual secondary structure. The differences in the extent of disorder are often linked with the function of these proteins, as some IUPs can become ordered upon binding to their partners, whereas others, such as entropic chains, are permanently unstructured under native conditions *(1–4)*.

The primary datasource for disordered proteins and regions longer than 30 residues is the DisProt database *(14)*. Its current release (release 3.4, Aug. 15, 2006) contains about 460 different proteins (1,100 disordered regions), shown to be disordered by various experimental techniques. Datasets of disordered regions were also compiled from x-ray structures, collecting residues missing from the electron density maps. Such regions are usually shorter than 30 residues in length *(15)*. Disorder of such short regions is more likely to depend on the structure of their flanking segments, that is, it is context dependent. The authors usually consider long disordered regions as the ones that lack a well-defined structure irrespective of their flanking regions. Although short disordered regions could be part of x-ray structures, and hence do not prevent crystallization, the presence of long disordered regions seems to be incompatible with structure determination. The complementarity of long disordered regions to globular proteins can explain that despite the lack of a generally accepted definition of disorder, and the uncertainties in databases of disordered proteins and segments (see Note 1), there is a pronounced difference between sequence determinants of ordered and disordered proteins.

1.2. Disorder Predictors

The group of Dunker pioneered the analysis and prediction of disorder, and showed that the lack of ordered structure, just like the three-dimensional structure in the case of globular proteins, is encoded in the sequence *(16)*.

Their classical predictor is PONDR® (the predictor of naturally disordered regions), which actually is a family of predictors (PONDR VLXT, VL3) *(17)* that rely on a neural network algorithm trained on various databases containing ordered and disordered sequences. They use sequence attributes such as the fractional composition of particular amino acids, hydropathy, or sequence complexity, which are averaged over windows of 9 to 21 amino acids and serve as input for the prediction.

In more general terms, the prediction of protein disorder can be phrased as a typical binary classification problem. Indeed, several predictors (Table 6.1) are

Table 6.1 Predictors for intrinsic disorder of proteins.

Predictor	URL	Principle	Download/ Script	PsiBlast profile	Ref.
PONDR® VSL2	http://www.ist.temple.edu/disprot/ predictorVSL2.php	SVM	Yes/Yes but limited	Optional	*(17)*
DISOPRED2	http://bioinf.cs.ucl.ac.uk/disopred/	SVM + NN for smooth- ing	Yes/No	Yes	*(6)*
IUPred	http://iupred.enzim.hu/index.html	estimated interac- tion energy	Yes/Yes	No	*(23)*
DisEMBL™	http://dis.embl.de	neural network	Yes/Yes	No	*(8)*
GlobPlot	http://globplot.embl.de/	single AA propensity	Yes/Yes	No	*(35)*
DISpro	http://www.ics.uci.edu/~baldig/dis- pro.html	1-D recur- sive NN with profiles	Yes/No	Yes	*(38)*
Spritz	http://protein.cribi.unipd.it/spritz/	SVM with non-linear kernel	No/No	Yes	*(39)*
FoldUnfold	http://skuld.protres.ru/~mlobanov/ ogu/ogu.cgi	single AA propensity	No/No	No	*(20)*
PreLink	http://genomics.eu.org/spip/PreLink	AA pro- pensity + hydropho- bic cluster analysis	No/No	No	*(21)*
FoldIndex©	http://bip.weizmann.ac.il/fldbin/fin- dex	AA propen- sities	No/Yes	No	*(40)*
RONN	http://www.strubi.ox.ac.uk/RONN	bio-basis function neural network	Yes/No	No	*(41)*
DRIP-PRED	http://www.sbc.su.se/~maccallr/dis- order/	self-organiz- ing maps	No/No	Yes	http://www. forcasp.org/ paper2127. html

based on this principle and use standard machine learning techniques such as feed-forward neural network (PONDR, DISEMBL), recursive neural network (DISpro), support vector machines (SVM) with linear or radial basis kernel functions (Spritz, DISOPRED2, VSL2), and biobasis neural network (RONN). This latter basically is a pattern recognition algorithm aiming to recognize the similarity to known examples of disordered regions. DRIP-PRED is based on clustering of sequence-profile windows of UniProt sequences, and sequence profiles that are not represented in the PDB are predicted as disordered. These methods usually consider information derived from sequence homologies in the form of profiles (see Note 2). They can also use predicted secondary structure, or accessibility, although the added value of these components has not been convincingly demonstrated *(6,15)*. As these machine-learning methods rely on training, their performance is limited by the uncertainties of the underlying dataset, as well as the simplification of disorder to a binary state. Most commonly, these methods rely on the definition of disorder, which is restricted to residues with missing atomic coordinates in x-ray structures. PONDR VSL2 is unique in its capacity to predict both short and long disorder by combining two specific predictors *(15)*.

A different rationale is utilized in predictors that extract relevant information from the amino acid composition of proteins. It has been suggested that the disorder of IUPs stems from their low average hydrophobicity and high net charge *(18)*. A more comprehensive model of disorder originates from the suggestion that IUPs lack a stable fold because they do not have the capacity to form sufficient stabilizing interresidue interactions that would be required to form a compact stable structure. IUPred estimates this potential energy by statistical interaction potentials *(19)*, whereas in another study the capacity of residues for interactions has been approximated by the average number of their contacting residue partners in globular proteins (FoldUnfold) *(20)*. PreLink recognizes disordered regions based on compositional bias and low hydrophobic cluster content *(21)*. These methods are generally more transparent, they involve significantly less parameters, and they rely on the known collection of disordered proteins to a much lesser extent. These factors might explain why these seemingly simple approaches are commensurate in performance with the more sophisticated machine learning methods.

Altogether, about 20 predictors have been described in the literature, 12 of which are available via the Internet (see Table 6.1). A comprehensive list with links can be found in the DisProt database at http://www.disprot.org/predictors.php and http://iupred.enzim.hu/IUP_links.html. As a preliminary guide among predictors, a good starting point is the recent CASP6 experiment *(22)*, in which the predictors were assessed and critically compared. Of course, the relative performance of predictors depends on the dataset used for testing, or more generally, the type of disorder studied; the CASP evaluations thus far have concentrated on residues with missing x-ray coordinates, and neglected blind testing for long disordered regions.

Another factor that should be taken into account when prediction methods are compared is the relative size of the two classes, ordered and disordered. Therefore, there are two separate measures commonly used to evaluate prediction accuracy, sensitivity, and specificity. Sensitivity is the fraction of disordered residues correctly identified, whereas specificity is the fraction of correctly identified ordered residues in a prediction. Good prediction methods should have both high specificity and sensitivity; however, there is a tradeoff between these parameters, and their optimal balance depends on the purpose of the prediction and cost of misclassification *(22)*. Although datasets usually contain more ordered residues than disordered, prediction of disordered residues is usually weighted more heavily. On the other hand, for the purpose of structural proteomics target selection, the more conservative prediction (predicting more order) is favored (see Note 3). From the CASP6 and other compilations, the best and most convenient methods have been selected and are summarized in Table 6.1. Overall, their performances at 90% specificity is above 50% sensitivity on missing residues, whereas nearly 80% of sensitivity is reported for long disordered regions *(15,23)*.

As a cautionary note, some other prediction algorithms are also in use for assessing protein disorder. For example, it is found recurrently in the literature that structural disorder correlates with regions of low complexity in sequences, and the algorithm developed to recognize such regions (SEG) *(24)* may be used for disorder prediction. However, low-complexity regions are not always disordered, and disordered regions are not always of low complexity in terms of their amino acid composition *(25)*. Another practice that may be subject to criticism comes from the simplification that disordered regions are devoid of regular secondary structure, and disorder can be predicted by an algorithm developed to recognize the lack of such regions (NORSp) *(26)*. Again, this view is not justified as IUPs may contain significant secondary structure *(27)* and proteins made up entirely of amino acids in coil conformation may have a well-defined three-dimensional structure (i.e., loopy proteins) *(28)*. Another concept related to protein disorder is flexibility. This property is usually defined through high B-factors in x-ray structures. Although sequence determinants for regions of high B factor and intrinsically disordered regions are correlated, they are significantly different *(29)*. Thus, these techniques generally would give an indirect and often imperfect description of protein disorder and are not recommended for predicting disorder.

2. Materials

Protein sequences should be prepared in the form of a plain sequence with no header (see Note 4). Some predictors prefer or at least can handle FASTA format as well. An example of the FASTA format is given below. In this format it may be critical that the long header line is not wrapped when copied and pasted from editors, because the servers would treat text from line 2 as part of the sequence.

>sp|P43532|FLGM_ECOLI Negative regulator of flagellin synthesis
MSIDRTSPLKPVSTVQPRETTDAPVTNSRAAKTTASTSTSVTLSDAQAKLMQP
GSSDINLERVEALKLAIRNGELKMDTGKIADALINEAQQDLQSN

3. Methods

Depending on the intended scale of the project, there are three possible ways of using the predictors. (1) Predictors come with convenient web interfaces, where the query sequence can be pasted and a disorder score pattern is returned both in a graphic format and as a text file. This mode of use, which can be conveniently applied in the case of up to a dozen or so sequences, requires no programming skills and is described in the following. Some predictors return the results promptly; others send it back by e-mail. All predictors that use PSI-BLAST profiles belong to the second category. (2) If a larger number of sequences are to be predicted, certain predictors allow automation of the process, when sequences may be sent to the server by a dedicated script. (3) For real genome-scale predictions, best is to download the source-file of the algorithm and install locally on your computer; speed of calculations may be a critical point under these circumstances (see Table 6.1 and Note 5). These two latter applications follow the logic of the first, but cannot be automated, and for their implementation one needs to follow descriptions given at the Internet sites of the predictors, not to be detailed here (see Table 6.1).

3.1. Predicting Mostly Disordered Proteins

At the present stage, most high-throughput structural genomics projects focus on protein domains with a large number of sequence homologues but without structural representatives. The domains are usually selected from existing databases of domain families such as PFAM *(30)*. Disorder prediction methods are often used to filter out or prioritize against those targets that are not amenable for structure determination. This requires to classify proteins, which have more disordered than ordered residues, as "mostly disordered" and distinguishing them in a binary classification from "mostly ordered" proteins, because they are the ones that are likely to cause trouble in expression, purification, and structure determination *(9,10)*.

3.1.1. Prediction by Charge-Hydropathy Analysis

The charge-hydropathy plot has been developed to provide an overall assessment of the disorder status of a polypeptide chain, and it is a very simple but only moderately accurate procedure for achieving the binary classification of a protein *(10,18)*. To carry out this analysis:

1. Determine the mean net charge $<R>$ of the sequence by calculating the absolute value of the difference of the number of positively and negatively charged amino acids, i.e., $|n_{Lys} + n_{Arg} - n_{Asp} - n_{Glu}|$, and divide it by the chain length.
2. Determine the mean hydrophobicity of the sequence. Hydrophobicity is to be defined by the Kyte-Doolittle scale, normalized to the range 0–1, and divided by the number of amino acids in the chain. A simple way to determine these numbers is to go to the ProtScale tool within "Primary structure analysis" at the ExPaSy server (http://us.expasy.org/tools/protscale.html), select the Hphob./Kyte-Doolittle scale, and normalize its scale from 0 to 1 at the bottom of the screen. Paste the sequence in the window, download the hydrophobicity score values in a text format, and calculate the mean for the entire chain ($<H>$).
3. Plot the values thus determined in a mean hydrophobicity-mean net charge plot, and also put on the plot a straight function corresponding to the equation $<R> = 2.743<H> - 1.109$ *(10)*. This line best separates mostly disordered proteins (upper left half, low mean hydrophobicity and high net charge) from mostly ordered proteins (lower right half, high mean hydrophobicity and low net charge).

3.1.2. Prediction by Position-Specific Scoring Schemes

Predicting proteins as mostly ordered or mostly disordered by definition has its limitations, as most proteins are not made up of clearly separated globular and disordered regions. Most advanced predictors provide a position-specific score to characterize order/disorder tendency, which offer a more detailed picture about order/disorder nature of proteins. These predictors assign a score to each position in the amino acid sequence. The scores usually range between 0 and 1, and residues with a score above a predefined threshold are considered disordered, whereas those below the threshold are considered ordered (see Note 6). This threshold is usually, but not always, 0.5. The output of the prediction is smoothed over a predefined window size (see Note 7).

Generally, positions with score close to 0 or 1 are confidently predicted as ordered or disordered, respectively, whereas residues close to the cutoff value represent borderline cases. Longer regions are usually predicted more confidently, and when only 1 to 3 consecutive residues are predicted either disordered or ordered, these should be treated very cautiously. In many cases mostly ordered and mostly disordered regions can be easily identified by visual inspection. Nevertheless, these regions quite often contain shorter regions with opposite characteristics. Although it could simply be due to some noise in the prediction, it often has structural or physiological significance. Within the context of mostly disordered regions, shorter regions with a pronounced tendency toward order were proposed as an indication of binding regions *(27,31,32)*. Short regions of disorder within an otherwise globular domain hint to increased flexibility, and are likely to correspond to regions with missing electron density. Regions with only a moderate tendency toward order could indicate a likely disorder-to-order transition upon binding to partner proteins or DNA. Figure 6.1 shows an example of the prediction for p53, a tumor suppressor transcription factor, taken from the IUPred server (http.//iupred.

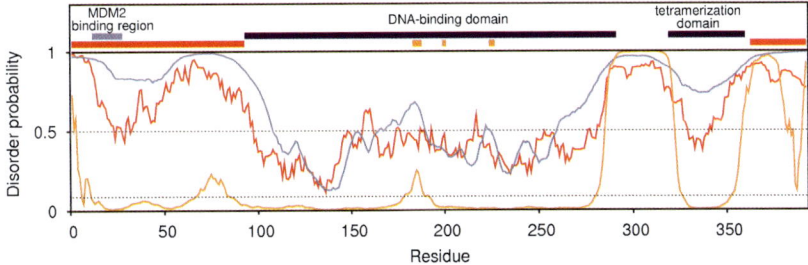

Fig. 6.1 Disorder score of human p53 by three different predictors. Disorder of p53 has been determined by DISPROT VSL2 *(blue)*, DISOPRED2 *(orange*, please note the cutoff value in this case is 0.08), and IUPred *(red)*. The figure demonstrates agreements and disagreements among the three predictors. The red bars correspond to disordered regions, the blue bars to the ordered regions with known coordinates. The light blue bar marks the MDM2 binding region. This region becomes ordered in the bound form, and the structure of this fragment could be solved in the complex. Short orange bars show some of the regions with missing electron density. It is of particular note that the predictors do not perfectly agree on the position of the border(s) separating ordered and disordered regions. The implications of this uncertainty for generating appropriate constructs for structural studies are discussed in the text.

enzim.hu). This example highlights some of these features. The N-terminal region (1–93) was shown to be disordered *(33)*, and in agreement with this, the region is predicted mostly disordered. Within this region, however, residues predicted close or even below the threshold correspond to a binding region, and the corresponding structure in complex with MDM2 could be solved *(34)*. The tetramerization domain is not confidently predicted ordered, but in this case the oligomeric form is essential, which could explain the lack of a strong signal for order. Note that individual predictions agree in certain regions, but disagree within the border regions that separate disordered and ordered segments. The structural status of this zone is uncertain and should be treated with caution.

To determine the binary disorder status of a protein, the following protocols can be used:

1. Determine the disorder score of the query sequence(s) by one of the predictors from Table 6.1. As a first approximation, determine what proportion of its amino acids are disordered (fall above the default threshold value). If the ratio is above 50%, the protein may be classified as mostly disordered, and should be withdrawn from the pipeline. However, this depends critically on the selection of the threshold value of the predictor (i.e., true positive vs. false positive rate) (see Note 6).
2. Alternatively, the score could be averaged over the whole target. If the average score is above the cutoff value (e.g., 0.5), the protein or protein region should not be considered for structure determination. This approach takes into account the extent of order or disorder predicted for the residues, and it depends on the cutoff value to a lesser extent.
3. The minimal size of globular proteins is 30–50 residues. Most of the smaller proteins contain disulfide bonds, or require metal ions such as Zn^{2+} for the stabilization of the structure, both associated with cysteine residues. If cysteine content is below 6%, the larger size should be used as a selection criterion. Targets that do not have a continuous stretch of ordered residues longer than this limit should also be removed from the pipeline.
4. Among globular proteins with known structure, the presence of at least 30 residues predicted as disordered is very rare. Thus, regions with a tendency for order encompassing such a long disordered sequence should not be considered as single folding units. If such a long disordered region is at the end of the chain, it suggests that construct should be further optimized, probably be removing potentially disordered regions (see Note 8).
5. To find reliably predicted ordered regions, use several predictors (see Note 9). Select the region that the predictors agree upon as ordered. As this approach may clip a little piece of the globular part of the protein off, which may impair folding and structure determination, as a second approach extend this construct up to the border were all predictors agree on disorder. This may allow a little disordered overhang on the globular domain, which, however, might be tolerated in structure determination.

3.2. Predicting Exact Ordered Segments in a Protein

The prediction of the exact location of disordered/ordered regions can lead to the generation of appropriate constructs for solving the structures of functional units even within a mostly disordered protein. This requires

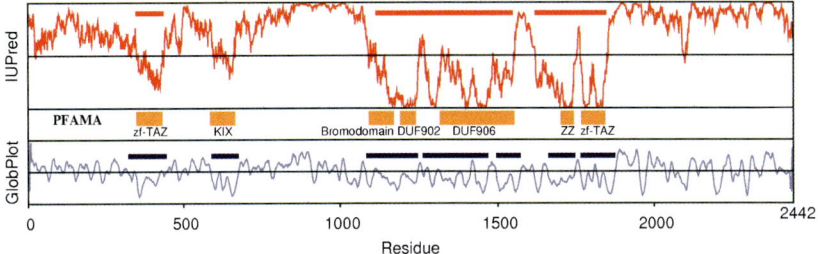

Fig. 6.2 Disorder score and suggested globular domains of human CBP by two different predictors. The suggested globular domains IUPred *(red)* and GlobPlot *(blue)*, and the underlying position specific scores for human CBP. Please note that the GlobPlot Web server shows the cumulative scores. The orange bars represent PFAM domains for this protein, which are likely to correspond to folded domains. There is a known structural homologue corresponding to, and/or functional assignment to these PFAM families, except for DUF902 and DUF906.

a specific algorithm that finds likely globular regions based on the position-specific prediction of disorder. At the practical level, the main source of difficulty is that an overall globular domain can contain shorter disordered regions and vice versa. One solution for this problem was first suggested for GlobPlot *(35)*, and later implemented also in IUPred *(23)*. The algorithm is the following:

1. Two globular regions are merged if they are separated by a short disordered region which is shorter than a predefined cutoff value (JOIN).
2. A globular region is merged into neighboring disordered regions if its length is below a threshold (DEL).
3. Follow this process for the whole sequences. This would eliminate short regions of disorder below length JOIN, and short ordered regions below length DEL.
4. To obtain such prediction, use GlobPlot or IUPred. Although such algorithm can be applied to any kind of position-specific prediction of disorder, other servers have not implemented such prediction option.

Figure 6.2 shows the suggested "constructs" for the human CREB binding protein, calculated by both IUPred and GlobPlot and compared with PFAM families *(30)*. In the selected example, likely globular regions are generally found, but neighboring globular domains can be merged into a single unit.

4. Notes

1. It is to be noted that due to several reasons, there are significant uncertainties in databases used for training and assessing disorder predictors. The major sources to emphasize are: (1) apparently we probably do not deal with one, but several kinds of disorder; (2) distinct physicochemical techniques are used in identifying individual IUPs; (3) erroneous entries occur in databases (both as false-positives and -negatives in both databases of order and disorder); (4) there is a noted principal difference between "short" and "long" disordered regions. These uncertainties variously affect current predictors

and make it impossible at the present to establish the "best" predictor for one, or any other, purpose.

2. As in many areas in protein structure prediction, evolutionary information in the form of profiles generated from homologous sequences can be incorporated into the prediction. However, the improvement over single sequences is relatively small, 1–2% *(6,15)*, compared with the 5% increase in accuracy observed in the case of secondary structure prediction. It should also be taken into account that evolutionary conservation and the validity of existing sequence search methods for disordered proteins are not well established. Generally, sequences with biased amino acid composition represent a serious problem for database searches because they obscure the assumptions made about the statistical significance of meaningful homology. Most often, the SEG *(24)* program is used to overcome this limitation, by masking regions with low sequence complexity. However, many disordered proteins and regions have very biased amino acid composition, without having low complexity segments. These segments, although they are not captured by the SEG algorithm, are still prone to produce false-positives in database searches that can mislead prediction methods *(36)*.

3. The users of predictors should be aware that differences among predictors could simply arise because of the different false prediction rates, manifest in the default cutoff values of the methods. Thus, PONDR methods usually predict more disorder than other methods, such as DISpro or DISOPRED2. It should be kept in mind, however, that these differences can be compensated to some extent by appropriately changing the cutoff values.

4. Most predictors take a single sequence as an input either in FASTA or plain sequence format (without any header) using the standard amino acids with single letters. Nonstandard amino acids or other characters are usually ignored; however, some predictors do not return any results if they are present. It is always the user's responsibility to check that the submitted sequence is a protein sequence and not DNA sequence, since the algorithm will interpret DNA sequences as a protein sequence composed of amino acids Ala, Thr, Cys, and Gly. Although minimal sequence length is usually not stated, it should be kept in mind the sequences shorter than 30 residues are highly unlikely to form a folded structure. Some programs (e.g., DISOPRED2) cannot handle too long sequences, i.e., sequence longer than 1,000 residues.

5. For large-scale studies the speed of calculations may represent a decisive factor. Disorder prediction methods that rely on a single (GlobPlot, FoldUnfold) or only a few (FoldIndex) amino acid propensities are very fast. The position-specific prediction of IUPred is also based on only a relatively small number of parameters and scales linearly with sequence length; hence, it is also a fast method. Other methods that rely on calculated PsiBlast profiles either directly or by using secondary structure predictions (DISOPRED2, DIPPRED, Dispro, Spritz, etc.), are significantly slower, and this can be a concern in large-scale genome-wide predictions. PONDR VSL2 can also be used in single sequence mode, with only small decrease in performance.

6. Position-specific scoring schemes usually define a threshold value, above which an amino acid is considered to be in a locally disordered state. It is to be noted that this threshold is not necessarily 0.5, the default value set in most predictors, but can attain other values, such as in the case of

DISOPRED2. Further, in several prediction methods this value is a function of the pre-set true-positive/false-negative ratio of prediction, which might not suit the given application.

7. To some extent, smoothing window size can also affect the accuracy of the predictions. For predicting short disordered regions (residues missing from electron density maps), a smaller window size is used (from 11 up to 21). For recognizing long disordered regions, the most common window sizes range from 21 to 41.

8. The ends of complete proteins or proteins fragments are often predicted disordered, especially by servers tuned for predicting short disorder (missing residues in x-ray structures). This is an attempt to model that most of the missing residues are from the chain termini. This could be partly explained by the increased flexibility of the ends, which are restrained by only one side of the chain. However, this effect might not be relevant in deciding about the overall tendency toward order or disorder.

9. In the current state of the art, no single prediction method is accurate enough that it could be entirely trusted. To increase the confidence of the judgment, it is wise to apply more than one predictor, and accept the assessment where most of them agree. This, however, is not automated at the moment. This actually could be more beneficial for this vibrant field in the long term. Many predictor methods have some strength; for example, IUPred is good for predicting long disordered regions, DISOPRED2 is good in recognizing short disorder, PONDR VSL2 is a good general predictor applicable of disorder at any length, PONDR VLXT is recommended for predicting functional sites within disordered regions, PreLink for linker regions *(37)*. Therefore, disagreement in predictors could also suggest the presence of specific types of disorder.

Acknowledgments

This work was supported by grants GVOP-3.2.1.-2004-05-0195/3.0, Hungarian Scientific Research Fund (OTKA) F043609, T049073, K60694, NKFP MediChem2, and the Wellcome Trust International Senior Research Fellowship ISRF 067595. We also acknowledge the Bolyai János fellowships for Z.D. and P.T.

References

1. Tompa, P. (2002) Intrinsically unstructured proteins. *Trends Biochem. Sci.* **27**, 527–533.
2. Dunker, A. K., Brown, C. J., Lawson, J. D., Iakoucheva, L. M., and Obradovic, Z. (2002) Intrinsic disorder and protein function. *Biochemistry* **41**, 6573–6582.
3. Tompa, P. (2005) The interplay between structure and function in intrinsically unstructured proteins. *FEBS Lett.* **579**, 3346–3354.
4. Dyson, H. J., and Wright, P. E. (2005) Intrinsically unstructured proteins and their functions. *Nat. Rev. Mol. Cell Biol.* **6**, 197–208.
5. Dunker, A. K., Obradovic, Z., Romero, P., Garner, E. C., and Brown, C. J. (2000) Intrinsic protein disorder in complete genomes. *Genome Inform. Ser. Workshop Genome Inform.* **11**, 161–171.

6. Ward, J. J., Sodhi, J. S., McGuffin, L. J., Buxton, B. F., and Jones, D. T. (2004) Prediction and functional analysis of native disorder in proteins from the three kingdoms of life. *J. Mol. Biol.* **337**, 635–645.

7. Tompa, P., Dosztányi, Z. and Simon, I. (2006) Prevalent structural disorder in E. coli and S. cerevisiae proteomes. *J. Proteome. Res.* **5**, 1996–2000.

8. Linding, R., Jensen, L. J., Diella, F., Bork, P., Gibson, T. J., and Russell, R. B. (2003) Protein disorder prediction: implications for structural proteomics. *Structure.* **11**, 1453–1459.

9. Oldfield, C. J., Ulrich, E. L., Cheng, Y., Dunker, A. K., and Markley, J. L. (2005) Addressing the intrinsic disorder bottleneck in structural proteomics. *Proteins.* **59**, 444–453.

10. Oldfield, C. J., Cheng, Y., Cortese, M. S., Brown, C. J., Uversky, V. N., and Dunker, A. K. (2005) Comparing and combining predictors of mostly disordered proteins. *Biochemistry.* **44**, 1989–2000.

11. Receveur-Brechot, V., Bourhis, J. M., Uversky, V. N., Canard, B., and Longhi, S. (2005) Assessing protein disorder and induced folding. *Proteins* **62**, 24–45.

12. Cortese, M. S., Baird, J. P., Uversky, V. N., and Dunker, A. K. (2005) Uncovering the unfoldome: enriching cell extracts for unstructured proteins by acid treatment. *J. Proteome Res.* **4**, 1610–1618.

13. Csizmok, V., Szollosi, E., Friedrich, P. and Tompa, P. (2006) A novel two-dimensional electrophoresis technique for the identification of intrinsically unstructured proteins. *Mol. Cell. Proteomics* **5**, 265–273.

14. Sickmeier, M., Hamilton, J. A., LeGall, T., Vacic, V., Cortese, M. S., Tantos, A., Szabo, B., Tompa, P., Chen, J., Uversky, V. N., Obradovic, Z. and Dunker, A. K. (2007) DisProt: the Database of Disordered Proteins. *Nucl. Acids Res.* **35**, D786–D793.

15. Peng, K., Radivojac, P., Vucetic, S., Dunker, A. K., and Obradovic, Z. (2006) Length-dependent prediction of protein intrinsic disorder. *BMC Bioinformatics* **7**, 208.

16. Xie, Q., Arnold, G. E., Romero, P., Obradovic, Z., Garner, E., and Dunker, A. K. (1998) The sequence attribute method for determining relationships between sequence and protein disorder. *Genome Inform. Ser. Workshop Genome Inform.* **9**, 193–200.

17. Obradovic, Z., Peng, K., Vucetic, S., Radivojac, P., and Dunker, A. K. (2005) Exploiting heterogeneous sequence properties improves prediction of protein disorder. *Proteins.* **61**, 176–182.

18. Uversky, V. N., Gillespie, J. R., and Fink, A. L. (2000) Why are "natively unfolded" proteins unstructured under physiologic conditions? *Proteins.* **41**, 415–427.

19. Dosztányi, Z., Csizmok, V., Tompa, P., and Simon, I. (2005) The pairwise energy content estimated from amino acid composition discriminates between folded and intrinsically unstructured proteins. *J. Mol. Biol.* **347**, 827–839.

20. Garbuzynskiy, S. O., Lobanov, M. Y., and Galzitskaya, O. V. (2004) To be folded or to be unfolded? *Protein Sci.* **13**, 2871–2877.

21. Coeytaux, K., and Poupon, A. (2005) Prediction of unfolded segments in a protein sequence based on amino acid composition. *Bioinformatics.* **21**, 1891–1900.

22. Jin, Y., and Dunbrack, R. L. Jr., (2005) Assessment of disorder predictions in CASP6. *Proteins* **61**, 167–175.

23. Dosztányi, Z., Csizmok, V., Tompa, P., and Simon, I. (2005) IUPred: web server for the prediction of intrinsically unstructured regions of proteins based on estimated energy content. *Bioinformatics* **21**, 3433–3434.

24. Wootton, J. C. (1994) Non-globular domains in protein sequences: automated segmentation using complexity measures. *Computers Chem.* **18**, 269–285.

25. Romero, P., Obradovic, Z., Li, X., Garner, E. C., Brown, C. J., and Dunker, A. K. (2001) Sequence complexity of disordered protein. *Proteins* **42**, 38–48.

26. Liu, J., and Rost, B. (2003) NORSp: predictions of long regions without regular secondary structure. *Nucleic Acids Res*. **31**, 3833–3835.

27. Fuxreiter, M., Simon, I., Friedrich, P., and Tompa, P. (2004) Preformed structural elements feature in partner recognition by intrinsically unstructured proteins. *J. Mol. Biol*. **338**, 1015–1026.

28. Liu, J., Tan, H., and Rost, B. (2002) Loopy proteins appear conserved in evolution. *J. Mol. Biol*. **322**, 53–64.

29. Radivojac, P., Obradovic, Z., Smith, D. K., Zhu, G., Vucetic, S., Brown, C. J., Lawson, J. D., and Dunker, A. K. (2004) Protein flexibility and intrinsic disorder. *Protein Sci*. **13**, 71–80.

30. Finn, R. D., Mistry, J., Schuster-Bockler, B., Griffiths-Jones, S., Hollich, V., Lassmann, T., Moxon, S., Marshall, M., Khanna, A., Durbin, R., et al. (2006) Pfam: clans, web tools and services. *Nucleic Acids Res*. **34**, D247–251.

31. Csizmok, V., Bokor, M., Banki, P., Klement, É., Medzihradszky, K. F., Friedrich, P., Tompa, K., and Tompa, P. (2005) Primary contact sites in intrinsically unstructured proteins: the case of calpastatin and microtubule-associated protein 2. *Biochemistry* **44**, 3955–3964.

32. Oldfield, C. J., Cheng, Y., Cortese, M. S., Romero, P., Uversky, V. N., and Dunker, A. K. (2005) Coupled folding and binding with alpha-helix-forming molecular recognition elements. *Biochemistry* **44**, 12454–12470.

33. Dawson, R., Muller, L., Dehner, A., Klein, C., Kessler, H., and Buchner, J. (2003) The N-terminal domain of p53 is natively unfolded. *J. Mol. Biol*. **332**, 1131–1141.

34. Kussie, P. H., Gorina, S., Marechal, V., Elenbaas, B., Moreau, J., Levine, A. J., and Pavletich, N. P. (1996) Structure of the MDM2 oncoprotein bound to the p53 tumor suppressor transactivation domain. *Science* **274**, 948–953.

35. Linding, R., Russell, R. B., Neduva, V., and Gibson, T. J. (2003) GlobPlot: exploring protein sequences for globularity and disorder. *Nucl. Acids Res*. **31**, 3701–3708.

36. Koonin, E. V., and Galperin, M. (2003) *Sequence-Evolution-Function: Computational Approaches in Comparative Genomics*. Kluwer Academic Publishers, New York.

37. Ferron, F., Longhi, S., Canard, B. and Karlin, D. (2006) A practical overview of protein disorder prediction methods. *Proteins* **65**, 1–14.

38. Cheng, J., Sweredoski, M., and Baldi, P. (2005) Accurate prediction of protein disordered regions by mining protein structure data. *Data Mining Knowledge Discovery* **11**, 213–222.

39. Vullo, A., Bortolami, O., Pollastri, G., and Tosatto, S. C. (2006) Spritz: a server for the prediction of intrinsically disordered regions in protein sequences using kernel machines. *Nucl. Acids Res*. **34**, W164–168.

40. Prilusky, J., Felder, C. E., Zeev-Ben-Mordehai, T., Rydberg, E. H., Man, O., Beckmann, J. S., Silman, I., and Sussman, J. L. (2005) FoldIndex: a simple tool to predict whether a given protein sequence is intrinsically unfolded. *Bioinformatics* **21**, 3435–3438.

41. Yang, Z. R., Thomson, R., McNeil, P. and Esnouf, R. M. (2005) RONN: the biobasis function neural network technique applied to the detection of natively disordered regions in proteins. *Bioinformatics* **21**, 3369–3376.

Chapter 7

Protein Domain Prediction

Helgi Ingolfsson and Golan Yona

Domains are considered to be the building blocks of protein structures. A protein can contain a single domain or multiple domains, each one typically associated with a specific function. The combination of domains determines the function of the protein, its subcellular localization and the interactions it is involved in. Determining the domain structure of a protein is important for multiple reasons, including protein function analysis and structure prediction. This chapter reviews the different approaches for domain prediction and discusses lessons learned from the application of these methods.

1. Introduction

1.1. How and When the *"Domain Hypothesis"* Emerged

Already in the 1970s, as data on protein sequences and structures started to accumulate, researchers observed that certain patterns tend to appear in multiple proteins, either as sequence motifs or structural substructures. The first studies that coined the term "domain" date back to the 1960s. One of the earliest studies that discovered protein domains is by Phillips *(1)*, who reported the existence of distinct substructures of lysozyme, one of the first protein structures that were determined. He noted that the substructures have a hydrophobic interior and somewhat hydrophilic surface. A later study by Cunningham et al. *(2)* separated immunoglobulins into structurally distinct regions that they call domains. They hypothesized that these regions evolved by gene duplication/translocation. A different explanation was suggested by Wetlaufer *(3)*, who was the first to examine multiple proteins and compile a list of their domains. Unlike the work of Cunningham et al., which suggested a separate genetic control for each region, he proposed that the structural independence is mostly due to rapid self-assembly of these regions. Later studies established the evolutionary aspect of domains and argued in favor of separate genetic control *(4)*; however, there is also evidence in support of Wetlaufer's approach *(5,6)*.

From: *Methods in Molecular Biology, Vol. 426: Structural Proteomics: High-throughput Methods*
Edited by: B. Kobe, M. Guss and T. Huber © Humana Press, Totowa, NJ

1.2. Domain Definition

Unlike a protein, a domain is somewhat of an elusive entity and its definition is subjective. Over the years several different definitions of domains were suggested, each one focusing on a different aspect of the domain hypothesis:

- A domain is a protein unit that can fold independently.
- It forms a specific cluster in three-dimensional (3D) space (depicted in Fig. 7.1).
- It performs a specific task/function.
- It is a movable unit that was formed early on in the course of evolution.

Most of these definitions are widely accepted, but some are more subjective than others. For example, the definition of a cluster in 3D space is dependent on the algorithm used to define the clusters and parameters of that algorithm. Furthermore, although the first two definitions focus on structural aspects, the other definitions do not necessarily entail a structural constraint. The structural definitions can also result in a domain that is not necessarily a continuous subsequence. However, this somewhat contradicts the evolutionary viewpoint that considers domains as elementary units that were put together to form multiple-domain proteins through duplication events. Although many of the proteins known today are single-domain proteins, it has been estimated *(7)* that the majority of proteins are multidomain proteins (contain several domains). The number of domains in multidomain proteins ranges from two to several hundred (e.g., titin protein, TrEMBL *(8)* Q10466_HUMAN; http://biozon.org/Biozon/Profile/68753). However, most proteins contain only a few domains (see Section 4.3.1.). Consistent with the evolutionary aspect of the domain hypothesis, more complex organisms have a higher number of multidomain proteins *(6,9)*.

Understanding the domain structure of proteins is important for several reasons:

- **Functional analysis of proteins.** Each domain typically has a specific function and to decipher the function of a protein it is necessary first to determine

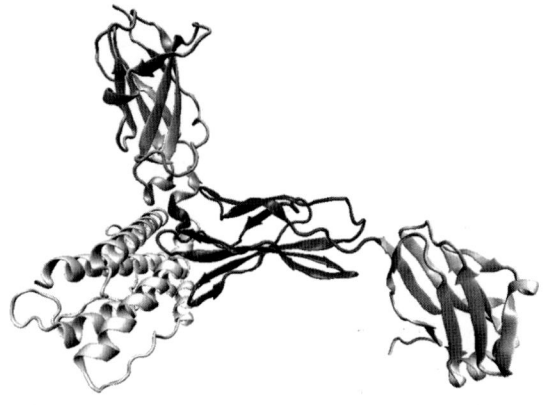

Fig. 7.1 Domain structure of a cytokine/receptor complex. Three-dimensional rendering of the asymmetric unit of a cytokine/receptor complex solved by x-ray diffraction (PDB 1i1r). This unit is composed of two chains: chain B is an all alpha helical bundle (light gray, *lower left*) and chain A consists of three SCOP domains colored dark gray (*d1i1ra1*, positions 2–101, *lower right*), black (*d1i1ra2*, positions 102–196, *middle*) and dark gray (*d1i1ra3*, positions 197–302, *upper left*).

its domains and characterize their functions. Since domains are recurring patterns, assigning a function to a domain family can shed light on the function of the many proteins that contain this domain, which makes the task of automated function prediction feasible. In light of the massive sequence data that is generated these days, this is an important goal.

- **Structural analysis of proteins.** Determining the 3D structure of large proteins using NMR or x-ray crystallography is a difficult task due to problems with expression, solubility, stability, and more. If a protein can be chopped into relatively independent units that retain their original shape (domains), then structure determination is likely to be more successful. Indeed, protein domain prediction is central to the structural genomics initiative.
- **Protein design.** Knowledge of domains and domain structure can greatly aid protein engineering (the design of new proteins and chimeras).

However, despite the many studies that investigated protein domains over the years and the many domain databases that were developed, deciphering the domain structure of a protein remains a nontrivial problem, especially if one has to predict it from the amino acid sequence of the protein without any structural information.

The goal of this chapter is to survey methods for protein domain prediction. In general the existing methods can be divided into the following categories: (1) experimental (biological) methods, (2) methods that use 3D structure, (3) methods that are based on structure prediction, (4) methods based on similarity search, (5) methods based on multiple sequence alignments, and (6) methods that use sequence-based features.

When discussing the modular structure of proteins it is important to make a distinction between motifs and domains. Motifs are typically short sequence signatures. As with domains, they recur in multiple proteins; however, they are usually not considered as "independent" structural units and hence are of less interest here. Another important concept related to the domain structure of a protein is the organization of domains into hierarchical classes. These classes group domains based on evolutionary (families), functional (superfamilies), and structural (folds) relationships, as discussed in Section 3.2.

The chapter starts with a review of the different approaches for domain prediction (Table 7.1) and the main domain databases (Table 7.2). It then discusses various computational and statistical issues related to domain assignments, and concludes with notes about the future of domain prediction.

2. Domain Detection Methods

2.1. Experimental Methods

Although the focus of this chapter is *computational* methods to predict domains, it is important to mention experimental methods that can "chop" a protein into its constituent domains. One such method is proteolysis, the process of protein degradation with proteases. Proteases are cellular enzymes that cleave bonds between amino acids. Proteases can only cleave bonds that are accessible; by carefully manipulating experimental conditions to make sure that the protein is in native or near-native state (not denatured), the proteases can only access relatively unstructured regions of the proteins to obtain "limited" proteolysis *(10)*. The method has

Table 7.1. Domain prediction methods.

Method	Reference	URL or corresponding author
Methods that use 3D structure		
Taylor	*(16)*	ftp://glycine.nimr.mrc.ac.uk/pub/
PUU	*(18)*	holm@embl-ebi.ac.uk
DOMAK	*(19)*	geoff@bio.ox.ac.uk
DomainParser	*(20)*	http://compbio.ornl.gov/structure/domainparser/
PDP	*(22)*	http://123D.ncifcrf.gov/pdp.html
DIAL	*(23)*	http://caps.ncbs.res.in/DIAL/
Protein Peeling	*(25)*	http://www.ebgm.jussieu.fr/~gelly/
Methods that use 3D predictions		
Rigden	*(28)*	daniel@cenargen.embrapa.br
SnapDragon	*(29)*	jhering@nimr.mrc.ac.uk
Methods based on similarity search		
Domainer	*(30)*	http://www.biochem.ucl.ac.uk/bsm/dbbrowser/ protocol/ prodomqry.htm/
DIVCLUS	*(32)*	http://www.mrc-lmb.cam.ac.uk/genomes/
DOMO	*(36,37)*	http://abcis.cbs.cnrs.fr/domo/
MKDOM2	*(38)*	http://prodes.toulouse.inra.fr/prodom/xdom/ mkdom2.html/
GeneRAGE	*(35)*	http://www.ebi.ac.uk/research/cgg//services/rage/
ADDA	*(43)*	http://ekhidna.biocenter.helsinki.fi/sqgraph/ pairsdb/
EVEREST	*(45)*	http://www.everest.cs.huji.ac.il/
Methods based on multiple sequence alignments		
PASS	*(50)*	kuroda@gsc.riken.go.jp
Domination	*(49)*	http://mathbio.nimr.mrc.ac.uk/
Nagarajan & Yona	*(42)*	http://biozon.org/tools/domains/
Methods that use sequence only		
DGS	*(51)*	ftp://ftp.ncbi.nih.gov/pub/wheelan/DGS/
Miyazaki et al.	*(52,53)*	ykuroda@cc.tuat.ac.jp
DomCut	*(54)*	http://www.bork.embl.de/~suyama/domcut/
GlobPlot	*(55)*	http://globplot.embl.de/
DomSSEA	*(56)*	
KemaDom	*(58)*	wangfei@fudan.edu.cn
Meta-DP	*(59)*	http://meta-dp.cse.buffalo.edu/

been applied to many proteins (e.g., on thermolysin by Dalzoppo et al. *(11)* and on streptokinase by Parrado et al. *(12))*. A related method is described by Christ and Winter *(13)*. To screen for protein domains, they cloned and expressed randomly sheared DNA fragments and proteolyzed the resulting polypeptides. The protease-resistant fragments were reported as potential domains.

Table 7.2. Domain databases.

Database	Reference	URL
Motif databases		
PROSITE	*(60)*	http://ca.expasy.org/prosite/
PRINTS	*(61)*	http://www.bioinf.man.ac.uk/dbbrowser/PRINTS/
Blocks	*(62)*	http://blocks.fhcrc.org/
Sequence-and MSA-based, automatically generated		
ProDom	*(39)*	http://www.toulouse.inra.fr/prodom.html/
Biozon	*(75)*	http://biozon.org/
MSA-based, manually verified		
Pfam	*(44)*	http://pfam.wustl.edu/
SMART	*(47)*	http://smart.embl-heidelberg.de/
TigrFam	*(48)*	http://www.tigr.org/TIGRFAMs/
CDD	*(65)*	http://www.ncbi.nlm.nih.gov/entrez/query. fcgi?db=cdd
3Dee	*(77)*	http://www.compbio.dundee.ac.uk/3Dee/
Integrated databases		
InterPro	*(63)*	http://www.ebi.ac.uk/interpro/
Biozon	*(75)*	http://biozon.org/
Structure-based		
SCOP	*(26)*	http://scop.berkeley.edu/
CATH	*(27)*	http://cathwww.biochem.ucl.ac.uk/
Dali/FSSP	*(76)*	http://www.ebi.ac.uk/dali/
DDBASE2	*(24)*	http://caps.ncbs.res.in/ddbase/
SBASE	*(67)*	http://www.icgeb.trieste.it/sbase/

2.2. Methods That Use Three-Dimensional Structure

The first methods that were developed for protein domain detection were based on structural information. All algorithms in this category are based on the same general principle that assumes domains to be structurally compact and separate substructures (with higher density of contacts within the substructures than with their surroundings). The differences lie in the algorithms employed to search for these substructures.

Early methods by Crippen *(14)* and Lesk and Rose *(15)* paved the way for methods that automatically define domains from 3D structure. These algorithms use a bottom-up agglomerative approach to cluster residues into domains. Crippen et al. used a binary tree clustering algorithm based on α-carbon distances, starting with small continuous segments with no "long-range" interactions and ending with a cluster containing all protein residues. They classified clusters with contact-density and radius of gyration lower than some cutoff values as domains. Lesk and Rose *(15)* presented an algorithm that identifies compact contiguous-chain segments by looking at segments of different lengths whose atoms fit into an ellipsoid with the smallest area.

Taylor's domain identification algorithm *(16)* is also based on a clustering method that uses only α-carbon spatial distances. The method starts by assigning all α-carbons an integer in sequential order from start to finish of the protein sequence. Then it increases or decreases the integers, depending on the average integer value of all its neighbors within a certain distance. This is done in an iterative manner until convergence. After post processing, smoothing and fixing β-sheets, all α-carbons with the same integer are assigned to the same domain, thus resulting in a partitioning of the protein into domains. A different approach is used in DETECTIVE *(17)*, a program that determines domains by identifying hydrophobic cores.

As an alternative to clustering, the following algorithms use a top-down divisive approach to split a protein into its domains. PUU *(18)* recursively splits proteins into potential folding units by solving an eigenvalue problem for the residue contact matrix. Finding optimal spatial separation is thus independent of the number of cuts in the protein sequence, allowing for multiple sequence cuts in linear time. DOMAK *(19)* cuts proteins into two structural subunits by selecting 1–4 cutting sites on the protein chain that result in subunits with a maximum "split value." The split value is defined as the ratio between the number of internal residues that are in contact (defined as residues that are within 5 Å of each other) in each subunit, and the number of residues that are in contact between the subunits. The cutting is iterated until a minimum split value or minimum sequence length is reached. DomainParser *(20)* represents protein 3D structure as a flow network, with residues as nodes and residue contacts as edges of capacity related to the distance between them. The algorithm iteratively locates bottlenecks in the flow network using the classical Ford-Fulkerson algorithm *(21)*. These bottlenecks represent regions of minimum contact in the protein structure and the protein is partitioned into domains accordingly. A similar approach is employed by Protein Domain Parser (PDP) *(22)*. PDP starts by defining a protein as a single domain and then recursively splits the domains into subdomains at sites that result in low 3D contact area between domains. In each iteration the algorithm uses a single or a double cut of the amino acid chain (where both cuts are required to be very close spatially). A cut is only kept if the number of contacts between the subdomains, normalized by subdomain size, is less than one half of the average domain contact density.

The organization of residues into secondary structures can also hint at the domain structure of a protein. DAIL *(23)* determines secondary structures from 3D data and then predicts domains by clustering the secondary structures based on the average distances between them. The algorithm was used to generate the domain database DDBASE2.0 *(24)*. Protein peeling, a method described in *(25)*, iteratively splits a protein into two or three subunits based on a "partition index" (PI), down to "protein units" that are intermediate substructures between domains and secondary structures. The PI is calculated from the protein's contact probability matrix and measures the ratio between the number of contacts within a subunit and the number of contacts between subunits.

There is no single objective definition of a cluster in 3D, and each algorithm in this category implies a slightly different definition of structurally compact substructures, depending on the criterion function it tries to optimize, thus resulting in different partitions into domains. Moreover, each method is sensitive to the parameters of the algorithm used to determine the clusters/ domains (the maximal distance, the threshold ratio value, etc.). However, to this day the most effective methods are still those that utilize the 3D information to define domain boundaries, as manifested in perhaps the two most important resources on protein domains: SCOP *(26)* and CATH *(27)* (see Section 3.2.). This is not surprising as the definition of domains relies most naturally on the 3D structure of a protein.

2.3. Methods That Use Three-Dimensional Predictions

Since structural data are available for only a relatively small number of proteins, several methods approach the problem of domain prediction by employing structure prediction methods first or by using other types of predicted 3D information. For example, Rigden *(28)* computes correlated mutation scores between different columns of a multiple sequence alignment to assess whether they are part of the same domain. Columns that exhibit signals of correlated mutations are more likely to be physically close to each other and hence are part of the same domain. The scores are predictive of the contacts between columns and are used to create a contact profile for the protein. Points of local minima along this graph are predicted as domain boundaries. Although conceptually appealing, signals of coevolution are often too weak to be detected. This method is also computationally intensive.

A similar approach is employed by the SnapDragon algorithm *(29)*, that starts by generating many 3D models of the query protein using hydrophobicity values (computed from multiple sequence alignments) and predicted secondary structure elements. These models are then processed using the method of Taylor *(16)* described in the preceding section. The domains are defined based on the consistency between the partitions of the different 3D models. The results reported by the authors suggest that this algorithm can be quite effective in predicting domains; however, it is also very computationally intensive since it requires generating many 3D models first.

2.4. Methods Based on Similarity Search

Given the computational complexity of structure prediction methods, methods that are based only on sequence information provide an appealing alternative, especially for large-scale domain prediction. Sequence information can be utilized in many ways, the most obvious of which is sequence similarity. Methods described in this subsection use homologous sequences detected in a database search to predict domains, most of which start with an all-vs.-all comparison of sequence databases. The similar sequences are then clustered and split into domains.

One of the first to be introduced in this category is Domainer *(30)*. The algorithm starts by generating homologous segment pairs (HSPs) based on the results of an all-vs.-all BLAST *(31)* similarity search. The HSPs are then clustered into homologous segment sets (HSSs) and if different parts of a sequence are split into different clusters then a link is formed between them, resulting in a large number of HSSs graphs. The domain boundary assignments are then made with N- or C-terminal information from within the HSSs and/or by analyzing the HSS graphs in search of repeats and shuffled domains. In the DIVCLUS program (32) there is a choice of running SSEARCH (an implementation of the Smith-Waterman algorithm (33)) or a faster but less accurate FASTA (34) search algorithm for the all-vs.-all sequence homology search. Pairs of similar sequences are linked using the single-linkage clustering algorithm to form clusters. If the sequences in a cluster are not matched in overlapping regions, then the cluster is split into smaller clusters, each representing a domain. The criteria used by these algorithms are somewhat ad hoc.

GeneRAGE (35) is based on a similar principle but uses a different algorithm. GeneRAGE attempts to identify domains by searching for problematic transitive relations. The algorithm starts with an all-vs.-all BLAST sequence similarity search and builds a binary relationship matrix from this data. To improve the quality of the data they check all significant similarities with the Smith-Waterman algorithm and symmetrize the matrix. To check for possible multidomain proteins they search the matrix for sites where transitivity fails to apply (that is, where protein A is related to B, B to C, but A is not related to C). However, this should be considered with caution, as transitivity can also fail when proteins are only remotely related.

To speed up the clustering phase, DOMO (36) uses amino acid and dipeptide compositions to compare sequences. One sequence is selected to represent each cluster, and the representative sequences are searched for local similarities using a suffix tree. Similar sequences are clustered and domain boundary information is extracted based on the positions of N- and C-terminals and repeats in the clusters using several somewhat ad hoc criteria (37). Another algorithm that attempts to speed up the process and avoid an all-vs.-all comparison is MKDOM2 (38). MKDOM2 is an automatic domain prediction algorithm (used to generate the ProDom (39) database) based on repeated application of the PSI-BLAST algorithm (40). The algorithm starts by pruning the database of all low-complexity sequences, with the SEG algorithm (41), and all sequences shorter than 20 amino acids (argued to be too short for genuine structured domains). Then, in an iterative manner, it finds the shortest repeat-free sequence in the database, which is assumed to represent a single domain. This sequence is used as input for a PSI-BLAST search against the database to locate all its homologous domains and form a domain family. The sequence and all its homologous sequences are removed from the database and the procedure is repeated until the database is exhausted. This approach, although fast, tends to truncate domains short of domain boundaries (42).

A more sophisticated approach is employed by ADDA (43). Like the others, it starts with an all-vs.-all BLAST search. The protein space is represented as a graph and edges are drawn between pairs that are significantly similar. Then, they iteratively decompose each sequence into a tree of putative domains, at positions that minimize the number of neighbors (in the similarity graph) that are aligned to both sides of the cut. The process terminates if the size of a putative domain falls below 30. Using a greedy optimization strategy, they traverse the putative hierarchical domain assignments and keep the partitions that maximize a likelihood model, which accounts for domain size, family size, residues not covered by domains and the integrity of alignments. In general, ADDA tends to overestimate the size of domains, but it is relatively fast and tends to agree with both SCOP and Pfam (44).

Finally, the EVEREST system *(45)* also starts with a pairwise all-vs.-all BLAST sequence comparison. The similar segments are extracted and internal repeats are removed. The segments are clustered into candidate domain families using the average linkage clustering algorithm. A subset of candidate domain families is then selected by feeding features of each family (cluster similarity, cluster size, length variance, etc.) to a regression function, which was learned with a boosting algorithm from a Pfam domain family training set. An HMM *(46)* is then created for each selected family and the profile-HMMs are used to scan the original sequence set and generate a new set of segments. The process (search, clustering, selection using a regression function, modeling) is repeated, and after three iterations overlapping HMMs are merged and a domain family is defined for each HMM. The use of powerful HMMs in an iterative manner is an advantage over a BLAST search, as it can detect remote homologies that are missed by BLAST. However, the method is complicated and needs another database of domain families to train its regression function.

2.5. Methods Based on Multiple Sequence Alignments (MSAs)

A database search provides information on pairwise similarities; however, these similarities are often erroneous (especially in the "twilight zone" of sequence similarity). Furthermore, many sequences in protein databases are fragments that may introduce spurious termination signals that can be misleading. A natural progression toward more reliable domain detection is to use MSAs. Not only do MSAs align sequences more accurately, but they can also detect remote homologies that are missed in pairwise comparisons.

MSA-based approaches are the basis of several popular domain databases, such as Pfam, SMART *(47)*, and TigrFam *(48)* that combine computational analysis and manual verification (see Section 3.1.). Other MSA-based methods start with a query sequence and run a database search to collect homologs and generate a MSA that is then processed to find signals of domain termination. For example, DOMAINATION *(49)* runs a PSI-BLAST database search, followed by SEG to filter low complexity regions. The MSA is then split at positions with multiple N- and C-terminal signals. This process is iterated with each subsequence as a new input, until there is no further splitting, and the remaining subsequences are predicted as domains. Another system that uses a similar procedure is PASS *(50)*, which is short for Prediction of AFU (autonomously folding units) based on Sequence Similarity. The program predicts domains (or AFUs) based on BLAST results. For each residue in the query protein PASS counts how many sequences were aligned to it. Then it scans the sequence and predicts domain start/end positions at residues where the count increased or decreased over a certain threshold. Another system is presented by Nagarajan and Yona in *(42)*, where 20 different scores derived from multiple sequence alignments (including sequence termination, entropy, predicted secondary structure, correlation and correlated mutations, intron/exon boundaries, and more) are used to train a neural network (NN) to predict domain vs. linker positions. The predictions are smoothed and the minima are fed to a probabilistic model that picks the most likely partition considering the prior information on domains and the NN output. Clearly, the quality of all these methods depends on the number and composition of homologs used to construct the MSA (for a discussion see Section 4.1.).

2.6. Methods that Use Sequence Only

In addition to homology-based and MSA-based methods, some methods utilize other types of sequence information to predict domain boundaries. For example, domain guess by size (DGS) *(51)* predicts domain boundaries based on the distribution of domain lengths in domain databases. Given a protein sequence DGS computes the likelihood of many possible partitions using the empirical distributions and pick the most likely one. This crude approach, however, tends to over-predict single-domain proteins (since these are the most abundant in domain databases). Miyazaki et al. *(52)* use a neural network to learn and predict linker regions between domains, based on their composition. In a follow-up study *(53)* they report that low-complexity regions also correlate with domain termini and combine both methods to predict domain boundaries. DomCut *(54)* also predicts domain linkers from amino acid composition. Using a large set of sequences with previously defined domain/linker regions, they first calculate a linker index for each amino acid that indicates its preference for domains vs. linkers. Given a query protein, the algorithm then computes the average linker index for a sliding window and if the average falls below a certain threshold it reports it as a linker region. Although simple and fast, predictions produced with these methods can be noisy.

Secondary structure information is utilized by methods such as GlobPlot *(55)*, which makes domain predictions based on the propensity of amino acids to be in ordered secondary structures (helices and strands) or disordered structures (coils). DomSSEA *(56)* also uses secondary structure information. It predicts the secondary structure of a protein using PSIPRED *(57)*, aligns the protein to CATH domains based on the secondary structure sequence and then uses the most similar domains to chop the protein.

Other methods combine several different sequence features. KemaDom *(58)* combines three support vector machines (SVMs) that are trained over different feature sets (including secondary structure, solvent accessibility, evolutionary profile, and amino acid entropy), derived from the protein sequence. Each residue is assigned a probability to be in a domain boundary, which is the maximum over the output of the three SVMs. These probabilities are then smoothed to give the final domain predictions. Finally, Meta-DP *(59)* is a domain prediction server that runs a variety of other domain predictors and derives a consensus from all the outputs.

3. Domain Databases

In parallel to the development of domain prediction algorithms, several very useful resources on protein domains were created (Table 7.2). In general, these resources can be categorized into sequence- and structure-based databases.

3.1. Databases Based on Sequence Analysis

Early databases of recurring patterns in proteins focused on short signature patterns or motifs. As mentioned, these patterns are typically shorter than domains and are not concerned with the structural aspects of domains. However, it is worth mentioning a few important resources on protein motifs, such as PROSITE *(60)*, PRINTS *(61)*, and Blocks *(62)*.

ProDom *(39)* is an automatically generated domain family database. It is built by applying MKDOM2 algorithm (see Section 2.4.) to a non redundant set of sequences gathered from SWISS-PROT *(8)* and TrEMBL. Additionally, ProDom provides links to a variety of other databases, including; InterPro *(63)*, Pfam, and PDB *(64)*.

Pfam *(44)* is a database of multiple alignments representing protein domains and protein families. Each family is represented by a profile HMM constructed from those alignments. Pfam is separated into two parts, Pfam-A and Pfam-B. Pfam-A is the manually curated version, constructed from high-quality manually verified multiple alignments. Pfam-B is the fully automated version over the remaining sequences, derived from ProDom. SMART *(47)* is a domain database that concentrates on signaling and extracellular proteins. As in Pfam-A, each domain is represented by a profile HMM that is constructed from a manually verified multiple alignment. TigrFam *(48)* is another database that uses HMMs generated with HMMER *(46)*. It is similar to Pfam-A, but while Pfam focuses on the sequence–structure relationship when defining domains, TigrFam focuses on the functional aspects and domain families are defined and refined such that sequences have homogeneous functions.

Several databases integrate domain definitions from multiple sources. For example, CDD (which stands for Conserved Domain Database) *(65)* is a domain database that is maintained by NCBI. It consists of two parts: one essentially mirroring Pfam, SMART, and conserved domains from COG (see Note 1), and the other containing curated definitions by the NCBI. SBASE *(67)* stores protein domain sequences and their annotations, collected from the literature and/or any of the following databases; Swiss-Prot, TrEMBL, PIR *(68)*, Pfam, SMART, and PRINTS. Domain sequences are assembled into similarity groups based on BLAST similarity scores. InterPro *(63)* also links together several resources on proteins and protein domains, including UniProt *(69)*, PROSITE, Pfam, PRINTS, ProDom, SMART, TIGR-FAMs, PIRSF *(70)*, SUPERFAMILY *(71)*, Gene3D *(72)*, and PANTHER *(73)*. To scan a protein for protein domains in InterPro one can use InterProScan *(74)*. Finally, Biozon *(75)* is another database that consolidates multiple databases, including domain databases. Biozon, however, is broader in scope and includes information also on protein sequences and structures, DNA sequences, interactions, pathways, expression data, and more.

3.2. Databases Based on Three-Dimensional Information

Domain databases that are based on structure analysis of proteins include SCOP, CATH, FSSP *(76)*, 3Dee *(77)*, and DDBASE2 *(24)*. SCOP *(26)* is based on an expert analysis of protein structures. Proteins are chopped into domains based on visual inspection and comparison of protein structures. The exact principles that are employed by Dr. Murzin are unknown, but one of the criteria is that the substructures defined as domains should recur in more than one structure. Once defined, the domains are organized into a hierarchy of classes. At the bottom level homologous domains are grouped into families based on sequence similarity. At the next level, structurally similar families are grouped into superfamilies based on functional similarity (common residues in functional sites). Next, structurally similar superfamilies are grouped into folds. Finally, folds are grouped into seven general classes: all-alpha, all-beta, alpha/beta, alpha+beta, multidomain proteins, membrane and cell surface proteins, and small proteins.

CATH *(27)* is based on a mixture of automated and manual analysis of protein structures. Proteins that have significant sequential and structural similarity to previously processed proteins inherit their classification. For other proteins, domain boundaries are manually assigned based on the output of a variety of different structural- and sequence-based methods and the relevant literature (see Note 2). Domains are hierarchically classified into four major levels: Homology (for domains that share significant sequence, structural and/or functional similarity), Topology (domains whose secondary structure shape and connectivity are similar), Architecture (similar secondary structure orientation), and Class (all domains with similar secondary structure composition).

Dali/FSSP *(76)* is another structural classification of proteins, however, unlike SCOP and CATH, it classifies complete PDB structures and it is fully automatic. It uses the structure comparison program Dali *(81)* to perform an all-vs.-all structure comparison of all entries in PDB. The structures are then clustered hierarchically based on their similarity score into folds, second cousins, cousins, and siblings.

3Dee *(77)* is another repository of protein structural domain definitions. Initially created with the automatic domain classifier DOMAK (see Section 2.2.), it has been subsequently manually corrected and updated through visual inspection and relevant literature. Additionally, 3Dee includes references to older versions of domain definitions and to alternative domain definitions, and allows multiple MSAs for the same domain family. DDBASE2 *(24)* is a small database of globular domains, automatically generated by applying the program DIAL (see Section 2.2.) to a set of non-redundant structures (with less than 60% sequence identity).

4. Domain Prediction-Lessons and Considerations

Despite the many domain databases and numerous domain prediction algorithms, the domain prediction problem is by no means solved for several reasons. The protein of interest might contain new domains that have not been characterized or studied yet, and therefore the protein might be poorly represented in existing domain databases with limited information about its domain structure or no information at all. One can attempt to predict the domain structure of a protein by applying any of the algorithms described in the previous sections, however, not all methods are applicable. For most proteins (and especially newly sequenced ones) the structure is unknown, thus ruling out the application of structure-based methods. Some methods might only work by clustering whole databases of proteins but do not work on individual proteins. If the protein has no homologues, then most of the prediction methods that are based on sequence similarity are ineffective. Other methods (e.g., those that are based on structure prediction) might be too computationally intensive.

Most importantly, existing domain prediction algorithms and domain databases can be very inconsistent in their domain assignments (Fig. 7.2). However, without structural information it is difficult to tell which ones (if any) are correct. Even when the structure is known, determining the constituent domains might not be straightforward (Fig. 7.3). Majority voting does not necessarily

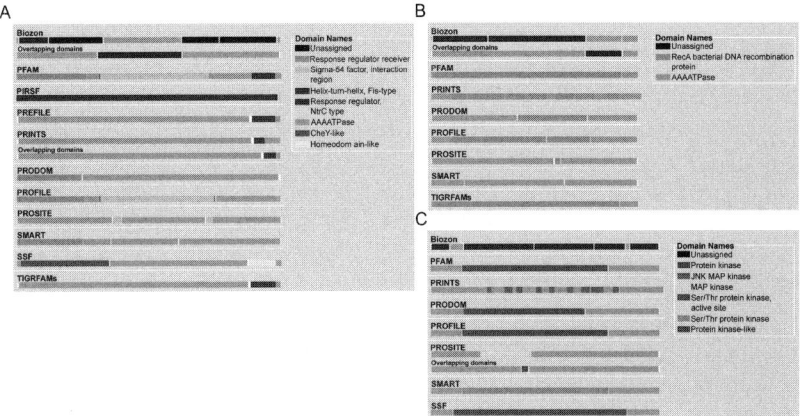

Fig. 7.2 Examples of disagreement between different domain databases. Figures are adapted from www.biozon.org. **A.** Nitrogen regulation protein ntrC (Biozon nr ID 004570000090). **B.** Recombinase A (Biozon nr ID *003550000123*). **C.** Mitogen-activated protein kinase 10 (Biozon nr ID *004640000054*). To view the profile page of a protein with Biozon ID *x* follow the link www.biozon.org/Biozon/ProfileLink/x. As these examples demonstrate, different domain databases have different domain predictions and domain boundaries are often ill-defined. Domain databases can also generate overlapping predictions. Overlapping domains are represented in separate lines. There are two possible reasons for overlaps: (1) Domain signatures in the source database overlap, where the longer one could be a global protein family signature and the shorter one a local motif. (2) Completely identical sequence entries that were analyzed independently by a source database (resulting in slightly different predictions) were mapped to the same nr entry in Biozon.

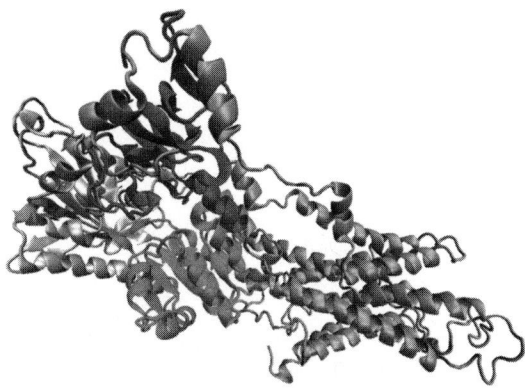

Fig. 7.3 Domain structure of calcium ATPase (PDB 1IWO_A). According to SCOP, this structure is composed of four domains: a transmembrane domain (*d1iwoa4*, positions 1–124, *colored cyan*), a transduction domain (*d1iwoa1*, positions 125–239, *colored red*), a catalytic domain (*d1iwoa2*, positions 344–360 and 600–750, *colored green*), and unlabeled fourth domain (*d1iwoa3*, positions 361–599, *colored gray*). Note that the domains are not well separated, structurally.

produce correct results, either. The methods are correlated and often use similar types of information (e.g., predicted secondary structures) and might provide similar but wrong predictions.

4.1. Multiple Sequence Alignment as a Source for Domain Termini Signals

The next subsections discuss methods to improve the reliability of domain predictions with similarity and MSA-based methods, which constitute the majority of the methods. MSA-based methods are especially appealing for several reasons. Well-constructed MSAs are a good source of information for domain predictions. They are much easier and faster to generate than 3D models and hence can be used in a large-scale domain prediction. Moreover, MSAs are usually more accurate than pairwise similarities and more sensitive, as they can be used to detect remote homologies (using a profile or a HMM that is built from the MSA). They can be used to predict certain structural features quite easily (e.g., secondary structures) and assess correlation and contact potential between different positions. Utilizing the fact that proteins typically do not start or end in the middle of domains and that various proteins have different domain organization, one can use the sequences' start/end positions as indicators for domain boundaries (Fig. 7.4). Indeed, this is the basis of methods such as DOMO, Domainer, DOMAINATION, PASS, and others.

However, there are several complicating factors to consider. If the MSA contains just highly similar sequences (which have the same domain organization), or it contains only remotely related sequences with no domains in common, then MSA-based approaches are ineffective. Moreover, MSAs are not necessarily optimally aligned with respect to domain boundaries. They might end prematurely, or include many spurious termination signals due to remotely related sequences that are loosely aligned. On the other hand, sequences that

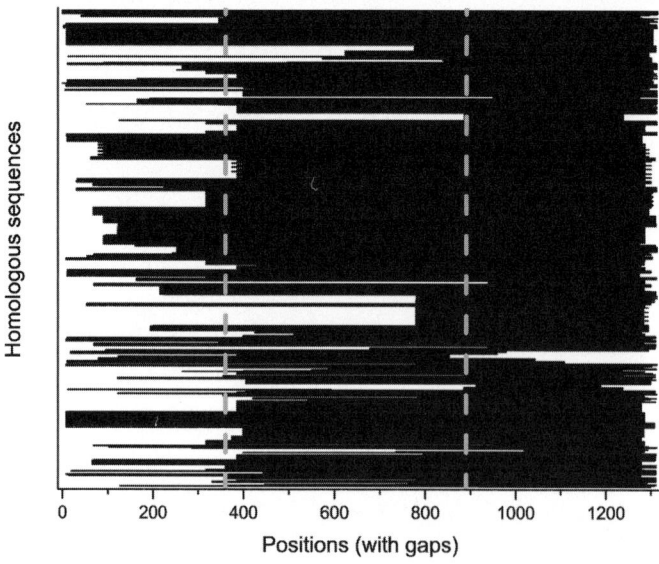

Fig. 7.4 A schematic representation of an MSA for Interleukin-6 receptor beta protein (PDB 1i1r). MSA was generated using PSI-BLAST. See *(42)* for details. Homologous sequences are in order of decreasing e-value (965 sequences in total). Domain boundaries according to SCOP are at positions 102, 197, and 302 (see Fig. 7.1). The *dashed gray lines* mark the positions where SCOP predicts domains after correction of gaps (1, 360, 890, and 1312). Note the correlation with sequence termini in the MSA.

are too similar usually contain little information on domain boundaries and bias predictions as they mask other equally important but less represented sequences. Defining the correct balance between distantly related sequences and highly similar sequences is not trivial and usually a weighting scheme is employed (see Section 4.3.4.). Additionally, many of the sequences in the alignment might not represent whole proteins but fragments, thus introducing artificial start/end signals and adding to the noise. Including gaps in the MSA further blurs the signal, especially in regions of low density (regions where most sequences in the alignment have a deletion or only a few sequences have an insertion, resulting in very low information for those regions).

One can try to mitigate these factors by using a more accurate multiple sequence alignment algorithm (e.g., T-coffee or POA) *(82,83)*, picking an appropriate e-value cutoff for the similarity search, removing or under-weighting highly similar sequences, and removing sequences that have been annotated as fragments. However, even after applying all these methods, MSAs will almost always have some noise, complicating domain prediction.

4.2. Detecting Signals of Domain Termination

To extract probable domain boundary signals from noisy MSAs it is necessary to assess the statistical significance of each signal. As an illustrating example, the focus here is on sequence termination information. Sequence termination has been shown to be one of the main sources of information on domain boundaries *(42)*. Through this example, the chapter also presents a useful framework for the integration of other sources of information on domain boundaries, such as domain and motif databases.

One way to extract plausible domain boundary signals from a MSA is to count the number of sequences that start or end at each residue position in the MSA and generate a histogram. However, these histograms tend to be quite noisy, especially for large MSAs with many sequences (Fig. 7.5A). The histogram can be smoothed by averaging over a sliding window of length l_w (see

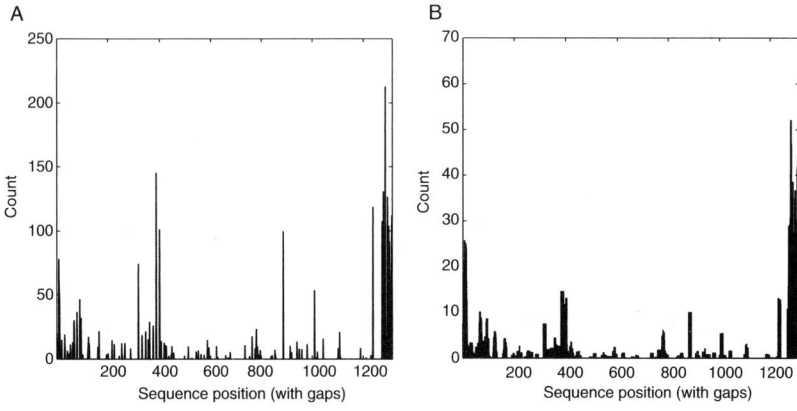

Fig. 7.5 Analysis of sequence termination signals in MSAs. Sequence termination signals for cytokine receptor signaling complex (PDB 1i1r). **A.** Original histogram of sequence termini counts. **B.** Smoothed histogram (window size $l_w = 10$).

Fig. 7.5B). However, even after smoothing it is still unclear which signals should be considered as representing true domain boundaries.

To evaluate each position and assess its statistical significance, one can check how likely it is to observe such a signal by chance. To estimate the probability that a sequence terminated at a certain position by chance, we assume a uniform distribution over all the MSA positions (see Note 3). Given a MSA of length M, the probability of a randomly aligned sequence to start or end at any histogram position is estimated as $p = \dfrac{1}{M}$. To assess the significance of a position with k termination signals, we evaluate the probability that k or more sequences terminate at the same position by chance. By assuming independence between the sequences (see Note 4), it is possible to use the binomial distribution with parameter $p = \dfrac{1}{M}$. The probability that k or more sequences (out of n sequences) terminate in one position is given by:

$$P(i \ge k) = \sum_{i=k}^{n} \binom{n}{i} p^{i}(1-p)^{n-i} \qquad k = 0,1,\dots,n \qquad (1)$$

Applying this procedure to the histogram of Fig. 7.5B results in the graph of Fig. 7.6. Peaks in this graph that are lower than a certain threshold T (here set to 0.05) are suspected as signals of true domain boundaries.

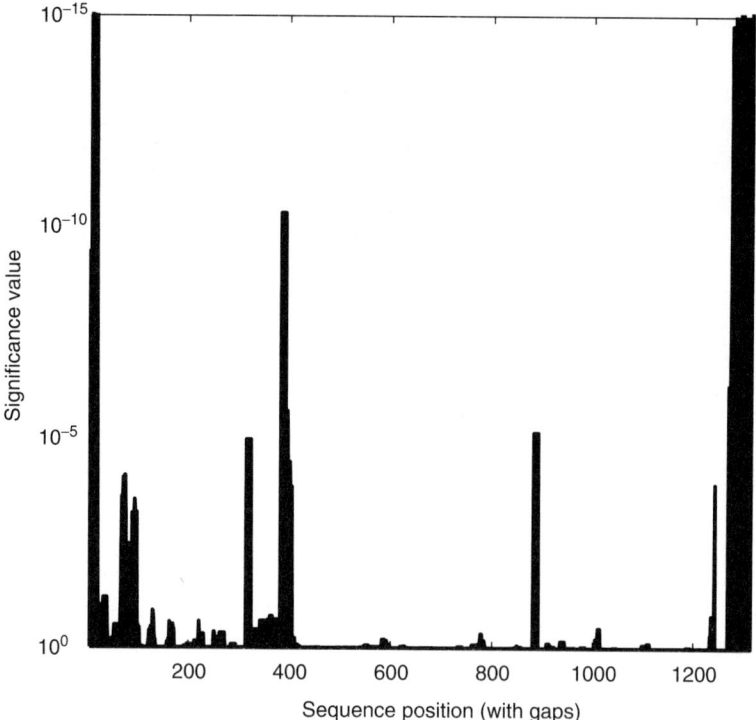

Fig. 7.6. Analysis of sequence termination signals in MSAs (PDB 1i1r). P-value of sequence termination signals computed using the binomial distribution. The p-value is plotted in logscale.

In the case of gapped MSAs (which contain gaps either due to deletions or, more likely, inserts in one sequence which introduce gaps in all others), some regions in the alignment can have very low density (where density at position i is defined as the number of non-gap elements at that position). This can be a problem in Equation (1) because proteins cannot start/end in gaps; therefore, one might be grossly overestimating p in positions with low density. To correct for this bias, we modify the probability above and define $p = \dfrac{Density_{(k)}}{Density_{total}}$ instead of $p = \dfrac{1}{M}$, where Density total is the total density of the alignment and Density$_{(k)}$ is the smoothed density at position k, defined as:

$$Density_{total} = \sum_{i=1}^{M} \sum_{j=1}^{n} \begin{cases} 0 & \text{if } seq_j(i) = gap \\ 1 & \text{else} \end{cases} \tag{2}$$

$$Density_{(k)} = \frac{1}{l_W} \sum_{i=k-(l_W/2)}^{k+(l_W/2)} \sum_{j=1}^{n} \begin{cases} 0 & \text{if } seq_j(i) = gap \\ 1 & \text{else} \end{cases} \tag{3}$$

The smoothing improves the statistical estimates; however, it also results in a large number of neighboring positions that could be deemed significant, since isolated peaks of low probability are replaced with small regions of low probability. To address that, one can pick a single point in each region of low probability (below the threshold T), the one with the minimal probability.

4.3. Selecting the Most Likely Domain Partition

Mis-aligned regions in the MSA and fragments might introduce noise that can lead to many erroneous predictions, often in the proximity of each other. Sometimes, different instances of the same domain in different proteins are truly of different lengths due to post-duplication mutations and truncations. This further complicates the task of predicting domains accurately.

The procedure described in the preceding section might still generate too many putative domains. Some might be consistent with existing domain databases. The others might still be correct, even if overlooked by other resources on protein domains, or erroneous (due to noise in the MSA). On the other hand, some of the true boundary positions between domains might be missed by the specific method used. These problems are typical of all domain prediction algorithms (Fig. 7.7).

If predicted boundary positions (transitions) are too close to each other (say, within less than 20 residues), then some can be eliminated by picking the most significant one in each region that spans a certain number of residues and eliminate every other putative transition in that region. A more rigorous approach is to use a global likelihood model that considers also prior information on domains. Once such a model is defined, the best partition is selected by enumerating all possible combinations of putative boundary positions and picking the subset of positions (and hence domain assignments) that results in maximum likelihood or maximum posterior probability; such an approach was used in *(42,43)*.

For example, the likelihood model of *(42)* defines a feature space over MSA columns that is composed of more than 20 different features, and assumes a

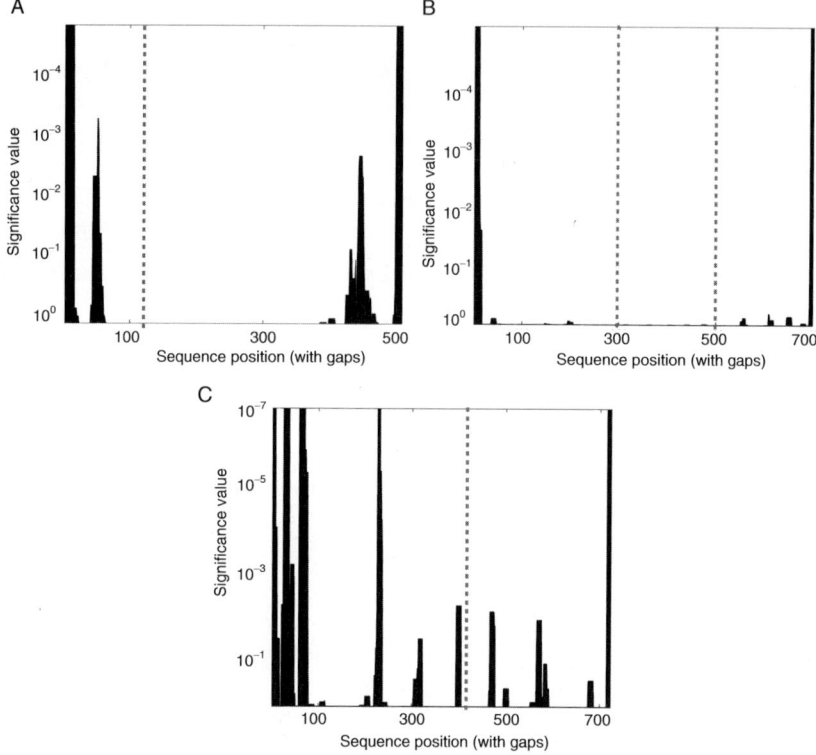

Fig. 7.7. Noisy signals of sequence termination in MSAs. Three examples where the graph is too noisy to predict domain boundaries. **A.** Dihydroorotate dehydrogenase B (PDB 1ep3_B). **B.** Erythroid membrane protein (PDB 1gg3_C). **C.** Tetanus toxin Hc (PDB 1fv3_B). SCOP domains were marked with dashed vertical lines at the corresponding positions (after correcting for gaps).

"domain generator" that cycles between domain and linker states. The domain generator emits a domain or linker while visiting each state according to a certain probability distribution over the MSA feature space. The probability distributions associated with each state are trained from known domains.

Given a sequence of length L, a multiple sequence alignment S (centered around the query sequence) and a possible partition D into n domains of lengths $l_1, l_2, \ldots, l_n$, the model computes the likelihood of the MSA given the partition $P(S/D)$ and prior probability to observe such a partition $P(D)$ (more details about these entities are given next). The best partition is selected by looking for the one that maximizes the posterior probability:

$$P(D/S) = \frac{P(S/D)P(D)}{P(S)} \tag{4}$$

Since $P(S)$ is fixed for all partitions, then $\max_D P(D/S) = \max_D P(S/D)P(D)$ and the partition that maximizes the product is picked.

4.3.1. Computing the Prior P(D)

To calculate the prior $P(D)$, we estimate the probability that an arbitrary protein sequence of length L will consist of n domains of the specific lengths $l_1, l_2, \ldots, l_n$. That is:

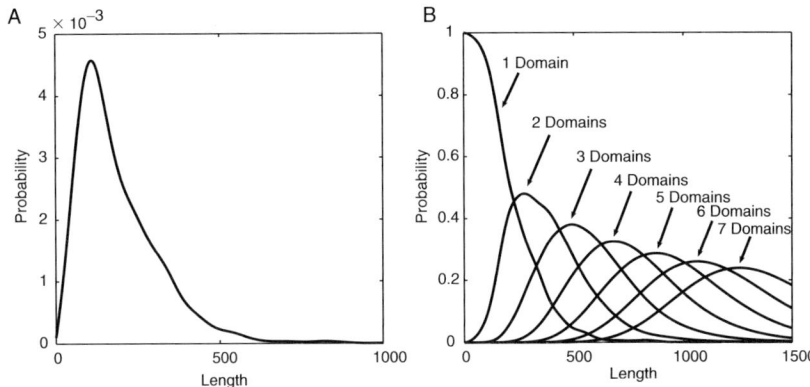

Fig. 7.8 Prior information on domain distributions. **A**. The empirical distribution of domain lengths in the SCOP database (smoothed). **B**. Extrapolated distributions of the number of domains n in proteins of a given length. Values are given for *n* from 1 to 7. For details see *(42)*.

$$P(D) = P((d_1, l_1)(d_2, l_2)\ldots(d_n, l_n) \text{ s.t. } l_1 + l_2 + \ldots + l_n = L) \quad (5)$$

Denote by $P_0(l_i)$ the prior probability to observe a domain of length l_i (estimated, for example, from databases of known domains, see Fig. 7.8A), and denote by $Prob(n/L)$ the prior probability to observe n domains in a protein of length L (see Fig. 7.8B). Then $P(D)$ can be approximated by:

$$P(D) \simeq Prob(n/L) \cdot \sum_{\pi(l_1, l_2, \ldots, l_n)} P_0(l_1/L) P_0(l_2/L - l_1) \ldots P_0(l_{n-1}/L - \sum_{i=1}^{n-2} l_i) \quad (6)$$

where the prior probabilities $P_0(l_i/L)$ are approximated by $P_0(l_i)$, normalized to the relevant range $[0\ldots L]$ *(42)*.

4.3.2. Computing the Likelihood P(S/D)

Computing the likelihood of the data given a certain partition $D = (d_1, l_1), (d_2, l_2), \ldots, (d_n, l_n)$ is more complicated, and depending on the representation/model used, is not always feasible. The model of *(42)* assesses the probability to observe each column in the MSA either in a domain state or in a linker state using a feature space with r different features. Each MSA column j is associated with certain feature values $f_{j1}, f_{j2}, \ldots, f_{jr}$ that are computed from the MSA, and by characterizing/training the distributions of these features using databases of known domains (e.g., SCOP) it is possible to estimate the probability $P(j/domain - state) = P(f_{j1}, f_{j2}, \ldots, f_{jr}/domain - state)$. $P(j/linker - state)$ can be estimated in a similar manner. The likelihood of a certain partition is given by the product over the MSA positions, using either $P(j/domain - state)$ or $P(j/linker - state)$ depending on the partition tested.

4.3.3. Generalization of the Likelihood Model to Multiple Proteins

Generalizing the likelihood model to multiple proteins is not trivial. For example, consider the problem of computing the prior probability. Start with a data set S with m proteins, and their partitions $D = D_1, D_2, \ldots, D_m$ as induced by one of the hypotheses generated by the pre-processing step. When computing the prior $P(D)$ for multiple proteins, we multiply the priors of each partition.

However, this assumes a certain random process in which every experiment is completely independent from the rest. This is clearly not the case, as the protein sequences that make up the multiple alignment are evolutionarily related. Moreover, the individual partitions are all induced by the same global partition for the multiple alignment as a whole. Therefore, the probabilistic setup is different and is more similar to the following.

Assume you are rolling a balanced die, with a probability of 1/6 for each facet. To compute the probability of events, such as "the number is even," we consider the complete sample space and add the mass probabilities associated with each realization of the die that is consistent with that event, i.e.:

$$P(\textit{the number is even}) = P(\textit{number} = 2) + P(\textit{number} = 4) + P(\textit{number} = 6) \quad (7)$$

The case of a domain prediction in multiple alignments is similar. One can think about each sequence as a different realization of the source, with uniform probability. The partition D (that induces all partitions $D_1, \ldots, D_m$) can be considered as the equivalent of the event "the number is even" and its probability.

$$P(D) = P(D_1) + P(D_2) + \ldots + P(D_m) \quad (8)$$

However, unlike the case of the die, the realizations are not independent. Rather, sequences are evolutionarily related, and although highly similar sequences are eliminated from the multiple alignment (see Section 4.3.4.), the representative sequences are still related. To account for this scenario, we need to estimate the effective probability mass associated with each sequence. This depends on the number of sequences that are closely similar to a given representative sequence, and the diversity among different representative sequences. Estimating these probabilities is difficult *(84,85)*; instead we use sequence weights as described in Section 4.3.4. as a rough measure for the probability mass of each representative sequence:

$$P(D) = w_1 \cdot P(D_1) + w_2 \cdot P(D_2) + \ldots + w_m \cdot P(D_m) \quad (9)$$

where $\sum_i w_i = 1$

4.3.4. Selecting Representative Sequences

MSAs often contain many sequences that are highly related. These over-represented sequences might bias the results. Moreover, even after filtering sequences that are annotated as fragments, the MSA might still contain un-annotated fragments. To try and mend for the overrepresentation of sequences and un-annotated fragments, one way would be to go through all possible pairs of sequences in the alignment and compute their similarity score. If the sequences match extremely well, it is likely that the smaller one is a fragment of the longer one. If the sequences have high similarity score, they might represent the same protein or a close homologue, and in that case we want to down weight the similar sequences to avoid overrepresentation. The problem with this method is that it cannot be applied to large MSAs, since it would entail too many pairwise comparisons. A more practical solution is to use the fact that the sequences are already ordered based on their similarity with the seed sequence of the MSA, and these similarity scores are not independent. That is to say, if sequences A,B and B,C have high similarity, then A,C are likely to have relatively high similarity score as well.

This procedure starts by marking the seed sequence as the first element in a so-called representative set. Each sequence in the representative set is considered a different realization of the source (protein family) and is associated with a group of highly similar sequences. The alignment is then processed from the top working our way down in order of decreasing similarity (increasing e-value). For each sequence in the MSA, compute its similarity with each key sequence in the representative set, from top to bottom. If the sequence similarity is higher than T_1 (where T_1 is a very high threshold, set to 0.95), and the current sequence is totally included in the alignment, and it is smaller than the key sequence being compared to, then this sequence is a fragment. If the sequence is not classified as a fragment, but the similarity score is higher than a second threshold T_2 (0.85 used in this procedure), this sequence is associated with that key sequence in the representative set. If the sequence does not have a similarity score of T_2 or greater to any of the key sequences in the representative set, then it is added at the bottom of the representative set as a new key sequence. Once all MSA sequences were exhausted, each key sequence is associated with a weight w that is computed using the method of (86), summed over all the sequences in its group. Finally, the weights of all key sequences are normalized such that they sum to 1. This simple and quick procedure produces grouping that can then be used to define more accurate priors (see Section 4.3.3.). Clearly, more elaborate clustering procedures can be applied to group the sequences.

4.4. Evaluation of Domain Prediction

Evaluating the quality of domain predictions is a difficult problem in itself. Since there is no single widely accepted definition of domains, there is no single yardstick or gold standard. Some studies focus on hand-picked proteins, but more typically, people use domain databases that are based on structural information, such as SCOP and CATH. However, even these two databases do not agree on more than 30% of their domains (42).

Once the reference set has been chosen, the next question is which proteins to use in the evaluation. Some use only single domain proteins, to simplify the evaluation (59). However, such results can be very misleading. With the multidomain proteins the evaluation is more difficult. Some test predictions by checking if the number of domains is accurate, however, this completely ignores that actual positions of domain boundaries. A more accurate approach is to test if the predicted domains are within a certain window from known domain boundaries, and vice versa (42).

5. Conclusions

The domain prediction problem is clearly a difficult problem. Typically, all one has is the protein sequence and the goal is to predict the structural organization of this protein. It is well known that structure prediction is one of the most challenging problems in computational and structural biology, and the domain prediction problem is closely related to it. Therefore, it is not surprising that success in this field is strongly driven from advances in structure prediction algorithms.

There are multiple factors that hinder algorithms for domain prediction. Probably the most important one is the lack of a consistent definition of domains.

A unified "correct" definition that everyone would agree on might not be on the immediate horizon, but more stringent and specific definitions of the ones that are already in use (independently foldable, structural subclusters, functional units, and evolutionary units), and analysis of how they differ would be a step in that direction. Another aspect is the lack of proper reference sets, that is, domain collections that can be considered "correct." So far most studies on domain prediction validated their results using one of the manually verified domain resources like SCOP, CATH, or Pfam-A as a gold standard. These are excellent resources that are based on expert analysis, but even these do not agree on domain definitions. For example, SCOP and CATH disagree in about 30% of the cases. The disagreement with Pfam-A is even higher. In some cases it is possible to resolve some of the ambiguity by using a "gradient of domains." That is, the reference set could have many levels of definitions. The highest level would have the fewest transition points that correspond to the most reliable and consistent domains that are documented in the literature. The middle level would have a higher number of transitions, representing also the "most likely" ones based on multiple algorithms, and the lowest level contains all transitions predicted with any algorithm.

Nevertheless, domain prediction algorithms have helped to advance the knowledge on the domain structure of proteins tremendously and there is a vast number of domain resources available. Although the authors tried to be comprehensive, the lists of algorithms and databases are by no means exhaustive and the authors apologize if they overlooked some resources about which they were not aware. In view of many available resources, efforts like InterPro, Meta-DP, SBASE, and Biozon, that integrate many resources become more and more important and further success in domain prediction is likely to rely on the consolidation of information from multiple sources and data types.

6. Notes

1. COG (*66*) is a database of groups of orthologous proteins from the complete genomes of several bacteria.
2. To assign domains through fold matching CATH mainly uses CATHEDRAL (*78*), an in-house algorithm that exploits a fast structure comparison algorithm based on graph theory (GRATH) (*79*). It also uses profile HMMs, SSAP scores (*80*) (a dynamic-programming based structure comparison algorithm) and relevant literature. In addition it uses structure-based domain detection algorithms such as PUU, DOMAK and DETECTIVE (see Section 2.2.).
3. A uniform distribution greatly simplifies the statistical analysis. A more accurate analysis would require more complex distributions.
4. This is clearly not the case, as the sequences are homologous and hence related to each other. However, one can take certain measures to reduce the dependency between sequences as described in Section 4.3.4.

References

1. Phillips, D. C. (1966) The three-dimensional structure of an enzyme molecule. *Sci. Am.* **215**, 78–90.
2. Cunningham, B. A., Gottlieb, P. D., Pflumm, M. N., and Edelman, G. M. (1971) Immunoglobulin structure: diversity, gene duplication, and domains, in (Amos, B., ed.), *Progress in Immunology.* Academic Press, New York, pp. 3–24.

3. Wetlaufer, D. B. (1973) Nucleation, rapid folding, and globular intrachain regions in proteins. *Proc. Natl. Acad. Sci. USA* **70**, 697–701.

4. Schulz, G. E. (1981) Protein differentiation: emergence of novel proteins during evolution. *Angew. Chem. Int. Edit.* **20**, 143–151.

5. Richardson, J. S. (1981) The anatomy and taxonomy of protein structure. *Adv. Protein Chem.* **34**, 167–339.

6. Branden, C., and Tooze, J. (1999) *Introduction to Protein Structure*. Garland Publishing, Inc., New York.

7. Liu, J., and Rost, B. (2004) CHOP proteins into structural domain-like fragments. *Proteins.* **55**, 678–688.

8. Boeckmann B., Bairoch A., Apweiler R., Blatter M. C., Estreicher A., Gasteiger E., Martin M. J., Michoud K., O'Donovan C., Phan I., Pilbout S., and Schneider M. (2003) The SWISSPROT protein knowledge base and its supplement TrEMBL in 2003. *Nucl. Acids Res.* **31**, 365–370.

9. Bornberg-Bauer, E., Beausart, F., Kummerfeld, S. K., Teichmann, S. A., and Weiner, J. (2005) The evolution of domain arrangements in proteins and interaction networks. *Cell. Mol. Life Sci.* **62**, 435–445.

10. Hubbard, S. J. (1998) The structural aspects of limited proteolysis of native proteins. *Biochim. Biophys. Acta.* **1382**, 191–206.

11. Dalzoppo, D., Vita, C., and Fontana, A. (1985) Folding of thermolysin fragments. Identification of the minimum size of a carboxyl-terminal fragment that can fold into a stable native-like structure. *J. Mol. Biol.* **182**, 331–340.

12. Parrado, J., Conejero-Lara, F., Smith, R. A., Marshall, J. M., Ponting, C. P., and Dobson, C. M. (1996) The domain organization of streptokinase: nuclear magnetic resonance, circular dichroism, and functional characterization of proteolytic fragments. *Protein Sci.* **5**, 693–704.

13. Christ, D. and Winter, G. (2006) Identification of protein domains by shotgun proteolysis. *J. Mol. Biol.* **358**, 364–371.

14. Crippen, G. M. (1978) The tree structural organization of proteins. *J. Mol. Biol.* **126**, 315–332.

15. Lesk, A. M. and Rose, G. D. (1981) Folding units in globular proteins. *Proc. Natl. Acad. Sci. USA.* **78**, 4304–4308.

16. Taylor, W. R. (1999) Protein structural domain identification. *Prot. Eng.* **12**, 203–216.

17. Swindells, M. B. (1995) A procedure for detecting structural domains in proteins. *Protein Sci.* **4**, 103–112.

18. Holm, L., and Sander, C. (1994) Parser for protein folding units. *Proteins* **19**, 256–268.

19. Siddiqui, A. S., and Barton, G. J. (1995) Continuous and discontinuous domains: an algorithm for the automatic generation of reliable protein domain definitions. *Protein Sci.* **4**, 872–884.

20. Xu, Y., Xu, D., and Gabow, H. N. (2000) Protein domain decomposition using a graph-theoretic approach. *Bioinformatics.* **16**, 1091–1104.

21. Ford, L. R., Jr., and Fulkerson, D. R. (1962) *Flows in Networks*. Princeton University Press, Princeton, NJ.

22. Alexandrov, N., and Shindyalov, I. (2003) PDP: protein domain parser. *Bioinformatics.* **19**, 429–430.

23. Pugalenthi, G., Archunan, G., and Sowdhamini, R. (2005) DIAL: a web-based server for the automatic identification of structural domains in proteins. *Nucleic Acids Res.* **33**, W130–132.

24. Vinayagam A., Shi J., Pugalenthi G., Meenakshi B., Blundell T. L., and Sowdhamini R. (2003) DDBASE2.0: updated domain database with improved identification of structural domains. *Bioinformatics.* **19**, 1760–1764.

25. Gelly, J. C., de Brevern, A. G., and Hazout, S. (2006) 'Protein Peeling': an approach for splitting a 3D protein structure into compact fragments. *Bioinformatics.* **22**, 129–133.

26. Murzin, A. G., Brenner, S. E., Hubbard, T., and Chothia, C. (1995) SCOP: a structural classification of proteins database for the investigation of sequences and structures. *J. Mol. Biol.* **247**, 536–540.

27. Orengo, C. A., Michie, A. D., Jones, S., Jones, D. T., Swindells, M. B., and Thornton, J. M. (1997) CATH-a hierarchic classification of protein domain structures. *Structure.* **5**, 1093–1108.

28. Rigden, D. J. (2002) Use of covariance analysis for the prediction of structural domain boundaries from multiple protein sequence alignments. *Protein Eng.* **15**, 65–77.

29. George, R. A., and Heringa, J. (2002) SnapDRAGON: a method to delineate protein structural domains from sequence data. *J. Mol. Biol.* **316**, 839–851.

30. Sonnhammer, E. L., and Kahn, D. (1994) Modular arrangement of proteins as inferred from analysis of homology. *Prot. Sci.* **3**, 482–492.

31. Altschul, S. F., Gish, W., Miller, W., Myers, E. W., and Lipman, D. J. (1990) Basic local alignment search tool. *J. Mol. Biol.* **215**, 403–410.

32. Park, J., and Teichmann, S. A. (1998) DIVCLUS: an automatic method in the GEANFAMMER package that finds homologous domains in single-and multidomain proteins. *Bioinformatics.* **14**, 144–150.

33. Smith, T. F., and Waterman, M. S. (1981) Identification of common molecular subsequences. *J. Mol. Biol.* **147**, 195–197.

34. Pearson, W. R., and Lipman, D. J. (1988) Improved tools for biological sequence comparison. *Proc. Natl. Acad. Sci. USA* **85**, 2444–2448.

35. Enright, A. J., and Ouzounis, C. A. (2000) GeneRAGE: a robust algorithm for sequence clustering and domain detection. *Bioinformatics.* **16**, 451–457.

36. Gracy, J., and Argos, P. (1998) Automated protein sequence database classification. I. Integration of compositional similarity search, local similarity search and multiple sequence alignment. *Bioinformatics.* **14**, 164–173.

37. Gracy, J., and Argos, P. (1998) Automated protein sequence database classification. II. Delineation of domain boundaries from sequence similarity. *Bioinformatics.* **14**, 174–187.

38. Gouzy, J., Corpet, F., and Kahn, D. (1999) Whole genome protein domain analysis using a new method for domain clustering. *Comput. Chem.* **23**, 333–340.

39. Servant, F., Bru, C., Carrere, S., Courcelle, E., Gouzy, J., Peyruc, D., and Kahn, D. (2002) ProDom: automated clustering of homologous domains. *Brief. Bioinform.* **3**, 246–251.

40. Altschul, S. F., Madden, T. L., Schaffer, A. A., Zhang, J., Zhang, Z., Miller, W., and Lipman, D. J. (1997) Gapped BLAST and PSI-BLAST: a new generation of protein database search programs. *Nucleic Acids Res.* **25**, 3389–3402.

41. Wootton, J. C., and Federhen, S. (1996) Analysis of compositionally biased regions in sequence databases. *Methods Enzymol.* **266**, 554–571.

42. Nagarajan, N., and Yona, G. (2004) Automatic prediction of protein domains from sequence information using a hybrid learning system. *Bioinformatics.* **20**, 1335–1360.

43. Heger, A., and Holm, L. (2003) Exhaustive enumeration of protein domain families. *J. Mol. Biol.* **328**, 749–767.

44. Bateman, A., Coin, L., Durbin, R., Finn, R. D., Hollich, V., Griffiths-Jones, S., Khanna, A., Marshall, M., Moxon, S., Sonnhammer, E. L., Studholme, D. J., Yeats, C., and Eddy, S. R. (2004) The Pfam protein families database. *Nucl. Acids Res.* **32**, D138–141.

45. Portugaly, E., Harel, A., Linial, N., and Linial, M. (2006) EVEREST: automatic identification and classification of protein domains in all protein sequences. *BMC Bioinformatics.* **7**, 277.

46. Eddy, S. R. (1998) Profile hidden Markov models. *Bioinformatics.* **14**, 755–763.

47. Schultz, J., Milpetz, F., Bork, P., and Ponting, C. P. (1998) SMART, a simple modular architecture research tool: identification of signaling domains. *Proc. Natl. Acad. Sci. USA* **95**, 5857–5864.

48. Haft, D. H., Loftus, B. J., Richardson, D. L., Yang, F., Eisen, J. A., Paulsen, I. T., and White, O. (2001) TIGRFAMs: a protein family resource for the functional identification of proteins. *Nucl. Acids Res.* **29**, 41–43.

49. George, R. A., and Heringa, J. (2002) Protein domain identification and improved sequence similarity searching using PSI-BLAST. *Proteins* **48**, 672–668.

50. Kuroda, Y., Tani, K., Matsuo, Y., and Yokoyama, S. (2000) Automated search of natively folded protein fragments for high-throughput structure determination in structural genomics. *Protein Sci.* **9**, 2313–2321.

51. Wheelan, S. J., Marchler-Bauer, A., and Bryant, S. H. (2000) Domain size distributions can predict domain boundaries. *Bioinformatics.* **16**, 613–618.

52. Miyazaki, S., Kuroda, Y., and Yokoyama, S. (2002) Characterization and prediction of linker sequences of multidomain proteins by a neural network. *J. Struct. Funct. Genomics.* **15**, 37–51.

53. Miyazaki, S., Kuroda, Y., and Yokoyama, S. (2006) Identification of putative domain linkers by a neural network -application to a large sequence database. *BMC Bioinformatics* **7**, 323.

54. Suyama, M., and Ohara, O. (2003) DomCut: prediction of inter-domain linker regions in amino acid sequences. *Bioinformatics* **19**, 673–674.

55. Linding, R., Russell, R. B., Neduva, V., and Gibson, T. J. (2003) GlobPlot: exploring protein sequences for globularity and disorder. *Nucleic Acids Res.* **31**, 3701–3708.

56. Marsden, R. L., McGuffin, L. J., and Jones, D. T. (2002) Rapid protein domain assignment from amino acid sequence using predicted secondary structure. *Protein Sci.* **11**, 2814–2824.

57. Jones D. T. (1999) Protein secondary structure prediction based on position-specific scoring matrices. *J. Mol. Biol.* **292**, 195–202.

58. Chen, L., Wang, W., Ling, S., Jia, C., and Wang, F. (2006) KemaDom: a web server for domain prediction using kernel machine with local context. *Nucleic Acids Res.* **34**, W158–163.

59. Saini, H. K., and Fischer, D. (2005) Meta-DP: domain prediction meta-server. *Bioinformatics.* **21**, 2917–2920.

60. Hulo, N., Bairoch, A., Bulliard, V., Cerutti, L., De Castro, E., Langendijk-Genevaux, P. S., Pagni, M., and Sigrist, C. J. A. (2006) The PROSITE database. *Nucleic Acids Res.* **34**, D227–D230.

61. Attwood, T. K., Bradley, P., Flower, D. R., Gaulton, A., Maudling, N., Mitchell, A. L., Moulton, G., Nordle, A., Paine, K., Taylor, P., Uddin, A., and Zygouri, C. (2003) PRINTS and its automatic supplement, prePRINTS. *Nucleic Acids Res.* **31**, 400–402.

62. Henikoff, J. G., Greene, E. A., Pietrokovski, S., and Henikoff, S. (2000) Increased coverage of protein families with the blocks database servers. *Nucleic Acids Res.* **28**, 228–230.

63. Mulder, N. J., Apweiler, R., Attwood, T. K., Bairoch, A., Bateman, A., Binns, D., Bradley, P., Bork, P., Bucher, P., Cerutti, L., Copley, R., Courcelle, E., Das, U., Durbin, R., Fleischmann, W., Gough, J., Haft, D., Harte, N., Hulo, N., Kahn, D., Kanapin, A., Krestyaninova, M., Lonsdale, D., Lopez, R., Letunic, I., Madera, M., Maslen, J., McDowall, J., Mitchell, A., Nikolskaya, A. N., Orchard, S., Pagni, M., Ponting, C. P., Quevillon, E., Selengut, J., Sigrist, C. J., Silventoinen, V., Studholme, D. J., Vaughan, R., and Wu, C. H. (2005) InterPro, progress and status in 2005. *Nucl. Acids Res.* **33**, D201–205.

64. Berman, H. M., Westbrook, J., Feng, Z., Gilliland, G., Bhat, T. N., Weissig, H., Shindyalov, I. N., and Bourne, P. E. (2000) The Protein Data Bank. *Nucl. Acids Res.* **28**, 235–242.

65. Marchler-Bauer, A., Anderson, J. B., Cherukuri, P. F., DeWeese-Scott, C., Geer, L. Y., Gwadz, M., He, S., Hurwitz, D. I., Jackson, J. D., Ke, Z., Lanczycki, C. J., Liebert, C. A., Liu, C., Lu, F., Marchler, G. H., Mullokandov, M., Shoemaker, B. A., Simonyan, V., Song, J. S., Thiessen, P. A., Yamashita, R. A., Yin, J. J., Zhang,

D., and Bryant, S. H. (2005) CDD: a Conserved Domain Database for protein classification. *Nucleic Acids Res.* **33**, D192–196.

66. Tatusov, R. L., Fedorova, N. D., Jackson, J. D., Jacobs, A. R., Kiryutin, B., Koonin, E. V., Krylov, D. M., Mazumder, R., Mekhedov, S. L., Nikolskaya, A. N., Rao, B. S., Smirnov, S., Sverdlov, A. V., Vasudevan, S., Wolf, Y. I., Yin, J. J., and Natale, D. A. (2003) The COG database: an updated version includes eukaryotes. *BMC Bioinformatics.* **4**, 41.

67. Vlahovicek, K., Kajan, L., Agoston, V., and Pongor, S. (2005) The SBASE domain sequence resource, release 12: prediction of protein domain-architecture using support vector machines. *Nucleic Acids Res.* **33**, D223–225.

68. George, D. G., Barker, W. C., Mewes, H. W., Pfeiffer, F., and Tsugita, A. (1996) The PIR-International protein sequence database. *Nucleic Acids Res.* **24**, 17–20.

69. Bairoch, A., Apweiler, R., Wu, C. H., Barker, W. C., Boeckmann, B., Ferro, S., Gasteiger, E., Huang, H., Lopez, R., Magrane, M., Martin, M. J., Natale, D. A., O'Donovan, C., Redaschi, N., and Yeh, L. S. (2005) The Universal Protein Resource (UniProt). *Nucleic Acids Res.* **33**, D154–159.

70. Wu, C. H., Nikolskaya, A., Huang, H., Yeh, L. S., Natale, D. A., Vinayaka, C. R., Hu, Z. Z., Mazumder, R., Kumar, S., Kourtesis, P., Ledley, R. S., Suzek, B. E., Arminski, L., Chen, Y., Zhang, J., Cardenas, J. L., Chung, S., Castro-Alvear, J., Dinkov, G., and Barker, W. C. (2004) PIRSF: family classification system at the Protein Information Resource. *Nucleic Acids Res.* **32**, D112–114.

71. Madera, M., Vogel, C., Kummerfeld, S. K., Chothia, C., and Gough, J. (2004) The SUPERFAMILY database in 2004: additions and improvements. *Nucleic Acids Res.* **32**, D235–239.

72. Yeats, C., Maibaum, M., Marsden, R., Dibley, M., Lee, D., Addou, S., and Orengo, C. A. (2006) Gene3D: modeling protein structure, function and evolution. *Nucleic Acids Res.* **34**, D281–284.

73. Mi, H., Lazareva-Ulitsky, B., Loo, R., Kejariwal, A., Vandergriff, J., Rabkin, S., Guo, N., Muruganujan, A., Doremieux, O., Campbell, M. J., Kitano, H., and Thomas, P. D. (2005) The PANTHER database of protein families, subfamilies, functions and pathways. *Nucleic Acids Res.* **33**, D284–288.

74. Quevillon, E., Silventoinen, V., Pillai, S., Harte, N., Mulder, N., Apweiler, R., and Lopez, R. (2005) InterProScan: protein domains identifier. *Nucleic Acids Res.* **33**, W116–120.

75. Birkland, A., and Yona, G. (2006) BIOZON: a system for unification, management and analysis of heterogeneous biological data. *BMC Bioinformatics.* **7**, 70.

76. Holm, L., and Sander, C. (1997) Dali/FSSP classification of 3D protein folds. *Nucl. Acids Res.* **25**, 231–234.

77. Siddiqui, A. S., Dengler, U., and Barton, G. J. (2001) 3Dee: a database of protein structural domains. *Bioinformatics* **17**, 200–201.

78. Pearl, F. M., Bennett, C. F., Bray, J. E., Harrison, A. P., Martin, N., Shepherd, A., Sillitoe, I., Thornton, J., and Orengo, C. A. (2003) The CATH database: an extended protein family resource for structural and functional genomics. *Nucl. Acids Res.* **31**, 452–455.

79. Harrison, A., Pearl, F., Sillitoe, I., Slidel, T., Mott, R., Thornton, J., and Orengo, C. (2003) Recognizing the fold of a protein structure. *Bioinformatics* **19**, 1748–1759.

80. Taylor, W. R., and Orengo, C. A. (1989) Protein structure alignment. *J. Mol. Biol.* **208**, 1–22.

81. Dietmann, S., Park, J., Notredame, C., Heger, A., Lappe, M., and Holm, L. (2001) A fully automatic evolutionary classification of protein folds: Dali Domain Dictionary version 3. *Nucl. Acids Res.* **29**, 55–57.

82. Notredame, C., Higgins, D. G., and Heringa, J. (2000) T-Coffee: a novel method for fast and accurate multiple sequence alignment. *J. Mol. Biol.* **302**, 205–217.

83. Lee, C., Grasso, C., and Sharlow, M. F. (2002) Multiple sequence alignment using partial order graphs. *Bioinformatics* **18**, 452–464.

84. Koehl, P., and Levitt, M. (2002) Protein topology and stability define the space of allowed sequences. *Proc. Natl. Acad. Sci. USA* **99**, 1280–1285.

85. Meyerguz, L., Kempe, D., Kleinberg, J., and Elber, R. (2004) The evolutionary capacity of protein structures. *In the Proceedings of RECOMB 2004.*

86. Henikoff, S., and Henikoff, J. G. (1994) Position–based sequence weights. *J. Mol. Biol.* **243**, 574–578.

Chapter 8

Protein Structure Modeling with MODELLER

Narayanan Eswar, David Eramian, Ben Webb, Min-Yi Shen, and Andrej Sali

Genome sequencing projects have resulted in a rapid increase in the number of known protein sequences. In contrast, only about one-hundredth of these sequences have been characterized using experimental structure determination methods. Computational protein structure modeling techniques have the potential to bridge this sequence-structure gap. This chapter presents an example that illustrates the use of MODELLER to construct a comparative model for a protein with unknown structure. Automation of similar protcols has resulted in models of useful accuracy for domains in more than half of all known protein sequences.

1. Introduction

The function of a protein is determined by its sequence and its three-dimensional (3D) structure. Large-scale genome sequencing projects are providing researchers with millions of protein sequences, from various organisms, at an unprecedented pace. However, the rate of experimental structural characterization of these sequences is limited by the cost, time, and experimental challenges inherent in the structural determination by x-ray crystallography and nuclear magnetic resonance (NMR) spectroscopy.

In the absence of experimentally determined structures, computationally derived protein structure models are often valuable for generating testable hypotheses *(1)*. Comparative protein structure modeling has been used to produce reliable structure models for at least one domain in more than half of all known sequences *(2)*. Hence, computational approaches can provide structural information for two orders of magnitude more sequences than experimental methods, and are expected to be increasingly relied upon as the gap between the number of known sequences and the number of experimentally determined structures continues to widen.

Comparative modeling consists of four main steps *(3)* (Fig. 8.1): (1) fold assignment that identifies overall similarity between the target and at least one known template structure (see Section on Materials for definitions of these terms); (2) alignment of the target sequence and the template(s); (3) building a

From: *Methods in Molecular Biology, Vol. 426: Structural Proteomics: High-throughput Methods*
Edited by: B. Kobe, M. Guss and T. Huber © Humana Press, Totowa, NJ

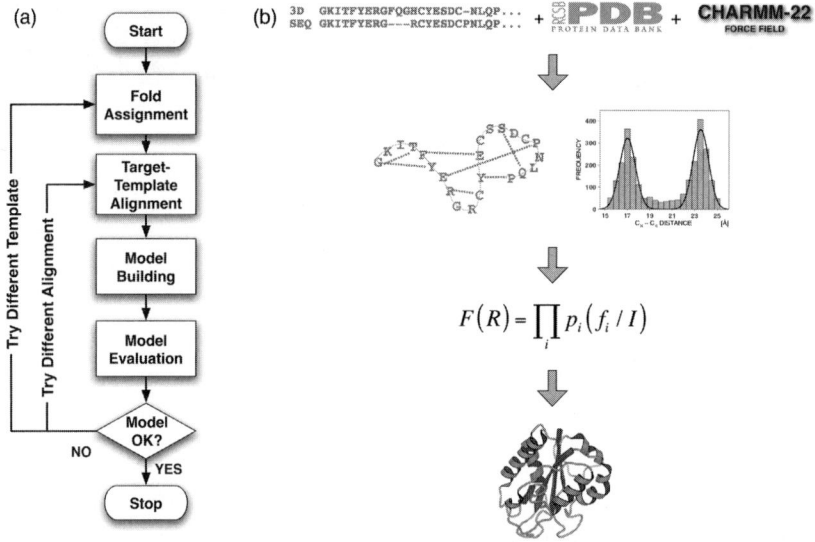

Fig. 8.1 Comparative protein structure modeling. (**a**) A flowchart illustrating the steps in the construction of a comparative model *(3)*. (**b**) Description of comparative modeling by extraction of spatial restraints as implemented in MODELLER *(5)*. By default, spatial restraints in MODELLER involve: (1) homology-derived restraints from the aligned template structures; (2) statistical restraints derived from all known protein structures; and (3) stereochemical restraints from the CHARMM-22 molecular mechanics force field. These restraints are combined into an objective function that is then optimized to calculate the final three-dimensional structure of the target sequence.

model based on the alignment with the chosen template(s); and (4) predicting the accuracy of the model.

MODELLER is a computer program for comparative protein structure modeling *(4,5)*. In the simplest case, the input is an alignment of a sequence to be modeled with the template structure(s), the atomic coordinates of the template(s), and a simple script file. MODELLER then automatically calculates a model containing all non-hydrogen atoms, without any user intervention and within minutes on a desktop computer. Apart from model building, MODELLER can perform auxiliary tasks such as fold-assignment *(6)*, alignment of two protein sequences or their profiles *(7,8)*, multiple alignment of protein sequences and/or structures *(9)*, clustering of sequences and/or structures, and *ab initio* modeling of loops in protein structures *(4)*.

MODELLER implements comparative protein structure modeling by satisfaction of spatial restraints that include: (1) homology-derived restraints on the distances and dihedral angles in the target sequence, extracted from its alignment with the template structures *(5)*; (2) stereochemical restraints such as bond length and bond angle preferences, obtained from the CHARMM-22 molecular mechanics force-field *(10)*; (3) statistical preferences for dihedral angles and non-bonded interatomic distances, obtained from a representative set of known protein structures *(11,12)*; and (4) optional manually curated restraints, such as those from NMR spectroscopy, rules of secondary structure packing, cross-linking experiments, fluorescence spectroscopy, image reconstruction from electron microscopy, site-directed mutagenesis, and intuition (see Fig. 8.1). The spatial restraints, expressed as probability density functions, are combined into an objective function that is optimized by a combination of conjugate gradients and molecular dynamics with simulated annealing. This model building procedure is similar to structure determination by NMR spectroscopy.

This chapter uses a sequence with unknown structure to illustrate the use of various modules in MODELLER to perform the four steps of comparative modeling. This is followed by a Notes section that highlights several underlying practical issues.

2. Materials

2.1. Hardware

1. A computer running Linux/Unix, Apple Mac OS X, or Microsoft Windows 98/NT/2000/XP; 512 MB RAM or higher; about 100 MB of free hard-disk space for the software, example and output files; and a connection to the internet to download the MODELLER program and example files described in this chapter (see Note 1).

2.2. Software

1. The MODELLER 8v2 program, downloaded and installed from http://salilab. org/modeller/download_installation.html. Instructions for the installation are provided as part of the downloaded package; they are also available over the internet at http://salilab.org/modeller/release.html#install.
2. The files required to follow the example described in this chapter, downloaded and installed from http://salilab.org/modeller/tutorial/MMB06-example.tar.gz (Unix/Linux/MacOSX) or http://salilab.org/modeller/tutorial/MMB06-example.zip (Windows).

2.3. Computer Skills

1. MODELLER uses Python as its control language. All input scripts to MODELLER are, hence, Python scripts. Although knowledge of Python is not necessary to run MODELLER, it can be useful to perform more advanced tasks.
2. MODELLER does not have a graphical user interface (GUI) and is run from the command-line by executing the input scripts; a basic knowledge of command-line skills on a computer is necessary to follow the protocol described in this chapter.

2.4. Conventions Followed in the Text

1. A sequence with unknown structure, for which a model is being calculated, is referred to as the "target."
2. A "template" is an experimentally determined structure, and/or its sequence, used to derive spatial restraints for comparative modeling.
3. Names of files, objects, modules, and commands to be executed are all shown in `monospaced` font.

4. Files with `.ali` extensions contain the alignment of two or more sequences and/or structures. Files with `.pir` extensions correspond to a collection of one or more unaligned sequences in the PIR format. Files with `.pap` extensions contain an alignment in a user-friendly format with an additional line indicating identical aligned residues with a *. All input scripts to MODELLER are Python scripts with the `.py` extension. Execution of these input scripts always produces a log file identified by the `.log` extension.

5. A typical operation in MODELLER would consist of: (1) preparing an input Python script; (2) ensuring that all required files (sequences, structures, alignments, etc.) exist; (3) executing the input script by typing `mod8v2 <input-script>`; and (4) analyzing the output and log files.

3. Methods

The procedure for calculating a three-dimensional model for a sequence with unknown structure is illustrated using the following example: A novel gene for lactate dehydrogenase (LDH) was identified from the genomic sequence of *Trichomonas vaginalis* (TvLDH). The corresponding protein had higher sequence similarity to the malate dehydrogenase of the same species (TvMDH) than to any other LDH *(13)*. Comparative models were constructed for TvLDH and TvMDH to study the sequences in a structural context and to suggest site-directed mutagenesis experiments to elucidate changes in enzymatic specificity in this apparent case of convergent evolution. The native and mutated enzymes were subsequently expressed and their activities compared *(13)*.

3.1. Fold Assignment

1. The first step in comparative modeling is to identify one or more template structure(s) that have detectable similarity to the target. This identification is achieved by scanning the sequence of TvLDH against a library of sequences extracted from known protein structures in the Protein Data Bank (PDB) *(14)*. This step is performed using the `profile.build()` module of MODELLER (file `build_profile.py`) (*see* Note 2). The `profile.build()` command uses the local dynamic programming algorithm to identify related sequences *(6,15)*. In the simplest case, `profile.build()` takes as input the target sequence (file `TvLDH.pir`) and a database of sequences of known structure (file `pdb_95.pir`) and returns a set of statistically significant alignments (file `build_profile.prf`). Execute the command by typing `mod8v2 build_profile.py`.

2. The results of the scan are stored in the output file called build_profile.prf. The first six lines of this file contain the input parameters used to create the alignments. Subsequent lines contain several columns of data. For the purposes of this example, the most important columns are: (1) the second column, containing the PDB code of the related template sequences; (2) the 11th column, containing the percentage sequence identity between the TvLDH and template sequences; and (3) the 12th column, containing the E-values for the statistical significance of the alignments.

3. The extent of similarity between the target-template pairs is usually quantified using sequence identity or a statistical measure such as E-value (*see* Notes 3 and 4). Inspection of column 11 shows that the template with the highest

sequence identity with the target is the 1y7tA structure (45% sequence identity). Further inspection of column 12 shows that there are six PDB sequences, all corresponding to malate dehydrogenases (1y7tA, 5mdhA, 1b8pA, 1civA, 7mdhA, and 1smkA) that show significant similarities to TvLDH with E-values of zero. Two variations of the model building procedure will be described below, one using a single template with the highest sequence identity (1y7tA; 3.2.1), and another using all six templates (3.2.2), to highlight their differences (see Note 5).

3.2. Sequence-Structure Alignment

Sequence-structure alignments are calculated using the `align2d()` module of MODELLER (see Note 6). Although `align2d()` is based on a global dynamic programming algorithm *(16)*, it is different from standard sequence-sequence alignment methods because it takes into account structural information from the template when constructing an alignment. This task is achieved through a variable gap penalty function that tends to place gaps in solvent exposed and curved regions, outside secondary structure segments, and between two positions that are close in space *(9)*. In the current example, the target-template similarity is so high that almost any method with reasonable parameters results in the correct alignment (see Note 7).

3.2.1. Single-Template

1. The input script `align2d-single.py` reads in the structure of the chosen template (1y7tA) and the target sequence (TvLDH) and calls the `align2d()` module to perform the alignment. The resulting alignment is written out to the specified alignment files (`TvLDH-1y7tA.ali` in the PIR format and `TvLDH-1y7tA.pap` in the PAP format).

3.2.2. Multiple-Template

1. The first step in using multiple templates for modeling is to obtain a multiple structure alignment of all the chosen templates. The structure alignment module of MODELLER, `salign()`, can be used for this purpose *(17)*. The input script `salign.py` contains the necessary Python instructions to achieve a multiple structure alignment. The script reads in all the six template structures into an alignment object and then calls `salign()` to generate the multiple structure alignment. The output alignment is written out to `TvLDH-salign.ali` and `TvLDH-salign.pap`, in the PIR and PAP formats, respectively.
2. The next step is to align the TvLDH sequence with the multiple structure alignment generated in the preceding step. This task is accomplished using the script file `align2d-multiple.py`, that again calls the `align2d()` module to calculate the sequence-structure alignment. Upon execution, the resulting alignments are written to `TvLDH-multiple.ali` and `TvLDH-multiple.pap` in the PIR and PAP formats, respectively.

3.3. Model Building

Two variations of the model building protocol are described, corresponding to the two alignments generated in the preceding: (1) modeling using a single template; and (2) modeling using multiple templates, followed by building and optimizing a consensus model. The files required for each of these protocols are present in separate subdirectories called `single/` and `multiple/`, respectively.

3.3.1. Single Template

1. The input script `model-single.py` lists the Python commands necessary to build the model of the TvLDH sequence using the information derived from 1y7tA structure. The script calls the `automodel` class specifying the name of the alignment file to use and the identifiers of the target (TvLDH) and template sequences (1y7tA). The `starting_model` and `ending_model` specify that 10 models should be calculated using different randomizations of the initial coordinates. The models are then assessed with the GA341 *(18,19)* and DOPE assessment functions *(12)*.

2. Upon completion, the 10 models for the TvLDH are written out in the PDB format to files called `TvLDH.B9990[0001-0010].pdb` (see Notes 8 and 9).

3.3.2. Multiple Templates with Consensus Modeling

1. The input script, `model-multiple.py`, is quite similar to `model-single.py`. The specification of the template codes to automodel now contains all six chosen PDB codes and additionally, the `cluster()` method is called to exploit the diversity of the 10 generated models via a clustering and optimization procedure to construct a single consensus model (see Note 10).

2 Upon completion, the 10 models for the TvLDH and the consensus model are written out to `TvLDH.B9990[0001-0010].pdb` and `cluster.opt`, respectively.

3.4. Model Evaluation

1. The log files produced by each of the model building procedures (`model-single.log` and `model-multiple.log`) contain a summary of each calculation at the bottom of the file. This summary includes, for each of the 10 models, the MODELLER objective function (see Note 11) *(5)*, the DOPE pseudo-energy value (see Note 12), and the value of the GA341 score (see Notes 13 and 14). These scores can be used to identify which of the 10 models produced is likely to be the most accurate model (see Note 15).

2. A residue-based pseudo-energy profile for the best scoring model, chosen as the one with the lowest DOPE statistical potential score, can be obtained by executing the `evaluate_model.py` script. This script is available in each of the subdirectories mentioned above. Such a profile is useful to detect local regions of high pseudo-energy that usually correspond to errors in the model (see Notes 16 and 17). Figure 8.2 shows the pseudo-energy profiles of the best scoring models from each procedure. It can be seen that some of the errors in the single-template model have been resolved in the model calculated using multiple templates.

4. Notes

1. Exactly the same job run on two different types of computers (e.g., Windows/Intel and a Macintosh) generally returns slightly different results. The reason for this variation is the difference in the rounding of floating point numbers, which may lead to a divergence between optimization trajectories starting at exactly the same initial conditions. Although these

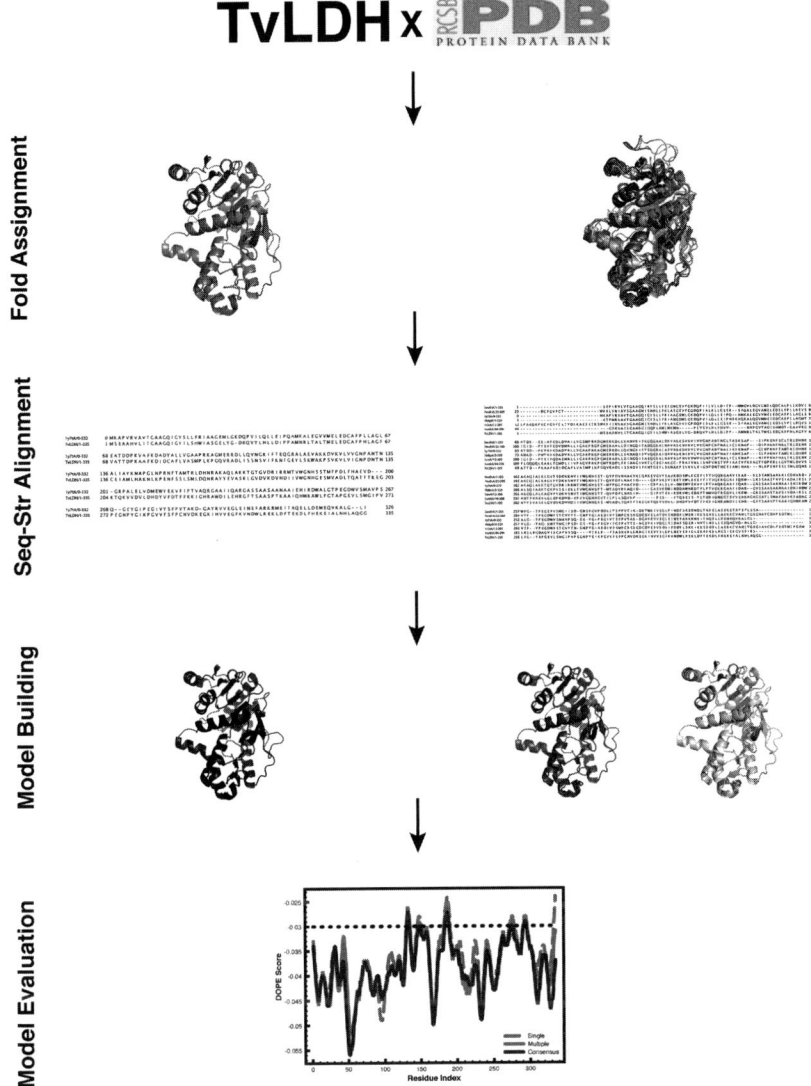

Fig. 8.2 The four steps of comparative modeling as applied to the described example: (1) Scanning the sequence of TvLDH against the sequences of PDB structures identifies 1y7tA as a single template with highest sequence identity; six other malate dehydrogenases are also identified with statistically significant E-values; (2) sequence-sequence structure alignments are generated using a variable gap penalty method; (3) 10 models are constructed per alignment and the best model is chosen using the DOPE statistical potential; in addition, a consensus model is calculated from the 10 models constructed using multiple templates. The model based on single template is shown in black, the one based on multiple templates is in dark gray and the consensus model is shown in light gray; (4) the residue-averaged DOPE scores are used to evaluate local regions of high, unfavorable score that tend to correspond to errors. The consensus model results in the best profile (black line).

differences are generally small, for absolute reproducibility, the same type of computer architecture and operating system need to be used.

2. As mentioned, knowledge of the Python scripting language is not a requirement for basic use of MODELLER. The lines in the script file are usually

self-explanatory and input/output options for each module are described in the manual. For the purpose of illustration, the various lines of the build_ profile.py script are described below (Fig. 8.3):

- log.verbose() sets the amount of information that is written out to the log file.
- environ() initializes the "environment" for the current modeling run, by creating a new environ object, called env. Almost all MODELLER scripts require this step, as the new object is needed to build most other useful objects.
- sequence_db() creates a sequence database object, calling it sdb, which is used to contain large databases of protein sequences.
- sdb.read() reads a file, in text format, containing non-redundant PDB sequences into the sdb database. The input options to this command specify the name of the file (seq_database_file="pdb_95.pir"), the format of the file (seq_database_format="pir"), whether to read all sequences from the file (chains_list="all"), upper and lower bounds for the lengths of the sequences to be read (minmax_db_seq_len=(30,3000)), and whether to clean the sequences of non-standard residue names (clean_ sequences=True).
- sdb.write() writes a binary machine-independent file (seq_data- base_format="binary") with the specified name (seq_database_ file="pdb_95.bin"), containing all sequences read in the previous step.
- The second call to sdb.read() reads the binary format file back in for faster execution.
- alignment() creates a new "alignment" object (aln).

```
log.verbose()
env = environ()

sdb = sequence_db(env)
sdb.read(seq_database_file='pdb_95.pir', seq_database_format='PIR',
        chains_list='ALL', minmax_db_seq_len=(30, 4000), clean_sequences=True)

sdb.write(seq_database_file='pdb_95.bin', seq_database_format='BINARY',
        chains_list='ALL')

sdb.read(seq_database_file='pdb_95.bin', seq_database_format='BINARY',
        chains_list='ALL')

aln = alignment(env)
aln.append(file='TvLDH.ali', alignment_format='PIR', align_codes='ALL')

prf = aln.to_profile()

prf.build(sdb, matrix_offset=-450, rr_file='${LIB}/blosum62.sim.mat',
        gap_penalties_1d=(-500, -50), n_prof_iterations=1,
        check_profile=False, max_aln_evalue=0.01)

prf.write(file='build_profile.prf')

aln = prf.to_alignment()

aln.write(file='build_profile.ali', alignment_format='PIR')
```

Fig. 8.3 The Python input script used for fold assignment. See Note 2 for details.

- `aln.append()` reads the target sequence TvLDH from the file `TvLDH.ali` and `aln.to_profile()` converts it to a profile object (`prf`). Profiles contain similar information as alignments, but are more compact and better suited for sequence database searching.
- `prf.build()` searches the sequence database (`sdb`) using the target profile stored in the `prf` object as the query. Several options, such as the parameters for the alignment algorithm (`matrix_offset, rr_file, gap_penalties,` etc.), are specified to override the default settings. `max_aln_evalue` specifies the threshold value to use when reporting statistically significant alignments.
- `prf.write()` writes a new profile containing the target sequence and its homologs into the specified output file (`file=build_profile.prf`).
- The profile is converted back to the standard alignment format and written out using `aln.write()`.

3. Sequence-structure relationships can be divided into three different regimes of the sequence similarity spectrum: (1) the easily detected relationships characterized by >30% sequence identity; (2) the "twilight zone" *(20)* corresponding to relationships with statistically significant sequence similarity, with identities generally in the 10–30% range; and (3) the "midnight zone" *(20)* corresponding to statistically insignificant sequence similarity. Hence, the sequence identity is a good predictor of the accuracy of the final model when its value is greater than 30%. It has been shown that models based on such alignments usually have, on average, more than approximately 60% of the backbone atoms correctly modeled with a root-mean-squared-deviation (RMSD) of less than 3.5 Å (Fig. 8.4).

   However, the sequence identity is not a statistically reliable measure of alignment significance and corresponding model accuracy for values lower than 30% *(20,21)*. During a scan of a large database, for instance, it is possible that low values occur purely by chance. In such cases, it is useful to quantify the sequence-structure relationship using more robust measures of statistical significance, such as E-values, that compare the score obtained for an alignment with an established background distribution of such scores.

4. One other problem of using sequence identity as a measure to select templates is that, in practice, there is no single generally accepted way to normalize it *(21)*. For instance, local alignment methods usually normalize the number of identically aligned residues by the length of the alignment, whereas global alignment methods normalize it by either the length of the target sequence or the length of the shorter of the two sequences. Therefore, it is possible that alignments of short fragments produce a high sequence identity but do not result in an accurate model. Measures of statistical significance do not suffer from this normalization problem because the alignment scores are always corrected for the length of the aligned segment before the significance is computed *(22,23)*.

5. After a list of all related protein structures and their alignments with the target sequence has been obtained, template structures are usually prioritized depending on the purpose of the comparative model. Template structures may be chosen based purely on the target-template sequence identity or a combination of several other criteria, such as the experimental accuracy of the structures (resolution of x-ray structures, number of restraints per residue for NMR structures), conservation of active site residues,

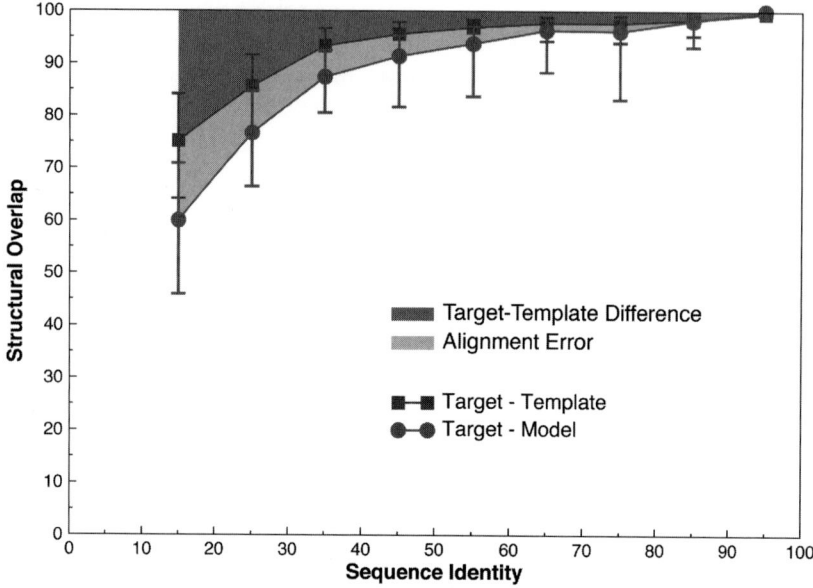

Fig. 8.4 Average model accuracy as a function of sequence identity *(40)*. As the sequence identity between the target sequence and the template structure decreases, the average structural similarity between the template and the target also decreases *(dark gray area, squares) (41)*. Structural overlap is defined as the fraction of equivalent C^α atoms. For the comparison of the model with the actual structure *(circles)*, two C^α atoms were considered equivalent if they belonged to the same residue and were within 3.5 Å of each other after least-squares superposition. For comparisons between the template structure and the actual target structure *(squares)*, two C^α atoms were considered equivalent if they were within 3.5 Å of each other after alignment and rigid-body superposition. The difference between the model and the actual target structure is a combination of the target-template differences *(dark gray area)* and the alignment errors *(light gray area)*. The figure was constructed by calculating ~1 million comparative models based on single template of varying similarity to the targets. All targets had known (experimentally determined) structures.

holo-structures that have bound ligands of interest, and prior biological information that pertains to the solvent, pH, and quaternary contacts.

6. Although fold assignment and sequence-structure alignment are logically two distinct steps in the process of comparative modeling, in practice almost all fold assignment methods also provide sequence-structure alignments. In the past, fold assignment methods were optimized for better sensitivity in detecting remotely related homologues, often at the cost of alignment accuracy. However, recent methods simultaneously optimize both the sensitivity and alignment accuracy. For the sake of clarity, however, they are still considered as separate steps in the current chapter.

7. Most alignment methods use local or global dynamic programming algorithms to derive the optimal alignment between two or more sequences and/or structures. The methods, however, vary in terms of the scoring function that is being optimized. The differences are usually in the form of the gap-penalty function (linear, affine, or variable) *(9)*, the substitution matrix used to score the aligned residues (20×20 matrices derived from

alignments with a given sequence identity, those derived from structural alignments, those incorporating the structural environment of the residues) *(24)*, or combinations of both *(25–28)*. There does not yet exist a single universal scoring function that guarantees the most accurate alignment for all situations. Above 30–40% sequence identity, alignments produced by almost all of the methods are similar. However, in the twilight and midnight zones of sequence identity, models based on the alignments of different methods tend to have significant variations in accuracy. Improving the performance and accuracy of methods in this regime remains one of the main tasks of comparative modeling *(29,30)*.

The single source of errors with the largest impact on comparative modeling is misalignments, especially when the target-template sequence identity decreases below 30%. It is imperative to calculate an accurate alignment between the target-template pair, as comparative modeling can almost never recover from an alignment error *(31)*.

8. Comparative models do not reflect the fluctuations of a protein structure in solution. That is, the variability seen in the structures of multiple models built for one set of inputs reflect different solutions to the molecular objective function, which do not correspond to the actual dynamics of the protein structure in nature.

9. If there are no large differences among the template structures (>2 Å backbone RMSD) and no long insertions or deletions (>5 residues) between the target and the template(s), building multiple models generally does not drastically improve the accuracy of the best model produced. For alignments to similar templates that lack many gapped regions, building multiple models from the same input alignment most often results in a narrow distribution of accuracies: the difference between the C^α RMSD values between each model and the true native structure is usually within a range of 0.5 Å for a sequence containing approximately 150 residues *(5)*. If, however, the sequence-structure alignment contains different templates with many insertions and/or deletions, it is important to calculate multiple models for the same alignment. Calculating multiple models allows for better sampling of the different templates segments and the conformations of the unaligned regions, and often results in a more accurate model than if only one model had been produced.

10. A consensus model is calculated by first clustering an ensemble of models and then averaging individual atomic positions. The consensus model is then optimized using the same protocol used on the individual models. Construction of a consensus model followed by optimization usually results in a model with a lower objective function than any of the contributing models; the construction of a consensus model can thus be seen as a part of an efficient optimization. When there are substantial variations in regions of the contributing models, due to the variation among the templates and the presence of gaps in the alignment, calculating the consensus using cluster averaging usually produces the most accurate conformation.

11. The MODELLER objective function is a measure of how well the model satisfies the input spatial restraints. Lower values of the objective function indicate a better fit with the input data and, thus, models that are likely to be more accurate *(5)*.

12. The Discrete Optimized Protein Energy (DOPE) *(12)* is an atomic distance-dependent statistical potential based on a physical reference state that accounts for the finite size and spherical shape of proteins. The reference state assumes a protein chain consists of non-interacting atoms in a homogeneous sphere of equivalent radius to that of the corresponding protein. The DOPE potential was derived by comparing the distance statistics from a non-redundant PDB subset of 1,472 high-resolution protein structures with the distance distribution function of the reference state. By default, the DOPE score is not included in the model building routine, and thus can be used as an independent assessment of the accuracy of the output models. The DOPE score assigns a score for a model by considering the positions of all non-hydrogen atoms, with lower scores corresponding to models that are predicted to be more accurate.

13. The GA341 criterion is a composite fold-assessment score that combines a Z-score calculated with a statistical potential function, target-template sequence identity, and a measure of structural compactness *(18,19)*. The score ranges from 0.0 for models that tend to have an incorrect fold to 1.0 for models that tend to be comparable to low-resolution x-ray structures. Comparison of models with their corresponding experimental structures indicates that models with GA341 scores greater than 0.7 generally have the correct fold with more than 35% of the backbone atoms superposable within 3.5Å of their native positions. Reliable models (GA341 score $\geq$ 0.7) based on alignments with more than 40% sequence identity, have a median overlap of more than 90% with the corresponding experimental structure. In the 30–40% sequence identity range, the overlap is usually between 75–90% and below 30% it drops to 50–75%, or even less in the worst cases.

14. The accuracy of a model should first be assessed using the GA341 score to increase or decrease our confidence in the fold of the model. An assessment of an incorrect fold implies that an incorrect template(s) was chosen or an incorrect alignment with the correct template was used for model calculation. When the target-template relationship falls in the twilight or midnight zones, it is usually difficult to differentiate between these two kinds of errors. In such cases, building models based on different sets of templates may resolve the problem.

15. Different measures to predict errors in a protein structure perform best at different levels of resolution. For instance, physics-based forcefields may be helpful at identifying the best model when all models are very close to the native state (<1.5 Å RMSD, corresponding to approximately 85% target-template sequence identity). In contrast, coarse-grained scores such as distance-based statistical potentials have been shown to have the greatest ability to differentiate between models in the approximately 3 Å C^{α} RMSD range. Tests show that such scores are often able to identify a model within 0.5 Å C^{α} RMSD of the most accurate model produced *(32)*. When multiple models are built, the DOPE score generally selects a more accurate model than the MODELLER objective function.

16. Segments of the target sequence that have no equivalent region in the template structure (i.e., insertions or loops) are among the most difficult regions to model *(4,33–35)*. This difficulty is compounded when the target and template are distantly related, with errors in the alignment leading to

incorrect positions of the insertions and distortions in the loop environment. Using alignment methods that incorporate structural information can often correct such errors *(9)*. Once a reliable alignment is obtained, various modeling protocols can predict the loop conformation, for insertions of less than 8–10 residues long *(4,33,36,37)*.

17. As a consequence of sequence divergence, the mainchain conformation of a protein can change, even if the overall fold remains the same. Therefore, it is possible that in some correctly aligned segments of a model, the template is locally different (<3 Å) from the target, resulting in errors in that region. The structural differences are sometimes not due to differences in sequence, but are a consequence of artifacts in structure determination or structure determination in different environments (e.g., packing of subunits in a crystal). The simultaneous use of several templates can minimize this kind of an error *(38,39)*.

References

1. Baker, D., and Sali, A. (2001) Protein structure prediction and structural genomics. *Science* **294**, 93–96.
2. Pieper, U., Eswar, N., Davis, F. P., Braberg, H., Madhusudhan, M. S., Rossi, A., Marti-Renom, M., Karchin, R., Webb, B. M., Eramian, D., Shen, M. Y., Kelly, L., Melo, F., and Sali, A. (2006) MODBASE: a database of annotated comparative protein structure models and associated resources. *Nucleic Acids Res.* **34**, D291–295.
3. Marti-Renom, M. A., Stuart, A. C., Fiser, A., Sanchez, R., Melo, F., and Sali, A. (2000) Comparative protein structure modeling of genes and genomes. *Annu. Rev. Biophys. Biomol. Struct.* **29**, 291–325.
4. Fiser, A., Do, R. K., and Sali, A. (2000) Modeling of loops in protein structures. *Protein Sci.* **9**, 1753–1773.
5. Sali, A., and Blundell, T. L. (1993) Comparative protein modelling by satisfaction of spatial restraints. *J. Mol. Biol.* **234**, 779–815.
6. Eswar, N., Madhusudhan, M.S., Marti-Renom, M.A., Sali, A. (2005) BUILD_PROFILE: a module for calculating sequence profiles in MODELLER. http://www.salilab.org/modeller
7. Marti-Renom, M. A., Madhusudhan, M. S., and Sali, A. (2004) Alignment of protein sequences by their profiles. *Prot. Sci.* **13**, 1071–1087.
8. Eswar, N., Madhusudhan, M.S., Marti-Renom, M.A., Sali, A. (2005) PROFILE_SCAN: A module for fold-assignment using profile-profile scanning in MODELLER. http://www.salilab.org/modeller
9. Madhusudhan, M. S., Marti-Renom, M. A., Sanchez, R., and Sali, A. (2006) Variable gap penalty for protein sequence-structure alignment. *Protein Eng. Des. Sel.* **19**, 129–133.
10. MacKerell, A. D., Jr., Bashford, D., Bellott, M., Dunbrack, R. L., Jr., Evanseck, J. D., Field, M. J., Fischer, S., Gao, J., Guo, H., Ha, S., Joseph-McCarthy, D., Kuchnir, L., Muczera, K., Lau, F. T. K., Mattos, C., Michnik, S., Nguyen, D. T., Ngo, T., Prodhom, B., reiher, W. E., III, Roux, B., Schlenkrich, M., Smith, J. C., Stote, R., Straub, J., Watanabe, M., Wiorkiewicz-Kuczera, J., Yin, D., and Karplus, M. (1998) All-atom empirical potential for molecular modleing and dynamics studies of proteins. *J.Phys.Chem.B.* **102**, 3586–3616.
11. Sali, A., and Overington, J. P. (1994) Derivation of rules for comparative protein modeling from a database of protein structure alignments. *Protein Sci* **3**, 1582–1596.
12. Shen, M. Y., and Sali, A. (2006) Statistical potential for assessment and prediction of protein structures. *Protein Sci.* **15**, 2507–2524.

13. Wu, G., Fiser, A., ter Kuile, B., Sali, A., and Muller, M. (1999) Convergent evolution of Trichomonas vaginalis lactate dehydrogenase from malate dehydrogenase. *Proc. Natl. Acad. Sci. USA* **96**, 6285–6290.

14. Deshpande, N., Addess, K. J., Bluhm, W. F., Merino-Ott, J. C., Townsend-Merino, W., Zhang, Q., Knezevich, C., Xie, L., Chen, L., Feng, Z., Green, R. K., Flippen-Anderson, J. L., Westbrook, J., Berman, H. M., and Bourne, P. E. (2005) The RCSB Protein Data Bank: a redesigned query system and relational database based on the mmCIF schema. *Nucleic Acids Res.* **33**, D233–237.

15. Smith, T. F., and Waterman, M. S. (1981) Identification of common molecular subsequences. *J. Mol. Biol.* **147**, 195–197.

16. Needleman, S. B., and Wunsch, C. D. (1970) A general method applicable to the search for similarities in the amino acid sequence of two proteins. *J. Mol. Biol.* **48**, 443–453.

17. Madhusudhan, M. S., Eswar, N., Marti-Renom, M.A., Sali, A. (2005) SALIGN: A module for multiple sequence/structure alignments in MODELLER. http://www.salilab.org/modeller

18. John, B., and Sali, A. (2003) Comparative protein structure modeling by iterative alignment, model building and model assessment. *Nucleic Acids Res.* **31**, 3982–3992.

19. Melo, F., Sanchez, R., and Sali, A. (2002) Statistical potentials for fold assessment. *Protein Sci.* **11**, 430–448.

20. Rost, B. (1999) Twilight zone of protein sequence alignments. *Protein Eng.* **12**, 85–94.

21. May, A. C. (2004) Percent sequence identity; the need to be explicit. *Structure* **12**, 737–738.

22. Altschul, S. F., Madden, T. L., Schaffer, A. A., Zhang, J., Zhang, Z., Miller, W., and Lipman, D. J. (1997) Gapped BLAST and PSI-BLAST: a new generation of protein database search programs. *Nucleic Acids Res.* **25**, 3389–3402.

23. Pearson, W. R. (1998) Empirical statistical estimates for sequence similarity searches. *J. Mol. Biol.* **276**, 71–84.

24. Henikoff, S., and Henikoff, J. G. (1992) Amino acid substitution matrices from protein blocks. *Proc. Natl. Acad. Sci. USA* **89**, 10915–10919.

25. Zhou, H., and Zhou, Y. (2005) Fold recognition by combining sequence profiles derived from evolution and from depth-dependent structural alignment of fragments. *Proteins* **58**, 321–328.

26. McGuffin, L. J., and Jones, D. T. (2003) Improvement of the GenTHREADER method for genomic fold recognition. *Bioinformatics* **19**, 874–881.

27. Karchin, R., Cline, M., Mandel-Gutfreund, Y., and Karplus, K. (2003) Hidden Markov models that use predicted local structure for fold recognition: alphabets of backbone geometry. *Proteins* **51**, 504–514.

28. Shi, J., Blundell, T. L., and Mizuguchi, K. (2001) FUGUE: sequence-structure homology recognition using environment-specific substitution tables and structure-dependent gap penalties. *J. Mol. Biol.* **310**, 243–257.

29. Dunbrack, R. L., Jr. (2006) Sequence comparison and protein structure prediction. *Curr. Opin. Struct. Biol.* **16**, 374–384.

30. Xiang, Z. (2006) Advances in homology protein structure modeling. *Curr. Protein Pept. Sci.* **7**, 217–227.

31. Sanchez, R., and Sali, A. (1997) Advances in comparative protein-structure modelling. *Curr. Opin. Struct. Biol.* **7**, 206–214.

32. Eramian, D., Shen, M. Y., Devos, D., Melo, F., Sali, A., and Marti-Renom, M. A. (2006) A composite score for predicting errors in protein structure models. *Protein Sci.* **15**, 1653–1666.

33. Jacobson, M. P., Pincus, D. L., Rapp, C. S., Day, T. J., Honig, B., Shaw, D. E., and Friesner, R. A. (2004) A hierarchical approach to all-atom protein loop prediction. *Proteins* **55**, 351–367.

34. Fernandez-Fuentes, N., Oliva, B., and Fiser, A. (2006) A supersecondary structure library and search algorithm for modeling loops in protein structures. *Nucleic Acids Res.* **34**, 2085–2097.

35. Zhu, K., Pincus, D. L., Zhao, S., and Friesner, R. A. (2006) Long loop prediction using the protein local optimization program. *Proteins* **65**, 438–452.

36. Coutsias, E. A., Seok, C., Jacobson, M. P., and Dill, K. A. (2004) A kinematic view of loop closure. *J. Comput. Chem.* **25**, 510–528.

37. van Vlijmen, H. W., and Karplus, M. (1997) PDB-based protein loop prediction: parameters for selection and methods for optimization. *J. Mol. Biol.* **267**, 975–1001.

38. Sanchez, R., Sali, A. (1997) Evaluation of comparative protein structure modeling by MODELLER-3. *Proteins* **1**, 50–58.

39. Srinivasan, N., and Blundell, T. L. (1993) An evaluation of the performance of an automated procedure for comparative modelling of protein tertiary structure. *Protein Eng.* **6**, 501–512.

40. Sanchez, R., Sali, A. (1998) Large-scale protein structure modeling of the Saccharomyces cerevisiae genome. *Proc. Natl. Acad. Sci. USA* **95**, 13597–13602.

41. Chothia, C., and Lesk, A. M. (1986) The relation between the divergence of sequence and structure in proteins. *Embo. J.* **5**, 823–826.

Section II

Protein Production

Chapter 9

High Throughput Cloning with Restriction Enzymes

Volker Sievert, Asgar Ergin, and Konrad Büssow

The systematic structural analysis of many target proteins involves generating expression clones in high throughput. This requires robust laboratory procedures and benefits from laboratory automation and data management systems. This chapter gives an overview of the Protein Structure Factory, a structural genomics project focusing on human proteins, and presents the authors' method for cloning bacterial expression clones with the restriction enzymes BamHI and NotI and compatible enzymes. PCR amplification, product purification and digestion and vector ligation were adapted to the 96-well microtiter plate format.

1. Introduction

The Protein Structure Factory (PSF) is a structural genomics project focusing on human proteins (1,2). To increase the throughput, standardized and automated procedures have been introduced to establish a pipeline from target selection to structure determination.

Targets were mainly selected according to biophysical criteria and availability of cDNA clones. Proteins were expressed as His-tag and StrepII-tag (3) fusion proteins in *Escherichia coli*, *Saccharomyces cerevisiae*, and *Pichia pastoris*. Procedures adapted to the 96-well microplate format were established for cloning and characterization of expression clones by small-scale expression and purification (4,5).

Protein production for crystallization is performed by a standardized procedure of affinity chromatography and His-tag removal by TEV protease digest, followed by ion exchange chromatography and gel filtration (6). Purified proteins are concentrated, to avoid aggregation, in a stepwise manner. The particle size is controlled by dynamic light scattering at each concentration step.

Protein crystallization at the PSF is highly automated (7). Crystallization screens are set up by nanoliter pipetting robots. The screening buffers are automatically prepared from stocks in deep well microplates by a pipetting robot. This robot allows setting up standard screens as well as fine screens for optimization of crystallization conditions. X-ray diffraction experiments are performed at the beamlines established by the PSF at the BESSY synchrotron storage ring.

From: *Methods in Molecular Biology, Vol. 426: Structural Proteomics: High-throughput Methods*
Edited by: B. Kobe, M. Guss and T. Huber © Humana Press, Totowa, NJ

The PSF pipeline consists of the following steps:

1. Target selection and design of expression constructs
2. High throughput cloning and clone sequence verification
3. Characterization of expression clones by small scale expression and purification
4. Automated protein crystallization
5. X-ray crystallography

Standard cloning methods based on restriction enzymes and DNA ligase can be adapted to the microplate format to achieve a moderately high throughput. Ligation-based cloning (LIC) using the T4 DNA polymerase to generate overhangs *(8)* or recombination-based methods such as the Invitrogen Gateway system are popular alternatives to the traditional approach *(9,10)*.

A large number of Gateway Entry clones available from academic and commercial sources allow to produce expression clones by a simple recombination reaction that avoids PCR amplification *(11)*. All other cloning approaches require a PCR step with primers that have tails of functional sequences, such as restriction or recombination sites. Restriction enzyme recognition sites are generally shorter than the sequences required for LIC and recombination methods. Therefore, restriction enzyme–based cloning allows for shorter PCR primers that are less expensive and less likely to contain mutated byproducts. A six–base pair recognition site only introduces two additional amino acids. This is advantageous when designing N-terminal fusion proteins. On the other hand, restriction enzyme–based cloning of PCR products involves more steps, including DNA purification, which makes it more difficult to set up in a robust way with high throughput. Moreover, restriction sites can occur within the target sequences to be cloned, demanding to choose alternative enzymes for these targets.

At the PSF, expression clones were generated with the restriction enzymes BamHI and NotI and other enzymes that generate compatible overhangs. The vectors used for cloning are pQTEV *(1)*, pQTEV2 (Fig. 9.1), pQLinkH, pQLinkG, pGEX-6P1 and pEntryTEV (Table 9.1). All these are *E. coli* expression vectors except for pEntryTEV, which is used to generate Entry clones with a TEV site for the Gateway recombination cloning system (Invitrogen). pQLinkH and pQLinkG plasmids can be easily combined into co-expression plasmids *(19)*.

For templates containing internal BamHI or NotI sites, compatible overhangs are produced by alternative enzymes or by the hetero-stagger cloning method *(12)*. Alternative enzymes are BglII for BamHI and the type IIs enzymes BpiI, Eco31I, and Esp3I, which can replace both BamHI and NotI. Type IIs enzymes cut outside their recognition sequence and can produce arbitrary overhangs.

The *E. coli* strain SCS1 carrying the helper plasmid pRARE was used for cloning of expression clones and protein expression experiments. pRARE is a plasmid of the Novagen Rosetta strain; it carries genes for overexpression of rare tRNAs and a chloramphenicol resistence marker *(13)*.

To control the correctness of new clones, their insert size was verified. Clones produced by PCR may contain mutations even if their insert size is correct. These mutations are mostly caused by byproducts of PCR primer synthesis in the authors' experience. It is advisable to test several transformants for protein expression in case some contain a mutation. Alternatively, clones can be verified by sequencing. This chapter describes how the Staden sequence analysis package can be used for this purpose.

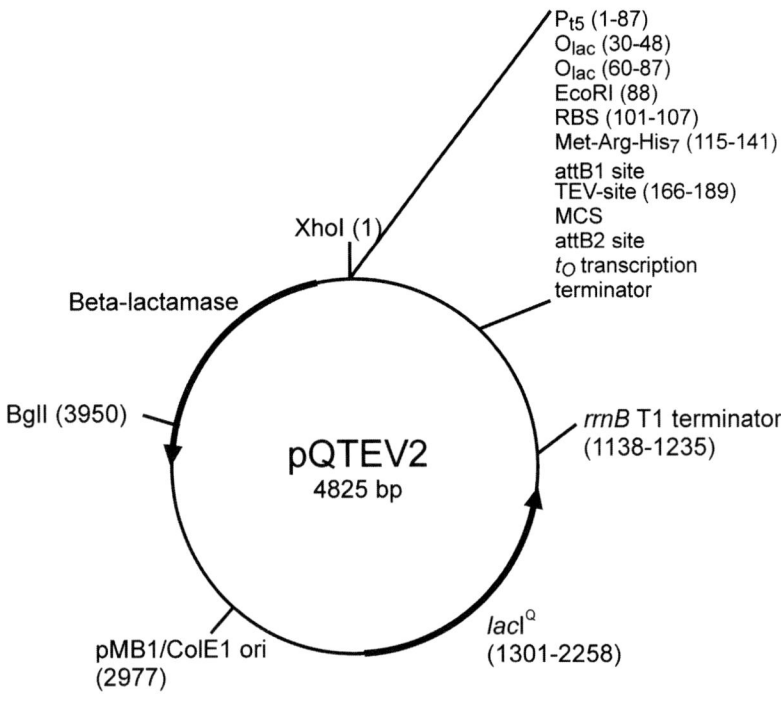

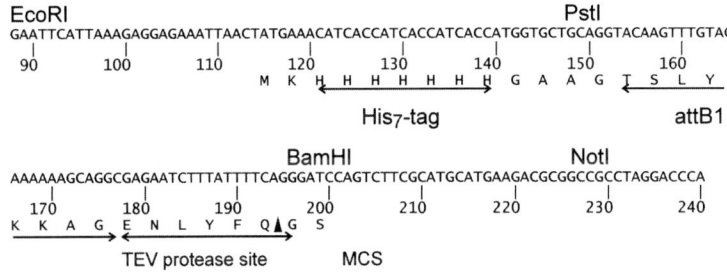

Fig. 9.1 Map and sequence of the multiple cloning site (MCS) of the pQTEV2 vector.

The generation of many clones in parallel requires electronic data management of some sort. A list of the constructs to be produced containing identifiers of the target proteins, the template cDNA clones, the start and end position of the construct with respect to the template cDNA clone sequence, and PCR primer identifiers should be available (see Note 1).

The aim of the high throughput cloning method described here is to adapt common laboratory procedures to allow processing of many cloning experiments in parallel. It uses a common restriction-ligation design. No special instrumentation is needed except for a 96-pin inoculation gadget. However, a pipetting robot may facilitate preparation of primer plates, purification of PCR products and reformatting of samples in microtiter plates (see Note 2). The procedure consists of the following steps:

Table 9.1 Vectors for cloning with BamHI and NotI.

Vector	N-terminal affinity tag, protease site	C-terminal affinity tag	Gateway att sites	Accession No.	Comment
pQTEV	His$_7$, TEV	—	—	AY243506	See ref. *(1)*
pQTEV2, pQLinkH	His$_7$, TEV	—	attB1, attB2	EF025688	Insert can be shuttled into a Gateway Donor vector (Fig. 9.1).
pQLinkG	GST, TEV	—	attB1, attB2	EF025689	
pGEX-6P1	GST, PreScission	—	—		
pEntryTEV	none, TEV	—	attL1, attL2		Gateway Entry clones

1. Preparation of template clone plates and diluted primer plates. Preparation of restriction digested and dephosphorylated vector (1–2 days).
2. PCR and evaluation of PCR products by gel electrophoresis. If necessary, rearraying of successful reactions (1–3 days).
3. Purification of PCR products and quantification with gel electrophoresis, set up of ligations (2 days).
4. Transformation and plating: 96 to 192 transformations can be processed in parallel (1 day).
5. Picking of transformed bacteria clones (0.5 day; needs overnight incubation).
6. Verification of clones by colony PCR (1–3 days).
7. Picking of correct clones for further analysis (0.5 day, needs overnight incubation).

2. Materials

2.1. Preparation of Competent Cells

1. SCS1 *E. coli* cells (Stratagene) carrying the pRARE plasmid (see Introduction).
2. DMSO.
3. SOB broth *(14)*: Dissolve 20 g bacto-tryptone, 5 g bacto-yeast extract, 0.5 g NaCl and 0.186 g KCl in 1 liter demineralised water and autoclave. Add 5 ml of sterile 2 M MgCl$_2$ just before use.
4. Inoue Transformation Buffer *(15)*: 10 mM PIPES, 55 mM MnCl$_2$, 15 mM CaCl$_2$, 250 mM KCl. Mix all components except for the MnCl$_2$. Adjust the pH with KOH and add the MnCl$_2$. Sterilize by filtration.

2.2. Preparation of Inserts

1. PCR primers to amplify the open reading frames of interest (*see* Note 3).
2. Polystyrene microtiter plates with lids for bacterial cultures and glycerol stocks (Nunc, No. 260895).

3. Expand High Fidelity PCR kit (Roche Applied Science).
4. 10 mM of each dATP, dCTP, dGTP, dTTP in TE buffer (Fermentas, R0151).
5. Thermowell 96 well plates (Costar).
6. 96-pinned replicators. Plastic and steel replicators are available from Genetix or Nunc.
7. Adhesive sealing sheets for microtiter plates (Abgene, Epsom, UK).
8. 96-well PCR product cleaning kit. We use an in-house magnetic beads procedure that is commercially available from Bruker (Genopure ds kit).
9. Restriction enzymes with buffers (NEB).

2.3. Preparation of the Vector

1. Chromaspin 1,000 size exclusion columns (Clontech, Mountain View, CA) for DNA purification.
2. Shrimp alkaline phosphatase (SAP, 1 unit/μl, Roche Applied Science, Indianapolis, IN).
3. NEBuffer BamHI: 10 mM Tris-HCl, 150 mM NaCl, 10 mM $MgCl_2$, 1 mM Dithiothreitol, pH 7.9 at 25°C.
4. Dephosphorylation buffer: 0.5 M Tris-HCl, 50 mM $MgCl_2$, pH 8.5 at 20°C.

2.4. Ligation and Transformation

1. T4 DNA ligase with buffer (NEB).
2. 2×YT broth: Add 16 g bacto-tryptone, 10 g bacto-yeast extract and 5 g NaCl per liter demineralised water, autoclave.
3. 2×YT agar: Add 16 g bacto-tryptone, 10 g bacto-yeast extract, 5 g NaCl and 15 g agar per liter, autoclave. Add supplements such as antibiotics and glucose after cooling to 50°C.
4. QTray with divider (Genetix, New Milton, UK) to prepare agar plates with 48 wells.
5. 10×HMFM: Dissolve 72 g KH_2PO_4 and 188 g K_2HPO_4 (Mw 174) in 800 ml water and autoclave. Dissolve 3.6 g $MgSO_4×7H_2O$, 18 g Na_3-citrate×2H_2O, 36 g $(NH_4)_2SO_4$ and 1,760 g glycerol (1,645 ml of 87% glycerol) in 3.2 L water, autoclave and add the 800 ml phosphate solution.
6. 40% (w/v) glucose: dissolve 400 g D+glucose monohydrate in distilled water to 1 liter and sterilize by filtration through a 0.2-μm pore size filter.
7. Polypropylene deep well plates (96 wells of 2.2 ml, non-sterile, Wertheim, Germany).

2.5. Colony PCR

1. Taq DNA polymerase, prepared according to Engelke et al. *(16)*.
2. 10× PCR buffer: 0.5 M KCl, 1%Tween 20, 15 mM $MgCl_2$, 350 mM Tris-base, 150 mM Tris×HCl.
3. 5 M betaine.

2.6. Clone Sequence Verification

1. The Staden package (http://staden.sourceforge.net).

3. Methods

3.1. Preparation of Competent Cells

The preparation of competent cells follows the method of Inoue et al. *(15)* with slight modifications.

1. Grow a 5-ml starter culture of *E. coli* SCS1/pRARE overnight at 37°C in SOB broth supplemented with 34 mg/l chloramphenicol.
2. Inoculate 250 ml of the same broth with 2 ml overnight culture and grow to OD_{600} = 0.6 at 18°C.
3. Cool on ice for 10 min.
4. Transfer to two sterile, prechilled 250-ml polypropylene centrifuge bottles and spin down at 2,500 g for 10 min.
5. Remove the supernatant and gently resuspend the pellet with 40 ml ice-cold Inoue Transformation Buffer.
6. Pool both suspensions and place them on ice for 10 min. Centrifuge again as above and resuspend the cells in 20 ml Inoue Transformation Buffer, incubate for 10 min on ice and add 1.4 ml DMSO.
7. Dispense the cell suspension into 900-µl aliquots, freeze them in liquid nitrogen and store at −80°C.
8. The quality of the cells is tested by transformation with 50 pg pUC19 DNA. The expected yield is around 10^7 colonies/µg DNA.

3.2. Preparation of Inserts

1. Arrange diluted primers (10 µM, see Note 3) and glycerol stocks of template cDNA clones in three different polystyrene microtiter plates, such that templates and their corresponding primers are located on corresponding positions. If different pairs of restriction enzymes are necessary for different targets, try to place targets with the same enzymes together.
2. Prepare master mixes of polymerase, buffer and dNTPs for 100 µl .Expand High Fidelity PCRs on ice according to the manufacturer's instructions. Dispense 100 µl per well of a Thermowell 96 well plate on ice. Dispense 1 µl of each primer. "Inoculate" the reactions with the template bacteria from cultures in a 96-well microtiter plate using a sterile 96-pinned replicator. Cover the Thermowell plate with an adhesive sealing sheet and run the following PCR program: 5 min at 96°C followed by 25 cycles of 1 min at 94°C, 0.5 min 55°C, and 1 min per 1,000 bp of maximum expected product size at 72°C followed by a final extension step of 15 min at 72°C.
3. Check a 5-µl aliquot of each reaction on an agarose gel. If all or most reactions are successful, proceed. Otherwise, consider repeating failed reactions using different parameters (Fig. 9.2).
4. Purify PCR products using a suitable 96-well PCR purification system (see Fig. 9.2).
5. Set up restriction digests with 30 µl of the purified DNA in a new Thermowell plate for 3 hours using 1 unit of each enzyme in 50 µl total volume according to the manufacturer's instructions. Purify again as above.
6. Separate 5 µl on a gel and estimate the concentration by comparison with control samples of known concentrations. Molar concentrations are obtained by automatically determination of product amounts with a gel image

PCR

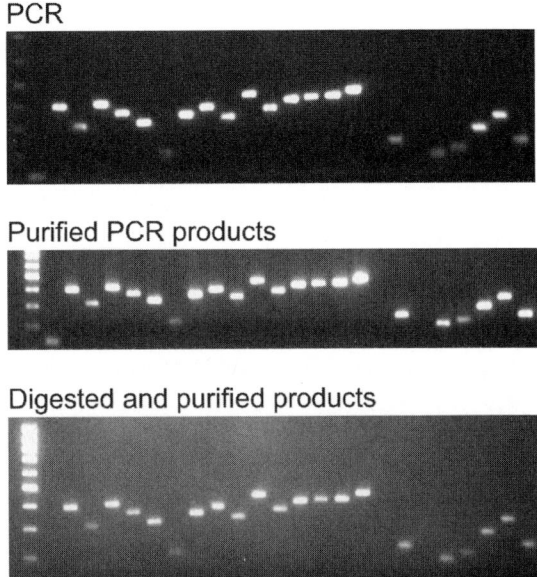

Purified PCR products

Digested and purified products

Fig. 9.2 PCR amplification, product purification, and digestion. The *upper panel* shows the results of PCR amplification of a number of targets on an agarose gel stained with ethidium bromide. The *middle panel* shows the same PCR products upon cleaning with magnetic beads. Note that the product amounts are more uniform after cleaning due to the limited amount of beads used for cleaning. The *lower panel* shows the PCR products after restriction digest and a second purification.

evaluation software and multiplication of DNA amounts with molecular weights stored in a database. We have used the software Phoretix 1D (now totallab, Nonlinear Dynamics, Inc., Newcastle upon Tyne, UK) and our in-house database for this purpose (see Note 1).

3.3. Preparation of the Vector

1. Linearize 5 µg vector DNA by digestion with 3 µl of BamHI (20 units/µl) and 3 µl NotI (10 units/µl) in 100 µl of NEBuffer BamHI containing 100 µg/ml bovine serum albumin (BSA) overnight at 37°C. Add one more microliter of each enzyme, mix gently and incubate for another 3 hours. Inactivate enzymes by heating to 65°C for 20 min. Purify the linearized vector with a Chromaspin 1,000 column.
2. Dephosphorylate 44 µl of the linearized vector DNA by adding 1 µl shrimp alkaline phosphatase and 5 µl 10× dephosphorylation buffer at 37°C for 1 hour. Inactivate the phosphatase by heating to 72°C for 20 min. Purify the linearized and dephosphorylated DNA with another Chromaspin 1,000 column.

3.4. Ligation and Transformation

1. Set up ligations according to PCR product concentration. Usually it is sufficient to distinguish low and high product concentrations and to use 6.5 or 2 µl product, respectively, for the ligation. In the wells of a Thermowell plate, add PCR product and water to 6.5 µl and a mix of 2 µl (5 ng) linearized

vector DNA, 0.5 µl T4 ligase (400 units/µl), 1 µl 10× ligase buffer. Cover the Thermowell plate with a sealing sheet and incubate the samples for 1 hour at 20°C in the PCR cycler. Heat inactivate the samples for 20 min at 65°C.

2. Thaw chemical competent E. coli SCS1/pRARE cells on ice. Meanwhile, transfer 5 µl of each ligation reaction to a well of a new prechilled Thermowell plate on ice. Add 100 µl bacteria to each well and incubate on ice for 5 min. Heat shock the cells at 42°C for 30 sec using a PCR cycler. Immediately put the plate back on the ice and incubate for 1–5 min. Recover cells with 500 µl 2×YT broth supplemented with 20 mM MgCl2 and 20 mM glucose for 45 min at 37°C in a polypropylene deep well plate. Spin down cells at 2,000 g for 10 min at 4°C. Discard 300 µl of the supernatant, resuspend the remainder, and plate on agar.

3. Agar plates are prepared in large QTray plastic dishes with 48 wells. The dishes are filled with 300 ml 2×YT agar supplemented with 100 mg/l ampicillin, 34 mg/l chloramphenicol and 2% glucose. 200 µl transformed bacteria are pipetted into each well. The liquid is spread by gently moving the dishes and dried in a sterile hood without the lid, followed by overnight incubation of the closed dishes at 37°C.

4. Fill a polystyrene microtiter plate with 200 µl of 2×YT broth containing 1×HMFM, antibiotics and 2% glucose. Pick six colonies of each agar dish well with sterile tooth picks into individual wells of the microtiter plate. Wrap the plate in plastic foil and incubate at 37°C until all wells are grown evenly (approx. 16 h).

3.5. Colony PCR

To check the presence of inserts of the new clones, a colony PCR is performed. Enhancers such as betaine should be added to the PCR to improve the amplification of difficult templates (17).

1. Prepare a PCR master mix for 30 µl reactions of 80 units/ml Taq polymerase, dNTPs (65.5 µM each), 1.25 M betaine, and vector primers pQE65 and pQE276 (0.3 µM of each) in PCR buffer. Dispense 30-µl volumes in a Thermowell 96-well plate. 'Inoculate' the reactions with the cultures of the picked transformants with a sterile 96-pinned replicator. Cover the Thermowell plate with sealing sheet and run the following PCR program: 5 min at 95°C followed by 30 cycles of 0.5 min at 95°C, 0.5 min at 60°C, and 1 min per 1,000 bp of maximum expected product size at 72°C followed by a final extension step of 15 min at 72°C.

2. Check an aliquot of each reaction on an agarose gel. Choose two clones per transformation that had PCR products of the expected size for further protein expression and purification. Again, a gel analysis software in combination with a database is useful, as it allows determination of product sizes and comparison with expected sizes stored in the database.

3. Inoculate two wells of another polystyrene microtiter plate with 2×YT broth and supplements with the selected clones using sterile tooth picks. A pipetting robot is helpful in this step to avoid the error prone reformatting by hand. Incubate the wrapped plate at 37°C as before and store at −80°C until testing the clones by small scale expression and purification (4,5).

3.6. Clone Sequence Verification

Mutation may be introduced by PCR during cloning. In the authors' experience, most errors are found in the sequences of the PCR primers and are probably caused by primer synthesis byproducts. The Staden package *(18)* is a useful tool to compare DNA sequencing results with the expected sequence. It is available for Unix/Linux, Windows and Mac computers. For each cloning experiment, a database has to be created with the *gap4* program of the package, according to the following procedure, that can be automated by a suitable script.

1. Create an empty database named "gap4" with the gap4 program.
2. Use pregap4 to create two configuration files. Choose the modules "Estimate Base Accuracies," "Initialize Experiment Files," and "Quality Clip" and deselect all other modules. Choose "Save all Parameters To": and save as pregap4-init.conf. Then select only the module "Gap4 Shotgun Assembly," set the database name to gap4, and save as pregap4-assembly. conf.
3. For each cloning experiment, copy the empty gap4 database files, the expected clone sequence in Staden text format, and the chromatogram files with the sequencing results into a new directory. The files should have the suffixes .sdn and .abi, respectively.
4. In the same directory, run pregap4 with the following parameters to create experiment files (.exp) for each chromatogram:

```
pregap4 -config [path]/pregap4-init.conf -nowin
- *.abi
```

5. To include all experiment files and the expected sequence into the gap4 database, run pregap4 again with the other configuration file:

```
pregap4 -config [path]/pregap4-assembly.conf -
nowin - *.exp *.sdn
```

6. The steps 3.–5. can be combined into a script to facilitate the analysis of many clones. The gap4 database are examined manually with *gap4* to verify agreement of sequencing results with the expected sequence.

4. Notes

1. Simultaneously handling larger amounts of samples requires an appropriate bookkeeping method. To record high-throughput cloning experiments, we implemented a sample management system, which is embedded into the authors' projectwide database structure. A brief description of this system can be found here:

   http://www.proteinstrukturfabrik.de/tp03page/lims.shtml

   The system contains a batch clone management module that allows creation, handling, and evaluation of large numbers of HTP cloning experiments simultaneously. For each cloning step, all necessary objects (PCR reactions to generate templates, microtiter plates for storage, agarose gels for evaluation, and new clones with respective data) are created automatically. Different kinds of experimental observations such as PCR results (product amount and size, success or failure of cleaning steps) can be entered interactively and are automatically assigned to the objects to which they belong. Results are always judged by the user, but the system supplies several kinds of support: Prediction of DNA and protein sequences of affinity tag fusion constructs and calculated fragment sizes.

2. Most steps of the cloning procedure are performed with a multichannel pipet but there are some reformatting and "cherry picking" steps which benefit from a pipetting robot. The authors' computer-based storage and sample management system (see Note 1) can create robot command files. A robot is especially useful to prepare the primer and template plates and to reformat successful candidate clones at the end of the procedure. The authors also use the robot to assemble restriction digests with different enzymes.

3. The primers should have a common annealing temperature of about 63°C. The authors' use the program ORFprimer to design primers. The program adjust the lengths of the primers to obtain similar annealing temperatures. Forward primers should have a BamHI site (GGATCC) tail encoding glycine, serine in frame with the amplified gene (see Fig. 9.1). Reverse primers should have a NotI site (GCGGCCGC) tail adjacent to the stop codon. For inserts including BamHI or NotI sites, choose alternative enzymes (see Introduction). For efficient cleavage, two nucleotides should be added to the 5' end before the BamHI site and four nucleotides before the NotI site. Here is an example:

Forward primer
5'-CAGGATCCGCTTGTGCTGAGTTTTCTTTTCATG -3'
G S A C A E F S F H
Reverse primer
5'-GACTGCGGCCGC<u>TCA</u>ATCTCGCCCAATTGAATGCG-3'
Stop

Acknowledgments

The authors thank Janett Tischer for excellent technical assistance and critical reading of the manuscript. This work was funded by the German Federal Ministry of Education and Research (BMBF) in the National Genome Network programme (NGFN FKZ 01GR0472) and the SFB 633 of the German Research Foundation.

References

1. Büssow, K., Scheich, C., Sievert, V., Harttig, U., Schultz, J., Simon, B., Bork, P., Lehrach, H., and Heinemann, U. (2005). Structural Genomics of human proteins—target selection and generation of a public catalog of expression clones. *Microb. Cell Fact.* **4**, 21.

2. Heinemann, U., Büssow, K., Mueller, U., and Umbach, P. (2003). Facilities and methods for the high-throughput crystal structure analysis of human proteins. *Acc. Chem. Res.* **36**, 157–163.

3. Voss, S., and Skerra, A. (1997). Mutagenesis of a flexible loop in streptavidin leads to higher affinity for the Strep-tag II peptide and improved performance in recombinant protein purification. *Protein Eng.* **10**, 975––982.

4. Scheich, C., Sievert, V., and Büssow, K. (2003). An automated method for high-throughput protein purification applied to a comparison of His-tag and GST-tag affinity chromatography. *BMC Biotechnol.* **3**, 12.

5. Berrow, N. S., Büssow, K., Coutard, B., Diprose, J., Ekberg, M., Folkers, G. E., Levy, N., Lieu, V., Owens, R. J., Peleg, Y., Pinaglia, C., Quevillon-Cheruel, S., Salim, L., Scheich, C., Vincentelli, R., and Busso, D. (2006). Recombinant protein expression and solubility screening: a comparative study. *Acta Cryst. Section D* **62**, 1218–1226.

6. Büssow, K., Quedenau, C., Sievert, V., Tischer, J., Scheich, C., Seitz, H., Hieke, B., Niesen, F. H., Götz, F., Harttig, U., and Lehrach, H. (2004). A catalogue of human cDNA expression clones and its application to structural genomics. *Genome Biol.* **5**, R71.

7. Mueller, U., Nyarsik, L., Horn, M., Rauth, H., Przewieslik, T., Saenger, W., Lehrach, H., and Eickhoff, H. (2001). Development of a technology for automation and miniaturization of protein crystallization. *J. Biotechnol.* **85**, 7–14.

8. Aslanidis, C., and de Jong, P. J. (1990). Ligation-independent cloning of PCR products (LIC-PCR). *Nucleic Acids Res.* **18**, 6069–6074.

9. Hartley, J. L., Temple, G. F., and Brasch, M. A. (2000). DNA cloning using in vitro site-specific recombination. *Genome Res.* **10**, 1788–1795.

10. Busso, D., Delagoutte-Busso, B., and Moras, D. (2005). Construction of a set Gateway-based destination vectors for high-throughput cloning and expression screening in *Escherichia coli*. *Anal. Biochem.* **343**, 313–321.

11. Rual, J. F., Hirozane-Kishikawa, T., Hao, T., Bertin, N., Li, S., Dricot, A., Li, N., Rosenberg, J., Lamesch, P., Vidalain, P. O., Clingingsmith, T. R., Hartley, J. L., Esposito, D., Cheo, D., Moore, T., Simmons, B., Sequerra, R., Bosak, S., Doucette-Stamm, L., Le Peuch, C., Vandenhaute, J., Cusick, M. E., Albala, J. S., Hill, D. E., and Vidal, M. (2004). Human ORFeome version 1.1: a platform for reverse proteomics. *Genome Res.* **14**, 2128–2135.

12. Li, C., and Evans, R. M. (1997). Ligation independent cloning irrespective of restriction site compatibility. *Nucleic Acids Res.* **25**, 4165–4166.

13. Novy, R., Drott, D., Yaeger, K., and Mierendorf, R. (2001). Overcoming the codon bias of E.coli for enhanced protein expression. *inNovations* **12**, 1–3.

14. Sambrook, J., Fritsch, E. F., and Maniatis, T. (1989). *Molecular Cloning: A Laboratory Manual*. 2 ed, Cold Spring Harbor Laboratory, Cold Spring Harbor, NY.

15. Inoue, H., Nojima, H., and Okayama, H. (1990). High efficiency transformation of Escherichia coli with plasmids. *Gene* **96**, 23–28.

16. Engelke, D. R., Krikos, A., Bruck, M. E., and Ginsburg, D. (1990). Purification of *Thermus aquaticus* DNA polymerase expressed in *Escherichia coli*. *Anal. Biochem.* **191**, 396–400.

17. Ralser, M., Querfurth, R., Warnatz, H. J., Lehrach, H., Yaspo, M. L., and Krobitsch, S. (2006). An efficient and economic enhancer mix for PCR. *Biochem. Biophys. Res. Commun.* **347**, 747–751.

18. Staden, R., Judge, D. P., and Bonfield, J. K. (2003). Analysing sequences using the Staden package and EMBOSS. Introduction to Bioinformatics. (Krawetz, S. A., and Womble, D. D., eds.), *A Theoretical and Practical Approach*, Humana Press, Inc., Totowa, NJ.

19. Scheich, C., Kümmel, D., Soumailakakis, D., Heinemann, U., and Büssow, K. (2007) Vectors for co-expression of an unrestricted number of proteins. *Nucleic Acids Research* **35**, e43.

Chapter 10

Automated Recombinant Protein Expression Screening in *Escherichia coli*

Didier Busso, Matthieu Stierlé, Jean-Claude Thierry, and Dino Moras

To fit the requirements of structural genomics programs, new as well as classical methods have been adapted to automation. This chapter describes the automated procedure developed within the Structural Biology and Genomics Platform, Strasbourg for performing recombinant protein expression screening in *Escherichia coli*. The procedure consists of parallel competent cells transformation, cell plating, and liquid culture inoculation, implemented for up to 96 samples at a time.

1. Introduction

During recent years, structural genomics programs have been initiated worldwide by large consortia as well as academic laboratories (http://www.isgo.org) with the aim of determining, as efficiently as possible, three-dimensional structures of proteins and their complexes in living organisms. Such initiatives led to technological and methodological developments using automation and robotics during each step of a structural study from gene cloning through structure determination *(1)*. Since soluble and homogeneous protein is required to carry on a structural study, some emphasis has been placed on screening expression conditions, resulting in soluble production of recombinant proteins. Although various expression systems are currently used to produce recombinant proteins, *Escherichia coli* is still the expression system of choice for initial expression screening due to fast growth, easy handling, and low cost *(2,3)*. Moreover, the recent advent of autoinducible methods *(4)* facilitates parallel expression screening in *E. coli* and is well adapted for automation.

The Structural Biology and Genomics Department (IGBMC, Illkirch, France) is involved in recent structural genomics programs through the SPINE (Structure Proteomics IN Europe) network and has implemented a platform in which a medium-throughput pipeline as well as tools for producing and characterizing protein samples suitable for structural biology studies have been developed *(5–7)*.

This chapter focuses on recombinant protein production in *E. coli* and describes the automated procedure the authors have developed from cell transformation through recombinant protein analysis. To validate the procedure, an experiment was run with 48 plasmid DNAs.

From: *Methods in Molecular Biology, Vol. 426: Structural Proteomics: High-throughput Methods*
Edited by: B. Kobe, M. Guss and T. Huber © Humana Press, Totowa, NJ

2. Materials

2.1. Competent Cell Preparation

1. Glycerol stock of BL21(DE3) *Escherichia coli* strain (see Note 1).
2. Luria Bertani (LB) medium (12780-052, Invitrogen, Carlsbad, CA): 20 g/L. Sterilize by autoclaving. Store at room temperature.
3. 10 m*M* NaCl, 75 m*M* CaCl$_2$ (see Note 2).
4. 87% Glycerol. Sterilize by autoclaving. Store at room temperature.
5. Control plasmid DNA: pUC19 (N3041S, New England Biolabs, Beverly, MA).
6. 50-mL sterile conical centrifuge tubes (352070, BD Biosciences, Franklin Lakes, NJ).
7. 250-mL sterile centrifuge bottles (Corning CLS430776, Sigma-Aldrich, St. Louis, MO).
8. 0.6-mL sterile PCR tubes for storage (964625901, TreffLab, Degersheim, Switzerland).

2.2. Automated Procedures

Automated procedures were performed on a Tecan Genesis Workstation robot (Tecan, Maennedorf, Switzerland).

2.2.1. Competent Cell Transformation

1. *E.* coli BL21(DE3) chemically competent cells (see Section 3.1.).
2. MicroAmpR optical 96-well PCR plate (N801-0560, Applied Biosystems, Foster City, CA).
3. 40-well bench top cooler rack for 0.6 mL PCR tubes (410098, Stratagene, La Jolla, CA).
4. Aluminum sealing tape (AB-0626, Abgene, Epsom, UK).
5. LB medium (*see* Section 2.1., Item 2.).

2.2.2. Transformation Reactions Plating

1. Transformation reactions (see Section 3.2.1.).
2. LB-agar medium (22700-025, Invitrogen): 32 g/L. Sterilize by autoclaving. Store at room temperature.
3. 24-well culture plates (3536, Costar-Corning, Corning, NY).

2.2.3. Liquid Culture Inoculation

1. ZYM-5052 medium: mix under sterile conditions 958 mL of ZY medium, 20 mL of 50 × 5,052 solution, 20 mL of 50 × M solution and 2 mL of 1 M MgSO$_4$ solution (*4*). Prepare fresh.
2. ZY medium: 10 g tryptone, 5 g yeast extract/L. Sterilize by autoclaving. Store at room temperature.
3. 50 × 5052 solution: 25% (w/v) glycerol, 2.5% (w/v) glucose, 10% (w/v) α-lactose. Sterilize by autoclaving. Store at room temperature.
4. 50 × M solution: 1.25 *M* Na$_2$HPO$_4$, 1.25 *M* KH$_2$PO$_4$, 2.5 *M* NH$_4$Cl, 0.25 *M* Na$_2$SO$_4$. Sterilize by autoclaving. Store at room temperature.
5. 24-deep well culture plate (7701-5102, Whatman, Middlesex, UK).
6. AirPore sealing tape (19571, Qiagen, Hilden, Germany).
7. Standard microtiter 96-well plate (655101, Greiner BioOne, Kremsmünster, Austria).

2.3. Protein Expression Analysis

2.3.1. Cell Disruption
1. Lysis buffer: $50\,\text{m}M$ Tris-HCl, pH 7.5, $150\,\text{m}M$ NaCl, 10% glycerol.
2. 24-probe head sonicator (Fisher Bioblock Scientific, Illkirch, France).
3. 96-well polypropylene PCR plate (652270, Greiner BioOne, Philadelphia, PA).
4. Aluminum sealing tape (Abgene).

2.3.2. SDS-Polyacrylamide Gel Electrophoresis (SDS-PAGE)
1. Separating buffer: $1.5\,M$ Tris-HCl, pH 8.8, 0.4% SDS. Store at room temperature.
2. Stacking buffer: $0.5\,M$ Tris-HCl, pH 6.8, 0.4% SDS. Store at room temperature.
3. Running buffer (5×): $125\,\text{m}M$ Tris, $1.25\,M$ glycine, 0.5% (w/v) SDS. Store at room temperature.
4. Staining solution: 0.25% (w/v) Coomassie brilliant blue R-250 (B0149, Sigma Aldrich), 45% (v/v) ethanol, 5% (v/v) acetic acid.
5. Destaining solution: 40% (v/v) ethanol, 10% (v/v) acetic acid.
6. Loading buffer (4×): $120\,\text{m}M$ Tris-HCl, pH 6.8, 4% (w/v) SDS, 16% (v/v) glycerol, 0.08% (w/v) bromophenol blue, $2.96\,M$ β-mercaptoethanol.
7. N,N,N',N'-Tetramethyl-ethylenediamine (TEMED) and 40% acrylamide/bis solution (29:1). Handle with care.
8. Ammonium persulfate solution: prepare 10% (w/v) solution in water and keep at 4°C.
9. Empty sealed Criterion cassette with 26-well comb (345-9903, Bio-Rad, Hercules, CA).
10. Prestained Precision Plus molecular weight marker (161-0373, Bio-Rad).
11. Impact2 eight channels pipettor (2130, Matrix, Hudson, NH).

3. Method

3.1. Competent Cells Preparation

Different procedures were tested to prepare the competent cells for chemical transformation of expression vectors suitable for automation (unpublished results). Good efficiency was obtained for single transformation, as well as for cotransformation with two expression vectors, by preparing the *E. coli* BL21(DE3), or its derivatives, competent cells using a calcium chloride procedure adapted from basic protocols previously described (8,9).

1. Add $150\,\mu\text{L}$ of a glycerol stock to $100\,\text{mL}$ of LB medium in a sterile 500-mL flask. Incubate the preculture for 16 hours at 37°C in a rotary shaker (Infors, Bottmingen, Switzerland) at 190 rpm.
2. Measure the optical density at $600\,\text{nm}$ of the overnight culture and inoculate $1\,\text{L}$ of 2 × LB medium in a 5-L flask at 0.05 $\text{OD}_{600\text{nm}}$ from the overnight culture (see Note 3). Grow cells at 37°C in a rotary shaker at 190 rpm until $\text{OD}_{600\text{nm}} = 0.4$ (see Note 4).
3. Transfer the cells aseptically into four 250-mL prechilled, sterile bottles. Leave the bottles on ice for 20 min.
4. Centrifuge cells at $2,500\,g$ for 15 min at 4°C.
5. Pour off the supernatant and resuspend the cells with $250\,\text{mL}$ of ice-cold $10\,\text{m}M$ NaCl and store for 10 min on ice (see Note 5).

6. Recover the cells by centrifugation as indicated in step 4. Discard the supernatant and resuspend cells in 250 mL of ice-cold 75 mM CaCl$_2$. Keep the resuspended cells on ice for 35 min.

7. Centrifuge cells as in step 4. Discard the supernatant and resuspend the cells in 15 mL of ice-cold 75 mM CaCl$_2$. Transfer the resuspended cells to a 50-mL prechilled, sterile conical tube. Add 3 mL of sterile 87% glycerol, mix gently and leave the cells on ice for 20 min.

8. Dispense cells into prechilled, sterile 0.6 mL PCR tubes (see Note 6) placed on crushed dry ice. Keep aliquots on ice until frozen.

9. Transfer tubes to a dedicated box to be placed in a −80°C freezer (see Note 7).

10. Check competent cells efficiency as well as viability (see Notes 8 and 9).

3.2. Automated Procedures

An automated and integrated protocol was established using a Tecan Workstation robot (Tecan) for competent cell transformation, spreading of transformation mix, and culture inoculation. The program allows users to enter different variables depending on their needs (i.e., number of samples, well positioning, volumes). The robot is equipped with four low-volume needles and four standard-volume needles used depending on the step of the protocol. Even though the program has been written to run the entire protocol for 1 to 96 samples, from cell transformation through culture inoculation, including cell plating, the program allows the different steps to be run independently of each other as well. For more clarity, each step is presented individually under the following subsections.

The robot's deck, designed for the integrated protocol, is presented in Fig. 10.1.

To illustrate this protocol, the procedure was performed with 48 samples (Table 10.1). The results obtained for recombinant protein analysis are displayed in Fig. 10.2.

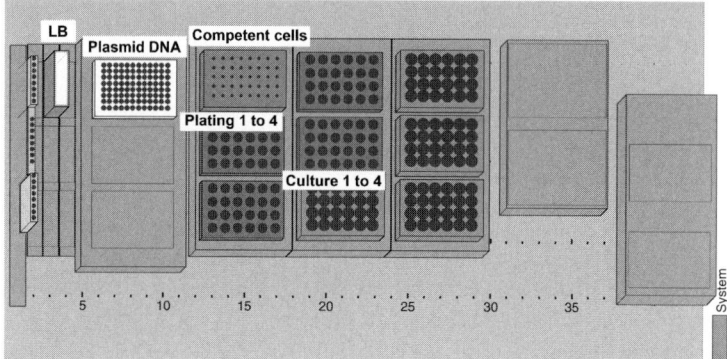

Fig. 10.1 Scheme of the robot's deck. The organization of the Tecan deck is displayed. Competent cells were dispensed into 0.6 mL PCR tubes and stored cold on the prechilled cooling rack (Competent cells, *purple*). Transformation reactions occurred in the MicroAmpR optical 96-well PCR plate (Plasmid DNA, *yellow*). The LB medium was stored in a 100-mL trough (LB, *green*). Transformation mixes were spread onto LB-agar + antibiotic(s) dispensed onto a 24-well culture plate (Plating 1 to 4, *red*) and used to inoculate 2 ml of ZYM-5052 medium + antibiotic(s) dispensed into a 24-deep well culture plate (Culture 1 to 4, *blue*).

Table 10.1 List of proteins selected for expression screening.

Lane	SBGP code	Mr (kDa)
Gel 1		
1	VE1196	109
2	VE1197	93
3	VE1249	18
4	VE1250	42
6	VE1251	58
7	VE1252	30
8	VE1255	15
9	VE1256	35
10	VE1257	55
11	VE1258	27
12	VE1308	40
13	VE1501	39
14	VE2861	62
15	VE2862	38
16	VE0338	31
17	VE1246	73
18	VE1441	73
19	VE1043	16
20	VE1044	56
21	VE1049	28
22	VE1050	68
23	VE1292	17
24	VE1293	42
25	VE1294	57
Gel2		
1	VE1295	72
2	VE1296	29
3	VE1297	41
4	VE1298	70
5	VE1299	27
6	VE1301	55
7	VE1302	70
8	VE1251	58
10	VE1304	41
11	VE1305	53
12	VE1306	83
13	VE1307	40
14	VE1290	63
15	VE1610	18

(continued)

Table 10.1 (continued)

Lane	SBGP code	Mr (kDa)
16	VE1611	61
17	VE0791	113
18	VE1512	46
19	VE1522	28
20	VE1752	110
21	VE0330	85
22	VE0454	85
23	VE2852	54
24	VE1290	63
25	VE1611	61

The proteins selected for the expression screen presented in the study are listed for each gel by lane number (see Fig. 10.2 for the corresponding gels).
The molecular mass (Mr) is approximate.
SBGP = Structural Biology and Genomics Platform.

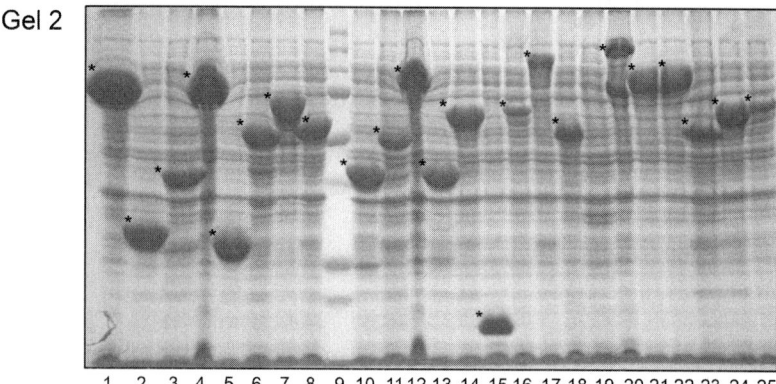

Fig. 10.2 Recombinant protein expression analysis. *E. coli* BL21(DE3) chemically competent cells were transformed with plasmid DNA. Cultures and protein sample preparations were performed as described in Methods. An aliquot (10 μl) of total protein of each sample was loaded onto either a 13.5% (Gel 1) or a 12% (Gel 2) SDS-PAGE. Stars indicate expressed fusion proteins at the expected molecular weight (see Table 10.1). The prestained molecular weight markers (Bio-Rad) consist of 250-, 150-, 100-, 75-, 50-, 37-, 25-, 20-, 15-, and 10-kDa bands from the top to the bottom (Gel 1, lane 5 and Gel 2, lane 9).

3.2.1. Competent Cell Transformation

1. Place the 40-well bench top cooling rack at −20°C for at least 20 min and place it at 4°C before starting the run (see Note 10).
2. Switch on the cooling block of the robot at least 20 min before starting the run.
3. Switch on the PCR machine (programmed to maintain a constant temperature at 45°C).
4. Place the 96-well PCR plate containing the plasmid DNA on the cooling block (see Note 11).
5. Thaw the appropriate volume of competent cells and distribute them into four 0.6 mL PCR tubes placed on the cooling rack (see Note 12). Open the lids.
6. Start the program and select the different variables.
7. The robot will extensively clean the inside and the outside of low volume needles (10 mL + 10 mL) and will operate the appropriate number of "Aspirate-Dispense-Wash" cycles for distributing competent cells (see Note 13).
8. After an incubation time of 20 min, a pop-up window appears asking the user to perform the heat shock step.
9. Seal the plate with aluminum tape and transfer manually the 96-well PCR plate containing the transformation reactions to the PCR machine (see Step 3). Leave the plate for 45 s and place it back on the robot's deck (see Note 14).
10. Place a 100-mL trough, containing at least 20 mL of LB medium, as indicated in Fig. 10.1.
11. Click "OK" on the pop-up window.
12. The robot will extensively clean the inside and the outside of the standard volume needles (10 mL + 10 mL) and will operate the appropriate number of "Aspirate-Dispense-Wash" cycles for distribution of LB medium (see Note 15).
13. Cover the PCR plate with aluminum tape and incubate at 37°C for 1 hour in a 2-mm orbital microplate rotary shaker (Infors) at 350 rpm.
14. Perform plating and/or culture inoculation (see Sections 3.2.2. and 3.2.3., respectively).

3.2.2. Plating of Transformation Reactions

1. Under sterile condition, dispense 1 mL of LB-agar + appropriate antibiotic(s) into each well of a 24-well culture plates and leave the plates at room temperature until the medium solidifies.
2. Place the 24-well culture plates onto the robot's deck (see Fig. 10.1).
3. After the 1-hour incubation at 37°C (see Section 3.2.1., Item 13), place the PCR plate containing the transformation reactions on the cooling rack and remove the aluminum tape.
4. Click "OK" on the pop-up window.
5. The robot will extensively clean the inside and the outside of the standard volume needles (10 mL + 10 mL) and operate the appropriate number of "Aspirate-Dispense-Wash" cycles for plating transformed competent cells (see Note 16).
6. Stack the four culture plates and perform a manual a gentle rotation.
7. Leave the plates at room temperature until the liquid has been adsorbed and invert the plates during incubation at 37°C for 16 hours (see Note 17).

3.2.3. Liquid Culture Inoculation

1. Under sterile conditions, dispense 2 mL of ZYM-5052 medium + antibiotic(s) into each well of the 24-deep well culture plates.
2. Place the 24-deep well culture plates onto the robot's deck (see Fig. 10.1).
3. After the 1-hour incubation at 37°C (see Section 3.2.1., Item 13), place the PCR plate containing the transformation reactions on the cooling rack, and remove the aluminum tape.
4. Click "OK" on the pop-up window.
5. The robot will extensively clean the inside and the outside of the low volume needles (10 mL + 10 mL) and will operate the appropriate number of "Aspirate-Dispense-Wash" cycles for liquid culture inoculation (see Notes 18 and 19).
6. Cover the culture plates with AirPore tape and incubate at 25°C for 42 hours in a microplate rotary shaker (Infors) at 350 rpm (see Note 20).
7. Before harvesting the cells, perform a 1:10 dilution of each culture (10 μL of culture + 90 μL of water) into a 96-well microtiter plate and measure OD_{600nm} using the GENios microplate reader (Tecan) installed on the robot's deck. This step can be automated.
8. Harvest the cells by centrifugation at 2,500 g for 15 min at 4°C. Cell pellets can be stored at −20°C until protein analysis (see Section 3.3.).

3.3. Protein Expression Analysis

3.3.1. Cell Disruption

1. Suspend thawed cell pellets into the 24-deep well culture plates with the appropriate volume of lysis buffer (see Notes 21 and 22).
2. Place 24 deep well culture plates containing the suspended pellet on ice and disrupt cells by sonication for 3 min (pulse 2/2) using a 24-probe head sonicator at 25% intensity.
3. Transfer 20 μL of disrupted cells into a 96-well PCR plate containing 5 μL of loading buffer. Seal the plate with aluminum tape and denature the samples by heating the plate for 5 min at 95°C using a PCR machine.
4. The 24-deep well culture plates can be centrifuged at 3,000 g for 1 hour at 4°C to obtain soluble proteins, which can be analyzed by SDS-PAGE and/or loaded on an affinity resin for initial high throughput protein purification (6).

3.3.2. SDS-PAGE

1. These instructions assume the use of a Criterion empty cassette and running apparatus (Bio-Rad).
2. Prepare a 1-mm thick 12.5% gel by mixing 2.5 mL of separating buffer with 3.75 mL of acrylamide/bis solution, 5.75 mL water, 80 μL ammonium persulfate solution and 13 μL TEMED. Pour the gel leaving space for the stacking gel and overlay with ethanol. The gel should polymerize in about 20 min.
3. Pour out the ethanol and rinse the top of the gel twice with water. Absorb the excess of water using Whatman paper.
4. Prepare the stacking gel by mixing 0.5 mL of stacking buffer with 0.5 mL of acrylamide/bis solution, 3.8 mL water, 50 μL ammonium persulfate solution and 13 μL TEMED. Pour the stacker and insert the comb. The stacking gel should polymerize in about 15 min.

5. Prepare 1× running buffer.
6. Once the stacking gel has set, install the gel in the running apparatus, add running buffer in the upper and lower chambers and carefully remove the comb. Use a 5-ml syringe equipped with a 22-gauge needle to rinse each well with running buffer.
7. Load total proteins previously prepared (see Section 3.3.1., Item 3) using a multichannel pipettor and include one well for the prestained molecular weight marker (Bio-Rad).
8. Complete the assembly of the apparatus and run the gel for about 45 min at 100 V.
9. After running the gel, separate both sides of the cassette using a spatula and stain the gel for 15 min in the staining solution at room temperature under gentle orbital shaking.
10. Remove staining solution and destain the gel for 1 hour at room temperature under gentle orbital shaking.
11. Place the gel in water and take a picture for archiving. Examples of recombinant protein expression profiles for 48 samples are shown in Fig. 10.2.

4. Notes

1. Glycerol stock is prepared by mixing 100 μL of a saturated *E. coli* culture grown in LB medium started from commercial strains (69450-3, Novagen-Merck Biosciences, Darmstadt, Germany) with 50 μL of sterile glycerol 87%. Glycerol stocks are kept at −80°C.
2. The 10 m*M* NaCl and 75 m*M* $CaCl_2$ solutions were prepared fresh by diluting with ultrapure water from a stock solution at 4 *M* and 1 *M*, respectively. After 0.22 μm filtration under sterile conditions, the solutions were kept on ice until used.
3. The culture inoculation at 0.05 OD_{600nm} means that for a 1-L culture, 50 OD_{600nm} (0.05 × 1,000 mL) were needed. Knowing the OD_{600nm} per mL of the overnight culture, it was possible to calculate the volume needed to start the culture. Usually, an OD_{600nm} = 2 to 3 was obtained for the overnight culture resulting in the addition of 25 to 17 mL of 2 × LB per liter, respectively.
4. Nakata et al. have described that high-efficiency transformation by calcium chloride was obtained by harvesting the cells at specific OD_{600nm} (*10*).
5. Resuspension should be performed very gently and cells should be kept on ice.
6. Aliquots of 50 μL are sufficient for one transformation. Nevertheless, for parallel transformations, 200- and 600-μL aliquots were prepared.
7. The competence of cells between aliquots gently frozen at −80°C and aliquots flash frozen into liquid nitrogen before storage were compared and higher efficiency was obtained by placing aliquots at −80°C without flash freezing.
8. Use 0.5 ng of pUC19 plasmid to transform 50 μL of competent cells. Resuspend transformed cells by adding 150 μL of LB medium. After a 1-h incubation at 37°C under shaking, to allow the bacteria to recover and to express the antibiotic resistance marker encoded by the plasmid, 1/5 of the transformation mix is plated onto LB-Agar plate plus ampicillin (0.1 μg/μL). Count the number of transformed colonies and calculate the transformation efficiency in cfu (colony forming unit)/μg DNA. Usually, transformation efficiency of 10^5–10^6 is obtained for *E. coli* BL21(DE3).

9. For the viability, a 50-µL aliquot was plated onto LB-Agar plate without antibiotic and incubated at 37°C for 16 hours. The day after, a uniformed layer of cells should be obtained, suggesting that the cells are viable and not infected by phage.

10. The 40-well bench top cooling rack is placed at −20°C to maintain the temperature during the process. Nevertheless, to avoid keeping the competent cells frozen, the cooling rack is placed at 4°C for 10 minutes before placing the tubes containing the thawed competent cells. The temperature of the cooling rack was followed during a full procedure for 96 samples, and the temperature was not higher than 8°C at the end of the process.

11. Plasmid DNA was dispensed to the bottom of a 96-well PCR plate using a multichannel pipettor. Usually, 1–2 µL (50 ng) of plasmid DNA is used.

12. Calculate the volume of competent cells (Vcc) needed per tube using the following formula:

$$\text{Vcc}\,(\mu L) = ((N+4) \times 25)/4 + 20 \text{ with } N = \text{number of transformation reactions}$$

The authors use 25 µL of competent cell per reaction and four individual tubes, one per needle. Adding four reactions to the total number of transformation reactions allows one to have enough volume. For instance, with 18 reactions, four cycles using the four needles and one cycle using the first two needles were required, meaning that an extra reaction for the first two tubes were necessary. Therefore, to avoid estimating if one, two, or three extra reactions were needed, it was decided to permanently add an extra reaction per tube. The additional 20-µL allowed the circumvention of the dead volume of needles.

13. During the "Aspirate" step, the robot aspirated 25 µL of competent cells from the PCR tubes placed on the cooling rack (see Fig. 10.1). During the "Dispense" step, the robot dispensed the competent cells into four wells of the PCR plate containing plasmid DNA (see Fig. 10.1). Since the needles dispensed the competent cells at the top of each tube avoiding contact with DNA, the robot operated a "Wash" step only after six "Aspirate-Dispense" cycles. During the "Wash" step, the robot cleaned the inside and the outside of each needle with 5 and 2 mL of water, respectively. No cross contaminations occurred.

14. Depending on the throughput, a PCR machine can be installed on the robot deck. This step can be easily automated using the robotic arm to move the plate.

15. During the "Aspirate" step, the robot aspirated 12 × 100 µL of LB medium per needle from the trough (see Fig. 10.1). During the "Dispense" step, the robot dispensed 12 times the LB medium into four wells of the PCR plate containing transformation reactions (see Fig. 10.1). Since the needles dispense the LB medium at the top of each tube, avoiding contact with the transformation reaction, the robot operates a "Wash" step after 12 "Dispense" cycles. During the "Wash" step, the robot cleans the inside and the outside of each needle with 5 and 2 mL of water, respectively.

16. The robot initially performs a mixing step to homogenize the cells. During the "Aspirate" step, 50 µL of transformation mix is taken from the PCR plate (see Fig. 10.1). During the "Dispense" step, the robot removed the cover of the 24-well culture plate and dispenses the transformation mix onto four wells of the 24-well culture plate containing LB-Agar medium + antibiotic(s) (see Fig. 10.1). During the "Wash" step, the robot cleans the

inside and the outside of each needle with 5 and 2 mL of water, respectively. Once the 24 wells of the plate have been completed, the cover is replaced and the procedure continued with the next plate.

17. Since the inoculation of liquid culture is performed directly from the transformation mix, the plating step is not required saving time and eliminating error prone colony picking *(11)*. Nevertheless, having plated colonies avoids retransformation for future cultures.

18. The robot initially performs a mixing step to homogenize the cells. During the "Aspirate" step, 10 μL of transformation mix is taken from the PCR plate (see Fig. 10.1). During the "Dispense" step, the robot dispenses the transformation mix into four wells of the 24-deep well culture plate containing ZYM-5052 medium + antibiotic(s) (see Fig. 10.1). During the "Wash" step, the robot cleans the inside and the outside of each needle with 5 and 2 mL of water, respectively. The protein expression profile was similar for cultures inoculated directly with the transformation mix compared to that of classical strategies, in which either an individual colony or a starter culture was used (*12*).

19. When plating and liquid culture inoculation are done at once, low-volume needles are used and the robot combined procedure described in Notes 16 and 18. Briefly, the robot initially performs a mixing step to homogenize the cells. During the "Aspirate" step, 60 μL of transformation mix is taken. During the "Dispense" step, 50 and 10 μL of the transformation mix are dispensed onto four wells of the 24 well culture plate containing LB-Agar medium + antibiotic(s) and into four wells of the 24 deep well culture plate containing ZYM-5052 medium + antibiotic(s), respectively. During the "Wash" step, the robot cleaned the inside and the outside of each needle with 5 and 2 mL of water, respectively.

20. Growing cells at 25°C lowers the metabolism rate of *E. coli* cells. We checked that recombinant protein expression was optimal at 25°C after 42 hours of cultivation (*12*).

21. Calculate the volume of lysis buffer (Vlb) to be added on cell pellet using the following formula:

$$\text{Vlb (μL)} = (OD_{600nm}/2) * Vc * 100 \text{ with } Vc = \text{volume of culture in mL}$$

22. This step can be automated using the results of OD_{600nm} measurement obtained in Section 3.2.3., Item 7, which are presented by the GENios microplate reader (Tecan) into an Excel spreadsheet, and used as variables for the robot.

Acknowledgments

The authors give special thanks to Dr. Rosalind Kim for critical reading of the manuscript and for useful discussions. The authors also thank David Rosé for his input in the initial stages of the protocol development and for conducting the comparison among different procedures to prepare competent cells for chemical transformation. The Structural Biology and Genomics Department is thanked for supplying plasmid templates. This work was supported by funds from RNG through the Genopole program and SPINE EEC QLG2-CT-2002-00988.

References

1. Hunt, I. (2005) From gene to protein: a review of new and enabling technologies for multi-parallel protein expression. *Protein Expr. Purif.* **40**, 1–22.
2. Baneyx, F. (1999) Recombinant protein expression in *Escherichia coli*. *Curr. Opin. Biotechnol.* **10**, 411–421.
3. Yokoyama, S. (2003) Protein expression systems for structural genomics and proteomics. *Curr. Opin. Chem. Biol.* **7**, 39–43.
4. Studier, F. W. (2005) Protein production by auto-induction in high density shaking cultures. *Protein Expr. Purif.* **41**, 207–234.
5. Busso, D., Thierry, J. C., and Moras, D. (in press) The Structural Biology and Genomics Platform in Strasbourg: an overview. *Methods Mol. Biol.*
6. Busso, D., Delagoutte-Busso, B., and Moras, D. (2005) Construction of a set Gateway-based destination vectors for high-throughput cloning and expression screening in *Escherichia coli*. *Anal. Biochem.* **343**, 313–321.
7. Busso, D., Poussin-Courmontagne, P., Rose, D., Ripp, R., Litt, A., Thierry, J. C., and Moras, D. (2005) Structural genomics of eukaryotic targets at a laboratory scale. *J. Struct. Funct. Genomics* **6**, 81–88.
8. Mandel, M., and Higa, A. (1970) Calcium-dependent bacteriophage DNA infection. *J. Mol. Biol.* **53**, 159–162.
9. Dagert, M., and Ehrlich, S. D. (1979) Prolonged incubation in calcium chloride improves the competence of *Escherichia coli* cells. *Gene* **6**, 23–28.
10. Nakata, Y., Tang, X., and Yokoyama, K. K. (1997) Preparation of competent cells for high-efficiency plasmid transformation of *Escherichia coli*. *Methods Mol. Biol.* **69**, 129–137.
11. Berrow, N. S., Büssow, K., Coutard, B., Diprose, J., Ekberg, M., Folkers, G. E., Levy, N., Lieu, V., Owens, R. J., Peleg, Y., Pinaglia, C., Quevillon-Cheruel, S., Salim, L., Scheich, C., Vincentelli, R., and Busso, D. (2006) Recombinant protein expression and solubility screening in *Escherichia coli*: a comparative study. *Acta Crystallogr.* **D62**, 1218–1226.
12. Busso, D., Stierlé, M., Thierry, J.C., and Moras, D. (submitted) A comparison of inoculation methods to simplify recombinant protein expression screening in *Escherichia coli*. *Biotechniques.*

Chapter 11

From No Expression to High-Level Soluble Expression in *Escherichia coli* by Screening a Library of the Target Proteins with Randomized N-Termini

Kyoung Hoon Kim, Jin Kuk Yang, Geoffrey S. Waldo,
Thomas C. Terwilliger, and Se Won Suh

For structural studies by x-ray crystallography and nuclear magnetic resonance it is important for the target protein to be available in large quantity and high purity. *Escherichia coli* expression systems remain the most versatile and convenient means to produce a large quantity of recombinant proteins. Unfortunately, some proteins fail to be expressed in *E. coli* or are expressed in an insoluble form. To overcome the difficulty of no expression or expression at a very low level, a simple and efficient approach of screening a library of variants of a target protein with randomized N-termini was devised. In this method, a few N-terminal residues are randomized by designing a mixture of oligonucleotides for the forward PCR primer and we fuse the library in front of green fluorescent protein, which serves as a reporter for the target protein expression level and folding yield. In favorable cases this approach can result in high-level soluble expression of recombinant proteins in *E. coli*. This chapter describes the results of a test of this approach with a bacterial protein (the HI0952 gene product) that is not well expressed in *E. coli*.

1. Introduction

Production of properly folded proteins or their isolated domains is a prerequisite for the detailed structural and functional characterization. For instance, structural studies of proteins by x-ray crystallography or nuclear magnetic resonance normally require significant amounts (5–50 mg) of a target protein in high purity. This is most often accomplished by expression of the target protein in soluble form using a suitable heterologous host. *Escherichia coli* is still the first choice for protein expression as it offers many advantages over other expression hosts in terms of speed, cost effectiveness, and convenience. Many expression vectors for heterologous expression in *E. coli* are already available and new vectors continue to be developed.

From: *Methods in Molecular Biology, Vol. 426: Structural Proteomics: High-throughput Methods*
Edited by: B. Kobe, M. Guss and T. Huber © Humana Press, Totowa, NJ

E. coli has some limitations as an expression host. One of the limitations is that some proteins fail to be expressed in *E. coli* or their level of expression or solubility is very low when they are expressed. Causes of such failure in protein expression in *E. coli* may include the toxicity of the target protein, improper protein folding, the codon usage problem, instability of the mRNA, susceptibility of the expressed protein to proteolysis, and the hairpin formation at the 5'-side of the mRNA. Several strategies are available to overcome some of these difficulties. Rosetta cells (Novagen) provide tRNAs corresponding to infrequently used codons. C41 and C43 cells, derivatives of BL21, were isolated for producing toxic proteins *(1)*. mRNA secondary structures at the 5'-side can be minimized by silent mutations of the gene *(2)*.

Alterations at the N-terminus of the target protein are well known to affect heterologous protein expression *(3–6)*. It was observed that NGG codons at positions +2, +3, and +5 downstream of the initiation codon lower the gene expression in *E. coli* at the translational level and it has been suggested that the low expression is not the result of mRNA secondary structure or a lowered intracellular mRNA pool *(7)*. As a specific example of altering the N-terminal region for enhancing the expression level, sequences within the first seven amino acid codons of the eukaryotic membrane protein (bovine cytochrome P450 17α-hydroxylase) were altered to optimize the expression in *E. coli* *(8)*. However, such methods are time consuming and usually require a large number of trials.

Here a simple and efficient approach was devised to reduce the problem of no expression or very low-level expression in *E. coli*. In this method, a few N-terminal residues of the target protein were randomized by employing a mixture of oligonucleotides as the 5'-primer for polymerase chain reaction (PCR) and fuse the library in front of green fluorescent protein, which is used as the reporter for the expression level and folding yield. Then the library of the target protein with randomized N-terminus is screened for the clones that express the target protein at high levels in a soluble form. In favorable cases this approach has the potential to enable expression of the recombinant protein in *E. coli* both at a sufficiently high level and in a soluble form. This approach was tested with a bacterial protein (HI0952 gene product) that was not well expressed in *E. coli*. GFP fusions can be expressed in a variety of cell types, so in principle this method is applicable to protein expression in other hosts, provided that methods for highly efficient transformation of host cells are available.

2. Materials

2.1. Enzymes and Proteins

1. T4 DNA ligase (Takara, Shiga, Japan).
2. NdeI (DCC Bionet, Sungnam, Korea).
3. BamHI (DCC Bionet, Sungnam, Korea).
4. High-fidelity thermophilic DNA polymerase (Phusion DNA polymerase from Finnzymes, Keilaranta, Finland).

2.2. Plasmids, Nucleotides, Gels, and Kits

1. Green fluorescent protein (GFP) folding reporter vector (X-FR) *(9)*.
2. pET-21a(+) (Merck, Darmstadt, Germany).

3. Oligonucleotide primers (Bioneer, Daejon, Korea) (see Note 1).
4. Template DNA for PCR: For cloning the HI0952 gene, the genomic DNA was obtained from ATCC.
5. 1% (w/v) Agarose gel (Bioworld, Dublin, OH).
6. 12.5% (w/v) SDS-PAGE gel.
7. Qiaquick spin gel extraction kit (Qiagen, Hilden, Germany).

2.3. Cells and Media

1. *E. coli* strain ElectroMAX DH10B (Invitrogen, Grand Island, NY).
2. *E. coli* strain BL21(DE3) (Invitrogen).
3. *E. coli* strain Rosetta II(DE3)pLysS (Invitrogen).
4. Luria-Bertani (LB) medium (BD Sciences, Franklin Lakes, NJ).
5. Kanamycin stock solution (35 mg/ml): 1.75 g of kanamycin in 50 ml of sterile water. Aliquot in 1 ml fractions and store at −20°C.
6. Isopropyl-β-D-thio-galactopyranoside (IPTG) stock solution (1.0 M): 4.8 g IPTG in 20 ml of sterile water. Aliquot in 1 ml fractions and store at −20°C.
7. SOC: 2% Bacto tryptone, 0.5% Bacto yeast extract, 10 mM NaCl, 2.5 mM KCl, 10 mM $MgCl_2$, 10 mM $MgSO_4$, 20 mM glucose.

2.4. Equipment

1. Microcentrifuge.
2. PCR machine (MJ Research Model PTC-200).
3. MicroPulser electroporation apparatus (Bio-Rad, Hercules, CA).
4. Gene Pulser cuvette: 0.2 cm electrode gap (Bio-Rad, catalog no. 165–2086).
5. Nitrocellulose membrane, polyester reinforced (GE Osmonics, Minnetonka, MN): 85 mm, 137 mm diameter.
6. Kirby-Bauer plate: 90 mm, 150 mm diameter.
7. Illumatool tunable lights system, 488 nm excitation filter and 520 nm emission filter (Lightools Research, Encinitas, CA).
8. Digital camera with a stand.

3. Methods

3.1. Library Construction of the Target Protein with Randomized N-Terminus

1. For the forward primer, design a mixture of oligonucleotides that contain NNY for the positions +2, +3, and +4 downstream of the translation initiation codon, where N is any base (A, T, G, or C) and Y is a pyrimidine base (C or T) (see Notes 1 and 2).
2. Amplify the target gene by running the PCR reaction (see Note 3).
3. The PCR product is digested with NdeI and BamHI.
4. The digested PCR product is inserted into the NdeI/BamHI-digested GFP folding reporter vector by T4 DNA ligase.

3.2. Amplification of Library Plasmid

The GFP folding reporter vector containing the library of the target protein with randomized N-terminus is transformed into the *E. coli* nonexpression strain DH10B by the electroporation method (see Note 4) and the amount of the plasmid library is amplified. The amplified plasmid library is then isolated for transformation into the *E. coli* expression strain BL21(DE3) by the chemical method. The following protocols for library transformation and recovery of the amplified library are a slight modification of those already described by Waldo *(10)*.

3.2.1. Library Transformation into DH10B by Electroporation

1. Thaw 100 µl aliquot of DH10B cells on ice (10–15 min).
2. Prechill 100 µl dd H_2O on ice (10 min) and three Gene Pulser cuvettes per library.
3. Combine 100 µl thawed DH10B and 50 µL pre-chilled dd H_2O, add the diluted DH10B to same tube on ice containing the 10 µL resuspended, ligated DNA, flick to gently mix, incubate 5 min on ice.
4. Transfer 50 µL of transformation mix in each Gene Pulser cuvette (see Note 4).
5. Recover each of the three transformations by immediately resuspending in 1 ml SOC in 12 ml culture tube with shaking at 37°C for 1.5 hours.
6. Centrifuge the recovered cultures in 1.5-ml Eppendorf tubes for 1 min at 14,000 rpm.
7. Leave 200 µL of supernatant over the pelleted cells (remove ~ 800 µl supernatant). Resuspend cells by pipetting.
8. Pool the three tubes into one 1.5-ml Eppendorf. The total volume will be about 600 µl.
9. Plate the library onto a large Kirby-Bauer plate (150 mm) selective media LB plate supplemented with 35 µg/ml kanamycin (LB-Kan).
10. On a separate counting agar plate, plate 1/300th of library (2 µl of the 600 µl pooled transformations into pool of ~ 200 µl SOC onto a standard LB/Kan selective plate to get the counts). Incubate 12–16 hours at 37°C (see Note 5).

3.2.2. Library Plasmid Recovery

1. Add 12 ml LB to the Kirby-Bauer DH10B library plate from plates in Section 3.2.1., resuspend with spreader.
2. Transfer suspension to 15-ml Falcon tube, vortex to suspend.
3. Perform QIAgen plasmid prep on 750 µl of cell suspension. Cell mass prepped should be equivalent to 3 ml of overnight LB culture, i.e., approx. 75–100 mg pellet.

3.3. Screening of the Clones on the Basis of Gfp Fluorescence

1. Prepare nitrocellulose membranes (137 mm).
2. Label then wet the membranes (see Note 6). Place on towels to dry until damp only.
3. Lay on an LB-Kan Kirby-Bauer plate (150 mm), excluding all air pockets.
4. Incubate at 37°C for at least 1 hour prior to plating.
5. The plasmid library (1 µl) is transformed into BL21(DE3) competent cells (50 µl) by the heat shock protocol (42°C, 1 min) (see Note 7).
6. The transformed cells are plated directly onto nitrocellulose membranes (137 mm) on an LB-agar plate (150 mm). After the membrane is dry, invert the plate, and incubate at 37°C for 10–15 hours until colonies are approximately 1 mm in diameter (see Note 7).

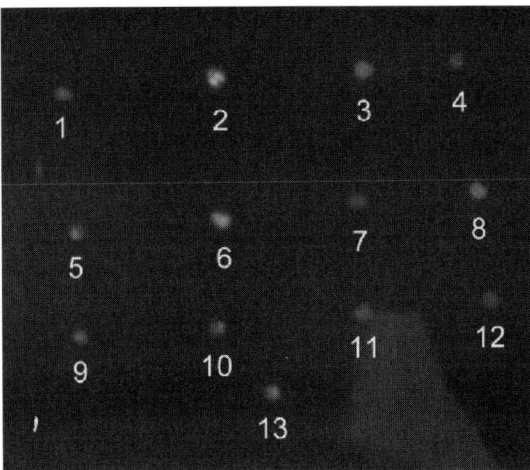

Fig. 11.1 Colonies of the selected HI0952 variants showing significant fluorescence improvements over the wild type, whose GFP fusion is almost nonfluorescent. Among these 13 colonies, colonies #2, #3, #5, #6, #8, and #13 were chosen for further characterization. The wild type colony is not included in this figure, because it is invisibly dark.

7. After incubation, the membrane is transferred onto an LB-agar plate (150 mm) containing 2 mM IPTG and is incubated for 5–6 hours for induction.
8. Pick the 30 brightest colonies and transfer them onto the master LB-agar plate (90 mm).
9. The master LB-agar plate (90 mm) is incubated at 37°C for 14–16 hours and its replica is made on a nitrocellulose membrane (85 mm).
10. The replica membrane is incubated on an LB-agar plate (90 mm) at 37°C for 8–10 hours, transferred onto an LB-agar plate containing 2 mM IPTG, and incubated for an additional 4–6 hours for induction.
11. Select colonies showing significant fluorescence improvements over the wild-type and the cell mass of the selected colonies on the plate is recovered. For HI0952, we selected 13 colonies (Fig. 11.1).

3.4. Test of Protein Expression as a Fusion with GFP

1. The selected variant colonies are overexpressed in BL21(DE3) cells.
2. Cells are grown at 37°C up to OD_{600} of 0.5 in LB medium containing 35 μg/ml kanamycin.
3. The protein expression is induced by 1.0 mM IPTG.
4. Cells are grown for additional 18 hours at 30°C.
5. The expression levels of the GFP fusion protein for the selected colonies are estimated by running SDS-PAGE on the total *E. coli* proteins.
6. Choose a few colonies that have a reasonably high expression level. As an example, six colonies were chosen for HI0952 (colonies #2, #3, #5, #6, #8, and #13 in Fig. 11.1).
7. The solubility of the GFP fusion protein for the selected colonies is estimated by running SDS-PAGE on the soluble and insoluble fractions of the total *E. coli* proteins (see Note 8).

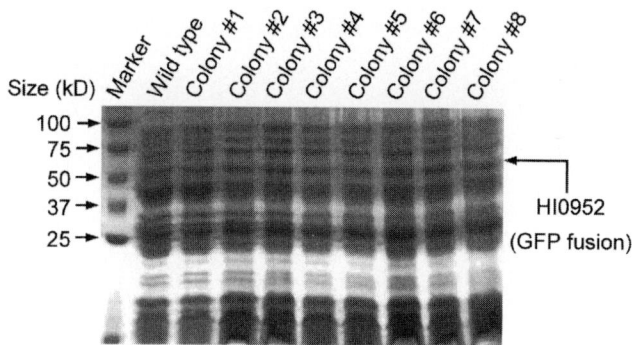

Fig. 11.2 SDS-PAGE analysis of HI0952 variants with C-terminally fused GFP. The colony #3 shows both bright fluorescence and a high expression level.

8. Choose a colony that has both a high degree of soluble expression and a reasonably high expression level. As an example, the colony #3 was chosen for HI0952 (Fig. 11.2).
9. Purify the GFP folding reporter plasmid containing the desired variant.

3.5. Test of Protein Expression Without GFP Fusion

1. The variant gene is amplified by high-fidelity PCR using the purified GFP plasmid as the template and the appropriate primers (see Note 9) for sub-cloning into an expression vector without GFP fusion.
2. The amplified PCR product is digested with NdeI and BamHI.
3. The digested PCR product is inserted into the pET-21a(+) expression vector, which has been already digested with the same restriction enzymes.
4. The plasmid pET-21a(+) carrying the desired variant gene is transformed into BL21(DE3) cells by the standard calcium chloride method.
5. The transformed cells are grown at 37°C up to OD_{600} of 0.5 in LB medium containing 35 μg ml^{-1} kanamycin.
6. The protein expression is induced by 1.0 mM IPTG.
7. Cells are grown for additional 18 hours at 30°C.
8. The expression level and solubility of the variant protein are assessed by running SDS-PAGE on soluble and insoluble fractions of the total *E. coli* protein. An example of SDS-PAGE analysis is shown in Fig. 11.3.
9. The mutations introduced to the N-terminus are determined by performing DNA sequencing (see Note 10).
10. If necessary, the conditions for soluble expression of the variant protein can be further optimized, for example, by expressing it in different *E. coli* cells (e.g., BL21(DE3)pLysS or Rosetta II (DE3)pLysS) or at lower temperatures (e.g., 15°C). The variant protein can be evolved for further improvement in solubility using the split-GFP *in vivo* solubility reporter system *(11)*.

4. Notes

1. Primer sequences designed for PCR (sub)cloning of the HI0952 gene are as follows. The underlined bases represent NdeI and BamHI digestion sites. The rare codon CGG was changed for Arg2 into CGT in the forward primer #1b for the wild-type to reduce the codon usage problem.

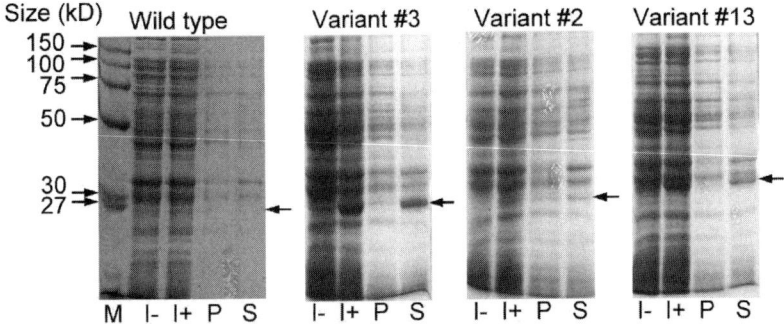

Fig. 11.3 SDS-PAGE analysis of HI0952 wild-type and variants (nonfusion), which are indicated by the arrows on the right side of each gel. The #3 variant shows the best solubility compared to other variants and the wild-type. I–, before IPTG induction; I+, after IPTG induction; P, pellet fraction; S, supernatant fraction.

- #1a (Forward primer for constructing a library of proteins with randomized N-terminus):
  G GAA TTC <u>CAT ATG</u> NNY NNY NNY TCT TTC TTA TTT TTC TTT TAT AAA TAT ATG (52-mer), where N is any base (A, T, G, or C) and Y is a pyrimidine base (C or T).
- #1b (Forward primer for wild type):
  G GAA TTC <u>CAT ATG</u> CGT TCA AAA TCT TTC TTA TTT TTC TTT TAT AAA TAT ATG (52-mer).
- #2 (Reverse primer for both wild type and the variant library; reverse primer for subcloning the selected variant into pET-21a(+)):
  CGC CGC <u>GGA TCC</u> TAA TAA ACA ATT CTC CGC GAA AGA A (37-mer).
- #3 (Forward primer for subcloning the selected variant into pET-21a(+)):
  CTA ATA CGA CTC ACT ATA GGG G (22-mer). This primer covers a T7 promoter region.

2. By limiting the third position of the codon to be either C or T we can avoid introducing stop codons and some of the rare codons in *E. coli*. That is, we exclude all codons for five amino acids (methionine, glutamine, glutamate, lysine, and tryptophan) and stop codons by limiting the codon to NNY. The designed primer mixture contains $(4 \times 4 \times 2)^3 = 32,768$ sequences, which correspond to $(15 \times 15 \times 15) = 3,375$ protein sequences. By limiting the codon to NNY, we also exclude the NGG codons at positions +2 and +3, which were found to lower the gene expression in *E. coli* at the translational level *(7)*. It is reasonable to assume that the structure and function of the recombinant protein will be affected only insignificantly by altering the second, third, and fourth amino acids in most cases. This is because the altered positions are generally variable in sequence.

3. For HI0952, we used the following protocol ("touchdown" PCR) programmed into a MJ Research Model PTC-200 PCR machine:

Step 1: 98°C, 30 seconds
Step 2: 98°C, 10 seconds

Step 3: 50°C, 20 seconds (Increase the temperature by 1°C per cycle.)
Step 4: 72°C, 20 seconds*
Step 5: Steps 2 through 4 are repeated 15 times
Step 6: 98°C, 10 seconds
Step 7: 58°C, 20 seconds
Step 8: 72°C, 20 seconds*
Step 9: Steps 6–8 are repeated 30 times.
Step 10: 72°C, 5 minutes
Step 11: 4°C

Details of the PCR reaction mixture:

35 µl ddH2O
10 µl 5 × High-fidelity buffer (Finnzymes)
1.0 µl 10 mM dNTPs (2.5 mM each dNTP)
0.5 µl Phusion DNA polymerase (Finnzymes)
0.5 µl Forward primer (100 pmol/µl stock)
0.5 µl Reverse primer (100 pmol/µl stock)
1.0 µl Template DNA (50 ng/µl)
1.5 µl DMSO
50 µl Total volume

4. Set electroporator to 2.5 kV. If arcing occurs and is outside the cuvette, recover mix and perform pulse in a fresh cuvette.
5. If fewer than 200 colonies are observed on the counting plate, the variant library may not be as complete as desired. In such a case, try to adjust the ratio of the GFP folding reporter plasmid to the PCR product to obtain more than 200 colonies.
6. It is necessary to preincubate the nitrocellulose membrane in distilled water for 5 minutes in order to avoid formation of air bubbles between the membrane and the plate, which adversely affect the growth of the colonies.
7. Depending on the competency of the BL21(DE3) strain, 1 µl of a "standard" plasmid prep could yield several hundred thousand colonies. One option is to make the BL21(DE3) expression library essentially a lawn, then washing the viable cells off the plate and diluting sequentially twice 350-fold and plating 1 ml of this per Kirby-Bauer plate (150 mm) for actually picking. This would add one extra step per day. Otherwise, one could do a set of dilution plates in step 6 of Heading 3.3, and use the one showing single colonies for picking. Since the library size is relatively small, this places an estimate on the number of clones needed to get good representation (see Note 5).
8. Fluorescence of the C-terminal GFP in the fusion context reflects the expression level of the fusion protein as well as the lack of misfolding by the upstream fusion partner protein. Therefore, increased fusion fluorescence is correlated with increased expression and solubility of the test protein in the nonfusion context. Note that GFP-fusions can have reduced solubility, so it is essential to test the solubility of the selected protein variants without the fused GFP.

---

*For Phusion DNA polymerase (Finnzymes), the supplier recommends an extension time of 15 seconds per 1 kb for low complexity DNAs (e.g., plasmid, lambda, or BAC DNA) and 30 seconds per 1 kb for high-complexity genomic DNAs. A longer extension time than recommended by the supplier is preferred for this PCR reaction.

9. Forward primer: #3 primer (see Note 1). Reverse primer: #2 primer (see Note 1).

10. Results of DNA sequencing of some variants of HI0952 are as follows.

Wild-type: ATG CGT TCA AAA (MRSK)
Colony #2: ATG ACC TGT GAC (MTCD)
Colony #3: ATG CGC TGT TGC (MRCC)
Colony #8: ATG GGC CTT GCC (MGLA)
Colony #13: ATG GGT CTT GAT (MGLD)

Acknowledgments

This work was supported by the KOSEF Basic Research Program grant funded by the Korea Ministry of Science and Technology (R01-2006-000-10311-0).

References

1. Miroux, B., and Walker, J. E. (1996) Over-production of proteins in *Escherichia coli*: mutant hosts that allow synthesis of some membrane proteins and globular proteins at high levels. *J. Mol. Biol.* **260**, 289–298.

2. Cèbe, R., and Geiser, M. (2006) Rapid and easy thermodynamic optimization of the 5'-end of mRNA dramatically increases the level of wild type protein expression in *Escherichia coli. Protein Expr. Purif.* **45**, 374–380.

3. Kis, M., Burbridge, E., Brock, I. W., Heggie, L., Dix, P. J., and Kavanagh, T. A. (2004) An N-terminal peptide extension results in efficient expression, but not secretion, of a synthetic horseradish peroxidase gene in transgenic tobacco. *Ann. Bot. (Lond)* **93**, 303–310.

4. Orchard, S. S., and Goodrich-Blair, H. (2005) An encoded N-terminal extension results in low levels of heterologous protein production in *Escherichia coli. Microb. Cell. Fact.* **21**, 22–31.

5. Sati, S. P., Singh, S. K., Kumar, N., and Sharma, A. (2002) Extra terminal residues have a profound effect on the folding and solubility of a *Plasmodium falciparum* sexual stage-specific protein over-expressed in *Escherichia coli. Eur. J. Biochem.* **269**, 5259–5263.

6. Sawant, S. V., Kiran, K., Singh, P. K., and Tuli, R. (2001) Sequence architecture downstream of the initiator codon enhances gene expression and protein stability in plants. *Plant Physiol.* **126**, 1630–1636.

7. Gonzalez de Valdivia, E. I., and Isaksson, L. A. (2004) A codon window in mRNA downstream of the initiation codon where NGG codons give strongly reduced gene expression in *Escherichia coli. Nucleic Acids Res.* **32**, 5198–5205.

8. Barnes, H. J., Arlotto, M. P., and Waterman, M. R. (1991) Expression and enzymatic activity of recombinant cytochrome P450 17α-hydroxylase in *Escherichia coli. Proc. Natl. Acad. Sci. USA* **88**, 5597–5601.

9. Waldo, G. S., Standish, B. M., Berendzen, J., and Terwilliger, T. C. (1999) Rapid protein-folding assay using green fluorescent protein. *Nature Biotechnol.* **17**, 691–695.

10. Waldo, G. S. (2003) Improving protein folding efficiency by directed evolution using the GFP folding reporter. *Methods Mol. Biol.* **230**, 343–359.

11. Cabantous, S., Pedelacq, J. D., Mark, B. L., Naranjo, C., Terwilliger, T. C., and Waldo, G. S. (2005) Recent advances in GFP folding reporter and split-GFP solubility reporter technologies. Application to improving the folding and solubility of recalcitrant proteins from *Mycobacterium tuberculosis. J. Struct. Funct. Genomics* **6**, 113–119.

Chapter 12

Application of High-Throughput Methodologies to the Expression of Recombinant Proteins in *E. coli*

Yoav Peleg and Tamar Unger

Despite the large body of knowledge accumulated on recombinant protein expression, production, primarily of eukaryotic proteins, remains a challenge. The biggest obstacle is in obtaining large amounts of a given protein in a correctly folded form. Several strategies are being used to increase both yields and solubility. These include expression with fusion proteins, co-expression with molecular chaperones or a protein partner, and use of multiple constructs for each protein. Any given method may help to increase expression and solubility for a given protein, but often more than one rescue strategy should be tried. To perform several different rescue strategies on multiple proteins, high throughout (HTP) methodologies are applied. This chapter presents HTP methodologies for DNA cloning in multiple expression vectors and expression screening to identify clones capable of producing soluble proteins.

1. Introduction

Three-dimensional (3D) structure determination of proteins is a crucial tool for studying protein function and, downstream, for designing lead compounds for drug development. The first bottleneck in the pipeline en route to structure determination is the production of adequate amounts of soluble and biologically active protein, viz. mg quantities. The *E. coli* expression system is the most widely used for production of proteins, whether for structure determination or functional characterization. However, researchers often encounter difficulties in expression in the *E. coli* system, primarily when using genes of eukaryotic origin. These include formation of inclusion bodies, degradation of the protein, low levels of expression, and production of a nonfunctional protein.

The principal goal of the Israel Structural Proteomics Center (ISPC) is to determine structures of proteins related to human health in their functional context *(1)*. Emphasis is being placed on solving structures of proteins complexed with their natural partner proteins. The ISPC is producing recombinant proteins for structural analysis by x-ray crystallography or low-resolution electron microscopy. In addition, it is producing proteins for use in a variety of biological, biochemical, and biophysical studies. Production of proteins for x-ray crystallography raises additional layers of complexity than mentioned in

From: *Methods in Molecular Biology, Vol. 426: Structural Proteomics: High-throughput Methods*
Edited by: B. Kobe, M. Guss and T. Huber © Humana Press, Totowa, NJ

the preceding; thus, crystal formation and/or crystal quality may be markedly affected by the characteristics of the protein preparation tested. Subjecting a soluble, monomeric, and monodispersed protein to crystallization trials does not guarantee crystal formation, nor that crystals, if obtained, will diffract well. Thus, obtaining a correctly folded and soluble protein in substantial amounts is certainly a prerequisite for crystallization; but in most cases, many expression trials, including variations such as changes in the construct, or even mutagenesis of the native protein, may be needed before a protein structure is solved.

Eukaryotic proteins are difficult targets for expression *(2)*. In certain cases obtaining soluble active protein requires correct processing of the produced protein (i.e., posttranslational modification). Therefore, the ISPC has adopted several strategies to cope with the obstacles encountered. These include expression in *E. coli* of domains of the intact protein and variation of the boundaries of the full-length as well as individual domains; coexpression of the protein with one or more molecular chaperones, and expression in eukaryotic hosts such as yeast and baculovirus.

Proteins often function in their natural environment via interaction with other proteins. Therefore, expression of an individual protein in a heterologous system may result in the production of an inactive protein (e.g., an intrinsically unfolded protein). By coexpression of a protein with its natural partner, a stable folded protein may form as part of a complex, which, subsequent to purification, will then be suitable for crystallization trials *(3)*.

The procedures described in the following outline HTP methodologies for expression and screening of histidine-tagged soluble recombinant proteins in *E. coli*. These HTP methodologies allow handling of a large number of samples in parallel, and facilitate the identification of clones expressing soluble proteins *(4–6)*; thus, permitting testing of a large number of protein variants in crystallization screens, thereby increasing the likelihood of obtaining well-diffracting crystals and, as a consequence, subsequent structure determination.

2. Materials

2.1. Strains

For such general procedures as subcloning and plasmid preparation, the DH5α strain of *E. coli* (Stratagene, La Jolla, CA) is used. For protein expression, the BL21(DE3) strain (Novagen San Diego, CA) is employed. For many proteins very little difference has been found in expression between various BL21-derived strains tested. For proteins containing several disulfide bonds, the Origami B(DE3) strain (Novagen) is used.

2.2. Expression Vectors

2.2.1. Expression Vectors with Fusion Partners
On a routine basis the authors are using a series of expression vectors based on the backbone of the pET vectors (Novagen). They have modified the pET28a(+) vector by introducing a linker sequence of a *Sac*II restriction site, and a tobacco etch virus (TEV) protease recognition site, followed by *Kpn*I and *Bam*HI restriction sites. This modified vector (designated

A

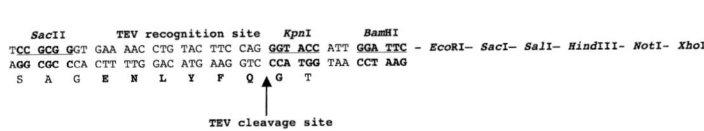

B

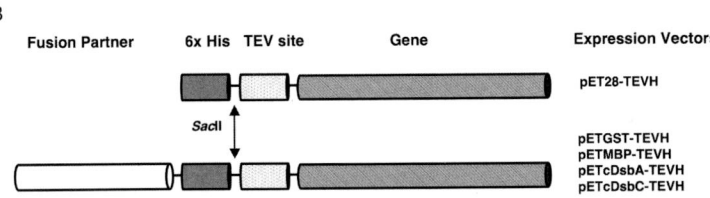

Fig. 12.1 pET-derived expression vectors harboring various fusion partners. **A.** Sequence of the *Sac*II-*Bam*HI cassette present in the various expression vectors. The cassette, including the TEV protease recognition site (amino acids marked in bold), was inserted down stream to the 6× His tag. The restriction sites listed downstream of the *Bam*HI site are part of the original pET vector, and are available for cloning of the PCR product. **B.** Schematic representation of the various fusion tags. The expression vectors used for HTP screening are listed at the right. The arrow indicates the *Sac*II site in the expression vectors.

pET28-TEVH) contains an N-terminal 6× His-tag followed by the newly introduced TEV cleavage site (Fig. 12.1). The same linker was introduced into several other expression vectors harboring various fusion tags (see Fig. 12.1). The fusion partners employed, viz. GST, MBP, DsbA, and DsbC, are intended to increase the solubility and/or to facilitate disulfide bond formation. Fusion proteins carrying DsbA and DsbC can be either expressed in the periplasm, using the pETDsbA-TEVH and pETDsbC-TEVH vectors, respectively, or in the cytoplasm, using the pETcDsbA-TEVH and pETcDsbC-TEVH vectors, respectively. In these latter expression vectors the periplasmic secretion signal is removed. For HTP applications, using the procedures described in the following, only cytoplasmic expression vectors are being used. Engineering the same restriction sites at corresponding positions in all the plasmids (see Fig. 12.1) permits parallel cloning of the PCR product into several different expression vectors. Kanamycin is used as a selection marker for all the expression vectors constructed.

2.2.2. Expression Vectors for Protein Co-expression

For co-expression, Duet vectors (Novagen) pACYDuet-1 and pETDuet-1 are being used. Two individual genes can be cloned into each expression vector. Each gene is under the control of the T7 promoter, resulting in multiple RNA transcripts from a single plasmid. Each vector harbors a 6× His-tag upstream from the first of the multiple cloning sites, thus permitting affinity purification of the two protein partners. The two vectors vary in their antibiotic resistance (chloramphenicol vs. ampicillin), copy number of the plasmids (low vs. medium) and in the origin of replication. The two vectors can be cotransformed into an appropriate bacterial strains (i.e., BL21(DE3)) for co-expression of up to four target genes.

2.3. Construction of Expression Vectors

1. Reagents, buffers, and high-fidelity thermostable DNA polymerase (Expand High Fidelity) kit for PCR, purchased from Roche (Pleasanton, CA).
2. Synthetic primers for PCR amplification (see Note 1).
3. Restriction enzymes and buffers suitable for DNA cleavage (purchased from New England Biolabs (NEB), Ipswich, MA, USA or Fermentas, Burlington, ON, Canada).
4. T4 DNA ligase, alkaline phosphatase (CIP), and suitable buffers (purchased from NEB or Takara).
5. PCR product purification kit (PCRquick, Intron Biotechnology, Daejoen, South Korea).
6. Plasmid DNA purification kit (DNA-Spin, Intron Biotechnology, Daejoen, South Korea).
7. Agarose gel extraction kit (MEGA-Spin, Intron Biotechnology, Daejoen, South Korea).
8. *E. coli* DH5α and BL21(DE3) competent cells ($CaCl_2$-treated cells prepared for heat shock transformation procedure).
9. Luria-Bertani (LB) medium and agar plates supplemented with kanamycin (30 μg/ml), ampicillin (100 μg/ml), and chloramphenicol (30 μg/ml).
10. Agarose gel electrophoresis apparatus; agarose; ethidium bromide (10 mg/ml stock); Tris/acetic acid/EDTA (TAE) buffer, 50× stock (purchased from Bio-Rad); microwave apparatus and visualization system (Gel Doc 2000, Bio-Rad). For most applications DNA can be analyzed on 1% (w/v) agarose gels in 0.5–1.0× TAE buffer.
11. 37°C incubator.
12. Ready-mix PCR master kit (Sigma, St. Louis, MO, or equivalent).

2.4. Protein Expression

1. Luria-Bertani (LB) medium, supplemented with kanamycin (30 μg/ml), ampicillin (100 μg/ml), and chloramphenicol (30 μg/ml).
2. 0.1*M* IPTG (isopropyl-thio-β-D-galactopyranoside).
3. 14 ml and 50 ml round bottom tubes (Falcon or equivalent).
4. Temperature-controlled shaker (15–37°C).

2.5. Protein Analysis

1. Cell-lysis reagent (BugBuster, Novagen).
2. Protein gel electrophoresis system (Mini-Protean 3 Cell from Bio-Rad or equivalent), protein sample buffer, gel staining solution (GelCode Blue Reagent, Pierce, Rockford, IL).
3. Ni-NTA agarose beads (Qiagen or equivalent).
4. Wash buffer for Ni-NTA agarose beads, containing 50 mM NaH_2PO_4, pH 8.0, 300 mM NaCl and 20 mM imidazole.
5. Elution buffer for Ni-NTA agarose beads, containing 50 mM NaH_2PO_4, pH 8.0, 300 mM NaCl, and 500 mM imidazole.

2.6. Automation

2.6.1. Equipment

1. A Tecan Freedom Evo robot is used for automated PCR clean up in 96-well format or in 8-strip format, for miniprep plasmid DNA purification and small-scale protein expression screening. The setup of the robot consists of a fixed, eight-needle pipetting arm (1-ml volume each needle), a handling arm for moving microplates, and shaking and vacuum units.

2. A Magnetight HT96 magnetic separation stand (Novagen) for magnetic-based affinity purification.

3. A PowerWave XS microplate reader (Bio-TEX Instruments, Winooski, VT, USA) for monitoring bacterial growth.

4. A Megafuge 1.0R or Heraeus benchtop centrifuge with adaptors for 96-well plates.

5. A benchtop orbital shaker.

6. Plate shakers:

   - A Titramax 1000 shaker (Heidolph) for up to six plates in parallel.
   - A Thermomixer comfort shaker (Eppendorf) for single plates.

2.6.2. Kits, Reagents, and Miscellaneous

1. A NucleoSpin Robot-96 extraction kit from Macherey-Nagel (MN) for PCR reaction purification.

2. A NucleoSpin Robot-96 Plasmid kit from MN for plasmid DNA purification.

3. Ni-NTA magnetic agarose beads (Qiagen) or Talon magnetic beads (Clontech, Mountain View, CA).

4. PCR plates (ABgene).

5. 96-deep well plate (2 ml volume) for growing bacteria; 96-deep well plate (1 ml volume) for expression with magnetic beads; round-bottom microplates for DNA or protein collection (Grenier, Philadelphia, PA).

6. His-Select iLAP plates (Sigma-Aldrich, catalogue number H9412).

7. Bacterial plates (150 mm).

8. Reagents for sample processing using the iLAP plates and for CoFi:

   - TBST buffer (wash buffer): 20 mM Tris- Cl, pH 7.5, 500 mM NaCl, and 0.05% (v/v) Tween 20.
   - Elution buffer: 50 mM sodium phosphate, pH 8.0, 300 mM NaCl, and 250 mM imidazole.

9. Reagents and materials for dot-blot analysis and CoFi:

   - 96-well format dot-blot apparatus (Bio-Rad).
   - Nitrocellulose transfer membrane BA85, 0.45 μm (Schleicher & Schuell)
   - Durapore membrane filters, 0.45 μm (Millipore, Billerica, MA, USA).
   - Filter paper 3MM (Whatman).
   - Wash buffers: TBST buffer (see item 8) and TBS buffer: 20 mM Tris- Cl, pH 7.5, 500 mM NaCl.
   - Blocking buffer: TBS buffer plus 10% (w/v) dry skim milk.
   - Monoclonal antipolyhistidine antibody conjugated with peroxidase (Sigma-Aldrich).
   - SuperSignal West Pico chemiluminescent substrate kit (Pierce).
   - X-ray film and exposure cassette and film developing apparatus.

3. Methods

3.1. Construction of Expression Vectors

Standard molecular biology procedures are described in detail elsewhere (7). The present protocols describe strategies and essential information required for performing parallel cloning into multiple-expression vectors and two-cloning steps for co-expression of protein partners.

3.1.1. Cloning into Modified pET Vectors

1. Design of specific primers for amplification of the gene of interest. The same PCR product can be cloned directly into the various pET-derivative expression vectors described in Section 2.2.1. (see Note 2).
2. Perform PCR on cDNA of the gene of interest in single tubes or in 96-well format PCR plates.
3. Analyze small aliquots from the reaction mixture on an agarose gel to verify that the DNA product obtained is of the correct size.
4. Purify the PCR product with PCR purification kit or separate the reaction mixture on an agarose gel, isolate the product of the correct size, and purify using an agarose gel extraction kit (see Note 3). If a 96-well format PCR plate is used, the products can be purified using the NucleoSpin Robot-96 Extract kit.
5. Double digest the PCR product using the appropriate restriction enzymes (see Note 4). Purify PCR product using the PCR purification kit.
6. Ligate the PCR product into the series of expression vectors digested with the same restriction enzymes and treated with alkaline phosphatase (see Note 5).
7. Transform the ligation mixture into $CaCl_2$-treated DH5α E. coli cells. Plate on LB-kanamycin plates (kanamycin is used as a selection marker for all the pET-derivative expression vectors described in Section 2.2.1.). Incubate plates at 37°C.
8. Optional step—On the following day, select single colonies from the transformation plate, and perform colony PCR to verify that the PCR fragment had been cloned (see Note 6). Analyze the PCR reactions on an agarose gel.
9. Select two individual colonies from each clone and grow at 37°C, for 16–20 hours, in LB medium supplemented with kanamycin (30 μg/mL). For HTP screening colonies are grown in 96-well format (Section 3.2.2.1.). For testing individual clones grow 5–10 ml cultures in 50 ml conical tubes.
10. Harvest cells and extract DNA using a plasmid DNA purification kit. If cells are grown in 96-well format a NucleoSpin Robot plasmid DNA purification kit is employed (Section 2.6.2. and Note 7).
11. Transform two plasmids from each clone to competent BL21(DE3) E. coli cells (or into any other E. coli strain used for expression of pET vectors), inoculate cells on LB agar plates supplemented with kanamycin and incubate at 37°C. Use 1–2 colonies from each plate for small-scale expression experiments.

3.1.2. Cloning into Duet Vectors

The principal methodologies for cloning genes into the Duet vectors are the same as for cloning of a single gene into the pET-derivatives vector described in Section 3.1.1. However, since two individual genes are cloned into each of the Duet expression vectors the following points need to be considered:

1. Decide which of the protein partners will be flagged with the 6× His-tag and as to whether the tag is to be placed at the C- or N-terminus of the protein (see Note 8).

2. Decide on the restriction sites to be used for cloning each of the two target genes. These restriction sites should be absence in the target gene sequence and should be part of the synthetic primer. Care should be taken to keep the sequence in frame while designing the primers.

3. DNA amplification by PCR, PCR purification, cloning, and transformation should be carried as described in Section 3.1.1. (steps 2–10).

4. Following cloning the first gene into the Duet vector, the second gene should be cloned into the second multiple-cloning site of the Duet vector using the guidelines described in Section 3.1.1.

5. Once the two genes have been cloned into the same expression vector, extract plasmid DNA from two individual DH5α colonies and transform into BL21(DE3) cells. Inoculate cells on LB agar plates containing the appropriate antibiotic and incubate at 37°C. Use 1–2 colonies from each plate for small-scale expression experiments.

3.2. Protein Expression

3.2.1. Screening for Protein Expression in Single Tubes

The procedure described in the following outlines steps for handling a small number of clones in parallel. For a larger number of samples (8 or more), the guidelines for HTP expression and screening described in Section 3.2.2 should be followed.

1. Inoculate a single colony into 4 mL of LB medium containing the appropriate antibiotic, in a 14-ml tube, followed by growth for 16–20 hours, at 37°C, with shaking.

2. Following growth (see Note 9) dilute the culture 1:100, into 4 ml of fresh LB medium containing the appropriate antibiotic, in a 14-ml tube. For each clone, two diluted cultures should be set up. Incubate cultures at 37°C with shaking until OD_{600} reaches 0.6–0.8.

3. Induce protein expression with 50–100 μM IPTG. For each clone, incubate one tube at 30°C for 3–4 hours, and a second one at 15°C for 16–20 hours (see Note 10).

4. Harvest cells by centrifugation. Chemically lysed cells using BugBuster reagent according to the manufacturer's specifications. Rotate extracts at room temperature for 30 minutes, followed by spinning for 20 minutes at 20,000g. The pellet fraction and supernatant fraction are retained for SDS-PAGE.

5. Purify the tagged proteins by batch affinity purification. For His-tagged proteins use Ni-agarose beads (Qiagen), following the manufacturer's recommendations.

6. Analyze proteins in the pellet and supernatant fractions as well as the Ni-agarose-purified proteins on SDS-PAGE.

3.2.2. HTP Screening for Protein Expression

3.2.2.1. Expression in 96-Deep-Well Format

1. Bacterial cultures are grown in 96-deep-well plates (2 ml volume). Each well is filled with 1 ml of LB medium containing the appropriate antibiotic. Single colonies are picked and inoculated into each well for 16–20 hours growth at 37°C with shaking.

2. Prepare two 96-deep-well plates (2-ml volume) with 1 ml LB medium containing the appropriate antibiotic in each well. Following 16–20 hours growth (see Note 9) dilute the culture 1:100 into the two 96-deep-well

plates using either a multichannel pipettor or a robot. Incubate both plates at 37°C with shaking until the OD_{600} reaches 0.6–0.8. Incubation makes use of either the Titramax 1000 (Heidolph) shaker or the Thermomixer Comfort (Eppendorf) shaker.

3. Induce protein expression with 50–100 μM IPTG. Transfer one plate to 30°C shaker for 3–4 hours, and the second one to a shaker maintained at 15°C for 16–20 hours induction (see Notes 10 and 11).

4. Harvest cells in the 96-well plates by centrifugation at 500g using a bench-top centrifuge. Chemically lysed cells using BugBuster reagent by gentle mixing at room temperature for 30 minutes. The extracts are then spun for 1 hour at 3,500g Supernatant and pellet are analyzed by SDS-PAGE.

5. Transfer supernatants using the Tecan robot, to a 1-ml 96-deep-well plate, positioned on the robot-shaker, after which 50 μL Ni-NTA magnetic beads (or Talon magnetic beads) are dispensed into each well. Shake plate at 1,500 rpm for 5 minutes. Move plate to magnetic separation stand and remove unbound proteins by aspiration of the supernatant.

6. Move plate to a robot-shaker and wash the magnetic beads with buffer containing 50 mM NaH_2PO_4, pH 8.0, 300 mM NaCl, and 20 mM imidazole. Shake plate at 1,500 rpm for 1 minute. Move plate to magnetic separation stand and remove washed unbound proteins by aspiration of the supernatant.

7. Wash the beads a second time.

8. The plate is moved to the robot-shaker and the bound proteins are eluted from the beads with 50 mM NaH_2PO_4, pH 8.0, 300 mM NaCl, and 500 mM imidazole, by shaking at 1,500 rpm for 1 minute. The plate is moved to the magnetic separation stand and the eluted proteins in the supernatant are transferred, by aspiration, to a microplate.

9. Analyze proteins in the pellet and the His-tagged purified proteins on SDS-PAGE (Fig. 12.2A).

3.2.2.2. HTP Screening for Protein Expression Using His-Select iLAP Plates: His-Select iLAP (Integrated Lysis and Affinity Purification) plates (Sigma-Aldrich), designed for combined cell lysis and capture of up to 96 His-tagged proteins in parallel, were employed (see Note 12). The His-Select iLAP procedure, combined with dot-blot analysis, provides a fast track for identification of even small amounts of soluble proteins expressed in *E. coli*.

The procedure described in the following can be performed either using a multichannel pipettor or instructing the Tecan Freedom EVO robot to perform the operations.

1. Cell growth and protein induction are performed as described in Section 3.2.2.1.

2. Remove aliquots of 200 μl from each of the IPTG-induced cultures in the 96-well plate and transfer to the same positions in the iLAP plate, followed by incubation at room temperature with shaking for approximately 3 hours (see Note 13).

3. Following the lysis/capture steps remove the lysates, and wash the wells four times with 0.3 ml per well of TBST (wash buffer).

4. Remove the wash buffer and wash four times with double-distilled water.

5. Add aliquots of 200 μl elution buffer to each well and incubate with slow agitation at room temperature for 1 hour (see Note 13).

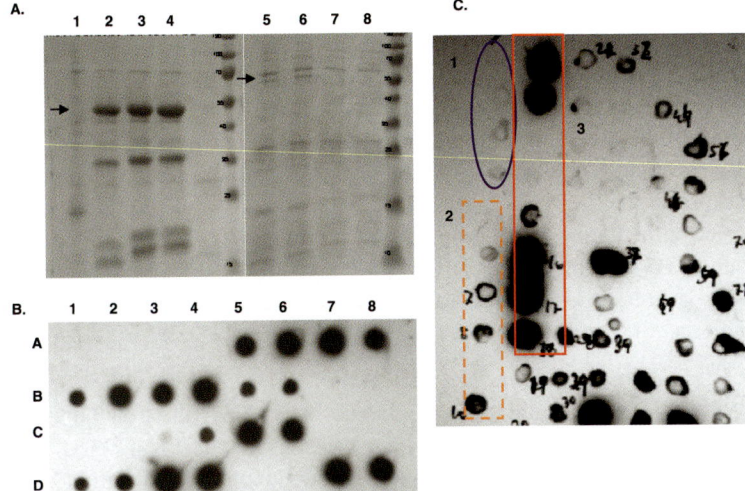

Fig. 12.2 HTP screening for soluble protein expression. **A.** Capture of soluble proteins on magnetic beads analyzed by SDS-PAGE. A representative example of variants of three different proteins. Four different variants of protein A were analyzed (lanes 1–4). Variant 1 showed no expression, whereas variants 2–4 showed high levels of expression. Two variants each of proteins B and C were tested (lanes 5–6 and 7–8, respectively). Both variants of protein B showed low levels of expression (lanes 5 and 6), whereas those of protein C revealed no expression of soluble protein (7 and 8). **B.** Dot-blot analysis of proteins eluted from a His-Select iLAP plate (see Section 3.2.2.2.). Samples A1–A4 are negative controls for cells devoid of an expression vector (A1 and A2) and for cells harboring the pET28-TEVH expression vector, but lacking an insert (A3 and A4). All other positions are for samples of different proteins displaying varying levels of expression. **C.** Colony filtration blot of protein variants. Variants of three different proteins are highlighted (numbered 1–3). Substantial differences are observed in expression levels. The four variants of protein 1 showed no expression of soluble protein, whereas the variants of protein 2 showed low expression levels, and those of protein 3 high expression levels. It should be noted that in this screening procedure lack of a positive signal can be due either to lack of soluble expression or to failure in the cloning procedure.

6. Transfer the eluates to a new 96-well plate. Remove 20–30 µl aliquots from each sample for SDS-PAGE analysis (see Note 13).

7. The remainder of the sample is subjected to dot-blot analysis.

Dot-blot analysis:

1. Prewet a nitrocellulose membrane (8.5 × 12 cm) in TBS buffer for 10 minutes.

2. Remove the membrane and let access of liquid drain on a filter paper.

3. Lay the membrane on the gasket of the 96-well dot-blot apparatus. Remove air bubbles (see Note 14).

4. Place the 96-well sample template on top of the membrane and tighten the four screws. An aliquot of 100 µl TBS buffer is added to each well, and the excess removed by vacuum.

5. Apply samples to the wells, and the entire sample is allowed to percolate through the membrane by gravity (30–40 minutes for a 100 µl aliquot).

6. Wash each well with 400 µl of TBST buffer, which is then removed by vacuum. Repeat the wash step.

7. Remove the membrane from the dot-blot apparatus, and incubate with 10 ml of blocking solution for 1 hour with gentle shaking at room temperature.

8. Remove the blocking solution and add a fresh blocking solution with anti-His antibodies diluted 1:3,000. Incubate for 1 hour at room temperature with slow shaking.

9. Remove the antibody solution and wash three times (10–15 minutes each) using TBST buffer.

10. View results following development of the membrane using the chemiluminescent kit and exposing to x-ray film (see Fig. 12.2B).

3.2.2.3. HTP Screening for Protein Expression Using the Colony Filtration Blot Procedure: The authors have adopted the colony filtration blot (CoFi) method developed by Nordlund and his colleagues (8). This method is used for parallel rapid screening of multiple clones expressing different soluble recombinant proteins. Only positive clones obtained by the CoFi method are subjected to DNA sequencing analysis and to small-scale expression (see Section 3.2.1.), followed by SDS-PAGE to check for production of protein of the expected size. The procedure is as follows:

1. On an LB agar plate (150-mm diameter) containing the appropriate antibiotic spot 96 clones harboring constructs of different proteins or of variants of the same protein and incubate 16–20 hours at 37°C.

2. The colonies are picked up by placing a 0.45 µm Durapore membrane on the surface of the plate. The membrane is then lifted and placed on an LB agar plate containing 0.1–0.2 mM IPTG with side bearing the colonies facing upward. Plates are incubated 16–20 hours at 15°C.

3. A "filter sandwich" is assembled. In the bottom of a 150-mm plate, a Whatman 3MM paper soaked in a lysis buffer (20 mM Tris pH 8.0, 100 mM NaCl, 0.1 mg/ml lysozyme, and 1 µg/ml DNAse I) is placed. On top of the Whatman paper a nitrocellulose membrane is positioned. The Durapore membrane with the colonies facing upward is then transferred on top of a nitrocellulose membrane. Incubate the plate at 22°C for 30 minutes. The proteins on the filters are freeze–thawed three times at −80°C for 10 minutes and then at 37°C for 10 minutes. The soluble proteins diffuse through the Durapore membrane and are captured on the nitrocellulose membrane. The nitrocellulose filter is removed and probed with monoclonal anti-polyhistidine peroxidase antibody as described for the iLAP procedure (see Section 3.2.2.2. and Fig. 12.2C).

4. Notes

1. Synthetic primers for PCR amplification are ordered from Sigma-Genosys. Primers of up to 50-mer are routinely ordered that have undergone basic desalting purification. Primers above 50-mer must be purified by other methods, such as HPLC or PAGE. Primers are designed so as to ensure that the Tm of the specific region in the primer (nucleotides that are part of the coding sequence of the gene of interest without flanking restriction sites or tag sequences) is at least 58–59°C. This helps to increase specificity and to obtain a uniform PCR product.

2. Designing a forward primer using a *Sac*II restriction site requires synthesis of a long primer that includes the TEV recognition site. In this case, only glycine remains at the N-terminus of the protein following TEV cleavage (see Fig. 12.1). The reverse primer should include one of the restriction sites available on the pET-derivative expression vectors (see Fig. 12.1), which is not present on the gene of interest (e.g., *Eco*RI, *Not*I).

3. The decision whether to use a PCR purification kit or to purify the product from an agarose gel depends on the pattern of the PCR product on the agarose gel. If the correct size DNA fragment is observed as the predominant product, a PCR purification kit can be used. However, if several PCR products appear, the reaction mixture should be run on an agarose gel, and the DNA fragment of the correct size isolated using a gel extraction kit.

4. It is recommended that the reaction conditions for double digestion be selected on the basis of the "double digest" tables available on the Web, such as the NEB (http://www.neb.com/nebecomm/DoubleDigestCalculator.asp) or Fermentas (http://www.fermentas.com/doubledigest) web sites.

5. Alkaline phosphatase treatment prior to ligation is needed in order to reduce the transformation background of a partially digested vector. A control ligation of the vector should be set up without an insert so as to assess the background transformation.

6. Performing colony PCR is an optional step. If a similar number of colonies are obtained both with and without the insert, colony PCR is performed to identify colonies carrying the insert. PCR is carried out using forward and reverse primers from sequences adjacent to the cloning sites in the vector. Alternatively, a combination of a specific primer from the insert and a primer from the expression vector are used.

7. Sequencing the DNA clones prior to the small-scale protein expression experiments is not essential. DNA sequencing can be performed only at a later stage, e.g., prior to large-scale production. In cases in which no protein is observed in small-scale experiments, or only the fusion protein is detected, DNA sequencing is recommended to verify the integrity of the gene.

8. When co-expressing two protein partners (or putative partners) either in different expression vectors or in the same expression vector (the Duet vectors) one of the protein partners is flagged with His-tag, thus permitting initial copurification of the complex using an affinity resin (e.g., Ni-agarose beads). If no prior experience is available, the decisions as to which of the partners should be tagged and as to its position (N- or C-terminal) are empirical. It should be kept in mind that interaction between the partners may be affected by placing the tag at a given position.

9. It is recommended to prepare from the culture grown overnight (16–20 hours) a glycerol stock (final glycerol concentration 20–25% v/v). The stock should be stored at −80°C until used.

10. Each culture is tested for protein expression at two temperatures. The authors observed that lowering the temperature to as low as 15°C increases in many cases the fraction of soluble protein.

11. The speed of the Thermomixer comfort shaker used for shaking the 96-well plates should not exceed 800 rpm.

12. The His-Select iLAP plates provide an alternative procedure for processing a large number of samples in parallel in a relatively short time of only

a few hours. A full-length description of the procedure can be found in the Sigma technical bulletin for product code H 9412.

13. Alternatively, the entire 1 ml of cells grown in the 96-well plate can be spun down, and most of the supernatant removed, leaving about 200 µl of medium. The cells are then resuspended, and transferred to iLAP plates. Applying a larger cell mass to the iLAP plate, and lowering the elution volume to 50 µl, should assist protein visualization following SDS-PAGE analysis.

14. Bio-Rad technical guidelines should be followed for operation of the dot-blot apparatus.

Acknowledgements

The authors acknowledge the support of the Divadol Foundation, the Newman Foundation, The Israel Ministry of Science, Culture and Sport grant for the ISPC, and the European Commission Sixth Framework Research and Technological Development Programme "SPINE2-COMPLEXES" Project under contract No. 031220. The authors thank Joel L. Sussman and Israel Silman for their comments on the manuscript.

References

1. Albeck, S., Burstein, Y., Dym, O., Jacobovitch, Y., Levi, N., Meged, R., Michael, Y., Peleg, Y., Prilusky, J., Schreiber, G., Silman, I., Unger, T., and Sussman, J. L. (2005) 3D structure determination of proteins related to human health in their functional context at the Israel Structural Proteomics Center (ISPC). *Acta Cryst.* **D61**, 1364–1372.

2. Aricescu, A. R., Assenberg, R., Bill, R. M., Busso, D., Chang, V. T., Davis, S. J., Dubrovsky, A., Gustafsson, L., Hedfalk, K., Heinemann, U., Jones, I. M., Ksiazek, D., Lang, C., Maskos, K., Messerschmidt, A., Macieira, S., Peleg, Y., Perrakis, A., Poterszman, A., Schneider, G., Sixma, T. K., Sussman, J. L., Sutton, G., Tarboureich, N., Zeev-Ben-Mordehai, T., and Jones, E. Y. (2006) Eukaryotic expression: developments for structural proteomics. *Acta Cryst.* **D62**, 1114–1124.

3. Romier, C., Ben Jelloul, M., Albeck, S., Buchwald, G., Busso, D., Celie, P. H., Christodoulou, E., De Marco, V., van Gerwen, S., Knipscheer, P., Lebbink, J. H., Notenboom, V., Poterszman, A., Rochel, N., Cohen, S. X., Unger, T., Sussman, J. L., Moras, D., Sixma, T. K., and Perrakis, A. (2006) Co-expression of protein complexes in prokaryotic and eukaryotic hosts: experimental procedures, database tracking and case studies. *Acta Cryst.* **D62**, 1232–1242.

4. Berrow, N. S., Bussow, K., Coutard, B., Diprose, J., Ekberg, M., Folkers, G. E., Levy, N., Lieu, V., Owens, R. J., Peleg, Y., Pinaglia, C., Quevillon-Cheruel, S., Salim, L., Scheich, C., Vincentelli, R., and Busso, D. (2006) Recombinant protein expression and solubility screening in *Escherichia coli*: a comparative study. *Acta Cryst.* **D62**, 1218–1226.

5. Vincentelli, R., Canaan, S., Offant, J., Cambillau, C., and Bignon, C. (2005) Automated expression and solubility screening of His-tagged proteins in 96-well format. *Anal. Biochem.* **346**, 77–84.

6. Moreland, N., Ashton, R., Baker, H. M., Ivanovic, I., Patterson, S., Arcus, V. L., Baker, E. N., and Lott, J. S. (2005) A flexible and economical medium-throughput strategy for protein production and crystallization. *Acta Cryst.* **D61**, 1378–1385.

7. Ausubel, F. M., Brent, R., Kingston, R. E., Moore, D. D., Seidman, J. G., Smith, J. A., and Struhl, K. (eds.) (1987) *Current Protocols in Molecular Biology*, John Wiley & Sons, New York.

8. Cornvik, T., Dahlroth, S. L., Magnusdottir, A., Herman, M. D., Knaust, R., Ekberg, M., and Nordlund, P. (2005) Colony filtration blot: a new screening method for soluble protein expression in *Escherichia coli*. *Nat. Methods* **2**, 507–509.

Chapter 13

A High Throughput Platform for Eukaryotic Genes

Yunjia Chen, Shihong Qiu, Chi-Hao Luan, and Ming Luo

The objective of structural proteomics is to determine the structures of all protein folds found in nature and develop a public resource to organize and analyze protein structures and fold families. High throughput (HTP) methods, which can process multiple samples in parallel, saving both time and cost, play important roles in achieving this goal. Using *C. elegans* and human as model organisms, a HTP cloning and expression pipeline was developed for structural proteomics that required production of a large number of recombinant proteins, applying the Gateway cloning/expression technology and utilizing a stepwise automation strategy on an integrated robotic platform. This system can process up to 384 unique samples in parallel and successfully automates most aspects of gene cloning and protein expression analysis, from PCR to protein solubility profiling.

1. Introduction

Proteome-scale studies of protein three-dimensional structures provide valuable information for both investigating basic biology and developing therapeutics. Naturally, gene cloning and protein expression are critical parts for these studies. The traditional cloning and expression approach, however, is inadequate for such a herculean task, and automated, parallelized HTP methods are clearly necessary *(1,2)*. To setup such a HTP cloning and expression system, the proper strategies that are amenable to HTP operation should be decided firstly. The Gateway system is an ideal choice for this endeavor since the Gateway technology employs *in vitro* site-specific recombination method, which is independent of the candidate target gene, to transfer gene of interest into multiple vector systems for function analysis and protein expression *(3–5)*. Miniaturization of basic cloning and expression protocols is the second step for constructing such a HTP system since this system can only perform liquid handling tasks with preferred volumes of 1 ml or less, at least at present. The third step for setting up a HTP system is automation of miniaturized protocols, whose objective is to process automatically multiple samples in parallel, saving both time and cost as well as generating consistent and reproducible data. Currently, major obstacles in automating molecular biology are the requirements of strict temperature control, and the long periods used for shaking, centrifuging and incubating.

From: *Methods in Molecular Biology, Vol. 426: Structural Proteomics: High-throughput Methods*
Edited by: B. Kobe, M. Guss and T. Huber © Humana Press, Totowa, NJ

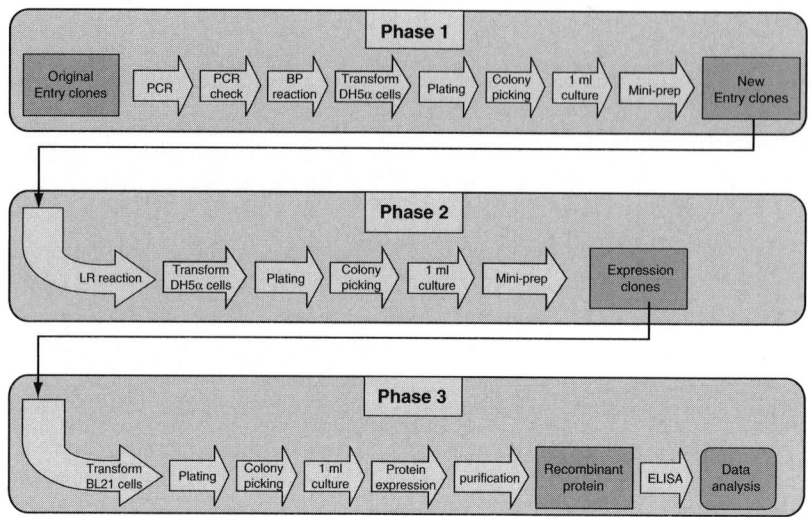

Fig. 13.1 Flowchart showing the overall strategy of the authors' HTP cloning and expression system. The first dark gray box represents the starting point and the light gray boxes represent stop points. Arrows represent process and the final box stands for resolution of data analysis, to address which proteins are soluble, insoluble, or negative for expression.

For this structural genomics project, a system to process multiple samples in parallel with *C. elegans (5)* and human *(6)* as the model organisms was designed, using the Gateway cloning and expression system in 96-well plate format. By optimizing and miniaturizing the basic protocols, all liquid handling steps in the HTP system are carried out in 96-well plates in 1 ml or less and many steps have been automated.

The authors' HTP system can be generalized with three phases, a total of 21 steps (Fig. 13.1). The first phase is to subclone selected regions (see Note 1) of ORFs (open reading frames) into new entry vectors. In this phase, a thrombin cleavage site is introduced between attB1 site and the selected region (see Note 2). The second phase is to obtain expression clones based on *in vitro* site-specific recombination, and the third phase is to express and purify proteins, and analyze corresponding expression and solubility results with automated enzyme-linked immunosorbent assay (ELISA).

After profiling expression level, solubility, and optimal expression conditions for each protein, individual clones are selected for large-scale production in six 1-L cultures using traditional methods. The soluble proteins are purified by use of the standard protocols with affinity, ion-exchange, and size exclusion chromatography to obtain homogenous protein preparations. The purified proteins are then concentrated and used in crystallization trials.

2. Materials

2.1. Apparatus for the High Throughput Pipeline

The authors' integrated robotic platform (Fig. 13.2) is centered on the Beckman/Sagian core system (see Note 3), including:

Fig. 13.2 Pictures of some apparatus for the HTP pipeline: Biomek FX *(left)*, Biomek 2000 *(middle)*, Beckman ORCA arm (*right*, **A**), and DNA Engine Tetrad Cycler (*right*, **B**).

1. Biomek FX and Biomek 2000 liquid handlers (Beckman Coulter, Fullerton, CA).
2. Beckman ORCA arm (Beckman).
3. DNA Engine Tetrad Cycler (MJ Research).
4. ELX-405UV plate washer (BioTek, Winooski, VT).
5. SpectraMax UV/Vis plate reader (Molecular Devices, Sunnyvale, CA).
6. Vortemp temperature-controlled shaker-incubators (National Labbet, Woodbridge, NJ).
7. Centrifuge with microplate carrier (Eppendorf, Westbury, NY).
8. BioRobot 9600 (Qiagen, Valencia, CA).

2.2. PCR and PCR Results Test

1. 96-well PCR plates (Denville, Metuchen, NJ).
2. AccuPrime™ Pfx DNA polymerase with 10× reaction mix (Invitrogen, Carlsbad, CA). Store at −20°C.
3. E-Gel® Pre-cast agarose electrophoresis system (Invitrogen): E-Gel® 96 mother base, 2% E-Gel® 96 agarose (store at room temperature), and E-Gel® low range quantitative DNA ladder (store at 4°C).
4. Primers: Four primers are used in each PCR reaction (see Note 4). Primer F1 contains a part of the thrombin cleavage site followed by the gene specific sequence of 5'-terminus:
   CCACGCGGCAGC- 5' gene specific sequence (see Note 5). Primer R1 contains a part of the attB2 site followed by the gene specific sequence of the 3'-terminal:
   CAAGAAAGCTGGGTTA-3' gene specific sequence. Primer F2 contains the attB1 and the thrombin cleavage site:
   GGGGACAAGTTTGTACAAAAAAGCAGGCTTGGTGCCACGCGGC AGC, and R2 contains attB2 and the termination codon:
   GGGGACCACTTTGTACAAGAAAGCTGGGTTA. Primer F1 and R1 are ordered in 96-deep-well plate from Illumina (San Diego, CA). Primer F2 and R2 were ordered in bulk from Invitrogen. Store at −80°C.
5. Templates: entry clones containing ORFs of *C. elegans*, human from *C. elegans* ORFeome version 3.1 and HORFeome versions 1.1. Store in 96-well plates at −80°C.

2.3. BP Reaction and LR Reaction

1. 96-well PCR plates (Denville).
2. BP Clonase enzyme mix with 5× BP Clonase Reaction Buffer (Invitrogen). Store at −80°C.
3. Donor vector: pDONR201 (Invitrogen). Store at −80°C.
4. LR Clonase enzyme mix with 5× LR Clonase Reaction Buffer (Invitrogen). Store at −80°C.
5. Destination vector: pET15G, an in-house vector based on pET15b (Novagen), which contains a His$_6$ tag before the N-terminal recombination site that could be used for Ni affinity purification. Store at −80°C.
6. TE buffer (1×): 10 mM Tris-HCl, 1 mM EDTA, pH 8.0. Store at room temperature (see Note 6).

2.4. Transformation, Bacterial Growth, Mini-prep, and Protein Expression

1. LB medium: Dissolve 25 g LB brother (Sigma, St. Louis, MO) in 1 L water, and then autoclave the medium. Store at 4°C.
2. 96-well PCR plates (Denville).
3. DH5α competent cells (Invitrogen), prepared in 96-well PCR plates, 50 μl per well. Store at −80°C.
4. BL21 AI competent cells (Invitrogen), prepared in 96-well PCR plates, 50 μl per well. Store at −80°C.
5. 12-well tissue culture plate (MIDSCI, St. Louis, MO), 1.5 ml LB agarose (Fisher) with 50 μg/ml kanamycin (Sigma) or 100 μg/ml ampicillin (Sigma) per well. Store at 4°C.
6. 96-well, 2.2 ml V-bottom block plate (Costar, Fisher), 1 ml LB medium (Sigma) with 50 μg/ml kanamycin or 100 μg/ml ampicillin per well (see Note 7).
7. Airport tape (Qiagen).
8. Qiagen 96-well turbo mini-prep kits (Qiagen, Valencia, CA), including: Solution1, Solution2, Solution3, 96-well, 2.4 ml culture plate, filter plate A, filter plate DB, 350 μl collection plate, plate seal, DNA Binding Buffer, Purification solution, molecular biology grade (MBG) water.
9. Arabinose (Sigma): 100×, 200 mg/ml.
10. IPTG (Denville): 100×, 100 mM.

2.5. Cell Lysis and Protein Purification

1. Native lysis buffer: 50 mM NaH$_2$PO$_4$, 300 mM NaCl, 10 mM imidazole, and 1 mg/ml lysozyme, pH 8.0. Store at room temperature (see Note 8).
2. Denaturing lysis buffer: 100 mM NaH$_2$PO$_4$, 10 mM Tris- Cl, and 8 M urea, pH 8.0. Wrap in Al foil and store at room temperature.
3. Wash buffer: 100 mM NaH$_2$PO$_4$, 10 mM Tris Cl, and 4 M urea, pH 6.3. Wrap in Al foil and store at room temperature.
4. Elution buffer: 100 mM NaH$_2$PO$_4$, 10 mM Tris- Cl, and 8 M urea, pH 4.5. Wrap in Al foil and store at room temperature.
5. Stripe buffer: 100 mM EDTA, 500 mM NaCl, 20 mM Tris- Cl, pH 8.0.
6. Charge buffer: 100 mM nickel sulfate.
7. 96-well Ni affinity chromatography: 96-well filters (Qiagen) and Ni-NTA resin (Qiagen).
8. 96-well 350 μl Collection Plate (Qiagen).

2.6. Enzyme-Linked Immunosorbent Assay

1. Enzyme-linked immunosorbent assay (ELISA) plates: EIA/RIA high binding plates (Costar).
2. PBSN buffer: 137 mM NaCl, 2.7 mM KCl, 4.3 mM NaH_2PO_4, 1.4 mM KH_2PO_4, 0.05% NaN_3, pH 7.3.
3. Blocking buffer: 17 mM $Na_2B_4O_7$ 120 mM NaCl, 0.05% (v/v) Tween20, 1 mM EDTA, 0.25% (w/v) protease-free bovine serum albumin, 0.05% NaN_3, pH 8.5.
4. pNPP: p-Nitrophenyl phosphate (ICN, Aurora, OH).
5. pNPP working buffer: 750 mM 2-amino-2-methyl-1.3-propandiol at pH 10.3.
6. Primary antibody: mouse penta-His IgG1 antibody (Qiagen).
7. Secondary antibody: rabbit anti-mouse IgG Fc alkaline phosphatase conjugate (Pierce, Rockford, IL).
8. 500 mM NaOH.
9. Airport tape (Qiagen).
10. Positive control: *C. elegans* UBC protein with His_6 tag (2.4 µg/µl).

3. Methods

3.1. PCR and PCR Results Test

1. PCR: PCR is the first step of the HTP method, which determines the final success rate of the whole HTP cloning process. To improve the success rate of PCR, the authors designed a multistep laddered PCR strategy (Fig. 13.3) of using two forward primers (F1, F2) and two backward primers (R1, R2) (Fig. 13.4).

   - F2 and R2 are dissolved manually with water to get a concentration of 20 pmol/µl, whereas F1 and R1 were dissolved with water to 100 pmol/µl and then diluted to 2 pmol/µl using a Beckman Biomek FX robot.
   - After optimizing, the amount of all compositions in each PCR reaction is decided (see Note 9) as following: AccuPrime™ Pfx, 0.7 µl; 10X AccuPrime™ Pfx reaction mix, 5 µl; Primer F1 (2 pmol/µl), 2 µl; Primer R1 (2 pmol/µl), 2 µl; Primer F2 (20 pmol/µl), 2 µl; Primer R2 (20 pmol/µl), 2 µl; Template DNA, 1 µl; water, 35.3 µl.
   - According to the proportion of compositions and the total reaction number, manually premix proper amounts of primer F2, R2, AccuPrime™ Pfx polymerase, 10X AccuPrime™ Pfx reaction mix and template (see Note 10).
   - Aspirate 2 µl of primer F1, 2 µl of primer R1, 35.3 µl of water and 10.7 µl of premixtures to respective wells of 96-well plates, then mix them using the Beckman Biomek FX robot.
   - PCR started on a tetrad PCR machine

2. PCR results check: After running the batch PCR protocol, E-Gel® Pre-cast Agarose Electrophoresis System was used to check PCR outcomes.

   - 3 µl of PCR sample in each well was diluted with 17 µl of water and the resulting 20 µl sample was loaded into the corresponding well of 2% E-Gel® 96 Agarose (see Note 11).

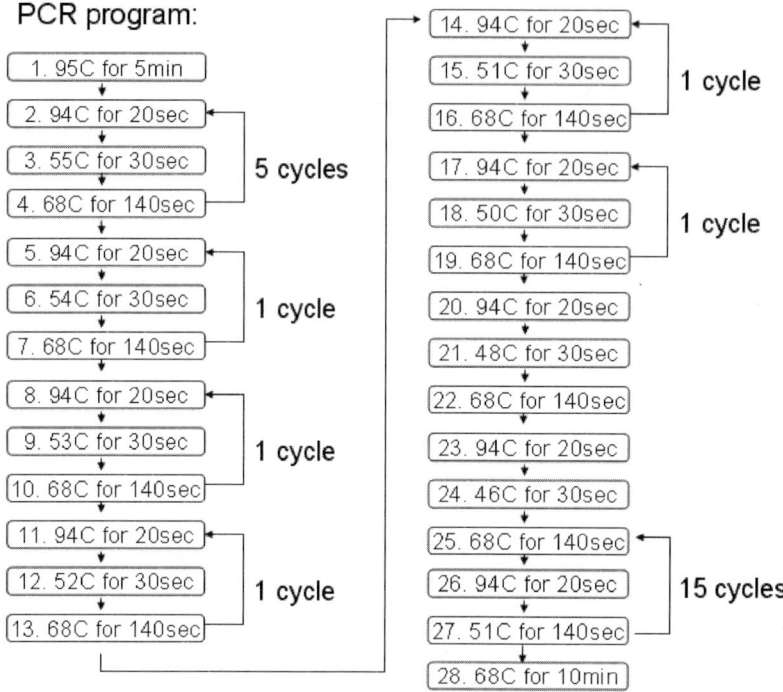

Fig. 13.3 A multi-step PCR protocol: Amplify template DNA for total 34 cycles with 5 minutes at 95°C for initial denaturation, 20 seconds at 94°C for denaturation, 30 seconds for annealing, 140 seconds at 68°C for extension, and 10 minutes at 68°C for final extension. Annealing temperature was variational: Start from a relatively high temperature (55°C), and then decreased 1–2 degrees each time, until the temperature reaches 46°C, and then increase and finally stabilize at 51°C.

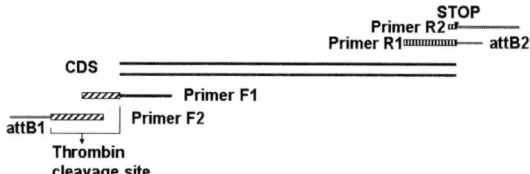

Fig. 13.4 The primer design strategy for engineering a thrombin cleavage site using the GATEWAY system and two pairs of primers. Primer F2 and R2 contained attB site and no gene specific region, which could be synthesized in bulk; Primer F1 and R1 contained gene specific sequences and an overlap region with Primer F2 and R2.

- Prepare 160 μl DNA ladder sample for each E-Gel® 96 gel by mixing 40 μl of E-Gel® Low Range Quantitative DNA Ladder and 120 μl of water, then load 20 ul into each marker well.
- Run E-Gel® 96 gel in the Mother E-Base device for 12 minutes using the pre-set 12 minute DNA program (see Note 12).

3.2. BP Reaction

1. After PCR products were confirmed by the E-Gel system, BP reaction was performed with pDONR201 as the donor vector.

2. The amount of all compositions in each BP reaction was decided as following: attB-PCR product, 2 μl; Donor vector, pDONR201 (150 ng/μl), 1 μl; 5× BP Clonase Reaction Buffer, 2 μl; TE buffer, pH 8.0, 3 μl; BP Clonase enzyme mix, 2 μl.

3. Manually premix proper amounts of donor vector, BP Clonase enzyme mix, 5× BP Clonase reaction buffer and TE buffer based on their proportions in each reaction and the total reaction number (also see Note 10).

4. With a Beckman Biomek FX robot, 8 μl of premixture and 2 μl of PCR products were added into the corresponding well of 96-well plates and mixed. The reaction was incubated overnight at the room temperature.

3.3. Generating Entry Clones

1. Transformation, colony picking, and cell culture:

- Unfreeze up to 4 plates of competent cells in a tetrad PCR machine at 4°C.
- Mix the BP reaction with competent cells using a Beckman Biomek 2000, and then incubate at 4°C for 30 minutes.
- Move the plates on a heat shocking station, which was created using two thermalizers, water-circulating blocks, holding at 4 and 42°C, with a Beckman ORCA arm.
- After heat shocking, the PCR plates were returned to 4°C for 2 minutes and 130 μl LB medium was added and incubated at 37°C for 1 hour with shaking.
- The cells were pelleted and resuspended in 50 μl LB and plated on 12-well LB agarose plates with 50 μg/ml kanamycin. Then incubate plates at 37°C overnight.
- Single colonies were selected and grown overnight in 1 ml LB with 50 μg/ml kanamycin in 96-well block plates covered with airport tape (Qiagen).
- Growth occurred in Vortemp shaking incubators with a small shaking radius of 3 mm and a speed of 1,000 rpm (see Note 13).

2. Mini-prep: Entry clones were produced using a Qiagen BioRobot 9600 and 96-well turbo mini-prep kits.

- Pellet bacteria by centrifuging 96-well block plates at 3,500 rpm for 5 minutes, and then pour off supernatants.
- Add 150 μl Solution 1 to each well and completely resuspend cells by vortexing 800 for 20 minutes (see Note 14).
- Add 150 μl Solution 2 to each well and vortex at 400 rpm for 5 minutes (see Note 15).
- Add 150 μl Solution 3 to each well and vortex at 600 rpm for 10 minutes.
- Transfer all of each lysate to respective wells of a filter plate A on top of a filter plate DB and short adaptor in the vacuum manifold. Then apply vacuum for 5 minutes.
- Release vacuum slowly. Discard the filter plate A. Remove the filter plate DB and short adaptor.
- Place used culture plate into manifold and place filter plate DB in lid of manifold.
- Add 300 μl DNA binding buffer to each filter plate DB well and apply vacuum for 2 minutes.
- Release vacuum and add 400 μl diluted purification solution to each filter plate DB well and apply vacuum for 5 minutes.

- Release vacuum and blot the bottom of the filter plate DB on a paper towel. Remove and discard the culture plate.
- Place a fresh 350 μl collection plate on top of the tall adaptor in the vacuum manifold and place the filter plate DB in the lid of the manifold.
- Add 50 μl MBG water and incubate 2 minutes.
- Apply vacuum for 5 minutes, and then release vacuum.
- Recover the collection plate from the manifold, and then seal it. Store plasmid DNA at −20°C.

3.4. LR Reaction

1. The amount of all compositions in each LR reaction was decided as following: entry clone, 2 μl; Destination vector, pET15G (150 ng/μl), 1 μl; 5 × LR Clonase Reaction Buffer, 2 μl; TE buffer, pH 8.0, 3 μl; LR Clonase enzyme mix, 2 μl.
2. Manually premix proper amounts of destination vector, LR Clonase enzyme mix, 5 × LR Clonase reaction buffer and TE buffer based on their proportions in each reaction and the total reaction number (also see Note 10).
3. Mix entry clones and pre-mixture using a Beckman Biomek FX robot.
4. Incubate the LR reaction overnight at room temperature.

3.5. Generating Expression Clones

Once the LR reaction was completed, transformation, colony picking, cell culture, and mini-prep were performed: Procedures are the same as those to get entry clones, except for substituting the LR reaction for the BP reaction and plates/LB medium containing 100 μg/ml ampicillin for those containing 50 μg/ml kanamycin.

3.6. Protein Expression and Purification

3.6.1. Expressing proteins

1. Transform *E. coli* strain BL21 AI with obtained expression vectors using the same transformation protocol described in Section 3.4.
2. Pick single colonies for recombinant protein expression.
3. After overnight growth, the bacteria were diluted 1:200 in 1 ml LB containing 100 μg/ml ampicillin in block assay plates and grown for 3–4 hours before induction.
4. Add 10 μl 100 mM IPTG and 10 μl 200 mg/ml arabinose to each well to induce protein expression.
5. Protein expression was carried out for 3 hours at 37°C or 20 hours at 18°C (see Note 16).

3.6.2. Preparing samples for protein purification:

1. After protein expression, spin plates at 4,000 rpm for 30 minutes and cell pellets were lysed by freezing overnight at −80°C and then thawed at room temperature for 15 minutes.
2. Cell lysis was continued by shaking for 30 minutes to 1 hour at 1,000 rpm in Vortemp shakers after the addition of 500 μl native lysis buffer.
3. Immediately after lysis, plates were spun at 4,000 rpm for 30 minutes and a Beckman Biomek FX robot was used to separate the supernatant from the pellet by slowly aspirating 300-μl lysis buffer from the top of the solution. This

300-µl supernatant was Sample1 for native purification, which contained only soluble proteins.

4. With Biomek 2000, 500 µl denature lysis buffer was added to the remaining solutions from native purification, followed by mixing and shaking at 1,000 rpm in Vortemp shakers for 30 minutes to dissolve the pellet. This cell lysate was Sample2 for denatured purification.

3.6.3. Protein purification:

Qiagen 96-well filter plates and Ni–NTA were used to purify proteins. All procedures of the Ni affinity purification were carried out on a Beckman Biomek FX robot using two vacuum manifolds at a high (about 800 mbar) and a low (about 200 mbar) vacuum setting as well as a 96-channel pipetting tool (see Note 17).

1. Suspend Ni-NTA resin and distribute 100 µl to each well of 96-well Qiagen Ni–NTA 96-well filter plates.
2. Add 600 µl of charge buffer to each well using Biomek FX robot. Then apply high vacuum for 5 minutes. Release vacuum.
3. Add 300 µl Sample1 or Sample2 into the Qiagen 96-well filter plates with Ni-NTA for native or denature purification and apply low vacuum for 5 minutes. Release vacuum.
4. Add 250 µl washing buffer into plates and apply high vacuum for 5 minutes. Release vacuum.
5. Repeat previous step.
6. Add 200 µl elution buffer to each well to elute purified proteins into prepared 350-µl 96-well plates. After purification, two new samples were obtained: one is from Sample1, named Sample3; another is from Sample2, named Sample4.
7. Add 300-µl strip buffer into filter plates and apply high vacuum for 5 minutes. Release vacuum (see Note 18).

3.7. Enzyme-Linked Immunosorbent Assay

After purification, each bacterial culture plate was separated into four plates for ELISA analysis, one for supernatant without purification (Sample1), one for supernatant with purification (Sample3), one for pellet without purification (Sample2), and one for pellet with purification (Sample4) (see Note 19). Indirect ELISAs were carried out on a Beckman/Sagian core system: an ORCA robotic arm for moving plates, a Biomek 2000 for handling liquid, a Biotek plate washer for washing plates, and a SpectraMax plate reader for recording and analyzing results.

3.7.1. Preparing a sufficient volume of antibody dilution and stain solution for ELISA:

1. Dissolve primary antibody in water to get a concentration of 1 µg/ul, and then dilute 1:500 in blocking buffer.
2. Dissolve secondary antibody in water to get a concentration of 1 µg/ul, and then dilute 1:1500 in blocking buffer.
3. Dissolve 120 mg pNPP in 600 µl of water. Then add to 120 ml of pNPP working buffer.

3.7.2. ELISA:

1. Add 50 µl of PBSN to each well of the ELISA plates using a Biomek 2000 liquid handler.

2. Add 50 μl of sample, 7.5 μl *C. elegans* UBC protein (positive control) (see Note 20) to respective wells, and mix with PBSN. Then seal plates with tape and incubate overnight at 4°C.

3. Wash plates three times with 250 μl washing buffer using an Elx-405 microplate washer.

4. Add 250 μl of blocking buffer and incubate for 30 minutes at room temperature.

5. Wash plates three times with 250 μl washing buffer.

6. Add 100 μl of primary antibody to each well and incubate for 2.5 hours at room temperature.

7. Discard antibody solution and rinse five times with water.

8. Add 250 μl of blocking buffer and incubate for 10 minutes at room temperature.

9. Discard blocking solution and rinse five times with water.

10. Add 100 μl of secondary antibody to each well and incubate for 2.5 hours at room temperature.

11. Discard antibody solution and rinse five times with water.

12. Add 250 μl of blocking buffer and incubate for 10 minutes at room temperature.

13. Discard blocking solution and rinse five times with water.

14. Add 100 μl of pNPP to each well and incubate overnight at room temperature.

15. Read absorbance at 405 nm using a SpectraMax Plus plate reader for 6 hours, with an interval of 30 minutes.

16. Stop hydrolysis with the addition of 50 μl of 0.5 NaOH.

17. With in-house software, the results were electronically compiled and automatically scored by comparison of the sample values to controls. Profiles for protein expression and solubility were generated through linking the gene information of a particular well to its corresponding ELISA values from total and soluble expression experiments.

4. Notes

1. The success rate of expressing soluble eukaryotic proteins is limited when the full-length ORF with Gateway tags was used to express the target protein, whereas a well-folded fragment/domain of the target protein is best suited for expression of a soluble recombinant protein in *E. coli (7–10)*. Therefore, based on available bioinformatics tools, a combined strategy was developed to find possible well-folded domains/fragments.

2. The recombination sequence attB1 used in Gateway system adds additional unwanted amino acids at the N-terminus since the insert is downstream from His6-tag and the additional amino acids in the recombinant protein may interfere with subsequent experiments, such as crystallization. Therefore, it is desirable to engineer a thrombin cleavage site after attB1 site (Fig. 13.4). After the protein is purified, amino acids corresponding to attB1 region can be removed by the thrombin.

3. The integrated robotic system is supported by its system software and in-house programs for specific applications, such as moving plates, shaking, washing, etc. All operations should be tested before carrying out experiments and operators should be trained before using the robots.

4. Since the thrombin cleavage site was included in the primer synthesis and the long primer would be costly and could increase the chance of errors, a PCR strategy was designed using two forward primers and two reverse primers. This strategy has two advantages: Only short primers are required, and primer F2, R2 could be synthesized in bulk. Such measures significantly reduce the cost and the error rate in 96-well operations.

5. Gene specific regions in primer F1 and R1 are designed by BatchPrimer, which was developed by the authors; it results in a pair of primers with a similar melting temperature by adjusting the oligo length. The length range of gene-specific oligos in the program was limited to between 20 to 30 bases according to previous experimental results.

6. Unless stated otherwise, all solutions should be prepared in autoclaved, distill water, which is referred to as "water" in the text.

7. The fresh LB medium containing 50 μg/ml kanamycin or 100 μg/ml ampicillin is added into plates on the day of small culture.

8. This buffer is prepared fresh on the day of purification and add lysozyme of required volume to get a concentration of 1 mg/ml before use.

9. A series of tests showed optimized PCR results could be obtained when proportions of primers F1:F2 and R1:R2 is 1:10 (11).

10. Given that some loss in the operation is unavoidable, the actual amount of premixture prepared was a little larger than that calculated one. For example, if the total reaction number was 96 and the volume of premixture for each reaction was 10.7 μl, the authors prepared 98 × 10.7, i.e., 1048.6 μl premixture, in which the total reaction number was not 96, but 98.

11. The concentration of E-Gel® 96 Agarose and DNA ladder should be chosen according to the length of PCR products: 1% gels resolves 1 kb to 10 kb DNA fragment, and 2% gels resolve 100 bp to 2 kb DNA fragments. For optimal results, load each E-Gel 96 gel within 30 minutes of removing it from the plastic pouch and run within 15 minutes of loading.

12. E-Gel results are checked by UV; used E-Gel 96 gels should be disposed of as hazardous waste since they contain ethidium bromide.

13. To identify growth conditions in 96-deep well block plates that most closely match routine bacterial growth, the *E. coli* strain DH5α was grown for 16 hours at various speeds, volumes, and rotation radii, and results demonstrate that a small shaking radius of 3 mm and a speed of 1,000 rpm reach a slightly higher optical density when compared with the control.

14. If the pellet is not resuspended completely, use the pipette to resuspend it manually.

15. The speed of shaking should not exceed 400 rpm in this step.

16. According the authors' test results, the maximum recovered soluble protein was seen at 20–25 and 3–5 hours postexpression for the 18 and 37°C expressions, respectively.

17. It is critical to program the correct timing for the vacuum steps to avoid residual liquid being left in some wells. Up to 10 minutes are used depending on the volume and pressure being applied to the 96-well filter plates. The final times for each vacuum step were determined by multiple mock runs and adding several minutes to average minimal time to clear all the wells. However, these times no doubt must be modified based on the robotic system, buffers, filter plates, and culture concentration being employed.

18. This step is used to release everything linked with Ni-NTA, and regenerated Ni-NTA could be reused in the next purification.

19. To increase the accuracy of the solubility profiling, which assists in determining optimal conditions for large-scale expression, each gene could be expressed at two temperatures, 37 and 18°C. At that condition, each gene plate was associated with eight ELISA data sets.

20. In these HTP experiments, at least two wells of a 96-well plate were left for the negative and positive control. Normally the well G12 was used, which contained only 1 ml LB medium, and was dealt with as a normal sample in all expression and purification procedures, as negative control for ELISA, and the well H12 contained nothing for the positive control.

Acknowledgment

This work was supported in part by the NIH grant 1P50-GM62407.

References

1. Service, R. F. (2001) Robots enter the race to analyze proteins. *Science* **292**, 187–188.

2. Stevens, R. C., Wilson, I. A. (2001) Industrializing structural biology. *Science* **293**, 519–520.

3. Hartley, J. L., Temple, G. F., and Brasch, M. A. (2000) DNA cloning using in vitro site-specific recombination. *Genome Res.* **10**, 1788–1795.

4. Walhout, A. J., Temple, G. F., Brasch, M. A., Hartley, J. L., Lorson, M. A., van den Heuvel, S., and Vidal, M. (2000) GATEWAY recombinational cloning: Application to the cloning of large numbers of open reading frames or ORFeomes. *Methods Enzymol.* **328**, 575–592.

5. Reboul, J., Vaglio, P., Rual, J.-F., Lamesch, P., Martinez, M., Armstrong, C. M., Li, S., Jacotot, L., Bertin, N., Janky, R., et al. (2003) *C. elegans* ORFeome version 1.1: experimental verification of the genome annotation and resource for proteome-scale protein expression. *Nat. Genet.* **34**, 35–41.

6. Rual, J. F., Hirozane-Kishikawa, T., Hao, T., Bertin, N., Li, S., Dricot, A., Li, N., Rosenberg, J., Lamesch, P., Vidalain, P. O., Clingingsmith, T. R., Hartley, J. L., Esposito, D., Cheo, D., Moore, T., Simmons, B., Sequerra, R., Bosak, S., Doucette-Stamm, L., Le Peuch, C., Vandenhaute, J., Cusick, M. E., Albala, J. S., Hill, D. E., and Vidal M. (2004) Human ORFeome version 1.1: a platform for reverse proteomics. *Genome Res.* **14**, 2128–2135.

7. Symersky, J., Zhang, Y., Schormann, N., Li, S., Bunzel, R., Pruett, P., Luan, C. H., and Luo, M. (2004) Structural genomics of *Caenorhabditis elegans*: structure of the BAG domain. *Acta. Crystallogr. D. Biol. Crystallogr.* **60**, 1606–1610.

8. Lu, S., Symersky, J., Li, S., Carson, M., Chen, L., Meehan, E., and Luo, M. (2004) Structural genomics of *Caenorhabditis elegans*: crystal structure of the tropomodulin C-terminal domain. *Proteins* **56**, 384–386.

9. Yoon, J., Kang, Y., Kim, K., Park, J., and Kim, Y. (2005) Identification and purification of a soluble region of BubR1: a critical component of the mitotic checkpoint complex. *Protein Expr. Purif.* **44**, 1–9.

10. Finch, D., and Webb, M. (2005) Identification and purification of a soluble region in the breast cancer susceptibility protein BRCA2. *Protein Expr. Purif.* **40**, 177–182.

11. Kagawa, N., Kemmochi, K., and Tanaka, S. (2004) One-step adapter PCR method for HTP Gateway technology cloning. *Quest* **1**, 53–55.

Chapter 14

High Throughput Production of Recombinant Human Proteins for Crystallography

Opher Gileadi, Nicola A. Burgess-Brown, Steve M. Colebrook, Georgina Berridge, Pavel Savitsky, Carol E. A. Smee, Peter Loppnau, Catrine Johansson, Eidarus Salah, and Nadia H. Pantic

This chapter presents in detail the process used in high throughput bacterial production of recombinant human proteins for crystal structure determination. The core principles are: (1) Generating at least 10 truncated constructs from each target gene. (2) Ligation-independent cloning (LIC) into a bacterial expression vector. All proteins are expressed with an N-terminal, TEV protease cleavable fusion peptide. (3) Small-scale test expression to identify constructs producing soluble protein. (4) Liter-scale production in shaker flasks. (5) Purification by Ni-affinity chromatography and gel filtration. (6) Protein characterization and preparation for crystallography. The chapter also briefly presents alternative procedures, to be applied based on specific knowledge of protein families or when the core protocol is unsatisfactory. This scheme has been applied to more than 550 human proteins (>10,000 constructs) and has resulted in the deposition of 112 unique structures. The methods presented do not depend on specialized equipment or robotics; hence, they provide an effective approach for handling individual proteins in a regular research lab.

1. Introduction

Protein biochemistry, particularly structural studies, has traditionally been a low throughput affair. Most projects start from a biological question and include extensive preliminary characterization of a protein, careful optimization of the source, expression strategy, purification and preservation of the protein, and continuous monitoring of its functional integrity. The demands and goals of structural genomics may be difficult to reconcile with the traditional approach. First, large sets of proteins are investigated, rather than individual proteins. Second, there is a premium on parallel and relatively uniform procedures. Third, there is a pragmatic focus on proteins that can be solved, even if this includes truncation or modification of the protein. Such a high throughput approach can easily become counterproductive, especially if a degree of flexibility in target selection offers the possibility of abandoning any protein that fails to be solved by straightforward procedures.

From: *Methods in Molecular Biology, Vol. 426: Structural Proteomics: High-throughput Methods*
Edited by: B. Kobe, M. Guss and T. Huber © Humana Press, Totowa, NJ

At the advent of structural genomics centers in the late 1990s, the best methods to treat large, diverse sets of proteins were not clear. It was crucial to identify core method(s) that would capture the largest number of proteins, and devise rational selection processes for second- and third-tier methods to rescue proteins that could not be produced by the standard methods. It was also important to consider which stages in the process would benefit most from parallel processing. Possible (nonexclusive) choices include generating very large numbers of subclones of every protein, testing multiple expression vectors, host cells, expression conditions, detection methods, or purification methods (1–10). Presently, the conclusions from several years of experience can be used to define a rational choice of core and alternative methods (10,11).

The present chapter presents an overview and a detailed protocol used in the Structural Genomics Consortium (SGC) in Oxford. The SGC investigated a predefined set of high-value human proteins: 900 distinct proteins implicated in a variety of metabolic and regulatory processes. This implied that a considerable effort must be devoted to each target to reach acceptable success rates in every step.

The process is based on the premise that the protein is the most important determinant of success (12): A well-behaved protein may be less sensitive to the specific details of expression, purification, and crystallization conditions. Specifically, a full-length native sequence may have disordered segments, internal flexibility, and hydrophobic patches or other surface properties, which may be detrimental to soluble expression or crystallization. Other protein characteristics, such as a preferred maturation pathway or dependence on interaction partners, may be absent when expressed in bacteria (13). Some of these problems may be overcome by using truncated or mutated versions of the protein (14,15). The authors' approach has been to generate 10 or more versions of each protein; the truncation products were tested in small scale, and the most promising versions were subjected to a standard production and purification protocol. This approach has been highly successful: The recovery of targets as soluble proteins and success in structure determination were both increased fivefold by the application of multiple constructs, compared with the recovery from full-length constructs. Not all proteins were optimally recovered from the standard procedure, but the initial tests of each protein identified the best clones for processing and indicated which alternative or follow-up conditions to be used.

The whole procedure requires very little specialized equipment. More than 10,000 clones corresponding to approximately 550 genes were generated and tested over 2 years using manual processes. Automation is incorporated during protein purification, but is not necessary when a lower throughput is required. Consequently, the process described here is applicable to any research lab investigating a limited number of proteins. Parallel processing of multiple versions of an individual protein can dramatically reduce the time and expense required to produce useful protein and determine a structure.

2. Materials

2.1. Enzymes (see Note 1)

1. Pfx Platinum kit (Invitrogen, Carlsbad, CA, see Note 2).
2. BioTaq red DNA polymerase (Bioline Bio-21041).
3. T4 DNA polymerase (Novagen or NEB, see Note 3).
4. Benzonase nuclease (Novagen, HC, 250 units/μl).

5. Restriction enzymes were from NEB.

6. TEV protease is available commercially; a recombinant, his$_6$-tagged enzyme in bacteria was produced in house.

2.2. Bacteria and Plasmids

General cloning is done in Mach1-T1R cells (Invitrogen); any general *recA*⁻ strain would do, but resistance to T1 bacteriophages is an advantage.

Expression is done in derivatives of BL21(DE3). After a serious outburst of bacteriophage infection, a phage-resistant derivative, BL21(DE3)-R3, was generated. This strain was then transformed with the plasmid pRARE2, extracted from the strain Rosetta 2 (Novagen). For unknown reasons, the resulting strain BL21(DE3)-R3-Rosetta has excellent growth and expression characteristics, which outperform the original Rosetta 2 strain.

The cloning vector pNIC28-Bsa4 (GenBank accession EF198106) is described in detail in Section 3.1.1 and Fig. 14.1. The vector and bacterial strains are available upon request.

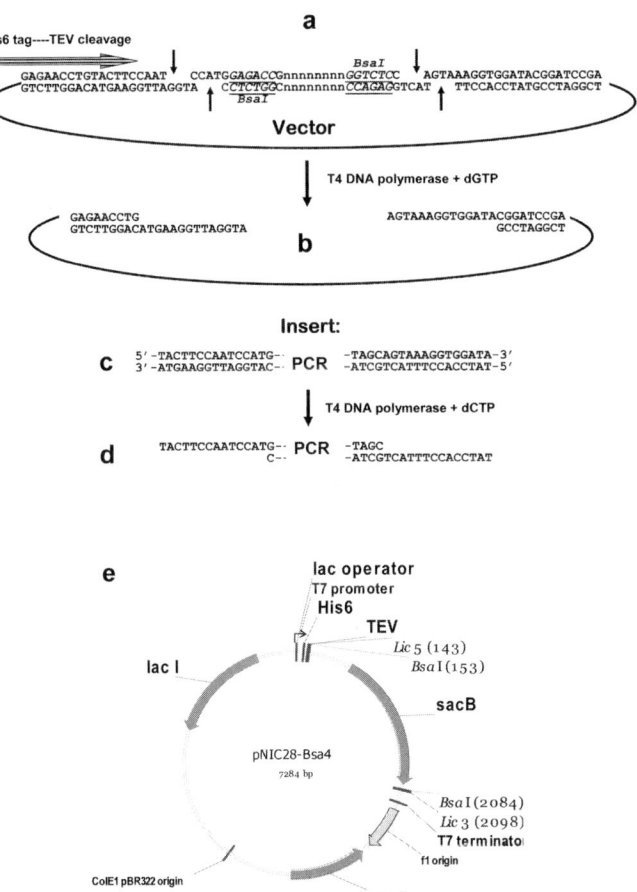

Fig. 14.1 Schematic view of LIC cloning. **A.** The cloning vector pNIC28-Bsa4 is cleaved by the non-palindromic enzyme BsaI. **B.** The resulting ends are trimmed by T4 DNA polymerase and dGTP. **C.** PCR fragments are generated with the indicated extensions. **D.** The PCR fragments are trimmed by T4 DNA polymerase and dCTP, generating single-stranded tails complementary to the vector tails. **E.** Map of pNIC28-Bsa4.

Chemically competent bacterial cells are prepared in house as described [TBF1/TFB2 protocol] *(16)* and stored at −80°C in 0.5- and 2.0-ml aliquots.

2.3. Human cDNA Clones

Entry clones from the Mammalian Gene Collection (IMAGE consortium) were purchased from Geneservice (Cambridge, UK). Other clones were purchased from Origene, Invitrogen, and FivePrime, or isolated in house by PCR from human cDNA. Synthetic DNA clones were purchased from GenScript (Piscatway, NJ; www.genscript.com); the coding sequence was adjusted to match codon frequencies in *E. coli*.

2.4. Chemicals

1. Protease inhibitors: complete, EDTA-free (Roche).
2. Bugbuster (10× concentrate) from Novagen.
3. Oligonucleotides were ordered from Invitrogen or MWG.
4. Ninety-six–well DNA purification kits for PCR products and plasmid minipreps were purchased from either Qiagen or Millipore. Single-cartridge DNA purification kits were from Qiagen.
5. SYBR-Safe DNA gel dye (Invitrogen): The stock solution is diluted 1:30,000 in agarose gels.
6. NiNTA-agarose (Qiagen) or Ni-Sepharose (GE Healthcare) were used with equal efficiency for manual affinity purification.
7. Standard primers for colony PCR:

   pLIC-for: TGTGAGCGGATAACAATTCC.
   pLIC-rev: AGCAGCCAACTCAGCTTCC.

8. Virkon (Antec), Precept (Johnson and Johnson), or bleach were used to decontaminate spent bacterial media and contaminated vessels and surfaces. Virkon is preferred because it does not release noxious gases. Plastic flasks should not be left in oxidizing solutions for more than 1–2 hours, otherwise it is difficult to remove the last traces.

2.5. Plasticware

1. 96-well thin-walled PCR plates (Abgene AB-0900, see Note 4).
2. Adhesive, heat resistant film (Abgene AB-0558).
3. 96 deep-well blocks, 2.2 ml (Thomson 951652B).
4. Disposable sterile inoculation loops (1 μl; Fisher).
5. 96-well filter plates (Chromacol Cat No: CLS-370011- or Thomson Instruments cat no: 921673B).
6. Porous adhesive film for 96-well blocks (Thomson 899410).
7. 2.5-L baffled flasks: Tunair or Ultra-yield (Thomson 931136B), with porous seals (Thomson Airotop seal, 899415).
8. E-gels (Invitrogen): 96-well agarose gels with ethidium bromide, used for rapid DNA analysis.
9. Serum acrodisc syringe filters, 0.2 μ (available as VWR 514–4119).

2.6. Instruments

1. 96-well thermocycler with heated lid.
2. Micro-Express Glas-Col shaker (8 × 96-well blocks; Glas-Col, Terre Haute, IN; see Note 5).

3. High-pressure cell homogenizer: The authors use either the Basic Z model cell disruptor (Constant Systems, Daventry, UK) or an Emulsiflex C5 high-pressure homogenizer (Glen Creston, Stanmore, UK).

4. Chilled shaker-incubators (Innova 4420 or Infors).

5. A visible light spectrometer for measuring OD600 of bacterial cultures. Optional: a 96-well plate reader for multiple, parallel measurements.

6. UV-spectrometer for measuring DNA and protein concentration (The Nanodrop spectrometer allows measurement from tiny volumes).

7. Gel electrophoresis apparatus for DNA and protein analysis. It is convenient to have a configuration that allows gel loading with multichannel pipettors. These are usually arranged so that alternate wells are loaded together.

8. Precast DNA minigels (E-gels, Invitrogen) for quick analysis of PCR products, a 2×52 lane agarose gel apparatus (BioRad) for full-length DNA gels, and 26-lane SDS-polyacrylamide gels (from BioRad or Invitrogen) for protein analysis.

2.7. General Stock Solutions

1. $0.5 \, M$ potassium phosphate, pH 8.0 (referred to as potassium phosphate buffer). This can be made by preparing separate solutions of $0.5 \, M$ K_2HPO_4 and $0.5 \, M$ KH_2PO_4; mix 940 ml of the K_2HPO_4 solution with approximately 50 ml of the KH_2PO_4 solution, measure the pH, and adjust if necessary. Autoclave, store at room temperature.

2. $5.0 \, M$ NaCl. Store at room temperature.

3. $3.0 \, M$ Imidazole, pH 7.5. Dissolve imidazole in 80% of the final volume, titrate with concentrated HCl to the desired pH, and then adjust the volume with water. Autoclave, and store at room temperature.

4. 60% (v/v) glycerol, autoclave, and store at room temperature.

5. $1.0 \, M$ HEPES, pH 7.5. Dissolve HEPES (free acid) in 80% of the final volume, titrate with $5 \, M$ NaOH to the desired pH, and then adjust the volume with water. Autoclave, and store at room temperature. Check for contamination, and discard if cloudy.

6. 5% PEI (polyethyleneimine, Sigma P3143), pH 7.5. Dilute 10 ml of the 50% stock in 80 ml water, and titrate to pH 7.5 with HCl. Adjust the volume to 100 ml.

7. $0.5 \, M$ TCEP (Tris[2-carboxyethyl] phosphine hydrochloride, MW = 286.7): Dissolve the solid in water, store in 1-ml aliquots at $-20°C$.

8. $1 \, M$ IPTG: Dissolve the solid in water, filter-sterilize and store in 1-ml aliquots at $-20°C$.

9. 25%(w/v) sucrose: dissolve completely in water, filter-sterilize.

10. Antibiotics: Kanamycin sulfate, 50 mg/ml in water (filter-sterilize and store in aliquots at $-20°C$). Chloramphenicol, 34 mg/ml in ethanol, store at $-20°C$.

2.8. Cloning

1. LB and SOC media are prepared according to standard formulations.

2. $2\times$ LB: LB medium prepared in 0.5 volume.

3. LB-agar plates (add antibiotics and sucrose as indicated after autoclave; pour 60-mm petri dishes).

   o LB + kanamycin (50 μg/ml) + sucrose (5%, from 25% stock).
   o LB + kanamycin (50 μg/ml).
   o LB + kanamycin (50 μg/ml) + Chloramphenicol (34 μg/ml).

- TB (terrific broth): can be made according to the original method (auto-claving separately the nutrient solution and the phosphate buffer) or using a premixed powder (Merck 1.10629.0500) plus 0.4% glycerol. Ensure the solids are completely dissolved before autoclaving. (Some formulations from other suppliers did not work well in the authors' hands).

2.9. Expression and Purification

1. Protease inhibitors: Complete® EDTA-free tablets from Roche are dissolved in a small volume (up to 4 tablets in 1–2 ml) of buffer by vigorous vortex-ing, and then added to the full amount of buffer. Other protease inhibitor formulations may be used, including 1 mM PMSF.

The following solutions are made by combining general stock solutions and water; no pH adjustment is needed, but it is good practice to check that the pH is between 7.5 and 8.1.

2. Bugbuster lysis buffer: 50 mM potassium phosphate buffer, pH 8.0, 0.5 M NaCl, 10 mM imidazole, 1/10 volume of 10× Bugbuster solution (Novagen), 10% (v/v) glycerol. Add: 1 mM TCEP, protease inhibitors (Complete, EDTA-free [Roche], 1 tablet/50 ml) and Benzonase (1 µl of HC per 5 ml, 50 U/ml) before use.
3. Wash buffer: 50 mM potassium phosphate buffer, pH 8.0, 0.5 M NaCl, 30 mM imidazole. TCEP added to 0.5 mM just before use.
4. Elution buffer: 50 mM potassium phosphate buffer, pH 8.0, 0.5 M NaCl, 0.3 M imidazole, 10% (v/v) glycerol, and 1 mM TCEP.
5. 2× Lysis buffer: 100 mM potassium phosphate buffer, pH 8.0, 1.0 M NaCl, 20% (v/v) glycerol, 1 mM TCEP, with protease inhibitors (Complete, EDTA-free [Roche], 1 tablet/25 ml) and Benzonase (3 µl of HC per 25 ml, 30 U/ml) added before use.
6. Affinity buffer: 50 mM potassium phosphate buffer, pH 8.0, 0.5 M NaCl, 10 mM imidazole, 10% (v/v) glycerol. TCEP is added to 1 mM just before use.
7. Gel filtration buffer: 10 mM HEPES, pH 7.5, 0.5 M NaCl, 5% glycerol, 0.5 mM TCEP added before use.
8. Buffers for protein gel electrophoresis (SDS-PAGE): Commonly available formulations.

3. Methods

This chapter does not include descriptions of standard microbiological tech-niques and gel electrophoresis, the use of instruments and commercial kits. The reader is referred to standard texts and to manuals and protocols provided with instruments and kits.

3.1. Cloning Vectors and Primer Design

3.1.1. Vector

The expression vector pNIC28-Bsa4 (GenBank EF198106) is based on the LIC vector pMCSG7 described in (17). The relevant features include (see Fig. 14.1):

1. pET28a-derived backbone: kanamycin resistance, and Lac-regulated T7 promoter.

2. 22-aa N-terminal fusion peptide, including a 6 × histidine tag and a TEV protease cleavage site. (Tag sequence: MHHHHHHSSGVDLGTENLYFQ* SM, TEV cleavage marked with *)

3. A stuffer fragment between the cloning sites which includes the *B. subtilis* sacB gene; this gene confers sensitivity to sucrose, and provides a selection against the uncut vector.

4. The cloning sites are generated by cleavage at two sites by the non-palindromic restriction enzyme BsaI, followed by digestion with T4 DNA polymerase and dGTP. The 3'-exonuclease activity exposes 14-base single stranded regions on both cleavage sites; complementary single stranded regions are generated in the inserts by T4 DNA polymerase and dCTP. The vector and insert are annealed briefly and introduced into *E. coli*. This cloning procedure is directional, independent of restriction sites, and highly efficient with very little background.

The vector is maintained as glycerol stocks in XL1-blue or other standard bacterial strains. Large-scale plasmid preps are made as required, 200–500-ml cultures, according to the kit instructions. Expect a modest yield of plasmid, 200–900 µg DNA from a 500-ml culture.

The authors have made several vectors suitable for LIC cloning, including bacterial expression vectors with different tags and baculovirus transfer vectors; these vectors, as well as pNIC28-Bsa4, are available upon request.

3.1.2. PCR Primer Design

Primers are designed for amplification of the entire protein coding sequence or smaller domains. Priming sites are chosen based on predicted domain boundaries and secondary structure elements, avoidance of potentially unfolded regions, and specific knowledge of the protein families investigated. Typically, three primers are designed around each of the predicted optimal boundaries, and 9–10 PCR fragments are generated for each target gene (3 × 3 internal primers plus the full-length construct). Primers are designed with the following 5' extensions: TACTTCCAATCCATG for the upstream primer and TATCCACCTTTACTGtca for the downstream primer. The lowercase "tca" in the downstream primer represents a termination codon, which could be replaced by other termination codons. The ATG (underlined) in the upstream primer is obligatory because of the LIC cloning method. Primers of length 35–45 are used without further purification.

Primers are dissolved at a standard concentration (100 µM) in 10 mM Tris-HCl, pH 7.5 or water and stored at −20°C.

3.2. Ligation-Independent Cloning

3.2.1. 96-Well PCR (see Note 6)

1. Design the PCR plate layout: Each template and primer may be used in several wells, so it is useful to prepare separate master plates of templates and primers in case some PCR reactions need to be repeated. This stage can take the better part of a day.

2. Prepare a template master plate: Estimate DNA concentration by A260 (test 1.5 µl on the nanodrop spectrometer) or by running 2 µl in an agarose gel. Dilute each template to 5 ng/µl in water. Distribute 10–20 µl into the appropriate wells of a 96-well plate, store at −20°C.

3. Prepare a primer master plate: Transfer 5 µl of each primer (100 µM stock) into wells of a 96-well plate (combine forward and reverse primers). Dilute the primers to 10 µM each by adding 40 µl of water.

4. Program the thermocycler as follows (see Note 7):

   - 95°C, pause (preheat lid)
   - 95°C, 1 minute
   - 30 cycles of 94°C, 30 seconds, 52°C, 30 seconds, 68°C, 1.5 minutes
   - 68°C, 5 minutes
   - 15°C, pause

   The length of the extension step (30 cycles of 94°C, 30 seconds, 52°C, 30 seconds, 68°C, 1.5 minutes, 68°C) depends on the expected fragment length, approximately 1 min/kb plus 30 seconds.

   Start the program, pausing at 95°C, pause (preheat lid).

5. Prepare the reaction mix, on ice. For a 96-well plate, mix:

	96-well	One reaction
Water	1420 µl	13.5 µl
Pfx amplification buffer	525 µl	5 µl
MgSO$_4$ (50 mM)	52.5 µl	0.5 µl
dNTP mix (10 mM each)	79 µl	0.75 µl
Pfx enzyme	21 µl	0.2 µl

6. Dispense 20 µl of the reaction mix to each well of a 96-well PCR plate on ice.

7. Using a multichannel pipettor, add 2.5 µl of the primer mix and 2.5 µl of diluted template to each well. Seal the plate with a heat-resistant adhesive film. Ensure all wells are sealed—especially along the edges.

8. Place the plate in the thermocycler, continue the run.

9. After completion, analyze 2–5 µl of each reaction by gel electrophoresis (96-well E-gels give results within 10 minutes, alternatively mix each aliquot with DNA sample buffer and run a standard 1% agarose/TAE gel). The E-gels only confirm the presence of bands with an approximate size, but the rapid answer leaves more time for repeating reactions, if necessary (see Note 7 and Fig. 14.2A).

10. Failed reactions may be repeated using different conditions; the reactions can be conveniently assembled using the pre-aliquoted templates and primers.

11. If the template plasmid and the cloning vector have the same antibiotic resistance (e.g., kanamycin), digestion of template DNA remaining after PCR with the methyl-specific restriction enzyme DpnI can reduce background colonies. Dilute 5 µl DpnI in 100 µl NEBuffer 2, add 1 µl to each PCR product and incubate at 37°C for 1 hour.

12. Purify the DNA fragments using a 96-well PCR purification kit. Elute in 50 µl of 10 mM Tris-HCl (pH 7.5–8.5).

3.2.2. Vector Preparation

1. Cut the vector (pNIC28-Bsa4) with restriction enzyme BsaI:
   100-µl reaction volume

   - Water (volume adjustable according to DNA concentration)
   - 10 µl NEB buffer 3
   - 1 µl BSA

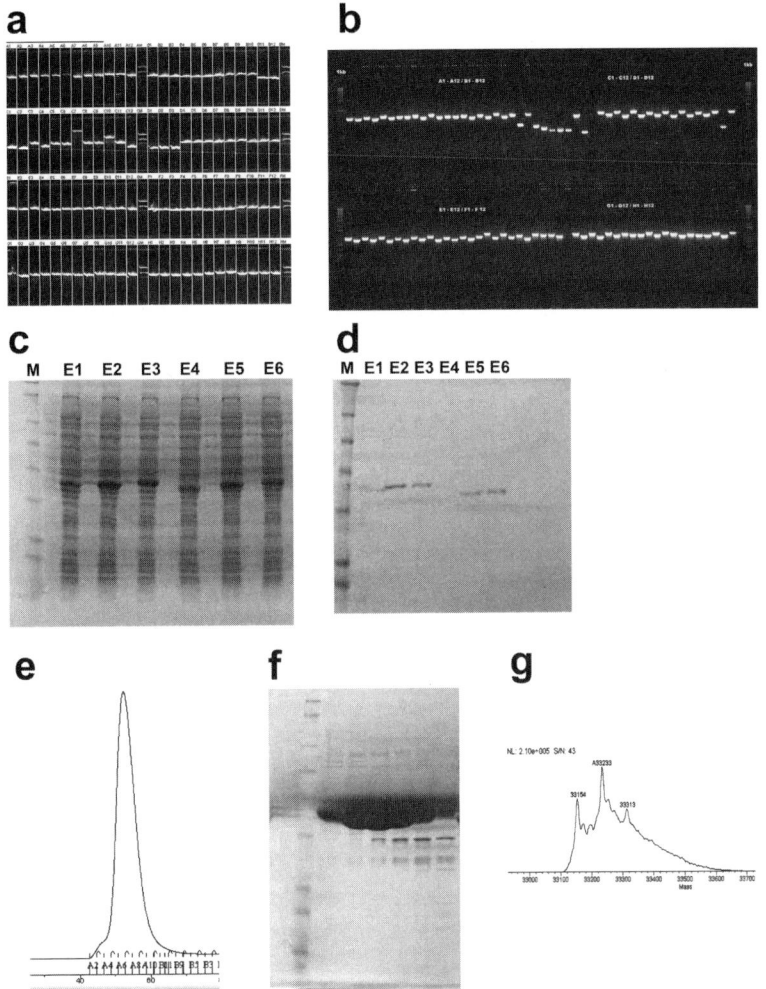

Fig. 14.2 Images of high throughput protein production. **A**. 96 PCR reactions examined using E-Gel precast agarose minigels. PCR results are obtained rapidly, albeit with limited size resolution. **B**. Identification of clones by colony PCR. These are run in full-length agarose gels, so size resolution is better. **C**. Total lysates of 1-ml test expression of six constructs of one gene (MAP3K5). The heavy band indicates high expression levels from all constructs. **D**. Eluted fractions from NiNTA chromatography of the fractions depicted in (c). Only 4 of the constructs yield soluble protein. **E**. Purification of construct E5: Gel filtration chromatogram. Fractions A5–A8 were pooled for further analysis and crystallization. **F**. A (overloaded) gel showing the fractions from gel filtration. **G**. The protein is multiply autophosphorylated in bacterial cells (not shown), treatment with phosphatase reduces the protein to a small number of phosphorylated forms (here, the unphosphorylated and monophosphorylated forms), as revealed by mass spectrometry. This protein (a catalytic domain of MAP3K5) crystallized and its structure was determined by x-ray crystallography (2CLQ).

- 5 μg plasmid (volume determined according to DNA concentration)
- 3 μl (30 units) BsaI restriction enzyme

Incubate 2–3 hours at 50°C.

Note: *BsaI digestion of the vector releases a 2-kb fragment containing the sacB gene, which does not need to be removed from the cut vector!*

2. Purify the DNA using a PCR purification column (Qiagen), elute in 50 μl EB.
3. Generation of cohesive ends (vector):

- In a PCR tube, mix:

Water	21.5 μl
BsaI-digested plasmid	50 μl
10 × T4 DNA polymerase buffer (Novagen or NEB buffer 2)	10 μl
dGTP (25 mM)	10 μl
BSA (NEB, 10 mg/ml)	1 μl
100 mM DTT	5 μl
T4 DNA polymerase (Novagen or NEB)	2.5 μl

4. Incubate 30 minutes at 22°C, and then 20 minutes at 75°C to inactivate the enzyme (preheat the thermocycler lid to 90°C).

3.2.3. Cohesive End Generation (Inserts)

1. Prepare a reaction mix (omitting the DNA):

	96-well plate	One reaction
DNA	—	5 μl
Water	237 μl	2.15 μl
10x T4 polymerase buffer	110 μl	1 μl
dCTP 25 mM	110 μl	1 μl
100 mM DTT	55 μl	0.5 μl
BSA (NEB, 10 mg/ml)	11 μl	0.1 μl
T4 DNA polymerase (Novagen or NEB)	27.5 μl	0.25 μ

2. Dispense 5 μl into each well of a 96-well PCR plate.
3. Using a multichannel pipettor, add 5 μl of each purified PCR fragment to the plate. Mix by pipetting gently up and down once. Seal with adhesive tape.
4. Incubate 30 minutes at 22°C, and then 20 minutes at 75°C to inactivate the enzyme (preheat the thermocycler lid to 90°C).

3.2.4. Annealing and Transformation

1. Prepare 96 60-mm LB-agar plates with 50 μg/ml kanamycin and 5% sucrose, mark each plate on the bottom with the well ID (A1, A2, etc.), and place in convenient stacks at room temperature.
2. Dispense 1 μl of the treated vector into each well of a 96-well PCR plate.
3. Using a multichannel pipettor, add 2 μl of T4-treated insert to each well; seal with adhesive tape (optional: centrifuge for a few seconds).
4. Incubate 10 minutes at room temperature, and then transfer to ice.
5. During the incubation of the DNA fragments, thaw 4 ml of frozen competent MACH1 cells on ice. When completely thawed, mix gently, and then add 40 μl of the cells to each well. Ensure there are no bubbles entrapped at the bottom of the tubes: these can be removed by gentle mixing with a sterile pipette tip.

6. Incubate on ice for 15–30 minutes. Transfer to a water bath or a thermocycler block preheated to 42°C, incubate for 45 seconds, and then return to ice for 1 minute.

7. Remove the plate from ice, add to each well 100 µl of SOC medium. Cover the plate with porous film, and incubate at 37°C for 1 hour (shorter incubation times are not recommended for kanamycin-resistance vectors).

8. Plate the entire transformation mix on 60-mm plates containing LB-agar with 5% sucrose and 50 µg/ml kanamycin. Incubate the plates at 37°C overnight until colonies appear.

3.2.5. Colony Screening (see Note 8 and Fig. 14.2b)

1. Individual colonies from each transformation are screened by placing bacteria directly in a PCR mix. Vector-specific primers are used, annealing to sites flanking the cloning site. The same colonies are used to inoculate 1-ml liquid cultures, which are subsequently used for plasmid purification. Normally, two colonies are picked for every transformation; most colonies will be positive, but the second set can serve as a back-up for failed colonies.

2. Program the thermocycler as follows:

   - 95°C, pause (preheat lid)
   - 95°C, 1 minute
   - 30 cycles of 94°C, 30 seconds, 52°C, 30 seconds, 72°C, 1.5 minutes (adjusted for long fragment size).
   - 72°C, 5 minutes
   - 15°C, pause
     Start the program.

3. For each PCR plate, prepare the following reaction mix on ice:

	96-well	One reaction
Water	1,530 µl	14.6 µl
Bioline 10× PCR buffer	210 µl	2 µl
50 mM MgCl$_2$	63 µl	0.6 µl
dNTP mix (10 mM each)	42 µl	0.4 µl
Vector-specific primers pLIC-for and pLIC-rev, 10 µM each	210 µl	2 µl
Bioline Red Taq DNA polymerase	42 µl	0.4 µl

4. Dispense 20 µl into each well of a PCR plate. Maintain the plate on ice throughout the procedure.

5. Dispense 1 ml of LB/kanamycin (no sucrose) into each well of a 96 deep-well block.

6. Using a sterile 1-µl inoculation loop, pick a well-isolated colony, dip into a PCR well, swirl to release bacteria, and then dip the same loop into the corresponding well in the deep-well block.

7. After inoculating all wells, cover the deep-well block with porous adhesive film, and seal the PCR plate with heat-resistant film.

8. Transfer the PCR plate directly from ice to the preheated block, and continue the program. Incubate the deep-well block overnight in a Glas-col shaker set to 37°C, 700 rpm.

9. Analyze 10 μl of each PCR reaction on a full-size 1% agarose gel (TAE buffer). No sample buffer/dye solution is required if using the Red Taq polymerase, otherwise add 2 μl of 10× DNA sample buffer to each well of the PCR plate before loading. Check carefully for the correct size of each DNA fragment—this is an important quality control. The fragments obtained from colony PCR are 131 bp longer than the original PCR fragments, because of the position of the vector-specific primers.

10. The next day, assemble (if necessary) the overnight cultures from the positive colonies into one 96-well block. Add one third volume of sterile 60% glycerol to each well. The block is then stored at −80°C as a glycerol stock. Inoculate a small amount (10 μl or a loopful) of each culture into a fresh 96 deep-well block containing 1 ml of 2× LB medium + kanamycin. Cover with porous film, and grow overnight in the Glas-col shaker for DNA minipreps.

11. Prepare DNA from all cultures in the block using a 96-well miniprep kit according to the manufacturer's instructions. Elute the DNA in 50 μl.

12. The yields from this vector are not very high, typically 50 μg/ml (2.5 μg total). This is usually sufficient for sequencing and transformation, but more DNA can be made from the glycerol stocks if needed.

3.3. Small-Scale Test Expression

3.3.1. Preparation of Glycerol Stocks (see Note 9)

1. Transform competent cells of BL21(DE3)-R3-Rosetta with the expression plasmids. Transformation efficiencies in this strain are relatively low, so use 7 μl of miniprep DNA (200–400 ng). Plate the transformation on the appropriate antibiotic plate, e.g. LB agar containing kanamycin (50 μg/ml) and chloramphenicol (34 μg/ml).

2. The next day, use an inoculation loop to scoop together 5–10 colonies from each plate, transfer to 1 ml of TB+ kana + chlp in wells of a deep-well block. Grow overnight with vigorous shaking at 37°C.

3. The next day, add one third volume of 60% glycerol, mix, and dispense the cells into a number of replicate microtiter plates. Seal each with adhesive film, and store at −80°C.

3.3.2. Cell Growth and Induction (see Notes 10 and 11)

1. Fill a 96 deep-well block with 0.5 ml of TB + kanamycin + chloramphenicol in each well. Inoculate each well with a lump of cells from the glycerol stock. Alternatively, thaw one of the glycerol stock plates, and use 20 μl of each well to inoculate the cultures; discard the glycerol plate.

2. Cover with porous film, grow overnight at 37°C in the Glas-col shaker at 700 RPM.

3. The next morning, prepare duplicate 96 deep-well blocks with 1 ml of TB + kanamycin. Do not use chloramphenicol at this stage. With a multichannel pipettor, transfer 20 μl of each overnight culture into each well of the new blocks, in duplicate. One block is used for OD measurements; the other block is used for induction. If necessary, additional replicate blocks can be inoculated in parallel, to explore different expression conditions.

4. Cover the blocks with porous film, place in the Glas-col shaker at 37°C, 650–700 rpm. Monitor the cell density by measuring OD600 of selected

wells from one block, starting 2 hours after inoculation (see Note 10). Select for these measurements 4–6 wells with representative densities; dilute 200 μl into 800 μl medium or higher dilutions, so that the measured OD is within the linear range (up to OD = 0.5).

5. When the average OD is 2, dilute 50 μl of each well (of the 96) into 150 μl of medium in flat-bottomed polystyrene plates (ELISA plates). Leave one well with medium only (omit one culture if necessary). Transfer the culture plates to 25°C (change the temperature in the shaker, see Note 11). Measure OD in a multiwell plate reader relative to the medium blank. Multiply all measurements by a constant factor to reach the OD values that would be measured in a 1-cm cuvette, and by 4 (the dilution factor).

6. After the cultures have been at 25°C for 30 min, add to each well 10 μl of 10 mM IPTG (dilute the 1 M stock 100-fold in sterile water; final concentration: 0.1 mM). Continue incubation overnight.

7. Dilute 25 μl of each culture into 175 μl of medium in a microtiter plate, measure ODs as before.

8. Centrifuge the entire block in a swing-out rotor for 20 minutes at 4,000 rpm. Pour off the supernatant into a beaker containing bleach or Virkon. Tap gently on paper towels to remove most of the medium. Seal the block with adhesive film. The block may be stored at −80°C or processed directly.

3.3.3. Small-Scale Purification and Analysis (see Note 12)

1. Prepare 2.6 ml (bed volume) of NiNTA beads by washing twice in water, and then three times in affinity buffer. Suspend in affinity buffer as a 50% slurry. Prepare 45 ml of lysis buffer by adding Benzonase, TCEP, and protease inhibitors.

2. Vortex the block extensively (2–3 minutes) to disrupt the cell pellets. Touch a different spot of the block on the vortex head every few seconds. If properly done, this can save a lot of handling in the next step.

3. Add 400 μl of Bugbuster lysis buffer to each well. Seal the block, and vortex to resuspend the cells. Complete resuspension is crucial; if cell clumps remain, resuspend each well by pipetting up and down using a 1-ml multichannel pipettor.

4. Seal the block, and incubate at room temperature for 30 minutes with occasional shaking. The block may be placed in the Glas-col shaker at 400 rpm.

5. Remove 10-μl aliquots of each well into a 96-well microtiter plate or PCR plate for later analysis as "total extract."

6. Centrifuge the block, 4,000 rpm, 20 minutes. Transfer the clear supernatants to a clean deep-well block using 1-ml multichannel pipette. Hold the block at an angle and take care not to touch the pellets when pipetting.

7. Remove 10-μl aliquots of each well into a 96-well microtiter plate or PCR plate for later analysis as "soluble fraction."

8. Add to each lysate 50 μl of a 50% slurry of NiNTA beads. Seal the block, place in the Glas-col shaker at room temperature, for 1 hour with 400 rpm shaking.

9. Place a 96-well filter block over a 96 deep-well block (drain block). Using a 1-ml multichannel pipettor, suspend the slurry from each well by pipetting up and down and transfer to the filter plate.

10. Centrifuge the filter block on top of the drain block in a swing-out rotor for 1 minute at 500*g*.

11. Discard the effluent from the drain block. Wash the beads by adding 800 µl of wash buffer to each well, and then centrifuging as above. Repeat the wash two more times.

12. Place the filter block on top of a microtiter plate with v-shaped wells and add 50 µl of elution buffer directly onto the beads in each filter well. Incubate 5 minutes, and then centrifuge the filter and microtiter plates for 2 minutes at 500*g*.

13. Analyze by SDS-PAGE the total, soluble and eluted fractions. Dilute 3 µl of the total and soluble lysate fractions with 9 µl water, and use 12 µl of the undiluted eluted fractions. Add to each 4 µl of 4× sample buffer, heat to 95°C for 2 minutes, and then load on the appropriate gels. Stain and de-stain the gels according to standard procedures.

3.4. Standard Scale-up Protocol

3.4.1. Culture Growth and Induction

1. Use glycerol stocks to inoculate fresh starter cultures: Scrape off a loopful of cells, transfer to a 50-ml tube containing 10 ml of TB + kanamycin + chloramphenicol. Grow overnight at 37°C in a shaker. If planning more than 1 L of expression culture, scale up the starter cultures proportionately.

2. Prepare 1 L of TB medium in a 2.5 L baffled flask (see Note 13). It is most convenient to prepare and autoclave the medium directly in the flask, covered with aluminium foil, Before use, add kanamycin (1:1,000 dilution of 50 mg/ml flask) to the cooled medium. During cell growth, cover the flask with an adhesive porous film.

3. Transfer the entire 10-ml culture into 1 L of TB + kanamycin. *The medium should not contain chloramphenicol* at this stage. This ensures more predictable growth, and the pRARE2 plasmid is not lost in the subsequent process.

4. Place in a 37°C shaker, grow with 200 rpm shaking. Monitor the OD600 by taking samples every hour. When taking samples, remove the flask from the shaker and use a 5-ml serological pipette to remove 1 ml of the culture, and then return the flask to the shaker for continued incubation. OD measurements are not linear above 0.5, so if your culture is at higher OD, dilute before measurement to reach the linear range.

5. When the OD reaches 2.00 ± 1, shift the temperature to 25°C. If you have two shaking incubators, prepare a low-temperature shaker in advance, and simply transfer the cultures when ready. If not, set the temperature to 25°C and leave the door open for a while to let the air cool down, and then close the door and continue.

6. After 30 minutes, add IPTG to induce expression. 0.1 mM final concentration is sufficient for most strains: add 1 ml of 0.1 M IPTG for each liter of culture. pLysS strains may require higher concentrations, 1–2 mM IPTG, for efficient induction.

7. Continue incubation for 5 hours to overnight at 25°C.

8. Transfer one or more 5-ml aliquots of the culture to 15-ml conical tubes. Dilute a 25-µl sample into 1 ml medium and measure OD600. Centrifuge the remainder of each aliquot 10 minutes at 4,000 rpm, and discard the supernatant. Remove traces of medium with a 1-ml pipettor. Store the pellets at −80°C for small-scale purification tests.

9. Harvest the cells from the liter-scale cultures by centrifugation, 20 minutes at 6000 rpm in a JA8.1 fixed-angle rotor. Pour the clear supernatant back into the original culture flask, and then decontaminate chemically (using Virkon or bleach) or by autoclave.

10. Scrape the cell pellet with a rubber spatula, transfer to a 50-ml conical polypropylene tube OR to a small ziplock bag. Weigh the cells, record the wet weight. This should be between 12–30 g per 1 L culture. Place in a −80°C freezer or process directly.

3.4.2. Test Purification from 5-ml Samples (optional but useful!)

1. Thaw the pellets from the 5-ml samples. Add 0.5 ml of Bugbuster lysis buffer, Resuspend the cells completely, and transfer to a 2.0-ml Eppendorf tube.

2. Incubate 30 minute at room temperature. Save a 20-μl aliquot for analysis ("total lysate"). Spin the remainder, 14,000 rpm, 15 minutes in chilled Microfuge.

3. Transfer the supernatant to a clean Eppendorf tube. Save a 20-μl sample as "soluble fraction."

4. Add 200 μl of 25% slurry of NiNTA beads, equilibrated in lysis buffer. Tumble for 30 minutes.

5. Spin 1 minute, remove as much of the supernatant as possible without disturbing the beads.

6. Wash the beads four times in wash buffer: each wash consists of resuspending in 10 ml, tumbling for 5 minutes, and then centrifuging and removing the supernatant.

7. Add 100 μl of elution buffer. Incubate with occasional mixing for 10–15 minutes. Spin, and transfer the supernatant carefully to a clean tube. This is the eluate.

8. For gel analysis, dilute the total and soluble fractions fivefold, and use undiluted eluate fraction. for example, mix 3 ul total (or soluble) fraction with 12 μl of water and 5 μl of LDS sample buffer (+DTT); and mix 15 μl of the eluted fraction with 5 μl of LDS sample buffer (+DTT).

9. *The 5-ml purifications provide an estimate of the yield expected from the large-scale cell pellets, allowing to decide how many pellets to process on a single His-trap column for optimal results. Furthermore, the protein from the 5-ml purification is often sufficient for mass spectrometry, to gain information on protein mass, post-translation modification, and proteolysis.*

3.4.3. Cell Extraction (see Note 14)

1. Prechill the cell breaker, centrifuge, and buffers. Add the supplements to the extraction buffers: TCEP, Benzonase, and protease inhibitors. To effectively dissolve the Complete protease inhibitor tables, place the tablets in a small volume (1–2 ml) of buffer in a 1.5- or 15-ml conical tube, and vortex intermittently until completely dissolved. Dilute this into the larger volumes of buffer as required.

2. Thaw the cell pellets at room temperature or (if in conical tube) by placing in a warm water beaker (20–37°C) until just thawed, and then transfer to ice.

3. Add 1 volume (i.e., 1 ml for every gram of cells) of 2× lysis buffer. Resuspend the cells thoroughly with a glass rod or similar implement. Add 2–3 more volumes of lysis buffer until the sample is manageably fluid with no lumps.

4. Transfer to the high-pressure cell homogenizer, and lyse the cells according to the instrument instructions. Continue to flush the homogenizer with lysis buffer until the effluent is clear (20–40 ml) and combine with the lysate.

5. Measure the volume. Add PEI from a pH-adjusted 5% stock solution to a final concentration of 0.15% (30 µl/ml of lysate), and mix thoroughly (see Note 15). The lysates should turn milky. Transfer the lysates to centrifuge tubes, *balance the tubes pairwise*, and centrifuge at 17,000 rpm for 30 minutes at 4°C.

6. There should be a firm pellet with a clear supernatant. Pour off the supernatant into a fresh tube. If the supernatant is considerably turbid or slimy or if some of the pellet dislodges, add further 0.05% (10 µl/ml) of the PEI solution, mix, and repeat the centrifugation. Transfer the clear supernatant to a fresh tube.

7. Filter the supernatant through an Acrodisc filter using a 30-ml syringe (a 10-ml syringe may be used, but take care not to rupture the filter or disengage the syringe at high pressure). For repetitive use of the same filter, remove the filter before pulling out the syringe plunger.

8. Proceed to affinity purification.

3.4.4. Protein Purification: Automated (see Notes 16 and 17)

1. Column preparation can be done the day before the extraction/purification. Set up an AKTA-express system in a cold room or cold cabinet with the desired number of affinity columns (HisTrap FF crude, 1 ml) and a gel filtration column (HiLoad 16/60 Superdex S75 or Superdex S200).

2. Wash the Histrap columns with 10 ml water, and then equilibrate with 10 ml of affinity buffer at 0.8 ml/min.

3. Wash the gel filtration column with two column volumes (CV) of water, and then 2 CV of gel filtration buffer, at 1.2 ml/min.

4. Set up an affinity-gel filtration purification method; change the default parameters as described in the notes. Steps 5–8 are performed automatically if using an AKTA-express system.

5. Load the filtered cell lysate on the affinity column at 0.8 ml/min.

6. After loading is complete, wash the column with 5–10 CV of affinity buffer until the A280 stabilizes. Wash with 10 CV of wash buffer. Elute with 5 CV of elution buffer; the eluted peak, identified by A280, is collected into the reinjection loop.

7. Inject the eluted peak onto the gel filtration column at 1.2 ml/min. Continue running 1.1 CV (132 ml) of gel filtration buffer at the same flow rate.

8. Collect 2-ml fractions based on A280 into a 96 deep-well block.

9. Analyze the factions by gel electrophoresis. Compare elution times with the elution pattern of molecular weight markers from the same column; avoid high-molecular-weight aggregates, and pool separately peaks corresponding to different oligomeric forms (monomer/dimers, etc.).

10. Proceed to Section 3.4.6.

3.4.5. Manual Affinity Purification

The main advantage of manual purification is the improved flexibility: It is possible to adjust the volume of resin to the amount of protein, and perform and collect several elution steps with gradual increases in imidazole concentration.

1. Equilibrate the NiNTA resin in affinity buffer (if originally in 20% ethanol, wash twice in water, and then 3–4 times in affinity buffer).
2. Bind the protein to the resin:
   - Add resin to the cell lysate in conical 50-ml tubes; use 0.25–1 ml of packed resin per litre of culture, depending on the estimated expression level. Close the tube tightly, and tumble for at least 1 hour at 4°C. Centrifuge for 10 minutes at 2,000 rpm, remove most of the supernatant (careful not to lose the resin!). Suspend the beads + remaining liquid in a few ml of affinity buffer, and then transfer to a column. Save the supernatant (unbound fraction). OR
   - Place 0.25–1 ml of NiNTA beads in an appropriate column at 4°C (e.g., a BioRad econocolumn equipped with a top funnel), and pour the cell lysate through the resin at gravity flow. Save the unbound fraction.
3. Wash with 10 column volumes of affinity buffer.
4. Wash with 20 column volumes of wash buffer.
5. At this stage, proceed to elution, either in one step or in several intermediate steps. Elution should be done slowly; it is easy to do by carefully adding small volumes of liquid on top of the resin (0.5 column volumes), waiting for the flow to stop before adding the next aliquot. If fractions are collected separately after each addition, the protein may be recovered in a more concentrated form.
6. Analyze fractions by SDS-PAGE for presence and purity of the recombinant protein.
7. Protein purified through one column is rarely pure enough for crystallization. The protein can be injected onto a gel filtration (after concentrating, if necessary).

3.4.6. Analysis, Further Purification, and Storage

1. Carefully select the column fractions to be pooled. Look for high protein concentration and low level of contaminating proteins. When looking at gel filtration, compare with the elution of MW standards, and collect separately peaks corresponding to monomers, dimers, etc. Take special care to eliminate aggregated proteins (eluting in the void volume): this can make the difference in crystallization! Look for well-formed, symmetric peaks: Avoid long tails that may represent some heterogeneity.
2. Inhomogeneity of the protein can result from contamination with other proteins, or from partial protein modification. In both cases, the protein can be further purified using a variety of chromatographic methods. Ion exchange columns are very effective; specific interactions may be utilized, such as binding to heparin, nucleotides, etc.
3. Tag cleavage: An overnight digestion with TEV protease at 4°C removes the N-terminal tag, including the hexahistidine sequence. The cleaved protein will not bind to a nickel column, whereas most of the contaminating proteins are expected to bind. Passing the cleaved protein through a small column with NiNTA beads (0.25–0.5 ml) removes most contaminating proteins. For this protocol to work, the protein solution must not contain imidazole: the

protein should undergo gel filtration or buffer exchange before attempting cleavage/purification.

4. Protein concentration: Even proteins that survived extraction and purification may aggregate at this stage. This becomes clearly apparent when attempting to concentrate by ultrafiltration: protein aggregates rapidly block the filter, and it becomes impossible to further concentrate the protein. Before committing the whole protein prep, it is best to test a small aliquot in a miniconcentrator. Measure the concentration of the protein, either by A280 (a nanodrop spectrometer is most useful) or by mini Bradford reaction. Transfer 0.2–0.5 ml of the protein solution to a concentration device that fits into a 1.5-ml microfuge tube. Centrifuge according to the manufacturer's instruction at 15°C: Take care not to exceed the recommended speed. Check the volume every 5–10 minutes. The retained protein solution should be concentrated quite rapidly (10–15 minutes) to a protein concentration of at least 5–10 mg/ml. If the process is stuck, with no apparent reduction in volume, it is likely that the protein is aggregating. At this stage, action should be taken to improve protein stability and solubility.

5. Storage: There is no universal agreement on whether to freeze proteins: the merits of freezing clearly depend on the protein at hand. Crystallization experiments should be done with identical protein batches if any reproducibility is desired: it is practically useless to try to follow up an initial hit observed with fresh protein, with a protein that waited on ice for a week or a protein from frozen storage. Hence, the best practice is to quick-freeze the protein in small aliquots, and then thaw the amount required for a crystallization experiment, avoiding repeated thawing. A significant minority of proteins do not survive freezing well. In some cases, a frozen batch of protein can be rescued by repeating the gel filtration, removing aggregated material and isolating monodisperse protein. Other proteins, unfortunately, must be prepared fresh for every crystallization experiment.

6. Freezing protocol: Concentrate the protein; it is useful to include glycerol in the buffer (the gel filtration buffer normally has 5% glycerol). Distribute aliquots of 35–100 µl in thin-wall PCR tubes (strips of eight tubes of different colors are most useful). Close the tubes, and plunge directly into liquid nitrogen. Wear protective face mask and gloves. Transfer the strips to −80°C for storage. When needed, thaw the protein rapidly by placing in a water bath at 25–27°C; watch carefully and transfer to ice when the tube contents have thawed. Transfer to a microfuge tube, and spin 10 minutes at 14,000 rpm to remove aggregated material; transfer the supernatant to a fresh tube. Measure protein concentration: It is important to check if the protein was lost and precipitated.

7. If available, mass spectrometric analysis of every purified protein is highly recommended. This confirms the identity of the protein, reveals mutations or cloning artefacts, and indicates posttranslational modifications. The protein is loaded in a small C3 HPLC column, for de-salting, and eluted onto an inline electrospray ionization—time of flight analyzer (Agilent). Any discrepancy needs to be explained—either by re-sequencing the DNA, by enzymatic removal of suspected modifications, or by MS/MS analysis of proteolytic fragments.

3.5. Quality Issues

1. Entry clones: Entry clones from the MGC collection are assumed to be sequence-verified. With the exception of a few plates that were switched and some missing clones, this has been true for all clones. Entry clones from other sources have been fully sequenced either by the supplier or ourselves; if only a fragment of a long gene is to be used, the relevant region was sequenced.

2. Cloning: In an ideal world, each subclone should be fully sequenced. However, this is expensive and time consuming. The common type of errors we found were primer errors, PCR errors, pipetting and mislabelling errors, and cross-contamination. Our verification of expression clones relies on the following data and assumptions: verified entry clone; correct combination of primers and template; correct transfer through the series of 96-well and agar plates; correct size of inserts; DNA sequencing; and precise molecular weight of the purified protein (by mass spectrometry). Sequencing all clones may be too costly; sequencing a subset of all clones (at least one clone from every gene) provides statistical verification and identification of entry clone errors. The authors also used this information to detect problems with primer quality: In some cases, a high incidence of errors within the primer-encoded sequences identified poor production runs. The main quality control tool is mass spectrometry of every protein batch. Mass deviations may indicate cloning problems or posttranslational modifications.

3. Avoiding bacteriophage infections: Bacteriophages can bring an entire lab or facility to a halt. Once widespread, it can be extremely difficult to completely remove. The authors are using bacterial strains resistant to T1-related phages, which were isolated after a massive phage infection. A lab-wide phage infection will first be seen as sporadic failed cultures: especially cultures that initially increased in density to OD 0.5–1.5 and subsequently decreased. Vigilance is critical: Check every such culture, verify that the correct strain and antibiotic were used, and test for bacteriophages using a standard plaque assay in top agar. If plaques are detected, decontaminate all lab surfaces and the interior of incubators with bleach or Precept, autoclave all vessels and solutions, and try to detect infected areas by swabbing a surface and plaque testing. An early response can save a lot of pain.

3.6. When the Standard Protocol Fails

A significant fraction of proteins are not effectively recovered using the standard procedures described above. Note 18 lists a hierarchy of alternative approaches that have worked in the authors' hands.

4. Notes

1. Reagents and instruments can be obtained from a variety of suppliers; unless stated otherwise, specific brands are listed as guidelines only.

2. For high throughput PCR amplification, it is advised to use a proofreading polymerase. Several products are available; the Pfx kit offers the further advantage of a hot start enforced by an antibody included in the enzyme mix. Comparing a limited number of alternatives, Pfx was found to be

effective, but other products may work as well (including home-made mixes of Taq polymerase with proofreading enzymes).

3. The quality of T4 DNA polymerase is critical; it was found that enzymes from Novagen (LIC-certified) and NEB performed equally.

4. Thin-walled, 96-well PCR plates are used for most reactions. The quality of the seal is critical during PCR reactions but not in most other incubations. The authors prefer the combination of plates and seals mentioned in the preceding for PCR, but cheaper plates and seals can be used for other incubations or storage.

5. The Glas-col shaker *(18)* is necessary for bacterial cultures in 96 deep-well blocks: the combination of 600–700 rpm vibration and the well geometry yield growth curves similar to those obtained in large baffled flasks (OD600 of 10–12 at saturation). Any other shaker/block combination should be tested for growth rate. Other workers have used 24-well plates with 1–3 ml of medium with shaking at 250–300 rpm; aeration may also be improved by adding a sterile 3-mm glass bead to each well.

6. Shifting from single-clone experiments to high throughput presents unexpected challenges. High throughput cloning and expression require, more than anything, a special discipline and well-designed routines to succeed.

   The principal rule is that nearly the whole pedigree and specificity of a clone or a cloning intermediate are defined by position in successive 96-well plates. It is critically important that all transfers from one plate to the next are carried out with utmost precision: it is not uncommon to misalign plates, transferring wells A1, A2, A3…into H12, H11, H10…. Another common error is pipetting a whole row from one plate into a different row of another plate. To avoid such errors, one needs a clean and organized workspace; full concentration on the work at hand; clearly marking a specific corner of each plate (position A1); using a whole tip box for a 96-well set, so that missing tips correspond to finished wells. Arrange all elements on the workspace (source and target plates, tip box, waste bin, etc.) so that all distances and angles allow convenient and smooth repetitive pipetting.

   Most of the work is done with multichannel pipettors (12- or 8-tip varieties). Invest in a good set of pipettors, and ensure the use of compatible tips. When using a multichannel pipettor, always remember that every tip counts: very often, 1 or 2 out of 12 tips will not be properly filled because of small bubbles in the wells or because the tip is not tight enough. ALWAYS look at all tips, after drawing up liquid, to ensure an equal level of liquid in each. If there are differences, return the liquid to the wells (or discard), identify the problem, and then try again. Do not compromise!

   One important step is not done in 96-wells: plating transformed cells for colony isolation. Both when plating the cells and when picking colonies into wells, double-check that the labeling on the agar plate matches the well coordinate.

7. The PCR program described is relatively quick, and gives satisfactory results in most cases. However, depending on the genes amplified and on the quality of the primers, several reactions may fail under those conditions. A more general reaction scheme involves touch-down PCR, which is longer but may be the method of choice in difficult cases.

- 95°C, 5 minutes
- (95°C, 60 seconds; 68°C, 180 seconds) × 5 cycles
- (95°C, 60 seconds; 60°C, 60 seconds; 68°C, 90 seconds) × 5 cycles
- (95°C, 60 seconds; 55°C, 60 seconds; 68°C, 90 seconds) × 5 cycles
- (95°C, 60 seconds; 50°C, 60 seconds; 68°C, 90 seconds) × 20 cycles
- 68°C, 10 minutes
- 15°C hold

When the templates are unusually GC-rich, it may help to add 6% glycerol and/or 5% DMSO to the reaction. The Enhancer component in the InVitrogen Pfx kit may perform a similar function.

8. The PCR fragments generated in this step will not be cloned, so the proofreading DNA polymerase can be replaced by cheaper Taq polymerase. The BioTaq red DNA polymerase used here has the (minor) advantage of a dye included in the reaction, so samples can be loaded directly on gels.

   The results of colony PCR are evaluated by full-length agarose gels in TAE buffer. DNA is visualized by SYBR-safe, a nonhazardous alternative to ethidium bromide. The authors use a gel tray and comb that can be loaded with a standard multichannel pipettor in alternate wells. Sample loading takes some care, but is very quick. The DNA bands are compared to a standard ladder, but more precisely to other samples: It is easier to see the differences between bands of 880, 890, and 900 than to evaluate the exact size of each separately.

   Make full use of this step to identify correctly sized clones.

9. It is widely found that individual colonies from a transformation into an expression strain yield different expression results. Standard protocols advise testing separately several colonies and selecting the best expressors. This is impractical in a high throughput experiment; instead, the authors combine 5–10 colonies to reduce the variability.

   Glycerol stocks are prepared in 5–6 replicates. Ideally, each should be used once and discarded. In practice, after scraping some cells from the top, the (unthawed) stock can be returned to the freezer and used a few more times.

   Finally, the authors have experienced a small number of clones that express recombinant proteins only when fresh transformants are used. In some cases, it may be advantageous to test fresh transformants without freezing and compare with cultures started from glycerol stocks of the same transformants.

10. Measuring culture densities by light scattering (as OD600) may be tedious, but is worthwhile. Growth curves are important characteristics of expression clones. Severe growth inhibition by the expression plasmid suggests using shorter growth/induction times and, possibly, the use of tightly repressed bacterial strains (e.g., pLysS or pLysE). Growth problems with a clone that grew well in previous experiments alerts to potential problems: deterioration of the glycerol stocks, problems with media, or even bacteriophage infection.

    Quick, parallel measurement of 96 culture densities can be done in plate readers. Ensure all wells are filled to a final volume of 200 µl, leaving one well with medium only. The measurement has to be calibrated with a few dilutions of a culture with known density, to derive a conversion constant.

11. The majority of human clones tested in *E. coli* produced high amounts of protein, often detected as a strong band in gels of total bacterial extracts. However, most of the expressed proteins are not soluble. Obtaining soluble proteins is the major challenge when working with most human proteins.

The temperature during inductions has the largest impact. Many proteins that are completely insoluble when expressed at 37°C yield a much larger soluble fraction at temperatures under 30°C. Temperatures in the range of 15–25°C have been used for induction (preinduction growth is done at 37°C). Generally, there is not much difference within this range, so it is often a matter of personal preference. An advantage of 25°C over lower temperatures is that the cells continue to grow at an appreciable rate. This can be convenient when growing several different cultures in parallel: Cultures may be shifted from 37°C to 25°C at the same time, and the slower cultures can be induced at later times than the cultures with high OD. However, a significant minority of clones produce more soluble protein at 15–18°C. This can be tested in small scale and the results implemented at larger scale.

Three more variables that are often modulated are the cell density at induction, the length of induction, and the growth medium. It has been found that growth in TB, induction at an OD600 between 1.5 and 4, and continued growth overnight, produce optimal results for most proteins. However, this may need to be optimized for individual proteins. Various proteins were found to express better in LB medium, at lower OD of induction, or with shorter induction; in this case, larger culture volumes may be needed to produce sufficient cell mass. Expression in LB and for shorter induction times may also help to reduce posttranslational modifications, such as autophosphorylation, which are more pronounced in vigorously growing cultures or in lengthy inductions.

12. The 1-ml test expression experiments have two purposes: identification of the clones with best expression of soluble protein, and parallel optimization of expression conditions.

Results from small scale expression can be good predictors of results at a larger scale *(18)*. There are three caveats: first, expression experiments are not absolutely reproducible, and even repetition of the same experiment can yield different results. Second, the results of small-scale experiments cannot be expected to give accurate quantitative results (i.e., absolute levels of expression). Third, a negative result in a 1-ml experiment cannot be taken as definitive: the sensitivity of these experiments is limited. Indeed, 30% of the structures solved in the SGC emerged from clones that did not produce detectable expression in 1-ml test expression.

Test expression in 50-ml cultures identifies all clones that were not seen in 1-ml cultures because of low sensitivity. When proceeding to protein production, clones that yielded visible bands in 1-ml tests can be grown at 1–4 L to obtain several mg of pure protein. Clones that gave visible results only at 50-ml scale often require 10–50 L to obtain sufficient protein for structure determination, but even this is not excessive in view of the modest price of bacterial cultivation.

Fifty-milliliter test expression is performed exactly as large-scale expression, using 250-ml baffled flasks. The cell pellets are lysed in 3 ml of Bugbuster lysis buffer or in 1× lysis buffer, using sonication to break the cells. The lysate is centrifuged to remove insoluble material, and the supernatant is purified on 100 μl of NiNTA resin as in the manual purification procedure (see Section 3.4.5.).

13. Aeration and stirring of cultures are important determinants of growth and protein production. Good aeration in shaker flasks can be achieved by a combination of flask geometry and by optimizing the ratio of liquid to the total volume. Baffled flasks improve mixing, and allow one to use larger liquid volumes. The flasks used here (Ultra Yield or Tunair) combine excellent mixing and a wide mouth, which provide good aeration for culture volumes up to 1 L in 2.5-L flasks. Conventional baffled Erlenmeyer flasks provide comparable aeration with lower culture to vessel ratios (typically 1:4). When choosing experimental parameters, it is important to grow several test cultures without induction to gauge the growth curves achieved with different culture volumes, vessels, and rotation speeds. An interesting alternative, using 2-L soft drink bottles *(19)*, may be more prone to contamination.

    The main problem with the wide-mouth Ultra-yield or Tunair flasks is the difficulty in maintaining absolute sterility. Acceptable results have been obtained by sealing the flask tops with porous adhesive films; care must be taken to avoid wetting the films, and they should be replaced if there are signs of wetting to puncturing.

    When working in large volumes there may be occasional spills or splashes into the shaker or on lab benches. It is essential to decontaminate such spills immediately.

14. Cell pellets that are not used immediately are stored frozen. The cells may be frozen directly, or after resuspension in a small volume of buffer. In the latter case, thawing may produce a very viscous solution which may be difficult to handle.

    Proteins used for crystallography are best extracted without the use of detergents, as these may be hard to remove. High-pressure cell disruptors provide a very efficient means to break bacteria. It is also possible to use sonication and/or lysozyme treatment.

    Cell suspensions can become extremely viscous if even a small amount of cellular DNA is released. It was found that adding a small amount of nuclease (Benzonase) improves the passage of the cell suspension through the cell disruptor.

15. Nucleic acids may be removed by precipitation with polycations [PEI *(20)*, protamine sulfate, etc.] in the presence of high salt. It is crucial to maintain more than 0.5 M NaCl during this step; otherwise some proteins may be lost by coprecipitation. An alternative method is passing the lysate through an anion exchange column (DEAE-cellulose, DE52) equilibrated in lysis buffer *(21)*. Subsequent filtration of the cleared supernatant is not essential, but can eliminate flow problems during affinity chromatography. The Acrodisc filters, designed for serum filtration, allow easy filtration, whereas regular 0.2–0.45 μ filters are rapidly blocked.

16. It is inherently difficult to find purification conditions that match a very diverse set of proteins. The purification scheme described here has been surprisingly general and effective. Affinity purification using immobilized

metal affinity chromatography (IMAC) is very general, and may be limited only by accessibility of the polyhistidine tag. The important parameter is to maintain a pH of 7.5–8.0 and moderately high salt, and avoid any chelators. High salt conditions (0.5 M NaCl) have proved to be most useful during elution and subsequent gel filtration; a significant number of proteins precipitated when we attempted to reduce the salt concentration. Glycerol has been found to improve protein solubility and stability. Phosphate buffers work best at the affinity chromatography stage. The buffer is later exchanged to HEPES during gel filtration, because phosphates may form salt crystal with many of the crystallization solutions. HEPES or Tris buffers should be used in the affinity step if metal ions (e.g., Mg^{2+}, Ca^{2+} or Zn^{2+}) are included, to avoid precipitation.

17. If using an AKTA-express system for purification, the detection parameters should be changed to accommodate varying protein loads. The authors recommend using the default parameters, with the following changes:

Affinity peak collection:

- Start: Watch level greater than 20 mAU, slope greater than 25 mAU/min.
- Stop: Peak max factor 0.5, watch level less than 20 mAU, watch stable plateau for 0.5 min, delta plateau 5 mAU/min

Gel filtration peak collection:

- Elution volume before fractionation: 0.3 CV
- Elution volume with fractionation: 0.8 CV
- Peak fractionation algorithm: level_OR_slope
- Start level 10 mAu, start slope 5 mAu/min
- Peak max factor 0.5, minimum peak width 0.5 min
- Stop level 10 mAU, slope 5 mAU/min

18. The standardized process, from gene to purified protein, is only the first step in production of a new protein. As a default process, it has yielded acceptable results for a majority of non-membrane proteins. Even in cases with a negative outcome, the standard protocol can provide an indication for selecting alternative approaches. Naturally, any prior knowledge of the protein can be incorporated into the procedure; for example, known stabilizing additives or conditions (e.g., salt, divalent cations, detergents, optimal pH, and nucleotides) can dictate different buffer systems; and the optimal expression conditions may be different from the standard method. The vast diversity of proteins precludes even a universal decision process. However, an outline of a general decision process can be suggested based on statistical impact of the various approaches tested in the authors' hands.

1. Design and generate approximately 10 expression constructs of the target gene in standard host strain.
2. Perform test expression at 1-ml scale, standard conditions (or, preferably, under 2–3 conditions in parallel). Identify constructs producing soluble proteins. If none are detected, proceed to 9.
3. Scale up a few constructs to 1–4 L. Test soluble protein production in 5-ml aliquots. If negative, proceed to 9.

4. Extract the proteins, purify by IMAC, followed by gel filtration, under standard conditions. If protein recovery is low, proceed to 8.

5. Concentrate small aliquots of the protein. If concentration proceeds well, concentrate the entire batch. If the protein does not concentrate, proceed to optimize buffer conditions and test for stabilizing ligands.

6. Concentrated protein should be tested for survival after freezing; if acceptable, freeze in aliquots, perform all crystallization screens from frozen material. Otherwise, crystallize freshly prepared protein.

7. It is good practice to cleave the tag using TEV protease, and test both tagged and cleaved versions in crystallization.

8. Purify the protein manually though affinity chromatography only. Test aliquots of the eluted protein by exchanging the buffer, identify stabilizing conditions. Change the extraction and chromatography buffers accordingly (return to step 3).

9. Perform test expression at 50-ml scale. If positive, scale up (5–50 L), proceed to step 3.

10. If very little soluble protein is recovered from 50-ml cultures, try varying expression conditions (three temperatures, 3 hour vs. longer induction, LB vs. TB).

11. Try other bacterial strains. The most effective, in our hands, has been chaperone coexpression (plasmids available from TaKaRa).

12. Clone the constructs into one or more different vectors. In the authors' hands, C-terminal his$_6$ tags have been very effective in recovery of some targets. Other, larger tags have been used; however, in many cases, subsequent tag removal results in a serious decrease in protein solubility.

13. If all attempts in bacteria fail, consider other recombinant expression systems (baculovirus, mammalian cells, yeast).

References

1. Makrides, S. C. (1996) Strategies for achieving high-level expression of genes in *Escherichia coli. Microbiol. Rev.* **60**, 512–538.

2. Christendat, D., Yee, A., Dharamsi, A., Kluger, Y., Gerstein, M., Arrowsmith, C. H., and Edwards, A. M. (2000) Structural proteomics: prospects for high throughput sample preparation. *Prog. Biophys. Mol. Biol.* **73**, 339–345.

3. Edwards, A. M., Arrowsmith, C. H., Christendat, D., Dharamsi, A., Friesen, J. D., Greenblatt, J. F., and Vedadi, M. (2000) Protein production: feeding the crystallographers and NMR spectroscopists. *Nat. Struct. Biol.* **7**, 970–972.

4. Gerstein, M., Edwards, A., Arrowsmith, C. H., and Montelione, G. T. (2003) Structural genomics: current progress. *Science* **299**, 1663.

5. Stevens, R. C. (2000) Design of high throughput methods of protein production for structural biology. *Structure* **8**, R177–185.

6. Stevens, R. C., Yokoyama, S., and Wilson, I. A. (2001) Global efforts in structural genomics. *Science* **294**, 89–92.

7. Vincentelli, R., Bignon, C., Gruez, A., Canaan, S., Sulzenbacher, G., Tegoni, M., Campanacci, V., and Cambillau, C. (2003) Medium-scale structural genomics: strategies for protein expression and crystallization. *Acc. Chem. Res.* **36**, 165–172.

8. Yakunin, A. F., Yee, A. A., Savchenko, A., Edwards, A. M., and Arrowsmith, C. H. (2004) Structural proteomics: a tool for genome annotation. *Curr. Opin. Chem. Biol.* **8**, 42–48.

9. Yee, A., Pardee, K., Christendat, D., Savchenko, A., Edwards, A. M., and Arrowsmith, C. H. (2003) Structural proteomics: toward high throughput structural biology as a tool in functional genomics. *Acc. Chem. Res.* **36**, 183–189.

10. Albeck, S., Burstein, Y., Dym, O., Jacobovitch, Y., Levi, N., Meged, R., Michael, Y., Peleg, Y., Prilusky, J., Schreiber, G., Silman, I., Unger, T., and Sussman, J. L. (2005) Three-dimensional structure determination of proteins related to human health in their functional context at The Israel Structural Proteomics Center (ISPC). This paper was presented at ICCBM10. *Acta Crystallogr. D Biol. Crystallogr.* **61**, 1364–1372.

11. Mehlin, C., Boni, E., Buckner, F. S., Engel, L., Feist, T., Gelb, M. H., Haji, L., Kim, D., Liu, C., Mueller, N., Myler, P. J., Reddy, J. T., Sampson, J. N., Subramanian, E., Van Voorhis, W. C., Worthey, E., Zucker, F., and Hol, W. G. (2006) Heterologous expression of proteins from *Plasmodium falciparum*: results from 1000 genes. *Mol. Biochem. Parasitol.* **148**, 144–160.

12. Dale, G. E., Oefner, C., and D'Arcy, A. (2003) The protein as a variable in protein crystallization. J. *Struct. Biol.* **142**, 88–97.

13. Baneyx, F. (1999) Recombinant protein expression in *Escherichia coli. Curr. Opin. Biotechnol.* **10**, 411–421.

14. Cornvik, T., Dahlroth, S. L., Magnusdottir, A., Flodin, S., Engvall, B., Lieu, V., Ekberg, M., and Nordlund, P. (2006) An efficient and generic strategy for producing soluble human proteins and domains in *E. coli* by screening construct libraries. *Proteins* **65**, 266–273.

15. Cornvik, T., Dahlroth, S. L., Magnusdottir, A., Herman, M. D., Knaust, R., Ekberg, M., and Nordlund, P. (2005) Colony filtration blot: a new screening method for soluble protein expression in *Escherichia coli. Nat. Methods* **2**, 507–509.

16. Hanahan, D., Jessee, J., and Bloom, F. R. (1991) Plasmid transformation of *Escherichia coli* and other bacteria. *Methods Enzymol.* **204**, 63–113.

17. Stols, L., Gu, M., Dieckman, L., Raffen, R., Collart, F. R., and Donnelly, M. I. (2002) A new vector for high throughput, ligation-independent cloning encoding a tobacco etch virus protease cleavage site. *Protein. Expr. Purif.* **25**, 8–15.

18. Page, R., Moy, K., Sims, E. C., Velasquez, J., McManus, B., Grittini, C., Clayton, T. L., and Stevens, R. C. (2004) Scalable high throughput micro-expression device for recombinant proteins. *Biotechniques* **37**, 364–368.

19. Millard, C. S., Stols, L., Quartey, P., Kim, Y., Dementieva, I., and Donnelly, M. I. (2003) A less laborious approach to the high throughput production of recombinant proteins in *Escherichia coli* using 2-liter plastic bottles. *Protein. Expr. Purif.* **29**, 311–320.

20. Burgess, R. (1991) Use of polyethylenimine in purification of DNA-binding proteins. *Methods Enzymol* **208**, 3–10.

21. Feaver, W. J., Gileadi, O., and Kornberg, R. D. (1991) Purification and characterization of yeast RNA polymerase II transcription factor b. J. *Biol. Chem.* **266**, 19000–19005.

Chapter 15

Assembly of Protein Complexes by Coexpression in Prokaryotic and Eukaryotic Hosts: an Overview

Anastassis Perrakis and Christophe Romier

Most functional entities within cells are composed of protein complexes. The actual challenge for structural biologists is to purify these complexes, or at least functional subcomplexes, in sufficiently large amounts for structural characterization. One major technique for assembling complexes is coexpression of complex components in the same host cell, as it combines the advantages of *in vivo* and *in vitro* techniques. Several hosts can be used for coexpression, including *Escherichia coli*, insect and mammalian cells. Strategies that enable high throughput combinatorial coexpression of many proteins are discussed. The simplicity, versatility, cost effectiveness and success of *E. coli* can only be rivalled by the sophistication of the eukaryotic cells, providing more complicated posttranslational processing of the complex components sometimes required for complex formation. The technique of coexpression can easily be integrated in semiautomated approaches for the high throughput characterization and structure determination of protein complexes.

1. Introduction

Protein complexes appear to play a central role in most cellular functional pathways *(1)*. As a consequence, the field of structural biology is moving to the challenging study of these complexes, making extensive use of the already implemented structural genomics platforms *(2,3)*. There are three major techniques for isolating protein complexes in sufficient quantities for *in vitro* studies. First, *in vitro* reconstitution from separately purified components has been used routinely for many—mostly binary—complexes, but this technique is dependent on the proper folding for each component on its own. Second, purification of endogenous complexes, notably through the use of the TAP-tagging method in yeast *(4)* and *in vivo* biotin tagging in mammalian cells *(5)* are giving promising results. This group of techniques, although excellent for the production of small amounts for general proteomic approaches, is inherently impaired by the low abundance of endogenous complexes.

The third major technique for producing protein complexes, especially at amounts suitable for structural studies, is coexpression. This method allows the heterologous coproduction of several proteins within a single cell and

From: *Methods in Molecular Biology, Vol. 426: Structural Proteomics: High-throughput Methods*
Edited by: B. Kobe, M. Guss and T. Huber © Humana Press, Totowa, NJ

combines several advantages: (1) the use of the efficient existing systems for overexpression of proteins can be directly used for coexpression; (2) *in vivo* reconstitution of soluble (sub)complexes through cofolding of protein partners in the same cell allows proper complex formation; (3) it facilitates the characterization of protein–protein interactions by combinatorial methods; and finally (4) enables the use of protein variants (constructs of various sizes or with point mutations) in an attempt to produce stable subcomplexes amenable to structural studies *(6–8)*.

Escherichia coli is the most common host for performing coexpression experiments, as its rapid growth allows the parallel set-up of a large number of experiments. However, in cases in which proteins are not correctly expressed in *E. coli*, other hosts, such as insect cells (baculovirus system) and mammalian cells can successfully be used *(9,10)*.

A major advantage of coexpression is its ability to be automated, making use of the already implemented platforms, thereby easing the assembly and structural characterization of protein complexes by HTP approaches.

2. Coexpression in *Escherichia coli*

2.1. Several Vectors

The simplest way of performing coexpression in *Escherichia coli* is to use several different vectors. In this case, each vector bears a single gene under the control of a single promoter. This approach is particularly flexible since it allows the screening of various parameters: different constructs for each target gene (see Note 1), different promoters, different tags/fusion proteins (see Note 2) *(6)*. The use of new proteins or new constructs within a study does not imply exhaustive modifications to the existing expression vectors; rather, it only requires the cloning of these proteins and/or constructs into the set of initially selected vectors (see Note 3).

For proper selection of cells having incorporated all plasmids used for cotransformation, each expression vector should harbour a specific antibiotic resistance gene (see Note 4). To avoid any potential competition and any other unwanted side effect during plasmid replication, the plasmids used should preferably have different origin of replications (e.g., ColE1, p15A). However, numerous examples have shown that plasmids with the same origin of replication can be used, without any apparent effect on complex stoichiometry and yield, as long as their genes for antibiotic resistance are different. More important, all vectors should have similar copy number to prevent stoichiometry and yield problems.

For coexpression experiments, *E. coli* cells are cotransformed, either through chemical transformation or electroporation (see Note 5), with the different plasmids chosen. Each protein construct to be coexpressed is encoded on one of these vectors. After transformation and incubation during 1 hour at 37°C in presence of LB, the cells are simply platted on petri dishes containing Agar and the antibiotics selecting for the different plasmids (see Notes 6) and incubated overnight at 37°C, using the conventional protocols. The colonies that grew are then used for inoculating either small (coexpression tests) or large (complex production) cultures (see Note 7), and complex production and purification is carried out using standard protocols as for single proteins.

Coexpression with multiple vectors is not only extremely flexible, but also robust since examples show that coexpression from several vectors can give higher yields than from a single vector. In the worst case, very few examples have shown no coexpression at all from a single vector approach *(8)*. However, the coexpression technique with multiple vectors suffers from some limitations. First, when a large number of proteins are to be coexpressed, the number of cells cotransformed by all plasmids decreases dramatically and growth rates can be extremely reduced in presence of many antibiotics. It is also unclear how the cell can replicate and maintain efficiently many plasmids at the same level, implying the risk of having stoichiometry and yield problems. Second, it is often interesting or even necessary to coexpress proteins with chaperones or rare tRNAs to obtain sufficient amounts of soluble complexes. This increases the number of vectors to be used and may cause compatibility problems between vectors having resistance against the same antibiotic. In these cases, a single vector strategy is better suited.

2.2. Single Vector–Single Transcript Strategies

One way of performing a single vector experiment is to use a vector in which all genes are under the control of a single promoter, each gene being preceded by a ribosome binding site and separated from the preceding gene by a linker region *(8,11)*. This strategy leads to a large single mRNA transcript. This approach is attractive since the size of the vector remains relatively small. However, several aspects have to be carefully considered. First, the order of the genes may have an effect on expression levels. Second, the size and composition of the linker region between two genes may also affect expression levels. Third, large mRNA transcripts may be less stable. However, much work remains to be done to quantify the importance, if any, of these problems. Especially, if the order of the genes was to be important, these kinds of vectors would generate a lot of cloning, unless: (1) one concatenates them through PCR, using long primers that would contain the linker region, or (2) design LIC adaptors as described in Section 2.3.

For coexpression of the genes, the same strategy as for single proteins production is followed, with the use of a single antibiotic. A strategy as described in Section 2.1. is to be carried out if other vectors coding for "helper" macromolecules (e.g., T7 lysozyme, chaperones, rare tRNAs) are used conjointly. Complex expression and purification follow the same requirements as for coexpression with multiple vectors.

2.3. Single Vector–Multiple Transcripts Strategies

An alternative single vector strategy is to have each gene on the same plasmid but under the control of its own promoter. This is exemplified, for instance, by the pET-Duet vectors (Novagen, Darmstadt, Germany). This strategy leads to multiple mRNA transcripts and is particularly interesting since various kinds of promoters can be used, as in the multi vectors approach, depending on the most appropriate for each gene. However, problems of plasmid maintenance within the host may occur due to the large size of these vectors. Clearly, cloning of several promoters within a single vector can also be cumbersome, unless one uses commercial strategies as the LIC adaptors (Novagen). Transformation, expression and purification are carried out as described in Section 2.2.

2.4. Combination Strategies

The various approaches discussed above all have positive and negative aspects. It is important to note that all these approaches are not mutually exclusive but can be nicely combined depending on the project carried out. For instance, when looking for putative partners of a known complex, it is sound to have all genes of the complex cloned onto a single vector (with either a single or multiple promoters), and the various proteins to be tested for interaction cloned independently on a second vector. This procedure can easily be automated and used in a HTP manner for characterizing and assembling protein complexes, as described in Section 5.

3. Coexpression in Insect Cells

Coexpression of proteins in baculovirus infected insect cells has been in use for a long time *(10)*. Typically, experiments are done with multiple viruses, each one bearing a single gene to be overexpressed. A major problem of this approach is that it gives partial as well as full coinfection, and often very careful quantification of the virus titre is required to obtain reasonable results. Typically a multiplicity of infection (MOI) of about 2–5 is needed for maximal efficiency: two to five viruses should infect one cell. When one attempts to coexpress five proteins, to ensure that at least one virus of each of the five components infect the same cell, large MOIs per virus need to be used; that can complicate considerably the standard protocols. However, large complexes have been obtained and purified using this technique, albeit in small amounts *(8)*.

The use of a single vector with multiple promoters appears to be a much more attractive alternative that has been successfully used in a number of cases *(12)*. The preceding strategy has been recently made considerably more efficient and compatible with HTP procedures. In *(13)* a modular system is presented, that is specifically designed for eukaryotic multiprotein expression (MultiBac). In this method any combination of genes, each cloned to an initial transfer vector, can be used to create multigene cassettes for protein coexpression in an *in vitro* Cre-loxP–mediated plasmid fusion procedure. The multigene expression cassette can be easily integrated in the MultiBac baculoviral genome, in an efficient *E. coli* mediated recombination based system. Baculovirus mediated coexpression in insect cells can be remarkably efficient, even if time consuming (see Note 8) and is a proven method for production of protein complexes.

4. Coexpression in Mammalian Cells

Protein expression for structural studies in mammalian cells is becoming an increasingly realistic strategy *(14)*, especially well suited for secreted proteins. Aricescu and coworkers have reported one coexpression trial with a receptor–ligand complex *(9)*, with a straightforward transient cotransfection (mixing two plasmids, each with a gene of interest). For human genes, owing to the homologous mammalian expression environment—as opposed to the truly heterologous cells of *E. coli* or even insect cells—it appears that protein production and correct folding are carried out very efficiently, e.g., in HEK293 cells. Combined with effective strategies for the limited glycosylation of secreted proteins by using genetically modified cell lines like HEK293GnTi *(15)*, expression in mammalian cells is being established as a method of choice when *E. coli* expression strategies fail.

A mammalian cell is typically transfected by a few hundreds of plasmids (unlike the insect cells getting infected by only a few viruses). Thus, cotransformation with many different plasmids will almost invariably achieve transformation of each cell by a few copies of each component. The Oxford Protein Production Facility (R. Aricescu., personal communication) has managed, using simple cotransfection protocols (effectively mixing plasmids together), to produce even a six-component complex in the HEK293 system and has successfully expressed and crystallized various other human protein complexes. In any case, the requirements for multiple constructs (see Note 1) remain; thus, the requirement for HTP platforms.

Transient transfection of cells is quick and efficient for protein production. In the case of proteins that are not toxic for the cell, stable transfection (genomic integration) of the gene of interest is a preferred strategy, especially in the light of the Flp-In™ system (Invitrogen, Carlsbad, CA), which alleviates the need of extensive screening for production-efficient transformants. Crucially, the vector that needs to be constructed for stable transfections with the Flp-In™ system is compatible with transient transfection strategies that can be used efficiently in small scale for high throughput screening of multiple constructs, prior to scale up and establishment of the stably transformed cell line.

For secreted proteins, coexpression (and protein expression in general) in mammalian cells appears to be a much preferred option to insect cells expression systems, when the trivial, efficient and cost-effective *E. coli* strategies fail. In this case, mammalian cell expression is faster, easier, and considerably less expensive than insect cells (see Note 9). Unlike virus construction for insect cells protein expression systems, the cloning steps for the production of plasmids ready for cell transfection (see Note 10), is identical to *E. coli* cloning and any existing HTP cloning platform (e.g., utilizing the LIC system) can be used for that. Further automation of the cell culture steps is feasible with existing of-the-shelf robotic solutions (e.g., Cytomat, Thermo Electron Co., Waltham, MA), but is not deemed necessary to achieve moderate to high throughputs. Finally, combinatorial techniques described for *E. coli* (see Section 5) are even more straightforward for the mammalian system: once a number of genes that correspond to the possible partners of a complex have been cloned into the transfection vector, any combination of these vectors can be tried to assess requirements for complex formation.

The expression of intracellular proteins in mammalian systems offers the additional challenge of producing a large mass of cells in suspension culture (see Note 11) and the intrinsic problem of the multitude of modifications that the mammalian hosts can achieve (e.g., phosphorylation or ubiquitination). Single protein expression has been reported in several cases, but no intracellular protein coexpression has been achieved to the authors' knowledge, although there is no concern fundamentally prohibiting it. The current experience does not allow concrete statements on yields, but initial data indicate they are comparable to the insect cells system in the worst cases.

5. Flowchart of High Throughput Protein Complexes Assembly by Coexpression

To maximize the advantages and avoid most of the pitfalls of the coexpression technique described in the previous sections, it is important to set up a strategy that requires the generation of a rather large set of expression vectors, and an even larger set of experiments, much more important than for single protein expression. A clear consequence is the requirement for parallel/HTP methodologies that allow the processing of many experiments, as well as the development of vectors that are compatible with these approaches.

To test complex formation *in vitro*, the untagged subunits are retained on the affinity column through their interaction with the tagged subunit and should copurify with it in the following purification steps, unless inadequate purification buffer is used. Although coexpression is extremely robust against false-positives in comparison to other techniques (in particular yeast two-hybrid, pull downs, and immunoprecipitation), care should be taken, when screening for potential interactions, to the nonspecific binding of untagged subunits to affinity columns, an issue that is otherwise often only detected at the large-scale purification step. Single expression of the untagged subunit should give an answer to this problem.

Complex characterization and assembly can be carried out using small volume coexpression experiments, prior to large production. In such an approach, initial low-complexity (typically dimeric, but also ternary and quaternary) subcomplexes are being characterized by a first cycle of coexpression experiments using multiple vectors. Once subcomplexes are identified, a second cycle is started to look for larger subcomplexes by coexpressing the known subcomplexes against the other subunits or other identified subcomplexes. In *E. coli*, and to a less extent with the baculovirus system, this strategy is further facilitated by the easy transfer of the known subcomplex genes onto a single vector, whereas with mammalian cells this transfer may not be necessary, although it is feasible. The whole procedure can be repeated as many times as necessary for assembly of the full complex, or at least a functional subcomplex. *E. coli* is the most used and convenient approach to identify stable subcomplexes when no other data on known interactions are specifically known by other techniques (e.g., yeast two-hybrid, immunoprecipitation). Moreover, well tested flowcharts and vector collection are already available. Typically, genes are initially cloned in two compatible vectors and therefore dimers can be identified. However, in some cases more than two proteins are necessary for forming an initial subcomplex. Up to four compatible vectors can be used, or several genes can be inserted in each vector, provided vectors are compatible, at least in terms of antibiotic resistance. Many vectors available for coexpression, either publicly or commercially, are compatible with this approach, although some of them may impose restrictions due to the number of genes that can be cloned onto a single vector, or may require many cloning steps. Strategies by concatenation using long primers can also be used, although they would require sequencing of the genes at each cycle. Two sets of vectors have been especially designed to fit with this procedure in *E. coli*: pET-MCN (Multi-Cloning and expressioN) and pET-MCP (Multi-Cloning

Promoter) *(8)*. For this, four compatible initial vectors harbouring single genes have been modified to allow the formation through a regenerative cut-and-paste strategy of vectors containing several genes under the control of a single promoter (pET-MCN) or several promoters (pET-MCP).

6. Notes

1. The choice of the boundaries for protein constructs, as well as internal deletions is of paramount importance for the coexpression of soluble complexes. Constructs that are too short, too long, or contain unstructured loops may give insoluble aggregates, unless coexpressed with additional partners. In this respect, a large number of constructs should be made in order to avoid false-negatives.

2. The position, the chemical/electrostatic character, and the size of the tags/fusion proteins are of paramount importance for the coexpression of soluble complexes. A tag or fusion protein can negatively affect the assembly of a complex. In this respect, different tags/fusion proteins should be tested at various positions (N- or C-terminal) for each construct. Unless specifically required (e.g., solubilization by fusion protein), coexpression should be performed with a single tagged subunit, the other subunits being untagged thereby decreasing the risk of false-negatives. For the same reason, the protein bearing the tag should also be varied. The coexpression of subunits bearing an identical tag or fusion protein that will be used for purification is to proscribe.

3. Due to the large number of cloning required for coexpression, it is highly advisable to choose a single or a limited set of restriction enzymes (e.g., NdeI/BamHI) that should be used for cloning of every construct into expression vectors. It is definitively less work to mutate the few internal restriction sites that may occur, than to choose different restriction sites for each protein. Alternatively, other cloning strategies (e.g., LIC or Gateway) are to be considered, but different primers should be ordered for each construct in order to obtain tagged and untagged versions.

4. For selection, the normal amount of each antibiotic should be used. However, to prevent slow growth in liquid cultures, the amount of each antibiotic can be reduced. Typically, antibiotic quantities can be halved for coexpression with two to three plasmids.

5. The number of cells cotransformed by all plasmids is decreasing upon increase of the number of plasmids used. Therefore, it is necessary to increase the quantity used for each plasmid to be cotransformed. Typically, quantities in the range 100 to 200 ng should be used for each plasmid in the case of a cotransformation with 2–3 vectors. Electroporation is a technique of choice for cotransformation, but it is not very well adapted to the large number of transformations to be made when performing large-scale coexpression experiments. In the latter case, chemical transformation should be preferred, but particular care must be taken to prepare highly competent cells. A more time-consuming but effective technique is to transform one plasmid, select by its antibiotic resistant the transformants, grow competent cells of the latter, and then transform them with the second plasmid, which has a new antibiotic marker.

6. For large scale coexpression tests that are performed in small volumes (typically 1–2 ml), transformed cells can be plated on 24-well culture plates, each well containing 500 µl of agar + antibiotics. A few colonies are necessary for inoculating the liquid cultures. For large cultures, the transformed cells should be plated on normal or large petri dishes that can be used directly as culture starters (see Note 7).

7. Complex yields are decreasing with the increasing number of subunits coexpressed. Several approaches can be used to prevent this decrease. First, use high-density media, such as the auto-induction media *(16)*, which is very well suited for performing large scale coexpression tests since there is no need to induce complex production at a certain time point that may vary from culture to culture. Second, use cells grown on Petri dishes as starter rather than precultures. Third, assemble only soluble subcomplexes, if possible, that can then be used for *in vitro* reconstitution.

8. The protocol for large scale virus production, to infect a sufficient amount of insect cells, requires many virus generations (P0, P1, P2) and requires several weeks. This step of virus production is not straightforward to integrate for HTP production. Reasonable throughput can be achieved though with the systems available (e.g., MultiBac), even in a manual manner, testing several possibilities within a few weeks. The growth of insect cells in large mass is straightforward in normal shaking incubators and using "flat" conical flasks, similar to the ones used for *E. coli* production. Since insect cells bear no antibiotic resistance, care must be given to avoid infections. Attempts to counter the typical infections by common antifungals (Fungizone, amphotericin) must be avoided at all costs since they interfere seriously with virus infection efficiency and subsequently protein production.

9. In the authors' hands, the expression of secreted proteins from mammalian cells is about 10 to 30 times more efficient (measured as expressed protein per volume of culture) than insect cells. This efficiency partially alleviates the cost concern for consumables (growth media) in the majority of cases. In a typical experiment, cells are grown in T175 cell culture flasks to 90% confluency, in growth media with 10% serum. Cells from one flask are then split to five 15 cm petri dishes and left to grow for 24 hours (to reach about 60–70% confluency), in 15 ml growth media containing 10% serum. At this point, cells are placed in serum-free growth medium and transfected.

10. An often overlooked step is the production of large quantities of DNA for transfection on scaled-up experiments and the amount of transfection agent needed. Some very efficient commercially available lipid-based transfection strategies are often prohibitively expensive; simple polyethylamine (PEI) protocols can however be easily optimized and are cost-efficient. Typically, each 15 cm dish (see the preceding) is treated with 100 µl of a 100 mg/ml PEI solution and transfected with 50 µg of plasmid DNA. Some basic cell culture infrastructure is needed. This includes a standard cell culture hood for cell manipulation, and an incubator for cell cultures (5% CO_2).

11. For intracellular protein expression, where a large cell mass is desired cells should grow unattached. HEK293 cells and specifically variants like the

HEK293Ebna1 cells are very well suited for that purpose. For unattached cell cultures, CO_2 shakers or some kind of well-controlled fermentation system is desirable.

References

1. Gavin, A. C., and Superti-Furga, G. (2003) Protein complexes and proteome organization from yeast to man. *Curr. Opin. Chem. Biol.* **7**, 21–27.
2. Banci, L., Bertini, I., Cusack, S., de Jong, R. N., Heinemann, U., Jones, E. Y., Kozielski, F., Maskos, K., Messerschmidt, A., Owens, R., Perrakis, A., Poterszman, A., Schneider, G., Siebold, C., Silman, I., Sixma, T., Stewart-Jones, G., Sussman, J. L., Thierry, J. C., and Moras, D. (2006) First steps towards effective methods in exploiting high throughput technologies for the determination of human protein structures of high biomedical value. *Acta Crystallogr. D Biol. Crystallogr.* **62**, 1208–1217.
3. Fogg, M. J., Alzari, P., Bahar, M., Bertini, I., Betton, J. M., Burmeister, W. P., Cambillau, C., Canard, B., Carrondo, M., Coll, M., Daenke, S., Dym, O., Egloff, M. P., Enguita, F. J., Geerlof, A., Haouz, A., Jones, T. A., Ma, Q., Manicka, S. N., Migliardi, M., Nordlund, P., Owens, R. J., Peleg, Y., Schneider, G., Schnell, R., Stuart, D. I., Tarbouriech, N., Unge, T., Wilkinson, A. J., Wilmanns, M., Wilson, K. S., Zimhony, O., and Grimes, J. M. (2006) Application of the use of high throughput technologies to the determination of protein structures of bacterial and viral pathogens. *Acta. Crystallogr. D Biol. Crystallogr.* **62**, 1196–1207.
4. Dziembowski, A., and Seraphin, B. (2004) Recent developments in the analysis of protein complexes. *FEBS Lett.* **556**, 1–6.
5. Rodriguez, P., Braun, H., Kolodziej, K. E., de Boer, E., Campbell, J., Bonte, E., Grosveld, F., Philipsen, S., and Strouboulis, J. (2006) Isolation of transcription factor complexes by in vivo biotinylation tagging and direct binding to streptavidin beads. *Methods Mol. Biol.* **338**, 305–323.
6. Fribourg, S., Romier, C., Werten, S., Gangloff, Y. G., Poterszman, A., and Moras, D. (2001) Dissecting the interaction network of multiprotein complexes by pairwise coexpression of subunits in E. coli. *J. Mol. Biol.* **306**, 363–373.
7. Jawhari, A., Uhring, M., Crucifix, C., Fribourg, S., Schultz, P., Poterszman, A., Egly, J. M., and Moras, D. (2002) Expression of FLAG fusion proteins in insect cells: application to the multi-subunit transcription/DNA repair factor TFIIH. *Protein Expr. Purif.* **24**, 513–523.
8. Romier, C., Ben Jelloul, M., Albeck, S., Buchwald, G., Busso, D., Celie, P. H., Christodoulou, E., De Marco, V., van Gerwen, S., Knipscheer, P., Lebbink, J. H., Notenboom, V., Poterszman, A., Rochel, N., Cohen, S. X., Unger, T., Sussman, J. L., Moras, D., Sixma, T. K., and Perrakis, A. (2006) Coexpression of protein complexes in prokaryotic and eukaryotic hosts: experimental procedures, database tracking and case studies. *Acta Crystallogr. D Biol. Crystallogr.* **62**, 1232–1242.
9. Aricescu, A. R., Lu, W., and Jones, E. Y. (2006) A time- and cost-efficient system for high-level protein production in mammalian cells. *Acta Crystallogr. D Biol. Crystallogr.* **62**, 1243–1250.
10. Kost, T. A., Condreay, J. P., and Jarvis, D. L. (2005) Baculovirus as versatile vectors for protein expression in insect and mammalian cells. *Nat. Biotechnol.* **23**, 567–575.
11. Tan, S., Kern, R. C., and Selleck, W. (2005) The pST44 polycistronic expression system for producing protein complexes in Escherichia coli. *Protein Expr. Purif.* **40**, 385–395.
12. Berger, I., Fitzgerald, D. J., and Richmond, T. J. (2004) Baculovirus expression system for heterologous multiprotein complexes. *Nat. Biotechnol.* **22**, 1583–1587.
13. Fitzgerald, D. J., Berger, P., Schaffitzel, C., Yamada, K., Richmond, T. J., and Berger, I. (2006) Protein complex expression by using multigene baculoviral vectors. *Nat. Methods* **3**, 1021–1032.

14. Aricescu, A. R., Assenberg, R., Bill, R. M., Busso, D., Chang, V. T., Davis, S. J., Dubrovsky, A., Gustafsson, L., Hedfalk, K., Heinemann, U., Jones, I. M., Ksiazek, D., Lang, C., Maskos, K., Messerschmidt, A., Macieira, S., Peleg, Y., Perrakis, A., Poterszman, A., Schneider, G., Sixma, T. K., Sussman, J. L., Sutton, G., Tarboureich, N., Zeev-Ben-Mordehai, T., and Jones, E. Y. (2006) Eukaryotic expression: developments for structural proteomics. *Acta Crystallogr. D Biol. Crystallogr.* **62**, 1114–1124.

15. Reeves, P. J., Callewaert, N., Contreras, R., and Khorana, H. G. (2002) Structure and function in rhodopsin: high-level expression of rhodopsin with restricted and homogeneous N-glycosylation by a tetracycline-inducible N-acetylglucosaminyl-transferase I-negative HEK293S stable mammalian cell line. *Proc. Natl. Acad. Sci. USA* **99**, 13419–13424.

16. Studier, F. W. (2005) Protein production by auto-induction in high density shaking cultures. *Protein Expr. Purif.* **41**, 207–234.

Chapter 16

Cell-Free Protein Synthesis for Analysis by NMR Spectroscopy

Margit A. Apponyi, Kiyoshi Ozawa, Nicholas E. Dixon, and Gottfried Otting

Cell-free protein synthesis offers fast and inexpensive access to selectively isotope labeled proteins thath can be measured by NMR spectroscopy in the presence of all the unlabeled proteins in the reaction mixture. No chromatographic purification is required. Using an extract from *Escherichia coli* in a simple dialysis system, the target protein can be prepared at a typical concentration of about 1 mg/ml, which is sufficient for subsequent analysis by NMR. This chapter describes in detail the protocol used in the authors' laboratory.

1. Introduction

Cell-free protein synthesis has won great interest in structural biology, since Kigawa et al. demonstrated the possibility of obtaining milligram-per-ml yields in a coupled cell-free transcription-translation system based on *E. coli* cell extract *(1)*. These concentrations are sufficient for analysis by nuclear magnetic resonance (NMR) spectroscopy on modern high-field NMR spectrometers, providing a convenient and rapid way to analyze proteins, which would be difficult or costly to make otherwise, e.g., proteins that are toxic or sensitive to proteolysis *in vivo (2)*, proteins enriched with non-natural amino acids *(3,4)* and proteins requiring binding partners for expression in soluble form *(5)*, including membrane proteins *(6)*. A number of recent reviews can be found in *(5,7–11)*.

Using NMR spectroscopy for the analysis of proteins made by cell-free protein synthesis is attractive, because of the efficiency with which selectively isotope-labeled amino acids can be incorporated into proteins *(5,8,12,13)*. Cell-free protein synthesis uses isotope-labelled amino acids much more economically than *in vivo* expression systems and is much less affected by isotope-scrambling by metabolic enzymes *(14)*. In addition, since the target protein is the only protein synthesized during the reaction, isotope-filtered NMR experiments allow selective observation of the target protein directly in the reaction mixture without prior purification of the protein by chromatography or concentration of the sample *(15)*. Cell-free protein synthesis coupled with selective isotope labeling and NMR analysis thus presents a powerful combination by which proteins can be made from DNA and structurally characterized within 24 hours *(13,15,16)*.

From: *Methods in Molecular Biology, Vol. 426: Structural Proteomics: High-throughput Methods*
Edited by: B. Kobe, M. Guss and T. Huber © Humana Press, Totowa, NJ

The multitude of potential applications of cell-free protein synthesis has been widely recognized and commercial kits for high-yield cell-free protein expression have become available (e.g., the RTS system by Roche, Indianapolis, IN, and the Expressway system by Invitrogen, Darmstadt, Germany). Yet, the costs of these systems tend to be prohibitive for routine applications. The following describes the *E. coli*–based cell-free protein synthesis protocol used in the authors' laboratory at the Australian National University. The process is very efficient once the cell extract and the stock solutions have been prepared, and well over 1,000 NMR-size reactions can be performed with a single 20-L fermenter preparation of cell extract.

Cellular extracts prepared in-house tend to be more active than commercial ribosomal extracts, which are prepared in view of shipping requirements. The methods used by different groups to obtain the cellular extract vary in details *(6,14,17,18)*. The authors' method has been evolved from the protocol by Kigawa et al. *(1,18,19)* in view of NMR applications with minimal sample handling. The preparation of S30 extract uses the protocol by Pratt *(20)* and the heat treatment described follows the protocol described by Klammt et al. *(6)*.

2. Materials

1. Spectrapor #2 and #4 dialysis tubing, 10 mm flat width (6.4 mm diameter), 12,000–14,000 MWCO (Spectrum Laboratories Inc., Rancho Dominguez, CA).
2. French press.
3. Centrifuge with rotors suitable for 50 ml, 500 ml and 1-L tubes.
4. Optical densitometer.
5. 10 ml ultrafiltration centrifugal filtration units, 10,000 MWCO, e.g., Amicon Ultra-4.
6. 10- and 50-ml disposable centrifuge tubes.
7. Shaking tray.
8. SDS-PAGE equipment.
9. Vortex mixer.
10. 20-L capacity fermenter.
11. Tris(hydroxymethyl)-aminomethane (Tris).
12. Potassium acetate.
13. Magnesium acetate.
14. Acetic acid.
15. Potassium dihydrogen phosphate.
16. Dipotassium hydrogen phosphate.
17. Yeast extract.
18. 1-Hydroxyethyl-piperazine ethane sulfonic acid (HEPES).
19. 1,4-Dithiothreitol (DTT).
20. Adenosine triphosphate (ATP).
21. Cytidine triphosphate (CTP).
22. Guanosine triphosphate (GTP).
23. Uridine triphosphate (UTP).
24. Cyclic adenosine monophosphate (cAMP).
25. Folinic acid.
26. Ammonium acetate.
27. tRNA (from *E. coli* MRE 600; Roche).
28. Creatine kinase.

29. Ribonuclease inhibitor (RNasin; Roche).
30. Potassium hydroxide.
31. Potassium glutamate.
32. Magnesium acetate.
33. Creatine phosphate.
34. All amino acids, optionally ^{15}N labeled.
35. Phenylmethanesulfonyl fluoride (PMSF).
36. Antifoam 289 (Sigma, St. Louis, MO).
37. Milli-Q (MQ) water.
38. 1 mg/ml thiamine.
39. 2 M glucose.
40. 1 M Tris- acetate pH 8.2.
41. 50% PEG 8000 w/w: 1000 g PEG 8000 and 1 L of 1× S30 buffer (total volume ~ 2L).
42. LBT: 10 g/L tryptone, 5 g/L yeast extract, 5 g/L sodium chloride, 2.5 ml/L 1 M sodium hydroxide, 0.5 mg/L thymine.
43. 4× Z-medium: 165 mM potassium dihydrogen phosphate, 664 mM potassium phosphate dibasic, 40 g/L yeast extract (20).
44. 10× S30 buffer: 100 mM Tris- acetate, pH 8.2, 160 mM potassium acetate, 140 mM magnesium acetate, pH 8.3 with potassium hydroxide.
45. S30 buffer α: 1× S30 buffer containing 0.5 mM PMSF, 1 mM DTT and 7.2 mM β-mercaptoethanol.
46. S30 buffer β: 1× S30 buffer containing 1 mM DTT.
47. S30 buffer γ: 1× S30 buffer containing 1 mM DTT and 400 mM sodium chloride.
48. Vector containing the gene of interest under the T7 promoter.
49. T7 RNA polymerase or T7 RNA polymerase vector pKO1166 (21).
50. E. coli strain BL21 DE3 star (invitrogen) or Rosetta (DE3) pRARE (Novagen).

3. Methods

The methods outlined in the following describe: (1) the preparation of the S30 extract, (2) the stock solutions required for cell-free protein production, (3) the preparation of the amino-acid mixtures for cell-free protein expression, (4) the protein synthesis reaction, and (5) the preparation of the samples for NMR analysis.

3.1. Preparation of the S30 Extract

S30 extracts can be prepared from E. coli grown in a flask or a fermenter. Fermenter preparations have the advantage of high ribosome concentrations. The protocol described here is for a fermenter preparation.

The stock solutions required for S30 extract preparation are listed in Section 2. The lifespan of the S30 extract when stored at −80°C is greater than 1 year, and a 20-L fermentation yields approximately 100 ml of concentrated extract. This volume of extract is sufficient to prepare over 1,000 NMR samples.

The extract can be heat-treated at 42°C to precipitate mRNA and some unnecessary proteins. This process acts to speed up initial expression of the desired proteins when the extract is used. It does not necessarily improve the final yield (6).

Concentration of the S30 fraction by dialysis against PEG 8000 (1) decreases the volume required to add to each reaction. This allows more room in the reaction mixture for other components.

3.1.1. Day One: Buffer Preparation and Overnight Culture

1. Prepare and autoclave stock solutions of glucose 2 M, (224 ml), thiamine (1 mg/ml, 200 ml), 10× S30 buffer (2 L) and 50% PEG 8000 (2 L) as listed in the materials section. Stir the mixture of 50% PEG 8000 at room temperature overnight. Autoclave the stock solutions of glucose, thiamine and 10× S30 buffer, 1 L of LBT medium and 20 L of MQ water for diluting the 10× S30 buffer stock. Autoclave a large glass funnel and two 250 ml measuring cylinders required for measuring the glucose and thiamine solutions to be added to the medium on day three.
2. Grow a 10-ml culture of BL21 DE3 star cells in LBT overnight at 37°C. If the BL21 DE3 Rosetta pRARE strain is used, include 33 μg/ml of chloramphenicol.
3. After autoclaving, store MQ water and 10× S30 buffer at 4°C.

3.1.2. Day Two: Z-Medium Preparation and Overnight 1 L culture

1. Prepare 5 L of 4× Z-medium buffer.
2. Add 5 L of Z-medium concentrate, 4.5 ml of antifoam and 14 L of MQ water into the fermenter. This gives a total volume of 19 L. When 1 L of overnight culture is added to this the following day, 20 L of 1× Z-medium will result.
3. Sterilize the medium at 121°C using the sterilization function of the fermenter.
4. Calibrate the pO_2 electrode of the fermentes to 0% after 25 min at 121°C.
5. Inoculate 1 L of LBT medium prepared the previous day with the 10 ml overnight culture, and incubate overnight at 37°C.
6. Autoclave 2 L of 50% PEG 8000 and store at 4°C.

3.1.3. Day Three: 20-L Culture and Processing of Cell Pellets

1. Equilibrate the fermenter at 37°C, with 100% air and mixing rate at 395 rpm.
2. Prior to addition of the overnight culture, remove a sample from the outlet at the base as a blank for later measurement of the absorbance of the culture at 595 nm (OD).
3. Add the 1-L overnight culture, 224 ml of 2 M glucose and 200 ml of 1 mg/ml thiamine to the Z-medium using a sterile glass funnel.
4. Allow these to mix before removing a time zero sample for OD measurement.
5. Remove samples for OD measurements every 30 minutes until the culture reaches an OD of 3. This density should be achieved after 3–4 hours.
6. Fill a large sink or tub with a large amount of crushed ice in preparation for the next steps.
7. Drain the culture into 5-L conical flasks and immediately place them on ice. Weigh empty 1-L tubes.
8. Harvest the cells by centrifugation at 4°C for 12 minutes at 10,000 g in 1-L tubes. Use preweighed tubes so that the pellet mass can be determined later.
9. Resuspend the cell pellets in 400 ml of 1× S30 buffer α. Centrifuge the resuspended cells for 10 minutes at 10,000 g and 4°C.
10. Remove the supernatant and pack the pellet by centrifugation for a further 5 minutes. Remove the remaining supernatant with a pipette.
11. Snap freeze the pellets in liquid nitrogen, then store them at −80°C.

3.1.4. Day Four: Preparation of the S30 Extract

1. Remove the pellets from the freezer and defrost slowly on ice for at least 1 hour prior to resuspension.
2. Resuspend the pellets in 400 ml of S30 buffer α, and centrifuge for 12 minutes at 10,000 g.

3. Remove the supernatant and pack the pellet by centrifugation for a further 5 minutes at 10,000 g.

4. Resuspend the pellets in 1.3 ml of S30 buffer α per gram of cells.

5. Subject the suspension to a single pass in a French press at a pressure of 6,000 psi.

6. Centrifuge the lysed cells at 30,000 g at 4°C for 30 minutes to produce the S30 fraction.

7. Prepare three batches of 3 L each of S30 buffer β. Dialyze the S30 fraction in Spectrapor # 4 dialysis tubing against each 3 L batch of S30 buffer β for 1 hour at 4°C, i.e., for 3 hours in total.

7a. Optional: take one third of the S30 extract before the following concentration step. Snap-freeze and store at –80°C.

8. Transfer the dialysis tubes containing the S30 fraction to a large evaporating dish and dialyse against two 1-L batches of 50% PEG 8000. Place the dish on a shaking tray to speed dialysis, and leave for 2 hours at 4°C.

9. Change the 50% PEG buffer and repeat until the volume of the extract has been reduced to approximately half the volume of the S30 extract (about 2 h). The volume is measured by visual comparison with dialysis tubing containing 60 ml of water.

10. The concentrated S30 fraction is then dialyzed against a further 3 L of 1× S30 buffer β for 15 minutes to remove the residual PEG 8000 buffer.

If the extract is to be heat-treated, continue with steps 13a onward. If the extract is not to be treated, continue with steps 11–13.

11. Measure the volume of the extract. Centrifuge the S30 extract (30 minutes at 30, 000 g) if the level of precipitation is high.

12. Dispense the extract into 1ml aliquots in 1.5-ml Eppendorf tubes.

13. Snap-freeze the aliquots in liquid nitrogen and store at –80°C.

11a. Place the dialysis bag containing the S30 extract into a further 2 L of S30 buffer γ for dialysis overnight at 4°C.

12a. Place the dialysis bag containing the S30 extract into a 250-ml Pyrex glass bottle, filled with the S30 buffer γ from the overnight dialysis. Prewarm the S30 buffer γ to 42°C.

13a. Heat the bottle in a 42°C water bath with gentle shaking for 45 minutes exactly.

14a. Dialyze the sample against 1 L of 1× S30 buffer β overnight. Dialysis for 4h is sufficient.

15a. Centrifuge the dialysate at 30,000 g for 30 minutes at 4°C.

16a. Divide the extract into 1ml aliquots and snap freeze in liquid nitrogen, and then store at –80°C.

For each new batch of S30 extract, the concentrations of magnesium acetate and the extract itself must be optimized. A series of reactions using between 9–25 μL S30 extract per 100 μL reaction mixture and between 15–25 mM magnesium acetate should be sufficient to find the optimal concentrations of each.

3.2. Stock Solutions for Cell-Free Protein Production

The stock solutions for cell-free protein production are prepared as listed in Table 16.1. The dry weight of each solute is specified for the corresponding stock volumes. The stock solutions stored at –80°C are removed from the freezer in the afternoon preceding the reaction preparation, and

Table 16.1 Stock solutions for cell-free protein synthesis.

Reagent	Stock conc.	MW	Dry weight (g)	Stock volume (ml)	Final conc. in reaction mixture	Supplier	Storage (°C)
HEPES	2.0 M pH 7.5	238.3	95.35	200	55 mM	Sigma	RT
DTT	0.5 M	154.25	3.85	50[a]	1.7 mM	Sigma	−20
ATP	96 mM	551.14	0.529	10	1.2 mM	Sigma	−80
rNTP	25 mM each	527.1	0.132		0.8 mM	Sigma	
		523.18	0.131		0.8 mM	Sigma	
		484.14	0.121		0.8 mM	Sigma	
Total				10	0.8 mM		−80
Cyclic AMP	100 mM	351.2	0.351	10	0.64 mM	Sigma	−80
Folinic acid	10 mM	511	0.0511	10	68 μM	Sigma	−80
Ammonium acetate (mono hydrate)	9.2 M	77.08	35.5	50	27.5 mM	Ajax Finechem	−80
tRNA	17.5 mg/ml		0.0175	1	0.175 mg/ml	Roche	−20
Creatine kinase	10 mg/ml	18 kDa	0.1	10[a]	250 μg/ml	Roche	−80
RNasin	40 U/μL	50 kDa			500 U/ml	Promega	−80
T7 RNA polymerase					93 μg/ml		−20
pKO1166					32 μg/ml		−20
Plasmid					16 μg/ml		−20
S30 extract					~225 μL/ml		−80
Potassium glutamate	4 M	203.23	40.65	50	208 mM	Sigma	−80
Magnesium acetate	1.07 M	214.45	11.47	50	~19.3 mM	Ajax Finechem	RT
Creatine phosphate	1 M	327.14	16.36	50	80 mM	Sigma	−80
L-Alanine	100 mM	89.09	0.089	10	2 mM	Sigma	−80

[a]In 100 μL aliquots.

placed on ice in a cold room overnight to defrost. To avoid degradation of creatine kinase, it should be taken from the freezer only immediately prior to sample preparation. Folinic acid is slower to defrost and should be left at 4°C overnight. Stocks stored at −20°C can be removed from the freezer at the commencement of sample preparation. The rNTP solution contains CTP, UTP, and GTP at 25 mM each in one stock.

3.3. Preparation of the Amino-Acid Stock Solutions

The amino-acid stock solutions are prepared in three groups, water-, acid-, and base-soluble. Each amino acid is present in the stock solutions at a 50 mM concentration. The acid- and base-soluble combinations are dissolved in 1 M hydrochloric acid and 1 M potassium hydroxide, respectively, such that when they are mixed, the pH of the solutions is close to neutral. The stocks can be kept on ice but the complete mixture must be mixed at room temperature to avoid precipitation at the final near-neutral pH.

For preparations with individual amino acids provided in isotope labelled form, the stocks are prepared minus the amino acid of interest. The labeled amino acid is added later to the final reaction mixture from a stock of a concentration adjusted for economical usage of the labeled amino acids (14). The concentrations of the individual 100× [15]N stock solutions are also listed in Table 16.2.

For combinatorial ^{15}N-labeling (13) each amino acid stock must be prepared in five sets, with different ^{15}N amino acids included in each set.

Table 16.2 Amino Acid Stock Solutions.

Water soluble mixture—dissolved in 10 ml water		
Amino acid	Weight (mg)	^{15}N stock (mM)
Ala	44.5	100
Arg	105.0	35
Gly	37.5	100
His	105.0	15
Lys	91.3	100
Pro	57.6	n/a
Ser	52.6	100
Thr	59.6	15
Val	58.6	100
Acid soluble mixture—dissolved in 10 ml 1 M HCl		
Asn	66.0	35
Asp	66.5	100
Cys	60.1	35
Glu	73.5	100
Gln	73.1	100
Leu	65.6	100
Met	74.6	100
Trp	102.0	5
Tyr	90.6	5
Base soluble mixture—dissolved in 10 ml 1 M KOH		
Ile	65.5	15
Phe	82.5	35

The same procedure is used to prepare each of the three types of amino acid stock solutions.

1. Weigh out the individual amino acids in the quantities listed in Table 16.2.
2. Place all of the amino acids into a 10-ml tube.
3. Add 10 ml of the appropriate solvent to the mixture and vortex until dissolved.
4. Store the amino acid stock at −80°C.

3.4. Carrying Out the Cell-Free Reaction

3.4.1. Preparation of the Reaction Chamber

1. Soak one 10 cm length of Spectrapor #2 dialysis tubing per reaction in MQ water for 5 minutes, then tie a knot as close as possible to one end. Trim the knot to have minimal overhang.
2. Using clean scissors, prepare one 500-µL Eppendorf tube per reaction by cutting off the lid and hinge sections. The lid is retained and the hinge is discarded. Removal of the hinge allows the tube to fit inside a 10-ml tube. The bottom 0.5 cm of the tube is then sliced off using a scalpel.
3. Insert each cut Eppendorf tube into the top of a pre-tied length of dialysis tubing, as shown in Fig. 16.1. The Eppendorf tube ensures the contents of the dialysis tube are readily accessible. Remove the dialysis tubing from the MQ water and insert the Eppendorf tube immediately prior to use, so that the tubing does not dry out before assembly of the complete reaction.

Fig. 16.1 Schematic drawing of the assembly of the cut Eppendorf tube in the knotted dialysis bag to provide a readily accessible container for the reaction mixture.

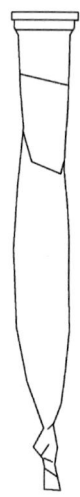

3.4.2. Preparation of the Reaction Mixtures

Working on ice at all times, mix the stock solutions in sequence according to Subheadings 3.4.2.1–3.4.2.3. The amino-acid mixture is prepared first, followed by the 10× B solution that contains nucleotides and other factors required in both the inner and outer chambers (Table 16.3). Finally, the reaction mixture and outside buffer are prepared (Table 16.4).

The quantities listed in this section are for one 500-µL reaction mixture with 5 ml outside buffer. An Excel spreadsheet is available online (http://rsc.anu.edu.au/~go) that facilitates scaling to different volumes, number of reactions or different vector concentrations. The quantities of vectors listed assume stock solutions with a concentration of 1 mg plasmid per ml to achieve final concentrations in the reaction mixture of 16 µg/ml, as per Table 16.1. If different volumes of vector solutions are to be added, adjust the quantity of water so that the final volume of the reaction is 500 µL. The pH of the outside buffer must be adjusted to 7.5 using 1 M KOH, and the volume of water added adjusted accordingly.

T7 RNA polymerase must be supplied if the expression of the target gene is under control of the T7 promoter. The pKO1166 vector is used in this example

Table 16.3 10× B Solution for a 500 µL Reaction with 5 ml Outside Buffer.

Stock solution	µL
rNTP	179.5
HEPES	162.5
ATP	70.0
Water	38.5
Folinic acid	38.0
Cyclic AMP	35.0
DTT	19.0
Ammonium acetate	17.0
Total	559.5

Table 16.4 Components of a 500-µL Reaction Mixture and 5-ml Outside Buffer.

Stock solution buffer (µL)	Reaction mixture (µL)	Outside
10× B	50.00	500
Creatine phosphate	40.00	400
Potassium glutamate	26.00	260
Magnesium acetate	7.00	90
L-Alanine	5.00	50
Amino acid mixture	35.00	335
RNAsin (optional)	6.25	
Creatine kinase	12.50	
T7 polymerase vector pKO1166	8.00	
S30 extract	112.50	
Vector (gene) of interest	8.00	
tRNA	10.00	
Water (adjust)	179.75	3,365
Total	500.00	5,000

rather than purified T7 RNA polymerase so that fresh polymerase is continuously generated during the reaction *(22)*. In most cases, pKO1166 delivers higher protein yields than purified T7 RNA polymerase, whereas experiments with incorporation of non-natural amino acids may depend on provision of the purified polymerase, which should be supplied at a final concentration of 32 µg/ml. When using the purified polymerase enzyme there is no need for additional L-alanine to be added to the reaction mixture. The water volume to be added must be adjusted accordingly *(21)*. The synthesis and purification of T7 RNA polymerase from a strain containing pKO1166 has been described in *(14)*.

3.4.2.1. Mixing the Amino-Acid Stock Solutions for Use in a 500-µL Reaction Mixture with 5 ml Outside Buffer

1. Pipette 37.5 µL water into a 1.5 ml Eppendorf tube.
2. Add 112.5 µL water soluble stock and pipette up and down to mix well.
3. Add 112.5 µL acid soluble stock and pipette up and down to mix well.
4. Add 112.5 µL base soluble stock and pipette up and down to mix well.

After mixing all amino acids, the solution must be kept at room temperature in order to avoid precipitation.

3.4.2.2. Preparation of the 10× B Mixture

Working on ice, add the stock solutions (described in Table 16.1) in the quantities listed in Table 16.3 in a 1.5-ml Eppendorf tube.

3.4.2.3. Preparation of the Reaction Mixture and Outside Buffer

Working on ice, mix the components listed in Table 16.4 for the reaction mixture in a 1.5-ml Eppendorf tube. Leave the reaction mixture on ice whilst the outside buffer is prepared in a 10-ml tube. Adjust the pH of the outside buffer using 1 M KOH prior to making up the final volume with water.

Fig. 16.2 Typical SDS-PAGE gel of S30 extract *(left lane)* and reaction mixture after expression of the protein τ_C16 *(right lane)*. Mobilities of molecular mass markers were as indicated.

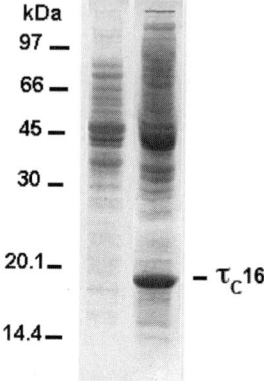

3.4.2.4. Assembly of the Reaction into the Reaction Chamber

1. Pipette the reaction mixture into the dialysis tubing and seal the Eppendorf tube with the pre-cut lid.
2. Place the dialysis tubing inside the 10-ml tube containing the outside buffer.
3. Place the entire assembly into a 37°C water bath, shaking at 200 rpm, for 6–8 hours or overnight.

At the end of the incubation period the reaction mixture may be placed on ice overnight or processed immediately. The sample is ready for initial SDS-PAGE electrophoresis after steps 1 and 2 of Section 3.5, which remove precipitated reaction components. Five µL of the supernatant is generally sufficient to give clear bands with Coomassie Brilliant Blue stain. Figure 16.2 shows a typical example from the expression of the protein τ_C16 *(13)*.

3.5. Preparation of the Samples for NMR Analysis

1. Pipette the reaction out of the dialysis tubing with a 200-µL pipette.
2. Place the mixture in an Eppendorf tube and centrifuge for 1 hour at 30,000–100,000 *g* at 4°C.
3. Prepare 2 L of the desired NMR buffer.
4. Remove the supernatant from the tube and place it into dialysis tubing (pure size depending on the molecular weight of the protein).
5. Dialyze the sample against the NMR buffer overnight.
6. Remove a 5–10 µL aliquot of the dialyzed sample for SDS-PAGE analysis.
7. Concentrate the dialyzed sample by centrifuging in an Amicon 10 K MWCO Ultra-4 tube (assuming a molecular weight of the protein greater than 10 kDa) for 10–15 minutes according to the manufacturer's instructions.
8. Adjust the concentrated sample to a total volume of 500 µL by adding 50 µL of D_2O and the NMR buffer.
9. Transfer the sample to an NMR tube for measurement.

The volume of the reaction can be scaled up or down. 200 µL reactions with 2 ml outside buffer are appropriate for optimization of the extract and magnesium acetate concentrations and also when a sample is only required for analysis by SDS PAGE.

Acknowledgment

The authors thank Nikki Moreland and Karin Loscha for a critical reading of the manuscript and Peter S. C. Wu for useful discussions. This work was supported by the Australian Research Council, including project grants (to G. O. and N. E. D.), a CSIRO-Linkage Fellowship (to K. O.) and a Federation Fellowship (to G. O.).

References

1. Kigawa, T., Yabuki, T., Yoshida, Y., Tsutsui, M., Ito, Y., Shibata, T., and Yokoyama, S. (1999) Cell-free production and stable-isotope labeling of milligram quantities of proteins. *FEBS Lett.* **442**, 15–19.
2. Renesto, P., and Raoult, D. (2003) In vitro expression of rickettsial proteins. *Ann. N.Y. Acad. Sci.* **990**, 642–652.
3. Kigawa, T., Yamaguchi-Nunokawa, E., Kodama, K., Matsuda, T., Yabuki, T., Matsuda, N., Ishitani, R., Nureki, O., and Yokoyama, S. (2001) Selenomethionine incorporation into a protein by cell-free synthesis. *J. Struct. Funct. Genomics* **2**, 29–35.
4. Ozawa, K., Headlam, M. J., Mouradov, D., Watt, S. J., Beck, J. L., Rodgers, K. J., Dean, R. T., Huber, T., Otting, G., and Dixon, N. E. (2005) Translational incorporation of L-3,4-dihydroxyphenylalanine into proteins. *FEBS J.* **272**, 3162–3171.
5. Ozawa, K., Dixon, N. E., and Otting, G. (2005) Cell-free synthesis of ^{15}N-labeled proteins for NMR studies. *IUBMB Life* **57**, 615–622.
6. Klammt, C., Löhr, F., Schäfer, B., Haase, W., Dötsch, V., Rüterjans, H., Glaubitz, C., and Bernhard, F. (2004) High level cell-free expression and specific labeling of integral membrane proteins. *Eur. J. Biochem.* **271**, 568–580.
7. Klammt, C., Schwarz, D., Löhr, F., Schneider, B., Dötsch, V., and Bernhard, F. (2006) Cell-free expression as an emerging technique for the large scale production of integral membrane protein. *FEBS J.* **273**, 4141–4153.
8. Ozawa, K., Wu, P. S. C., Dixon, N. E., and Otting, G. (2006) ^{15}N-Labelled proteins by cell-free protein synthesis: Strategies for high-throughput NMR studies of proteins and protein–ligand complexes. *FEBS J.* **273**, 4154–4159.
9. Shimizu, Y., Kuruma, Y., Ying, B. W., Umekage, S., and Ueda, T. (2006) Cell-free translation systems for protein engineering. *FEBS J.* **273**, 4133–4140.
10. Staunton, D., Schlinkert, R., Zanetti, G., Colebrook, S. A., and Campbell, I. D. (2006) Cell-free expression and selective isotope labelling in protein NMR. *Magn. Reson. Chem.* **44**, S2–S9.
11. Vinarov, D. A., Newman, C. L. L., and Markley, J. L. (2006) Wheat germ cell-free platform for eukaryotic protein production. *FEBS J.* **273**, 4160–4169.
12. Kainosho, M., Torizawa, T., Iwashita, Y., Terauchi, T., Ono, A. M., and Güntert, P. (2006) Optimal isotope labelling for NMR protein structure determinations. *Nature* **440**, 52–57.
13. Wu, P. S. C., Ozawa, K., Jergic, S., Su, X.-C., Dixon, N. E., and Otting, G. (2006) Amino-acid type identification in ^{15}N-HSQC spectra by combinatorial selective ^{15}N-labelling. *J. Biomol. NMR* **34**, 13–21.
14. Ozawa, K., Headlam, M. J., Schaeffer, P. M., Henderson, B. R., Dixon, N. E., and Otting, G. (2004) Optimization of an *Escherichia coli* system for cell-free synthesis of selectively ^{15}N-labelled proteins for rapid analysis by NMR spectroscopy. *Eur. J. Biochem.* **271**, 4084–4093.
15. Guignard, L., Ozawa, K., Pursglove, S. E., Otting, G., and Dixon, N. E. (2002) NMR analysis of in vitro-synthesized proteins without purification: a high-throughput approach. *FEBS Lett.* **524**, 159–162.
16. Keppetipola, S., Kudlicki, W., Nguyen, B. D., Meng, X., Donovan, K. J., and Shaka, A. J. (2006) From gene to HSQC in under five hours: high-throughput NMR proteomics. *J. Am. Chem. Soc.* **128**, 4508–4509.

17. Kang, S. H., Oh, T. J., Kim, R. G., Kang, T. J., Hwang, S. H., Lee, E. Y., and Choi, C. Y. (2000) An efficient cell-free protein synthesis system using periplasmic phosphatase-removed S30 extract. *J. Microbial. Meth.* **43**, 91–96.

18. Kigawa, T., Yabuki, T., Matsuda, N., Matsuda, T., Nakajima, R., Tanaka, A., and Yokoyama, S. (2004) Preparation of *Escherichia coli* cell extract for highly productive cell-free protein expression. *J. Struct. Funct. Genomics* **5**, 63–68.

19. Kigawa, T., Muto, Y., and Yokoyama, S. (1995) Cell-free synthesis and amino acid-selective stable isotope labeling of proteins for NMR analysis. *J. Biomol. NMR* **6**, 129–134.

20. Pratt, J. M. (1984) Coupled transcription-translation in prokaryotic cell-free systems. in (Hames, B. D., and Higgins, S. J., eds.), *Transcription and Translation*, pp. 179–209, IRL Press, Oxford, UK.

21. Ozawa, K., Jergic, S., Crowther, J. A., Thompson, P. R., Wijffels, G., Otting, G., and Dixon, N. E. (2005) Cell-free protein synthesis in an autoinduction system for NMR studies of protein-protein interactions. *J. Biomol. NMR* **32**, 235–241.

22. Nevin, D. E., and Pratt, J. M. (1991) A coupled in vitro transcription-translation system for the exclusive synthesis of polypeptides expressed from the T7 promoter. *FEBS Lett.* **291**, 259–263.

Chapter 17

A Medium or High Throughput Protein Refolding Assay

Nathan P. Cowieson, Beth Wensley, Gautier Robin, Gregor Guncar, Jade Forwood, David A. Hume, Bostjan Kobe, and Jennifer L. Martin

Expression of insoluble protein in *E. coli* is a major bottleneck of high throughput structural biology projects. Refolding proteins into native conformations from inclusion bodies could significantly increase the number of protein targets that can be taken on to structural studies. This chapter presents a simple assay for screening insoluble protein targets and identifying those that are most amenable to refolding. The assay is based on the observation that when proteins are refolded while bound to metal affinity resin, misfolded proteins are generally not eluted by imidazole. This difference is exploited here to distinguish between folded and misfolded proteins. Two implementations of the assay are described. The assay fits well into a standard high throughput structural biology pipeline, because it begins with the inclusion body preparations that are a byproduct of small-scale, automated expression and purification trials and does not require additional facilities. Two formats of the assay are described, a manual assay that is useful for screening small numbers of targets, and an automated implementation that is useful for large numbers of targets.

1. Introduction

Many of the proteins that are overexpressed in *Escherichia coli* form insoluble inclusion bodies *(1,2)*. A typical outcome from a high throughput project expressing recombinant proteins in *E. coli* might be that 80% of cloned genes result in some protein product detectable by gel electrophoresis, and this 80% can be subdivided into 20% soluble protein suitable for further analysis and 60% insoluble protein *(3–6)*. Despite this major attrition and despite the development of other high throughput systems such as cell-free *(7)* and yeast expression *(8)*, the vast majority of high throughput structural biology projects use *E. coli* fermentation exclusively as a source of protein.

In many cases proteins can be recovered from the insoluble fraction of an *E. coli* expression and folded into a native conformation via a denaturing step *(9)*. This process is known as protein refolding and has the potential to increase the number of protein targets taken through to structural studies in high throughput projects *(9,10)*. This chapter describes a protein refolding assay suitable for high throughput applications. The assay was designed primarily so that it could be easily integrated into a high throughput structural biology pipeline.

In a standard refolding experiment one protein of special interest is screened against a variety of refolding buffers to identify the condition in which refolding yield is optimal *(11)*. In a high throughput setting screening many buffers can increase sample numbers to a level that is intractable. Although it is true that the choice of refolding buffer can have a dramatic effect on refolding yield, the use of a single standard buffer maximizes throughput. An assumption is made that a protein that can be refolded to some extent in one buffer is worth optimizing by screening other buffers. If no refolded protein can be recovered after refolding in the standard buffer, then optimization is unlikely to be fruitful.

The assay is based on the observation that most 6×His tagged proteins, solubilized from inclusion bodies in a denaturing buffer, then bound to metal affinity resin and equilibrated in a renaturing buffer, will only elute from the resin with an imidazole gradient if they are correctly folded *(12)*. It is hypothesized that misfolded proteins do not elute because they interact strongly with the chromatography resin through hydrophobic interactions. There is a good correlation between the amount of protein that is eluted from the metal affinity resin after refolding and the yield of protein with native conformation from larger-scale fermentations *(12)*. A strength of this assay is that it employs a similar procedure to a standard small-scale protein purification, using equipment, reagents, and protocols common to high throughput laboratories.

Two methods are described—a low throughput manual implementation suitable for several dozens of proteins and a high throughput, automated implementation suitable for large numbers of proteins.

2. Materials

2.1. Low Throughput, Manual Refolding Assay

2.1.1. Protein Resolubilization and Resin Binding
1. Cell pellets from 2-5 ml *E. coli* culture, expressing 6×His tagged protein.
2. Lysis buffer: 8 M urea, 50 mM HEPES pH 7.5, 10 mM imidazole.
3. 50% TALON (Clontech, Mountain View, CA) metal affinity resin preequilibrated in 8 M urea.
4. Sonicator with microprobe.
5. 15-ml conical tubes (Falcon).
6. Benchtop centrifuge capable of spinning conical tubes at speeds of 4,500*g*.

2.1.2. Refolding and Elution of Solubilized Protein
1. Wash buffer: 50 mM HEPES pH 7.5, 300 mM NaCl, 10 mM imidazole.
2. Elution buffer: 50 mM HEPES pH 7.5, 300 mM NaCl, 200 mM imidazole.
3. Benchtop centrifuge capable of spinning conical tubes at speeds of 4,000*g*.

2.2. High Throughput, Automated Refolding Assay

1. 96-well microtiter plate of inclusion body pellets.
2. Resolubilization buffer: 8 M urea, 50 mM HEPES pH 7.5.
3. Wash buffer: 50 mM HEPES pH 7.5, 300 mM NaCl, 10 mM imidazole.

4. Elution buffer: 50 mM HEPES pH 7.5, 300 mM NaCl, 200 mM imidazole.
5. 50% nickel magnetic resin (Promega, Madison, WI) preequilibrated in 8 M urea.
6. Biomek 2000 liquid handling workstation (Beckman, Fullerton, CA) with plate shaker, "Magnabot" (Promega), gripper tool and P200 8-channel pipette head.
7. Automated MagneHis™ Protein Purification System Protocol #EP011 (Promega) running on Biomek workstation (see Note 1).
8. One box of 200-µl Biomek pipette tips (Beckman).
9. One round-bottomed 96-well microtiter plate (Greiner, Philadelphia, PA).
10. One 200-µl 96-well skirted PCR plate (Greiner).

3. Methods

3.1. Low Throughput, Manual Implementation

This method is for using a manual approach to test a small number of protein samples for their ability to refold. The method begins with cell pellets from small-scale (2–5 ml *E. coli* culture in which a 6×His tagged protein has been overexpressed. The method could alternatively begin with inclusion body preparations made from similar cultures. The advantage of using inclusion body preparations is that it is an efficient purification step. However, binding protein to metal affinity resin and washing generally results in protein with better than 90% purity and so the extra steps involved in making inclusion bodies are generally not necessary.

3.1.1. Protein Resolubilization and Resin Binding
1. Aspirate 5 ml of lysis buffer to each cell pellet (see Note 2).
2. Sonicate each sample 2 minutes with microprobe on med-high setting (see Note 3).
3. Incubate 30 minutes at room temperature spinning end over end on a wheel (see Note 4).
4. Pellet remaining insoluble material 20 min at 4,500*g* in a benchtop centrifuge (see Note 5).
5. Aspirate 100 µl each 50% TALON metal affinity resin to fresh 15-ml conical tubes (see Note 6).
6. Aspirate cell lysates onto metal affinity resin.
7. Incubate 20 minutes at room temperature spinning end over end on a wheel.
8. Pellet metal affinity resin 10 min at 2,500*g* (see Note 7).
9. Aspirate cell lysate from resin (see Note 8).
10. Resuspend resin in 15 ml wash buffer.
11. Pellet metal affinity resin in a benchtop centrifuge for 10 min at 2,500*g*.
12. Aspirate wash buffer from resin.
13. Repeat steps 10 to 12 a further two times.

3.1.2. Refolding and Elution
1. Resuspend resin in 15 ml of ice cold wash buffer (see Note 9).
2. Incubate on ice for 20 minutes (see Note 10).
3. Pellet metal affinity resin in a benchtop centrifuge for 10 minutes at 2,500*g* and 4°C.

4. Aspirate refolding buffer from resin.
5. Resuspend resin in 0.5 ml of ice-cold elution buffer.
6. Incubate on ice for 10 minutes; resuspend regularly by flicking.
7. Pellet metal affinity resin in a benchtop centrifuge for 10 min at 2,500g and 4°C.
8. Aspirate exactly 0.5 ml of elution buffer from resin and reserve in labeled microfuge tubes on ice (see Note 11).
9. Aspirate 0.5 ml of wash buffer onto resin and keep on ice.

3.1.3. Analysis of Refolding Yield

1. Mix a sample of each elution buffer samples with SDS-PAGE gel loading dye and boil, this is the refolded protein sample.
2. Thoroughly resuspend resin in wash buffer by flicking and mix a sample with SDS-PAGE gel loading dye and boil; this is the misfolded protein sample (see Note 12).
3. Run folded and misfolded samples for each protein side by side on an SDS-PAGE gel.
4. Refolding efficiency can be determined by comparing the ratio of protein in the folded and misfolded samples (see Note 13).

3.2. High Throughput, Automated Implementation

The starting material of the high throughput assay is inclusion body pellets in a 96-well, microtiter plate. In high throughput structural biology projects small-scale *E. coli* "test expression" cultures are grown in 96-well deepwell block plates to determine solubility and purification properties and migration on SDS-PAGE. Often bacterial cells are lysed using a detergent-based cell lysis reagent for metal affinity purification, and typically both of these steps are automated using a liquid handling workstation. Crude inclusion body preparations can be obtained simply by pelleting the insoluble material from these detergent cell lysis reactions. This automated high throughput refolding assay uses a protocol analogous to a small-scale protein purification protocol but the cell pellets are replaced with inclusion body preparations and the cell lysis reagent is replaced by 8 M urea.

3.2.1. Automated Protein Refolding Assay

1. Set up work surface of Biomek workstation with Biomek hardware, microtiter plates and tips according to the layout described in the Automated MagnaHis™ protocol (see Note 1).
2. Replace the cell lysis reagent in the reservoir with resolubilization buffer
3. Place 96-well plate of inclusion body preparations on the workstation surface in place of the cell pellets.
4. Run protocol.

3.2.2. Analysis of Refolding Yield

1. Mix eluted protein in the "elution plate" from the Biomek protocol with SDS-PAGE loading dye and boil (see Note 14).
2. Analyze by SDS-PAGE, proteins with good yield from refolding are those with strong bands on the gel.

4. Notes

1. At the time of writing the Automated MagneHis™ Protein Purification System Protocol #EP011 (Promega) can be downloaded from the Promega web site (www.promega.com) in formats for the Biomek 2000, 3000, and FX.

2. There is no need to resuspend the *E. coli* cell pellets at this step, because sonication in the following step takes care of cell suspension.

3. The sonicator produces some heat in the sample with these low sample volumes. In addition, some of the following steps are incubated at room temperature. This is not a problem, because protease is largely inactive in 8 M urea. The urea also protects from temperature-induced aggregation.

4. This half-hour incubation step is to allow the dense inclusion body granules time to dissolve in the chaotropic agent. Ambient temperatures allow this process to happen more quickly than in refrigerated samples.

5. Following this relatively low speed spin the cell lysates very often remain cloudy. This is not a major problem, because the fine particles in suspension also are not pellets during the resin washing and are diluted out.

6. When aspirating resin in 200-µl pipette tips it is helpful to use a sharp scalpel blade to remove the last 4 or 5 mm from the end of the tip. This widens the opening and prevents resin clogging in the end. Failure to do this can lead to lower resin percentages in the pipette tip than the solution.

7. Resin is spun at low speed to prevent compaction and fragmentation.

8. Resin does not form a compact pellet in the tube and care must be taken when aspirating supernatants away. Leaving too much of the previous wash behind reduces the efficacy of the wash steps but overstringent removal of buffers can result in a removal of resin with each successive wash step.

9. After removal of the final 8 M urea wash buffer there should be 50–100 µl of resin and residual buffer remaining in the tubes. Resuspension in 15 ml of wash buffer will result in a final urea concentration of approximately 50 mM. This should not interfere with refolding. However, the amount of residual buffer left after the wash steps should not be greater than necessary. Moving proteins from a denaturing buffer to a renaturing one by a rapid dilution step is generally thought to improve refolding yield, because the protein spends less time as aggregation prone folding intermediates when compared with slower dialysis methods. Consequently, the renaturing buffer should be added to the resin in a rapid motion and the resin resuspended immediately. Once the renaturing buffer has been added it must be assumed that proteases are renatured into an active form and that the refolded protein is now sensitive to temperature-dependent aggregation. All further steps are at 4°C.

10. This 20-minute incubation step is to allow refolding to take place and proteins to reach equilibrium. It is likely that the refolding reaction is relatively rapid, but in some cases proteins eluted from the resin without this incubation step can fall out of solution in the hours afterward, suggesting that it takes this long for the refolded proteins to reach equilibrium.

11. To accurately compare folded and misfolded protein the same quantity of each must be analyzed by SDS-PAGE. To achieve this, the same quantity of wash buffer should be added to the resin as refolding buffer was removed. These are then directly comparable if similar quantities are loaded onto the gel.

12. Misfolded protein will remain bound to the surface of the resin. To take a sample for SDS-PAGE that can be compared to the elution samples it is important that a representative sample of the resin is taken. The resin should be resuspended to homogeneity. The end of the pipette tip used for aspiration should be removed to prevent tip clogging (see Note 6). Misfolded protein is removed from the surface of the resin by SDS in the gel-loading dye. Therefore when loading the boiled sample onto a gel it is not necessary to resuspend and load the resin that settles to the bottom of the sample tube. Resin loaded into the wells of the gel along with sample will remain there and will not effect the migration of proteins through the gel matrix.

13. To make this technique more quantitative gel densitometry can be used to compare percentages of protein in the folded and misfolded samples.

14. In the manual, low-throughput implementation of this protocol, misfolded and folded protein are analyzed to estimate refolding efficiency. In the high throughput implementation only the eluted samples are analyzed to reduce the sample number; comparison across all of the proteins analyzed gives an estimate of the relative yield of protein that can be expected from a larger-scale fermentation and refolding. This will be a function of both the refolding efficiency and the expression yield. In practice this is all the information required to choose those proteins suitable for refolding, although information is lost about how best to optimize yield. This assay requires that a large number of samples be analyzed by SDS-PAGE. A microfluidic gel analyser capable of processing 96-well microtiter plates such as the AMS 90 (Caliper, Newton, MA) can be useful.

Acknowledgements

The authors thank Mareike Kurz, Christine Gee, Thomas Huber, Timothy Ravasi, Munish Puri, and Lynn Pauron for research support. This work was supported by a University of Queensland Postdoctoral Fellowship and an Australian Synchrotron Research Program Fellowship (to N. C. P.) and by an Australian Research Council (ARC) grant to J. L. M. and B. K. B. K. is an ARC Federation Fellow and a National Health and Medical Research Council (NHMRC) Honorary Research Fellow.

References

1. Huang, R. Y., Boulton, S. J., Vidal, M., Almo, S. C., Bresnick, A. R., and Chance, M. R. (2003) High throughput expression, purification, and characterization of recombinant Caenorhabditis elegans proteins. *Biochem. Biophys. Res. Commun.* **307**, 928–934.

2. Christendat, D., Yee, A., Dharamsi, A., Kluger, Y., Gerstein, M., Arrowsmith, C. H., and Edwards, A. M. (2000) Structural proteomics: prospects for high throughput sample preparation. *Prog. Biophys. Mol. Biol.* **73**, 339–345.

3. Gilbert, M., and Albala, J. S. (2002) Accelerating code to function: sizing up the protein production line. *Curr. Opin. Chem. Biol.* **6**, 102–105.

4. Edwards, A. M., Arrowsmith, C. H., Christendat, D., Dharamsi, A., Friesen, J. D., Greenblatt, J. F., and Vedadi, M. (2000) Protein production: feeding the crystallographers and NMR spectroscopists. *Nat. Struct. Biol.* **7**, 970–972.

5. Cowieson, N. P., Listwan, P., Kurz, M., Aagaard, A., Ravasi, T., Wells, C., Huber, T., Hume, D. A., Kobe, B., and Martin, J. L. (2005) Pilot studies on the parallel production of soluble mouse proteins in a bacterial expression system. J. *Struct. Funct. Genomics* **6**, 13–20.

6. Christendat, D., Yee, A., Dharamsi, A., Kluger, Y., Savchenko, A., Cort, J. R., Booth, V., Mackereth, C. D., Saridakis, V., Ekiel, I., Kozlov, G., Maxwell, K. L., Wu, N., McIntosh, L. P., Gehring, K., Kennedy, M. A., Davidson, A. R., Pai, E. F., Gerstein, M., Edwards, A. M., and Arrowsmith, C. H. (2000) Structural proteomics of an archaeon. *Nat. Struct. Biol.* **7**, 903–909.

7. Endo, Y., and Sawasaki, T. (2004) High throughput, genome-scale protein production method based on the wheat germ cell-free expression system. *J. Struct. Funct. Genomics* **5**, 45–57.

8. Holz, C., Hesse, O., Bolotina, N., Stahl, U., and Lang, C. (2002) A micro-scale process for high throughput expression of cDNAs in the yeast Saccharomyces cerevisiae. *Protein Expr. Purif.* **25**, 372–378.

9. Maxwell, K. L., Bona, D., Liu, C., Arrowsmith, C. H., and Edwards, A. M. (2003) Refolding out of guanidine hydrochloride is an effective approach for high throughput structural studies of small proteins. *Protein Sci.* **12**, 2073–2080.

10. Tresaugues, L., Collinet, B., Minard, P., Henckes, G., Aufrere, R., Blondeau, K., Liger, D., Zhou, C. Z., Janin, J., Van Tilbeurgh, H., and Quevillon-Cheruel, S. (2004) Refolding strategies from inclusion bodies in a structural genomics project. *J. Struct. Funct. Genomics* **5**, 195–204.

11. Willis, M. S., Hogan, J. K., Prabhakar, P., Liu, X., Tsai, K., Wei, Y., and Fox, T. (2005) Investigation of protein refolding using a fractional factorial screen: a study of reagent effects and interactions. *Protein Sci.* **14**, 1818–1826.

12. Cowieson, N. P., Wensley, B., Listwan, P., Hume, D. A., Kobe, B., and Martin, J. L. (2006) An automatable screen for the rapid identification of proteins amenable to refolding. *Proteomics* **6**, 1750–1757.

Chapter 18

Structural Proteomics of Membrane Proteins: a Survey of Published Techniques and Design of a Rational High Throughput Strategy

Melissa Swope Willis and Christopher M. Koth

Approximately one third of the proteins encoded in prokaryotic and eukaryotic genomes reside in the membrane. However, membrane proteins comprise only a minute fraction of the entries in protein structural databases. This disparity is largely due to inherent difficulties in the expression and purification of sufficient quantities of membrane targets. To begin addressing the challenges of membrane protein production for high throughput structural proteomics efforts, the authors sought to develop a simple strategy that would permit the standardization of most procedures and the exploration of large numbers of proteins. Successful methods that have yielded membrane protein crystals suitable for structure determination were surveyed first. A number of recurrent trends in the expression, solubilization, purification, and crystallization techniques were identified. Based largely on these observations, a robust strategy was then developed that rapidly identifies highly expressed membrane protein targets and simplifies their production for structural studies. This method has been used to express and purify intramembrane proteases to levels sufficient for crystallization. This strategy is a paradigm for the purification of many other membrane proteins, as discussed.

1. Introduction

Many unique factors limit the high throughput (HTP) production of membrane proteins for structural studies (1–6). Foremost among these is the availability of sufficient amounts of purified protein. This is a considerable challenge for several reasons. First, overexpression and membrane insertion in both endogenous and heterologous hosts is often toxic (6,7). Second, many membrane proteins are not produced in functional form in heterologous expression systems (7,8). This is due, for example, to an inability to undergo critical post-translational modifications or to differences in the membrane composition of recombinant versus endogenous hosts. The solubilization and purification of membrane proteins that *can* be expressed poses a third problem: detergents, which may render the target protein unstable or inactive, must be used for extraction from the lipid membrane and are necessary throughout the course

From: *Methods in Molecular Biology, Vol. 426: Structural Proteomics: High-throughput Methods*
Edited by: B. Kobe, M. Guss and T. Huber © Humana Press, Totowa, NJ

of purification and crystallization *(5,9,10)*. Rational, generally applicable strategies that focus on methods to express, solubilize, and purify membrane proteins could have considerable benefit to structural proteomics efforts and are the focus of this chapter.

The methods presented are organized as follows. The chapter begins with an in-depth survey of practical strategies that have already been used for successful membrane protein crystal structures. Based on key observations from this study, a general method is then proposed that aims to serve as an optimistic starting point for HTP membrane protein production. It is difficult to generalize the purification of any protein, let alone those in the cell membrane. Nevertheless, the strategy presented here works for many targets, with only minor variations in technique *(11–13)*.

For simplicity, the Materials and Methods sections detail a strategy for the production of Rhomboid intramembrane proteases, using the *Pseudomonas aeruginosa* Rhomboid, *PA3086*, as a specific example. The same purification scheme is used for several other Rhomboids *(13)*. Rhomboids are the only class of intramembrane serine proteases known and are among the most evolutionarily conserved membrane proteins. They cleave membrane-bound substrates, catalyzing the release of soluble factors that are involved in growth factor signaling *(14)*, mitochondrial fusion *(15)*, apoptosis *(16)*, bacterial quorum sensing *(17)*, and parasite invasion *(18)*. Rhomboids were examined as part of a HTP membrane protein platform, reasoning that structural studies of various members could provide significant insight into their function *(11,13)*.

1.1. Practical Observations Regarding Membrane Protein Production

The aim was to establish a general starting point for the production of membrane proteins that was based on an analysis of successful strategies to date. Further, to sample relevant membrane protein methods for structural biologists, the expression, solubilization, purification, and crystallization conditions that have been used to successfully determine membrane protein structures by x-ray crystallography were examined. Targets were identified by surveying the White Lab's Membrane Proteins of Known 3D Structure Database (http://blanco.biomol.uci.edu/Membrane_Proteins_xtal. htmL), which comprises a list of all membrane protein structures determined to approximately 4.5 Å resolution, or better. The analysis was limited to structures of unique, non-monotopic membrane proteins solved by x-ray crystallography (as of July 2006). From this pool, only two additional exclusionary criteria were implemented. First, targets crystallized in lipid cubic phases were excluded. The lipid cubic phase technique has been applied to only a very limited number of unique membrane proteins and is extensively detailed in several reviews *(19,20)*. Second, to remove a bias when multiple structures were solved of the same protein (e.g., when additional structures with bound ligands/inhibitors, or subsequent structures of the same protein solved at a higher resolution), all but the first structure of each unique protein were excluded from this study. These criteria left a total of 101 unique membrane protein crystal structures for analysis (data not shown). For each, the following information was collected: type (α-helical or β-sheet), source (prokaryotic or eukaryotic, and native or recombinant), solubilization detergent(s), purification strategy, purification detergent(s), method of detergent exchange, crystallization detergent(s), protein concentration for crystallization, and crystallization conditions. Parameters were derived primarily from the respective publications, and also from the PDB of the Research Collaboratory for Structural Bioinformatics (RCSB; http://www.pdb.org) *(21)* and the Membrane Protein Data Bank (MPDB; http://www.lipidat.chemistry.ohio-state.edu/MPDB/index.asp) *(22)*.

Figures 18.1 to 18.3 outline key results of the survey. Of 101 unique targets, 65.3% are classified as α-helical and 34.7% as β-sheet proteins. The majority of crystal structures are of prokaryotic proteins (84.2%), whereas 15.8% are of eukaryotic origin. A clear majority of crystal structures are of proteins produced by recombinant methods (61.3% vs. 38.7% from native sources). When examining strategies for solubilization, purification, and crystallization, some remarkable trends are observed. The detergent *n*-dodecyl-β-D-maltopyranoside (DDM) is used for the solubilization of almost one quarter of all targets (23.9%) (see Fig. 18.1A). In fact, just six different detergents could be used to solubilize 67.3% of all unique targets (DDM, 23.9%; LDAO, 13%; C8POE, 8.7%; DM, 8.7%; OG, 6.5%; and Triton-X100, 6.5%). (See Note 21 for a list of the full detergent names.) This trend is very similar to that of detergents used for purification (see Fig. 18.1B). Here, just six different detergents could be used to purify 59.2% of all targets (DDM, 22.8%; LDAO, 14.1%; C8POE, 7.6%; OG, 6.5%; C8E4, 4.3%; and DM, 4.3%). Again, DDM is by far the most commonly used detergent for purification (22.8%). The methods for purification are very common among the protocols for each target. Since the number of purification steps was often quite varied, just the statistics for the last purification step are illustrated here (see Fig. 18.2A). Unexpectedly, size exclusion chromatography (SEC) is the final purification step for more than half of all targets (51.1%). A further 27.7% is covered by just one additional purification strategy, ion exchange. It should also be noted that almost all recombinant targets are produced with a hexahistidine tag to facilitate initial purification by immobilized metal affinity chromatography (IMAC, data not shown). Many targets are solubilized with one particular detergent and then exchanged to a different detergent over the course of purification. This is often necessary to obtain well-diffracting crystals. Since this is a frequent step, the different methods used for detergent exchange were surveyed (see Fig. 18.2B). SEC is the most prevalent method (45.7%), followed by ion exchange (24.5%) and dialysis (13.8%).

Several common themes were also noted in the various crystallization parameters. Only six different detergents are used in 60% of the crystallization conditions (OG, 18%; DDM, 11%; C8E4, 11%; LDAO, 10%; NG, 6%; and DM, 4%) (see Fig. 18.1C). OG is the most successful detergent for crystallization, in contrast to the prevalent use of DDM for solubilization and purification. The concentration of membrane protein for crystal trials is varied, but typically in the range of 5–20 mg/ml (see Fig. 18.3A). The method of crystallization is overwhelmingly dominated by the vapor diffusion technique (this includes targets that crystallize by hanging drop and sitting drop), which makes up 90.5% of total entries (see Fig. 18.3B).

The results of this survey suggest that many of the steps involved in the production of membrane proteins, such as detergent selection, can be reduced to just a few possible parameters while still maintaining a reasonable level of success. The following sections outline an HTP membrane protein production strategy that is based on some of the most significant observations from

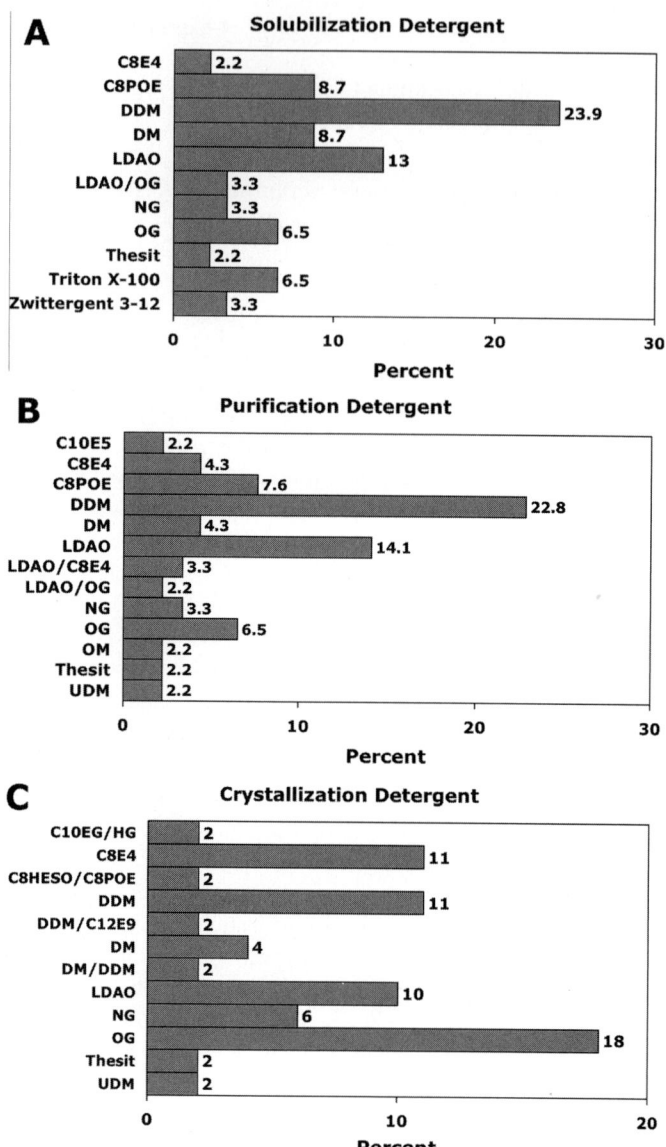

Fig. 18.1 Detergents used in the production of membrane proteins. Detergents used for the solubilization (*panel* A, n = 92), purification (*panel* B, n = 92), and crystallization (*panel* C, n = 100) of each protein were identified. Note that for some targets, certain steps were not required or were not reported (therefore, n is not the total number of targets, 101). Collected data for this and other figures was analyzed using JMP v6.0 software, SAS Institute. For clarity, detergents that have been used for only one structure were omitted from the graphs. In the case of solubilization detergent, these were: αDDM, C12DAO, C12E8, C12E9, Deoxycholate, DHPC, Elugent, Fos-choline 14, LAPAO, Sodium cholate, Sodium cholate/OG, NG/DGDG, SDS, Triton X-100/ OG, Tween 20/Brij 35, UDM, and Zwittergent 3–14. For purification detergent, these were: αDDM, C12DAO, C12E8, C12E9, C8POE/C8E4, Cholate/Tween 20, DDM/ DM, DDM/DM/Lipid, DHPC, DM/tri-DM, Elugent/C8E4, Fos-choline 14, LAPAO, LDAO/UDM, Sodium cholate, NG/DGDG, SDS/C8POE, tri-DM, Triton X-100, Triton X-100/C8POE, and Zwittergent 3–12. For crystallization detergent, these were αDDM, C10DAO, C10E5, C12E8, C12E9, C8E4/C-HEGA-10, C8E5, Cymal-5, DDG/ GxG/HG/OG, DHPC, DM/Lipids, DMG/SPC, Fos-choline 14, LAPAO, LDAO/C8E4, LDAO/Deoxy-BigCHAP, LDAO/OG, NM/DM, OG/C6DAO/HG, OG/C8E4, OM, OTG, tri-DM, tri-DM/HEGA-10, UDAO, UDM/DOPC, and Zwittergent 3–12. See Note 21 for a complete list of detergent abbreviations.

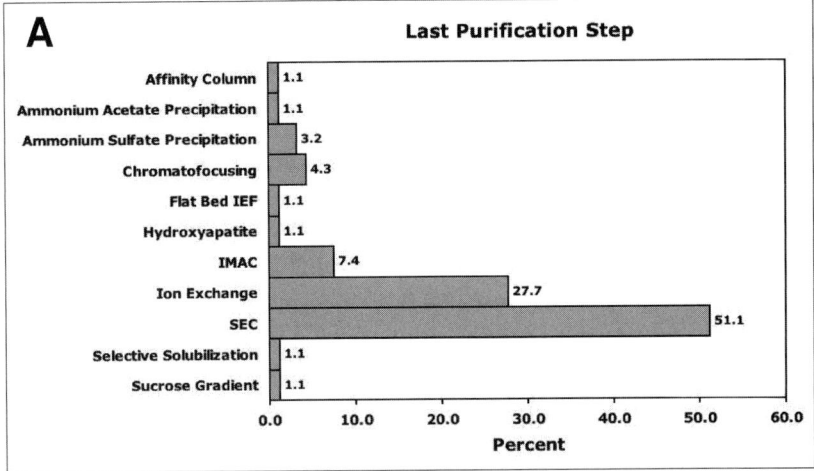

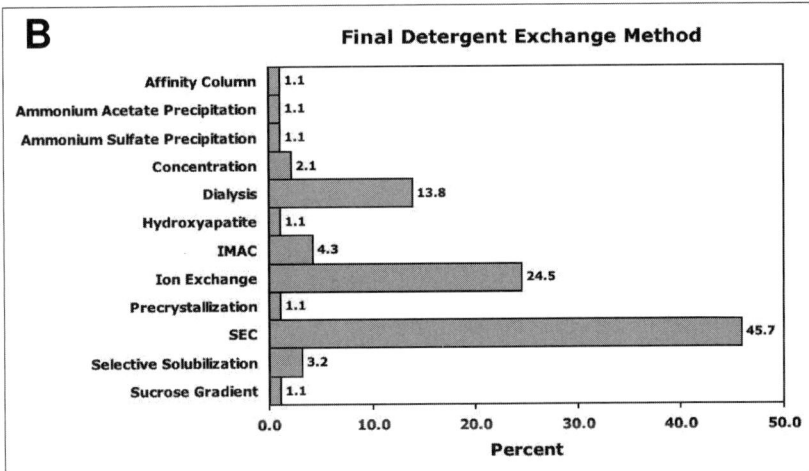

Fig. 18.2 Purification strategies used in the production of membrane proteins. **A**. The purification protocol for each protein was examined and the last step immediately prior to crystallization was scored ($n = 94$). **B**. The method of detergent exchange was scored for each protein ($n = 94$). Note that some targets did not require purification or detergent exchange. For others, the purification strategy was not reported. Therefore, n is not the total number of targets, 101.

the survey and also adopts some successful strategies employed for soluble protein HTP efforts (Fig. 18.4). This strategy has been used successfully for many membrane targets, including the Rhomboid proteases discussed here, ion channels, transporters, and sensor kinases (*11–13*). Membrane proteins are expressed in a recombinant host, *E. coli*, as fusions to a hexahistidine tag. Whenever possible, multiple orthologues of a given target are also screened. These are identified from sequenced genomes and the literature. A single detergent, DDM, is used for solubilization, purification, and crystallization. Vapor diffusion techniques are used for crystallization, and detergent exchange is facilitated by SEC if necessary. Alternative courses of action are also highlighted that could be explored if the strategy fails at any point for a given target. The alternative courses are largely based on the results of the above survey.

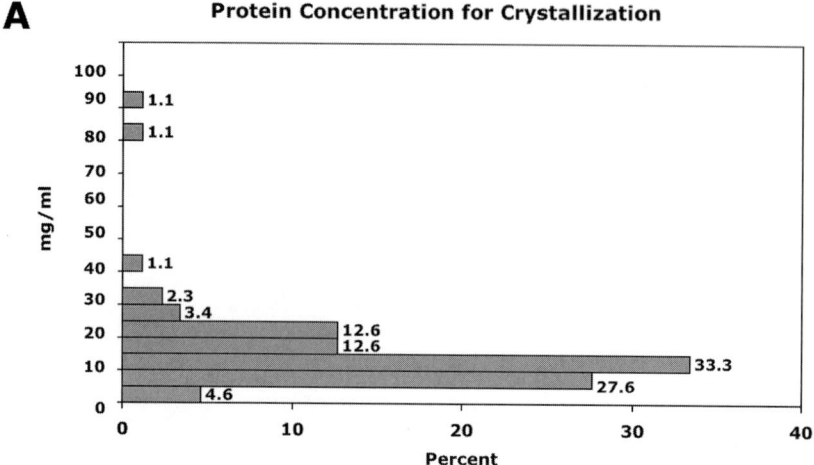

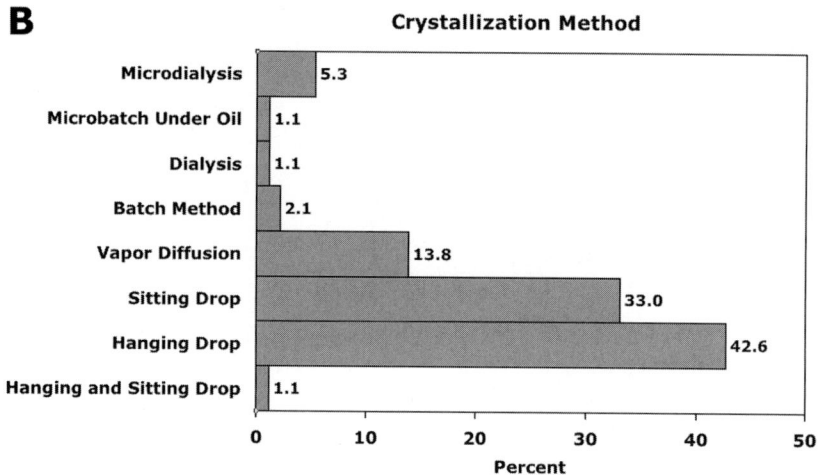

Fig. 18.3 Crystallization statistics of membrane proteins. The protein concentration used for crystallization (**A**, $n = 87$) and the method of crystallization (**B**, $n = 94$) was identified for each protein. Note that for some targets, information on the concentration or method of crystallization was not reported (therefore, n is not 101).

2. Materials

2.1. Target Identification and Cloning

1. Source DNA (typically genomic DNA from ATCC, Manassas, VA). Clones of several Rhomboids (in pCDNA3.1, Invitrogen, Carlsbad, CA) were also obtained from Matthew Freeman *(14,27)*.
2. Zero Blunt TOPO PCR Cloning Kit (Invitrogen).
3. pET25b(+) T7 polymerase expression vector (Novagen, San Diego, CA), modified to contain a C-terminal TEV protease cleavage site and hexahistidine tag (see below).
4. *Pfu* polymerase, Quickchange mutagenesis kit (Stratagene, La Jolla, CA).
5. Restriction enzymes *NdeI, KpnI*, and *NheI* (Fermentas, Burlington, ON, Canada).

Fig. 18.4 Outline of a HTP membrane protein strategy. Steps involved in the HTP production of both soluble and membrane proteins are outlined. Procedures that are unique to membrane proteins are highlighted in gray.

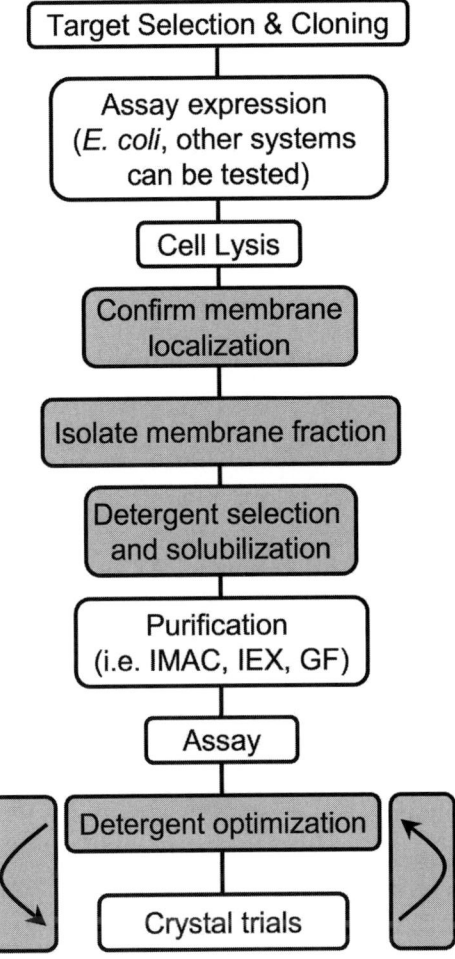

2.2. Cell Growth, Lysis and Expression Analysis

1. BL21-Gold (DE3) competent cells (Stratagene).
2. pRARE2 plasmid (Novagen).
3. Luria-Bertani media (LB), per L: 10 g Bacto Tryptone, 5 g yeast extract (Sigma, St. Louis, MO), 5 g NaCl. Autoclaved as 1 L media per 2 L Erlenmeyer flask.
4. LB Amp/Chlor plates: as above but with 15 g Bacto Agar (Sigma) added. Autoclaved and cooled to approximately 50°C; 100 μg/ml ampicillin and 34 μg/ml chloramphenicol is added prior to pouring plates.
5. Isopropyl-6-D-thiogalactopyranoside (IPTG, Anatrace, Maumee, OH); prepared as a 1 M stock in water.
6. n-Dodecyl-β-D-maltopyranoside (DDM, Anatrace) (see Note 1), prepared as a 20% stock in water.
7. Buffer A: 20 mM HEPES/NaOH pH 7.4, 100 mM NaCl, 5 mM $MgCl_2$, 10% glycerol, 1 × protease inhibitors (1 × PI; note that the inclusion of protease inhibitors in this and subsequent buffers is indicated by the "1 × PI" abbreviation. This refers to the use of 1 Complete EDTA-free Protease Inhibitor Cocktail tablet (Roche, Pleasanton, CA) per 50 ml of buffer).
8. Buffer B: 50 mM HEPES/NaOH pH 7.4, 100 mM NaCl, 15% glycerol, 1 × PI.

9. Buffer C: 50 mM HEPES/NaOH pH 7.4, 250 mM NaCl, 15% glycerol, 1.5% DDM, 1 × PI.

10. Avestin EmulsiFlex C3 High Pressure Homogenizer (Avestin, Ottawa, ON, Canada).

11. 2× SDS-PAGE Membrane Sample Buffer: 200 mM Tris-HCl pH 6.8, 20% glycerol, 2% SDS, 200 mM DTT, 0.2% bromophenol blue, 4 M urea. Filter solution through a 0.2 μm filter and store at −20°C.

12. α-Hexahistidine tag mouse monoclonal antibody (Novagen).

13. Goat α-mouse HRP conjugated secondary antibody, SuperSignal West Femto Maximum Sensitivity Substrate (Pierce, Rockford, IL).

2.3. Solubilization and Protein Purification

1. Buffer D: 50 mM HEPES/NaOH pH 7.3, 200 mM NaCl, 1 mM $MgCl_2$, 1 mM $CaCl_2$, 10% glycerol, 5 mM imidazole, 1% n-dodecyl-β-D-malto-pyranoside (DDM), 1 × PI.

2. Buffer E: 50 mM HEPES/NaOH pH 7.3, 1 mM $MgCl_2$, 1 mM $CaCl_2$, 10% glycerol, 30 mM imidazole, 0.1% DDM.

3. Buffer F: 50 mM HEPES/NaOH pH 7.3, 200 mM NaCl, 1 mM $MgCl_2$, 1 mM $CaCl_2$, 10% glycerol, 400 mM imidazole, 0.1% DDM.

4. Buffer G: 15 mM HEPES/NaOH pH 7.4, 0.05% DDM.

5. Buffer H: 15 mM HEPES/NaOH pH 7.4, 40 mM NaCl, 0.05% DDM.

6. Buffer I: 15 mM HEPES/NaOH, pH 7.4, 1 M NaCl, 0.05% DDM.

7. IMAC Superflow resin (Qiagen, Valencia, CA).

8. Tris(2-carboxyethyl)phosphine hydrochloride (TCEP, Bioshop).

9. UNO Q-1 ion exchange column (Bio-Rad, Hercules, CA).

10. Vivaspin 30K cut-off centrifugal filter device (Fisher Scientific, Pittsburgh, PA)

2.4. Crystal Trials

1. NeXtal MBClass I and II Suites, EasyXtal 24-well Hanging-Drop Crystal Support (for hanging-drop vapor diffusion) (Qiagen).

2. Crystal Screen I and II, Greiner CrystalQuick 96 well-sitting drop format plates (for sitting-well vapor diffusion) (Hampton Research, Aliso Viejo, CA).

2.5. Optimization Strategies

1. Detergent Screen I, II and III (Hampton).

2. Superdex 200 16/60 SEC column (GE Healthcare, Piscataway, NJ).

3. Methods

3.1. Target Identification and Cloning

Structural proteomics efforts with soluble proteins have shown that screening several orthologues within a protein family can increase the probability of obtaining at least one sample suitable for structural studies (23). This approach is also advantageous for membrane proteins (12,24–26). Rhomboid orthologues are identified from the literature (14,27) and through searches of various sequenced genomes. In the interests of throughput, a hexahistidine tag fusion strategy is used. This offers a simple, robust purification, has been used for many membrane protein crystal structures, and has proved successful for structural proteomics efforts with soluble proteins *(23,28,29)*. It should be noted that some, but not all, membrane proteins contain an amino-terminal membrane targeting signal sequence that is cleaved *in vivo*. With this in mind, a cloning strategy is used that generates fusions to a C-terminal hexahistidine tag (see Note 2). A cleavage site for the tobacco etch virus (TEV) protease is also incorporated, upstream of the hexahistidine tag. The general cloning strategy is as follows:

1. PCR: Amplify rhomboid coding sequences by PCR from genomic DNA or pcDNA3.1 parent clones *(14,27)* using high-fidelity *Pfu* polymerase. The natural stop codons are not incorporated into the PCR products. Design primers to include the restriction site *NdeI* (CATATG), which includes an ATG start codon, and the restriction site *NheI* (or *KpnI* if there is an internal *NheI* site) to the 5' and 3' end of PCR products, respectively. In the case of the *P. aeruginosa PA3086* rhomboid, the primer pairs are:

   Forward: 5' ctacatatgagcgcggtgcaggtc 3'
   Reverse: 5' ctagctagcacggcgacgccgcgc 3'

2. Digest the amplified fragments with the appropriate restriction enzymes *(NdeI-NheI* or *NdeI-KpnI)*.
3. Clone digested PCR products into the corresponding sites of a modified pET25b(+) T7 polymerase expression vector in which the nucleotide sequence between the *NcoI* and *NheI* restriction enzyme sites is replaced, via Quickchange mutagenesis, with the sequence: 5' ccatggaattcggatccgg taccaagcttgctagcctcgagagagaaaacttgta tttccagggcagcagccaccaccaccaccac-cactgactagt 3'. Insertion of this sequence results in the expression of cloned targets that contain a C-terminal hexahistidine tag preceded by a TEV protease recognition site (ENLYFQ^G).
4. Confirm positive clones by sequencing.

3.2. Cell Growth, Lysis, and Expression Analysis

In the interests of throughput, the authors rely on a single low-temperature induction protocol that has been optimized for the expression of soluble proteins in a general sense *(11,28)*. Also, to avoid incorporating a large, potentially *n*-dimensional and expensive detergent screen, solubilization is tested with only one detergent: DDM (see Note 3). As detailed in the above survey of successful membrane protein crystal structures, DDM is one of the most effective, widely used detergents for solubilizing membrane proteins. It is also among the most successful of all detergents for membrane protein purification and crystallization. Many targets simply will not be solubilized with DDM. However, from a purely HTP point of view, this is offset by the simple fact that standardizing protocols with one detergent permits the exploration of considerably more targets.

1. Transform pET25b Rhomboid clones into the *E. coli* strain BL21-Gold (DE3), which harbors an additional vector (pRARE2) that encodes transfer RNAs for rare *E. coli* codons (AUA, AGG, AGA, CUA, CCC, CGG, and GGA) (see Note 4). Plate cells on LB Amp/Chlor plates overnight at 37°C.
2. Transfer single colonies from the overnight plates to 30 ml starter cultures of LB media containing 100 µg/ml ampicillin and 34 µg/ml chloramphenicol (for selection of the pET25b and pRARE2 plasmids, respectively). Incubate cultures overnight at 30°C, with shaking at 100 rpm.

3. The following morning, transfer starter cultures to 200 ml LB containing antibiotics. Grow cells at 30°C, 225 rpm, to an OD_{600} of 0.3–0.4 (see Note 5). IPTG is then added to a final concentration of 0.5 mM. Grow cells for an additional 12 hours at 16°C before harvesting (see Note 6).

4. This and subsequent steps are performed at 4°C. Harvest cells by centrifuging at 7,000g for 7 minutes.

5. Resuspend the cell pellet in 30 ml buffer A.

6. Lyse cells by passing through an Avestin EmulsiFlex C3 High Pressure Homogenizer at a pressure setting of 15,000 psi (see Note 7). Centrifuge lysed cells at 50,000g for 30 minutes and discard the supernatant.

7. Resuspend the membrane-containing pellet in 5 ml of buffer B. Then add 10 ml of buffer C, which contains the detergent DDM. Incubate the resulting suspension at 4°C for 1–2 hours (see Notes 8 and 9).

8. Centrifuge the suspension at 100,000g for 1 hour (see Note 10) and retain the DDM-solubilized supernatant. Mix a sample of the solubilized fraction with an equal volume of 2× SDS-PAGE Membrane Sample Buffer and immediately resolve by SDS-PAGE (see Note 11). For most Rhomboids, expression is obvious after Coomassie Blue staining (Fig. 18.5A).

9. Protein expression can also be monitored by western blotting. Resolve supernatants from step 8 by SDS-PAGE, transfer to nitrocellulose, and probe using an α-hexahistidine tag mouse monoclonal antibody (1:2000), followed by a goat α-mouse HRP conjugated secondary antibody (1:50,000). Develop blots using SuperSignal West Femto Maximum Sensitivity Substrate and expose to film.

3.3. Solubilization and Protein Purification

The following protocol is used for purifications from 2 L of original culture. Expression and lysis is performed as outlined in Section 3.2., steps 1–5, except that volumes are increased proportionately. All procedures are carried out at 4°C. As discussed, the detergent DDM is used throughout purification. Although not detailed here, the functional properties of purified Rhomboids can be analyzed through the use of an *in vitro* protease assay *(13)* (see Note 12).

3.3.1. Solubilization

1. Resuspend membrane-containing pellets from 2 L of culture (Section 3.2., step 6) in 25 ml buffer D and stir gently for 1–2 hours (see Notes 8, 9, and 13).

2. Centrifuge the suspension at 100,000g for 1 hour and retain the DDM-solubilized supernatant.

3.3.2. Nickel Affinity Chromatography

1. Load the supernatant (Section 3.3.1., step 2), containing DDM-solubilized membrane protein, onto a 1 × 2.5 cm IMAC Superflow gravity column, preequilibrated in buffer D.

2. Collect the IMAC flow-through and load onto the same column a second time (see Note 14).

3. Wash the column with 10 volumes of buffer E containing 500 mM NaCl, followed by 10 volumes of buffer E containing 200 mM NaCl (see Note 15).

4. Elute bound protein with buffer F. Collect 1-ml fractions and analyze by SDS-PAGE. Pool fractions containing purified Rhomboid protein and determine the concentration by monitoring the absorbance at 280 nm (see Note 16).

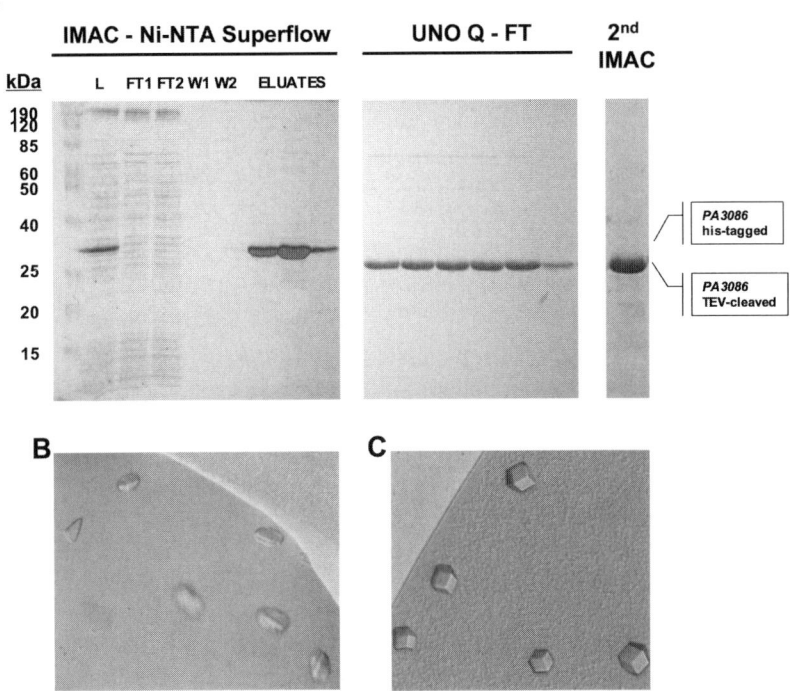

Fig. 18.5 Solubilization, purification, and initial crystallization of *P. aeruginosa* Rhomboid PA3086. **A.** Expressed, DDM-solubilized protein was subject to IMAC chromatography, TEV protease digestion, UNO Q ion exchange, and a second IMAC polishing step. *L*, load; *FT*, flow-through fractions, *W*, wash fractions (see text). The migration of standard molecular weight markers is indicated to the left of the gels. **B.** Initial *PA3086* crystals, as outlined in the text. **C.** Optimization around the initial crystallization condition yields cube-shaped crystals (hanging drop, protein:reservoir ratio of 1:1, using a reservoir solution of 10% PEG 2000 MME, 0.5 M ammonium acetate, 0.56 mM CYMAL-6, 0.1 M MES pH 6.0). Crystal improvement was monitored both qualitatively, in this case as a sharpening of crystal edges, and quantitatively, as better diffraction (30 Å, data not shown).

3.3.3. TEV Protease Digestion

TEV protease is highly specific, active in many detergents *(30)* and commercially available in a noncleavable hexahistidine-tagged form. This latter feature facilitates its removal on a second IMAC column, following the digestion of target proteins. When expressed from the modified pET25(b) vector, TEV-cleaved Rhomboids contain seven additional residues at their C-termini (ASENLYFQG). Cleavage is performed as follows:

1. Add 0.5 mM TCEP to pooled fractions eluted from the IMAC Superflow column.
2. Then add 0.05 mg TEV protease per mg of eluted membrane protein and incubate samples for 12 hours at 4°C. Monitor the digestion by Coomassie Blue-stained SDS-PAGE.

3.3.4. Ion-Exchange Chromatography

1. Dilute TEV-digested samples with 4 volumes of buffer G.
2. Immediately resolve samples on an UNO Q-1 anion exchange column using a 60-ml linear gradient from buffer H to buffer I and a flow rate of 1.0 ml/min. The elution profile of various Rhomboids is varied, as expected. For instance, the *E. coli* rhomboid *GlpG* elutes between 25–30% buffer I and the *B. subtilis* rhomboid *YqgP* between 55–60% buffer I, whereas the *P. aeruginosa* rhomboid *PA3086* does not bind the UNO Q-1 resin (although most remaining contaminant proteins do).
3. Collect 1-ml fractions and resolve by Coomassie Blue-stained SDS-PAGE.
4. Pool fractions containing purified protein.

3.3.5. Second Nickel Affinity Column Chromatography

1. Load pooled fractions from the UNO Q-1 anion exchange purification onto a second 1 × 2.5 cm IMAC Superflow gravity column, preequilibrated in buffer H.
2. Collect purified Rhomboids in the flow-through. Any remaining TEV protease, cleaved hexahistidine tag, undigested Rhomboid, and/or other "contaminating" proteins bind to the column and are effectively removed (see Fig. 18.5A).
3. Concentrate purified Rhomboid protein to approximately 10 mg/ml using a centrifugal filter device (Vivaspin, 30K cutoff) (see Note 17).
4. Use concentrated Rhomboid protein immediately in crystal trials (see the following) or flash-freeze in liquid nitrogen and store at −80°C (see Note 18).

3.4. Crystal Trials

The authors' general crystallization strategy for any membrane protein is to test samples after every key purification step. Thus, in the case of Rhomboids, crystal trials are set up: (1) after the first IMAC purification (Section 3.3.2., step 4), (2) after TEV digestion (Section 3.3.3., step 2), (3) after UNO Q-1 anion exchange chromatography (Section 3.3.4., step 4), and (4) on the flow-through from the second IMAC column (Section 3.3.5., step 4). Initial crystal trials are performed using commercially available sparse matrix screens. If protein is limiting, the authors' preference for screens, in order, is the following:

1. NeXtal MBClass I Suite (Qiagen, 96 conditions)
2. NeXtal MBClass II Suite (Qiagen, 96 conditions)
3. Crystal Screen I (Hampton Research, 50 conditions)
4. Crystal Screen II (Hampton Research, 50 conditions)

Note that this order is based on a limited sampling of membrane proteins and does not reflect the propensity of any particular screen to yield crystals. When possible, trials are performed at 22°C and 4°C, with protein samples at the following concentrations (concentrations are those suggested by the above survey of successful crystallization conditions):

1. 10 mg/ml
2. 5 mg/ml
3. 15–20 mg/ml

The general crystallization strategy is as follows:

1. Pipette 100 μl of each screen solution into the corresponding wells of Greiner sitting drop trays. For each condition, pipette 1 μl of purified protein into

the crystallization well (there are three wells, so three concentrations of protein can be tested at once). Using the same pipette tip, gently apply 1 μl of reservoir solution to the protein drop. Do not further mix the two solutions.

2. Seal trays with clear pressure tape and incubate at the desired temperature. Check trays for crystals after 24 hours and then every 48 hours for the first week. After that, check trays on a weekly basis.

3.5. Optimization Strategies

Following the aforementioned strategy, initial crystals of the *P. aeruginosa* Rhomboid, *PA3086*, can be obtained on sample screened after the second IMAC chromatography step (Section 3.3.5., step 4, and see Fig. 18.5B). Initial crystals form in 0.4 M ammonium acetate, 10% PEG 2000 MME, pH 7.0, although these are of very poor diffraction quality (data not shown). Many parameters can be varied in an effort to optimize crystals. These include varying the protein:precipitant ratio, pH, salt concentration, additives, and temperature (see Note 19). As one example, Fig. 18.5C shows the results of some optimization around the initial *PA3086* crystal condition.

Importantly, the above survey suggests that sampling even just a small number of other detergents would dramatically increase the likelihood of obtaining crystals or of improving upon initial "hits." There are many examples where this is indeed the case *(9)*. One simple strategy is to add small amounts of other detergents, typically near their CMC, to the crystal trials. Hampton Research sells several convenient kits (Detergent Screens 1, 2, and 3) designed for this purpose. A second strategy is also suggested by the survey and used by many groups: Exchange the protein into a different detergent by SEC. The initial sparse matrix crystal screens are then repeated with the exchanged sample (i.e., it should be treated as an entirely new sample). SEC is recommended because it is the most common method for detergent exchange of membrane proteins that have yielded crystal structures (see Fig. 18.2B). A method for this strategy is detailed in the following:

1. Concentrate protein purified as described (Section 3.3.5., step 4, or an earlier step) to 5–10 mg/ml (Section 3.3.5., step 3).
2. Resolve protein on a Superdex-200 (10/30) gel filtration column at 0.5 ml/ min. Buffer H is the running buffer except that the DDM in buffer H is replaced with an alternative detergent, i.e., one suggested by the results of our survey (Fig. 18.1C and see Notes 9 and 20). For example, OG, C8E4, LDAO, NG, and DM have all been used repeatedly for membrane protein crystallization and, together with DDM, are represented in 60% of unique crystal structures.
3. Pool protein-containing fractions and concentrate (Section 3.3.5., step 3).
4. Perform crystal trials as described earlier (Section 3.4.).

3.6. Conclusion

The production of membrane proteins for x-ray structural studies often presents enormous challenges and uncertainties. However, a thorough examination of the various methods that have been used for all successful membrane protein structures indicates that, in many cases, remarkably similar expression, solubilization, purification, and crystallization strategies have been used. This suggests that many of the procedures involved in the production of membrane proteins can be reduced to just a few possible parameters while still maintaining a reasonable level of success. With this in mind, the most recurrent membrane protein trends have been applied to the development of a robust HTP strategy. The approach detailed in this chapter serves as a foundation for the rapid production of large numbers of prokaryotic membrane proteins and is a reasonable first-start in the emerging field of membrane protein structural genomics.

4. Notes

1. Anatrace sells two grades of many popular detergents (Sol-grade and Anagrade). For solubilization and purification purposes, the more economical Sol-grade grade detergent usually can be used. Detergent stocks (for DDM, 20% w/v) can be prepared in distilled, deionized water and stored at −20°C. For crystal trials, the protein should be exchanged into Anagrade detergent using freshly prepared buffers. For reproducibility, it was found helpful to use the same batch of detergent throughout all crystal trials of a particular protein. This may require purchasing larger quantities of expensive detergent, but can prove critical if reproducibility is an issue.

2. For many other membrane proteins, it was also found that an N-terminal hexahistidine tag is compatible with expression *(11)*. One can rarely predict the effect a particular tag will have on expression, so it is advantageous to test both N- and C-terminal tag fusion strategies *(31,32)*.

3. The use of only one detergent for the production of *every* membrane protein is not advocated. However, in the interests of throughput, the strategy necessarily avoids detergent screening until an initial solubilization and purification with the detergent DDM has been attempted. If unsuccessful, alternative detergents should be tested, perhaps others widely used in membrane protein production (see Fig. 18.1). Almost all expressed membrane proteins can be solubilized with the zwitterionic detergent FC12 (FOS-CHOLINE 12, Anatrace, Maumee, OH). Eshaghi et al. also report this in a HTP membrane protein expression screening strategy *(32)*. However, Rhomboids have considerably reduced activity in FC12 (data not shown), and the survey does not suggest that this detergent is widely used in the production of membrane proteins for x-ray crystallography.

4. Plasmid pRARE2 can be obtained from a DNA miniprep of BL21 Rosetta2 cells (Novagen, San Diego, CA). Testing of other *E. coli* expression strains, including various BL21 derivatives and the so-called C41 and C43 "Walker" strains, is likely to increase the number of expressed targets *(32)*. The C41 and C43 strains have been selected for increased expression of membrane proteins *(6)*.

5. Despite an increase in the final mass of cell paste, we have observed no improvement in overall protein yield if cells are induced at an OD_{600} of approximately 0.8 versus 0.3–0.4.

6. It has been generally observed that a low temperature induction (<20°C) has a positive effect on both soluble and membrane protein expression *(11,28,33)*. Various concentrations of IPTG can also be tested, although lower concentrations (<0.5 mM) also tend to improve expression.

7. Efficient lysis can also be achieved by passing the suspension once through a Microfluidics M-100EH microfluidizer at a pressure setting of 15,000 psi, or by the use of a French press. The use of high-pressure disruption devices, although costly, is *highly* recommended. Lysis by sonication is contraindicated because of membrane-damaging effects and local heating near the sonication probe *(4,34)*.

8. Longer incubation times may increase the overall yield of solubilized protein. However, keep in mind that it is advantageous to purify membrane proteins as quickly as possible. Therefore, the solubilization time should be kept at a minimum.

9. The final concentration of DDM for solubilization is 1%. It may be possible to increase the overall yield by varying the amount of detergent. However, concentrations that are too high can denature the target protein. Also, always work above the CMC of the detergent being used! For a detailed rationale on determining the concentration of detergent that should be used for solubilization, purification and crystallization, the reader is strongly advised to read the reviews from M. Weiner *(4)* and M. Keyes *(10)*.

10. By definition, proteins are considered solubilized if they remain in solution after subjecting to 100,000g for 1 hour.

11. Some membrane proteins aggregate irreversibly when heated in the presence of SDS. As a result, they may get trapped in the sample wells during SDS-PAGE. The addition of 4 M urea to the sample buffer and immediate loading of the samples (without prior heating) helps to reduce the aggregation of some targets (unpublished observations).

12. Although a detergent-compatible assay for Rhomboid activity was developed *(13)*, an unavoidable issue with many membrane proteins is that they are vectorial in nature, transferring a signal or substrate from the outside to the inside of the cell or *vice versa* (i.e., ion channels, transporters, porins, etc.). This property is obviously lost in detergent, often rendering functional assays extremely difficult. The gold standard for such targets is usually to perform functional assays on purified protein that has been reconstituted into lipid bilayers.

13. The use of a pipette to resuspend the sticky membrane–containing pellets will usually result in some degree of frothing. A simple method to solubilize while avoiding excessive frothing is to remove the pellet using a thin metal spatula, transfer to a small beaker containing solubilization buffer and stir gently at 4°C. Any part of the pellet still sticking to the centrifugation tube can then be easily washed off with solubilization buffer.

14. Loading the sample a second time may increase the overall yield. This step has been found to be helpful for a number of membrane proteins.

15. Washing the column with buffer containing 500 mM NaCl will remove some proteins that bind nonspecifically to the resin.

16. Many companies also sell detergent-compatible kits for determining protein concentration. For example, the EZQ Protein Quantitation Kit (Invitrogen) is compatible with almost all detergents, even at high concentrations.

17. Some membrane proteins (and, indeed, some soluble proteins) may bind tightly to the concentrator membrane or housing. If this is suspected, it is recommended that concentrators with a polypropylene housing be tested (i.e., Macrosep, PALL). Also, it is important to remember that detergent

micelles can often be greater than 10–20 kDa in size and may not pass through a concentrator membrane (information on the micelle size of most popular detergents can be found in the Anatrace catalogue, available online and at www.anatrace.com). As a result, as the protein is concentrated, the detergent concentration also increases, often to a point where it can negatively affect activity and/or the outcome of crystal trials. For crystal trials, the detergent concentration should be above the CMC, at a concentration sufficient to provide 1.5–2 micelles per protein molecule *(4)*. Reproducibility is also an issue since it is usually difficult, if not impossible, to determine the detergent concentration accurately. In practice, there are really only a few ways of addressing this problem. One should use a concentrator with a molecular weight cutoff near the expected size of the protein–detergent complex, which is often much greater than the protein or detergent micelle alone. This will permit the removal of at least some detergent into the filtrate. The starting volume to be concentrated should also be as low as possible. It is sometimes possible to remove excess detergent by dialysis or the use of polystyrene beads (i.e., Biobeads, BioRad), but this invariably results in some loss of protein and/or activity. Of course, the most ideal solution is to avoid concentrating altogether by, say, eluting the protein off an ion exchange column in a very small volume. Still, from an extensive review of the literature, it is clear that further concentration of membrane proteins is almost always necessary prior to crystallization.

18. The final yield of purified Rhomboid is typically in the range of 0.3–0.5 mg per L of original culture. At this expression level, induced protein is obvious in the membrane fraction when analyzed by Coomassie Blue-stained SDS-PAGE (see Fig. 18.5A) and can also be readily detected by Western Blotting. Using the expression strategy outlined here, one rarely observes induction levels yielding greater than 0.6 mg of purified protein per L of original culture (data not shown).

19. With rare exceptions, the optimization of membrane protein crystals is an exceedingly difficult and time-consuming process. For a very specific and rational strategy to improve upon initial membrane protein crystal "hits," the reader is strongly advised to read Iwata's review *(35)*.

20. Variations in SEC retention time and peak shape are used as relative indicators of membrane protein stability. Typically, the aim is to identify detergents that promote monodispersity of a membrane protein target or complex. By comparing SEC profiles, detergents that may cause the protein(s) to form oligomers or aggregates can be readily identified. In the absence of a functional assay, as is often the case for detergent-solubilized proteins, this strategy is one of few methods available to quickly identify detergents that favor protein stability and that are suitable for crystal trials.

21. Detergent abbreviations are as follows:

αDDM, alpha n-Dodecyl-β-D-maltopyranoside
Brij 35, Polyoxyethylene glycol(23)monododecyl ether
C10DAO, n-Decyl-N,N-dimethylamine-N-oxide
C10E5, Pentaethyleneglycolmonodecyl ether
C12DAO, Dodecyldimethylamine oxide
C12E8, Octaethyleneglycol mono-n-dodecyl ether
C12E9, Polyoxyethylene(9)dodecyl ether
C6DAO, Hexyl-(dimethyl)-amine oxide

C8E4, Tetraethyleneglycolmonooctyl ether
C8E5, Pentaethyleneglycolmonooctyl ether
C8HESO, N-octyl-2-hydroxy-ethylsulfoxide
C8POE, Octyl-polyoxyethylene
C-HEGA-10, Cyclohexylbutanoyl-N-hydroxyethylglucamide
Cymal-5, 5-Cyclohexyl-1-pentyl-β-D-maltoside
DDG, n-Dodecyl-β-D-glucoside
DDM, n-Dodecyl-β-D-maltopyranoside
Deoxy-BigCHAP, N,N'-bis(3-D-gluconamidopropyl)deoxycholamide
DGDG, Digalactosyl diglyceride
DHPC, 1,2-diheptanoyl-sn-glycero-3-phosphocholine
DM, n-Decyl-β-D-maltoside
DMG, Decanoyl-N-methyl-glucamide
DOPC, 1,2-dioctanoyl-sn-glycero-3-phosphoethanolamine
Elugent, Elugent
Fos-choline 14, N-Tetradecylphosphocholine
HEGA-10, Decanoyl-N-hydroxyethylglucamide
HG, n-Heptyl-β-D-glucoside
HxG, n-Hexyl-β-D-glucoside
LAPAO, 3-Dodecylamido-N,N'-dimethylpropylaminoxide
LDAO, n-Dodecyl-N,N-dimethylamine-N-oxide
Lipids: 1-palmitoyl-2-oleoyl-sn-glycero-3-phosphocholine, 1-palmitoyl-2-oleoyl-sn-glycero-3-phosphoethanolamine, and 1-palmitoyl-2-oleoyl-sn-glycero-3-[phospho-rac-(1-glycerol)]
Na Cholate, Sodium cholate
NG, n-Nonyl-β-D-glucoside
NM, n-Nonyl-β-D-maltoside
OG, n-Octyl-β-D-glucoside
OM, n-Octyl-β-D-maltoside
OTG, n-Octyl-β-D-thioglucoside
SDS, Sodium dodecyl sulfate
SPC, Soybean-phosphatidylcholine
Thesit, Polyethylene glycol 400 dodecyl ether
tri-DM, Tri-decyl-β-D-maltoside
Triton X-100, 4-(1,1,3,3-Tetramethylbutyl)phenyl-polyethylene glycol
Tween 20, Polyoxyethylene (20)sorbitan monolaurate
UDAO, Undecylamine-N,N-dimethyl-N-oxide
UDM, n-Undecyl-β-D-maltoside
Zwittergent 3–12, n-Dodecyl-N-N-dimethyl-3-ammonio-1-propanesulfonate
Zwittergent 3–14, n-Tetradecyl-N-N-dimethyl-3-ammonio-1-propanesulfonate

Acknowledgments

The authors gratefully acknowledge many stimulating discussions with fellow researchers at Vertex Pharmaceuticals, the Ontario Centre for Structural Proteomics and the Structural Genomics Consortium. The authors are also grateful to J. Moore, A. Edwards, and D. Doyle for much advice and to M. Lemberg,

J. Menendez, A. Misik, and M. Garcia for their work on Rhomboids. The authors are especially indebted to M. Freeman for the generous gift of several Rhomboid clones. C. M. K. conducted experiments on Rhomboids while at the Ontario Centre for Structural Proteomics (University of Toronto). This work was funded by Genome Canada and the Ontario Genomics Institute.

References

1. Grisshammer, R., and Tate, C. G. (1995) Overexpression of integral membrane proteins for structural studies. *Q. Rev. Biophys.* **28**, 315–422.
2. Loll, P. J. (2003) Membrane protein structural biology: the high throughput challenge. *Journal of Structural Biology* **142**, 144–153.
3. Tate, C. G., and Grisshammer, R. (1996) Heterologous expression of G-protein-coupled receptors. *Trends Biotechnol.* **14**, 426–430.
4. Wiener, M. C. (2004) A pedestrian guide to membrane protein crystallization. *Methods* **34**, 364–372.
5. Selinsky, B. S. (ed.) (2003) Membrane Protein Protocols: Expression, Purification, and Characterization, Humana Press, Totowa, NJ.
6. Miroux, B., and Walker, J. E. (1996) Over-production of proteins in Escherichia coli: mutant hosts that allow synthesis of some membrane proteins and globular proteins at high levels. *J. Mol. Biol.* **260**, 289–298.
7. Sarramegna, V., Talmont, F., Demange, P., and Milon, A. (2003) Heterologous expression of G-protein-coupled receptors: comparison of expression systems from the standpoint of large-scale production and purification. *Cell. Mol. Life Sci.* **60**, 1529–1546.
8. Tate, C. G. (2001) Overexpression of mammalian integral membrane proteins for structural studies. *FEBS Lett.* **504**, 94–98.
9. Lemieux, M. J., Song, J., Kim, M. J., Huang, Y., Villa, A., Auer, M., et al. (2003) Three-dimensional crystallization of the Escherichia coli glycerol-3-phosphate transporter: a member of the major facilitator superfamily. *Protein Sci.* **12**, 2748–2756.
10. Keyes, M. H., Gray, D. N., Kreh, K. E., and Sanders, C. R. (2003) Solubilizing detergents for membrane proteins, in (Iwata, S., ed.), *Methods and Results in Crystallization of Membrane Proteins*, International University Line, La Jolla, CA.
11. Dobrovetsky, E., Lu, M. L., Andorn-Broza, R., Khutoreskaya, G., Bray, J. E., Savchenko, A., et al. (2005) High throughput production of prokaryotic membrane proteins. *J. Struct. Funct. Genomics* **6**, 33–50.
12. Lunin, V., Dobrovetsky, E., Khutoreskaya, G., Rongguang, Z., Joachimiak, A., Doyle, D. A., et al. (2006) Crystal structure of the CorA Mg^{2+} transporter. *Nature* **440**, 833–837.
13. Lemberg, M. K., Menendez, J., Misik, A., Garcia, M., Koth, C. M., and Freeman, M. (2005) Mechanism of intramembrane proteolysis investigated with purified rhomboid proteases. *EMBO J.* **24**, 464–472.
14. Urban, S., Lee, J. R., and Freeman, M. (2001) *Drosophila* rhomboid-1 defines a family of putative intramembrane serine proteases. *Cell* **107**, 173–182.
15. McQuibban, G. A., Saurya, S., and Freeman, M. (2003) Mitochondrial membrane remodelling regulated by a conserved rhomboid protease. *Nature* **423**, 537–541.
16. Cipolat, S., Rudka, T., Hartmann, D., Costa, V., Serneels, L., Craessaerts, K., et al. (2006) Mitochondrial rhomboid PARL regulates cytochrome c release during apoptosis via OPA1-dependent cristae remodeling. *Cell* **126**, 163–175.
17. Gallio, M., Sturgill, G., Rather, P., and Kylsten, P. (2002) A conserved mechanism for extracellular signaling in eukaryotes and prokaryotes. *Proc. Natl. Acad. Sci. USA* **99**, 12208–12213.

18. Brossier, F., Jewett, T. J., Sibley, L. D., and Urban, S. (2005) A spatially localized rhomboid protease cleaves cell surface adhesins essential for invasion by *Toxoplasma. Proc. Natl. Acad. Sci. USA* **102**, 4146–4151.

19. Nollert, P., Navarro, J., and Landau, E. M. (2002) Crystallization of membrane proteins in cubo. *Methods Enzymol.* **343**, 183–199.

20. Nollert, P. (2004) Lipidic cubic phases as matrices for membrane protein crystallization. *Methods* **34**, 348–353.

21. Berman, H. M., Westbrook, J., Feng, Z., Gilliland, G., Bhat, T. N., Weissig, H., et al. (2000) The Protein Data Bank. *Nucleic Acids Res.* **28**, 235–242.

22. Raman, P., Cherezov, V., and Caffrey, M. (2006) The Membrane Protein Data Bank. *Cell. Mol. Life Sci.* **63**, 36–51.

23. Savchenko, A., Yee, A., Khachatryan, A., Skarina, T., Evdokimova, E., Pavlova, M., et al. (2003) Strategies for structural proteomics of prokaryotes: quantifying the advantages of studying orthologous proteins and of using both NMR and x-ray crystallography approaches. *Proteins* **50**, 392–399.

24. Locher, K. P., Lee, A. T., and Rees, D. C. (2002) The E. coli BtuCD structure: a framework for ABC transporter architecture and mechanism. *Science* **296**, 1091–1098.

25. Jiang, Y., Lee, A., Chen, J., Ruta, V., Cadene, M., Chait, B. T., et al. (2003) X-ray structure of a voltage-dependent K+ channel. *Nature* **423**, 33–41.

26. Chang, G., Spencer, R. H., Lee, A. T., Barclay, M. T., and Rees, D. C. (1998) Structure of the MscL homolog from *Mycobacterium tuberculosis*: a gated mechanosensitive ion channel. *Science* **282**, 2220–2226.

27. Wasserman, J. D., Urban, S., and Freeman, M. (2000) A family of rhomboid-like genes: *Drosophila* rhomboid-1 and roughoid/rhomboid-3 cooperate to activate EGF receptor signaling. *Genes Dev.* **14**, 1651–1663.

28. Christendat, D., Yee, A., Dharamsi, A., Kluger, Y., Savchenko, A., Cort, J. R., et al. (2000) Structural proteomics of an archaeon. *Nat. Struct. Biol.* 7, 903–909.

29. Moreland, N., Ashton, R., Baker, H. M., Ivanovic, I., Patterson, S., Arcus, V. L., et al. (2005) A flexible and economical medium-throughput strategy for protein production and crystallization. *Acta Crystallogr. D Biol. Crystallogr.* **61**, 1378–1385.

30. Mohanty, A. K., Simmons, C. R., and Wiener, M. C. (2003) Inhibition of tobacco etch virus protease activity by detergents. *Protein Expr. Purif.* **27**, 109–114.

31. Tucker, J., and Grisshammer, R. (1996) Purification of a rat neurotensin receptor expressed in *Escherichia coli. Biochem. J.* **317**, 891–899.

32. Eshaghi, S., Hedren, M., Nasser, M. I., Hammarberg, T., Thornell, A., and Nordlund, P. (2005) An efficient strategy for high throughput expression screening of recombinant integral membrane proteins. *Protein Sci.* **14**, 676–683.

33. Wang, D. N., Safferling, M., Lemieux, M. J., Griffith, H., Chen, Y., and Li, X. D. (2003) Practical aspects of overexpressing bacterial secondary membrane transporters for structural studies. *Biochim. Biophys. Acta* **1610**, 23–36.

34. Hopkins, T. R. (1991) Physical and chemical cell disruption for the recovery of intracellular proteins, in (R. Seetharam, R. and Sharma, S. K., eds.), *Purification and Analysis of Recombinant Proteins*, Marcel Dekker, New York.

35. Iwata, S. (ed.) (2003) *Methods and Results in Crystallization of Membrane Proteins*. International University Line, La Jolla, CA.

Section III

Biophysical and Functional Characterization of Proteins

Chapter 19

Methods for Protein Characterization by Mass Spectrometry, Thermal Shift (ThermoFluor) Assay, and Multiangle or Static Light Scattering

Joanne E. Nettleship, James Brown, Matthew R. Groves, and Arie Geerlof

Mass spectrometry (MS) is widely used within structural and functional proteomics for a variety of tasks including protein quality assessment, identification, and characterization. MS is used routinely for the determination of the total mass of proteins, including N-glycosylated proteins, analysis of selenomethionine incorporation, crystal content verification, and analysis of N-glycosylation site occupancy. Protocols for sample preparation, data collection, and analysis are given.

A recent development is the fluorescence-based thermal shift (ThermoFluor) assay. It uses an environmentally sensitive dye, Sypro Orange, to monitor the thermal stability of a protein and investigate factors (e.g., buffers, additives, and ligands) affecting this stability. This chapter describes the application of this method using a 96-condition in-house screen. The measurements are performed on a commercially available real-time PCR machine.

Multiangle or static light scattering (SLS) is a very powerful technique to determine the conformational state of proteins in solution, especially when used in combination with size exclusion chromatography (SEC). In the authors' experimental set-up the SLS detector is connected in-line to a standard protein purification machine (e.g., the Äkta Purifier) equipped with an analytical SEC column. The data collection and analysis are performed using commercial software.

1. Introduction

Protein characterization plays an important role in two key aspects of proteomics. The first is the quality assessment of the produced protein preparations. Obtaining protein of sufficient quality for structural and/or functional studies is one of the major bottlenecks of most proteomics projects. This is reflected in the overall statistics for the major structural genomics consortia (for an overview, see http://sg.pdb.org/target_centers.html), which show that so far the structures of less than 4% of the cloned genes could be determined. Hence, it is essential to perform an extensive quality assessment of the protein preparations prior to their application and use the results in the evaluation

From: *Methods in Molecular Biology, Vol. 426: Structural Proteomics: High-throughput Methods*
Edited by: B. Kobe, M. Guss and T. Huber © Humana Press, Totowa, NJ

of the production process *(1)*. The second is the determination of protein properties such as domains, oligomeric state, posttranslational modifications and protein–protein and protein–ligand interactions.

In most laboratories SDS-PAGE and dynamic light scattering (DLS) are routinely used to determine the purity and monodispersity of a protein. However, the application of these methods is well documented in the literature; therefore, it is not considered here. This chapter gives protocols for the sample preparation and data collection and analysis of three other protein characterization methods that are routinely used in the authors' laboratories.

1.1. Protein Characterization by Mass Spectrometry

MS is a useful analytical technique and is widely used within structural proteomics consortia *(2–6)*. Analysis of both intact proteins and protein digests allow for the verification of construct along with any modifications such as selenomethionine incorporation or methylation as well as giving information on posttranslational modifications.

The protocols in this section describe how to analyze intact proteins using minimal sample preparation and automated procedures (Section 3.1.1.); remove N-glycans prior to automated LC-MS (Section 3.1.2.); and prepare proteins for LC-MS from crystals (Section 3.1.3.). A protocol for preparation of peptides by digestion of proteins followed by automated LC-MS or LC-MS/MS is given in Section 3.1.4., followed by methodology for the determination of site occupancy of N-glycosylated proteins (Section 3.1.5.).

1.2. ThermoFluor Assay

Thermal shift (ThermoFluor) assays offer a rapid and simple technique for assessing the thermal stability of proteins and to investigate factors affecting this stability *(1,7,8)*. An environmentally sensitive fluorescent dye is used to monitor protein unfolding with respect to temperature. Melting curve analysis determines the melting temperature, T_m (the midpoint of the unfolding transition); a shift in T_m under different conditions indicates a change in stability. Commercially available real-time PCR machines have sensitive thermal control and fluorescent detection capabilities, allowing the assay to be routinely performed using low amounts of protein. Thermal shift is amenable to drug screening and screening of buffer conditions, additives, ligands, and cofactors to indicate promising crystallization or storage conditions or to assign function *(1,9,10)*.

1.3. Protein Characterization by Static Light Scattering

SLS is a noninvasive technique whereby an absolute molecular mass of a protein sample in solution may be experimentally determined to an accuracy of better than 5% through exposure to low-intensity laser light (690 nm). The intensity of the scattered light is measured as a function of angle and may be analyzed to yield the molar mass, root mean square radius, and second viral coefficient (A_2). The results of an SLS experiment can be used as a quality control in protein preparation (e.g., for structural studies) in addition to the determination of solution oligomeric state (monomer/dimer, etc.). SLS experiments may be performed in either batch or chromatography modes. However, as the measurement yields the volume-averaged molecular weight of the sample within the laser beam, it is more powerful to utilize the technique in combination with protein purification. As the measurements are performed in a flow cell, there is no loss of sample and the SLS detector can be integrated easily into standard protein purification equipment. Due to the necessity of obtaining good baselines in both 280 nm absorption measurements (UV) and light scattering (LS) measurements, SEC represents a good choice of separation media, due to the use of only a single buffer system for the entire purification. Since the light scattering and concentration are measured for each eluting fraction, the mass and size can be determined independently of the elution position. This is particularly important for protein species with nonglobular shapes, which may elute at positions distant from that predicted by the calibration curve for the column.

2. Materials

2.1. Mass Spectrometry

2.1.1. Preparation and Automated Mass Spectrometry of Intact Protein Samples

1. Protein solution(s) to be analyzed.
2. 20 µM myoglobin solution (Sigma, Gillingham, UK). Dissolve protein in water and store at −80°C in 15-µl aliquots.
3. Ultimate HPLC with autosampler (Dionex, Camberley, UK) coupled to an electrospray ionization Q-Tof Micro mass spectrometer (Waters, Manchester, UK) (see Note 1).
4. C4 PepMap 300 µ-precolumn cartridge (Dionex) (see Note 2).
5. Wash solvent: 97.3% water, 2% acetonitrile, 0.5% formic acid, 0.2% trifluoroacetic acid.
6. Solvent A: 95% water, 5% acetonitrile with 0.1% formic acid.
7. Solvent B: 20% water, 80% acetonitrile with 0.1% formic acid.
8. Skirted PCR plate (Abgene, Epsom, UK) and Pierceable Power Seals (Greiner Bio-One, Stonehouse, UK).

2.1.2. Preparation of N-Glycosylated Samples

1. Peptide-N-glycosidase F (PNGase F, Sigma). Reconstitute the PNGase F according to the manufacturer's instructions, make 5-µl aliquots, and store at −80°C.

2.1.3. Preparation of Samples from Crystals (11)

1. Extra fine long paper wicks (Hampton Research).
2. Acetonitrile.
3. Resuspension buffer: 20 mM Tris-HCl pH 7.5, 200 mM NaCl, 8 M urea.
4. Skirted PCR plate (Anachem, Luton, UK) and Pierceable Power Seals (Greiner Bio-One).

2.1.4. Preparation and Automated Mass Spectrometry of Peptide Samples

1. Sequencing grade trypsin (Promega, Southampton, UK). Make 5-µl aliquots and store at −80°C to cut down on degradation of the enzyme due to freeze–thaw cycles.
2. Ultimate HPLC with autosampler (Dionex) coupled to an electrospray ionization Q-Tof Micro mass spectrometer (Waters) (see Note 1).
3. Ubiquitin (6 µM in water).

4. Glufibrinopeptide B (3 µM in water).

5. Jupiter™ 4µ Proteo 90 Å column (Phenomenex, Macclesfield, UK).

6. Wash solvent: 97.3% water, 2% acetonitrile, 0.5% formic acid, 0.2% trifluoroacetic acid.

7. Solvent A: 95% water, 5% acetonitrile with 0.1% formic acid.

8. Solvent B: 20% water, 80% acetonitrile with 0.1% formic acid.

9. Skirted PCR plate (Anachem) and Pierceable Power Seals (Greiner Bio-One).

2.1.5. Preparation of Samples for N-Glycosylation Site Analysis and Interpretation of Results

1. Sequencing grade trypsin (Promega). Make 5-µl aliquots and store at −80°C. to cut down on degradation of the enzyme due to freeze–thaw cycles.

2. Peptide-N-glycosidase F (PNGase F). Reconstitute the PNGase F according to the manufacturer's instructions, make 5-µl aliquots and store at −80°C.

3. Zwitterion Chromatography-Hydrophilic Interaction Chromatography (ZIC-HILIC) resin (HiChrom, Berkshire, UK).

4. Isopropanol.

5. Rainin Finepoint P20 filtered tips (Anachem, catalogue number: RT-20F).

6. Wash solvent: 20% water, 80% acetonitrile with 0.1% formic acid.

7. Resuspension solvent: 100% acetonitrile with 0.1% formic acid.

8. Elution solvent: 100% water with 0.1% formic acid.

9. 1 M Tris-HCl pH 7.5, 0.5 M NaCl.

10. Skirted PCR plate (Anachem), Pierceable Power Seal (Greiner Bio-One) and Adhesive PCR seal (ABgene).

11. ProteoMass™ Guanidination Kit (Sigma) (optional depending on quality of results).

2.2. ThermoFluor Assay

1. Solution of purified protein at 1 mg/ml (see Note 11).

2. Sypro Orange dye (Molecular Probes); 5,000× stock in DMSO diluted to 10× working stock in water and stored at 4°C (see Note 12).

3. 96-well thin-walled PCR plate (e.g., Thermo-Fast 96 skirted, Abgene) (see Note 13).

4. Optically clear plate seals (e.g., Microseal "B" Film, Bio-Rad).

5. Plate sealer (not essential).

6. Buffer/additive/ligand screen. For example, buffers at 20–100 mM, salts at 25–500 mM, ligands at suitable concentrations compared to protein and other additives at concentrations comparable to those used in the Hampton Research Additive Screen (see Note 14).

2.3. Static Light Scattering

1. Protein solution(s) to be analyzed (see Notes 22 and 23).

2. Äkta Purifier (GE Heathcare) applied with a size exclusion chromatography column (see Notes 24 and 25).

3 Static light scattering detector (Wyatt Technology miniDAWN Tristar) (see Note 26).

4. BSA solution (10 mg/ml in water).

3. Methods

3.1. Mass Spectrometry

3.1.1. Preparation and Automated Mass Spectrometry of Intact Protein Samples

1. Prepare 15 µl of a 20-µM solution for every protein to be analyzed by mass spectrometry in buffer solution containing only ionic salts, for example, 20 mM Tris-HCl pH 7.5, 200 mM NaCl. Buffer components such as detergents, polyethylene glycol (PEG) or high imidazole are to be avoided for best results, since components of this nature are favorably ionized within the mass spectrometer and therefore mask the protein signal (see Fig. 19.1).
2. Pipette 15 µl 20 µM myoglobin into the first well of the PCR plate. This is used for the calibration of the instrument (see Note 3).
3. Aliquot 15 µl water into the second well to wash the precolumn, followed by the sample(s) of interest, which are pipetted into the proceeding wells. Preparation protocols for some specialized samples are given in the following (see Sections 3.1.2. and 3.1.3.).
4. The PCR plate is sealed before automated running of the samples, which can take place overnight.

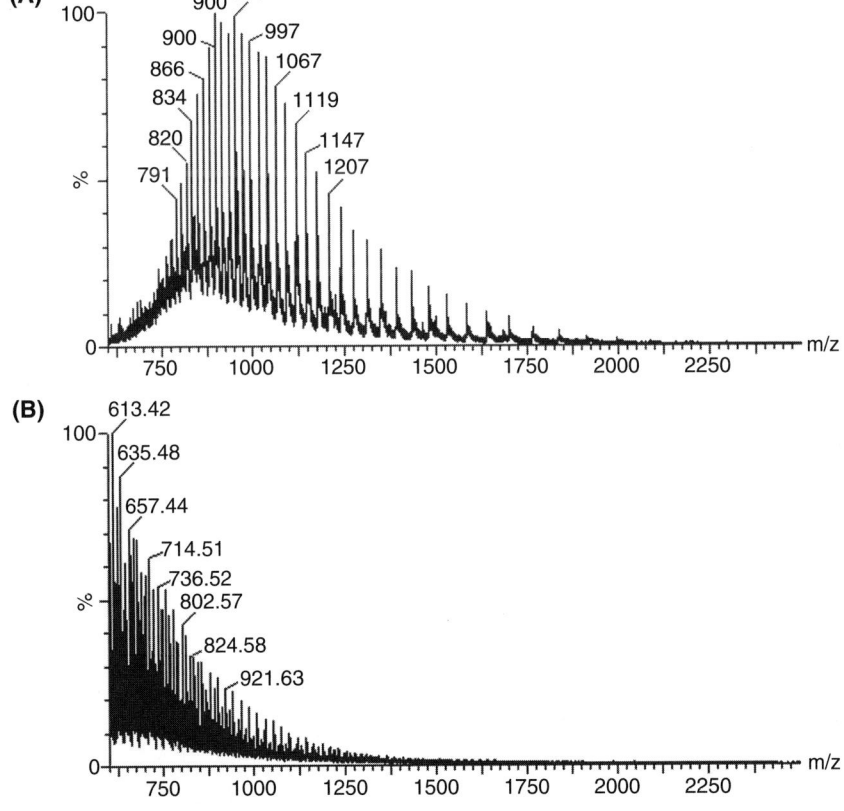

Fig. 19.1 (continued)

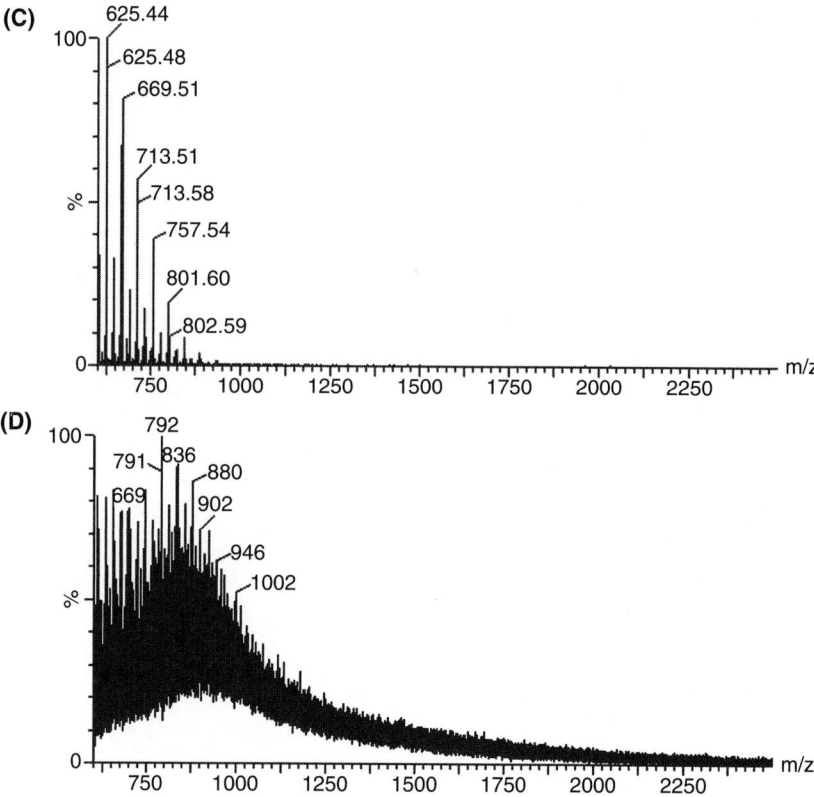

Fig. 19.1 Examples of raw data of good quality (**A**); with 10% PEG3350 contamination (**B**); with 1% Triton X100 (detergent) contamination (**C**); and with 0.5 M imidazole contamination (**D**).

5. Each sample is automatically loaded onto the C4 precolumn through the autosampler and washed with wash solvent at 5 μl/min for 5 minutes.

6. The sample is then eluted at 5 μl/min in the reverse direction directly into the mass spectrometer using a gradient of 5% to 80% solvent B over 1 minute. The concentration of mobile phase is held at 80% Solvent B for 10 minutes before re-equilibration of the column in 95% Solvent A, 5% Solvent B.

7. Analysis of samples is by MS (as opposed to MS/MS). Data is processed by combining the data under the peak on the chromatogram (Fig. 19.2A) to give a raw data file showing % ions against mass/charge (see Fig. 19.2B). This raw data is then deconvoluted to give a single mass peak using the MaxEnt algorithm (see Fig. 19.2C). The MaxEnt algorithm is part of the MassLynx software used to control the mass spectrometer.

3.1.2. Preparation of N-Glycosylated Samples

1. Boil 15 μl of a 20-μM protein sample in buffer solution containing only ionic salts for 10 minutes to denature the protein. This can be done either in a closed microcentrifuge tube or a sealed PCR plate.

2. After the solution has cooled to room temperature, add 1 μl of PNGase F and incubate at 37°C for over 3 hours (see Note 4).

3. Perform LC-MS and data analysis as described in Section 3.1.1.

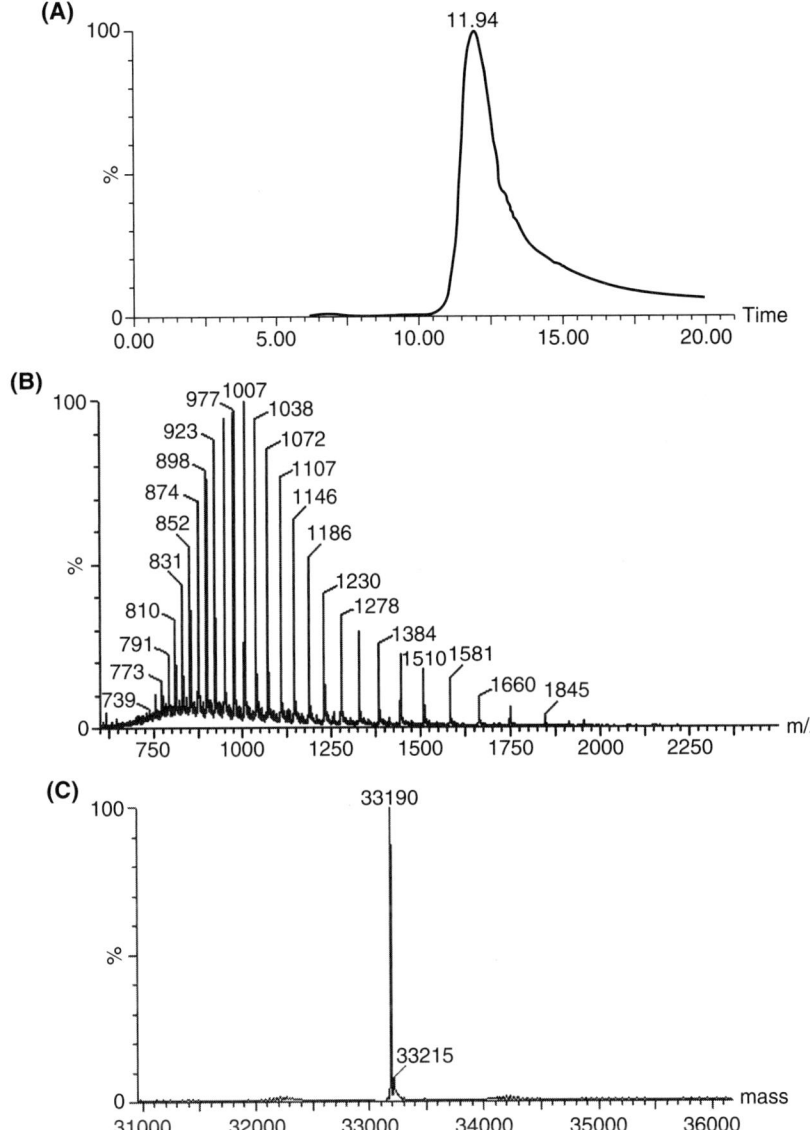

Fig. 19.2 Example data from a standard run showing the peak on the chromatogram (**A**); the raw combined data (**B**); and the deconvoluted mass peak (**C**).

3.1.3. Preparation of Samples from Crystals (11)

1. Open the well containing the crystal(s) of interest and place the crystallization plate under a microscope. This technique is limited by the concentration of protein in the crystal(s) therefore larger crystal(s) will give better data.
2. Wick away the mother liquor from around the crystal(s) using an extra fine long paper wick.
3. Carefully wash the crystal(s) in 15 µl of reservoir solution from the crystallization experiment.
4. Remove all solution from around the crystal(s) using a wick.

5. Carefully wash the crystal(s) twice with 15 μl acetonitrile, wicking away any solution in between washes. The washes remove PEG and other contaminants from the crystal(s). If any contaminants remain, this will result in less satisfactory data.

6. Dissolve the crystal(s) in 15-μl resuspension buffer and transfer to a skirted PCR plate and seal ready for mass spectrometry.

7. Analyze the sample using LC-MS as described in Section 3.1.1. (see Note 5).

3.1.4. Preparation and Automated Mass Spectrometry of Peptide Samples

1. To 15 μl of a 10-μM protein solution, add 1 μl trypsin solution and incubate at room temperature for 30 minutes (see Note 4). For glycosylation site analysis, follow the sample preparation protocol given in Section 3.1.5.

2. Aliquot calibrants into the first wells of a PCR plate. For this technique, ubiquitin (6 μM) and glufibrinopeptide B (3 μM) are used as they can be loaded into a well of the PCR plate and automatically injected onto the column as part of the run. However, accurate calibration can also be performed by direct injection using 10 μl of 0.05 μg/μl CsI in 50:50 isopropanol:water (see Note 3).

3. Pipette the sample(s) of interest into the proceeding wells and seal the plate.

4. Each sample is automatically loaded onto the Jupiter™ Proteo column through the autosampler and washed with wash solvent at 8 μl/min for 5 minutes.

5. The sample is then eluted at 10 μl/min into the mass spectrometer using a gradient of 95% solvent A:5% solvent B to 55% solvent A:45% solvent B over 40 minutes. The concentration of Solvent B is then increased to 100% over 1 minutes and held at 100% for 10 minutes before re-equilibration of the column into 95% Solvent A:5% Solvent B.

6. Peptides can be detected using either MS or MS/MS modes as required (see Note 6).

7. Analyze MS Data by combining the data under a peak of interest. MS/MS data are analyzed using the ProteinLynx Global Server™ program, which automatically processes the data and searches the online databases for proteins containing the peptides in the sample of interest.

3.1.5. Preparation of Samples for N-Glycosylation Site Analysis and Interpretation of Results

1. Pipette five samples of 20 μl of a 10-μM protein solution into five wells of the skirted PCR plate.

2. Incubate the different samples as described in step 4 and in the summary given in Fig. 19.3A. Timing of incubations can be combined as in Fig. 19.3A to allow parallelization of the experiments in the PCR plate.

3. The following types of peptide are generated by these different samples:

  • Sample 1: Nonglycosylated peptides
  • Samples 2 and 3: Both glycosylated and nonglycosylated peptides
  • Sample 4: Negative control for sample 5
  • Sample 5: Glycosylated peptides only

4. Seal the plate using an adhesive PCR seal during incubations (see Notes 4 and 7).

  • *For sample 1*: Add 1 μl of trypsin and incubate for 30 minutes at room temperature.

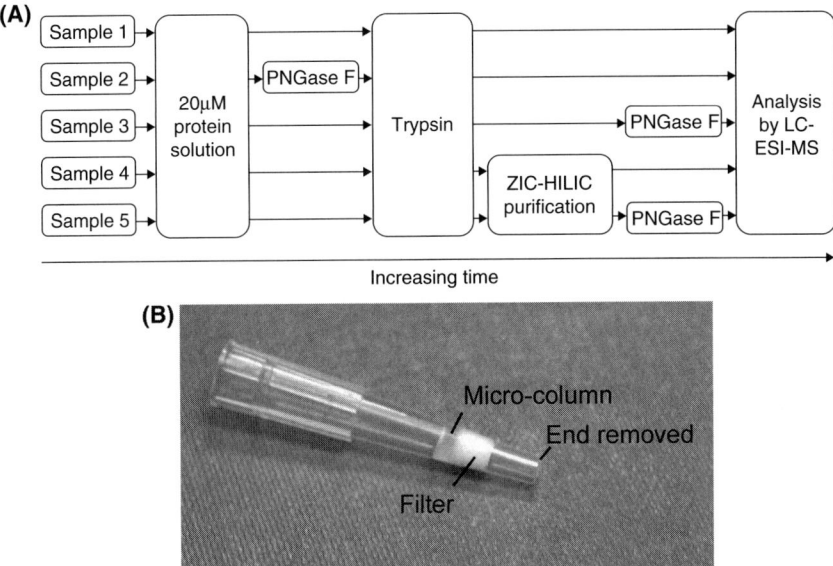

Fig. 19.3 Scheme showing the workflow for glycosylation site analysis (**A**); and construction of the ZIC-HILIC micro-column (**B**).

- *For sample 2*: Add 1 μl of PNGase F and incubate at 37°C for 3 hours. After the reaction is complete, add 1 μl of trypsin and incubate for 30 minutes at room temperature.
- *For sample 3*: Add 1 μl of trypsin and incubate for 30 minutes at room temperature. After the reaction is complete, add 1 μl of PNGase F and incubate at 37°C for 3 hours.
- *For sample 4*: Add 1 μl of trypsin and incubate for 30 minutes at room temperature. Then perform the ZIC-HILIC purification (see the following).
- *For sample 5*: Add 1 μl of trypsin and incubate for 30 minutes at room temperature, followed by ZIC-HILIC purification (see the following). Afterward, add 4 μl of 1 M Tris-HCl pH 7.5, 1 M NaCl to increase the pH before adding 1 μl of PNGase F. Incubate the sample for 3 hours at 37°C.

ZIC-HILIC purification:

- Resuspend the ZIC-HILIC resin in isopropanol to make a 50% (w/v) solution.
- Cut the end off of a filtered P20 tip about 5 mm after the filter and add 20 μl of resin solution above the filter to form a micro-column (see Fig. 19.3B and Note 8).
- Equilibrate the micro-column twice in 100 μl of wash solvent, using a P200 Pipetman to force the liquid through.
- Add 80 μl of resuspension solvent to the sample, which will give a final acetonitrile concentration of 80% and load the sample through the tip.
- Wash the micro-column three times in 100 μl wash solvent.
- Ensure all solvent is removed from the end of the tip before elution of the sample (see Note 9).
- Elute the sample in 20 μl of elution solvent.

5. Remove the adhesive PCR seal and replace it with a Pierceable Power Seal prior to analysis by LC-MS, which is run as in Section 3.1.4 using the MS mode of operation.

6. For analysis, perform an *in silico* digest of your protein using the Peptide Cutter tool (http://www.expasy.org/tools/peptidecutter/). Map the potential glycosylation sites onto the cleaved sequence. Glycosylation site occupancy prediction is performed using NetNGlyc (http://www.cbs.dtu.dk/services/NetNGlyc/) and NetOGlyc (http://www.cbs.dtu.dk/services/NetOGlyc/) for N- and O-glycosylation, respectively.

7. Using the calculated monoisotopic masses of the peptides containing potential glycosylation sites search for these in the MS data.

8. If the correct mass is found, this indicates the site is unoccupied. If the mass is found to be 1Da larger than calculated, the site is occupied. This is due to the PNGase F converting the asparagine to an aspartic acid during the deglycosylation reaction.

9. As indicated, the following types of peptide should be detected in the samples

   Sample 1: Nonglycosylated peptides.
   Samples 2 and 3: Both glycosylated and nonglycosylated peptides.
   Sample 4: Negative control for sample 5 (see Note 10).
   Sample 5: Glycosylated peptides only (see Note 10).

10. A consensus of the results from the five samples allows each potential site to be assigned as "glycosylated," "not glycosylated," or "partially glycosylated." The "partially glycosylated" category is for sites that appear as both glycosylated and not glycosylated in the different experiments (Fig. 19.4). It is these sites that may be mutated out to allow for secretion of a homogeneous sample.

11. If results are unsatisfactory, the signal of peptides ending in lysine can be enhanced using the ProteoMass™ Guanidination Kit *(12)*. This converts the lysine to homoarginine thus allowing the peptide to be more easily ionized.

Guanidination:

12. Add 10 µl of the Base Reagent into each well containing the digest sample and mix well.

13. Add 10 µl of Guanidination Reagent to each well and mix. Incubate the reactions at 65°C for 30 minutes.

14. Add 4 µl of 100% formic acid per well to stop the reaction instead of the 30–60 µl Stop Solution mentioned in Step 7 of the Sigma protocol. This keeps the sample volume to a minimum.

15. Run the samples by LC-MS as described previously (Section 3.1.4.).

3.2. ThermoFluor Assay

1. The thermal shift assay can be performed in any commercially available real-time PCR machine. Although certain elements of the protocol described here refer specifically to Opticon Monitor software version 3.1.32 (running a BioRad DNA Engine Opticon 2 real-time PCR machine), the general principles are applicable to other systems.

2. The software must be programmed with the necessary details for running the protocol, namely, which dye filters (see Note 12) and thermal parameters to

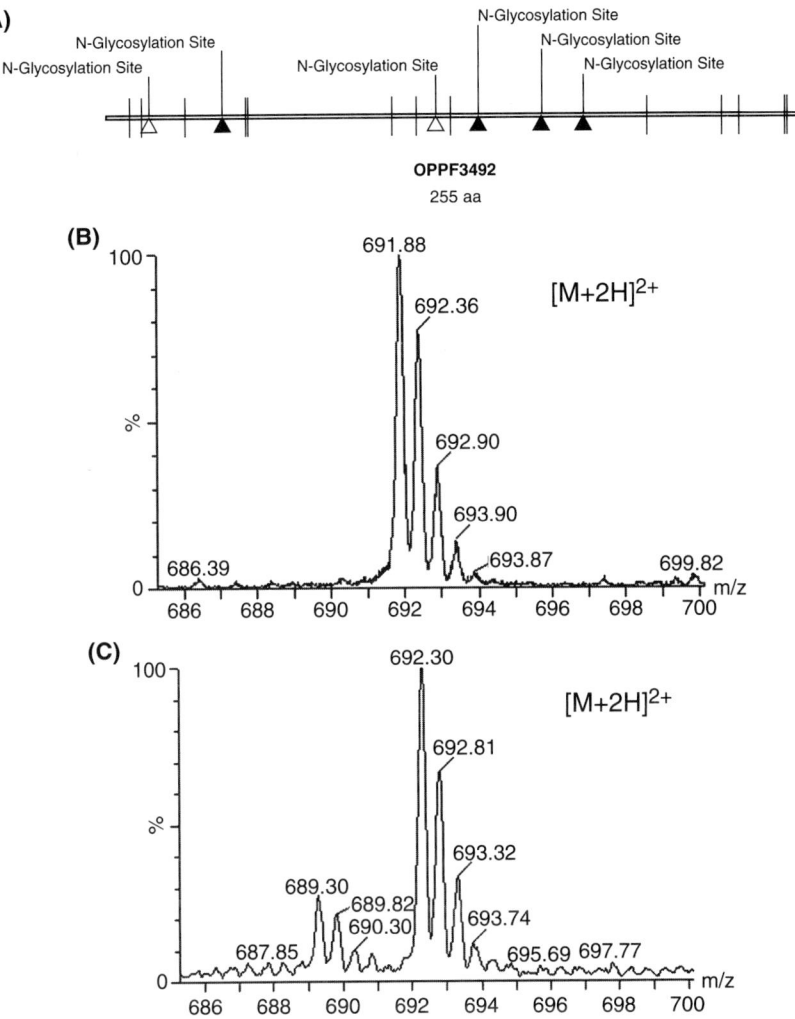

Fig. 19.4 The assigned glycosylation sites mapped onto the amino acid sequence with tryptic cleavage sites indicated by a perpendicular line, glycosylated sites by a black triangle and partially glycosylated sites by a white triangle (**A**). Example of ESI-LC-MS data for the peptide containing the third potential glycosylation site (LSNLDPGNYSFR) showing the peak with no shift (**B**) and with a +1 Da shift (**C**).

use. For the Opticon Monitor software, this involves specifying a plate template and running method. In the "Plate Setup" section, a new template should be programmed, choosing the appropriate plate type (clear/white) and selecting all wells to be read as samples using the SYBR Green filter (SBG1: see Note 12). In the "Protocol Setup" section, a new running method should be entered as a melting curve, with heating from 20°C to 95°C and a 15-second hold every 0.5°C, followed by a fluorescence reading. The assay volume should be entered here to allow the software to calculate the actual temperature in the reaction mixture. Also in this window, the authors set the temperature control at "Sample Calculation" with "Lid Settings" set at "Constant 101°C (shut off <20°C)." Save the templates with appropriate filenames.

3. Prepare the assay plate using a multichannel and/or repeater pipette to reduce pipetting errors. Total reaction volume is 50 µl (see Note 15). To each well, add 5 µl protein solution, 5 µl 10× Sypro Orange solution and 40 µl of whichever screen is under investigation. The order of addition is unimportant. Up to 96 additives or buffers can be tested on each plate; replicates can be used for more reliable evaluation of fewer conditions.

4. Seal the plate carefully (see Note 16) and centrifuge for 1 minute at 500g to mix components.

5. Place the plate into the PCR machine, close the lid and set the program running. For the Opticon Monitor software, select the appropriate plate and method, click "Run" and give a filename when prompted. After about 1 hour the assay is complete.

6. Assay progress can usually be followed in real-time. In the Opticon Monitor software, this is done via the "Status" window where the "Optical Read Status" tab allows the user to highlight wells and follow the melting curve(s).

7. When the assay is complete, the melting curves can be analyzed and T_ms calculated. Some software packages perform these calculations automatically while others require that the data be exported into statistical analysis software where curve fitting can be performed to extract T_ms. The "Melting Curve" window of the Opticon Monitor software calculates the T_m from the maximum value of the first derivative curve of the melting curve (Fig. 19.5). In the "Graphs" tab, selecting a well displays the data for that well, with radio buttons allowing display of the intensity curve, the first derivative curve or both. The "Calculations" tab at the bottom of the window displays the T_ms. An ideal melting curve would be sigmoidal but this is not always the case (see Notes 17–19) and adjustment of the "Peak Location Boundaries" in the "Graphs" tab might be necessary to define the correct region for the calculation. Care must be taken with the interpretation and it is advisable to visually inspect the melting curves before accepting the calculated T_ms. A "smooth" function which removes noise from the curves can also affect calculated T_ms and should be used with care (see Note 20).

8. Upon completion of the protocol, all melting curves can be compared and the results correlated with the various screens (see Note 21). Running multiple samples allows comparison of data from well-to-well; while large thermal shifts are obvious, such direct comparison can aid in the detection of smaller shifts. If necessary, raw data can usually be exported for curve fitting and further analysis.

9. Ideally, follow-up experiments should be performed using titrations of ligand or additive and assaying in replicate.

3.3. Static Light Scattering

3.3.1. Experimental Set-up and Calibration

1. For SLS measurements in the chromatography mode the miniDAWN Tristar (SLS detector) should be placed in the flow path of the Äkta Purifier. The instrument is typically equipped with a Superdex 200 10/300 GL SEC column. The SLS detector uses the UV signal from the Äkta Purifier to measure protein concentrations and care needs to be taken to ensure that a suitable delay volume is programmed into the ASTRA software (Wyatt Technology, Santa Barbara, CA) to correct for the difference in flow path between the UV

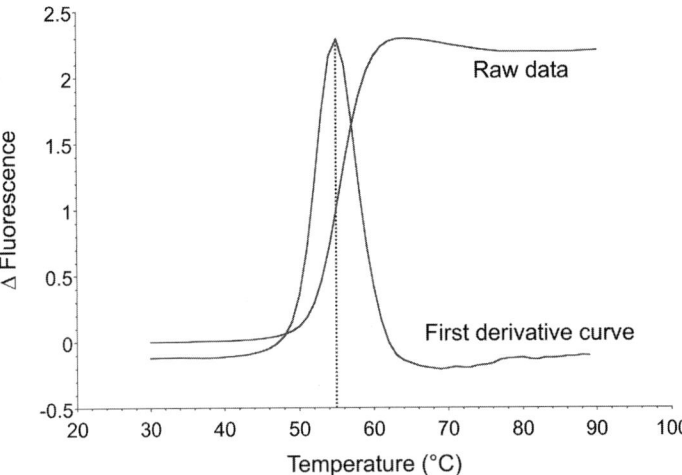

Fig. 19.5 Example thermal shift data showing the raw data curve alongside the first derivative curve. The temperature at the peak of the first derivative curve is the melting temperature, T_m *(dotted line)*.

and SLS measurements. This can be calculated from the difference in elution volume between the UV and SLS signals. A simple way to ensure the correct determination of the delay volume is available through the analysis software (see "Alignment" in the "View" menu). An overlaid display is given of both the SLS and UV signals and a right mouse click and drag between the two respective peaks allows the user to manually overlay the two signals. High precision is achievable by zooming in on the peaks (Control-left mouse to drag a zoom area). Once this delay volume is defined it should only need to be redefined if tubing length between the UV and SLS detectors is altered.

2. A 0.22-µm prefilter should be placed immediately upstream of the SLS detector in order to remove large particles (e.g., produced by the pumps), which will disturb the measurements (see Note 27).

3. The system is equilibrated with 2–3 column volumes of an appropriate buffer (see Note 28).

4. Once the system has been correctly set up and equilibrated a standard protein is used for calibration purposes. Bovine serum albumin (BSA) represents a good choice of calibration sample, as it forms a number of known oligomeric states in solution (monomer, dimer, and tetramer of 66-kDa subunits).

The calibration is performed as follows:

1. Inject 100 µl BSA solution and start data collection (see Section 3.3.2.).

2. Adjust the value of the "AUX1 calibration constant" in the "system set-up" window of the "Collect" menu during data analysis until the correct mass for BSA is obtained (see Section 3.3.3.). This value needs to be determined periodically to correct for changes in the intensity of the UV light.

However, any protein of "known" molecular weight and specific absorption coefficient may be used and after initial installation and calibration it is recommended that the user confirms the calibration using, e.g., lysozyme (molecular weight of 14 kDa).

3.3.2. Data Collection

1. Start a new experiment by selecting "New" from the "File" menu of the ASTRA software. A window appears with two screens monitoring the UV and LS signals.
2. Set-up the experimental parameters in the "Collect" menu. Enter in the "System set-up" window "solvent" (typically water), "flowrate" and "AUX1 calibration constant." Correct setting of the flow rate is essential as the device uses timers to define data collection windows over the required volume ranges. Hence incorrect flow rates will result in miscollection of the data. The new "AUX1 calibration constant" is calculated by correcting the value determined during the BSA calibration (see Section 3.3.1.) for the difference in the absorption coefficients of the sample protein and BSA:

$$(\text{AUX1})_{new} = (\text{AUX1})_{BSA\ calibration}\ \text{X}\ A_{280,\ BSA}^{0.1\%}\ /\text{X}\ A_{280,\ sample}^{0.1\%}$$

   Where $(\text{AUX1})_{BSA\ calibration}$ is the AUX1 calibration constant determined during BSA calibration; $A_{280,\ sample}^{0.1\%}$ is the absorption coefficient at 280 nm for a 1 mg/ml sample solution (see Note 29); $A_{280,\ BSA}^{0.1\%}$ is the absorption coefficient at 280 nm for a 1 mg/ml BSA solution (0.68 cm^{-1}).

   Enter in the "Collection set-up" window the following parameters: "operator," "sample ID," "injection-to-collect-delay volume," "collection duration," and "collection interval" (see Note 30). Save the experiment.
3. Fill the injection loop with 100 µl of the sample protein solution (see Note 22).
4. Start the chromatography method on the Äkta Purifier (see Note 31).
5. The data collection on the SLS detector has to be started manually: choose "single injection" in the "Collect" menu. Press "OK" to start data collection at the same time the sample is injected on the SEC column (indicated by the shift of the injection valve).

3.3.3. Data Analysis

The data obtained from the SLS detector is analyzed using the ASTRA software.

1. Choose "Baselines" in the "View" menu. Draw the baselines in the UV and the three LS signals by clicking on the traces in the linear parts to either side of the peak(s). It is worthwhile optimizing the baselines since they are crucial for the outcome of the data analysis. An option is available to define a baseline only in the second LS window, through "Auto baseline" in the "Options" menu, but the user is strongly recommended to visually check the first and third windows manually.
2. Select "Peaks" in the "View" menu. To check if the peaks in the UV and LS signals overlay choose two traces (AUX1 and one LS) by clicking on the "Data" button in the window. After a second click on "Data" both traces appear in the window. This can be repeated for the other LS signals. When the peaks overlay well select the area of the peak(s) to be analyzed by left mouse clicking and dragging over the area to be analyzed. The selected area is temporarily visualized by a gray bar and can be adjusted if needed.
3. Select "Report" in the "View" menu and choose "Summary." This window summarizes the results of the data analysis. A graphical display of the

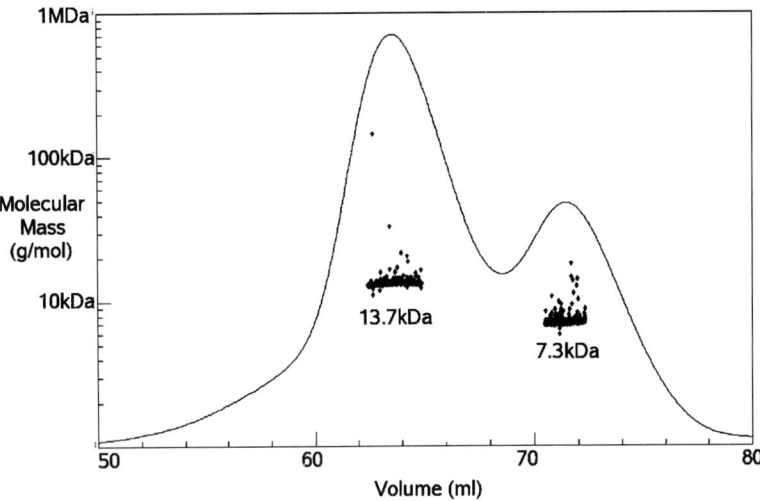

Fig. 19.6 Analysis of the solution oligomeric state of a 7.0-kDa protein by static light scattering in combination with size exclusion chromatography. The separation of the oligomers was performed on a HiLoad 16/60 Superdex 75 column (GE Healthcare). The light scattering signal *(dots)* is shown as the mass distribution in a slice in each of the two peaks in the elution profile monitored by the absorbance at 280 nm *(solid line)*. The molecular masses were calculated to be 7.3 and 13.7 kDa, respectively, which correlates well with the monomeric and dimeric protein forms.

masses present in each data slice is available through "MM vs. Volume" in the "Distribution" menu (Fig. 19.6). In order to measure only regions of the chromatogram in which a single species is eluting it may be necessary to adjust the analysis region (see Step 2) such that a horizontal line is obtained. The presence of nonhorizontal sections within the analysis area indicates the presence of two species, whose relative concentrations are not constant—leading to a change in the mass averaged molecular weights of these sections.

4. Notes

1. All protocols can be adapted for use with LC-MS systems from other manufacturers.
2. C4 precolumns can be reused, however some samples "stick" to the column and re-elute in all subsequent samples. On average each precolumn lasts for around 2 months.
3. Calibration can be performed either before running the samples or retrospectively and then applied to relevant data during analysis.
4. As PNGase F reactions and tryptic digests need to go to completion and protein denaturation is not an issue, longer incubation times—up to overnight—can be used. This will ensure a complete reaction without jeopardizing the sample.

5. When running samples prepared from crystals, it is essential to either use a new precolumn on the LC-MS or to make sure the precolumn is clean as the concentration of protein is likely to be below ideal.

6. MS analysis gives the accurate mass of a peptide, whereas MS/MS analysis gives both the accurate mass and fragmentation data. The fragmentation data can be used to give information on the sequence of the peptide. MS/MS data for a digested protein provide a mass fingerprint, which can be used to identify a protein by BLASTing the fingerprint against databases such as SwissProt.

7. This protocol can be used with digestion enzymes other than trypsin, which may lead to more informative results depending on the positioning of the cleavage sites in relation to putative glycosylation sites in the protein of interest. Other enzymes used routinely are chymotrypsin, glutamyl endopeptidase, thermolysin, and lysyl endopeptidase, although any digestion enzyme is appropriate for this method.

8. The ZIC-HILIC resin will migrate into the filter of tips from some manufacturers; however, Rainin Finepoint tips have proved successful for this method.

9. Removal of wash solvent or sample from the end of the tip can be performed using a P20 pipetman.

10. The ZIC-HILIC resin binds glycosylated peptides; however, some non-glycosylated peptides do bind such as a peptide containing a His_6 tag. Purification of the glycosylated peptides away from non-glycosylated ones increases the chance of detection

11. The authors' standard protocol uses protein at 1 mg/ml, with each well containing 5 μg protein. This amount was selected to ensure that most proteins give an adequate signal when first assayed. As little as 1.5 μg have been used per well with reproducible results.

12. The dye generally used is Sypro Orange, which fluoresces strongly when bound in the hydrophobic regions exposed in unfolded proteins. Its fluorescent properties are similar to those of SYBR Green, commonly used in real-time PCR applications, and hence the SYBR Green filters on real-time PCR machines can be used for excitation and detection.

13. Various types of plate are available depending on which PCR machine is being used (white or clear plates, skirted or non-skirted, and so on). Although the authors prefer opaque white plates, they have used clear plates with success.

14. The authors use a 96-conditions in-house screen (Table 19.1) for standard thermal shift assays which is stored as a master block at −20°C. As well as testing protein stability in several buffer types at different pHs and varying salt concentrations, the screen also assesses an assortment of metal salts, nucleotide cofactors, sugars, reducing agents, and other chemicals. When required, the block is defrosted and 40 μl are taken from each well for assaying. Each block contains enough material for five thermal shift assays.

15. The basic experimental conditions are widely applicable to different proteins. Assay volume and protein concentration can be lower (as low as 15 μl with as little as 1.5 μg protein/well), but this may not give an adequate signal for all proteins.

16. Care should be taken not to touch the seals with ungloved hands. Wear gloves to avoid smearing across the surface, which may affect optical properties.

Table 19.1 The 96 conditions of the in-house screen, which is routinely used for the thermal shift assay

	1	2	3	4	5	6	7	8	9	10	11	12
A	50 mM Sodium acetate pH 4.0	50 mM Sodium acetate pH 4.4	50 mM Citric acid pH 5.0	50 mM Citric acid pH 5.4	50 mM Sodium cacodylate pH 6.0	50 mM Sodium cacodylate pH 6.4	50 mM HEPES pH 7.0	50 mM HEPES pH 7.4	50 mM Tris-HCl pH 8.0	50 mM Tris-HCl pH 8.4	50 mM CAPSO pH 9.0	50 mM CAPSO pH 9.4
B	50 mM Sodium acetate pH4.0 100 mM NaCl	50 mM Sodium acetate pH4.4 100 mM NaCl	50 mM Citric acid pH5.0 100 mM NaCl	50 mM Citric acid pH5.4 100 mM NaCl	50 mM Sodium cacodylate pH 6.0 100 mM NaCl	50 mM Sodium cacodylate pH 6.4 100 mM NaCl	50 mM HEPES pH 7.0 100 mM NaCl	50 mM HEPES pH 7.4 100 mM NaCl	50 mM Tris-HCl pH 8.0 100 mM NaCl	50 mM Tris-HCl pH 8.4 100 mM NaCl	50 mM CAPSO pH 9.0 100 mM NaCl	50 mM CAPSO pH 9.4 100 mM NaCl
C	50 mM Sodium acetate pH4.0 200 mM NaCl	50 mM Sodium acetate pH4.4 200 mM NaCl	50 mM Citric acid pH5.0 200 mM NaCl	50 mM Citric acid pH5.4 200 mM NaCl	50 mM Sodium cacodylate pH 6.0 200 mM NaCl	50 mM Sodium cacodylate pH 6.4 200 mM NaCl	50 mM HEPES pH 7.0 200 mM NaCl	50 mM HEPES pH7.4 200 mM NaCl	50 mM Tris-HCl pH 8.0 200 mM NaCl	50 mM Tris-HCl pH8.4 200 mM NaCl	50 mM CAPSO pH 9.0 200 mM NaCl	50 mM CAPSO pH 9.4 200 mM NaCl
D	50 mM Sodium acetate pH4.0 500 mM NaCl	50 mM Sodium acetate pH4.4 500 mM NaCl	50 mM Citric acid pH5.0 500 mM NaCl	50 mM Citric acid pH 5.4 500 mM NaCl	50 mM Sodium cacodylate pH 6.0 500 mM NaCl	50 mM Sodium cacodylate pH 6.4 500 mM NaCl	50 mM HEPES pH 7.0 500 mM NaCl	50 mM HEPES pH 7.4 500 mM NaCl	50 mM Tris-HCl pH 8.0 500 mM NaCl	50 mM Tris-HCl pH 8.4 50 mM NaCl	50 mM CAPSO pH 9.0 500 mM NaCl	50 mM CAPSO pH 9.4 500 mM NaCl
E	2 mM ATP	2 mM ADP	2 mM AMP	2 mM AMPCPP	2 mM AMPCCP	2 mM AMPcPP	2 mM UTP	2 mM GTP	2 mM GDP	2 mM GTP-γS	2 mM TMP	2 mM FAD
F	2 mM β-NAD	2 mM β-γ MethylGTP	2 mM dCMP	2 mM dGMP	ssDNA 7mer	ssDNA 9mer	1% glycerol	5% glycerol	10% glycerol	20% glycerol	1 mM DTT	5 mM DTT
G	10 mM $CaCl_2$	10 mM $MgCl_2$	10 mM $MnCl_2$	10 mM $ZnCl_2$	10 mM $FeCl_3$	100 mM KCl	100 mM LiCl	200 mM NaThyiocyanate	10 mM L-Proline	10 mM Phenol	3% DMSO	10 mM $NiCl_2$
H	100 mM Glycine	10 mM Spermidine	10 mM Urea	5% PEG 400	3% D(+)-Glucose	3% D-Galactose	10 mM L-Alanine	10 mM L-Methionine	10 mM L-Serine	10 mM L-Arginine	0.5% n-Octyl-β-D-glucoside	0.5% n-Dodecyl-b-D-maltoside

The screen has been composed by Erika Mancini and Christian Siebold (Oxford, UK).

17. Not all proteins give good melting curves with a defined transition between folded and unfolded. This does not necessarily reflect instability, since the authors have seen poor melting curves from otherwise well-behaved and crystallizable proteins.

18. Following the complete unfolding of the protein (where fluorescence peaks), it is common to see a drop off in fluorescence, possibly due to exclusion of the dye during aggregation of denatured protein. Also, some proteins display a slight initial drop in fluorescence before the true transition begins, perhaps due to dye binding weakly in hydrophobic surface pockets and dissociating at low temperatures.

19. Sometimes, other interactions can be observed in the thermal shift profile, e.g., a plateau partway along the melting curve indicating melting of oligomers or complexes before the transition corresponding to unfolding of the individual proteins.

20. Thermal shifts are frequently significant enough to see clearly and there is often no need for accurate T_m determination. Following visual inspection, the authors generally take T_m values as calculated by the Opticon Monitor software. If necessary, data can be exported to other software for potentially more accurate T_m determination via curve fitting.

21. We have observed thermal shifts in excess of 12°C, but it is possible to see much smaller shifts of less than 1°C, especially when melting curves are compared side by side. For very low shifts, although they might turn out to be significant, experimental error must be considered as a factor. If necessary, this could be investigated by further testing in replicate.

22. The nature of the protein affects the upper and lower amounts that can be accurately analyzed. Proteins with a relatively high absorption coefficient ($A_{280}^{0.1\%} > 0.7\,\mathrm{cm^{-1}}$) will produce significantly more deflection in the UV measurements than those with low absorption coefficients ($A_{280}^{0.1\%} < 0.5\,\mathrm{cm^{-1}}$) for identical sample loading. It is possible to overload the UV signal used by the SLS detector, so care must be taken that sample concentrations do not become excessive. Conversely, insufficient sample concentration in the elution fractions will result in weak LS and UV signals. Such weak signals then result in large errors in the analysis. As a general rule of thumb, 100 μg to 1 mg of protein sample applied to a Superdex 200 10/300 GL column (GE Healthcare, Piscataway, NJ) result in reasonable LS and UV signals. When a preparative SEC column is used [e.g., the HiLoad 16/60 Superdex 200, GE Healthcare] 5 to 10 mg of sample should be applied.

23. The results of the SLS experiment are volume averaged. Hence, the best results are obtained when highly pure samples are analyzed by SEC since it is relatively easy to separate different oligomeric species of one protein from one another. However, when impure samples are applied increased errors appear due to insufficient separation of different protein species.

24. Other protein purification equipment can be used provided the UV signal of the instrument can be used by the SLS detector.

25. Both analytical and preparative SEC columns can be used for the analysis (see Note 22).

26. This protocol is based on the authors' experience with the Wyatt Technology mini-DAWN Tristar detector connected in-line with an Äkta Purifier. Other equipment combinations are also possible in order to

increase the accuracy of the method (e.g., the installation of a refractive index concentration measurement device). Such additional equipment will improve the accuracy of the technique, with an attendant increase in the cost of the installation. There are currently at least two companies which provide SLS detectors: Wyatt Technology (http://www.wyatt.com/) and Viscotek GmbH (http://www.viscotek.com). The reader is strongly recommended to browse the company Web pages for further information.

27. The addition of a prefilter immediately upstream of the SLS detector is a sensible step to prevent particles entering the device and affecting measurements. A standard 0.22-μm paper filter supported by a metal frit is sufficient for this purpose, although the backpressure of the system needs to be monitored. Increases in the backpressure may indicate that the filter has become clogged and needs replacing. Care also needs to be taken that the column itself suffers no damage due to increases in backpressure at standard operating flow rates. In the authors' experience it is permissible to add the measured backpressure of the filter and SLS detector to the maximum recommended backpressure rating of the column(s) used. This value can be determined by selecting a column bypass and noting current pump pressure (typically 0.3 MPa). Thus the Superdex 200 10/300 GL column (rated for a maximum backpressure of 1.5 MPa) can now be run at 1.8 MPa.

28. The SLS measurement is highly sensitive to baseline errors in both the UV and LS signals and as a result a thorough equilibration is needed for precise SLS measurements. Typically 2–3 column volumes of buffer represents a suitable equilibration volume for size exclusion columns. The components of the buffer should also not present too strong a background in either absorption of the UV signal or the LS signal. High concentrations of reducing agents (e.g., dithiothreitol) or glycerol should also be avoided if possible. If essential for the experiment these buffers should be prepared immediately prior to the experiment. As is true for all size exclusion experiments the buffer should be well filtered and degassed prior to use.

29. Inaccurate absorption coefficients represent the major source of errors in the data analysis. While absorption coefficients generated from the linear sequence are, in the main, sufficiently accurate it is recommended that UV absorption spectra are recorded of the sample in both the size exclusion buffer and a denaturing buffer in order to experimentally establish the extinction coefficients of the folded proteins. The presence of cofactors (e.g., nucleotides), which have significant absorption at 280 nm, also results in inaccurate concentration estimates from the UV signal. Additionally, samples that show significant absorption at the wavelength used for the SLS measurements (690 nm) result in errors in the LS measurements.

30. The maximal number of data points that can be collected per experiment is 14,400. Hence the minimal collection interval that can be chosen depends on the collection duration. For an experiment performed with a Superdex 200 10/300 GL column a typical collection duration is 30 minutes, which allows a collection interval of 0.125 seconds.

31. To allow simultaneous sample injection and start of the data collection it is advisable to program a short column equilibration step (e.g., 0.1 column volumes for the Superdex 200 10/300 GL column) in the chromatography method.

Acknowledgments

The work described here was supported by the MRC and the European Commission as SPINE, contract-no. QLG2-CT-2002-00988.

References

1. Geerlof, A., Brown, J., Coutard, B., Egloff, M.-P., Enguita, F. J., Fogg, M. J., Gilbert, R. J. C., Groves, M. R., Haouz, A., Nettleship, J. E., Owens, R. J., Ruff, M., Sainsbury, S., Svergun, D. I., and Wilmanns, M. (2006) The impact of protein characterization in structural proteomics. *Acta Crystallogr. D Biol. Crystallogr.* **62**, 1125–1136.

2. Yokoyama, S., Hirota, H., Kigawa, T., Yabuki, T., Shirouzu, M., Terada, T., Ito, Y., Matsuo, Y., Kuroda, Y., Nishimura, Y., Kyogoku, Y., Miki, K., Masui, R., and Kuramitsu, S. (2000) Structural genomics projects in Japan. *Nat. Struct. Biol.* **7**, 943–945.

3. Jeon, W. B., Aceti, D. J., Bingman, C. A., Vojtik, F. C., Olson, A. C., Ellefson, J. M., McCombs, J. E., Sreenath, H. K., Blommel, P. G., Seder, K. D., Burns, B. T., Geetha, H. V., Harms, A. C., Sabat, G., Sussman, M. R., Fox, B. G., and Phillips, G. N., Jr. (2005) High-throughput purification and quality assurance of *Arabidopsis thaliana* proteins for eukaryotic structural genomics. *J. Struct. Funct. Genomics* **6**, 143–147.

4. Pantazatos, D., Kim, J. S., Klock, H. E., Stevens, R. C., Wilson, I. A., Lesley, S. A., and Woods, V. L., Jr. (2004) Rapid refinement of crystallographic protein construct definition employing enhanced hydrogen/deuterium exchange MS. *Proc. Natl. Acad. Sci. USA* **101**, 751–756.

5. Jawhari, A., Boussert, S., Lamour, V., Atkinson, R. A., Kieffer, B., Poch, O., Potier, N., van Dorsselaer, A., Moras, D., and Poterszman, A. (2004) Domain architecture of the p62 subunit from the human transcription/repair factor TFIIH deduced by limited proteolysis and mass spectrometry analysis. *Biochemistry* **43**, 14420–14430.

6. Potier, N., Billas, I. M., Steinmetz, A., Schaeffer, C., van Dorsselaer, A., Moras, D., and Renaud, J. P. (2003) Using nondenaturing mass spectrometry to detect fortuitous ligands in orphan nuclear receptors. *Protein Sci.* **12**, 725–733.

7. Pantoliano, M. W., Petrella, E. C., Kwasnoski, J. D., Lobanov, V. S., Myslik, J., Graf, E., Carver, T., Asel, E., Springer, B. A., Lane, P., and Salemme, F. R. (2001) High-density miniaturized thermal shift assays as a general strategy for drug discovery. *J. Biomol. Screen.* **6**, 429–440.

8. Lo, M. C., Aulabaugh, A., Jin, G., Cowling, R., Bard, J., Malamas, M., and Ellestad, G. (2004) Evaluation of fluorescence-based thermal shift assays for hit identification in drug discovery. *Anal. Biochem.* **332**, 153–159.

9. Carver, T. E., Bordeau, B., Cummings, M. D., Petrella, E. C., Pucci, M. J., Zawadzke, L. E., Dougherty, B. A., Tredup, J. A., Bryson, J. W., Yanchunas, J., Jr., Doyle, M. L., Witmer, M. R., Nelen, M. I., DesJarlais, R. L., Jaeger, E. P., Devine, H., Asel, E. D., Springer, B. A., Bone, R., Salemme, F. R., and Todd, M. J. (2005) Decrypting the biochemical function of an essential gene from *Streptococcus pneumoniae* using ThermoFluor technology. *J. Biol. Chem.* **280**, 11704–11712.

10. Matulis, D., Kranz, J. K., Salemme, F. R., and Todd, M. J. (2005) Thermodynamic stability of carbonic anhydrase: measurements of binding affinity and stoichiometry using ThermoFluor. *Biochemistry* **44**, 5258–5266.

11. Nettleship, J. E., Walter, T. S., Aplin, R., Stammers, D. K., and Owens, R. J. (2005) Sample preparation and mass-spectrometric characterization of crystal-derived protein samples. *Acta Crystallogr. D Biol. Crystallogr.* **61**, 643–645.

12. Brancia, F. L., Oliver, S. G., and Gaskell, S. J. (2000) Improved matrix-assisted laser desorption/ionization mass spectrometric analysis of tryptic hydrolysates of proteins following guanidination of lysine-containing peptides. *Rapid Commun. Mass Spectrom.* **14**, 2070–2073.

Chapter 20

High Throughput Methods for Analyzing Transition Metals in Proteins on a Microgram Scale

Anelia Atanassova, Martin Högbom, and Deborah B. Zamble

Transition metals are among the most common ligands that contribute to the biochemical and physiological properties of proteins. In the course of structural proteomic projects, the detection of transition metal cofactors prior to the determination of a high-resolution structure is extremely beneficial. This information can be used to select tractable targets from the proteomic pipeline because the presence of a metal often improves protein stability and can be used to help solve the phasing problem in x-ray crystallography. Recombinant proteins are often purified with substoichiometric amounts of metal loaded, so additional metal may be needed to obtain the homogeneous protein solution crucial for structural analysis. Furthermore, identifying a metal cofactor provides a clue about the nature of the biological role of an unclassified protein and can be applied with structural data in the assignation of a putative function. Many of the existing methods for transition metal analysis of purified proteins have limitations, which include a requirement for a large quantity of protein or a reliance on equipment with a prohibitive cost.

The authors have developed two simple high throughput methods for identifying metalloproteins on a microgram scale. Each of the techniques has distinct advantages and can be applied to address divergent experimental goals. The first method, based on simple luminescence and colorimetric reactions, is fast, cheap, and semiquantitative. The second method, which employs HPLC separation, is accurate and affords unambiguous metal identification.

1. Introduction

The structural genomic initiatives have provided incentives to generate high throughput (HTP) methods for the rapid, sensitive, and accurate determination of the metal content of purified proteins. A transition metal can contribute extensively to the stability, activity, and function of a protein and enhance the success of high-resolution structural analysis, so methods for identifying metalloproteins are valuable. Various protocols for transition metal analysis have been developed based on principles related to the

From: *Methods in Molecular Biology, Vol. 426: Structural Proteomics: High-throughput Methods*
Edited by: B. Kobe, M. Guss and T. Huber © Humana Press, Totowa, NJ

physicochemical properties of the analyzed metals, such as the atomic signals produced in a plasma state *(1–3)*, complexation with chromophores or fluorophores *(4–7)*, or catalytic properties *(8)*. If the molecules under investigation are biomolecules, such as proteins, then some additional manipulation may be required to ensure metal release into solution to perform quantitative analysis.

The main advantages of the two methods described here include the use of small amounts of protein (microgram quantities), simple practical requirements, and adaptability to HTP analysis *(7,9)*. Both methods are focused on identifying the six most biologically common transition metals: manganese, iron, cobalt, nickel, copper, and zinc *(10,11)*, although the methods are not limited to these elements. The first method is based on the two established reactions of transition metals with luminol and 4-(2-pyridylazo)resorcinol (PAR) *(7)*. Many metal ions enhance the reaction of luminol with hydrogen peroxide at alkaline pH, producing an excited-state intermediate that emits light upon relaxation to the ground state *(8,12–14)*. A coincident set of metals forms a complex with the colorimetric indicator PAR that can be detected by electronic absorption spectroscopy *(4–6,15)*. This straightforward method to detect if a protein contains metals can be performed over a short timeframe (several hours) in a multi-well plate format, but in the analysis of unknown proteins it is not appropriate for metal identification or quantification *(7)*. The second method is based on the separation of the metal ions by mixed-bed ion exchange chromatography *(16–19*, and Application Note 108 at www.dionex.com), followed by postcolumn detection upon complexation with PAR *(9)*. This technique allows for precise metal identification and quantification in 100 samples over the course of about 24 hours, but the use of a metal-free HPLC instrument equipped with a postcolumn mixer is strongly recommended.

2. Materials

2.1. Multi-Well Plate Method

1. Chelex-100 resin (Bio-Rad, Hercules, CA) (see Note 1).
2. Chelex-treated 500 mM Na_2CO_3, stable for at least a week when refrigerated (see Note 2).
3. Luminol solutions: Freshly dissolve luminol in 500 mM Na_2CO_3 to prepare an 11 mM stock and add 30% H_2O_2 to a final concentration of 230 mM (see Note 3).
4. PAR solution: Prepare a 2 mM solution in water and store over Chelex. Stable refrigerated in the dark for 1–2 weeks.
5. Urea solution: Prepare 8 M in water and store over Chelex. Stable at room temperature for several months (see Note 4).
6. Metal standards: Individual metal salts are dissolved in water to prepare 1 mM stocks (see Note 5).
7. Protein standards (see Note 6).
8. White, nontransparent multi-well plates (see Note 7) with optical bottom (Nunc, Rochester, NY).
9. Fluor-S Multimager (Bio-Rad) or any instrument that detects luminescence. The settings will be optimized for the instrument.

10. Fluostar Galaxy plate reader (BMG Labtech, Offenburg, Germany) or any plate reader that measure absorbance at 492 nm. The settings will be optimized for the instrument.

2.2. HPLC Assay

1. Concentrated metal-free hydrochloric acid (cHCl) (Baseline, Seastar Chemicals, Sidney, Canada).
2. Metal standards (0.1 mg/ml) in 5% nitric acid (SCP Science, Baie D'Urfé, Canada), stable at room temperature for a year (see Note 8).
3. Pyridine-2,6-dicarboxylic acid (PDCA): 7 mM PDCA in 66 mM KOH, 5.6 mM K_2SO_4, and 74 mM HCOOH. The filtered solvent is stable at room temperature for several months.
4. PAR solution: 0.2 mM PAR in 0.5 M NH_4OH and 0.3 M CH_3COOH (see Note 9). The filtered solvent is stable at room temperature in a dark bottle for 2 weeks.
5. Commercial proteins with known amounts of transition metals bound (see Note 6).
6. Metal-free HPLC instrument, Model AS50 (Dionex), including a gradient pump GS50, a post-column pneumatic controller for post-column reagent addition, an AS50 auto-sampler, a PDA-100 detector with variable wavelength absorbance set at 530 nm, and software for data analysis (see Note 10).
7. An IonPac CS5A (Dionex, Oakville, Canada) analytical column (250 × 4 mm ID) for the metal separation, connected with a 50 × 4-mm I.D. IonPac CG5A guard column (see Note 11).
8. Vacuum system for drying protein samples, including a trap for toxic gases.

3. Methods

3.1. Multi-Well Plate Method

This method can be broken down into three sequential steps (Fig. 20.1), analysis of the luminescence signal before and after the addition of PAR and detection of the colorimetric signal produced by the PAR complexes. It has been optimized for the analysis of the six transition metals manganese through zinc, but it is not limited to these elements (see Note 12). A sample containing 0.1 nmol of a metal is sufficient for the full analysis, although the luminescent steps are even more sensitive (Fig. 20.2).

Based on the analysis of the individual metals the identity of each transition metal could be determined except for the distinction between nickel and zinc (see Figs. 20.1 and 20.2). However, combinations of two or more metals can provide misleading results (7), so it is not possible to assign the metals in unknown proteins with confidence. Furthermore, urea is used to denature the proteins, but metal release into solutions may not be complete (see Note 13). For both of these reasons, this method can only be used for metal quantification in proteins for which the identity of the metals is known and controls can be performed to confirm that the metal release is quantitative. In this case, serial dilutions of metals can be used to prepare standard curves for semiquantitative analysis.

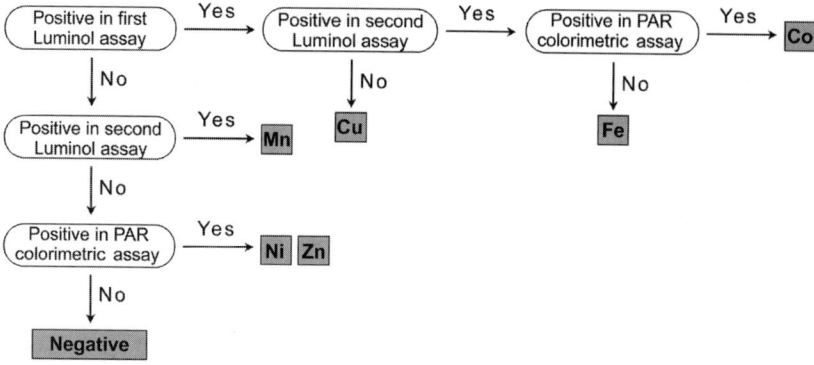

Fig. 20.1 Flow chart of the diagnostic signals of the metals based on the results in the three parts of the assay. In the simple case with only one metal in the sample, all metals except nickel and zinc can be distinguished. Taken from *(7)*.

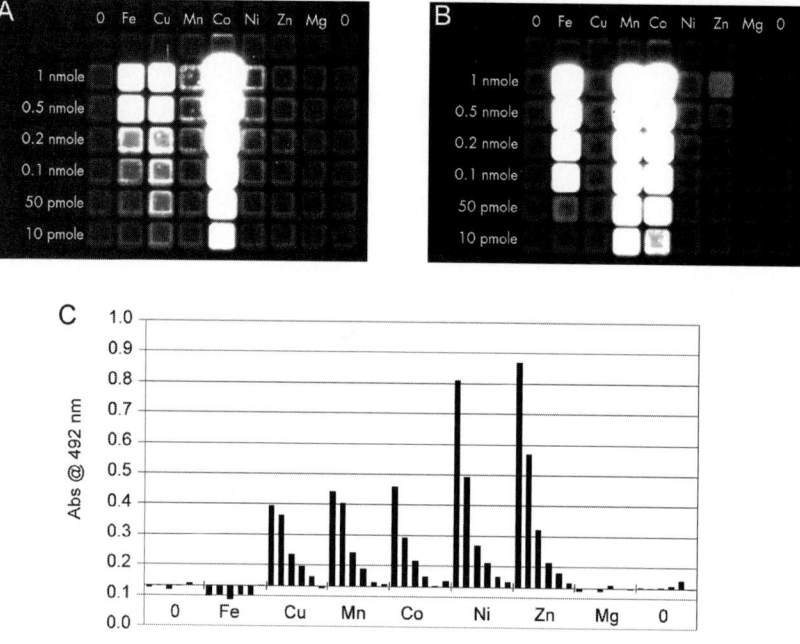

Fig. 20.2 Serial dilutions of divalent metal salt solutions. **A.** Luminescence test. **B.** Luminescence test after incubation with PAR. Note especially the disappearance of the copper signal and the appearance of the manganese signal in comparison with (**A**). **C.** absorbance at 492 nm of the same samples as in (**B**); nickel and zinc produce the strongest signals. Taken from *(7)*.

3.1.1. Analysis of Standards for Multi-Well Plate Method

1. Add 4 μl of 8 M urea to the wells of a 384-well plate plate. Add 1 μl of an individual standard solution followed by 10 μl of a freshly prepared luminol solution.

2. Measure the luminescence of the samples after 10 minutes on a Bio-Rad Fluor-S Multimager with an exposure time of 10–100 seconds (see Note 14).

3. Add 1 μl of a 2 mM PAR solution to each well, and repeat the luminescence reading.
4. Incubate the plate at room temperature for 2 hours.
5. Centrifuge the plate at 3,000g for 2 minutes to minimize interference from N_2 bubbles produced by the luminol reaction (see Note 15).
6. Measure the absorbance at 492 nm on a Fluostar Galaxy plate reader (see Note 16).
7. The results of analysis of serial dilutions of divalent metal salts are shown in Fig. 20.2. The addition of 1 nmol of iron, copper, or cobalt to luminol produces very bright signals, whereas manganese, zinc, and nickel do not (see Fig. 20.2A). The weak light that appears in the top of the manganese and nickel rows is due to light contamination from the extremely bright cobalt signal of the well next to these positions. Fortuitously, the subsequent addition of PAR has an effect on the luminescence signal of the different metals: it quenches the copper signal, but activates manganese luminescence, allowing both of these metals to be clearly identified (see Fig. 20.2B). Spectroscopic analysis at 492 nm reveals a strong absorption signal for copper, manganese, cobalt, nickel, and zinc but not iron (see Fig. 20.2C).

3.1.2. Analysis of Proteins for Multi-Well Plate Method

1. Add 4 μl of 8 M urea solution to the wells of the plate. Add 1–2 μl of protein (10 μg) to each well (see Note 17) followed by the addition of 10 μl of a freshly prepared luminol solution.
2. The same volume of protein buffer should also be analyzed as a control (see Note 18).
3. Follow steps 2–6 from Section 3.1.1.

3.2. HPLC Determination of Transition Metal Content in Proteins

For this method the proteins are degraded by acid hydrolysis to release the metal ions into solution. The metals are separated by ion exchange chromatography and as they elute from the column they form a complex with PAR, in a solution in a postcolumn mixer, which can be detected by absorption spectroscopy. The six transition metals, iron, copper, nickel, zinc, cobalt, and manganese can be separated with baseline resolution (Fig. 20.3). Furthermore, Fe^{3+} and Fe^{2+} can be resolved, although if quantification of these two states is desired, then rigorous anaerobic techniques for sample preparation and analysis should be used. During the development of this method all initial results were compared with analysis by inductively coupled plasma-atomic emission spectrometry (ICP-AES), and the data were comparable (9). The method is sensitive and accurate, requiring only micrograms of protein (Fig. 20.4), which is at least an order of magnitude less than required for ICP-AES. The limitation of the method is that a metal-free HPLC is recommended to limit contamination, and a postcolumn mixer and autosampler are required for HTP experiments.

3.2.1. Preparation of Standards and Samples for HPLC Assay

1. Dilute protein standards and protein samples to 20–30 mg/ml in the working buffer (see Note 19).

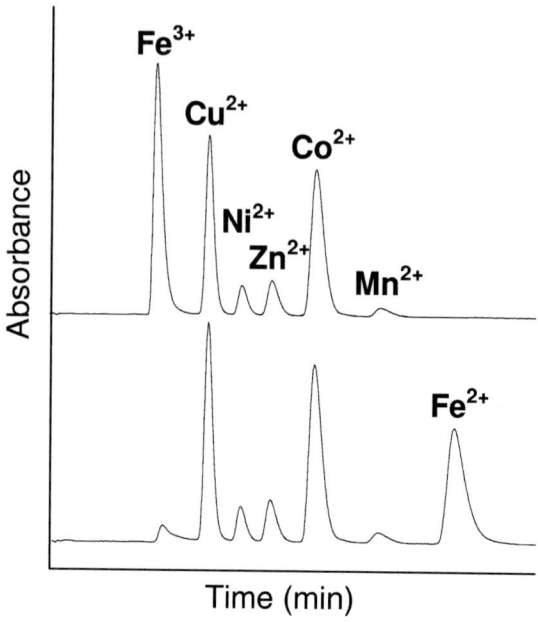

Fig. 20.3 HPLC chromatograms of metal standards. Equimolar amounts of metal stocks prepared in water (5 nmol) were injected under oxidizing conditions *(top)* or under reducing conditions *(bottom)*. The metals were separated in a 15-minute isocratic run. Adapted from *(9)*.

2. Dry 10–50 μg of the proteins in small volume microcentrifuge tubes (50 or 200 μl) by centrifugation under vacuum at room temperature (see Note 20).
3. Resuspend the dried samples in 2.5 μl cHCl and hydrolyze overnight at 100°C (see Note 21).
4. Dry the hydrolyzed mixture under vacuum at room temperature.
5. Dissolve the dried samples in 75 μl Milli-Q water.
6. Prepare serial dilutions of the standards in the working buffer (see Note 22). Mix 100 μl of Fe^{3+}, Cu^{2+}, Co^{2+}, each with 300 μl Ni^{2+}, Zn^{2+}, and 500 μl Mn^{2+} (see Note 23) and use this stock to prepare serial dilutions. Prepare at least six standard dilutions, up to 20-fold diluted.

3.2.2. HPLC Assay

1. Purge the HPLC solvents (PDCA buffer and H_2O) with nitrogen for at least 30 minutes before use (see Note 24).
2. Start equilibration of the system at 1.2 ml/min for at least 30 minutes.
3. Develop a program: isocratic separation for 15 minutes at a 1.2 ml/min, with detection at 530 nm.
4. Run the highest metal standard concentration at least two times to check the reproducibility of the system and to ensure that the peak intensities are below saturation (see Note 25).

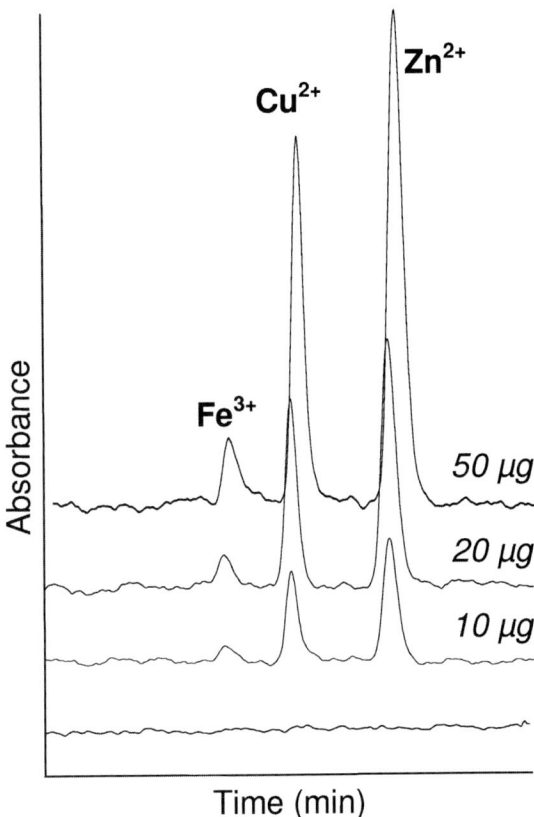

Fig. 20.4 HPLC chromatograms of commercial bovine carbonic anhydrase. The indicated amounts of protein were digested with HCl, dried, resuspended in water, and injected onto the HPLC. The bottom chromatogram is from the analysis of a buffer blank prepared in the same manner as the protein samples. There is a linear correlation between the peak area for each metal and the amount of protein analyzed, and the percent metal bound is in agreement with data from ICP-AES. The metals were separated in a 15-minute isocratic run. Adapted from *(9)*.

5. If the two consecutive runs produce reproducible data, then proceed to the analysis of the rest of the standard dilutions. An example of an analysis of all six metals is shown in Fig. 20.3.
6. Generate standard curves by using the HPLC software. If the deviations of the standard curves are in an acceptable range, then proceed to the next step (see Note 26).
7. Run the protein standards. Estimate the amount of the bound metal by using the generated standard curves. Proceed to the next step.
8. Run the protein samples (see Note 27). Estimate the amount of the bound metal by using the generated standard curves.

4. Notes

1. All buffer solutions for both methods should be treated with Chelex resin or prepared from Chelex-treated components to minimize background from metal impurities. Solutions should be stirred with Chelex for at

least 1 hour and then filtered or stored over Chelex as it settles to the bottom of the container. Alternatively, solutions can be run through a column prepared with Chelex. The amount of Chelex needed is specified by the manufacturer, although in practice significantly less resin can be used unless high background levels of metals become apparent in the analysis. It is possible, but not optimal, to omit the Chelex treatment if precise quantification is not required. In such a case, it is best to dilute the metal standards in the same buffer as the protein samples and analyze the buffer as a separate sample so that trace contaminations can be subtracted.

2. All solutions should be prepared in Milli-Q water, 18.2 MΩ-cm resistance (Millipore, Billerica, MA). Unless otherwise stated, all reagents are a minimum of analytical grade and purchased from Sigma-Aldrich. A metal-free grade is preferable; however, if the cost is prohibitive, then treatment with Chelex is usually sufficient.

3. Do not treat this solution with Chelex, use Chelex-treated 500 mM Na_2CO_3.

4. Guanidine hydrochloride was tested and resulted in a high background, probably because of impurities in the reagent used.

5. Serial dilutions of the standards can be prepared (see Fig. 20.2) if quantification is desired. However, this is not the strength of this method, so in practice the analysis of individual metals needs to be performed only once to confirm the diagnostic signal.

6. Commercial metalloproteins are dissolved in Chelex-treated 20 mM HEPES, pH 7.5, 100 mM NaCl, to a final concentration of about 60 mg/ml and aliquots are stored at −80°C. Select at least two metalloproteins that can be run as standards in parallel with the protein samples. A metal-free protein can also be used as a negative control. Analyze one aliquot of protein for each experiment to verify the success of the assay. Initially, it is a good idea to determine the metal content of the protein standards by using another method, such as ICP-AES. Several commercial metalloproteins that can be used as standards include bovine carbonic anhydrase (zinc, as well as traces of copper and iron); superoxide dismutase containing iron, manganese, or copper/zinc; cytochrome c (heme iron); and transferrin (iron).

7. The number of wells/plate depends on the capabilities of the plate reader used. The volumes described here are suitable for a 384-well format.

8. Metals initially should be analyzed separately to assign the retention time of each metal under investigation. Following this assignment, a stock metal mix can be prepared and used as a routine standard.

9. PAR dissolves faster in basic solutions. Dissolve the powder in 0.5 M NH_4OH and neutralize with 0.3 M CH_3COOH.

10. A metal-free HPLC is recommended and can be purchased from several suppliers. This instrument can also be used for any standard analytical HPLC methods.

11. The IonPac SC5A column is available in the 2-mm format for microbore separations that reduce the amount of sample needed and the operating costs. The current protocol is developed on a 4-mm column format, therefore optimization is required if the 2-mm column is used.

12. Previous studies suggest that several other metals may produce signals in this assay; for example gold, rhodium, chromium, vanadium, niobium, tantalum, and titanium have been reported to enhance the luminol reaction *(8,12–14,20–23)*, and vanadium, cadmium, and lead are also known to give a color change with PAR *(4,15)*. Under the present conditions, the oxidation states of the metal do not seem to influence the signals obtained with this method. Thus, distinguishing Fe^{2+} from Fe^{3+} and Mn^{2+} from Mn^{7+} is not possible.

13. Acid treatment of the protein samples can be included as an efficient step to release the metals into solution. In such a case, neutralization of the samples is required before the luminol assay. The same (acid-base) treatment of both metal and protein standards is required to verify compatibility and performance. This extra step was not optimized because the strength of this method is the identification of metalloproteins from a large number of unknowns. If accurate quantification of the metal content is required, the second (HPLC) method is recommended.

14. The acquisition time may vary with the settings and sensitivity of the instrument so optimization is required.

15. Centrifuge at room temperature to avoid condensation. Inspect the plate for bubbles and remove by gentle tapping.

16. The maximum absorption for the PAR-metal complexes varies slightly between 490 and 510 nm *(6)*. The most suitable wavelength is 492 nm because it is optimal for complexes of nickel and zinc, the two metals that are not detected in the luminol tests. Visual inspection of the samples in this test is also possible due to the reddish color of the PAR-metal complexes compared to the pale yellow PAR solution.

17. In the optimal case about 1 µg protein is sufficient for a clear positive signal. However, the larger proteins require analysis of a larger mass because an equimolar amount of bound metal is less compared to the same amount in a smaller protein. In addition, many proteins may be purified with substoichiometric amounts of metals loaded. Typically 5–10 µg of protein will provide a more dependable response.

18. Common components of protein solutions have been examined, and analysis of 1 µl of the following solutions did not interfere with the analysis: 10 mM $CaCl_2$, 10 mM $MgCl_2$, 20 mM DTT, 50 mM β-mercaptoethanol, 20% Me_2SO, 200 mM ammonium sulfate, 100 mM Tris, 100 mM HEPES, 100 mM Na_2HPO_4, 100 mM imidazole, and 100 mM citrate *(7)*. EGTA and EDTA affect the analysis and should be removed.

19. The buffer solution that we recommend is 20 mM Hepes, pH 7.5, 0.1 M NaCl. Metal standards should be diluted in the same solution as the proteins because the type of buffer can modulate retention times and intensities of the signals *(9)*. Avoid EDTA, EGTA, Ca^{2+}, and high salt concentrations. The presence of 10% glycerol or 50 mM glycine has no effect, and 1 mM reducing agents such as DTT or TCEP only influences the relative amounts of Fe(II) and Fe(III).

20. It is critical to determine the protein concentrations to calculate the amount of the bound metal. Prepare the samples at least in triplicate. If only a small fraction of the protein molecules are metal-loaded, higher concentrations may be necessary.

Table 20.1 Detection limit of metal standards (50 μl injected) in water at a signal-to-noise ratio of 3:1

Metal ion	Detection limit (mg/L)
Fe^{3+}	0.11
Cu^{2+}	0.13
Ni^{2+}	0.43
Zn^{2+}	0.43
Co^{2+}	0.20
Mn^{2+}	0.86
Fe^{2+}	0.50

21. A convenient method for the hydrolysis is to use a PCR machine with a heated lid. This method ensures uniform heating and minimizes volume loss due to condensation.
22. The detection limits for the metals differ (Table 20.1), so the dilutions should be prepared in ranges that allow comparable area of the chromatographic peaks. A good working range to start with is 10–200 μM.
23. If analysis of other metals, such as lead or cadmium, is desired, the PDCA can be replaced with oxalic acid (see Application Note 108 at www. Dionex.com).
24. Traces of metals in the chromatographic system can be removed by flushing with 0.2 M oxalic acid for 2 hours at 1 ml/min, followed by extensive rinsing with Milli-Q water. When Fe^{2+} is analyzed, the column is flushed with 0.1 M Na_2SO_3 for 2 hours at 1 ml/min, followed by prolonged equilibration.
25. The retention times of the individual metal standards should be almost identical for two consecutive runs. Usually, separation of the metals requires about 15 minutes. Each batch of PDCA solution may result in small changes in retention time due to variations in the pH. If necessary, include corrections of the retention times for the standards in the software.
26. The standard curves should have correlation coefficients higher than 0.98 to produce an accurate estimation of the metal content.
27. The system can be used to run 50 samples overnight.

Acknowledgments

The authors thank Ulrika B. Ericsson, Robert Lam, M. Amin Bakali, H. Ekaterina Kuznetsova, and Pär Nordlund for their participation in the development of these assays; Ivan Ho for the technical assistance; Dionex Canada for the technical support; and Aled Edwards for advice. This work was supported by the Natural Sciences and Engineering Research Council of Canada (NSERC) and the Canada Research Chairs Program.

References

1. Szpunar, J. (2000) Bio-inorganic speciation analysis by hyphenated techniques. *Analyst* **125**, 963–988.
2. Sanz-Medel, A., Montes-Bayón, M., and Luisa Fernández Sánchez, M. (2003) Trace element speciation by ICP-MS in large biomolecules and its potential for proteomics. *Anal. Bioanal. Chem.* **377**, 236–247.
3. Jakubowski, N., Lobinski, R., and Moens, L. (2004) Metallobiomolecules. The basis of life, the challenge of atomic spectroscopy. *J. Anal. At. Spectrom.* **19**, 1–4.
4. Jezorek, J. R., and Freiser, H. (1979) 4-(Pyridylazo)resorcinol-based continuous detection system for trace levels of metal ions. *Anal. Chem.* **51**, 373–376.
5. Hunt, J. B., Neece, S. H., and Ginsburg, A. (1985) The use of 4-(2-pyridylazo) resorcinol in studies of zinc release from *Escherichia coli* aspartate transcarbamoylase. *Anal. Biochem.* **146**, 150–157.
6. McCall, K. A., and Fierke, C. A. (2000) Colorimetric and fluorometric assays to quantitate micromolar concentrations of transition metals. *Anal. Biochem.* **284**, 307–315.
7. Högbom, M., Ericsson, U. B., Lam, R., Bakali, H. M., Kuznetsova, E., Nordlund, P., and Zamble, D. B. (2005) A high throughput method for the detection of metalloproteins on a microgram scale. *Mol. Cell. Proteomics* **4**, 827–834.
8. Huang, B., Li, J.-J., Zhang, L., and Cheng, J.-K. (1996) On-line chemiluminescence detection for capillary ion analysis. *Anal. Chem.* **68**, 2366–2369.
9. Atanassova, A., Lam, R., and Zamble, D. B. (2004) A high-performance liquid chromatography method for determining transition metal content in proteins. *Anal. Biochem.* **335**, 103–111.
10. Finney, L. A., and O'Halloran, T. V. (2003) Transition metal speciation in the cell: insights from the chemistry of metal ion receptors. *Science* **300**, 931–936.
11. Gray, H. B. (2003) Biological inorganic chemistry at the beginning of the 21st century. *Proc. Natl. Acad. Sci. USA* **100**, 3563–3568.
12. Parejo, I., Petrakis, C., and Kefalas, P. (2000) A transition metal enhanced luminol chemiluminescence in the presence of a chelator. *J. Pharmacologic. Toxicologic. Methods* **43**, 183–190.
13. Xiao, C., King, D. W., Palmer, D. A., and Wesolowski, D. J. (2000) Study of enhancement effects in the chemiluminescence method for Cr(III) in the ng/l range. *Anal. Chim. Acta* **415**, 209–219.
14. Liu, E.-B., Liu, Y.-M., and Cheng, J.-K. (2001) Separation of niobium(V) and tantalum(V) by capillary electrophoresis with chemiluminescence detection. *Anal. Chim. Acta* **443**, 101–105.
15. Lu, Q., and Collins, G. E. (2001) Microchip separations of transition metal ions via LED absorbance detection of their PAR complexes. *Analyst* **126**, 429–432.
16. Motellier, S., and Pitsch, H. (1996) Simultaneous analysis of some transition metals at ultra-trace level by ion-exchange chromatography with on-line preconcentration. *J. Chromatogr. A* **739**, 119–130.
17. Cardellicchio, N., Cavalli, S., Ragone, P., and Riviello, J. M. (1999) New strategies for determination of transition metals by complexation ion-exchange chromatography and post column reaction. *J. Chromatogr. A* **847**, 251–259.
18. Sarzanini, C. (1999) Liquid chromatography: a tool for the analysis of metal species. *J. Chromatogr. A* **850**, 213–228.
19. Nesterenko, P. N., and Haddad, P. R. (2000) Zwitterionic ion-exchangers in liquid chromatography. *Anal. Sci.* **16**, 565–574.
20. Alwarthan, A. A., and Townshend, A. (1987) Chemiluminescence determination of iron(II) and titanium(III) by flow injection analysis based on reactions with and without luminol. *Anal. Chim. Acta* **196**, 135–140.

21. Imdadullah, Fujiwara, T., and Kumamaru, T. (1991) Chemiluminescence from the reaction of chloroauric acid with luminol in reverse micelles. *Anal. Chem.* **63**, 2348–2352.

22. Imdadullah, Fujiwara, T., and Kumamaru, T. (1994) Catalytic effect of rhodium(III) on the chemiluminescence of luminol in reverse micelles and its analytical application. *Anal. Chim. Acta* **292**, 151–157.

23. TheingiKyaw, Kumooka, S., Okamoto, Y., Fujiwara, T., and Kumamaru, T. (1999) Reversed micellar-mediated luminol chemiluminescence reaction with bis(acetylac etonato)oxovanadium(IV). *Anal. Sci.* **15**, 293–297.

Chapter 21

High Throughput Screening of Purified Proteins for Enzymatic Activity

Michael Proudfoot, Ekaterina Kuznetsova, Stephen A. Sanders,
Claudio F. Gonzalez, Greg Brown, Aled M. Edwards, Cheryl H. Arrowsmith,
and Alexander F. Yakunin

Understanding the functions of every protein in the proteome is one of the great challenges of the postgenomic era. Global genome sequencing efforts revealed that in any genome 30–50% of genes encode proteins with unknown function (hypothetical proteins). To directly test purified hypothetical proteins for catalytic activity, the authors have designed a series of general and specific enzymatic screens. The described screens are designed to detect hydrolases (phosphatases, phosphodiesterases, proteases, and esterases), and oxidoreductases (dehydrogenases and oxidases). The general screens use either general chromogenic substrates or pools of substrates. The positive hits with the model substrates are then tested in the secondary screens with a set of potential natural substrates, or the substrate pools can be deconvoluted to identify the preferred *in vitro* substrate. The identification of a biochemical activity of a hypothetical protein helps to determine its cellular role.

1. Introduction

There are at the time of writing, 413 sequenced genomes and 2,132 genomes in the process of being sequenced. There are also numerous metagenomic sequencing efforts underway, which are supplying sequences for millions of potential gene products (www.genomesonline.org/). These efforts are creating an accumulation of hypothetical proteins (proteins with no known function), many of them conserved over many genomes *(1)*. One of the major goals of postgenomic biology will be to develop parallel high throughput methods to discover what all these proteins do. Bioinformatic analysis of sequences is the least expensive and often best way to annotate functions for proteins, but it still fails to assign a function to many proteins in any genome, or can often only assign a broadly defined function to proteins (i.e., putative hydrolase). There are also other numerous experimental large-scale approaches being employed to assign cellular functions to proteins *(2)*, such as the varied protein–protein interaction networks being created *(3–5)*, gene expression experiments using DNA microarray approaches *(6,7)*, or protein localization *(8,9)*. These approaches are somewhat indirect as they identify relationships to other proteins and pathways that require confirmation through activity assays.

From: *Methods in Molecular Biology, Vol. 426: Structural Proteomics: High-throughput Methods*
Edited by: B. Kobe, M. Guss and T. Huber © Humana Press, Totowa, NJ

The authors are developing general enzymatic screening of purified proteins as a direct experimental approach to identify the biochemical function of a hypothetical protein. While these general screens were designed to identify new enzymes and assign them to a subclass or sub-subclass (phosphatase, dehydrogenase, etc.), the authors have developed secondary screens with broad ranges of naturally occurring potential substrates for each of the general screens. These secondary screens can identify biologically relevant catalytic activities for any enzymes identified in the general screens. They can also be used to identify specific activities of proteins that have been placed in enzyme groups by bioinformatics analysis. There are also many known enzyme activities identified through biochemical experimentation for which there are no known sequences *(10)*. The authors' general screens can play an integral part in discovering which proteins are responsible for these activities. With the numerous large-scale structural proteomics efforts underway worldwide, there are many highly purified proteins in large quantities suitable for analysis using these screens. The authors' screens have been used to biochemically characterize 36 unknown proteins *(11)*.

2. Materials

Unless otherwise noted, all chemicals were purchased from Sigma (St. Louis, MO).

2.1. General

1. 1 M HEPES-K, pH 7.5.
2. 1 M HEPES-K, pH 8.0.
3. 1 M Tricine-K, pH 8.5.
4. 1 M $MgCl_2$.
5. 100 mM $MnCl_2$.
6. 100 mM $NiCl_2$.
7. 100 mM $ZnCl_2$.
8. 100 mM $CaCl_2$.
9. Dilution buffer: 50 mM HEPES-K pH 7.5, 10% glycerol.
10. Malachite green reagent: 1.1 g of malachite green dissolved in 900 ml of 3 M sulfuric acid.
11. 7.5% ammonium molybdate.
12. 10% v/v TWEEN® 20.

2.2. General Screens

1. 0.2 M *p*NPP (4-nitrophenyl-phosphate).
2. 33 mM Bis- *p*NPP (Bis(p-nitrophenyl) phosphate). Precipitates during storage; resuspend by heating to 37°C and vortexing.
3. 50 mM NAD (β-nicotinamide adenine dinucleotide).
4. 50 mM NADP (β-nicotinamide adenine dinucleotide phosphate).
5. 4 mM NADH (reduced β-nicotinamide adenine dinucleotide).
6. 4 mM NADPH (reduced β-nicotinamide adenine dinucleotide phosphate).
7. Amino acids mix: 5 mM of each of the 20 L-amino acids.
8. Acids mix: 0.48 mM each of acetic acid, fumarate, L-malate, DL-lactate, DL-isocitrate, succinate, oxaloacetate, and α-ketoglutarate.

9. Sugars mix: 0.8 mM each of D-glucose, D-galactose, D-mannitol, D-fructose, D-arabinose, D-sorbitol, and D-arabitol.

10. Alcohol mix: 0.48 mM each of methanol, ethanol, 1-hexanol, 1-decanol, and benzyl alcohol.

11. Aldehyde mix: 0.16 mM each of hexanal, decanal, benzaldehyde, and 2-naphthaldehyde.

12. 1 mg/ml peroxidase from horseradish, type II.

13. 3 mg/ml fast blue (*o*-dianisidine).

14. 100 mM EDTA (ethylenediaminetetraacetic acid), pH 8.0.

15. 1 mM palmitoyl-CoA.

16. 6 mM DTNB (5,5'-dithiobis(2-nitrobenzoic acid).

17. Lipase assay buffer: 50 mM Tris-HCl, pH 8.0, with 4.44 g/L Triton X-100, and 1.11 g/L gum arabic.

18. 8 mg pNP-palmitate dissolved in 2 ml isopropanol. Prepare fresh just before use, heat and vortex to dissolve.

19. Protease mix: 4 mM each of BAPNA (*N*α-benzoyl-DL-arginine *p*-nitroanilide), Leu-pNA (L-leucine-*p*-nitroanilide), Pro-pNA (L-proline *p*-nitroanilide), Suc-AAA-pNA (N-succinyl-Ala-Ala-Ala-*p*-nitroanilide), and Suc-Phe-pNA (N-succinyl-L-phenylalanine-*p*-nitroanilide) dissolved in dimethyl sulfoxide (DMSO). These substrates were chosen to detect a wide variety of proteases.

20. Prepare the phosphatase mix by combining 1 ml of 1 M HEPES-K pH 7.5, 100 µl 1 M $MgCl_2$, 200 µl 100 mM $MnCl_2$, 100 µl 100 mM $NiCl_2$, and 2 ml 0.2 M *p*NPP, and add water to a final volume of 18 ml.

21. Prepare the phosphodiesterase mix by combining 1 ml of 1 M Tricine-K, pH 8.5, 100 µl 1 M $MgCl_2$, 200 µl 100 mM $MnCl_2$, 100 µl 100 mM $NiCl_2$, and 6 ml 33 mM bis-*p*NPP, and add water to a final volume of 18 ml.

22. Prepare the five dehydrogenase mixes by combining 1 ml of 1 M Tricine-K, pH 8.5, 100 µl 1 M $MgCl_2$, 200 µl 100 mM $MnCl_2$, 100 µl 100 mM $NiCl_2$, 100 µl 100 mM $ZnCl_2$, 200 µl 50 mM NAD, 200 µl 50 mM NADP, and 12.5 ml of one of the pre-made substrate mixes (amino acids, acids, sugars, alcohols, aldehydes), and add water to a final volume of 18 ml.

23. Prepare the NADH/NADPH oxidase mix by combining 1 ml of 1 M HEPES-K pH 8.0, 100 µl 1 M $MgCl_2$, 200 µl 100 mM $MnCl_2$, 100 µl 100 mM $NiCl_2$, 600 µl 4 mM NADH, and 600 µl 4 mM NADPH, and add water to a final volume of 18 ml.

24. Prepare the five oxidase mixes by combining 1 ml of 1 M HEPES-K pH 8.0, 100 µl 1 M $MgCl_2$, 200 µl 100 mM $MnCl_2$, 100 µl 100 mM $NiCl_2$, 200 µl 1 mg/ml type II peroxidase from horseradish, 200 µL 3 mg/ml Fast Blue and 12.5 ml of one of the pre-made substrate mixes (amino acids, acids, sugars, alcohols, aldehydes), and add water to a final volume of 18 ml.

25. Prepare the lipase mix by pipetting 18 ml of the lipase assay buffer into a 50-ml beaker with a stirbar, and slowly dripping 2 ml of 4 mg/ml *p*NP-palmitate in isopropanol into the beaker with stirring to avoid precipitation. *p*NP-palmitate is used as a representative substrate because of its relative stability (see Note 1). If there is a positive reaction in this screen, other *p*NP-carboxylates should be assayed (*p*NP-butyrate, *p*NP-caprate, etc.).

26. Prepare the thioesterase mix by combining 1 ml 1 M HEPES-K pH 7.5, 400 µl 100 mM EDTA, 800 µl 1 mM palmitoyl-CoA, and add water to a volume of 17 ml. Palmitoyl-CoA is used as a representative substrate of

other acyl-CoAs. If there is a positive reaction in this screen, other acyl-CoA should be tested to find out which one is the best substrate.

27. Prepare the protease mix by combining 1 M HEPES-K pH 7.5, 100 μl 1 M MgCl$_2$, 200 μl 100 mM MnCl$_2$, 100 μl 100 mM NiCl$_2$, 100 μl 100 mM ZnCl$_2$, 100 μl 100 mM CaCl$_2$, 4 ml of protease mix, and add water to 18 ml.

2.3. Natural Phosphatase

1. 4 mM stocks of each substrate to be screened, in a 96-well block, with some exceptions due to high backgrounds, outlined in Table 21.1. Store at −20°C, and thaw just before use (see Note 2).

Table 21.1 List of all 89 substrates used in the natural phosphatase screen, stored in a 96-well block at 4 mM

AMP	dCTP	Phytic acid
CMP	dGTP	Imidodiphosphate[b]
GMP	dTTP	Thiamine-monophosphate
IMP	Inorganic pyrophosphate	Thiamine-pyrophosphate
UMP	NADP	P-Serine
XMP	FMN	P-Threonine
2'AMP	PEP	P-Tyrosine
2'CMP	CoA	P-Arginine
2'UMP	Phosphocholine	P-Glycolate
3'AMP	α-Glucose-1P	2P-Ascorbate
3'CMP	β-Glucose-1P	DHAP[b]
dAMP	Glucose-6P	PLP[a]
dCMP	Fructose-1P	Poly-P[b]
dGMP	Fructose-6P	5'PRPP[b]
dIMP	Ribose-5P	Phosphono acetate
dTMP	Mannose-1P	Phosphono formate
dUMP	Mannose-6P	(Phosphono-methyl) glycine
ADP	Galactose-1P	bis(Phosphono-methyl) glycine
CDP	Fructose-1,6 bisP	AMP Ramidate
GDP	Erythrose-4P[a]	UMP morpholidate
IDP	Trehalose-6P	Diethyl phosphono-ramidate
TDP	Glucose-1,6 bisP	a-Lactose-1P
UDP	Sucrose-6P	Phosphoryl ethanolamine
ATP	2-deoxyglucose-6P	Phosphatidic acid
CTP	2-deoxyribose-5P	Glycero-1P
GTP	Glucosamine-6P	Glycero-2P
ITP	6P-Gluconate	Glycero-3P
TTP	2P-Glycolate	GMP morpholidate
UTP	3P-Glycolate	Ribulose-5P
dATP	3P-Glyceraldehyde	

[a]These substrates are at 1.3 mM.
[b]These substrates are at 2 mM.

2.4. Phosphodiesterase

1. 10 mM stocks of all substrates, with some exceptions, outlined in Table 21.2. Store at −20°C, thaw just before use (see Note 2).
2. SAP buffer: 12.64 ml water, 3.2 ml 1M diethanolamine pH 9.8, 160 μl 1 M MgCl$_2$. Can be stored at room temperature, place on ice before use so that it is cool upon addition to the reaction mixture.
3. SAP: shrimp alkaline phosphatase (Fermentas, Burlington, Canada). Keep at −20°C.

2.5. Nucleotide Pyrophosphatase Assay

1. 10 mM stocks of all ATP, CTP, GTP, ITP, UTP, XTP, dATP, dCTP, dGTP, dITP, dTTP, dUTP, ADP, CDP, GDP, IDP, UDP, dADP, dCDP, and dGDP. Store at −20°C, thaw just before use (see Note 2).
2. Inorganic pyrophosphatase from baker's yeast. Keep at −20°C.

3. Methods

3.1. General Screens

The general screens are useful for determining general enzymatic activities for proteins of unknown function, or proteins comprising domains of unknown function. The substrates chosen for the assays have very broad reactivity, so as not to miss identifying any new enzymes; for instance, most phosphatases cleave the phosphate group from *p*NPP forming the chromophore *p*NP, which is easily detected at 410 nm. The dehydrogenase and oxidase assays use pools of electron donors designed to cover the most prevalent among currently known enzymes. As enzymes almost always work with more than one similar substrate (an octanol dehydrogenase usually also works with decanol as its electron donor), the entire substrate space is not fully covered; instead a representative set of electron donors is used.

The pHs selected for these assays were chosen based on their overlap among known enzymes in the category. Although the pH might not be optimum for the specific enzyme, there is still enough activity to be detected. For instance, a pH of 10.0 may be ideal for an alkaline phosphatase, and this screen would certainly identify alkaline phosphatases were one to do the

Table 21.2 List of all 22 substrates used in the natural phosphodiesterase screen, stored in tubes at 10 mM

cAMP	2'3'-cGMP	CDP-glycerol
cCMP	3-dephospho-CoA[a]	CDP-ethanolamine
cGMP	GDP glucose	UDP-galactose
cIMP	UDP-n-acetyl glucosamine	ADP-glucose
cTMP	GDP-mannose	ADP-ribose
cUMP	UDP-glucaronic acid[a]	UDP-glucose[a]
2'3'-cAMP	CDP-choline	FAD[a]
2'3'-cCMP		

[a]These substrates are at 5 mM.

reactions at that pH, but any possible acid phosphatases in the tested proteins would be missed; so do the assays at pH 7.5 to get activity for alkaline, acid, and neutral phosphatases. Somewhat trickier is the selection of divalent metal cations to add to the reaction mixtures. For many enzymes one is needed for activity, but at the same time certain metals are inhibitory. Mg^{2+} seems to be the most prevalent cation needed among enzymes, and Mn^{2+} and Ni^{2+} are also important to many enzymes. These three metals are also not common inhibitors. For dehydrogenases, enough instances in the literature were found in which Zn^{2+} is important to warrant adding it to the reaction mixtures. For the protease assay, there were enough known proteases requiring Zn^{2+} or Ca^{2+} to necessitate their addition in the reaction mixture. This screen is described for a 96-well format, but the mixtures can be scaled up or down depending on the number of proteins to be assayed.

1. Prepare the proteins for screening by pipetting 180 µg of each protein to be assayed into a well in a 96-well block with 1-ml wells. Keep this block on ice and dilute each well to 360 µl using dilution buffer for a final protein concentration of 0.5 mg/ml. Remember to keep at least one well with just dilution buffer as a negative control (see Note 3).
2. Using a multichannel pipette, pipette 20 µl of protein from each well of the block into the corresponding wells of 16 96-well plates suitable for your UV/vis plate spectrophotometer.
3. Also using a multichannel pipette, dispense 180 µl of each of the mixes into each well of one of the plates. For the thioesterase mix, only dispense 170 µl.
4. Place the plates in an incubator at 37°C for 1 hour.
5. Using a multichannel pipette, add 10 µl per well of 6 mM DTNB to the thioesterase assay plate. The reason that this is done now as opposed to in the original reaction mixture is that DTNB has been shown to inhibit thioesterase activity in some enzymes (see Note 4).
6. Read the phosphatase, phosphodiesterase, protease, lipase, and thioesterase plates at 410 nm, and note the increase in absorbance above the negative control.
7. Read the increase in absorbance at 340 nm for the dehydrogenase plates (see Note 5) and decrease in absorbance at 340 nm for the NADH/NADPH oxidase plate.
8. Read the increase in absorbance at 460 nm for the oxidase plates (see Note 5).

3.2. Phosphatase Screen Using Natural Substrates

1. Thaw the prepared block of phosphatase substrates by floating it in a water bath at 37°C. Once thawed, place the block on ice to keep cool.
2. While thawing, prepare the reaction mixture. Combine 800 µl 1M HEPES-K pH 7.5, 80 µl 1 M $MgCl_2$, 160 µl 100 mM $MnCl_2$, 80 µl 100 mM $NiCl_2$, and add water to 15 ml. Make one reaction mixture for the blank plate, and one for every protein to be assayed (see Note 6).
3. Using a multichannel pipette, dispense 10 µl from each well into the corresponding well of one 96-well plate suitable for your UV/vis plate spectrophotometer as a blank, and one 96-well plate for every protein to be assayed. Place these plates in the fridge until they are to be used.
4. For the blank plate, simply add 150 µl of the reaction mixture to each well by multichannel pipette.

5. For each protein, add 200 μg to its reaction mixture, mix quickly, and add 150 μl of this reaction mixture into each well in its plate. The final protein concentration will be 2 μg/well; less protein can be used if it is suspected that the protein is particularly active.

6. Place each plate in an incubator at 37°C for 30 minutes. Use 60°C for proteins from thermophilic organisms (any higher produces higher backgrounds due to substrate stability problems at high temperatures).

7. While the plates are in the incubator prepare fresh malachite green development reagent. For each plate, combine 4 ml of malachite green reagent with 1 ml of 7.5% ammonium molybdate, and 80 μl 10% v/v TWEEN 20. Scale up as needed. This solution is good for a few hours, and should be prepared fresh for each assay, or at least each day *(12)*.

8. Remove the plates from the incubator and add 40 μl of the fresh malachite green detection reagent to each well using a multichannel pipette.

9. Wait 5 minutes, and measure the absorbance at 630 nm.

3.3. Phosphodiesterase Screen Using Natural Substrates

1. Thaw the 10 mM substrates, and then place them on ice.

2. Prepare reaction mixtures by combining 120 μl 1 M HEPES-K pH 7.5, 12 μl 1 M $MgCl_2$, 24 μl 100 mM $MnCl_2$, 12 μl 100 mM $NiCl_2$, and add water to 2.34 ml. Make one of these mixtures for the blank and one more for every protein to be screened (see Note 6).

3. Spot 2 μl of each substrate into up to four wells in a 96-well plate. Up to three proteins can be screened per plate, and it is most efficiently done by spotting the substrates in rows such that rows A and B would be all the substrates as a blank, and rows C and D would have the same substrate positions and contain protein X, and so on.

4. Using a multichannel pipette, dispense 78 μl of one reaction mixture into each well of one set of substrates.

5. Add 60 μg of protein to a reaction mixture, and quickly swirl it. Then, using a multichannel pipette, dispense 78 μl of one reaction mixture into each well of one set of substrates.

6. Place the plates in an incubator at 37°C for 30 minutes. Use 60°C for proteins from thermophilic organisms (any higher produces higher backgrounds due to substrate stability problems at high temperatures).

7. Add 7.5 μl of SAP to 2.4 ml of cold SAP buffer for each set of substrates (for one protein being assayed, add 15 μl of SAP to 4.8 ml of SAP buffer) and place on ice while the plates are in the incubator.

8. Add 80 μl of the SAP mixture to each well in the plate, and incubate for ten more minutes at 37°C.

9. Follow steps 7–9 of Section 3.2. (Fig. 21.1).

3.4. Nucleotide Pyrophosphatase

This screen does not have an associated general screen, but is especially useful for identifying and characterizing nucleotide pyrophosphatases that have been identified through a bioinformatics approach.

1. Thaw out the 10 mM stocks of the NTPs, dNTPs, NDPs, and dNDPs.

2. While the substrates are thawing, prepare the reaction mixes; again, one for the blank and one for each protein to be screened.

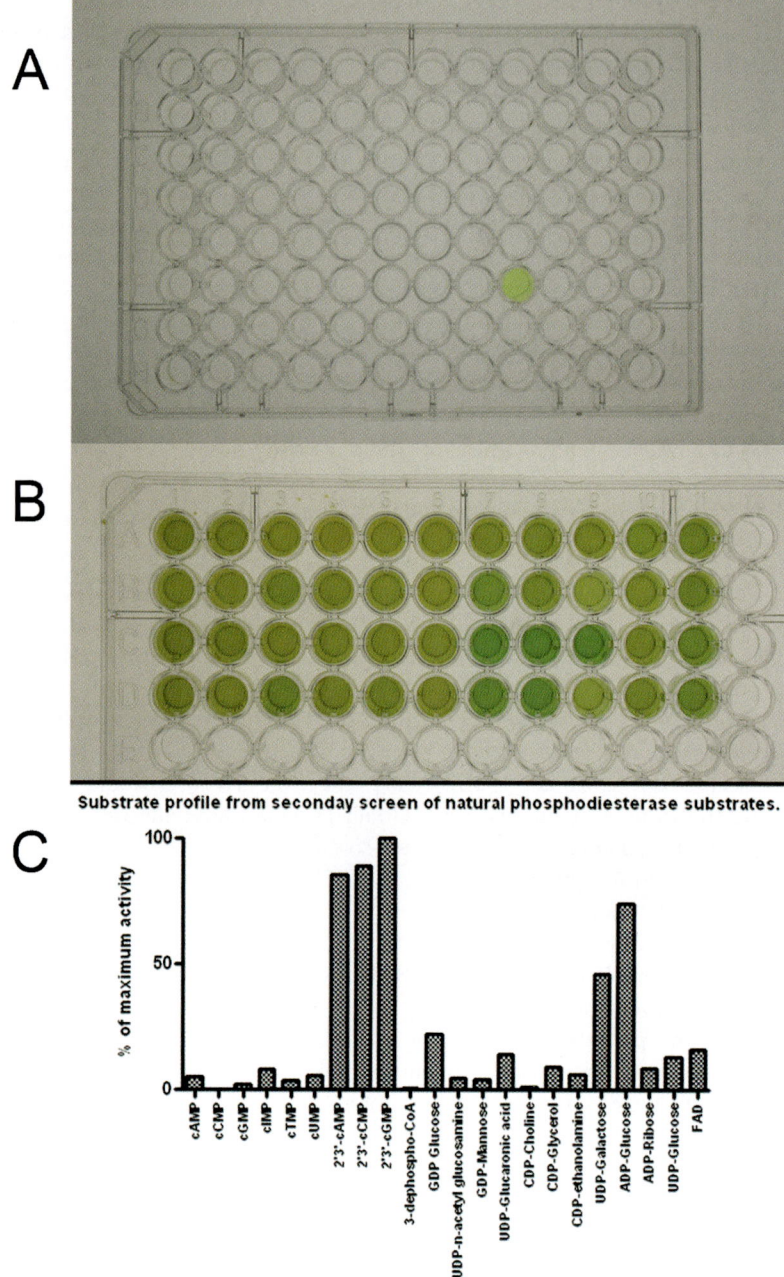

Substrate profile from seconday screen of natural phosphodiesterase substrates.

Fig. 21.1 The screening data of a discovered phosphodiesterase. **A**. A phosphodiesterase general screen using the model substrate; the positive reaction can be seen in well F9 by the yellow color. **B**. The phosphodiesterase secondary screen using the 22 substrates outlined in Table 21.2. Rows A and B are the blanks, whereas rows C and D are with 2 ug/well of the protein identified in the general screen. **C**. The results of the experiment pictured in **(B)** shown graphically.

3. Prepare reaction mixtures by combining 240 μl 1 M HEPES-K pH 7.5, 24 μl 1 M MgCl$_2$, 48 μl 100 mM MnCl$_2$, 24 μl 100 mM NiCl$_2$, and 0.25 units inorganic pyrophosphatase from baker's yeast, and add water to 4.74 ml.

4. Spot 2 μl of each substrate into a 96-well plate in the same fashion as in step 3 of the phosphodiesterase screen method.

5. Add 158 μl of reaction mix to each well of the blank rows.

6. Add 60 μg of each protein to be assayed to their mix.

7. Add 158 μl of these reaction mixes to a set of spotted substrates.

8. Place in an incubator at 37°C for 30 minutes. Use 60°C for proteins from thermophilic organisms (any higher produces higher backgrounds due to substrate stability problems at high temperatures). If higher than 37°C is used, the inorganic pyrophosphatase must be added as a second step after the first incubation and the plate must then be incubated for 10 minutes at 37°C.

9. Follow steps 7–9 of Section 3.2.

4. Notes

1. In the pNP-palmitate based esterase assay, high imidazole concentrations (0.1–0.3 M) present in the protein samples reaction mixture result in nonenzymatic degradation of the substrate and increased absorbance at 410 nm, producing a positive result. For this reason, any proteins purified by immobilized metal affinity chromatography (IMAC) using imidazole as eluent must be dialyzed prior to being screened for lipase activity.

2. One of the problems with the assays using malachite green to detect phosphate is high endogenous background for many of the substrates. The screens are done at relatively low substrate concentrations (0.25 mM) due to background issues (high absorbance in the blank), and some substrates have to be used in the screens at even lower concentrations (0.08–0.2 mM) than the rest. Some substrates, notably the NTPs, have higher backgrounds the more times they are thawed and refrozen, and it may be necessary to make fresh stocks of these more often than some of the other chemicals. Other chemicals have varying backgrounds depending on the batch from Sigma; this is usually only the case for substrates with high backgrounds such as erythrose-4-phosphate. It is unclear if the background issues are due to phosphate contamination in the chemicals or substrate degradation during assay or incubation with malachite green (although it is probably a combination of these). β-Glucose-1-phosphate for instance increases in absorbance fairly quickly after the addition of malachite green, indicating the degradation of the substrate. To avoid the problems this causes, the plates should be scanned 3 minutes after the addition of the malachite green development reagent. For secondary screens in which high backgrounds become problematic, the authors have found that the phosphate detection method of Saheki et al. is suitable (13), although this method is less sensitive.

3. The protein samples should be at least 90% pure, but the higher the purity the better. Often if a contaminant in the protein sample is an enzyme, there are positive hits in the screen that are not due to the protein intended. This problem seems to occur most often with the phosphodiesterase screen, with many somewhat dirty enzymes showing low but detectable reactivity with bis-pNPP. Another concern is protein precipitation. There are many cases in which in certain assays the well become turbid, and the spectrophotometer

incorrectly shows a positive reaction due to light scattering caused by protein precipitation.

4. The thioesterase/esterase assay uses DTNB to detect the formation of a free thiol group. Therefore, proteins with exposed and reduced cysteine residues may create false-positive reactions. Nothing much can be done except to check the protein and DTNB together without the substrate to determine if absorbance increase is due to exposed cysteines. Also, DTT must be kept out of the reaction mixture because it will also result in a sharp increase in absorbance at 410 nm.

5. For the dehydrogenase and oxidase assays, once a positive reaction has been detected, one should deconvolute the substrate mixture to determine the electron donor or donors responsible for the activity.

6. The reaction mixtures listed are the ones that should be used for potential enzymes with no other information. Normally, once a potential enzyme has been discovered from the general screens, further experiments are done with the model substrate to determine the optimal pH and divalent metal for that enzyme. This can be done by performing the general screen with the same reaction buffer and model substrate concentration, but having separate wells for each divalent metal. The authors use Mg^{2+}, Mn^{2+}, Ni^{2+}, Co^{2+}, Cu^{2+}, Zn^{2+}, Ca^{2+} (all metals are at 0.5 mM, except Mg^{2+} [5 mM] and Mn^{2+} [1 mM]) and have one well with no metal, as there are some enzymes that do not require one. Then perform another experiment with varying buffers and pH in the reaction wells, using the best metal from the previous experiment. The ideal buffer and metal should then be used in the secondary screen for that enzyme, although it is still a good idea to include Mg^{2+} in the reaction mixture as the best metal or buffer for the model substrates are not necessarily the best for the natural substrates. So for further characterization, metal and buffer optimization must be repeated for the natural substrate or substrates that are identified from the secondary screens.

References

1. Galperin, M. Y., and Koonin, E. V. (2004) "Conserved hypothetical" proteins: prioritization of targets for experimental study. *Nucleic Acids Res.* **32**, 5452–5463.
2. Phizicky, E. M., and Grayhack, E. J. (2006) Proteome-scale analysis of biochemical activity. *Crit. Rev. Biochem. Mol. Biol.* **41**, 315–327.
3. Phizicky, E. M., and Fields, S. (1995) Protein-protein interactions: methods for detection and analysis. *Microbiol. Rev.* **59**, 94–123.
4. Ito, T., Tashiro, K., Muta, S., Ozawa, R., Chiba, T., Nishizawa, M., Yamamoto, K., Kuhara, S., and Sakaki, Y. (2000) Toward a protein-protein interaction map of the budding yeast: a comprehensive system to examine two-hybrid interactions in all possible combinations between the yeast proteins. *Proc. Natl. Acad. Sci. USA* **97**, 1143–1147.
5. Uetz, P., Giot, L., Cagney, G., Mansfield, T. A., Judson, R. S., Knight, J. R., Lockshon, D., Narayan, V., Srinivasan, M., Pochart, P., Qureshi-Emili, A., Li, Y., Godwin, B., Conover, D., Kalbfleisch, T., Vijayadamodar, G., Yang, M., Johnston, M., Fields, S., and Rothberg, J. M. (2000) A comprehensive analysis of protein-protein interactions in *Saccharomyces cerevisiae*. *Nature* **403**, 623–627.
6. Brown, P. O., and Botstein, D. (1999) Exploring the new world of the genome with DNA microarrays. *Nat. Genet.* **21**, 33–37.
7. DeRisi, J. L., Iyer, V. R., and Brown, P. O. (1997) Exploring the metabolic and genetic control of gene expression on a genomic scale. *Science* **278**, 680–686.

8. Cai, Y. D., and Chou, K. C. (2004) Predicting 22 protein localizations in budding yeast. *Biochem. Biophys. Res. Commun.* **323**, 425–428.

9. Dreger, M. (2003) Proteome analysis at the level of subcellular structures. *Eur. J. Biochem.* **270**, 589–599.

10. Karp, P. D. (2004) Call for an enzyme genomics initiative. *Genome Biol* **5**, 401.

11. Kuznetsova, E., Proudfoot, M., Sanders, S. A., Reinking, J., Savchenko, A., Arrowsmith, C. H., Edwards, A. M., and Yakunin, A. F. (2005) Enzyme genomics: application of general enzymatic screens to discover new enzymes. *FEMS Microbiol. Rev.* **29**, 263–279.

12. Baykov, A. A., Evtushenko, O. A., and Avaeva, S. M. (1988) A malachite green procedure for orthophosphate determination and its use in alkaline phosphatase-based enzyme immunoassay. *Anal. Biochem.* **171**, 266–270.

13. Saheki, S., Takeda, A., and Shimazu, T. (1985) Assay of inorganic phosphate in the mild pH range, suitable for measurement of glycogen phosphorylase activity. *Anal. Biochem.* **148**, 277–281.

Section IV

Structural Characterization of Proteins

Chapter 22

Strategies for Improving Crystallization Success Rates

Rebecca Page

The production of crystals suitable for high-resolution structure determination is still one of the major bottlenecks in the structure determination process. This is especially true in structural genomics (SG) consortia, where the implementation of protein-specific purification and optimization strategies is not readily implemented into the structure determination workflow. This chapter describes four strategies that have been implemented by a number of SG groups to increase the number of protein targets that resulted in atomic resolution structures: (1) orthologue screening; (2) the use of 1D ^{1}H NMR spectroscopy to screen for the folded state of a protein prior to crystallization; (3) deletion constructs generation, in which regions of the target protein predicted to be disordered are omitted from the construct, to maximize the likelihood of crystal formation; and (4) crystallization optimum solubility screening to identify more suitable buffers for a given protein. The implementation of these strategies can lead to a substantial increase in the number of protein structures solved. Finally, because these strategies do not require the implementation of expensive robotics, they are highly applicable not only for the SG community but also for academic laboratories.

1. Introduction

For both academic laboratories and structural genomics consortia, the major bottlenecks in the crystal structure determination process are the production of protein samples in amounts suitable for structural studies and the subsequent formation of diffraction quality crystals. This is especially true for high throughput (HT) SG consortia, in which the efficient identification of protein *targets* (constructs) most suitable for high-resolution structure determination results in a substantial savings in cost and effort. Numerous strategies have been developed over the years to maximize both the production of soluble protein and the formation of diffraction quality crystals. This includes methods such as the use of solubility-enhancing tags to improve the production of soluble protein in *E. coli (1,2)*, surface entropy reduction (SER) to improve crystallization behavior of a protein *(3)*, and crystal dehydration to improve crystal diffraction quality and resolution *(4,5)*. However, it is not always obvious how

From: *Methods in Molecular Biology, Vol. 426: Structural Proteomics: High-throughput Methods*
Edited by: B. Kobe, M. Guss and T. Huber © Humana Press, Totowa, NJ

to prioritize these strategies. This is even more challenging in SG consortia, in which the implementation of protein-specific purification and crystal optimization strategies are not readily incorporated into the structure determination workflow.

The Joint Center for Structural Genomics (JCSG) developed and implemented high throughput structure determination pipelines *(6,7)* to obtain structural coverage of a bacterial (*Thermotoga maritima*, TM) and eukaryotic *(Mus musculus)* proteome. One of the JCSG pipelines was dedicated to processing bacterial and yeast homologues of mouse proteins (MH proteins, hereafter referred to as mouse homologue pipeline; all MH proteins expressed in *E. coli*). When compared with the TM proteins, these targets typically required additional care to optimize the production of diffraction-quality crystals and NMR-amenable protein samples. In 2003, the MH pipeline consisted of four basic processing steps, cloning, macroexpression, purification, and crystallization. However, because of the additional care required for the production of protein samples suitable for high-resolution structure determination, it was essential that the JCSG, like many other SG consortia, incorporate additional steps into the workflow that would allow them to more readily determine which targets were most likely to lead to structures.

Over a 15-month period, the MH pipeline was expanded to include six new processing steps: (1) microfermentation, (2) tailored and HT crystal fine screening, (3) orthologue screening, (4) 1D ^{1}H NMR fold screening, (5) deletion construct generation, and (6) crystallization optimum solubility screening. In the MH pipeline, the Vertiga microfermentation device (GlasCol, Thomson Instruments, Carlsbad, CA; 750-µl culture volume); was used to verify target expression and solubility prior to large-scale fermentation and purification *(8)*. It was found that the nearly 95% of the targets that expressed (or did not express) in microfermentation trials expressed (or did not express) again in macro-fermentation trials. In addition, microfermentation was also used to produce protein in sufficient quantities for subsequent purification and biophysical studies, such as 2D [^{1}H, ^{15}N] HSQC NMR screening *(9)*. This type of microexpression screening prior to scale-up allowed the MH group and others to not only identify those targets that were highly expressed and soluble, but also to carry out initial biophysical studies that allowed the proteins' likelihood of forming diffraction quality crystals to be better predicted *(10–14)*. Efforts to increase the speed and decrease the cost of microexpression screening are ongoing. For example, one of the most recently developed microexpression screening procedures is carried out on the *E. coli* colony itself, using the colony filtration blot (Co-Fi) *(15)*, and thus does not require small-scale expression cultures.

The second new processing step, crystal optimization, especially using standard crystal fine screening methods *(16)*, significantly increased the rate at which diffraction-quality crystals were obtained. In fact, more than 90% of the crystals that led to high-resolution structures in the MH pipeline during this period were produced from high throughput or tailored fine screening (data not shown). In addition, HT fine screening, described by DiDonota et al. *(17)* has also been important for the successful structural determination of TM proteins within the JCSG *(18)*. The importance of fine screening for the production of diffraction quality crystals has even led to the development of new robotics, such as the ScreenDeveloper Workstation (Hamilton Robotics, Reno, NV) or

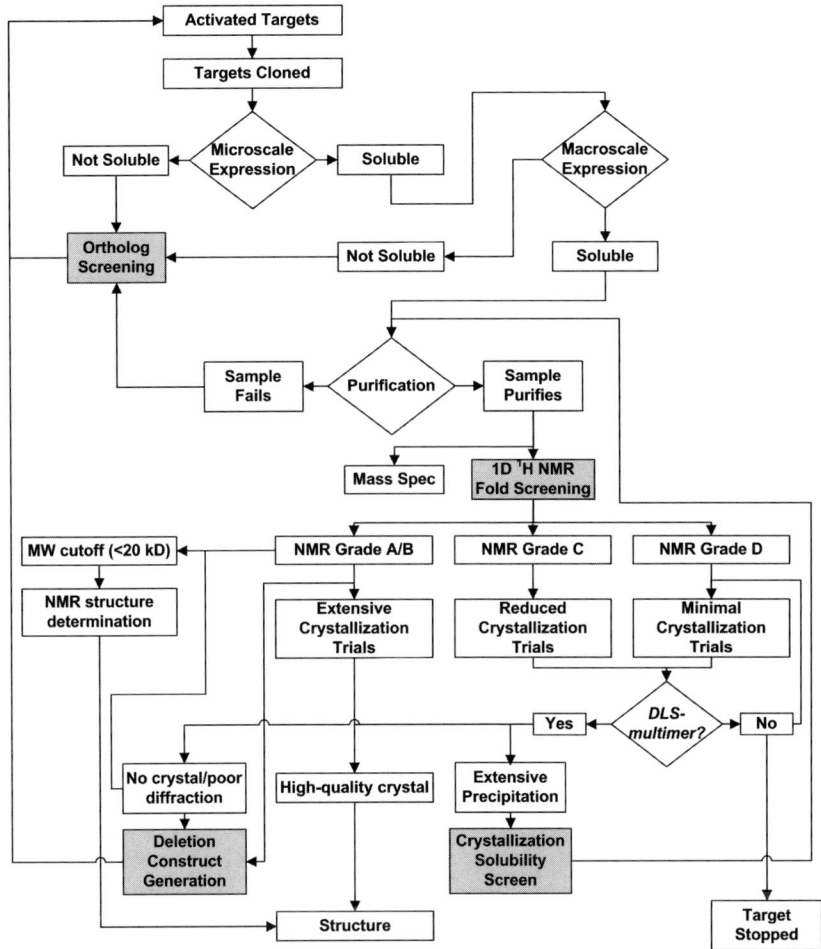

Fig. 22.1 Integration of several strategies into a structural genomics pipeline. Implementation of strategies highlighted in gray led to a 10-fold increase in the number of structures solved to high resolution within the MH pipeline. Currently at Brown University, the author regularly uses a subset of these strategies for academic projects *(42)*.

the Matrix Maker (Emerald Biosystems, Bainbridge Island, WA), which allow tailored crystallization screens to be designed by the user.

The last four methods, orthologue screening, 1D ^{1}H NMR fold screening, deletion construct generation, and "crystallization" optimum solubility screening, are discussed in more detail in this chapter. Orthologue screening, a method employed by multiple structural genomics consortia, was implemented to maximize the likelihood a protein from a given family would crystallize. 1D ^{1}H NMR fold screening was shown to be a suitable method for *prioritizing* crystallization efforts. Deletion constructs, in which regions of the protein N- and C-termini predicted to be disordered were omitted, were processed to maximize the likelihood a protein, or a shortened version of it, formed crystals suitable for structure determination. Finally, the optimum solubility buffer exchange protocol developed at the Berkeley Structural Genomics Center *(19)* was adapted for the MH high throughput structure determination pipeline to improve protein stability and, in turn, crystallization success. The implementation of these six strategies to the JCSG MH pipeline resulted in a 10-fold increase in the number of structures solved over these 15 months. The workflow ultimately established for the MH pipeline is shown in Fig. 22.1. Although many of these methods are highly applicable, and widely used within academic laboratories, the current challenge for all structural biology efforts is to determine which strategy to apply to their project, and in which order, to most rapidly produce a high-resolution structure. Thus, this chapter also discusses how these strategies have been adapted for protein target-focused projects at Brown University.

1.1. Orthologue Screening

One frequently used strategy to maximize the production of diffraction quality crystals of a protein family of interest is to use multiple "orthologue screening." Orthologues of a "parent" protein (the protein target of interest) often behave very differently from both the parent and one another during the different stages of the structure determination process. For example, although a parent target may express but fail to crystallize the small differences in sequence and length of a close orthologue may enable it to both express *and* crystallize. Thus, a strategy to maximize the likelihood that at least one version of a target sequence will express, purify, crystallize, and form crystals suitable for structure determination is to process multiple orthologues of a target sequence in parallel. It is important to point out that although this type of screening is typically referred to within the SG community as "orthologue screening," the proteins typically identified as orthologues may also be paralogues of one another. The formal definition of an orthologue is that orthologues are genes in different organisms that are direct evolutionary counterparts of one another. Paralogues, in contrast, are genes in the same organism that evolve by gene duplication. Thus, although both orthologues and paralogues are assumed to have similar biochemical functions, orthologues are expected to be part of similar regulatory pathways and thus be subjected to similar evolutionary pressures. In the studies described in the following, two criteria, similarity in sequence and functional annotation, were used to identify a target as an orthologue. Since the additional requirement of verifying that these proteins also functioned in the same biochemical pathways was not carried out, these "orthologues" may also be "paralogues" of one another. This distinction, however, does not substantially change the conclusions of these studies.

A comprehensive two-organism orthologue study carried out by Savchenko and colleagues *(20)* showed that the inclusion of even one orthologue increased the total number of samples suitable for structural studies by a factor of two. Specifically, they compared the expression, purification, crystallization, and NMR properties of 68 orthologous protein pairs from *E. coli* and *T. maritima*. Although 20 of the 136 proteins either did not express or expressed insolubly, in no case did *both* orthologues fail to express or express insolubly. Thus, the inclusion of even one additional orthologue significantly increased the likelihood of obtaining soluble protein, the first step toward high-resolution structure determination. In a similar study, Burley and colleagues identified orthologues for 262 gene families from 40 different organisms. They found that the use of orthologues led to the identification of at least one protein that expressed solubly from 80% of the gene families and was thus suitable for high-resolution structural studies (S. K. Burley, personal communication). Similarly, in the small study described here, the inclusion of orthologues increased the likelihood that the structure of a protein from a protein family of interest could be determined to high resolution. Although orthologue screening is not new, it has become more easily and widely applicable with the commercial availability of hundreds of cDNA libraries (www.atcc.org) and the availability of their corresponding genome sequences.

1.2. 1D ^{1}H NMR Fold Screening for Prioritization of Crystallization Efforts

The biophysical characterization of protein samples prior to crystallization has become increasingly important for success in obtaining high-resolution structures of multiple targets within structural genomics consortia. A variety of methods are now regularly used to characterize the structural, oligomeric, and dynamic behavior of proteins prior to crystallization trials, including limited proteolysis coupled with mass spectrometry *(21,22)*, protein stability by thermal shift *(23)*, static/dynamic light scattering *(19)*, size exclusion chromatography *(24)*, and small-angle x-ray scattering *(25)*, among numerous others. The author tested whether 1D ^{1}H NMR spectroscopy, which can be used to determine the folded state of a protein, could be used to predict which proteins were most likely to form diffraction quality crystals. It was found that it was highly successful, but, as additional comparative studies have elegantly demonstrated *(26,27)*, only when the *same* buffer that is used for NMR screening is also used for subsequent crystallization trials *(28)*.

1.3. Deletion Construct Generation

Deuterium exchange mass spectroscopy (DXMS) is a powerful method for experimentally identifying the boundaries of ordered protein domains suitable for crystallization trials *(29)*. Currently, however, DXMS is labor intensive and requires significant training before it can be efficiently used to accurately identify domain boundaries for large numbers of protein targets. Alternatively, a high throughput adaptation of limited proteolysis coupled with mass spectrometry has also been developed showing that proteins and protein domains that form diffraction quality crystals are typically resistant to proteolysis *(21)*. An alternative to the experimental approaches is to use a computational approach to identify domain boundaries. Within the MH pipeline, two programs, PONDR *(30,31)* and DisEMBL *(32)* were used to predict regions of potential disorder based on the target protein sequence. These results were then used to generate alternative constructs of the parent target in which these disordered regions, when located at the N- or C-termini, are genetically removed from the parent target.

1.4. Crystallization Optimum Solubility Buffer Screening

Maximizing protein solubility prior to crystallization trials can often improve crystal growth *(33)*. This was elegantly shown by Jancarik and coworkers, who, using buffer screening coupled with dynamic light scattering (DLS) developed a straightforward screening protocol to identify the optimal buffer for protein stability prior to crystallization trials *(19)*. Briefly, proteins are screened against 24 distinct buffers, set up in drops similar to those used for crystallization trials, and allowed to equilibrate overnight at room temperature. Those drops that are clear the following day are then diluted with an additional 15 µl of buffer and the resulting samples screened by dynamic light scattering for the aggregated state of the protein. Those buffer(s) in which the protein is monodisperse is then selected as the optimal buffer for crystallization trials of the protein. A variation of this method has been developed for the MH pipeline *(34)*. Instead of screening for optimal buffers as described, the results of the initial crystallization trials are analyzed for the presence of *clear* drops and the buffers of these drops identified. When a trend exists and a single buffer can readily be identified, the protein sample is transferred into this buffer (by dialysis or spin concentration buffer exchange) and the new sample rescreened for crystal formation.

2. Materials

2.1. Orthologue Screening

The materials required are simply a set of genomes to "blast" (www.ncbi. nlm.nih.gov/blast/) against for orthologue identification and the corresponding cDNA libraries (www.atcc.org) so the identified orthologues can be subcloned. To verify the identification of bona fide orthologues, refer to methods described by Storm and Sonnhammer *(35)*.

2.2. 1D ^{1}H NMR Fold Screening

2.2.1. Protein Sample

1. Concentration: 0.1–1 mM (preferably ~0.5 mM, as close to the same concentration to be used for crystallization trials as possible) and a volume of either 500 or 300 µl.
2. Protein buffer: 10 mM Tris, pH 7.8, 100 mM NaCl, and, as needed, 0.25 mM TCEP.
3. D$_2$0: Immediately prior to adding samples to NMR tubes, add 10% (v/v) D$_2$0
4. NMR tubes:

   • 500-µl Samples: disposable NMR tubes (NE-RG5-7, New Era, Vineland, NJ).
   • 300 µl Samples: reusable shigemi tubes (Shigemi Inc., Allison Park, PA).

2.3. Deletion Construct Generation

Protein sequence of the target under investigation; computer; access to a protein disorder prediction programs such as DisEMBL (dis.embl. de) *(32)*, PONDR (www.pondr.com; note, this is proprietary software)

(30,31) or GlobPlot (globplot.embl.de) *(36)* and the FFAS03 server (ffas. ljcrf.edu) *(37)*.

2.4. Crystallization Optimum Solubility Buffer Screening

2.4.1. Protein Sample
Purified protein concentrated to the appropriate concentration for initial crystallization trails

2.4.2. Crystallization Solutions
A broad, sparse matrix crystallization screen that samples a wide range of buffers and precipitants, for example, the JCSG crystallization screen *(38,39)*. The author typically uses this method for proteins that are poorly behaved during purification and concentration, although it is applicable for any protein under investigation.

3. Methods

3.1. Orthologue Screening

3.1.1. Orthologue Identification
For the MH study, a set of 22 protein targets were selected for orthologue identification. The 22 parent sequences were then "blasted" against 22 different bacterial genomes, which included 5 hyperthermophilic bacterial genomes. These hyperthermophilic genomes were included to capitalize on the crystallization success that has been observed for the *T. maritima* proteome screen, i.e., that 84% of the *T. maritima* proteins screened crystallized *(6)*. Between 1 and 7 orthologues were selected per parent sequence, with an average of 3–4 each, and orthologues from hyperthermophilic organisms, for the reason stated, were preferentially selected over those from mesophilic organisms. A total of 56 orthologue targets for the 22 parent sequences were identified, for 78 targets total (Table 22.1). In the MH pipeline, a broad sequence identity was used for the definition of an orthologue, with a cutoff range of 22–80% for sequence identity and 38–88% for sequence similarity. These values are similar to those used by Savchenko for their two-organism orthologue study, in which the *E. coli* and *T. maritima* protein pairs had sequence identities of 24–52%.

3.2. 1D ¹H NMR Fold Screening

3.2.1. Preparation of Protein Sample
Protein samples are purified using standard methods. Typically, the proteins are buffered in crystallization buffer and concentrated. Immediately prior to the addition of D_2O, the samples are spun for 15–20 minutes in a microcentrifuge at maximum speed to pellet any precipitate that forms during concentration. Once centrifuged, D_2O is added to the sample to a final concentration of 10% (v/v). At this point, the sample is added to the NMR tube (500 or 300 µl, as appropriate for the final sample volume).

3.2.2. NMR Spectroscopy
1D ¹H NMR solvent presaturation experiments are recorded (presaturation of 2.0 during the relaxation delay) with a 9,000-Hz sweep width. The measurement

Table 22.1 Orthologue target list: results of initial orthologue screen

Parent target		Results of orthologue screen			
Accession number	Function	Total orthologues	Soluble	Crystallized	Traced/ solved
6458035	Hypothetical protein DR0358	2	2	1	0
6460006	Hypothetical protein DR2203	4	2	2	1
17130350	Alanine–glyoxylate aminotransferase	2	1	1	1
17130499	Anthranilate phosphoribosyltransferase 2	3	2	1	1
2648890	Ornithine cyclo-deaminase (ARCB).	2	1	1	1
15156516	Hypothetical protein Atu1441	4	1	1	0
2649935	Hypothetical protein AF0690	3	1	1	0
15159614	Hypothetical protein Atu3699	2	2	2	2[1]
10173275	Response regulator aspartate phosphatase	4	2	2	0
10175988	Spore coat polysaccharide synthesis	4	3	1	1
10174364	Hypothetical protein BH1746.	3	2	2	0
10176030	NADH dehydrogenase	5	3	1	0
10174951	Hypothetical protein BH2331.	2	1	1	0
2632906	Hypothetical protein ydid	6	3	3	0
1788364	GDP-mannose mannosyl hydrolase	2	1	1	0
1787595	Fumarate and nitrate reduction regulatory protein	5	3	1	0
1787158	Hypothetical protein ycbL	4	1	1	0
YDR280W	Exosome complex exonuclease RRP45	4	4	3	1
YGL067W	Hypothetical protein	4	4	3	1
YGL213C	Antiviral protein SKI8	6	1	1	0
YDL236W	4-nitrophenylphosphatase	5	4	3	0
YGL221C	NGG1-interacting factor 3	2	2	1	0
Totals		78	46	34	9

Mutliple orthologues per target were selected and processed in the MH pipeline. In spite of the fact that these proteins are functionally related, many behave very differently, with some failing to express, some expressing and crystallizing, and others producing crystals sufficient for high-resolution structure determination.
Number includes parent targets.

times varied from 20 seconds to 20 minutes, depending on the concentration of the sample. Measurements were performed at 285 K on a Bruker Avance600 spectrometer (Bruker, Billerica, MA) equipped with a triple resonance probe (TXI-HCN with z-gradient). Immediately following data collection, the protein samples are removed, dispensed into an Eppendorf tube, and used immediately for crystallization trials or flash-frozen in liquid nitrogen and stored at −80°C until needed. Importantly, it was found that the presence of D_2O in the protein sample neither inhibits nor substantially changes the crystallization behavior of the protein; thus, the same sample used for the NMR screening is also used for the initial crystallization trials.

3.2.3. Spectra Evaluation

In the 1D ^{1}H NMR spectra, the dispersion of the NMR signals in three spectral regions provides the main indicators of folded globular proteins: (1) methyl protons (−0.5 to 1.5 ppm), (2) α-protons (3.5–6 ppm), and (3) amide protons (6–10 ppm). The author used a qualitative assessment of the line widths in the 1D ^{1}H NMR spectra recorded with water presaturation as a further criterion to characterize the state of the protein. Spectra are divided into four groups, A to D, which reflect the quality of the spectra and thus, the folded state of the protein.

- *A or B proteins* generate 1D ^{1}H NMR spectra with significant dispersion of resonance lines in the three spectral regions where ring current-shifted methyl resonances, downfield-shifted α-proton resonances, and downfield-shifted amide protein resonances are observed. For A-type proteins, the resonances are well separated with sharp lines and their intensities represent the entire population of protein molecules. For B-type proteins, the shape and intensities of the dispersed resonance lines do not quite meet the stringent criteria required for A-type proteins. A- and B-type proteins are characterized as well folded.
- *C-type proteins* have spectra that show evidence of chemical shift dispersion, especially in the region of −2.0 to 1.0 ppm, but also show evidence of line broadening throughout the spectrum. Possible causes of this line broadening may be intrinsic large size of the protein under investigation, nonspecific aggregation, or higher-order multimer formation.
- *D-type proteins* have spectra with no resonance lines in the spectral regions from −2.0 to 0.5 and from 5.0 to 6.5 ppm. They also have narrow chemical shift dispersion throughout the spectrum, especially in the amide protein region. These D-type proteins are characterized as unfolded.

3.2.4. Prioritization of Crystallization Efforts

Once the spectrum is collected and evaluated, the quality (grade) of the spectrum is used to determine the efforts that will be expended to crystallize the protein (prioritization).

3.3. Deletion Construct Generation

3.3.1. Disorder Analysis

Global disorder prediction programs, such as DisEMBL, PONDR, and GlobPlot are used to predict regions of potential disorder based on the protein sequence. Secondary structure prediction programs such as PsiPred (bioinf.cs.ucl.ac.uk/psipred/psiform.html) *(40)* are used to provide complementary information to that determined using the global disorder prediction programs. Finally, the program FFAS, which uses profile–profile alignments and fold recognition to detect remote homologies, is used to determine similar proteins whose structures have been solved to high resolution to provide an independent, experimental parameter to define the appropriate boundary domains for the protein under investigation.

3.3.2. Example: B. halodurans hypothetical protein BH2331 (10174951)

B. halodurans hypothetical protein BH2331 (10174951) crystallized as a full-length protein (265 amino acids), but the crystals were never suitable for diffraction screening. Therefore, this protein was selected as a candidate for deletion construct generation. DisEMBL predicts that the N-terminal 1–17 residues and C-terminal 218–229/234-265 residues of the *BH2331* sequence are unstructured. Similarly, PONDR predicts that residues 1–9 and 246–265 are disordered. Finally, FFAS predicts that PDBID 1RIL (mRNA capping enzyme) to be most similar to that of BH2331, in spite of an only 11% sequence identity between the protein sequences. The residues in the 1RIL structure correspond to the full-length BH2331 protein, suggesting that the large N- or C-terminal deletions predicted by the disorder programs may not necessarily be required for crystallization and structure determination of BH2331. These results were combined to generate 17 BH2331 deletion constructs. The nine constructs that were processed by the MH pipeline (full-length and eight deletion constructs; the structure was solved before the rest were processed) are reported in Table 22.2, column 1.

4.4. Crystallization Optimum Solubility Buffer Screening

A protein sample that precipitated heavily during purification was identified. This protein was subjected to crystallization trials against 192 different crystallization conditions at one temperature, and the crystallization results were then examined for clear drops. The chemical components of the clear drops were then compared to determine if a trend existed. In this case, 5 of the 10 clear drops contained 100 mM CHES. The protein was then re-expressed and purified,

Table 22.2 Deletion constructs of hypothetical protein BH2331 (10174951)

Construct	Residues	Residues D deleted from N-terminus	Residues deleted from C-terminus	Soluble expression
Parent construct (full length)	1–265	0	0	Y
1	19–265	18	0	Y
2	1–255	0	10	Y
3	17–255	16	10	Y*
4	1–217	0	48	N
5	1–245	0	20	N
6	1–211	0	54	N
7	1–240	0	25	N
8	17–200	16	65	N
9	17–240	16	25	N

*This construct produced crystals sufficient for structure determination.
The difference in expression behavior with the number of residues eliminated from the truncation domain constructs is reported. Only 4 of the 10 domain truncation constructs were soluble. Notably, the full-length construct had a 1D ^{1}H NMR spectra grade of C, whereas that of construct 3 had a grade of B. The full-length protein never produced crystals sufficient for diffraction screening. However, deletion construct 3 (residues 19–265) did produce diffraction quality crystals and the data collected were used to solve the structure to 1.9 Å (PDBID 1VL5).

except that this time the purified protein was immediately exchanged into a new CHES protein buffer (100 mM CHES, pH 9.25, 100 mM NaCl, and 0.25 mM TCEP) prior to crystallization trials.

4. Notes and Results

4.1. Orthologue Screening

All 78 parent and orthologous targets selected were processed through the MH pipeline. As expected, these proteins behaved very differently from one another during the different stages of the structure determination process. Of the 78 proteins tested, 46 were soluble, 34 crystallized, and 9 produced crystals suitable for high-resolution structure determination. Notably, of the 34 proteins that crystallized, the crystallization success varied dramatically from protein to protein within a set of orthologues. Some targets crystallized in just a few conditions, whereas other crystallized in many conditions, with two proteins crystallizing in 10 or 15% of all the conditions tested (61 or 92 of 576). Two examples that highlight the differences in orthologue behavior within the pipeline are shown in Fig. 22.2. Parent target 15159614 (hypothetical protein Atu3699 from *A. tumefaciens*) diffracted to high resolution, but the data was difficult to process and unable to be phased. However, its orthologue, 13421216 (putative DNA-binding protein from *C. crescentus;* 49% identical), crystallized in four different conditions, one of which produced crystals of exceptional quality and were used for structure determination to 1.62 Å (PDBID 1VJF). (The parent target, 15159614, was later solved by MR replacement

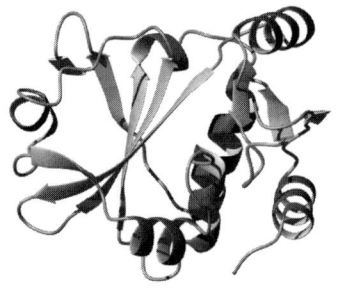
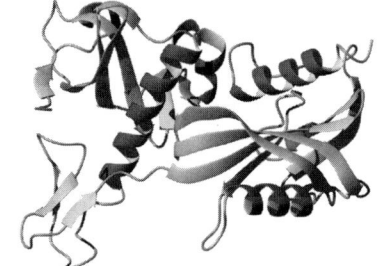

1.8 Å
13421216
putative DNA binding protein

2.1 Å
1790429
NADH pyrophosphatase

| **Parent Target:** *C. cresentus* | **poor diffraction** |
| **Ortholog 1:** *A. tumafaciens* | **4 crystal hits** |

Parent Target: *S. cerevisiae*	**No crystals**
Ortholog 1: *A. tumafaciens*	**1 crystal hit**
Ortholog 2: *C. cresentus*	**No expression**
Ortholog 3: *E. coli*	**60 crystal hits**

Fig. 22.2 Orthologues screening maximizes the likelihood that structures of target sequences will be obtained. The structures of two orthologues of two parent target sequences, a putative DNA binding protein (13421216; PDBID 1VJF) and NADH pyrophosphatase (1790429; PDBID 1VK6), were solved to high resolution. The orthologues of these parent sequences behaved very differently, with some orthologues failing to express, and others expressing and crystallizing in multiple conditions.

using the 13421216 structure as a search model [PDBID 1VKI]. Crystals used for final data collection were produced from 15159614 protein in which the N-terminal expression/purification tag had been proteolytically removed prior to crystallization trials.)

Similarly, parent target YGL067W (protein with no annotation from *S. cerevisiae*) expressed solubly but failed to crystallize. Of the three YGL067W orthologues processed, only two crystallized. The first, 15154967 (predicted MutT/nudix family protein from *A. tumefaciens*; 31% identity), crystallized in only one condition, whereas the second, 1790429 (hypothetical protein from *E. coli*; 31% identity) crystallized in 60 different crystallization conditions. The structure of 1790429 was ultimately solved to 2.2 Å resolution (PDBID 1VK6). Seven of the proteins (either the parent targets or the orthologues), which corresponded to eight different parent sequences, were determined to high resolution (1.8–2.2 Å), whereas an additional two orthologue proteins selected from two additional parent sequences have been traced (3–3.5 Å resolution). This is a success rate of 36% (8 of 22) for parent target sequences solved via orthologous proteins and clearly demonstrates the utility of using orthologues to obtain structures for a particular class of proteins.

4.2. 1D ¹H NMR Fold Screening

79 proteins were screened by 1D ¹H NMR spectroscopy and subjected to extensive crystallization trials to determine whether 1D ¹H NMR screening could be used to predict which proteins are most likely to form diffraction quality crystals and thus produce high-resolution structures. It was shown that 1D ¹H screening could, in fact, be successfully used for this prediction and the author used it to develop a crystallization prioritization strategy. The primary conclusions from the initial study, which were used to establish the prioritization strategy shown in Fig. 22.1, are the following:

- *A and B proteins have a substantially higher likelihood of forming crystals suitable for high-resolution structure determination.* Eighty-seven percent (13) of the structures determined (15) from the initial set of 79 proteins were characterized as either A or B proteins, indicating that A and B proteins have a higher likelihood of forming crystals suitable for high-resolution structure determination. This was in spite of the fact that the percentage of proteins that formed crystals per category was nearly identical from A through D, indicating that the potential to form crystals does not necessarily predict that a structure ultimately will be obtained.
- *Not all A and B proteins crystallize.* Although A and B proteins are more likely to produce crystals that will lead to high-resolution structures, not all proteins classified as A and/or B crystallized. In the present study, 12 of the 50 proteins classified as A or B proteins did not crystallize. Therefore, these samples are more highly suited for structure determination using NMR spectroscopy.
- *Some C proteins form crystals suitable for high-resolution structure determination.* Of the 22 proteins classified as C proteins, two formed crystals suitable for high-resolution structure determination.
- *No D proteins form crystals suitable for high-resolution structure determination.* Although six of the seven D proteins crystallized, none could ever be sufficiently optimized to provide a high-resolution structure.

The prioritization strategy that developed out of these observations was to expend the majority of the crystallization efforts on those proteins that had A or B quality spectra, yet allow all proteins to be screened for crystal formation (see Fig. 22.1). First, since the presence of D_2O does not alter protein crystallization behavior, and since the author recovers the screened protein samples following 1D ^1H NMR screening, *all* proteins are subjected to initial crystallization trials, regardless of their NMR spectra quality. The C and D proteins that fail to crystallize and do not form monodisperse oligomers, as determined using DLS, are not pursued further. The C and D proteins that do crystallize, but also do not form monodisperse oligomers, as determined using DLS, are subjected only to limited crystallization optimization when protein is available, but when those attempts also fail, crystallization of the protein target is stopped. The A and B proteins, however, are subjected to extensive crystallization trials and optimization. This includes methods such as fine screening, buffer optimization, and cofactor/ligand stabilization. Finally, for those A and B proteins that fail to crystallize, their sequences are then analyzed for deletion construct generation (see the following) or their structures determined using NMR spectroscopy *(41)*.

4.3. Deletion Construct Generation

Within the MH pipeline, 19 targets were analyzed for domain truncation. Eight of these targets crystallized, but failed to produce diffraction quality crystals, whereas 11 of the targets were determined by 1D ^1H NMR screening to be well folded (spectra of A or B quality) and subjected to analysis in case their full-length sequences failed to crystallize. Alternative constructs of these domains were generated by first analyzing the sequence of each target for the existence of any disordered domains using the PONDR and DisEMBL programs and then using these results to generate alternative domain constructs of each protein with the disordered domains removed. Two to 16 truncated constructs were generated for the 19 parent targets, with an average of four to five each for a total of 81.

During this experiment, 42 of the truncated constructs were processed, representing 16 different parent targets. Notably, 69% of the targets either failed to express or expressed insolubly. This percentage is much higher than was expected since each of the parent targets had expressed well and was soluble. This indicates that many of the truncated versions either ameliorated protein expression or led to a sufficient change in structure/folding kinetics that the truncated protein was insoluble during expression.

Sixteen truncated versions of eight targets, however, did express solubly. Two of these targets were solved to high resolution using truncated versions of the parent target. In the first case, the parent target (hypothetical protein BH2331, 10174951) crystallized as a full-length protein but failed to produce diffracting crystals. Because the purification gels indicated the presence of the triple proteolysis product, it was suspected that this protein would be a good candidate for domain truncation. Seventeen constructs were generated, of which nine were processed (see Table 22.2). Of these, six were insoluble whereas three were soluble. One of the soluble constructs (BH2331 residues 19–265), was purified and subjected to initial coarse screen crystallization trials. The truncated construct crystallized at a rate five times that of the full-length parent target and the crystals produced after fine screening about one of these conditions diffracted to high resolution. Ultimately, the structure was solved to 1.95 Å (PDBID 1VL5). In a second case, the parent target (hypothetical protein Alr5027, 17134165) diffracted to 2.2 Å, but the structure could not be solved. A crystal from a truncated version of the target produced high resolution, and more interpretable data, and the structure was solved to 1.5 Å (PDBID 1VL7).

4.4. Crystallization Optimum Solubility Buffer Screening

In this optimized buffer (100 mM CHES, pH 9.25, 100 mM NaCl, 0.25 mM TCEP), AF2059 was easily concentrated to 0.5 mM and no precipitation was observed. The newly buffered sample was then screened for crystal formation a second time against 192 conditions. This time, 25% of the drops were clear and 6% of the drops (24 conditions) produced crystals. Optimization of one of these conditions ultimately resulted in crystals suitable for data collection and structure determination.

5. Applicability for an Academic Laboratory

5.1. Orthologue Screening

While orthologue screening is highly suitable for most structural genomics efforts, in which the aim is obtain complete structural coverage of protein sequence space, the use of orthologue screening is not always applicable for projects that are focused on investigating a *specific* protein. In this latter case, if no structural information is already known about the protein family, orthologue screening, using highly orthologous proteins, can be used to increase the likelihood that the structure of a family member will be solved. This structure can then guide future experiments for the protein of interest.

5.2. 1D ^{1}H NMR Fold Screening

The use of 1D ^{1}H NMR screening is straightforward to implement in an academic laboratory if an NMR spectroscopist is at hand. It is most useful when coupled with either gel filtration or dynamic light scattering (DLS) to verify the oligomeric state of the protein under investigation. Proteins with molecular weights higher than 50 kDa have been screened, but the reader should find the method most informative for proteins with MW less than or equal to 40 kDa.

5.3. Deletion Construct Generation

Deletion construct generation is often used in the academic laboratory for protein optimization prior to expression and crystallization. A complementary, experimental approach is to use limited proteolytic digestion to identify stable folded domains of the protein under investigation. In the author's laboratory at Brown University, four to eight constructs (full-length and deletion, identified as described) are cloned in parallel *(42)*. If these constructs are shown by 1D ^{1}H NMR screening to be well folded, yet fail to crystallize, one should both re-analyze the sequence to identify additional deletion constructs that might provide samples suitable for high-resolution structure determination (as shown in Fig. 22.1) and, in parallel, perform limited proteolysis coupled with mass spectrometry on the purified constructs to identify the smallest stable structural domains.

5.4. Crystallization Optimum Solubility Buffer Screening

Optimum solubility (OS) screening, using either the Jancarik or crystallization method, is easily incorporated within academic-sized laboratories. The Jancarik Optimum Solubility screen coupled with dynamic light scattering is a powerful tool for identifying buffers and/or additives that result in a homogenous, monodisperse protein sample, i.e., those more likely to form diffraction quality crystals *(43)*. This type of buffer screening is also highly useful for optimizing samples for NMR spectroscopy *(44)*. The crystallization trial method can be used to complement the Jancarik method and is especially useful for those laboratories that do not have access to DLS instrumentation. Alternative solubility screening methods, in which the solubility of a protein is analyzed not by DLS, but by its ability to be precipitated out of a particular buffer using polyethylene glycols, have also been shown to successfully identify new protein buffers that lead to improved rates and quality of crystal formation *(45)* when screened against multiple crystallization solutions.

6. Summary

Orthologue screening, 1D ^{1}H NMR spectroscopy screening, deletion construct generation, and optimum solubility screening are four strategies that can lead to significant increases in the crystallization success rates of protein samples pursued for high-resolution structural studies. Each of these strategies plays an important role at distinct steps of the structure determination process. Orthologue screening is most effective when used at the inception of a structural project. Deletion construct generation, on the other hand, is effective at both the project inception and following 1D ^{1}H NMR screening and/or crystallization. Typically, the author's laboratory initially clones four to eight constructs (full-length and deletion) of a protein of interest (project inception) and processes the constructs in parallel. If one or more of the constructs is determined to be an A or B protein by 1D ^{1}H NMR screening, yet fails to crystallize, the sequence will be reanalyzed to determine if additional deletion constructs should be generated, or limited proteolysis coupled with mass spectrometry will be carried out to identify stable, structural domains. Once protein constructs have been purified, 1D ^{1}H NMR screening is one of a number of methods that can be used to characterize the biophysical characteristics of the samples prior to crystallization. 1D ^{1}H NMR spectroscopy is the most useful of these methods because it is fast, provides more structural information than other commonly used methods (e.g., CD spectroscopy), and one can recover the sample following screening and use it for initial crystallization trials. Finally, optimum solubility screening is applicable immediately prior to, or, if using the crystallization OS method, immediately following initial crystallization trials. The author prefers to use initial crystallization trials to provide indications of the optimal buffer for a protein under investigation, and then screen these buffers using traditional OS screening coupled with dynamic light scattering.

Acknowledgment

The author thanks members of the laboratory, especially D. Critton, for careful reading of the manuscript and the past and current members of the JCSG.

References

1. Nallamsetty, S., and Waugh, D. S. (2006) Solubility-enhancing proteins MBP and NusA play a passive role in the folding of their fusion partners. *Protein Expr. Purif.* **45**, 175–182.
2. Waugh, D. S. (2005) Making the most of affinity tags. *Trends Biotechnol.* **23**, 316–320.
3. Derewenda, Z. S. (2004) The use of recombinant methods and molecular engineering in protein crystallization. *Methods* **34**, 354–363.
4. Heras, B., and Martin, J. L. (2005) Post-crystallization treatments for improving diffraction quality of protein crystals. *Acta Crystallogr. D Biol. Crystallogr.* **61**, 1173–1180.
5. Bowman, G. D., O'Donnell, M., and Kuriyan, J. (2004) Structural analysis of a eukaryotic sliding DNA clamp-clamp loader complex. *Nature* **429**, 724–730.
6. Lesley, S. A., Kuhn, P., Godzik, A., Deacon, A. M., Mathews, I., Kreusch, A., Spraggon, G., Klock, H. E., McMullan, D., Shin, T., Vincent, J., Robb, A., Brinen, L. S., Miller, M. D., McPhillips, T. M., Miller, M. A., Scheibe, D., Canaves, J. M., Guda, C., Jaroszewski, L., Selby, T. L., Elsliger, M. A., Wooley, J., Taylor, S. S., Hodgson, K. O., Wilson, I. A., Schultz, P. G., and Stevens, R. C. (2002) Structural genomics of the Thermotoga maritima proteome implemented in a high-throughput structure determination pipeline. *Proc. Natl. Acad. Sci. USA* **99**, 11664–11669.
7. Lesley, S. A. (2001) High-throughput proteomics: protein expression and purification in the postgenomic world. *Protein Expr. Purif.* **22**, 159–164.
8. Page, R., Moy, K., Sims, E., Velasquez, J., McManus, B., Grittini, C., and Stevens, R. C. (2004) Scalable high-throughput microliter expression device for *Escherichia coli* recombinant proteins. *Biotechniques* **37**, 364–370.
9. Peti, W., Page, R., Moy, K., O'Neil-Johnson, M., Wilson, I. A., Stevens, R. C., and Wuthrich, K. (2005) Towards miniaturization of a structural genomics pipeline using micro-expression and microcoil NMR. *J. Struct. Funct. Genomics* **6**, 259–267.
10. Knaust, R. K., and Nordlund, P. (2001) Screening for soluble expression of recombinant proteins in a 96-well format. *Anal. Biochem.* **297**, 79–85.
11. Dummler, A., Lawrence, A. M., and de Marco, A. (2005) Simplified screening for the detection of soluble fusion constructs expressed in *E. coli* using a modular set of vectors. *Microb. Cell Fact.* **4**, 34.
12. Nguyen, H., Martinez, B., Oganesyan, N., and Kim, R. (2004) An automated small-scale protein expression and purification screening provides beneficial information for protein production. *J. Struct. Funct. Genomics* **5**, 23–27.
13. Moy, S., Dieckman, L., Schiffer, M., Maltsev, N., Yu, G. X., and Collart, A. F. (2004) Genome-scale expression of proteins from Bacillus subtilis. *J. Struct. Funct. Genomics* **5**, 103–109.
14. Sugar, F. J., Jenney, F. E., Jr., Poole, F. L., 2nd, Brereton, P. S., Izumi, M., Shah, C., and Adams, M. W. (2005) Comparison of small- and large-scale expression of selected *Pyrococcus furiosus* genes as an aid to high-throughput protein production. *J. Struct. Funct. Genomics* **6**, 149–158.
15. Cornvik, T., Dahlroth, S. L., Magnusdottir, A., Herman, M. D., Knaust, R., Ekberg, M., and Nordlund, P. (2005) Colony filtration blot: a new screening method for soluble protein expression in *Escherichia coli*. *Nat. Meth.* **2**, 507–509.
16. McPherson, A. (1994) *Crystallization of Biological Macromolecules*, Cold Spring Harbor Laboratory Press, Cold Spring Harbor, NY.

17. DiDonato, M., Deacon, A. M., Klock, H. E., McMullan, D., and Lesley, S. A. (2004) A scaleable and integrated crystallization pipeline applied to mining the *Thermotoga maritima* proteome. *J. Struct. Funct. Genomics* **5**, 133–146.

18. Page, R., Deacon, A. M., Lesley, S. A., and Stevens, R. C. (2005) Shotgun crystallization strategy for structural genomics II: crystallization conditions that produce high resolution structures for T. maritima proteins. *J. Struct. Funct. Genomics* **6**, 209–217.

19. Jancarik, J., Pufan, R., Hong, C., Kim, S.-H., and Kim, R. (2004) Optimum solubility (OS) screening: an efficient method to optimize buffer conditions for homogeneity and crystallization of proteins. *Acta Crystallogr. D Biol. Crystallogr.* **60**, 1670–1673.

20. Savchenko, A., Yee, A., Khachatryan, A., Skarina, T., Evdokimova, E., Pavlova, M., Semesi, A., Northey, J., Beasley, S., Lan, N., Das, R., Gerstein, M., Arrowmith, C. H., and Edwards, A. M. (2003) Strategies for structural proteomics of prokaryotes: quantifying the advantages of studying orthologous proteins and of using both NMR and X-ray crystallography approaches. *Proteins* **50**, 392–399.

21. Gao, X., Bain, K., Bonanno, J. B., Buchanan, M., Henderson, D., Lorimer, D., Marsh, C., Reynes, J. A., Sauder, J. M., Schwinn, K., Thai, C., and Burley, S. K. (2005) High-throughput limited proteolysis/mass spectrometry for protein domain elucidation. *J. Struct. Funct. Genom.* **6**, 129–134.

22. Dokudovskaya, S., Williams, R., Devos, D., Sali, A., Chait, B. T., and Rout, M. P. (2006) Protease accessibility laddering: a proteomic tool for probing protein structure. *Structure* **14**, 653–660.

23. Ericsson, U. B., Hallberg, B. M., Detitta, G. T., Dekker, N., and Nordlund, P. (2006) Thermo fluor-based high-throughput stability optimization of proteins for structural studies. *Anal. Biochem.* **357**, 289–298.

24. McMullan, D., Canaves, J. M., Quijano, K., Abdubek, P., Nigoghossian, E., Haugen, J., Klock, H. E., Vincent, J., Hale, J., Paulsen, J., and Lesley, S. A. (2005) High-throughput protein production for X-ray crystallography and use of size exclusion chromatography to validate or refute computational biological unit predictions. J. Struct. Funct. Genomics **6**, 135–141.

25. Geerlof, A., Brown, J., Coutard, B., Egloff, M. P., Enguita, F. J., Fogg, M. J., Gilbert, R. J., Groves, M. R., Haouz, A., Nettleship, J. E., Nordlund, P., Owens, R. J., Ruff, M., Sainsbury, S., Svergun, D. I., and Wilmanns, M. (2006) The impact of protein characterization in structural proteomics. *Acta Crystallogr D Biol Crystallogr* **62**, 1125–1136.

26. Yee, A. A., Savchenko, A., Ignachenko, A., Lukin, J., Xu, X., Skarina, T., Evdokimova, E., Liu, C. S., Semesi, A., Guido, V., Edwards, A. M., and Arrowsmith, C. H. (2005) NMR and X-ray crystallography, complementary tools in structural proteomics of small proteins. *J. Am. Chem. Soc.* **127**, 16512–16517.

27. Snyder, D. A., Chen, Y., Denissova, N. G., Acton, T., Aramini, J. M., Ciano, M., Karlin, R., Liu, J., Manor, P., Rajan, P. A., Rossi, P., Swapna, G. V., Xiao, R., Rost, B., Hunt, J., and Montelione, G. T. (2005) Comparisons of NMR spectral quality and success in crystallization demonstrate that NMR and X-ray crystallography are complementary methods for small protein structure determination. *J. Am. Chem. Soc.* **127**, 16505–16511.

28. Page, R., Peti, W., Wilson, I. A., Stevens, R. C., and Würthrich, K. (2005) NMR screening and crystal quality of bacterial expressed prokaryotic and eukaryotic proteins in a structural genomics pipeline. *Proc. Natl. Acad. Sci. USA* **102**, 1901–1905.

29. Pantazatos, D., Kim, J. S., Klock, H. E., Stevens, R. C., Wilson, I. A., Lesley, S. A., and Woods, V. L., Jr. (2004) Rapid refinement of crystallographic protein construct definition employing enhanced hydrogen/deuterium exchange MS. *Proc. Natl. Acad. Sci. USA* **101**, 751–756.

30. Li, X., Romero, P., Rani, M., Dunker, A. K., and Obradovic, Z. (1999) Predicting protein disorder for N-, C-, and internal regions. *Genome Inform. Ser. Workshop Genome Inform.* **10**, 30–40.

31. Romero, P., Obradovic, Z., Li, X., Garner, E. C., Brown, C. J., and Dunker, A. K. (2001) Sequence complexity of disordered protein. *Proteins* **42**, 38–48.

32. Linding, R., Jensen, L. J., Diella, F., Bork, P., Gibson, T. J., and Russell, R. B. (2003) Protein disorder prediction: implications for structural proteomics. *Structure (Camb)* **11**, 1453–1459.

33. Collins, B. K., Tomanicek, S. J., Lyamicheva, N., Kaiser, M. W., and Mueser, T. C. (2004) A preliminary solubility screen used to improve crystallization trials: crystallization and preliminary X-ray structure determination of *Aeropyrum pernix* flap endonuclease-1. *Acta Crystallogr. D Biol. Crystallogr.* **60**, 1674–1678.

34. Collins, B., Stevens, R. C., and Page, R. (2005) Crystallization Optimum Solubility Screening: using crystallization results to identify the optimal buffer for protein crystal formation. *Acta Crystallograph. Sect. F Struct. Biol. Cryst. Commun.* **61**, 1035–1038.

35. Storm, C. E., and Sonnhammer, E. L. (2003) Comprehensive analysis of orthologous protein domains using the HOPS database. *Genome Res.* **13**, 2353–2362.

36. Linding, R., Russell, R. B., Neduva, V., and Gibson, T. J. (2003) GlobPlot: exploring protein sequences for globularity and disorder. *Nucleic Acids Res.* **31**, 3701–3708.

37. Jaroszewski, L., Rychlewski, L., Li, Z., Li, W., and Godzik, A. (2005) FFAS03: a server for profile–profile sequence alignments. *Nucleic Acids Res.* **33**, W284–W288.

38. Page, R., Grzechnik, S. K., Canaves, J. M., Spraggon, G., Kreusch, A., Kuhn, P., Stevens, R. C., and Lesley, S. A. (2003) Shotgun crystallization strategy for structural genomics: an optimized two-tiered crystallization screen against the *Thermotoga maritima* proteome. *Acta Crystallogr. D Biol. Crystallogr.* **59**, 1028–1037.

39. Page, R., and Stevens, R. C. (2004) Crystallization data mining in structural genomics: using positive and negative results to optimize protein crystallization screens. *Methods* **34**, 373–389.

40. McGuffin, L. J., Bryson, K., and Jones, D. T. (2000) The PSIPRED protein structure prediction server. *Bioinformatics* **16**, 404–405.

41. Johnson, M. A., Peti, W., Herrmann, T., Wilson, I. A., and Wuthrich, K. (2006) Solution structure of Asl1650, an acyl carrier protein from *Anabaena* sp. PCC 7120 with a variant phosphopantetheinylation-site sequence. *Protein Sci.* **15**, 1030–1041.

42. Peti, W., and Page, R. (2007) Strategies to maximize heterologous protein expression in *Escherichia coli* with minimal cost. *Protein Expr. Purif.* **51**, 1–10.

43. Bergfors, T. M. (1999) Protein crystallization: techniques, strategies and tips, International University Line, La Jolla, CA.

44. Lepre, C. A., and Moore, J. M. (1998) Microdrop screening: a rapid method to optimize solvent conditions for NMR spectroscopy of proteins. *J. Biomol. NMR* **12**, 493–499.

45. Izaac, A., Schall, C. A., and Mueser, T. C. (2006) Assessment of a preliminary solubility screen to improve crystallization trials: uncoupling crystal condition searches. *Acta Crystallogr. D Biol. Crystallogr.* **62**, 833–842.

Chapter 23

Protein Crystallization in Restricted Geometry: Advancing Old Ideas for Modern Times in Structural Proteomics

Joseph D. Ng, Raymond C. Stevens, and Peter Kuhn

In the structural genomics period traditional methods for protein crystallization have been eclipsed by automation using batch or vapor diffusion equilibration to find conditions conducive for protein crystal growth. Although many globular and soluble proteins predominantly from prokaryotes have been crystallized and their structures solved by high throughput approaches, the remaining difficult proteins require more systematic and reflective methods combining miniaturization and integration of modern and traditional crystallography techniques. One of these conventional methods is growing crystals in restricted geometry, which is a historically well-known concept and a practical technique under-used by today's crystallographers. This chapter presents practical guidelines to use capillaries for microbatch crystallization screening and counter-diffusion crystallization as valuable techniques to obtain protein crystals in confined volumes. The emphasis in the authors' application is to perform broad-based screening with a microgram amount of protein, optimize crystal growth in a supersaturation gradient, and undergo *in situ* x-ray data analysis for x-ray crystallography without invasive manipulation. Applications and concepts presented here bring to light future prerequisites for the next generation of automation for structural genomics.

1. Introduction

Protein crystallization in the structural proteomics (or often referred to as structural genomics) era has utilized sophisticated automation to obtain protein crystals for structure determination by x-ray crystallography. The most common methods used in high throughput mode are batch or vapor diffusion equilibration of a purified protein against commercialized or custom screening reagents using robotic or semiautomated preparations *(1–11)*. Each protein is screened against hundreds of conditions per day in

From: *Methods in Molecular Biology, Vol. 426: Structural Proteomics: High-throughput Methods*
Edited by: B. Kobe, M. Guss and T. Huber © Humana Press, Totowa, NJ

very small volumes (hundreds of nanoliters) in most proteomic pipelines
(12–18). Although initial small volume crystals from coarse screens can
be used directly for data collection, often those initial conditions found to
produce crystals are further screened either in fine screening experiments
using similar volumes, or in larger volumes (microliter quantity) to optimize
crystal growth conditions and obtain superior crystals. Despite the advanced
capabilities to screen large number of crystallization conditions, only about
50% of recombinant proteins expressed and purified in structural genomic
centers were able to be crystallized and less than 10% of the crystallized
proteins were useful for structure determination by x-ray crystallography
(19,20). The reasons for the low success rate, in part, are inadequate optimization of crystal growth conditions and destruction of the crystal from physical manipulation and chemical treatment, such as heavy atom incorporation
or cryogenic preparation.

In recent years, crystal growth has been achieved in a controlled volume
under nonconvective conditions for structural genomics *(21–23)*. Proteins that
can be crystallized in a fixed volume, such as in a microfluidic or capillary
format, have the advantage of being analyzed *in situ* without disturbing the
crystal. In particular, if the volumes are small and fixed, convection is also
diminished, resulting in more diffusion limited crystallization and consequently increasing the chances of acquiring higher-quality crystals. Most of
today's interesting proteins are very limited, chemically sensitive, and expensive and time consuming to prepare. This chapter describes the methodology
to screen for initial protein crystallization conditions in nanoliter quantity in
a restricted volume, to optimize crystal growth conditions by counter-diffusion
equilibration and evaluate the crystal quality by *in situ* x-ray diffraction. Of
particular interest these days are the more difficult human proteins, which
often have large flexible regions, large macromolecular complexes, and
membrane proteins. Therefore, exploring fine supersaturation regimes of a
particular protein and post crystal treatment are much more important to either
provide access to a different part of crystallization space or refine crystallization
conditions in more controlled way.

Protein crystal growth in restricted geometry has been an effective approach
to obtain protein crystals for *in situ* x-ray crystallography *(22,24–26)*.
Crystallization by this method is not unique to modern approaches. In fact,
the principles of growing crystals in a constrained volume were developed
over a century ago *(27–29)*. In terms of applying old technology to respond to
current demands, the following steps describe its application for today's need
in structural proteomics:

1. Initial screening for crystallization conditions using nanoliter plugs in
   microcapillary tubing (glass, quartz, COC polymer, COP polymer)
2. Optimization and preparation of protein crystals for *in situ* x-ray diffraction by counter-diffusion crystallization in capillaries (glass, quartz, COC
   polymer, COP polymer)
3. Systematic evaluation of crystal quality
4. Preparation of heavy atom soaking and cryogenic preservation
5. Room temperature data collection using home source x-ray and synchrotron
   radiation

2. Materials

2.1. Chemical Preparation of Precipitants, Heavy Atoms, and Cryosolutions

1. Buffers for precipitating agents: 100 mM sodium acetate pH 5.0, 100 mM cacodylic acid pH 6, 100 mM HEPES pH 7.0, 100 mM Tris-HCl pH 8.0, 100 mM Tris-HCl pH 9, 100 mM CAPS (pH 10.0 and 11.0).
2. Precipitants: 90% saturated sodium chloride, 90% saturated ammonium sulfate, 50% PEG 8000, and 50% PEG 2000, 50% MPD.
3. Additives: 1 M magnesium chloride.
4. Heavy atoms: potassium iodide, sodium bromide (Table 23.1).
5. Commercial sparse-matrix screens (e.g., Hampton Research, Qiagen, Jena Bioscience, Emerald Structure, Molecular Dimensions).
6. Cryoprotectants: glycerol, MPD (Table 23.2).

2.2. Microcapillary Crystallization in Nanoliter Volume

1. PHD 2000 syringe pumps (Harvard Apparatus, Holliston, MA) with 10- and 50-µl Hamilton Gastight syringes (1700 series, TLL; Hamilton, Reno, NV).
2. Polydimethylsiloxane (PDMS) microchannels with $150 \times 100\,\mu m^2$ cross-sectional dimensions (Sylgard Brand 184 Silicone Elastomer; Dow Corning, Midland, MI).

Table 23.1 Common heavy atoms used for counter-diffusion soaking

cis-Dichlorodiamine platinum	Mercury acetate
Dichloro(ethylenediamine)platinum	Platinum potassium iodide
Dipotassium tetriodomercurate	Platinum potassium thiocyanate
Ethyl mercurithiosalicylate	Potassium hexachloroiridate
Ethylmercuryphosphate	Potassium hexachlorplatinate
Ethylmercuryphosphate	Potassium iodide
Gold chloride	Potassium tetrachloroaurate hydrate
Gold potassium bromide	Potassium tetrachloroplatinate
Gold potassium bromide cyanide	Potassium tetranitroplatinate trihydrate
Lead acetate trihydrate	Sodium bromide
Mecury chloride	Trimethyleadacetate

Table 23.2 Common cryoprotectants that work well with counter-diffusion equilibration

Glycerol
Ethylene glycol
2-methyl-2,4-pentadiol (MPD)
Dimethylsulfoxide (DMSO)
PEG 400
(2R, 3R)-(–)-butane 2,3-diol

3. Fluorocarbon oils, FC (40, 70, 3283) (3M, St. Paul, MN).
4. Plastic tubing composed of Zeonor Cyclic-Olefin Polymers (Paradigm Optical, British Columbia, Canada) with inner diameters of 0.1 mm.
5. Glass syringe (50–500 μL) (Hamilton).
6. Microsyringe pump (Stoelting, Wood Dale, IL).

2.3. Counter-Diffusion Crystallization

1. Laboratory gloves.
2. Borisilicated glass capillary with length, outer diameter, and inner diameters of 80, 0.40, and 0.3 mm, respectively (Hilgenberg GmbH, Malsfeld, Germany).
3. Soft temperature resistant Bol-Wax no. 1 (Fluidmaster, San Juan Capistrano, CA) or equivalent toilet sealing wax.
4. Diamond Shine nail polish (Sally Hansen Del Laboratories, Uniondale, NY) or equivalent hard base enamel nail polish.
5. Sterile autoclaved graduated 2-ml microcentrifuge tubes that are protease and nuclease free. The tubes preferably should be clear and non-colored for easy viewing.
6. Sterile syringe needle 23G.

2.4. Visualization and X-Ray Analysis

1. Modeling clay.
2. Microscope slides.
3. Color camera–mounted light stereomicroscope.
4. Laboratory x-ray generator and detector system (e.g., MSC R-AXIS IV; Rigaku, The Woodlands, TX) with 300-μm radius x-ray beam.
5. Crystallographic software for data processing and reduction (e.g., HKL2000) *(30)*.

3. Methods

3.1. Microcapillary Protein Crystallization for Initial Screening

Microcapillary-based microbatch screening allows crystallization trials to be performed with high efficiency using protein volume in nanoliter quantity. Using x-ray–transparent plastic tubing with diameters of 0.1 mm or less, crystallization plugs of nanoliter-sized are formed using micorfluidic devices fabricated by soft lithography *(31,32)*. The following procedure describes the initial screening of any protein against hundreds of precipitating conditions with less than 100 μg of purified protein (Fig. 23.1).

1. Load stock solutions containing different buffers, precipitants, additives, and protein into syringes. Four syringes are to be filled at a time, each containing one type of buffer at a single pH, a precipitating salt or alcohol, an additive, and protein. A typical screen would mix different combination of precipitating agents and additives at pH 5, 6, 7, 8, 9, 10, and 11. Commercialized sparse matrix screen solutions can also be used. Even though commercial screens contain similar chemical compositions, they often produce drastically different results *(33)*.
2. Connect the syringes to the convening channels of the microfluidic device. In the current system, up to six pumps can be used at a time.

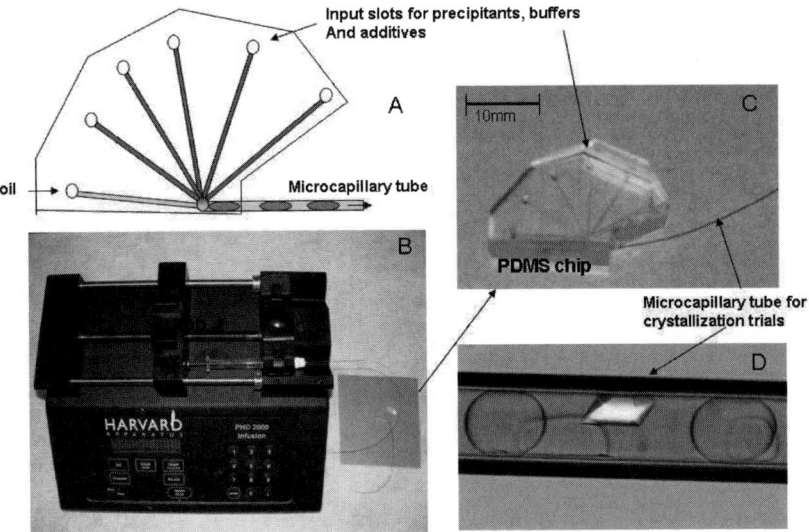

Fig. 23.1 Microcapillary system for nanoliter crystallization screening in restricted geometry. **A.** PDMS device is illustrated to have a maximum of six inlet channels and one outlet. Each inlet can accommodate microtubes to feed in oil, precipitants, buffers, and additives. The inlet channels merge into one to form plug formation output in a microcapillary tube. **B.** A Hamilton syringe can be mounted on each pump for fine injection of nanoliter volumes into the PDMS device. In general, there is one pump for each inlet, even though only one is shown for simplicity. A maximum of six pumps can be used for the preparation. **C.** PDMS device is shown close up displaying the input and out put channels. The microcapillary tube is indicated, containing consecutive nanoliter batch experiments. **D.** Example of a protein crystal grown in one crystallization plug flanked by oil spacers (courtesy of Rustem Ismagilov, University of Chicago).

3. Connect a fifth syringe containing water-immiscible fluorinated oil to a perpendicular channel.
4. Load the syringes to PHD 2000 syringe pumps (Harvard Apparatus) and program each pump such that the volume of each aqueous solution injected into a droplet is directly proportional to volumetric flow rate of that solution at the time when a droplet is formed in the microcapillary tube. In practice, over 500 conditions of microbatch screening plugs (10 nl each) can be prepared with about 10 µl of concentrated protein with the highest concentration that the protein can remain soluble.
5. Seal the end of plastic tubing when all plugs are loaded and observe results within 2–10 days under a light microscope. Record the results by imaging the plugs.
6. Conditions that gave rise to protein are further optimized by the counter-diffusion crystallization method. In some cases, crystals in the tubing can be directly analyzed by x-ray diffraction as described in the following for crystals grown by counter-diffusion.

3.2. Counter-Diffusion Crystallization

The principle of counter-diffusion crystallization has been well known and explained for over 100 years. Liesegang (27) first remarked a self-organizing pattern formed by the complex interplay of diffusion, chemical reaction, and salt precipitation termed as "Liesegang rings" or "bands" in circular and linear geometries, respectively. This unique formation was later explained by Ostwald *(28)* as an action-coupling chemical precipitation and diffusion in which the most stable nucleating particles in a supersaturated environment would dominate and grow at the expense of their surrounding products. Consequently, a decreasing supersaturation gradient is formed during the equilibration process and an integrated area of the protein's solubility curve can be covered *(24,26)* and simulated *(34)*. Garcia-Ruiz was first to show counter-diffusion application for proteins *(35)* and Gavira et al. demonstrated that protein crystals can be screened with heavy atoms and cryogenic solutions and analyzed *in situ* within the length of a capillary by x-ray diffraction and its complete three-dimensional structure solved without any physical manipulation *(25)*. The crystallographic application using this method has been reviewed *(22,26,36)*. The following steps require purified protein within the concentration of 15–20 mg/ml. Each capillary chamber requires about 5 µl.

1. Wear gloves while handling the capillary. The oil and other contaminants from the skin can block the entry ports of the capillary as well as hurt the protein samples.
2. Fill 0.5 ml of precipitating agent into a clean graduated microcentrifuge tube. Close the cap tight and gently pierce a hole through the cap with a 23-gauge syringe needle.
3. Carefully remove a glass capillary from its container and fill the length of the capillary simply by touching one end into the protein solution. By virtue of capillary action, the protein solution should fill up the capillary with ease. One can turn the capillary horizontal to facilitate the uptake.
4. Seal the opposite end with soft wax (toiling sealing wax works best) by gently piercing the capillary tube directly into the wax back and forth two or three times. If the seal is not proper, the protein solution will expel out during subsequent handling.
5. Cautiously, put the capillary filled with protein through the pierced hole in the microcentrifuge tube containing the precipitant. Make sure the open end is in contact with the precipitating agent. The capillary should be erected in the microcentrifuge tubes and one should start to see slight precipitation at the precipitant–protein interface.
6. Coat the wax-sealed end with enamel nail polish to assure a tight seal.
7. Multiple experiments can be conducted using the same precipitating agent by piercing multiple holes on the top of the microcentrifuge cap and inserting up to six protein-filled capillary into the precipitating agents as previously described.
8. Store the capillary setups in a humid environment. If the crystallization trials are kept in an enclosed incubator, installing an open beaker filled with distilled water will prevent the risk of drying up the capillary or precipitating agent.

Inexpensive plastic ware is presently being explored and expected to be available for routine use as an alternative material used in restricted geometry configuration for both micro- and capillary counter–diffusion crystallization.

3.3. Visualization and X-Ray Analysis

Three general methods are considered when assessing the quality of a crystal *(37)*. First, direct visualization with an optical microscope can reveal the immediate presence of defects in the crystalline habit such as step or multicluster formation. Be careful not to over-discriminate against what appears to be imperfect crystal, as sometimes the ugliest crystalline features provide the most useful diffraction data. Similarly, crystals that may appear to be large and perfect may not diffract well at all. Visualization may give you an idea of the quality of the crystals but is by no means a sufficient indicator as to how well it may diffract. A significant advantage of *in situ* diffraction is the fact that one can directly evaluate the most important characteristic of a crystalline sample, the diffracting power. No other measure or technique is able to evaluate this diffraction quality of a crystal more directly than x-rays. An additional benefit is that crystals screened *in situ* of their growth have not been perturbed by mounting or freezing damage. The second method to evaluate crystal quality is a more accurate assessment as it relies on the actual diffraction quality of the crystal by examining the intensity over the noise ratio of the diffraction spots (average $I/\sigma I$) as a function of resolution. Also, the spread of the reflection spots is evaluated and referred to as the mosaicity. The tighter and more circular the spots are, the lower the mosaicity indicative of high intrinsic lattice order. On the other hand, broad reflections spots are often associated with disordered lattices and measured as having high mosaicity. Typically, crystals that display significant values of average I/σ to greater than 3 Å resolution, with acceptable mosaicity, are good candidates for complete data collection.

1. The progress of the counter-diffusion equilibration can be monitored almost immediately after setup by viewing directly through the clear microcentrifuge tube under a light microscope. One should see almost immediately slight precipitation near the precipitant–protein interface, indicative of a precipitation zone. As time progresses, a gradient of high to low supersaturation can be observed along the length of the capillary, revealing precipitation, and high nucleation of crystal formation to a single crystal. Depending on the viscosity of the protein solution and precipitating media, a significant supersaturation gradient can be established to see crystals within 3–7 days. If required, the capillary can be easily taken out and examined under the microscope and replaced back into the precipitating agent without any ill effects (see Note 1).

2. When crystals are large enough for x-ray diffraction (>0.2 mm), remove the capillary from the microcentrifuge tubes, and seal the open end with wax and enamel nail polish as described.

3. Use modeling clay to support the capillary from the end of the precipitant–protein interface. Carefully mount the capillary on a goniometer head by adhering the clay holding the capillary to the goniometer base. Center and align the targeted crystals as one would normally do when preparing for x-ray diffraction. It is preferred to use an x-ray system with a horizontal phi axis instead of one that is vertical. Occasionally, there is crystal slippage due to inadequate attachment of the crystal to the inner capillary wall. Conduct initial x-ray diffraction analysis at room temperature or under a non-freezing cryostream (e.g., 4–22°C) (see Note 3).

4. Expose crystals to x-rays for 5 minutes at the home laboratory source at 0 and 90 degrees with 1° oscillation. If a large unit cell is expected such as for virus crystals, oscillation angles of 0.25 to 0.50° can be used. The detector to crystal distance should be initially adjusted so that the maximum detection resolution is 2.0–3.0 Å.

5. Diligently examine the reflection spots in terms of resolution limit and mosaicity. Index the reflections and determine the Bravais lattice type and unit cell parameters using conventional data processing software (e.g., HKL2000).

6. Translate the goniometer along the direction of the capillary and center the next well-separated crystal to perform the same diffraction analysis. The region of the capillary that appears to have well-separated crystals should be assessed in the same way by x-ray diffraction. The quality of each crystal based on diffraction limit and mosaicity will be slightly different along the length of the capillary, as they are grown in slightly different conditions within the supersaturation gradient (Fig. 23.2).

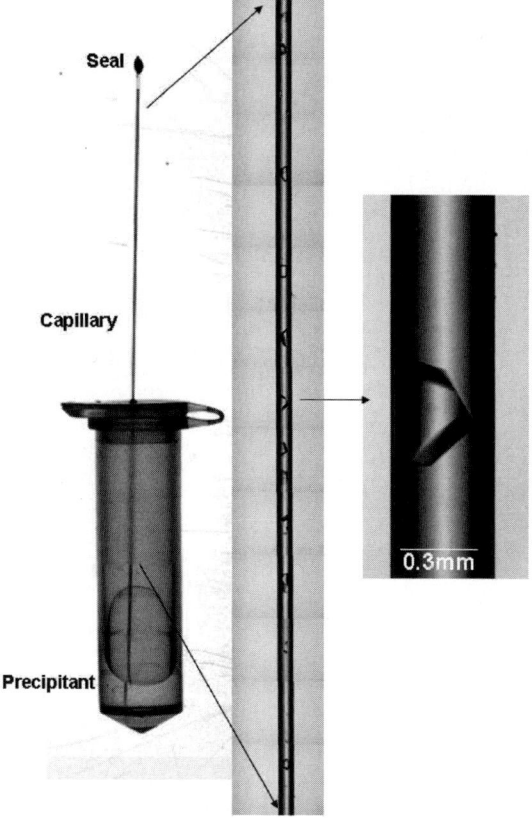

Fig. 23.2 Counter-diffusion crystallization in microcentrifuge tubes. Crystallization conditions are optimized using starting conditions from nanoliter screens that produced crystals. Proteins are loaded along the length of a glass capillary tube (≤ 0.3 mm) and sealed at the top with wax and enamel seal. The open end is immersed into 500 μl of precipitating agent with the option of containing heavy atoms, cryoprotectants, or specific additives. After 7 days of equilibration, the length of the capillary is viewed under a light microscope showing the well-separated crystals grown in a supersaturation gradient. Each protein crystal is grown out of a slightly supersaturation state and thus will be of different quality. The best quality crystal is shown in the far right panel, giving diffraction images that have the least mosaicity and maximum resolution. In this case, the protein is a hypothetical protein derived from a hypothermophilic archaeon. It diffracted to 2.2 Å resolution *in situ* using a home laboratory x-ray source at room temperature.

7. Pick the best crystal and prepare for a complete data set collection. The overall angle collected should acquire at least 90% completeness according to the space group type (e.g., at least 180° for monoclinic system).

3.4. Preparation of Heavy Atom Soaking and Cryogenic Preservation

Counter-diffusion crystallization has great advantages for heavy atom soaking and cryogenic preservation. Supplemental solutions can be diffused into the capillary chamber, and selecting the best derivatized and cryogenically prepared crystal can be achieved directly by *in situ* x-ray diffraction analysis. Historically, early crystallographic techniques in data collection were all performed in capillaries at room temperature *(38)*. Even though mounting the crystals into the capillary is problematic, there are no concerns about damages to the crystal caused by flash freezing, and the diffraction of the crystal is indicative of the crystal quality. As the x-ray beam becomes more focused and intense (e.g., at a synchrotron), radiation damage prevents complete data sets from one single crystal without flash cooling. However, with more complex proteins leading to more unstable crystals, room temperature data collection may prove to be more advantageous than rendering the protein crystal to drastic temperature changes. Therefore, early screening to obtain the best crystal for structure determination is best performed in the least invasive manner. The following steps describe the soaking and screening of heavy atom derivatives and cryogenic conditions if the crystal absolutely requires cryogenic cooling.

1. Prepare a series of 2-ml microcentrifuge tubes containing high scattering atoms in the same precipitant solution that gave rise to crystals. Halide salts such as KI or NaBr are examples that work very well in terms of successful derivatization for single anomalous scattering (SAS) phasing *(39–41)*. In these cases, 500 mM halide salts are generally prepared in the precipitating agent in a volume of 500 µl. Commercial collections of heavy atom compounds are now readily available for thorough screening. Table 23.1 lists some examples of commonly used heavy atom compounds that can be used. Usually, a concentration range of 100–500 mM is tried. Close the cap tightly and pierce a hole on top with a 23-gauge syringe needle as described.

2. Remove capillary containing the protein crystals of interest from the original precipitant solution. Do not seal the end of the capillary. Immediately insert the capillary containing the already grown crystals into the new microcentrifuge tubes with the heavy atom solutions. Allow 24 hours for the diffusion of the heavy atom salts (see Note 2).

3. Proceed to screen the crystals by x-ray diffraction analysis, as described, targeting the crystals along the capillary that do not appear to have been destroyed by the heavy atom soaking. The best native crystals are not necessarily the best derivatized crystals.

4. Collect a few consecutive frames of diffraction data for a few potential derivatized crystals. If a crystal has significant intensity differences between the x-ray data of the native crystals and those from the heavy atom soak, continue collecting a complete data set. Use the most suitable phasing strategy (e.g., SAS, MIR, SIRAS) for the data set.

5. Cryogenic preservation can be prepared in a similar fashion. Prepare a new precipitant solution containing 50% cryoprotectant mixed in the same

precipitant solution that gave rise to the original crystals. The cryoprotectant is mixed with the original precipitant solution in a volume of about 500 µl. Some common cryoprotectants are listed in Table 23.2. Take a capillary that already contains crystals (native or derivatized), remove it from its original precipitant, and place it directly into the microcentrifuge containing the candidate cryogenic solutions. Allow for the counter-diffusion process to proceed for 3–10 days. Screen the crystals again by x-ray diffraction as described.

6. The sequence of screening for heavy atoms and cryogenic solutions is a personal choice and depend on the sensitivity of the protein. In some cases crystals exposed to the cryogenic solution uptake heavy atoms more isomorphously with less crystal damage.

7. Protein crystallization can also be performed in the capillary against the precipitating agent already containing the heavy atom derivatives and cryogenic solution. Since the diffusion rate of each component is different, several chemical gradients are imposed such that nucleation, crystal growth, heavy atom incorporation, and cryogenic soaking all can be done simultaneously *(25)*.

8. Crystals can be flash-frozen directly on the goniometer platform. Gently wipe the glass capillary with ethanol to remove any dust particle or contaminants that may serve as a source for nucleation of water. Cover the cryostream (precooled for cryogenic data collection) with a hard plastic card (e.g., used credit cards are perfect) and mount the capillary with the targeted crystal roughly centered in the flow of the cryostream. Remove the plastic card quickly. Center the crystal in the normal fashion and collect data as described.

3.5. Synchrotron Radiation

In situ data analysis on crystals grown in capillary and plastic tubing can be performed with synchrotron radiation. The Stanford Synchrotron Radiation Laboratory (SSRL) beamline 1–5 using a 345-mm MAR research imaging plate scanner is ideal for room temperature crystal analysis and data collection. The intensity of the beam source can be easily attenuated to accommodate the sensitivity of the crystal. The detector is also sensitive enough to perform SAS phasing with halide soaks and even with sulfur intrinsic to the protein.

1. Tune the x-ray wavelength appropriate for the particular SAS or MAD experiments (e.g., S at 1.75 Å)

2. Enter the beamline hutch when authorized and carefully withdraw the cryostream nozzle with the fine adjustment. Cover the nozzle with aluminum foil such that it does not block the x-ray beam.

3. Place the capillary on the goniometer as described and center the crystal.

4. At the synchrotron, it is very important to know your space group and unit cell parameters with as little x-ray exposure as possible to minimize radiation damage. Therefore, the preliminary crystals analysis should include x-ray exposition times of a few seconds (1–5 seconds). Be sure to attenuate the beam intensity to its minimum.

5. After obtaining the crystal orientation and knowing the space group, strategize the shortest data collection time to obtain at least 90% data completeness. If the crystal decays before completing the projected data collection, translate the capillary along the phi axis and diffract a different part of the crystal or separate

crystal juxtaposition to it. The beam width should be much finer than the home source such that focusing on different parts of the crystal is possible.

6. Proceed with data processing, reduction, and structure determination by conventional crystallographic manipulation.

3.6. Perspectives

The future tasks for structural biology will include the necessity to produce proteins that are not easy to purify, limiting in yield, more sensitive to environmental conditions outside of their native states, and higher in complexity. Since the crystallographic structures of most proteins in this category remains unsolved to date there will be a continued demand to efficiently and reproducibly explore all chemical space with very little material to crystallize thousands of future macromolecular targets. Restricted geometry crystallization offers significant advantages to explore initial crystallization conditions with minute material as well as to refine crystal growth in an integrated supersaturation environment. Although it is still not clear what is really necessary in obtaining crystals to become "diffraction quality," it is evident that rapid and efficient methods must be developed to distinguish crystals that may be suitable for crystallographic analysis from those that need to be further refined. Therefore, crystallographic techniques such as those described here will be invaluable in high throughput mode as much as it has been historically used in traditional applications. The success of these early concept experiments will provide opportunities to fabricate widely accessible, cost-effective, and user-friendly instrumentation for all structural biologists. Future automation devices for next-generation structural genomic tasks are worth valuable consideration, as is the continued development of novel technologies and methods in this area.

4. Notes

1. Do not leave a capillary under the warm lamp of a microscope too long. The liquid from the open end will start to evaporate, leaving a space at the bottom. In most cases, this will not pose a problem in continuing the experiment if the capillary is put back into the tube. If the concern is great, then one can simply use a razor blade and excise the small end of the capillary containing the air gap and immediately place the capillary back into the precipitating solution.

2. Salts can be diffused out into a reservoir containing a lower concentration of salt. This is particularly useful when one wants to backwash over derivatized proteins. Also in cases in which proteins are less soluble in lower salt concentrations, one can reverse the salt diffusion where the higher salt concentrations is in the capillary and allowed to diffuse out into the reservoir solution. This forms a reverse solubility gradient.

3. If the goniometer does not have enough movement to scan the entire capillary within the x-ray beam, one can carefully excise the capillary in sections without drastically disturbing the crystals. Find empty spaces in the capillary, cautiously cut the glass with a razor blade or glass cutter, and immediately seal the ends with wax and nail polish. The shorter segments of the capillary can be mounted again with modeling clay for diffraction analysis.

Acknowledgments

The techniques reported here are derived from the collaborative effort between the University of Alabama in Huntsville and The Scripps Research Institute. Proof of concept of counter-diffusion crystallization for *in situ* crystallography was developed with Juan Manual García-Ruiz and José A. Gavira. The authors thank NSF (Alabama Structural Biology Consortium NSF-EPSCoR) and NIH (NIH Roadmap award GM073197) for their support.

References

1. Abola, E., Kuhn, P., Earnest, T., and Stevens, R. C. (2000) Automation of X-ray crystallography. *Nat. Struct. Biol.* **7**, 973–977.
2. Page, R., Moy, K., Sims, E. C., Velasquez, J., McManus, B., Grittini, C., Clayton, T. L., and Stevens, R. C. (2004) Scalable high-throughput micro-expression device for recombinant proteins. *Biotechniques* **37**, 364, 366, 368 passim.
3. Miyatake, H., Kim, S.-H., Motegi, I., Matsuzaki, H., Kitahara, H., Higuchi, A., and Miki, K. (2005) Development of a fully automated macromolecular crystallization/observation robotic system, HTS-80. *Acta Crystallogr. D Biol. Crystallogr.* **61**, 658–663.
4. Kuhn, P., Wilson, K., Patch, M. G., and Stevens, R. C. (2002) The genesis of high-throughput structure-based drug discovery using protein crystallography. *Curr. Opin. Chem. Biol.* **6**, 704–710.
5. Fu, Z.-Q., Rose, J., and Wang, B.-C. (2005) SGXPro: a parallel workflow engine enabling optimization of program performance and automation of structure determination. *Acta Crystallogr. D Biol. Crystallogr.* **61**, 951–959.
6. Berry, I. M., Dym, O., Esnouf, R. M., Harlos, K., Meged, R., Perrakis, A., Sussman, J. L., Walter, T. S., Wilson, J., and Messerschmidt, A. (2006) SPINE high-throughput crystallization, crystal imaging and recognition techniques: current state, performance analysis, new technologies and future aspects. *Acta Crystallogr. D Biol. Crystallogr.* **62**, 1137–1149.
7. Puri, M., Robin, G., Cowieson, N., Forwood, J. K., Listwan, P., Hu, S.-H., Guncar, G., Huber, T., Kellie, S., and Hume, D. A. (2006) Focusing in on structural genomics: the University of Queensland structural biology pipeline. *Biomol. Eng.* **23**, 281–289.
8. Pusey, M. L., Liu, Z. J., Tempel, W., Praissman, J., Lin, D., Wang, B. C., Gavira, J. A., and Ng, J. D. (2005) Life in the fast lane for protein crystallization and X-ray crystallography. *Prog. Biophys. Mol. Biol.* **88**, 359–386.
9. Page, R., Grzechnik, S. K., Canaves, J. M., Spraggon, G., Kreusch, A., Kuhn, P., Stevens, R. C., and Lesley, S. A. (2003) Shotgun crystallization strategy for structural genomics: an optimized two-tiered crystallization screen against the Thermotoga maritima proteome. *Acta Crystallogr. D Biol. Crystallogr.* **59**, 1028–1037.
10. Lesley, S. A., Kuhn, P., Godzik, A., Deacon, A. M., Mathews, I., Kreusch, A., Spraggon, G., Klock, H. E., McMullan, D., Shin, T., Vincent, J., Robb, A., Brinen, L. S., Miller, M. D., McPhillips, T. M., Miller, M. A., Scheibe, D., Canaves, J. M., Guda, C., Jaroszewski, L., Selby, T. L., Elsliger, M. A., Wooley, J., Taylor, S. S., Hodgson, K. O., Wilson, I. A., Schultz, P. G., and Stevens, R. C. (2002) Structural genomics of the Thermotoga maritima proteome implemented in a high-throughput structure determination pipeline. *Proc. Natl. Acad. Sci. USA* **99**, 11664–11669.
11. Santarsiero, B. D., Yegian, D. T., Lee, C. C., Spraggon, G., Gu, J., Scheibe, D., Uber, D. C., Cornell, E. W., Nordmeyer, R. A., Kolbe, W. F., Jin, J., Jones, A. L., Jaklevic, J. M., Schultz, P. G., and Stevens, R. C. (2002) An approach to rapid protein crystallization using nanodroplets. *J. Appl. Crystallogr.* **35**, 278–281.
12. Liu, Z. J., Tempel, W., Ng, J. D., Lin, D., Shah, A. K., Chen, L., Horanyi, P. S., Habel, J. E., Kataeva, I. A., Xu, H., Yang, H., Chang, J. C., Huang, L., Chang, S. H., Zhou,

W., Lee, D., Praissman, J. L., Zhang, H., Newton, M. G., Rose, J. P., Richardson, J. S., Richardson, D. C., and Wang, B. C. (2005) The high-throughput protein-to-structure pipeline at SECSG. *Acta Crystallogr. D Biol. Crystallogr.* **61**, 679–684.

13. Wang, B. C., Adams, M. W., Dailey, H., DeLucas, L., Luo, M., Rose, J., Bunzel, R., Dailey, T., Habel, J., Horanyi, P., Jenney, F. E., Jr., Kataeva, I., Lee, H. S., Li, S., Li, T., Lin, D., Liu, Z. J., Luan, C. H., Mayer, M., Nagy, L., Newton, M. G., Ng, J., Poole, F. L., 2nd, Shah, A., Shah, C., Sugar, F. J., and Xu, H. (2005) Protein production and crystallization at SECSG—an overview. *J Struct Funct Genomics* **6**, 233–243.

14. DiDonato, M., Deacon, A. M., Klock, H. E., McMullan, D., and Lesley, S. A. (2004) A scaleable and integrated crystallization pipeline applied to mining the Thermotoga maritima proteome. *J. Struct. Funct. Genomics* **5**, 133–146.

15. Li, F., Robinson, H., and Yeung, E. S. (2005) Automated high-throughput nanoliter-scale protein crystallization screening. *Anal Bioanal Chem* **383**, 1034–1041.

16. DeLucas, L. J., Hamrick, D., Cosenza, L., Nagy, L., McCombs, D., Bray, T., Chait, A., Stoops, B., Belgovskiy, A., and William Wilson, W. (2005) Protein crystallization: virtual screening and optimization. *Prog. Biophys. Mol. Biol.* **88**, 285.

17. Fenglei, L., Howard, R., and Edward, S. Y. (2005) Automated high-throughput nanoliter-scale protein crystallization screening. *Anal. Bioanal. Chem.* **383**, 1034.

18. Rebecca, P., Ashley, M. D., Scott, A. L., Raymond, C. S. (2005) Shotgun crystallization strategy for structural genomics ii: crystallization conditions that produce high resolution structures for *T. maritima* proteins. *J. Struct. Funct. Genomics* **6**, 209.

19. *Symposia of Protein Crystal Growth and Structural Genomics*, Quebec City, August 17–18, 2006.

20. Chandonia, J.-M., and Brenner, S. E. (2006) The impact of structural genomics: expectations and outcomes. *Science* **311**, 347–351.

21. Hansen, C. L., Skordalakes, E., Berger, J. M., and Quake, S. R. (2002) A robust and scalable microfluidic metering method that allows protein crystal growth by free interface diffusion. *Proc. Natl. Acad. Sci. USA* **99**, 16531–16536.

22. Garcia-Ruiz, J. M., and Ng, J. D. (2006) Counter-diffusion capillary crystallization for high throughput applications, in (Chayen, N., ed.), *Protein Crystallization Strategies for Structural Genomics*, International University Line, La Jolla, CA.

23. Carter, D. C., Rhodes, P., McRee, D. E., Tari, L. W., Dougan, D. R., Snell, G., Abola, E., and Stevens, R. C. (2005) Reduction in diffuso-convective disturbances in nanovolume protein crystallization experiments. *J. Appl. Crystallogr.* **38**, 87–90.

24. Garcia-Ruiz, J. M. (2003) Counterdiffusion methods for macromolecular crystallization. *Methods Enzymol.* **368**, 130–154.

25. Gavira, J. A., Toh, D., Lopez-Jaramillo, J., Garcia-Ruiz, J. M., Ng, J. D. (2002) *Ab initio* crystallographic structure determination of insulin from protein to electron density without crystal handling. *Acta Crystallogr. D Biol. Crystallogr.* **58**, 1147–1154.

26. Ng, J. D., Gavira, J. A., Garcia-Ruiz, J. M. (2003) Protein crystallization by capillary counterdiffusion for applied crystallographic structure determination. *J. Struct. Biol.* **142**, 218–231.

27. Liesegang, R. (1897) Chemische Fernwirkung. *Photographisches Arch.* **800**, 305–309.

28. Ostwald, W. (1897) Besprechung der Arbeit von Liesenganga A-Linien. Z. *Phys. Chem.* **23**, 365.

29. Ostwald, W. (1899) Lehrb. D. allgem. Chem., 2nd ed., Leipzig, Germany, 778.

30. Otwinowski, Z., and Minor, W. (1997) Processing of X-ray diffraction data collected in oscillation mode. *Methods Enzymol.* **276**, 307–326.

31. Zheng, B., Roach, L. S., and Ismagilov, R. F. (2003) Screening of protein crystallization conditions on a microfluidic chip using nanoliter-size droplets. *J. Am. Chem. Soc.* **125**, 11170–11171.

32. Yadav, M. K., Gerdts, C. J., Sanishvili, R., Smith, W. W., Roach, L. S., Ismagilov, R. F., Kuhn, P., and Stevens, R. C. (2005) In situ data collection and structure refinement from microcapillary protein crystallization. *J. Appl. Crystallogr.* **38**, 900–905.

33. Wooh, J. W., Kidd, R. D., Martin, J. L., and Kobe, B. (2003) Comparison of three commercial sparse-matrix crystallization screens. *Acta Crystallogr. D Biol. Crystallogr.* **59**, 769–772.

34. Otalora, F., and Garcia-Ruiz, J. M. (1996) Computer model of the diffusion/reaction interplay in the gel acupuncture method. *J. Cryst. Growth* **169**, 361–367.

35. Garcia-Ruiz, J. M. (1991) Uses of crystal growth in gels and other diffusing-reacting systems. *Key Eng. Mater.* **88**, 87–106.

36. Ng, J. D., and Garcia-Ruiz, J. M. (2006) Counter-diffusion capillary crystallization for structural genomics. *Screening-Trends in Drug Discovery* **3**, 36.

37. Ng, J. D. (2002) Space-grown protein crystals are more useful for structure determination. *Ann. NY Acad. Sci.* **974**, 598–609.

38. McPherson, A. (1999) *Crystallization of Biological Macromolecules.* Cold Spring Harbor Laboratory Press, Cold Spring Harbor, NY.

39. Dauter, Z. (2004) Phasing in iodine for structure determination. *Nat. Biotechnol.* **22**, 1239–1240.

40. Dauter, Z., Dauter, M., and Rajashankar, K. R. (2000) Novel approach to phasing proteins: derivatization by short cryo-soaking with halides. *Acta Crystallogr. D Biol. Crystallogr.* **56**, 232–237.

41. Dauter, Z., Li, M., and Wlodawer, A. (2001) Practical experience with the use of halides for phasing macromolecular structures: a powerful tool for structural genomics. *Acta Crystallogr. D Biol. Crystallogr.* **57**, 239–249.

Chapter 24

Fluorescence Approaches to Growing Macromolecule Crystals

Marc Pusey, Elizabeth Forsythe, and Aniruddha Achari

Trace fluorescent labeling, typically less than 1%, can be a powerful aid in macromolecule crystallization. Precipitation concentrates a solute, and crystals are the most densely packed solid form. The more densely packed the fluorescing material, the brighter the emission from it; thus, fluorescence intensity of a solid phase is a good indication of whether or not one has crystals. The more brightly fluorescing crystalline phase is easily distinguishable, even when embedded in an amorphous precipitate. This approach conveys several distinct advantages: one can see what the protein is doing in response to the imposed conditions, and distinguishing between amorphous and microcrystalline precipitated phases is considerably simpler. The higher fluorescence intensity of the crystalline phase led the authors to test if they could derive crystallization conditions from screen outcomes that had no obvious crystalline material, but simply "bright spots" in the precipitated phase. Preliminary results show that the presence of these bright spots, not observable under white light, is indeed a good indicator of potential crystallization conditions.

1. Introduction

Structural genomics projects, rational drug design, and understanding of structure–function relationships for proteins all require crystals for x-ray structure determination. In response to this expanded need for macromolecule crystals, a number of automated systems have been developed for setting up crystal screens *(1–14)*. These systems are generally based on high-density "standard format" 96-, 384-, or 1536-well plates, and are designed to use a minimal amount of protein per screen condition.

Setting up large numbers of crystal screens places a large burden on the subsequent data analysis process. The response to this has been the development of automated analysis systems by the commercial sector. The advantage of these systems is that one can look at the results, as stored images, from the relative comfort of a desktop computer, and not hunched over a microscope. The systems typically provide rudimentary image analysis, to let one know if a given drop is clear, precipitated, or possibly has a crystal. Although rapidly

From: *Methods in Molecular Biology, Vol. 426: Structural Proteomics: High-throughput Methods*
Edited by: B. Kobe, M. Guss and T. Huber © Humana Press, Totowa, NJ

eliminating clear or precipitated drops from subsequent consideration can dramatically reduce the overall analysis workload, the price may be that some possible crystallization conditions are missed.

Judge et al. have shown the utility of using a protein's intrinsic fluorescence for finding crystals (15). Subsequently, the authors' laboratory reported that use of covalent modification of a subpopulation of macromolecules with a fluorescent probe would eliminate some of the disadvantages of using intrinsic fluorescence; specifically, having to work in the UV, the relatively higher cost of UV optics, and the possibility that the protein of interest may not have a tryptophan (16). As the procedure involves generating a microheterogeneous subpopulation, it was important to also show that at sufficiently low levels the data quality obtained by x-ray diffraction was not affected by the presence of the probe. This work is extended herein, and it is shown that it is possible to use the appearance of "bright spots" in a precipitated phase as a starting point for developing subsequent crystallization conditions. These signals would not be observed under white light illumination. The ramifications of this are that the possible lead conditions space may be greatly expanded, further enhancing one's chances of deriving crystallization conditions from a given screen.

2. Materials

2.1. Protein Fluorescent Derivatization

1. Xylanase (Hampton Research, Aliso Viejo, CA), MW = 21 kDa. Used as a methodological example/control.
2. Carboxyrhodamine 6G succinimidyl ester (CR, Molecular Probes, Eugene, OR, cat.#C-6157), 5-mg bottle dissolved in 1 ml of dimethylformamide. Unused solution may be stored at −20°C for several months (see Note 1).
3. 0.025 M Hepes buffer, pH 7.5. If the protein is not soluble in this buffer, other components or buffers may be used to increase solubility as long as they do not have any primary or secondary amines.
4. G-50 column (1.5 × 30 cm), equilibrated in Hepes buffer, to place the protein into an appropriate reaction buffer and to separate unbound fluorescent dye from the modified protein.
5. Centrifugal desalting columns (Pierce, Rockford, IL).
6. Centrifugal ultrafiltration cells 10 kDA MWCO (Vivaspin; Sartorius Stedim Biotech, Aubagne, France); the MWCO used should be appropriate for your protein.
7. Fraction collector with drop counter (model 80; Gilson, Middleton, WI).

2.2. Protein Crystallization Screening

1. Greiner Bio One 96-well plates, having 3 droplet positions for each precipitant well (available from Hampton Research).
2. Crystal Screen HT screening solutions (Hampton Research, cat.# HR2-130).
3. PEG 4,000 (cat.# 33136, Serva, Heidelberg, Germany), 50% solution in distilled water.
4. Repeating pipettor (Gilson Distriman with 12.5 ml Distritip).

5. Multichannel pipettor, 8 channel (Finnpipette BioControl, 0.5–10 µL, Thermo Fisher Scientific, Waltham, MA).

2.3. Fluorescence Observations

1. Low magnification, long working distance, fluorescence microscope, compatible with crystallization plates and able to be used in either fluorescence or white light mode. A CR-specific filter set is not necessary, one suitable for GFP or fluorescein will be adequate.

3. Methods

3.1. Preparation of CR-Derivatized Protein

1. Xylanase is supplied in a concentrated aqueous solution. To place it in a reaction buffer suitable for N-terminal amine derivatization (see Note 2) a 1-ml aliquot is passed down the G-50 column under gravity feed, and 50 drop fractions are collected. The absorbance of the fractions is determined at 280 nm, and the first protein peak fractions are pooled. The total mass of protein in the pooled fractions is determined, and the moles of protein present calculated. As an example:

Pooled fraction	$= 2.2$ mg/ml, total volume $= 12$ ml
Total mass of protein	$= 26.4$ mg $= 26.4 \times 10^{-3}$ g
Moles of protein	$= 26.4 \times 10^{-3}$ g$/21 \times 10^3$ g/mol $= 1.25 \times 10^{-6}$ mol

If the entire 5-mg bottle of CR is taken up in 1 ml of dimethylformamide, then the CR concentration will be 5×10^{-3} g*ml^{-1}/555.6 g*mol^{-1} = 8.99×10^{-6} mol*ml^{-1}. The target derivatization level is 0.5–1.0% of the N-terminal amines present. The authors assume an approximately 25% yield, and desire a 0.75% derivatization level, which means they add dye to $(1/0.25) * 0.75 = 3\%$ of the protein concentration. So the total amount of CR solution to be added is:

$$1.25 \times 10^{-6} \text{ moles protein}/8.99 \times 10^{-6} \text{ moles *ml}^{-1} \text{ CR} = 0.139 \text{ ml (for 100\%)}$$
$$\rightarrow 0.139 * 0.03 \text{ (for 3\%)} = 0.00417 \text{ ml CR, or 4.2 µl}$$

2. The CR solution is added to the protein solution, preferably while stirring. Once added the protein solution is stored in a cool, dark location for at least half an hour, and overnight if possible.
3. The derivatized protein solution is concentrated by centrifugal ultrafiltration to a minimal volume. This serves to remove much of the unbound CR. The remainder is removed by passing the concentrated derivatized protein solution down a short gel filtration column. Alternatively, centrifugal desalting columns may be used (see Note 3), which have the added advantages of being much faster and resulting in less dilution of the protein sample. The unbound dye removal process also serves to put the protein into suitable buffer conditions for the subsequent crystallization screening trials.
4. Once the protein has been obtained free of unbound dye, the protein and dye concentrations are determined. It is not strictly necessary to correct the protein absorbance reading for the dye concentration. If this is done,

the CR absorbance is determined at 530 nm, and the molar concentration determined using $\varepsilon = 92,000 \, cm*mole^{-1}$. The 530 nm absorbance reading is multiplied by 0.26, the absorbance ratio 280 nm/530 nm for free CR, and this value is subtracted from the protein absorbance reading at 280nm, giving the corrected protein absorbance. However, assuming 1% of the protein molecules are covalently modified with dye, and assuming a protein absorptivity of $1 \, ml*mg^{-1}$, the error in protein concentration by not making this correction would only be 0.26%. The molar ratio of bound dye to protein (the percentage of protein molecules that are modified) is calculated. If too much dye is bound the solution can be diluted with non-derivatized protein to obtain a lower ratio. If insufficient protein is derivatized (0.1% or less for a protein of 20 kDa or larger), then the process can be repeated.

3.2. Setting Up Crystallization Plates

1. The Distriman repeating pipettor is charged with 50% PEG 4000 solution, and 100 μL aliquots are dispensed into each reservoir of the Greiner BioOne plate (Greiner, Philadelphia, PA) (see Note 4).
2. The 8-channel pipettor is set to dispense 10 × 1 μL aliquots, and the protein taken up. There are three protein crystallization wells, here identified as wells a, b, and c, for each precipitant reservoir, and the protein solution is dispensed in 1-, 1-, and 2-μL volumes, respectively, into these wells.
3. This process is repeated until all wells are filled.
4. The multichannel pipettor is reset to dispense 4 × 0.5μL, and is loaded with precipitant from the first column of solutions in the 96-well block.
5. Precipitant is dispensed onto the protein in volumes of 1, 0.5, and 0.5 μL into wells a, b, and c, respectively, in the first column of the crystallization plate.
6. This results in three droplets per reservoir, having protein:precipitant ratios of a = 1:1, b = 2:1, and c = 4:1. The used tips are ejected, new tips put on, and the process is repeated for the next column of precipitant and corresponding protein solutions until all have been dispensed. The plate is then covered with clear adhesive film and set in an incubator.

3.3. Fluorescent Observations

The authors have found it good practice to take a first look at the plate as soon as possible after setting it up. When using a strong, uniform dehydrant solution such as 50% PEG 4000, the crystallization droplets appear to go to equilibrium much faster than they would when using the precipitant solution, so additional observations should be made on a regular basis for the first week or so.

1. After turning on the fluorescent lamp and allowing it to warm up (~10 minutes), put the appropriate set of excitation, dichroic, and emission filters in place and begin observations at low magnification, such that all three protein droplets for a given precipitant solution can be seen at one time. Doing the first pass in this manner enables one to quickly determine if any crystals are present, as they can be easily seen even at low resolution, as shown in Fig. 24.1.
2. Clear solutions have uniform background fluorescence, whereas amorphous precipitate has a dull muddy fluorescence, with some nonuniformity in distribution, as shown in Fig. 24.2A. More structured precipitate has brighter

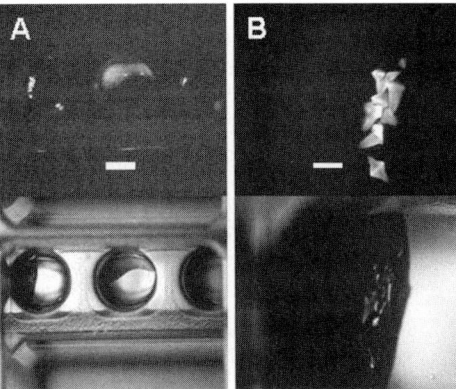

Fig. 24.1 Trace fluorescent tagging facilitates rapid surveying of crystal screening results. Upper images under fluorescent illumination, lower under white light. **A**. Low-resolution image of the three droplets for canavalin at condition B9 (scale bar = 1 mm). **B**. Magnified image of the first well in (**A**) (scale bar = 200 μm).

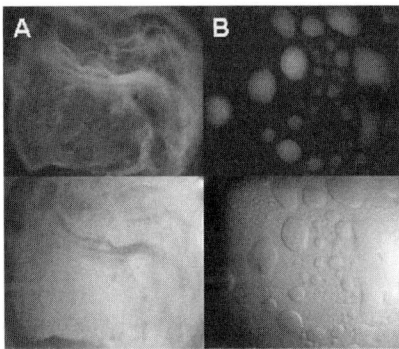

Fig. 24.2 Precipitated protein under fluorescent *(upper images)* and white light *(lower images)* illumination. **A**. Amorphous precipitate. **B**. Oiled-out protein.

fluorescent spots. "Oiled out" protein shows up as spherical droplets, as shown in Fig. 24.2B, which may fluoresce or not, depending on whether the protein or some other solution component has phase separated.

3. Nonprotein crystalline material does not fluoresce, and do not show up, except possibly as a blacker hole in the view, under fluorescent illumination (Fig. 24.3A and B). Crystalline material is easily distinguished, whether the crystals are in clear solution (Fig. 24.4A) or obscured within a precipitated phase (Fig. 24.4B). Needle-shaped crystals typically have a light pipe effect, and even if the body of the crystal does not fluoresce brightly, one can see a bright spot at the end of the needle if it is correctly aligned, as shown in Fig. 24.4C.

3.4. Bright Fluorescent Regions as Potential Lead Conditions

After testing the use of trace fluorescent labeling with several proteins, it was observed that some of the wells with precipitated protein had regions of

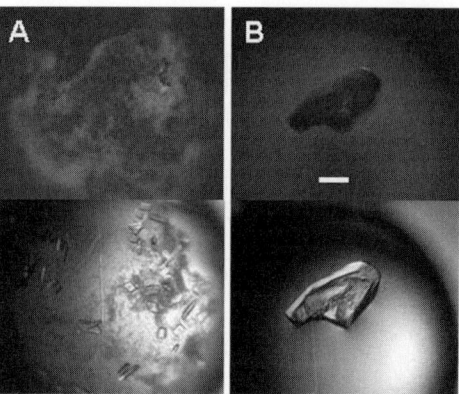

Fig. 24.3 Salt crystals under fluorescent *(upper images)* and white light *(lower images)* illumination. **A**. Canavalin + ionic liquid at screen condition D10. **B**. Chymotrypsinogen + ionic liquid at screen condition A2. The ionic liquids crystallized under the precipitant conditions (scale bar = 200 μm).

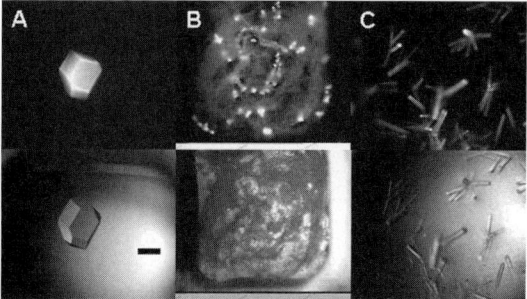

Fig. 24.4 Crystalline protein under fluorescent *(upper images)* and white light *(lower images)* illumination. **A**. Glucose isomerase, screen condition H3. **B**. Canavalin, screen condition B11. **C**. Glucose isomerase, screen condition E1. (**A** and **C** to same scale, bar = 200 μm).

bright fluorescence within the precipitate that did not correspond to any obvious crystalline material. Following the logic that intensity is proportional to density of packing, and crystallographic packing would be the densest, it was hypothesized that these might represent either failed crystallizations or partially crystalline material, either of which could be considered as potential lead conditions. This hypothesis was tested by selecting several of these conditions, then performing screens around those precipitant conditions.

1. Figure 24.5 shows the screen results for the protein canavalin at condition H1 (10% PEG 8K, 0.1 M sodium Hepes, pH 7.5, 8% ethylene glycol). Figure 24.5A shows a microgranular or microcrystalline precipitate, which has a noticeably different textural appearance under fluorescent illumination when compared with the amorphous precipitates shown in Figs. 24.2A and 4B. A sitting drop screen was set up, varying the ammonium sulfate and PEG 8K concentrations, using the same buffer. The results, shown in Fig. 24.5D,E,F indicate that these conditions could be optimized to obtain crystals.

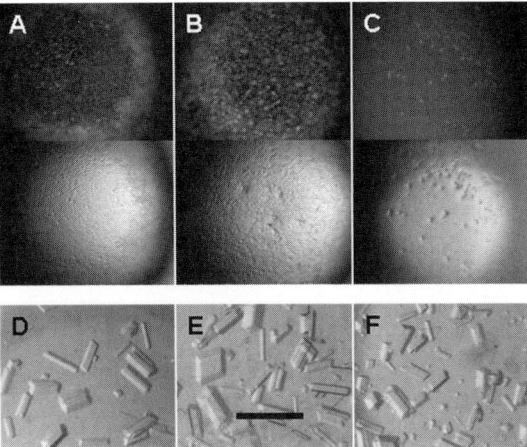

Fig. 24.5 Optimization of a canavalin precipitate. **A.–C.** The results *(upper under fluorescent, lower under white light illumination)* for screen condition H1. Protein:precipitant ratios are 1:1, 2:1, and 4:1 for **(A)**, **(B)**, and **(C)**, respectively. **D.–F.** Optimization screen results, at 0.1 M Na Hepes, pH 7.5, 5.4% PEG 8K, and **(D)** 2%, **(E)** 3%, and **(F)** 4% ethylene glycol, respectively. (**D, E,** and **F** to same scale, bar = 300 μm).

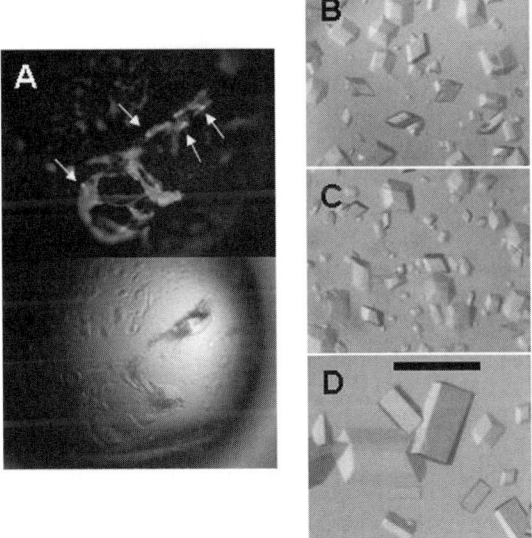

Fig. 24.6 Optimization of canavalin from precipitated screen results. **A.** *(upper image under fluorescent, lower under white light illumination)* shows the screen results for canavalin at condition F1 (30% PEG MME 2000, 0.1 M Na acetate pH 4.6, 0.2 M ammonium sulfate). The arrows point to bright spots in the precipitated protein. **B.–D.** show the outcomes of a screen set up at 0.1 M Na cacodylate, pH 6.5, 10% PEG MME 2000, and 0.1 M **(B)**, 0.15 M **(C)**, and 0.2 M **(D)** ammonium sulfate (scale bar = 300 μm).

2. The precipitate in Fig. 24.5 was clearly of an at least partially crystalline nature. A more difficult test is shown in Fig. 24.6A, again for canavalin. The screen conditions in this case are F1 (30% PEG MME 2000, 0.1 M sodium acetate pH 4.6, 0.2 M ammonium sulfate). In this case the precipitate

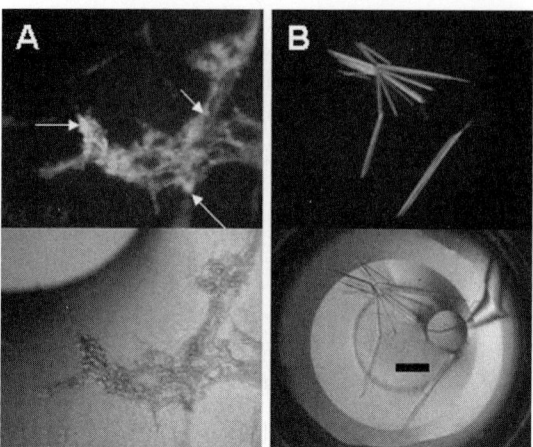

Fig. 24.7 Optimization of β-lactoglobulin B precipitate. Upper images are under fluorescent and lower under white light illumination. **A**. The initial screen results for screen solution D9 (18% PEG 8K, 0.2 M zinc acetate, 0.1 M Na cacodylate, pH 6.5), with arrows pointing to some of the bright spots observed under fluorescent illumination. **B**. Optimized crystals obtained at 10% PEG 8K, 0.01 M zinc acetate, 0.1 M Na cacodylate, pH 6.5 (scale bar = 500 μm).

was a clear gummy mass. However, a number of regions, some indicated by arrows in the fluorescent image, were considerably brighter than the body of the mass, suggesting possibly more densely packed protein. Subsequent screening was carried out in two steps. The first step involved varying the pH of the precipitant solution, from 3.5 to 8.5, with two different combinations each of PEG MME 2000 and ammonium sulfate. The results of this screen indicated that higher pHs would be most likely to give success, and a second screen was set up with 0.1 M sodium cacodylate pH 6.5 as the buffer, varying the PEG MME 2000 and ammonium sulfate concentrations. Again, crystals were obtained, as shown in Fig. 24.6B–D.

Precipitates such as those in Figure 24.6 had been observed for other proteins as well, and one such for β-lactoglobulin B is shown in Fig. 24.7A. The screen conditions in this case were 18% PEG 8K, 0.2 M zinc acetate, 0.1 M sodium cacodylate, pH 6.5. Again, as indicated by arrows, bright spots were noticed within the precipitated mass. A grid screen was set up with the starting conditions as the limiting concentrations, and the results shown in Fig. 24.7B were obtained. These crystals were obtained at the lowest zinc acetate concentration used in the grid screen, 0.01 M. Attempts at growing crystals without the zinc acetate were unsuccessful.

4. Notes

1. Alternate fluorescent probes may also be used. CR was chosen for its high absorptivity and quantum yield, relatively low sensitivity to pH, and emission maximum at the peak sensitivity for the human eye.
2. Higher pH solution may be used if one prefers instead to randomly label amines on the protein surface. If labeling of other groups or use of other

reactive species is desired, then the buffer conditions should be adjusted appropriately.

3. Centrifugal desalting columns come in several sizes, depending on the sample volume to be processed. For use, the column storage solution is first removed by centrifugation, after which the columns are equilibrated in the desired final protein solution buffer by filling the reservoir, spinning the column for the recommended time and g forces (for the Pierce columns, 2 minutes at $1,000g$, repeated two times), overlaying the column with the protein solution, then repeating the centrifugation process. This method may also be used for the initial protein processing, rather than the G-50 column.

4. The reservoirs can be filled with precipitant solution instead of 50% PEG 4,000. Alternatively, the reservoirs can be filled with 1.5 M NaCl or 1.0 M ammonium sulfate, each of which may have some advantage *(17)*. The approach of filling all reservoirs with the same dehydrant solution considerably speeds up the rate at which a 96-well plate can be set up.

Acknowledgments

This work was supported by grant GM071581 from the NIGMS.

References

1. Oldfield, T. J., Ceska, T. A., and Brady, R. L. (1991) A flexible approach to automated protein crystallization *J. Appl. Cryst.* **24**, 255–260.
2. Soriano, T. M. B., and Fontecilla-Camps, J. C. (1993) ASTEC: an automated system for sitting drop protein crystallization *J. Appl. Cryst.* **26**, 558–562.
3. Sadaoui, N., Janin, J., and Lewit-Bentley, A. (1994) TAOS: an automatic system for protein crystallization *J. Appl. Cryst.* **27**, 622–626.
4. Luft, J. R., Wolfley J., Collins, R., Bianca, M., Weeks, D., Jurisica, I, Rogers, P., Glasgow, J., Fortier, S., and DeTitta, G. T. (2000) Gearing up for structural genomics: the challenge of hundreds of proteins and thousands of crystallization experiments per year. *Acta Cryst A*, **56**, s57.
5. Luft, J. R., Wolfley, J., Jurisica, I., Glasgow, J., Fortier, S., and DeTitta, G. T. (2001) Macromolecular crystallization in a high throughput laboratory—the search phase *J. Cryst. Growth* **232**, 591–595.
6. Mueller, U., Nyarsik, L., Horn, M., Rauth, H., Przewieslik, T., Saenger, W., Lehrach, H., and Eickhoff, H. (2001) Development of a technology for automation and miniaturization of protein crystallization *J. Biotechnol.* **85**, 7–14.
7. Krupka, H. I., Rupp, B., Segelke, B. W., Lekin, T. P., Wright, D., Wu, H. C., Todd, P., and Azarani, A. (2002) The high-speed Hydra-Plus-One system for automated high throughput protein crystallography *Acta Cryst. D* **58**, 1523–1526.
8. Rupp, B., Segelke, B. W., Krupka, H. I., Lekin, T. P., Schafer, J., Zemla, A., Toppani, D., Snell, G., and Earnest, T. (2002) The TB structural genomics consortium crystallization facility: towards automation from protein to electron density *Acta Cryst. D* **58**, 1514–1518.
9. Santarsiero, B. D., Yegian, D. T., Lee, C. C., Spraggon, G., Gu, J., Scheibe, D., Uber, D. C., Cornell, E. W., Nordmeyer, R. A., Kolbe, W. F., Jin, J., Jones, A. L., Jaklevic, J. M., Schultz P. G., and Stevens, R. C. (2002) An approach to rapid protein crystallization using nanodroplets *J. Appl. Cryst.* **35**, 278–281.
10. Sulzenbacher, G., Gruez, A., Roig-Zamboni, V., Spinelli, S., Valencia, C., Pagot, F., Vincentelli, R., Bignon, C., Salomoni, A., Grisel, S., Maurin, D., Huyghe, C., Johansson, K., Grassick. A., Roussel, A., Bourne, Y., Perrier, S., Miallau, L.,

Cantau, P., Blanc, E., Genevois, M., Grossi, A., Zenatti, A., Campanacci, V., and Cambillau, C. (2002) A medium-throughput crystallization approach *Acta Cryst. D* **58**, 2109–2115.

11. Brown, J., Walter, T. S., Carter, L., Abrescia, N. G. A., Aricescu, A. R., Batuwangala, T. D., Bird, L. E., Brown, N., Chamberlain, P. P., Davis, S. J., Dubinina, E., Endicott, J., Fennelly, J. A., Gilbert, R. J. C., Harkiolaki, M., Hon, W.-C., Kimberley, F., Love, C. A., Mancini, E. J., Manso-Sancho, R., Nichols, C. E., Robinson, R. A., Sutton, G. C., Schueller, N., Sleeman, M. C., Stewart-Jones, G. B., Vuong, M., Welburn, J., Zhang, Z., Stammers, D. K., Owens, R. J., Jones, E. Y., Harlos, K., Stuart, D. I. (2003) A procedure for setting up high-throughput nanolitre crystallization experiments. II. Crystallization results *J. Appl. Cryst.* **36**, 315–318.

12. DeLucas, L. J., Bray, T. L., Nagy, L., McCombs, D., Chernov, N., Hamrick, D., Cosenza, L., Belgovskiy, A., Stoops, B., and Chait, A. (2003) Efficient protein crystallization *J. Struct. Biol.* **143**, 188–206.

13. Hosfield, D., Palan, J., Hilgers, M., Scheibe, D., McRee, D. E., and Stevens, R. C. (2003) A fully integrated protein crystallization platform for small-molecule drug discovery *J. Struct. Biol.* **142**, 207–217.

14. Walter, T. S., Diprose, J., Brown, J., Pickford, M., Owens, R. J., Stuart, D. I., and Harlos, K. (2003) A procedure for setting up high-throughput nanolitre crystallization experiments. I. Protocol design and validation *J. Appl. Cryst.* **36**, 308–314.

15. Judge, R. A., Swift, K., and Gonzalez, C. (2005) An ultraviolet fluorescence-based method for identifying and distinguishing protein crystals *Acta Cryst. D* **61**, 60–66.

16. Forsythe, E. L., Achari, A., and Pusey, M. L. (2006) Trace fluorescent labeling for high throughput crystallography, *Acta Cryst. D* **62**, 339–346.

17. Newman, J (2005) Expanding screening space through the use of alternative reservoirs in vapor-diffusion experiments *Acta Cryst. D* **61**, 490–493.

Chapter 25

Efficient Macromolecular Crystallization Using Microfluidics and Randomized Design of Screening Reagents

Andrew P. May and Brent W. Segelke

Microfluidic technologies enable a relatively new approach to macromolecular crystallization, but offer several significant advantages over more traditional techniques. Microfluidic devices provide significant savings in the amount of material required to complete a set of experiments, although recent innovations with vapor diffusion and microbatch methods have also greatly reduced their material requirements. When compared with these other methods, microfluidic approaches still consume 5–100× less material. In addition, comparisons in one set of experiments suggest that microfluidic free-interface diffusion may also offer substantially higher success rates than sitting drop vapor diffusion. Microfluidic methods also provide opportunities for experimental strategies involving testing multiple samples in parallel. When combined with randomized design of screening reagents, microfluidic devices provide a highly efficient method for sampling crystallization space. Commercial microfluidic crystallization chips have been in circulation for a number of years now and stable protocols for their use, tips and tricks, and data on their success and failure are now available.

1. Introduction

Recently, a number of microfluidic platforms have been reported that miniaturize and parallelize macromolecular crystallization using a variety of methods *(1–4)*. A number of approaches to macromolecular crystallization are used today for structural genomics efforts worldwide. Vapor diffusion, both sitting and hanging drop, are the most prevalent methods, although microbatch and free-interface diffusion (FID) have their proponents and each offer unique advantages *(5–9)*. Several of these approaches have been adapted to microfluidic applications. Microfluidic systems provide for random access reagent combination *(1,3)*, providing a direct substitute for commonly used crystallization labware, or alternatively, provide the ability to mix reagents on the fly and enabling highly systematic screening across a range of conditions *(2,10,11)*. Although innovation with vapor diffusion

From: *Methods in Molecular Biology, Vol. 426: Structural Proteomics: High-throughput Methods*
Edited by: B. Kobe, M. Guss and T. Huber © Humana Press, Totowa, NJ

and microbatch methods have also greatly reduced their material requirements, microfluidic systems still provide tremendous savings in the amount of material required to screen crystallization conditions or crystallization phase space and present new opportunities for experimental strategy. When combined with efficient sampling of reagent space through the use of randomized reagents, these devices greatly increase the efficiency with which a crystallization campaign can be conducted.

Fluidigm Corporation (San Francisco, CA) has commercialized the TOPAZ® system for microfluidic crystallization, which uses FID within microfluidic chips. The TOPAZ system has evolved through several generations of chip designs along with supporting hardware and software. By now the system has been widely distributed and significant practical experience has been gained with its use. The latest versions of TOPAZ chips (termed X.96 chips, where X is the number of sample inlets, 1, 4, or 8) are simple to use and require only 1 µl of protein stock solution (or ~10–20 µg of protein) to set up 96 crystallization experiments. This chapter describes the use of TOPAZ X.96 integrated fluidic circuits for high-throughput macromolecular crystallization. It also describes the design of randomized crystallization reagents for use in combination with TOPAZ chips for efficient sampling of crystallization space.

2. Materials

2.1. Crystallization Screens

1. Randomized reagent design. Approximately 70 crystallization stock reagents, including precipitating agents, buffers, detergents, and additives, are made well in advance of the setup of crystallization trials (see Notes 1–5). Each is sterile filtered before use and stored at room temperature. Screens are designed using CrysTool (12) and mixed crystallization cocktails are made on the fly using a MultiProbe liquid handling robot (PerkinElmer, Waltham, MA). Screens designed usign CrysTool can be purchased commercially from Axygen Biosciences (Union City, CA).
2. Preformulated crystallization reagents. Reagent screens can be purchased from a number of vendors (Fluidigm Corp., San Francisco, CA; Qiagen, Valencia, CA; Hampton Research, Aliso Viejo, CA; Emerald BioSystems, Bainbridge Island, WA; Molecular Dimensions, Apopka, FL; Jena Bioscience, Jena, Germany).
3. The recipes for initial screens used to pursue statistical sampling of phase space are designed with the CrysTool design engine (Axygen) using the default random template.
4. For initial screening experiments using commercially available reagents, Fluidigm OptiMix-1, Hampton Research Index, Qiagen PEGS, and Molecular Dimensions PACT are recommended as a general first-pass screen.
5. Optimization screens can be generated in a number of ways. Custom templates in CrysTool can be generated that limit the reagents to those giving rise to initial leads. Optimization and translation reagents can be designed using the TOPAZ Experiment Manager or the workbooks designed in Microsoft Excel that are available at www.fluidigm.com/guides.htm#workbook.

2.2. Sample Preparation and Analysis

1. Loading buffer, gels, and apparatus for SDS-PAGE analysis (Bio-Rad, Hercules, CA).
2. Gel filtration buffer: 50 mM Hepes, pH 7.5, 150 mM NaCl.
3. Analytical size exclusion column, Superdex S200 HR10/30 (GE Healthcare, Pittsburg, PA).
4. Minimal crystallization buffer (e.g., 25 mM Hepes pH 7.5, 50 mM NaCl).
5. HisTrap FF crude Column (GE Healthcare).
6. Nickel affinity loading buffer: 50 mM NaH_2PO_4, 300 mM NaCl, 10 mM imidazole, adjusted to pH 8.0 with NaOH.
7. Nickel affinity wash buffer: 50 mM NaH_2PO_4, 300 mM NaCl, 20 mM imidazole, adjusted to pH 8.0 with NaOH.
8. Nickel affinity elusion buffer: 50 mM NaH_2PO_4, 300 mM NaCl, 250 mM imidazole, adjusted to pH 8.0 with NaOH.
9. Preparative size exclusion column, HiLoad 16/60 Superdex 75 pg (GE Healthcare).
10. Ni-NTA Spin Columns (Qiagen).
11. Prescreen buffers: 15%, 30%, and 45% isopropanol in minimal crystallization buffer; 10%, 20%, and 30% PEG 4k in minimal crystallization buffer; and 0.5 M, 1.25 M, and 2 M ammonium sulfate in minimal crystallization buffer.
12. Vivaspin 500 concentrators (Sartorius, Aubagne, France) various molecular weight cutoffs (see Note 6).
13. Thin-walled 0.2-ml PCR tubes (VWR).

2.3. Microfluidic Crystallization

1. TOPAZ 1.96, 4.96, and 8.96 Screening chips and FID Crystallizer (Fluidigm).
2. TOPAZ Hydration Fluid (Fluidigm).

   - One part hydration fluid concentrate and two parts sterile D.I. water, for use with most sparse-matrix and PEG-based crystallization screens (e.g., Fluidigm Optimix-1, Qiagen PEGS, Hampton Research Index).
   - One part hydration fluid concentrate and one part sterile D.I. water, for use with high concentration (>1.5 M) salt-based reagents (e.g., Fluidigm OptiMix-3, Hampton Research SaltRx, Qiagen ammonium sulfate)

3. 1–10 μl 8-channel pipettor (e.g., Eppendorf, Hamburg, Germany).
4. 0.2–2 μl single channel pipettor (e.g., Eppendorf).
5. 2–20 μl extended length pipette tips (E&K Scientific, Santa Clara, CA).
6. 1,000 μl single channel pipette.

2.4. Drop-Based Crystallization

1. Intelliplate crystallization plates (Art Robbins Instruments, Sunnyvale, CA) for sitting drop vapor diffusion experiments.
2. Greiner Impact crystallization plates (with reservoir) for microbatch experiments (Hampton Research).
3. ClearSeal Film (Hampton Research).
4. Al's Oil (Hampton Research) for microbatch experiments.

3. Methods

Obtaining diffraction-quality crystals from a sample is an empirical process that is tied to the nature and quality of the sample being studied. There are a number of practical considerations that can increase the likelihood of success as well as emerging methods to characterize crystallizability, such as self-affinity chromatography, of a given sample. However, at the time of this writing, no single method is as good a predictor of crystallizability as crystallization screening itself. Use of microfluidic FID along with careful attention paid to experimental design, sample preparation, characterization of samples, and interpretation of results all maximize the chance of success. The reduction in material required by microfluidic chips and the apparent enhanced success rate compared with other methods (Fig. 25.1) enables more comprehensive empirical screening to be carried out early in the structure determination pipeline. Crystallizability of a sample can be assessed through actual crystallization experiments rather than orthogonal biophysical measurements such as dynamic light scattering.

3.1. Experimental Strategy

For traditional crystallization approaches, an iterative strategy of crystallization screening and optimization is commonly pursued following the expression and purification of a single construct or complex. If no crystals are obtained after exhaustive crystallization screening, then a new sample or variant is prepared and the cycle is repeated. It is estimated that, for most samples, if no crystals are obtained within 500 statistically independent reagent conditions, then the likelihood of that sample preparation yielding crystals is so low that that sample should be shelved and new constructs/sample preparations should be

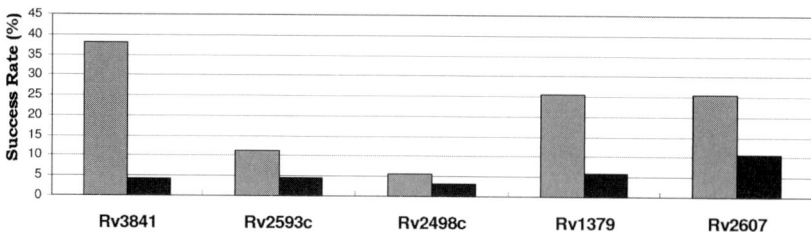

Fig. 25.1 Success rate comparison of TOPAZ versus sitting drop vapor diffusion. A large set of experiments were performed for each of five proteins using TOPAZ 1.48 chips and sitting drop vapor diffusion in IntelliPlate crystallization plates using 400 nl + 400 nl drops. Crystallization conditions were generated by random combination using CrysTool (12). Microfluidic experiments were monitored over a period of 1 month and sitting drop experiments were monitored over a period of 4 months. Success rate is determined by dividing the total number of experiments leading to crystallization by the total number of experiments performed. Microfluidic free-interface diffusion using TOPAZ chips (gray bars) led to a two- to eightfold increased success rate compared to sitting drop vapor diffusion (black bars).

tried. Alternative strategies involving parallel processing of multiple sample variants become possible with the low sample volumes required for microfluidic crystallization experiments. A number of these strategies are described in the following sections.

3.1.1. Focusing on a Single Sample Preparation

The simplest variant on the standard screening strategy enabled by microfluidic devices is the exploration of a much larger number of crystallization conditions (see Notes 7 and 8). Very little sample is required for crystallization experiments using X.96 chips (1 µl for 96 experiments) allowing a more comprehensive search of crystallization phase space to be explored. It is best to prepare novel reagent sets such as those generated by CrysTool *(12)* to conduct such a comprehensive search.

3.1.2. Varying the Sample Preparation

A novel and likely more efficient approach enabled by microfluidic crystallization involves increasing the number of sample variations that are subjected to crystallization experiments and pursuing them in parallel. Current molecular biology techniques and purification methods enable the parallel production of multiple constructs. Therefore, it is possible to approach any project now as a campaign in which construct space is sampled in parallel as part of the initial preparation prior to initiating crystallization experiments. An important example of a method for generating sample variations involves the generation of a number of surface mutations that are targeted to reduce the surface entropy of a sample *(13)*. Further sample variations can be generated by combining samples with endogenous or artificial small molecule binding partners. In many cases, the addition of small molecules (in the form of ligands, substrates, or substrate analogues) has resulted in changes in protein stability that have a positive effect on crystallization.

3.2. Preparation of Samples

3.2.1. Purification of Samples

3.2.1.1. Crude Purification

Samples can be tested for crystallizability without purification to homogeneity using single step affinity chromatography. Purify His6-tagged samples using nickel affinity purification.

1. Equilibrate HisTrap nickel affinity column mounted on an FPLC with loading buffer.
2. Load the sample containing the protein of interest, usually a crude cell extract.
3. Wash the column with protein bound using wash buffer.
4. Elute with a single step gradient.
5. Concentrate the collected elution fractions exchanging buffer to minimal crystallization buffer during concentration or simply dialyze against minimal crystallization buffer.

3.2.1.2. High Purity

To obtain better-quality crystals, samples can often be purified to near homogeneity after affinity purification by size exclusion chromatography.

1. Equilibrate preparative size exclusion column mounted on an FPLC with minimal crystallization buffer
2. Load the affinity purified sample containing the protein of interest and run with minimal crystallization buffer until the sample elutes as assessed by A280.
3. Collect the fractions containing the protein of interest and concentrate.

3.2.1.3. Parallel Purification

When pursuing a crystallization screening campaign with multiple constructs of a sample, crystallizability is assessed using single step, affinity purified material. Samples are purified using a protocol similar to crude purification but in purification apparatus that enables parallel processing. This is a brief description of purification using NickelNTA spin columns.

1. Equilibrate spin columns with loading buffer and centrifuge with a microfuge for 2 minutes at 700g.
2. Load 600 µl of 50× concentrated cleared lysate containing the His-tagged into the spin column and centrifuge for 2 minutes at 700g.
3. Wash the spin column twice with wash buffer.
4. Elute the protein with 200 µl elution buffer and repeat combining the elutions.
5. Concentrate the collected elution fractions exchanging buffer to minimal crystallization buffer during concentration.

3.2.2. Concentration of Samples

Higher sample concentrations are often required for successful microfluidic crystallization experiments, as with nanovolume drop-based experiments.

1. Measure the start concentration of sample (use absorption at 280 nm or a colorimetric assay such as the Bradford assay)
2. Wash the concentrator in either gel filtration or minimal crystallization buffer.
3. Add sample to the concentrator and concentrate (target 20 mg/ml for <30 kDa proteins and 15 mg/ml for >30 kDa proteins). After each concentration step pipette protein solution up and down to remove any concentration gradients that have built up during process of concentration. If the sample shows signs of precipitation during the concentration step, stop concentrating, and add buffer to reduce the concentration by one third.
4. Measure the final concentration of the sample. If the concentration is less than the target concentration, add back to the concentrator and repeat steps 1–4.

3.2.3. Components of the Sample Buffer

Ligands should be present in 3×–5× excess of protein.

3.3. Analysis of Samples Prior to Crystallization

3.3.1. SDS-PAGE

1. Add concentrated sample to 1× SDS-PAGE sample buffer following manufacturer's instructions.
2. Load several micrograms (up to 10 µg) of sample and run SDS-PAGE.

3. If the sample contains a single protein, the stained gel lane should contain greater than 90% of the stained protein within a single band. If the target sample is a complex and contains more than one protein, the combined bands of the components should comprise greater than 90% of the total stained protein.

4. If the sample is less than 90% pure by SDS-PAGE, further purification steps should be considered (see Section 3.2. and Notes 9–31).

3.3.2. Analytical Gel Filtration

1. Equilibrate gel filtration column with two column volumes of gel filtration buffer.

2. Pipette 1 μl concentrated sample into a clean microfuge tube and make volume up to 10 μl with gel filtration buffer.

3. Load sample into injection loop and run 1.5 column volumes of buffer through column.

   • The column trace should show no evidence of aggregation, as indicated by a peak at the exclusion volume of the column. The sample should elute as a single, symmetrical peak at a volume consistent with the sample molecular weight or integer multiples of the sample molecular weight (indicating oligomerization). If the sample contains multiple isotropic peaks, it should be subjected to preparative gel filtration before proceeding to crystallization. If the sample consists of multiple overlapping peaks, then significant purification will be required before proceeding with crystallization. New constructs may need to be prepared, or new initial buffer conditions may be required.

3.3.3. Sample Stability

1. Rerun SDS-PAGE and analytical gel filtration over time to assess sample stability.

2. Prepare fresh sample or store frozen samples if stability is a concern.

3.3.4. Sample Storage

1. Aliquot 30μl concentrated sample in to thin walled PCR tubes and flash freeze in liquid nitrogen or an ethanol dry ice bath.

2. Store at −80°C.

3.4. Setting up a Microfluidic Crystallization Experiment

A detailed description of how to set up TOPAZ screening chips can be found in the user guides and quick reference cards found at www.fluidigm. com/guides.htm. A brief description of the process is included in the following:

1. Hydrate the TOPAZ screening chip (Fig. 25.2) according to the following protocol using the appropriate hydration reagent (see Materials).

   • Remove the hydration chamber lid and pipette 0.75 ml of the hydration solution onto each of the two sponges in the hydration frame.
   • Replace the hydration chamber lid and press down firmly to close.

2. Run the Prep script on the FID Crystallizer. This closes the containment valves in the chip before reagents are dispensed and prevents unwanted wicking of reagents into the chip.

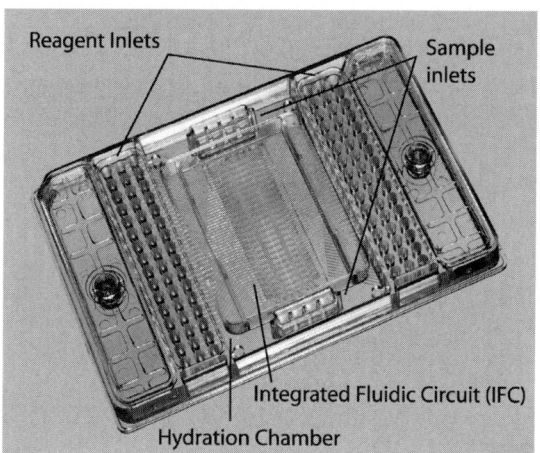

Fig. 25.2 Image of a TOPAZ 8.96 crystallization chip and carrier, indicating major components of device. Reagents are pipetted into the reagent inlets and samples into the sample inlets. The samples and reagents are loaded into the integrated fluidic circuit (IFC) using an FID crystallizer. The humidity level in the chip is controlled by the addition of hydration fluid to the reservoirs in the hydration chamber.

3. Dispense crystallization reagents into the chip carrier. When loading crystallization reagents in to the TOPAZ chip use a 1–10 µl eight-channel pipettor with extended length tips or a liquid-dispensing robot to introduce 10 µl reagent into each reagent inlet. Touch the tips to the bottom of the reagent well and dispense until the first stop of the pipettor. This reduces the chances of introducing bubbles, which can cause load failures.

4. Dispense sample into the protein inlet(s) on the carrier immediately after pipetting reagents has been completed. When loading protein in to the TOPAZ chip use a 0.2- to 2-µl pipettor to dispense 1.4 µL sample into the protein inlet. Do not press through to the second stop of the pipettor. Skilled users can use as little as 1 µl protein during this step.

5. Load protein and reagents into the chip on the FID crystallizer using the appropriate Load script (1.96, 4.96, or 8.96) immediately after the protein has been pipetted.

6. Initiate the free-interface diffusion experiment either by running an "FID Control" script, or the FID start script on the FID crystallizer. For samples with MW greater than 200 K and for membrane protein samples, running the FID start script, and leaving the interface valves open is recommended.

3.5. Interpreting the Results of a Microfluidic Crystallization Experiment

The results obtained from a microfluidic crystallization experiment appear different from those obtained within a standard drop-based experiment. Less precipitation is observed than in drop-based experiments. Patterns of crystal formation related to the relationship between the diffusion of samples and reagents are often observed. The information from these patterns can be used to guide follow-on experiments.

TOPAZ X.96 chips are inspected using the AutoInspeX workstation or an inverted light microscope on days 1, 2, 4, and 7. Beyond 7 days, the chance of false-positive crystals forming increases significantly. At 4°C, however, the lifetime of the chip experiment can be extended out to at least 14 days.

3.5.1. Little to No Precipitation or Crystals

If no crystals are obtained and no evidence of precipitation is observed in an experiment run in a X.96 chip over the entire chip, then the protein concentration should be doubled and the experiment repeated.

3.5.2. Crystals

Crystals of sample are usually observed in less than 7 days from the start of a chip-based experiment. Crystals that grow within the first 24–48 hours typically demonstrate gradients within the protein chamber of the chip (Fig. 25.3A,D,E). The direction of this gradient can be highly informative with respect to scaling up conditions. Crystals that grow after this time are typically distributed evenly across the protein chamber, indicating that they grew after the chamber had fully equilibrated, and once dehydration of the experiment has started to take place.

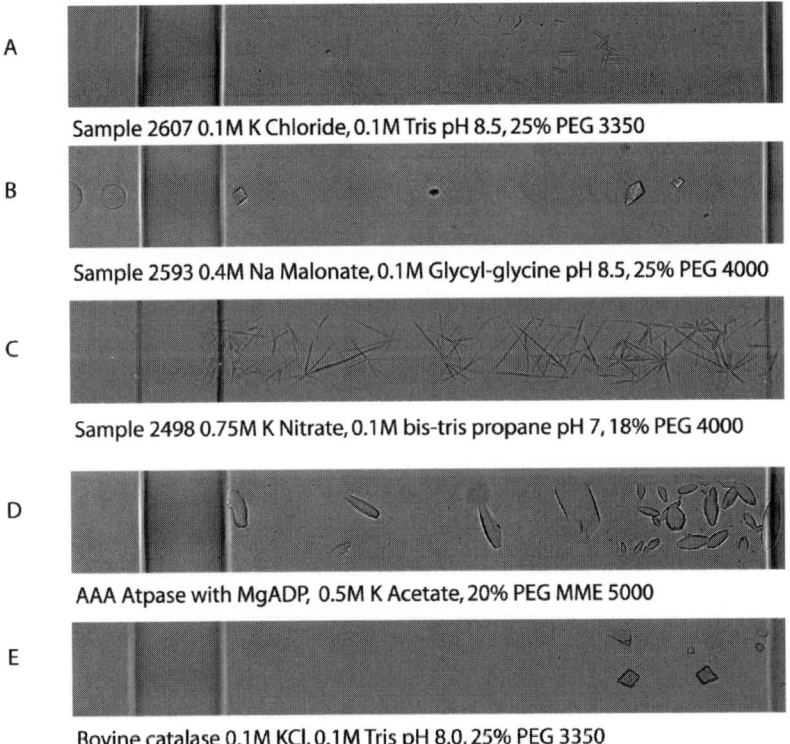

A

Sample 2607 0.1 M K Chloride, 0.1 M Tris pH 8.5, 25% PEG 3350

B

Sample 2593 0.4 M Na Malonate, 0.1 M Glycyl-glycine pH 8.5, 25% PEG 4000

C

Sample 2498 0.75 M K Nitrate, 0.1 M bis-tris propane pH 7, 18% PEG 4000

D

AAA Atpase with MgADP, 0.5 M K Acetate, 20% PEG MME 5000

E

Bovine catalase 0.1 M KCl, 0.1 M Tris pH 8.0, 25% PEG 3350

Fig. 25.3 Images of crystals grown in TOPAZ X.96 crystallization chips. **A.** Crystallization gradient with smaller crystals closer to the interface valve (and reagent) and larger crystals closer to the containment valve (furthest from reagent). **B.** Small number of high-quality, larger crystals distributed evenly across chip. **C.** Crystals grown with even distribution along protein well. **D.** Crystallization gradient with more, smaller crystals furthest from reagent well, and larger, single crystals closer to the interface valve (and reagent). **E.** Crystals grown with a few, single crystals grown furthest from the interface valve (and reagent).

Table 25.1 Reagent components likely to generate nonprotein crystals when present crystallization experiments.

First component	Second component
Mg^{2+} salts	pH > 8.5
	Phosphate salts
	Tartrate salts
Ca^{2+} salts	pH > 8.0
	Phosphate salts
	Tartrate salts
Zn acetate	Concentration > 25 mM
Ca acetate	Concentration > 100 mM
Transition metal (Cd^{2+}, Ni^{2+}, Cu^{2+}) salts	Sulfur-containing reducing agents (DTT, β-mercaptoethanol)
	Imidazole
Nitrate salts	
Tartrate salts	pH <5

3.5.3. Reagent Components Likely to Give Salt Crystals: False-Positives

Table 25.1 contains a list of reagent components that have a high likelihood of generating nonprotein crystals when present in reagents used in crystallization experiments.

3.6. Scaling up Hits from a Microfluidic Crystallization Experiment

Once crystals have been obtained from a TOPAZ X.96 chip, the conditions must be translated to conventional format experiments to obtain crystals suitable for harvesting and diffraction data collection. A guide describing how to approach translation can be found at www.fluidigm.com/PDF/trans.39VB. pdf. Comparisons between experiments run in X.96 chips and vapor diffusion have shown that in 40–50% of cases, setting up the identical reagent in drop-based experiments, also yield crystals. This rate of translation increases to approximately 70% for high-quality crystals that grow with 3–10 crystals per protein chamber. As a first approach to obtaining diffraction-quality crystals, translation should be attempted with the exact reagent using different protein: reagent ratios.

3.6.1. Translation Using the Original Reagent (Microbatch)

1. Identify the reagent that results in successful crystal growth.
2. Prepare the reagent at 50% concentration by pipetting 5 μl reagent and 5 μl MilliQ (or equivalent) water into a clean microfuge tube.
3. Add 40 μl Al's Oil (Hampton) to six wells in a microbatch plate (Hampton).
4. Pipette 1.5, 1, 0.5 μl 100% reagent into three wells in the plate. Add 0.5, 1, and 1.5 μl protein to each of those wells, respectively.
5. Pipette 1.5, 1, 0.5 μl 50% reagent into three wells in the plate. Add 0.5, 1, and 1.5 μl protein to each of those wells.
6. Incubate plates at the same temperature as the original chip-based crystallization experiment for up to 14 days.

3.6.2. Translation Using a Reagent Grid Based upon the Chemistry of the Original Condition

An alternative method for producing crystals in a drop-based experiment, which can be used in place of or in addition to the direct translation method, is to generate new reagents based upon variations in the components found in the original crystallization reagent. In some cases, the concentrations and relative proportions of the components of the crystallization condition used in the X.96 screening chip need to be varied during translation to conventional methods. The "Guide to Translating Screening Conditions" (www.fluidigm.com/PDF/trans.39VB.pdf) describes how to generate reagents for translation. In brief, this protocol describes preparing reagents for translating high-quality hits. For lower-quality hits, optimization in chip is highly recommended before attempting translation. Protocols for optimization are provided in the "Guide to Translating Screening Conditions." A workbook designed in Microsoft Excel® is also available for designing optimization reagents (www.fluidigm.com/guides.htm#workbook). It is assumed that for high-quality hits, the pH of the crystallization reagent is optimal. The recommended reagents should be prepared by reducing the concentrations of the principal components of the reagent (e.g., PEG and salt) as described in the protocol. This protocol is most successful for crystals that grow within the first 2 days. However, experience gained since the protocol was published indicates that it is also important to vary the pH of the translation solutions to sample pH values between the reagent pH and the pH of the protein buffer.

1. Crystals have grown within 2 days and show a gradient of size along the protein chamber. Prepare reagents as described in the "Guide to Translating Screening Conditions from TOPAZ Screening Chips to Macroscopic Methods." Recommendations for reagent grid design are given for different types of reagents.
2. Crystals grew in 2–4 days. Design a reagent grid that varies the precipitant and salt concentrations from 130% to 80% of the concentrations in the original reagent. The reagents should be prepared at the pH of the original reagent, and at at least one pH value intermediate between that of the original reagent and the sample buffer.
3. Crystals grew in 4–7 days. Reagents should be designed as in step 2, and the sample concentration should be increased by 30%.

If crystals first appear over 7 days after the chip was originally set up, the composition of the solution, and possible also the integrity of the sample will no longer be well defined, and more extensive exploration of reagent space will be required.

4. Notes

1. A list of suggested stock reagents for use with CrysTool and recipes to make them is listed at porter.llnl.gov. Additional usable stock reagents can be identified by reviewing the stock reagents listed in commercially available screens.
2. An alternative to making screens from stock reagents using CrysTool and a liquid handling robot is to buy pre-made screens. Reagents for

crystallization obtained from the commercial vendors in the materials section may be used for crystallization experiments in TOPAZ chips. However, these reagents are typically highly redundant and are less efficient for sampling reagent space than reagents designed using the CrysTool design engine.

3. The gel filtration buffer described is the most basic form of gel filtration buffer, and may need to be altered for specific projects by the addition of appropriate components. For example, for samples containing unpaired cysteine residues, 10 mM DTT should be added.

4. The affinity chromatography wash buffer can be modified to contain 45 mM imidazole if the protein binds strongly to the affinity media. This will improve the purity achieved.

5. The minimal crystallization buffer should represent the minimal buffer in which the sample will remain soluble. Increasing the salt concentration or adding glycerol to the buffer often helps with protein solubility, although this may reduce the total phase space that can be explored in crystallization experiments.

6. Vivaspin 500 concentrators provide a consistent means for concentrating samples down to 15 μl. The MW cutoff of the concentrator should be chosen to be less than 50% of the sample molecular weight.

7. Analytical crystallization. One mode of using TOPAZ chips is as an analytical tool to assess the crystallizability of a sample before committing the resources to a full-scale screening experiment using drop-based crystallization. In this approach, TOPAZ chips are used as a positive filter for the experiment: any samples that yield greater than four crystal hits from 192 reagents will be selected for larger-scale screening. Any samples that do not yield crystals will be lowered in priority, and pursued in a full-scale crystallization effort only if the other biophysical data still strongly indicate likelihood to crystallize or if the sample is of such significant interest that it still merits further pursuit.

8. Rescue strategy. TOPAZ X.96 screening chips employ free-interface diffusion to mix samples and reagents, which samples a broader area of sample and reagent space compared to other methods. FID can lead to hits where other methods have failed, so TOPAZ can be used as a fallback or a final alternative when other methods have been exhausted. The structure of cytochrome P450 2D6 *(14)* provides a good example of crystals that could only be obtained using TOPAZ screening chips.

9. The sample is often highly concentrated right off of the nickel affinity column and may not need further concentration. The high salt buffer needs to be exchanged for minimal crystallization buffer, however. Pay careful attention that the sample does not precipitate during buffer exchange or dialysis.

10. It is not known how purity impacts success rate, although it is often true that higher purity can lead to better-quality crystals.

11. Protein purification and optimization of purification protocols can be an involved subject. Presented here is a basic procedure for His-tagged proteins. A variant of the protocol can be used with other types of affinity tags.

12. For proteins expressed without affinity tags, weak ion exchange or a combination of ion exchange and size exclusion chromatography can be substituted for affinity chromatography. Purification of proteins from

natural sources or proteins that express in low abundance is beyond the scope of this chapter.

13. There are a number of other products on the market for parallel affinity purification.

14. When using the nickel-NTA spin columns the cleared lysate is concentrated 50× prior to loading to maximize protein yield. If the protein of interest expresses very well the lysate does not need to be concentrated as much. The goal is to recover approximately 100 μg of protein for crystallization screening.

15. If the sample shows signs of precipitation while concentrating, stop the concentration step and reduce the concentration by one third before setting up crystallization experiments. If more than one half of the wells in the chip precipitate within 24 hours of setting up the chip, reduce the starting protein concentration by one third.

16. In principle there is a wider usable range of concentration with TOPAZ FID compared with other methods. The concentration of protein varies in the course of a FID crystallization experiment in a different way to that in a traditional crystallization experiment. In a vapor diffusion experiment, the protein concentration is reduced suddenly (typically by half) during the initial drop mixing, and then returns to initial concentration gradually over the course of the experiment as the drop equilibrates with the reservoir. In a FID experiment carried out in a TOPAZ chip, the protein concentration drops slightly during the initial stage of the experiment, as a small amount of protein diffuses toward the reagent reservoir at a rate dependent on the molecular weight of the protein. After the interface valve has been closed, the mixture of protein and reagent gradually increases as the chip dehydrates.

17. One alternative approach to determining a protein concentration that is suitability for crystallization is prescreening. Combine 0.2 μL of concentrated protein to an equal volume of each of the nine prescreen solutions described in the Materials section. Monitor the resulting drops with a light microscope. If all of the drops precipitate rapidly the protein concentration is too high and the protein concentration should be reduced by one third with minimal crystallization buffer. If none of the drops precipitate the protein concentration is too low and should be concentrated 2× as described in the Methods section.

18. The nature of the FID crystallization experiment means that unbound, low molecular weight ligand that is present in the protein–ligand mixture may diffuse away from the higher molecular weight protein sample during the FID part of the experiment. For example, if the initial concentration of the ligand in the protein sample is 1 mM, given the protein-to-reagent ratio of 1:3.4 in a X.96 chip, the equilibrium concentration will be approximately 225 μM. The effect of this reduction in concentration depends on the strength of the interaction between the sample and ligand. However, it is recommended that ligands are introduced at 3–5× according to our protocol sample concentration where possible. If this concentration is not achievable, then it is recommended that the sample is purified and concentrated in the presence of the ligand.

19. As with all macromolecular crystallization experiments, sample homogeneity is often critical for successful crystallization. There are many methods for assessing sample homogeneity, most of which rely on bulk measurements

of purity. Depending on the experimental approach chosen different levels of assessment of the sample purity and homogeneity may be required.

20. It has been suggested by some that crystallizability can be assessed from less pure samples (<90% pure). A viable strategy for purification and crystallization may be rapid purification by affinity purification only followed immediately by crystallization screening. Given promising initial lead crystallization further purification may lead to better quality or larger diffraction quality crystals. This is particularly compelling when pursuing multiple constructs in parallel.

21. Many other methods are also available for assessing sample homogeneity. These include analytical ion-exchange chromatography, mass spectrometry, dynamic light scattering, isoelectric focusing gel electrophoresis, HSQC NMR, and proteolytic susceptibility. These are useful measures of sample homogeneity, but none provides a direct assessment of the propensity of a sample to crystallize. Ultimately, the best measure of the crystallizability of a sample is its behavior in crystallization trials.

22. The sample should be assayed for amenability to flash freezing. Flash freeze one aliquot and thaw after freezing. A number of samples will precipitate immediately after thawing and are not amenable to storage by flash freezing.

23. An alternative approach is to dispense sample directly into a liquid nitrogen bath, retrieve the frozen drops and place the frozen drops into a cryo-vial.

24. The default random template generates novel random screens with a modest preference for reagents empirically discovered to be most likely to give rise to protein crystals (unpublished data). This is appropriate for most samples for which little or no crystallization information is known.

25. The default template is commonly modified to take into account prior information such as pI, solubility limits as might be observed during the prescreen procedure, or reagent sensitivity, e.g., instability in alcohol.

26. There may be cross-talk between experiments with dramatically different solute concentration in close proximity in the crystallization chip. Ideally, the output of the design engine is sorted to match solute concentrations of proximal experiments.

27. The design of templates for optimization can involve considerable sophistication. If only a few marginal initial leads are discovered from initial screening, then the optimization template should not be restricted to just the reagents and concentration range from initial lead conditions, but rather should be slightly biased toward these conditions. If strong leads are obtained from initial screening or optimization, then scale up experiments should be pursued.

28. It is often useful to continue with *de novo* random screening for crystallization conditions even after initial lead conditions are discovered. Further screening can help to define the crystallization habits of the protein of interest much better and often leads to the discovery of multiple crystal forms.

29. The FID Control scripts cause the interface valves in the chip to be opened for a defined period of time and then closed again for the remainder of the experiment. Closing the interface valves prevents the sample from diffusing out of the sample chamber and keeps the effective sample concentration

higher. The equilibration times used in the FID Control scripts were chosen on the basis of the time taken for PEG 3350 to reach 90% of the equilibrium concentration. The FID Control RT script initiates diffusion and then terminates FID after 60 minutes. The FID Control 4C script initiates diffusion and then terminates FID after 100 minutes.

30. An often-used rule of thumb for drop-based experiments is that if less than one third of experiments produce precipitation, the concentration of the sample should be increased. However, in microfluidic crystallization experiments run in TOPAZ chips, the same rule does not appear to apply. This is due in part to the diffusion mediated mixing of sample and reagent, which is a gradual process and results in less dramatic crashing of samples from solution. It is also due to the different appearance of precipitate in the low-profile protein chambers in each experiment in the chip. Heavy precipitate is still observed as such in the chip, but conditions that would normally result in light or flocculent precipitate in drops may only lead to one or two precipitate particles, or insufficient material to scatter light effectively from the experiment.

31. Both vapor diffusion and microbatch methods have been shown to be effective in scaling up hits from TOPAZ chips. Microbatch using semipermeable oil is a particularly successful approach *(7)*.

Acknowledgments

This work was carried out at the Lawrence Livermore National Lab and was performed under the auspices of the U.S. Department of Energy by University of California, Lawrence Livermore National Laboratory under contract W-7405-Eng-48. The authors thank Kevin Farrell and Yong Yi for comments on the manuscript. The authors also thank the TB structural genomics consortium for support for some of the work described.

References

1. Hansen, C. L., Skordalakes, E., Berger, J. M., and Quake, S. R. (2002) A robust and scalable microfluidic metering method that allows protein crystal growth by free interface diffusion. *Proc. Natl. Acad. Sci. USA* **99**, 16531–16536.

2. Zheng, B., Roach, L. S., Rustem, F., and Ismagilov, R. F. (2003) Screening of protein crystallization conditions on a microfluidic chip using nanoliter-size droplets. *J. Am. Chem. Soc.* **125**, 11170–11171.

3. Yamada, M., Sasaki, C., Isomura, T., and Seki, M. (2003) Microfluidic reactor array for high-throughput screening of protein crystallization conditions. 7th International Conference on Miniaturized Chemical and Biochemical Analysts Systems October 5–9, 2003, Squaw Valley, CA.

4. Chao, W.-C., Collins, J., Bachman, M., Lia, G. P., and Lee, A. P. (2004) Droplet arrays in microfluidic channels for combinatorial screening assays. Solid-State Sensor, Actuator and Microsystems Workshop Hilton Head Island, SC, June 6–10, 2004.

5. Segelke, B. W., Schafer, J., Coleman, M. A., Lekin, T. P., Toppani, D., Skowronek, K. J., Kantardjieff, K. A., and Rupp, B. (2004) Laboratory scale structural genomics. J. Struct. Funct. Genom. 5, 147–157.

6. D'Arcy, A., Sweeney, A. M., and Haber, A. (2004) Practical aspects of using the microbatch method in screening conditions for protein crystallization. *Methods* **34**, 323–328.

7. D'Arcy, A., MacSweeney, A., and Habera, A. (2004) Modified microbatch and seeding in protein crystallization experiments. *J. Synchrotron Radiat.* **11**, 24–26.

8. Luft, J. R., Collins, R. J., Fehrman, N. A., Lauricella, A. M., Veatch, C. K., and DeTitta, G. T. (2003) A deliberate approach to screening for initial crystallization conditions of biological macromolecules. *J. Struct. Biol.* **142**, 170–179.

9. Stevens, R. C. (2000) High-throughput protein crystallization. *Curr. Opin. Struct. Biol.* **10**, 558–563.

10. Hansen, C. L., Sommer, M. O. A., and Quake, S. R. (2004) Systematic investigation of protein phase behavior with a microfluidic formulator. *Proc. Natl. Acad. Sci. USA* **101**, 14431–14436.

11. Chen, D. L., Gerdts, C. J., and Ismagilov, R. F. (2005) Using microfluidics to observe the effect of mixing on nucleation of protein crystals. *J. Am. Chem. Soc.* **127**, 9672–9673.

12. Segelke, B. W. (2001) Efficiency analysis of screening protocols used in protein crystallization. *J. Crystal Growth* **232**, 553–562.

13. Derewenda, Z. S., and Vekilov, P. G. (2006) Entropy and surface engineering in protein crystallization. *Acta Crystallogr. D Biol. Crystallogr.* **62**, 116–124.

14. Rowland, P., Blaney, F. E., Smyth, M. G., Jones, J. J., Leydon, V. R., Oxbrow, A. K., Lewis, C. J., Tennant, M. G., Modi, S., Eggleston, D. S., Chenery, R. J., and Bridges, A. M. (2006) Crystal structure of human cytochrome P450 2D6. *J. Biol. Chem.* **281**, 7614–7622.

Chapter 26

Increasing Protein Crystallization Screening Success with Heterogeneous Nucleating Agents

Anil S. Thakur, Janet Newman, Jennifer L. Martin, and Bostjan Kobe

The crystallization step is considered to be a major bottleneck in the process of structure determination by x-ray crystallography. Successful crystallization requires both the formation of nuclei that are capable of supporting crystal growth, and a subsequent crystal growth. Nucleation can occur spontaneously in a supersaturated solution. However, in the commonly used sparse matrix crystallization screens, protein and precipitant concentrations are not extensively sampled, so that suitable supersaturation conditions for nucleation are often missed. This chapter describes a simple method for enhancing nucleation and subsequent crystallization by the addition of heterogeneous nucleating agents.

1. Introduction

Crystallization occurs in two steps: (1) nucleation and (2) growth of nuclei to macroscopic crystals. The term "nucleus" as used here is defined as the smallest solid-phase aggregate of molecules formed during precipitation and capable of spontaneous growth. When nucleation occurs spontaneously in a homogeneous solution, it is termed homogenous nucleation; when it occurs on the surface of solids, it is termed heterogeneous nucleation *(1)*.

In typical protein crystallization experiments, crystals grow from a supersaturated aqueous solution by homogenous nucleation. In homogenous nucleation, the nuclei are formed by protein aggregates of critical size, which are in a state of equilibrium with the mother liquor. When there is a high enough supersaturation for the nuclei to overcome the free energy barrier, they can grow into a crystal (Fig. 26.1). The nucleation event does not occur if the protein concentration does not reach the required supersaturation level, and in that case the crystallization drop remains clear *(2,3)*.

Heterogeneous nucleation involves a solid material, hereby referred to as a "heterogeneous nucleating agent," with nucleation-inducible properties. Nucleation occurs on the surface of this material (Fig. 26.2). The heterogeneous nucleating agent reduces the activation energy for nucleation by means of reducing the interfacial free energy between the surface and the nucleus, when the solid particles are immersed in solution *(4)*. A lower level of supersaturation

From: *Methods in Molecular Biology, Vol. 426: Structural Proteomics: High-throughput Methods*
Edited by: B. Kobe, M. Guss and T. Huber © Humana Press, Totowa, NJ

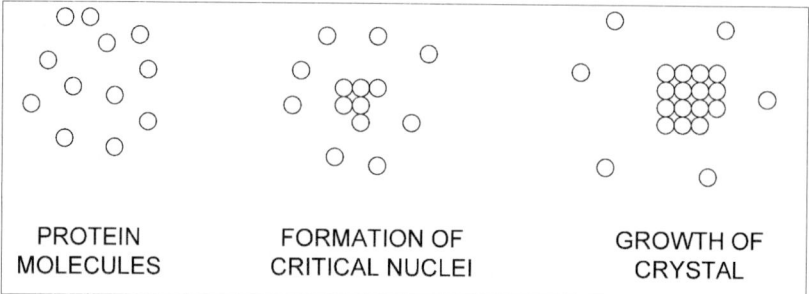

Fig. 26.1 A schematic depiction of homogenous nucleation in protein crystallization, which occurs in a typical crystallization experiment under supersaturated conditions. The protein molecules form nuclei that subsequently lead to ordered crystal growth. Adapted from *(20)*.

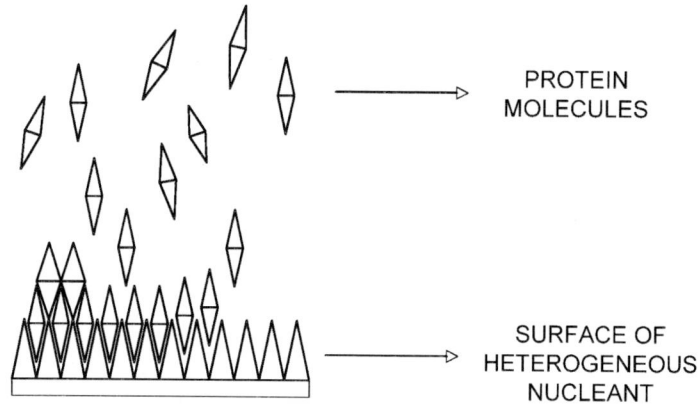

Fig. 26.2 A schematic depiction of heterogeneous nucleation. The protein molecules come in contact with the surface of a heterogeneous nucleate and form nuclei, leading to crystal growth.

is required under such circumstances for the nucleation step to occur, compared to homogenous nucleation. Crystals grown at a lower level of supersaturation may also have improved diffraction quality *(5)*. Epitaxial nucleation is a specific type of heterogeneous nucleation event that requires that there is a correlation between the lattice of the heterogeneous nucleating agent and the nascent protein crystal *(6)*.

The activation energy for nucleation can also be reduced by the process of seeding, in which crystal seeds are introduced into a slightly supersaturated solution. With this method, crystal seeds must be generated and then placed into the crystallization drop in the "metastable" zone (Fig. 26.3). This zone is preferred for seeding as it prevents the crystal seeds from dissolving, enables crystal growth, and it also prevents spontaneous nucleation (Fig. 26.3). By contrast, insoluble heterogeneous nucleating agents can be included in the crystallization experiments in any area of the phase diagram (including addition at the very beginning of a vapor diffusion crystallization experiment) *(7)*.

There are many examples of protein crystals that have grown as a consequence of fortuitous impurities in the drop, including dust particles and fibers.

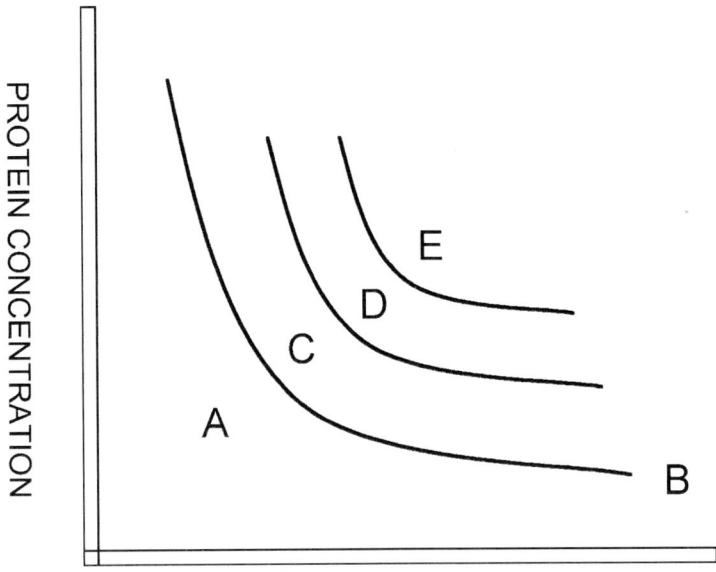

Fig. 26.3 A schematic diagram showing phase behavior in the crystallization drop as a function of protein and precipitant concentrations. Protein concentration is varied on the y axis and precipitant concentration is varied on the x axis. **A**. Undersaturated zone. **B**. Limit of solubility. **C**. Metastable zone. **D**. Nucleation zone. **E**. Precipitation zone. Adapted from *(7)*.

Furthermore, several studies have shown the benefits of including heterogeneous nucleation agents in protein crystallization using more systematic approaches. Examples include the growth of crystals on the surface of minerals *(8)*, lipids *(9)*, poly-L-lysine–modified glass substrate *(4)*, polyvinylidene *(10)*, porous silicone *(11–13)*, other porous substrates *(14)*, and polymeric films containing ionizable groups *(15)*. Hairs such as cat whiskers have also been used widely to introduce crystal seeds into the crystallization solution *(16,17)*.

In a typical high-throughput crystallization experiment, initial crystallization screening is carried out using sparse-matrix screens *(18)*. In such screens, protein and precipitant concentrations are not extensively sampled; therefore, suitable supersaturation conditions for nucleation are often missed. The authors reasoned that by including heterogeneous nucleating agents crystallization could be induced in a larger number of screen conditions. A study was performed of the effect of 10 different nucleating agents (fumed silica, cellulose fibers, titanium dioxide, horse hairs, sand particles, glass wool, dried seaweed, hydroxyapatite fines, Sephadex fibers, and combinations of these) on the crystallization of 10 different proteins in a sparse matrix screen *(19)*. It was observed that several nucleating agents induced crystallization in conditions that did not yield crystals in the absence of nucleating agents (Fig. 26.4). The best results were observed with dried seaweed as a single nucleating agent, and a combination of dried seaweed, horse hairs, cellulose fibers, and hydroxyapatite fines as a combined nucleating agent.

A

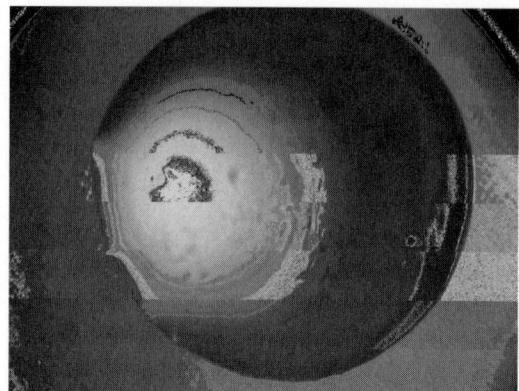

B:

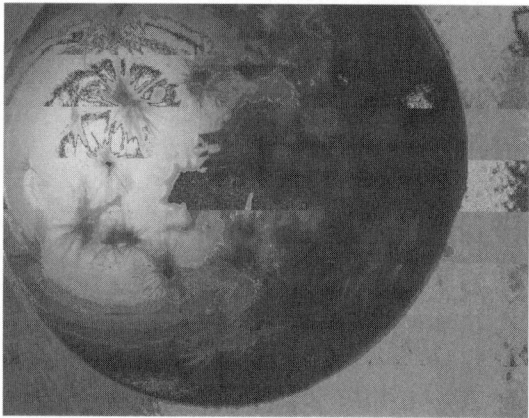

Fig. 26.4 Induction of crystallization by a heterogeneous nucleating agent. **A**. Control drop with no heterogeneous nucleating agent added. **B**. Drop with identical crystallization conditions as in (**A**), but with horse hairs added as a heterogeneous nucleating agent. A large crystal and needle crystal clusters were observed with the addition of the nucleating agent.

This chapter describes a simple method for including heterogeneous nucleating agents in crystallization experiments. This approach often results in more crystallization hits in a sparse matrix crystallization screen than in the absence of the heterogeneous nucleating agent. The protocol can be used with any common crystallization technique, such as hanging drop or sitting drop vapor diffusion or batch crystallization.

2. Materials

1. Heterogeneous nucleating agents:

- Dried seaweed (purchased from an Asian grocery store and dried before use; see Note 1).
- Horse hairs (obtained from a violin repair shop).
- Cellulose fibers (Sigma, St. Louis, MO).
- Hydroxyapatite fines (Sigma).

2. Liquid nitrogen.
3. 96-Condition sparse matrix crystallization screen (see Note 2).
4. 96-Well crystallization plates.
5. Mortar and pestle.
6. Sonicator.

3. Methods

3.1. Preparation of 96-Well Plates

1. Dispense the 96 conditions from the sparse matrix screen (for example from a deep-well-block) into the wells of a 96-well crystallization plate.
2. Cover the plate with sealing tape.
3. Store the plate at 4°C until further use.

3.2. Preparation of Heterogeneous Nucleating Agent

1. Weigh 1 g of the heterogeneous nucleating agent and place into the mortar (see Note 3).
2. Pour liquid nitrogen in the mortar to one fourth level of the size of the mortar.
3. Crush the heterogeneous nucleating agent with pestle.
4. Repeat steps 2 and 3 until a fine powder is obtained (see Note 4).

3.3. Addition of Heterogeneous Nucleating Agent to Protein Solution

1. Weigh out the heterogeneous nucleating agents in the ratio of 0.5 µg to 1 µl of the protein solution.
2. Add the heterogeneous nucleating agent to the protein solution and mix it gently by tapping the tube (see Note 5).
3. Store on ice.

3.4. Setting up of the 96-Well Crystallization Plate

1. Gently shake the tube to resuspend the heterogeneous nucleating agent into solution.
2. Use the protein sample to which the heterogeneous nucleating agent has been added to setup the 96-well crystallization plates with the sparse matrix screen conditions.

3.5. Monitoring Crystallization Plates

1. Monitor the plates for the next few weeks.
2. If any conditions with crystals are observed, proceed with optimization (see Note 6).

4. Notes

1. The seaweed was purchased from a local Asian grocery store in the form of fresh dry seaweed. The seaweed was then washed with Milli-Q water to remove any surface contaminants and dried in a drying oven at 60°C for 36 hours (until it became completely dry).
2. A wide range of sparse matrix screens can be used for initial screening in protein crystallization; for example, these are available from Hampton

Research, Emerald BioSystems, Jena Bioscience, Molecular Dimensions, and Qiagen.

3. Heterogeneous nucleates can also be combined. Prepare them by following Section 3.2. After obtaining a fine powder for each heterogeneous nucleating agent, add an equal amount (e.g. 0.5 mg) of each heterogeneous nucleate to an Eppendorf tube and mix well by vortexing before proceeding to the next step.

4. Prepare sufficient heterogeneous nucleating agent to make sure that the same batch can be used for further experiments.

5. In this protocol, an additional step can be added in preparing a homogenous mixture of heterogeneous nucleating agents in Milli-Q water or buffer, by sonicating the suspended heterogeneous nucleating agent solution. Do not sonicate the mixture after addition of heterogenous nucleate to the protein sample. The addition of the heterogeneous nucleating agent in solution form to the protein sample will dilute the protein concentration; this should be taken into account for crystallization.

6. In the optimization steps, once a heterogeneous nucleating agent has been identified, a series of dilutions of the heterogeneous nucleating agent can be performed if necessary, to reduce the amount of nucleation in the crystallization drop.

References

1. McPherson, A. (2004) Introduction to protein crystallization. *Methods* **34**, 254–265.
2. McPherson, A. (1999) *Crystallization of Biological Macromolecules*, Cold Spring Harbor Laboratory Press, Cold Spring Harbor, NY.
3. Asherie, N. (2004) Protein crystallization and phase diagrams. *Methods* **34**, 266–272.
4. Rong, L., Komatsu, H., Yoshizaki, I., Kadowaki, A., and Yoda, S. (2004) Protein crystallization by using porous glass substrate. *J. Synchrotron Radiat.* **11**, 27–29.
5. Yoshizaki, I., Sato, T., Igarashi, N., Natsuisaka, M., Tanaka, N., Komatsu, H., and Yoda, S. (2001) Systematic analysis of supersaturation and lysozyme crystal quality. *Acta Crystallogr. D Biol. Crystallogr.* **57**, 1621–1629.
6. McPherson, A., and Schlichta, P. (1988) Heterogeneous and epitaxial nucleation of protein crystals on mineral surfaces. *Science* **239**, 385–387.
7. Bergfors, T. (2003) Seeds to crystals. *J. Struct. Biol.* **142**, 66–76.
8. McPherson, A., and Shlichta, P. J. (1987) Facilitation of the growth of protein crystals by heterogeneous/epitaxial nucleation. *J. Cryst. Growth* **85**, 206–214.
9. Edwards, A. M., Darst, S. A., Hemming, S. A., Li, Y., and Kornberg, R. D. (1994) Epitaxial growth of protein crystals on lipid layers. *Nature Struct. Biol.* **1**, 195–197.
10. Punzi, J. S., Luft, J., and Cody, V. (1991) Protein crystal growth in the presence of poly(vinylidene diflouride) membrane. *J. Appl. Cryst.* **24**, 406–408.
11. Chayen, N. E., Shaw Stewart, P. D., Maeder, D. L., and Blow, D. M. (1990) An automated system for micro-batch protein crystallization and screeing. *J. Appl. Crytallogr.* **23**, 297–302.
12. Pechkova, E., and Nicolini, C. (2002) Protein nucleation and crystallization by homologous protein thin film template. *J. Cell Biochem.* **85**, 243–251.
13. Chayen, N. E., Saridakis, E., El-Bahar, R., and Nemirovsky, Y. (2001) Porous silicon: an effective nucleation-inducing material for protein crystallization. *J. Mol. Biol.* **312**, 591–595.
14. Hench, L. L. (1998) Biomaterials: a forecast for the future. *Biomaterials* **19**, 1419–1423.
15. Fermani, S., Falini, G., Minnucci, M., and Ripamonti, A. (2001) Protein crystallization on polymeric film surfaces *J. Cryst. Growth* **224**, 327–334.

16. Leung, C. J., Nall, B. T., and Brayer, G. D. (1989) Crystallization of yeast iso-2-cytochrome c using a novel hair seeding technique. *J. Mol. Biol.* **206**, 783–785.

17. Stura, E. A., and Wilson, I. A. (1992), Seeding techniques in (Ducruix, A., and Giegé, R., eds.), *Crystallization of Nucleic Acids and Proteins: A Practical Approach*, pp. 99–126, Oxford University Press, Oxford, UK.

18. Jancarik, J., and Kim, S.-H. (1991) Sparse matrix sampling: a screening method for crystallization of proteins. *J. Appl. Cryst.* **24**, 409–411.

19. Thakur, A. S., Robin, G., Guncar, G., Saunders, N. F., Newman, J., Martin, J. L., and Kobe, B. Improved success of protein crystallization sparse matrix screening with heterogeneous nucleating agents. PLoS ONE.

20. Chernov, A. A. (2003) Protein crystals and their growth. *J. Struct. Biol.* **142**, 3–21.

Chapter 27

High Throughput pH Optimization of Protein Crystallization

Ran Meged, Orly Dym, and Joel L. Sussman

Most high throughput structural proteomics centers use the sitting-drop method to obtain diffracting crystals for three-dimensional (3D) structure determination of biological macromolecules by x-ray crystallography. Although several robotic systems are available for dispensing the initial sitting-drop screening conditions, generally they are not used for optimization of crystallization conditions. This chapter describes a protocol for such automated systems, which permits easy construction of pH optimization grids using any desired fixed buffer set with varying ionic strengths directly dispensed into the crystallization plate.

1. Introduction

Despite technical and methodological advances, obtaining diffracting crystals remains a major obstacle to the 3D structure determination of biological macromolecules by x-ray crystallography *(1)*. Several methods are commonly used for crystallizing proteins, including vapor diffusion, micro-batch under oil, free interface diffusion, and microdialysis. The most commonly used method is vapor diffusion, and the majority of high throughput (HTP) crystallizations centers use the sitting-drop variant *(2)*.

Various robotic systems are available for dispensing the initial set of sitting-drop screening conditions. However, the sitting-drop method poses serious challenges to these systems:

1. Two different solution volumes are used, that of the reservoir (50–500 μL), and that of the drop (0.05–2 μL) (Fig. 27.1). Most dispensing systems for crystallization robots are adapted to utilizing very small drop volumes, so as to minimize protein consumption. For higher volumes these robots have limited capabilities, which can be overcome by the use of a more general-purpose liquid handling robot specifically designed to handle such volumes.
2. In many cases the chemicals used for crystallization, primarily the precipitants, are rather viscous (Table 27.1), rendering construction of fine and accurate gradients difficult for automated systems in terms of efficiency and accuracy *(3)*.

From: *Methods in Molecular Biology, Vol. 426: Structural Proteomics: High-throughput Methods*
Edited by: B. Kobe, M. Guss and T. Huber © Humana Press, Totowa, NJ

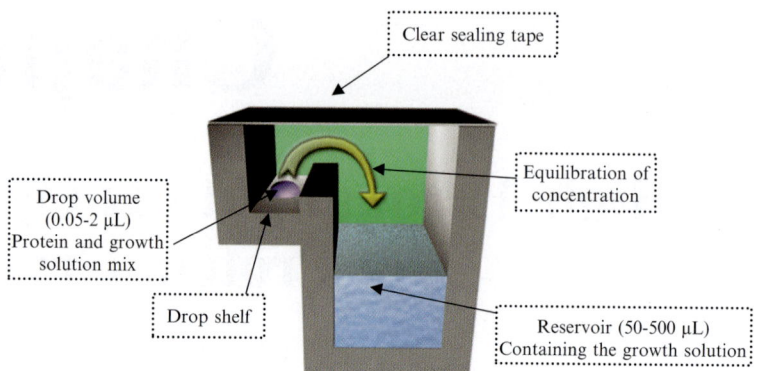

Fig. 27.1 Closed chamber for crystallization using the sitting-drop procedure: side view.

Table 27.1 Examples of liquid viscosities.

Chemical	Concentration	Viscosity, cP at 25°C
Water	100%	~1
Ethylene glycol	100%	~16
MPD (2-methyl-2,4-pentanediol)	100%	~36
PEG 200 (polyethylene glycol)	100%	~50
PEG-1000	50%	~20
PEG-4000	50%	~100

Once a protein crystal is grown from a given set of screening conditions, whether commercial or "in house," its diffraction pattern is measured, and an x-ray data set at cryogenic temperature is normally collected. For poorly diffracting protein crystals, with a resolution, e.g., worse than approximately 3Å, both crystal growth and freezing conditions need to be optimized *(4)* to improve diffraction quality. Furthermore, optimization can produce crystals of larger dimensions, and may also increase the probability of obtaining single crystals. Many variables that affect crystal quality can potentially be optimized, including:

1. The protein sequence (mutations, truncations, etc.), purity, concentration, or presence of inhibitors or cofactors
2. Optimization of the initial *"hit solution" (5)*, i.e., the crystallization conditions for the initial crystal *hits*, which include a precipitant, a buffer, and often a salt, an additive, and a detergent
3. Thermodynamic parameters, including, in particular, temperature *(2,3,6,7)*

The most convenient ways for optimizing the initial "hit solution" are either varying the pH, or varying the precipitant concentration *(8)* (which can lead to the construction of the phase diagram) (Fig. 27.2).

The authors have developed a pH optimization protocol designed for automated systems that is easy to use, thus making it very suitable for HTP optimization following initial screening.

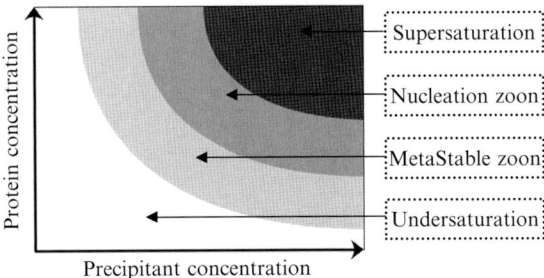

Fig. 27.2   Phase diagram plotting the solubility curve of a protein. The horizontal axis shows the parameter varied, usually the precipitant concentration. The vertical axis shows the protein concentration.

2. Materials

2.1. Apparatus

1. Liquid-handling robot with an eight-tip dispensing arm, suitable for use with either fixed or disposable tips.
2. Robotic stage capable of accepting at least four SBS plates (see Note 1).
3. Three polypropylene 96-well master blocks with a well volume of 2 ml.
4. One SBS sitting-drop crystallization plate.
5. Transparent sealing tape.
6. An eight-channel multi-pipettor (optional).

2.2. Stock Solutions

1. Suggested buffer solutions (Table 27.2) with two extreme ionic strengths.

3. Methods

3.1. Preparation of Buffer Stocks

1. Select specific buffers (see Table 27.2).
2. Prepare two solutions of minimal and maximal pH.
3. Set up the pH gradient according to the crystallization plates being used (see Table 27.2 for suggestions).
4. Generate the desired pH by mixing appropriate volumes of the two extreme pH buffers (see Note 2). A liquid-handling robot can construct such plates quite easily because the viscosities of all the buffers are close to that of water.

3.2. Optimization of the Hit Growth Solution

1. Prepare a solution for optimizing crystal growth, i.e., a *hit growth solution* (Fig. 27.3A,B).
2. Select buffers with pH values that are most compatible with the hit solution, and dispense them, using appropriate gradients closest to the hit value, as illustrated in Table 27.1 and Fig. 27.3C.

Table 27.2 Suggested buffer compositions, gradients, concentrations and configurations.

Deep Well 1:

Buffer	Min	Max	Conc (M)	pKa (at ~20°)
Na-acetate	4.5	5.5	1	4.76
	5.5	6.5	1	4.76
Imidazole-maleate	4.5	5.5	1	7.05
	5.5	6.5	1	7.05
	6.5	7.5	1	7.05
	7.5	8.5	1	7.05
Na-cacodylate	4.5	5.5	1	6.27
	5.5	6.5	1	6.27
	6.5	7.5	1	6.27
Na-citrate	4.5	5.5	1	4.76
	5.5	6.5	1	4.76
	6.5	7.5	1	4.76

	1	2	3	4	5	6	7	8	9	10	11	12
A	4.5	5.5	4.5	5.5	6.5	7.5	4.5	5.5	6.5	4.5	5.5	6.5
B	4.6	5.6	4.6	5.6	6.6	7.6	4.6	5.6	6.6	4.6	5.6	6.6
C	4.8	5.8	4.8	5.8	6.8	7.8	4.8	5.8	6.8	4.8	5.8	6.8
D	4.9	5.9	4.9	5.9	6.9	7.9	4.9	5.9	6.9	4.9	5.9	6.9
E	5.1	6.1	5.1	6.1	7.1	8.1	5.1	6.1	7.1	5.1	6.1	7.1
F	5.2	6.2	5.2	6.2	7.2	8.2	5.2	6.2	7.2	5.2	6.2	7.2
G	5.4	6.4	5.4	6.4	7.4	8.4	5.4	6.4	7.4	5.4	6.4	7.4
H	5.5	6.5	5.5	6.5	7.5	8.5	5.5	6.5	7.5	5.5	6.5	7.5
	Na-acetate		Imidazole-maleate				Na-cacodylate			Na-citrate		

	1	2	3	4	5	6	7	8	9	10	11	12
A	6.5	7.5	8.5	6.5	7.5	7.0	7.5	7.5	8.5	8.5	8.5	10.0
B	6.6	7.6	8.6	6.6	7.6	7.2	7.6	7.6	8.6	8.6	8.7	10.1
C	6.8	7.8	8.8	6.8	7.8	7.4	7.8	7.8	8.8	8.8	8.9	10.3
D	6.9	7.9	8.9	6.9	7.9	7.6	7.9	7.9	8.9	8.9	9.1	10.4
E	7.1	8.1	9.1	7.1	8.1	7.9	8.1	8.1	9.1	9.1	9.4	10.6
F	7.2	8.2	9.2	7.2	8.2	8.1	8.2	8.2	9.2	9.2	9.6	10.7
G	7.4	8.4	9.4	7.4	8.4	8.3	8.4	8.4	9.4	9.4	9.8	10.9
H	7.5	8.5	9.5	7.5	8.5	8.5	8.5	8.5	9.5	9.5	10.0	11.0
Buffer	BisTriPro			HEPES		TRIS-HCl	Na-Tricine	HEPPSO	CAPSO	AMPSO	CHES	CAPS

(continued)

Deep Well 2:

Buffer	Min	Max	Conc (M)	pKa (at ~20°)
BisTriPro	6.5	7.5	1	6.8
	7.5	8.5	1	6.8
	8.5	9.5	1	6.8
HEPES	6.5	7.5	1	7.55
	7.5	8.5	1	7.55
TRIS-HCl	7.0	8.5	1	8.1
Na-Tricine	7.5	8.5	1	8.16
HEPPSO	7.5	8.5	1	7.8
CAPSO	8.5	9.5	0.5	9.6
AMPSO	8.5	9.5	1	9
CHES	8.5	10.0	1	9.55
CAPS	10.0	11.0	1	10.56

	1	2	3	4	5	6	7	8	9	10
A	4.5	5.5	4.5	5.5	6.5	5.5	6.5	7.5	5.5	5.5
B	4.6	5.6	4.6	5.6	6.6	5.6	6.6	7.6	5.6	5.7
C	4.8	5.8	4.8	5.8	6.8	5.8	6.8	7.8	5.8	5.9
D	4.9	5.9	4.9	5.9	6.9	5.9	6.9	7.9	5.9	6.1
E	5.1	6.1	5.1	6.1	7.1	6.1	7.1	8.1	6.1	6.4
F	5.2	6.2	5.2	6.2	7.2	6.2	7.2	8.2	6.2	6.6
G	5.4	6.4	5.4	6.4	7.4	6.4	7.4	8.4	6.4	6.8
H	5.5	6.5	5.5	6.5	7.5	6.5	7.5	8.5	6.5	7.0
	TRIS-maleate		Na-succinate				Na-K-phosphate		MES	BisTRIS

Table 27.2 (continued)

Deep Well 3:

Buffer	Min	Max	Conc (M)	pKa (at ~20°)
TRIS-maleate	4.5	5.5	1	6.26
	5.5	6.5	1	6.26
Na-succinate	4.5	5.5	1	4.19
	5.5	6.5	1	4.19
	6.5	7.5	1	4.19
Na-K-phosphate	5.5	6.5	1	7.21
	6.5	7.5	1	7.21
	7.5	8.5	1	7.21
MES	5.5	6.5	1	6.16
BisTRIS	5.5	7.0	1	6.5

(A)

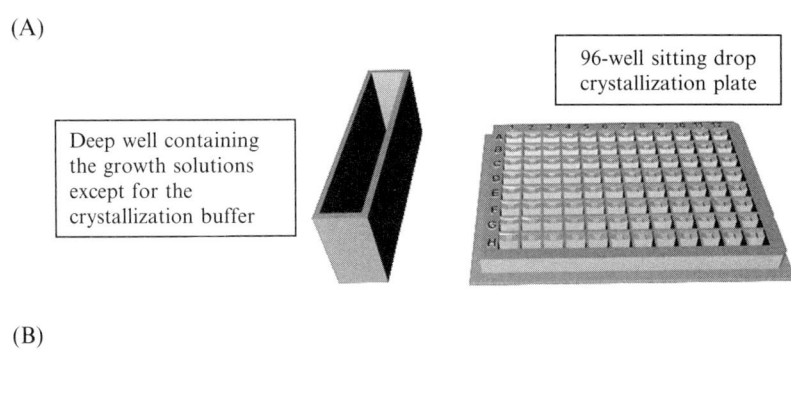

Deep well containing
the growth solutions
except for the
crystallization buffer

96-well sitting drop
crystallization plate

(B)

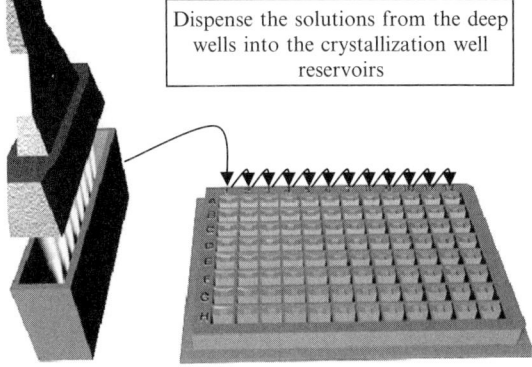

Dispense the solutions from the deep
wells into the crystallization well
reservoirs

(C)

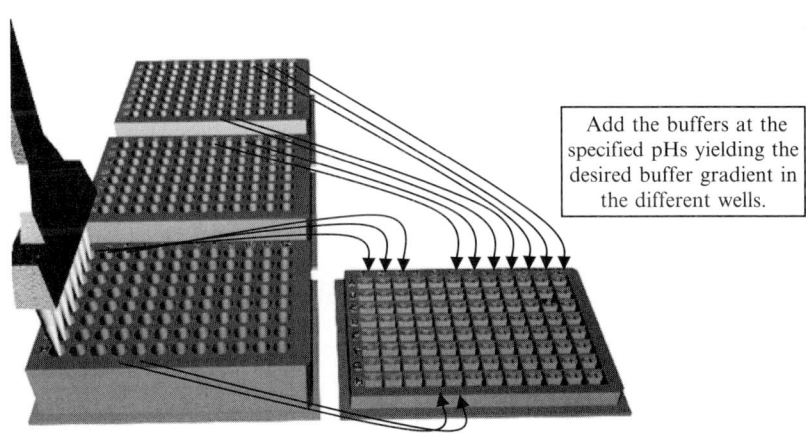

Add the buffers at the
specified pHs yielding the
desired buffer gradient in
the different wells.

Fig. 27.3 Schematic illustration of the use of a liquid-handling robot to prepare the optimized crystallization growth solutions. **A.** Growth solutions dispensed into the deep well with the exception of the crystallization buffer. **B.** Dispense the solutions from the deep wells (**A**) into the crystallization well reservoirs. **C.** Add the buffers at the specified pHs yielding the desired buffer gradient in the different wells.

3. Dispense the protein and the reservoir solution together onto the drop shelf (see Fig. 27.1), cover the plate with sealing tape, and store at the desired temperature.

4. Notes

1. The standards dimensions and tolerance for crystallization plates are compiled by the Microplate Standards Working Group of the Society for Biomolecular Sciences and can be found at www.sbsonline.org/msdc/approved.php
2. Ratio of the volumes of the two buffers

The ratio of the volumes of the two buffers needed to achieve a given pH value is calculated using the Henderson-Hasselbach equation:

$$pH = pka + \log 10([salt]/[acid])$$

which can be rewritten:

$$[salt][H] - Ka[acid] = 0$$

Acknowledgments

The authors acknowledge a grant from The Israel Ministry of Science, Culture and Sport to the ISPC, and the support of the Divadol Foundation, the Newman Foundation, and the European Commission Sixth Framework Research and Technological Development Programme SPINE2-COMPLEXES Project under contract No. 031220. The authors thank Israel Silman for his critical reading of the manuscript. J. L. Sussman is the Morton and Gladys Pickman Professor of Structural Biology.

References

1. Bränden, C., and Tooze, J. (1999) *Introduction to Protein Structure*, Garland Publishing, New York.
2. McPherson, A. (1990) Current approaches to macromolecular crystallization. *Eur. J. Biochem.* **189**, 1–23.
3. Ducruix, A., and Giegé, R. (Eds.) (1999) *Crystallization of Nucleic Acids and Proteins: A Practical Approach*, Oxford University Press, Oxford, UK.
4. Chayen, N. E., and Saridakis, E. (2002) Protein crystallization for genomics: towards high throughput optimization techniques. *Acta Cryst.* **D58**, 921–927.
5. Leulliot, N., Tresaugues, L., Bremang, M., Sorel, I., Ulryck, N., Graille, M., Aboulfath, I., Poupon, A., Liger, D., Quevillon-Cheruel, S., Janin, J., and van Tilbeurgh, H. (2005) High throughput crystal-optimization strategies in the South Paris Yeast Structural Genomics Project: one size fits all? *Acta Cryst.* **D61**, 664–670.
6. Giegé, R., and Mikol, V. (1989) Crystallogenesis of proteins. *Trends Biotechnol* **7**, 277–282.
7. Lorber, B., and Giegé, R. (1992) A versatile reactor for temperature controlled crystallization of biological macromolecules. *J. Cryst. Growth* **122**, 168–175.
8. Shaw Stewart, P. D., and Baldock, P. F. M. (1999) Practical experimental design techniques for automatic and manual protein crystallization. *J. Cryst. Growth* **196**, 665–673.

Chapter 28

Automated Structure Solution with the PHENIX Suite

Peter H. Zwart, Pavel V. Afonine, Ralf W. Grosse-Kunstleve, Li-Wei Hung, Thomas R. Ioerger, Airlie J. McCoy, Erik McKee, Nigel W. Moriarty, Randy J. Read, James C. Sacchettini, Nicholas K. Sauter, Laurent C. Storoni, Thomas C. Terwilliger, and Paul D. Adams

Significant time and effort are often required to solve and complete a macromolecular crystal structure. The development of automated computational methods for the analysis, solution, and completion of crystallographic structures has the potential to produce minimally biased models in a short time without the need for manual intervention. The PHENIX software suite is a highly automated system for macromolecular structure determination that can rapidly arrive at an initial partial model of a structure without significant human intervention, given moderate resolution, and good quality data. This achievement has been made possible by the development of new algorithms for structure determination, maximum-likelihood molecular replacement (PHASER), heavy-atom search (HySS), template- and pattern-based automated model-building (RESOLVE, TEXTAL), automated macromolecular refinement (phenix. refine), and iterative model-building, density modification and refinement that can operate at moderate resolution (RESOLVE, AutoBuild). These algorithms are based on a highly integrated and comprehensive set of crystallographic libraries that have been built and made available to the community. The algorithms are tightly linked and made easily accessible to users through the PHENIX Wizards and the PHENIX GUI.

1. Introduction

The worldwide efforts of many structural genomics projects, in particular the NIH Protein Structure Initiative have lead to many new technological advances in robotized cloning, sample expression and purification, screening of crystallization conditions *(1)*, data collection at synchrotron sources *(2)*, and structure solution *(3)*. These have made high throughput structure determination possible, and achievable for a number of the structures solved at structural genomics centers. More recently these technologies have started to be adopted by many structural biology research laboratories, where they are being applied to challenging systems such as large molecular complexes, and membrane proteins.

The demand for better software for crystallographic structure solution is increasing as it becomes possible for researchers to study more systems using

From: *Methods in Molecular Biology, Vol. 426: Structural Proteomics: High-throughput Methods*
Edited by: B. Kobe, M. Guss and T. Huber © Humana Press, Totowa, NJ

high throughput methods. This increased demand will need to be at least partially met by automated software for structure solution. Automation cannot rely on the availability of manual input from a trained crystallographer; the software must be able to make many complex decisions itself. Individual investigator research groups face a related problem as more biologists and biochemists make use of crystallography purely as a technique to better understand their biological system. Often there is insufficient time to obtain a very detailed expert crystallographic knowledge. A significant amount of this knowledge must therefore be built into the software. Furthermore, automated processes avoid possible subjective interpretation from manual interpretation of complex numerical data that can lead to delays or even inhibit reaching a high quality, final structure. Automated methods have the potential to produce minimally biased models in an efficient manner.

Current software packages such as SOLVE *(4)*, SHARP *(5)*, ACrS *(6)*, SHELX-C/D/E *(7)* CRANK *(8)*, Elves *(9)*, Auto-Rickshaw *(10)*, and BnP *(11)* are capable of automatic structure solution using MAD, SAD, or other sources of experimental phases. Molecular replacement can be carried out in an automated fashion by software including Amore *(12)*, PHASER *(13)*, EPMR *(14)*, and MOLREP *(15)*. Model-building can be carried out automatically by several algorithms including those in ARP/wARP *(16)*, RESOLVE *(17, 18)*, TEXTAL *(19)*, and MAID *(20)*. Manual model building programs such as O *(21)*, XtalView *(22)*, COOT *(23)*, and MAIN *(24)* have also incorporated an increasing number of tools that help automate complex tasks such as validation and model building. However, there still remain serious computational bottlenecks in structure determination. Truly automated structure solution is still limited to routine structures for which high-quality experimental data are available, typically at 2.5 Å or better.

Current shortcomings of automated algorithms are unlikely to be overcome by simply combining current software packages into automated "pipelines." Rather, new algorithms must be developed and combined with new approaches to decision making. The PHENIX software *(25,26)* has been developed with the needs for automation and complex decision making in mind.

2. Materials

PHENIX builds upon Python, the Boost.Python Library, and C++ to provide an environment for automation and scientific computing. Many of the fundamental crystallographic building blocks, such as data objects and tools for their manipulation are provided by the Computational Crystallography Toolbox (CCTBX) *(27)*. The computational tasks which perform complex crystallographic calculations are then built on top of this. Finally, there are a number of different user interfaces available in PHENIX. In order to facilitate automated operation there is the Project Data Storage (PDS) that is used to store and track the results of calculations.

2.1. User Interfaces

Different user interfaces are required depending on the needs of a diverse user community. There are currently three different user interfaces, each described in the following.

2.1.1. Command Line Interface

For a number of applications a command-line interface is most effective. This is particularly the case when rapid results are required, such as data quality assessment and twinning analysis, or substructure solution at the synchrotron beam line. Tools that facilitate the ease of use at the early stages of structure solution, such as data analyses (phenix.xtriage), substructure solution (phenix.hyss) and reflection file manipulations such as the generation of a test set, reindexing and merging of data (`phenix.reflection_file_converter`) are available via simple command line interfaces. Another major application that is controlled via the command line interface is phenix.refine.

To illustrate the command line interface, the command used to run the program that carries out a data quality and twinning analyses is: `phenix.xtriage data.sca (options)`

Further options can be given on the command line, or can be specified via a parameter file: `phenix.xtriage parameters.def`

A similar interface is used for macromolecular refinement: phenix.refine model.pdb data.mtz

Although SCALEPACK and MTZ formats are indicated in the above example, reflection file formats such as D*TREK, CNS/XPLOR or SHELX can be used, as the format is detected automatically.

Help for all command line applications can be obtained by use of the flag –help: `phenix.refine —help`

2.1.2. Tasks and Strategies

The PHENIX strategy interface provides a way to construct complex networks of tasks to perform a higher-level function (Fig. 28.1). For example, the steps required to go from initial data to a first electron density map in a SAD experiment can be broken down into well-defined tasks (available from the task window in the graphical user interface) that can be reused in other procedures. Instead of requiring the user to run these tasks in the correct order, they are connected together by the software developer, and thus can be run in an automated way. However, because the connection between tasks is dynamic, if problems occur they can be reconfigured or modified and new tasks introduced as necessary. This provides the flexibility of user input and control while still permitting complete automation when decision-making algorithms are incorporated into the environment. The tasks and their connection into strategies rely on the use of plain text task files written using the Python scripting language. This enables the computational algorithms to be used easily in a nongraphical environment. The PHENIX GUI permits strategies to be visualized and manipulated. These manipulations include loading a strategy distributed with PHENIX, customizing, and saving it for future recall.

Current tasks and strategies available include:

- Density modification; carries out a single run of RESOLVE.
- Substructure solution; runs phenix.hyss *(28)*.
- Molecular replacement; computes rotation and translation functions with PHASER.
- Model building; using TEXTAL or RESOLVE.

2.1.3. Wizards

The decision making in strategies is local, with decisions being made at the end of each task to determine the next path in the network.

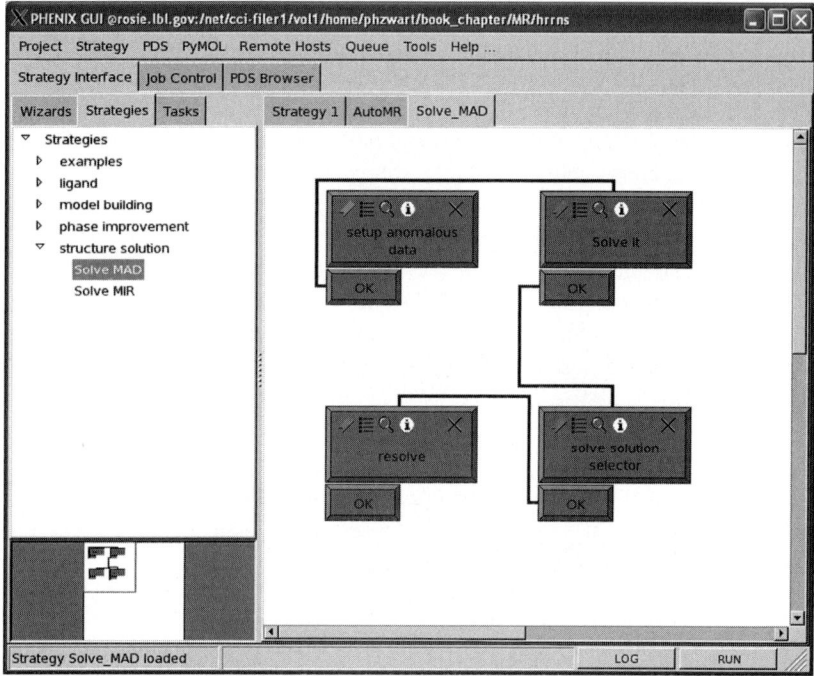

Fig. 28.1 Example of the PHENIX strategy interface, showing a substructure and phasing strategy for MAD data. The tasks are connected by lines, which indicate the flow of program execution dependant upon the outcome of each task (in this case the two possible outcomes are OK or Fail, the latter being implicit).

Crystallographers typically make decisions in a very similar way during structure solution; a program is run, the outputs are manually inspected, and a decision is made about the next step in the process. By contrast, a wizard provides a user interface that can make more global decisions, by considering all of the available information at each step in the process. Wizards are designed to lead the users through the process of setting up a desired task, making automatic decisions when possible, but prompting the user for additional information when necessary. The wizard interface uses the same graphical environment as the strategies, but consists of only a single input/output area (Fig. 28.2).

Currently available wizards perform the following tasks:

• Structure solution using experimental phasing approaches such as SAD/MAD and SIR.
• Structure solution via molecular replacement.
• Automated model building, structure completion, and refinement of structures
• Automated ligand building.

2.2. Common Crystallographic Computations

The following paragraphs are a brief description of a number of common tasks that can be performed within the PHENIX framework.

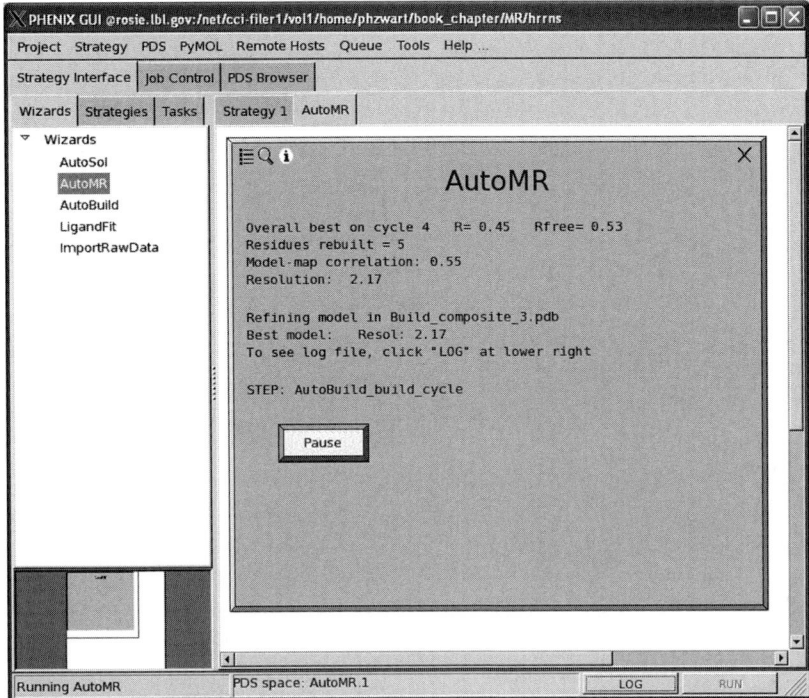

Fig. 28.2 Example of the PHENIX wizard interface. Shown is the AutoBuild cycle invoked after a potential molecular replacement solution has been found.

2.2.1. Automated Structure Solution Using Experimental Phasing Techniques

Structure solution via SAD, MAD, or SIR(AS) can be carried out with the AutoSol wizard. The AutoSol wizard performs heavy atom location, phasing, density modification, and initial model building in an automated manner.

The heavy atoms are located with substructure solution engine also used in phenix.hyss *(28)*, a dual space method similar to SHELXD *(7)*, and Shake and Bake *(29)*. Phasing is carried out with PHASER for SAD cases and with SOLVE for MAD and SIR(AS) cases. Subsequent density modification is carried out with RESOLVE. The hand of the substructure is determined automatically on the basis of the quality of the resulting electron density map. It is noteworthy that the whole process is not necessarily linear but that the wizard can decide to step back and (for instance) try another set of heavy atoms if appropriate.

In the resulting electron density map, a model is built (currently limited to proteins). Further model completion can be carried out via the AutoBuild wizard. The AutoBuild wizard iterates model building and density modification with refinement of the model in a scheme similar to other iterative model building methods, for example ARP/wARP *(16)*.

2.2.2. Automated Structure Solution via Molecular Replacement

Structure solution via molecular replacement is facilitated via the AutoMR wizard. The AutoMR wizard guides the user through setting up all necessary parameters to run a molecular replacement job with PHASER.

The molecular replacement carried out by PHASER uses a likelihood-based scoring function *(13,30)*, improving the sensitivity of the procedure and the ability to obtain reasonable solutions with search models that have a relatively low sequence similarity to the crystal structure being determined. Besides the use of likelihood-based scoring functions, structure solution is enhanced by detailed bookkeeping of all search possibilities when searching for more than a single copy in the asymmetric unit or the choice of space group is ambiguous.

When a suitable molecular replacement solution is found, the AutoBuild wizard is invoked and rebuilds the molecular replacement model given the sequence of the model under investigation.

2.2.3. Automated Model Building

Automated model building given a starting model or a set of reasonable phases can be carried out by the AutoBuild wizard. A typical AutoBuild job combines density modification, model building, macromolecular refinement, and solvent model updates (water picking) in an iterative manner.

Various modes of building a model are available. Depending on the availability of a molecular model, model building can be carried by locally rebuilding an existing model *(rebuild in place)* or building in the density without any information of an available model. The rebuilding in place model building is a powerful building scheme that is used by default for molecular replacement models that have a high sequence similarity to the sequence of the structure that is to be built.

A fundamental feature of the AutoBuild wizard is that it builds various models, all from slightly different starting points. The dependency of the outcome of the model building algorithm on initial starting conditions provides a straightforward mechanism to obtain a variety of plausible molecular models. It is not uncommon that certain sections of a map are built in one model, whereas not in another. Combining these models allows the AutoBuild wizard to converge faster to a more complete model than when using a single model building pass for a given set of phases.

Dedicated loop fitting algorithms are used to close gaps between chain segments. This feature, together with the water picking and side chain placement, typically results in highly complete models of high quality that need minimal manual intervention before they are ready for deposition.

2.2.4. Refinement

The refinement engine used in the AutoBuild and AutoSol wizards can also be run from the command line with the phenix.refine command. The phenix. refine program carries out likelihood based refinement and has the possibility to refine positional parameters, individual or grouped atomic displacement parameters, individual or grouped occupancies. The refinement of anisotropic displacement parameters (individual or via a TLS parameterization) *(31,32)* is also available. Positional parameters can be optimized using either traditional gradient-only based optimization methods or simulated annealing protocols *(33,34)*.

The command line interface allows the user to specify which part of the model should be refined in what manner. It is in principle possible to refine half of the molecule as a rigid group with grouped B values, whereas the other half of the molecule has a TLS parameterization. The flexibility of specifying the level of parameterization of the model is especially important for the refinement of low-resolution data or when starting with severely incomplete atomic models. Another advantage of this flexibility in refinement strategy is that a user can perform a complex refinement protocol that carries out simulated annealing, isotropic B refinement, and water picking in "one go."

Another main feature of phenix.refine is the way in which the relative weights for the geometric and ADP restraints with respect to the x-ray target are determined. Considerable effort has been put into devising a good set of defaults and weight determination schemes that results in a good choice of parameters for the data set under investigation. Of course, if the user chooses to, defaults can be overwritten.

Besides being able to handle refinement against x-ray data, phenix.refine can refine against neutron data or against x-ray and neutron data simultaneously.

2.2.5. Ligands

Automated fitting of ligands into the electron density is facilitated via the LigandFit wizard. The ligand building is performed by finding an initial fit for the largest rigid domain of the ligand and extending the remaining part of the ligand from this initial "seed." Besides being able to fit a known ligand into a difference map, the LigandFit wizard is capable of identifying ligands on the basis of the difference density only. In the latter scheme, density characteristics for ligands occurring frequently in the PDB (35,36) are used to provide the user with a range of plausible ligands.

Stereochemical dictionaries of ligands whose chemical description is not available in the supplied monomer library (37) for the use in restrained macromolecular refinement can be generated with the electronic ligand builder and optimization workbench (eLBOW). eLBOW generates a 3D geometry from a number of chemical input formats, including MOL2 or PDB files and SMILES strings (38). SMILES is a compact, chemically dense description of a molecule that contains all element and bonding information and optionally other stereo information, such as chirality. To generate a 3D geometry from an input format that contains no 3D geometry information, eLBOW uses a Z-Matrix formalism in conjunction with a table of bond lengths calculated using the Hartree-Fock method with a 6-31G(d,p) basis set to obtain a Cartesian coordinate set. The geometry is then optionally optimized using the semi-empirical quantum chemistry method AM1. The AM1 optimization provides chemically meaningful and accurate geometries for the class of molecule typically complexed with proteins. eLBOW outputs the optimized geometry and a standard CIF restraint file that can be read in by phenix.refine and can also be used for real space refinement during manual model building sessions in the program COOT (23). An interface is also available to use eLBOW within COOT.

2.2.6. Twinned Data

The presence of twinning can severely delay structure solution, model completion, and refinement if not explicitly taken into account. Detection of twinning on the basis of intensity statistics only is facilitated via the program phenix.xtriage. This command line–driven program analyzes an experimental data set and provides diagnostics that aid in the detection of other common idiosyncrasies such as the presence of pseudotranslational symmetry or certain data processing problems. Other sanity checks, such as a Wilson plot sanity check (39) and an algorithm that tries to detect the presence of ice rings from the merged data are performed as well.

If twin laws are present for the given unit cell and space group, a Britton plot *(40)* is computed, an H-test *(41,42)* is performed and a likelihood-based method is used to provide an estimate of the twin fraction. Twin laws are deduced from first principles for each data set, avoiding the danger of overlooking twin laws by incomplete lookup tables. If a model is available, more efficient twin detection tools are available. The RvsR statistic *(43)* is particularly useful in the detection of twinning in combination with pseudo-rotational symmetry. This statistic is computed by phenix.xtriage if calculated data are supplied together with the observed data. A more direct test for the presence of twinning is by refinement of the twin fraction given an atomic model. The command line utility phenix.twin_map_utils provides a straightforward way to refine a twin fraction given an atomic model and an x-ray data set and also produces "detwinned" $2F_o$-F_c and gradient maps. Restrained macromolecular structure refinement of a model against twinned data is possible in phenix.refine via an amplitude based, least squares target function.

The routines in Xtriage can also detect the presence of higher intensity symmetry than specified by the space group of the data. If higher intensity symmetry is detected, the user is advised to consider reprocessing the data.

3. Methods: Worked Examples

A few examples are given here to highlight many of the points mentioned in the previous sections. The results shown here have been generated with PHENIX version 1.26b-d2 (December 2006).

3.1. Structure Solution via S-SAD Phasing

An x-ray data set of insulin measured at a wavelength of 1.54 Å, was input to the AutoSol wizard for substructure solution and phasing (see Note 1). Seven anomalous sites are found and refined with PHASER. Phasing is carried out for both choices of the hand. The quality of the electron density is used to determine the correct hand. The solution that produces the best map is used for further density modification and model building.

Initial phasing of the sites produces a map with a mean figure of merit equal to 0.38. The experimental phases are of such a quality that large aromatic side chains such as tyrosine can be recognized in the map even before any density modification is applied (Fig. 28.3). The AutoSol wizard generated an almost complete model lacking only five N-terminal residues in weak density. Subsequent manual building of these residues in COOT and automated placement of waters and additional refinement by phenix.refine resulted in R-values (work/free) of 18%/20%.

3.2. Molecular Replacement

To illustrate a typical structure solution with molecular replacement, the structure of epsin *(44)* was solved using the structure of 1XGW as a search template. The sequence identity of the search template is 54% over the length of the alignment. The x-ray data extend to 1.84 Å. A Matthews analysis suggested 1 molecule per asymmetric unit with an approximate solvent content of 44%.

The rotation function reveals two significant peaks with Z-scores of 5.7 and 6.2, differing by approximately 5 degrees in orientation. A subsequent translation search yields two similar solutions with Z-scores of 10 and 13.

(A)

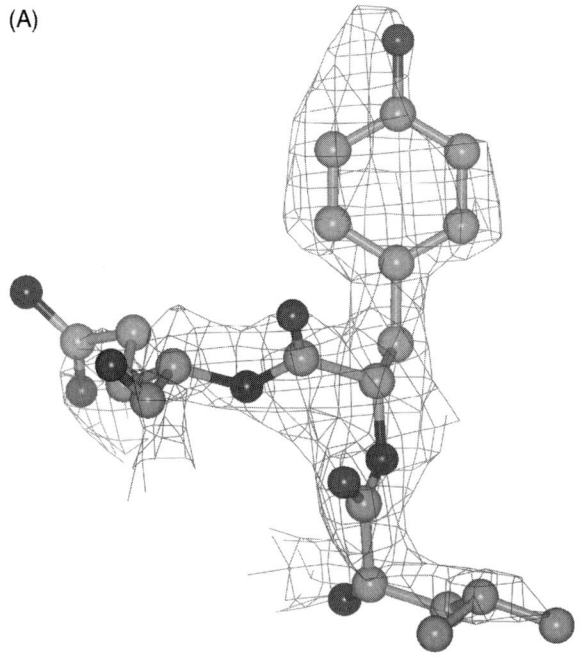

(B)

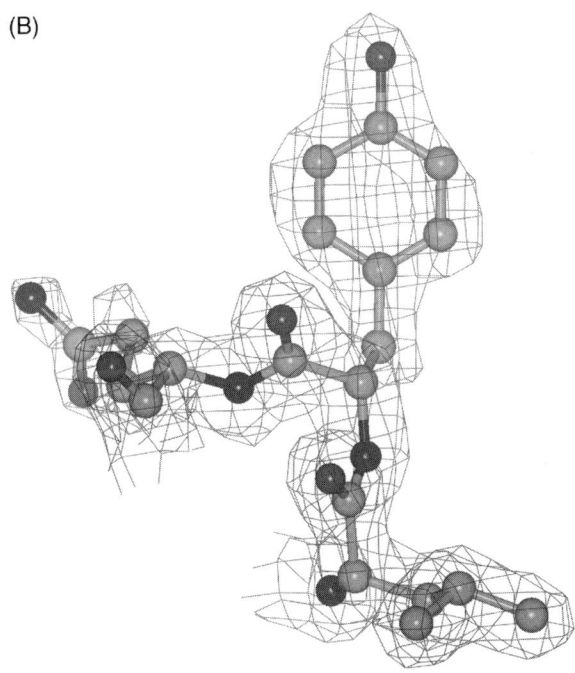

Fig. 28.3 A. Experimental S-SAD phases generate by PHASER during the execution of the AutoSol wizard for cubic insulin, before any density modification. The density is of a quality that it can be readily interpreted. The figure was prepared with CCP4MG *(51)*. **B**. The electron density map corresponding to the refined cubic insulin model after manual model completion and refinement with phenix.refine. The figure was prepared with CCP4MG *(51)*.

After rigid body refinement the two solutions are recognized automatically as equivalent, so that only a single unique solution remains. This solution can be used as a starting point in automated model building. Maps with virtually no model bias can be obtained (see Note 2) in a subsequent run of the AutoBuild wizard.

The space group of this particular data set is $P3_121$. The AutoMR wizard can be instructed to search in space groups with the same point groups as well. In this particular case, the translation function only gives a satisfactory result in space group $P3_121$. Possible solutions in $P3_221$ and in $P321$ have low Z-scores and multiple clashes with symmetry related copies.

The molecular replacement shown in this example is relatively straightforward. Molecular replacement attempts with low similarity search templates and multiple copies in the unit cell can be particularly challenging (see Note 3), but are often successful with the likelihood algorithms implemented in PHASER.

3.3. Ligand Building

The ligand building capabilities of the LigandFit wizard are illustrated by fitting NADH and cholic acid into the structure of SS_LADH *(45)*. The x-ray data extends to 1.54 Å, and the difference density is rather clear.

Atomic models for NADH and cholic acid were constructed from SMILES strings using elbow.builder. The command used to obtain the cholic acid model is:

```
phenix.elbow -smiles="CC(CCC(O)=O)C1CCC2C3C(O)CC4C
C(O)CCC4(C)C3CC(O)C12C" -opt
```

SMILES strings can be constructed with molecular editors such as JME (http://www.molinspiration.com/jme/) or can be obtained directly from MSDChem *(46)*.

The automated ligand building procedure uses a protein model (without ligands) and the x-ray data to compute a difference map in which the ligand is built. Two copies of NADH and two copies of cholic acid were built. The quality of the model is shown in Fig. 28.4.

3.4. Refinement

A typical refinement with phenix.refine is initiated with the following command: `phenix.refine data.mtz model.pdb`

The refinement program will try to determine which columns in the MTZ file to use for refinement and which column contains the test flags for cross validation purposes.

An example of the application of TLS refinement and its effects on the R-values is illustrated by refinement of the synaptotagmin structure *(47)*. The available x-ray data extended to 3.2 Å and is over 97% complete. Standard refinement (positional parameters and individual atomic displacement parameters; ADPs) results in R-values of 24.6% and 27.7% for the work and test set, respectively. At this resolution, ADPs are often refined in groups by applying constraints to the ADP values for all atoms within a residue. The refinement of ADPs in this manner resulted in R-values of 24.7% and 28.9% for the work and test set, respectively. The application of a TLS model to the atomic displacement parameters that models the displacement of rigid groups within a crystal reduced the R-values to 22.7% and 25.9% for the work and test set respectively. An ORTEP diagram of the anisotropic ADPs is shown in Fig. 28.5.

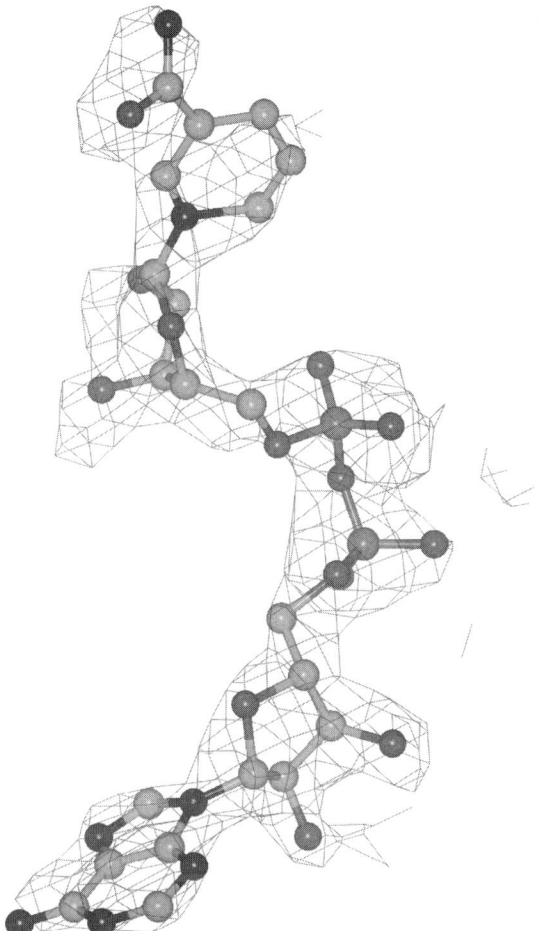

Fig. 28.4 The difference density and the model of NADH build in an automated manner by the LigandFit Wizard. The figure was prepared with CCP4MG *(51)*.

If only a TLS parameterization is used to model the ADPs (see Note 4), local variations in ADPs due to increased or decrease flexibility cannot be taken into account. A more complete ADP model includes both a TLS and individual ADP parameterization. The refinement of both the TLS parameters and the individual ADPs result in R-values of 20.7% and 24.4% for the work and test sets, respectively.

The command that is needed to perform this last refinement is straightforward:
phenix.refine scale.hkl synaptotagmin.pdb tls.param

Note that besides the experimental data and the atomic model and extra parameter file is specified. This parameter file has the following content:

```
refinement.strategy{
 strategy = *individual_sites *individual_adp *tls
 adp{
  tls="(chain A and resid :421)"
  tls="(chain A and resid 422:430)"
  tls="(chain A and resid 431:)"
 }
}
```

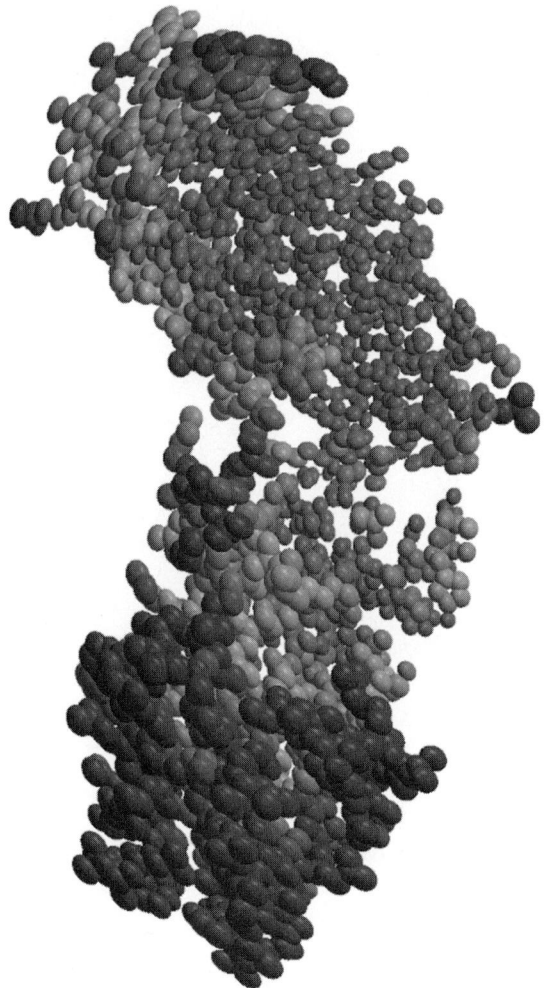

Fig. 28.5 An ORTEP-style diagram with anisotropic displacement parameters shown as three-dimensional ellipsoids, color coded by the magnitude of the total displacement for each atom (darker gray indicates a higher total ADP for an atom). The lower domain of the protein clearly shows significant rigid-body displacements, which are well modeled by the TLS formalism with a small number of parameters. This figure was made using RASTEP *(52)*.

The line containing the keyword *strategy* specifies that the positional parameters for individual sites should be refined, as well as a TLS model and individual ADPs. The TLS groups are defined by the *adp* scope. In this case, three TLS domains are specified within chain A.

Although most parameters for the refinement are set automatically, defaults (such as weights) can be set manually if desired (see Notes 5 and 6).

3.5. Twinning

The deposited x-ray dataset of PDB ID 1GH7 was analyzed by phenix. xtriage for the presence of twinning. A single twin law was found (-h-k, k,-l). Analysis of the intensity statistics by phenix.xtriage, indicates that the data is twinned:

```
Statistics independent of twin laws
```

- `<I^2>/<I>^2:1.795`
- `<F>^2/<F^2>:0.843`
- `<|E^2-1|>:0.658`
- `<|L|>, <L^2>: 0.396, 0.219`
- `Multivariate Z score L-test: 8.104`

```
The multivariate Z score is a quality measure of the
given spread in intensities. Good to reasonable data
is expected to have a Z score lower than 3.5. Large
values can indicate twinning, but small values do not
necessarily exclude it.
   The results of the L-test indicate that the inten-
sity statistics are significantly different then is
expected from good to reasonable, untwinned data.
   As there are twin laws possible given the crystal
symmetry, twinning could be the reason for the depar-
ture of the intensity statistics from normality. It
might be worthwhile carrying out refinement with a
twin specific target function.
```

An H-test *(41,42)* and Britton analyses *(40)* indicate a twin fraction of approximately 7%. Refinement of the twin fraction and bulk solvent and scaling parameters reveals that the data is 16% twinned, a fact not recognized during the original structure solution *(48)*.

4. Notes

1. *The effect of the data quality on the ability to solve the substructure.*

   The quality of the anomalous signal has a large impact on the ability to solve the substructure. The AutoSol wizard analyses the anomalous signal in a data set by computing either a correlation coefficient between the anomalous differences or a signal to noise ratio for SAD data. On the basis of these statistics, resolution limits for substructure solution are chosen.

   The quality of the anomalous data can be checked manually with phenix. reflection_statistics. It computes correlation coefficients between anomalous differences and a statistic known as the measurability *(49)* for SAD data sets. Correlation coefficients larger than 30% indicate significant anomalous signal in a MAD data set. For SAD data sets, measurabilities larger than 6% indicate the presence of significant anomalous signal.

   Although the AutoSol wizard analyses the signal to noise level of the anomalous data and makes appropriate default resolution cutoffs, it can be worthwhile running phenix.hyss with various resolution cutoffs if the AutoSol wizard fails to find the substructure with weak anomalous diffraction data.

2. *Bias removal in molecular replacement maps.*

The presence of bias in molecular replacement phases can make the interpretation of the electron density difficult or misleading. This bias can be removed by computing a *Full Omit* map in the AutoBuild wizard. The Full Omit procedure is reminiscent of the composite omit maps of the central nervous system *(50)* but provides a means to remove nearly all the bias, at the cost of computing time.

3. *Difficult molecular replacement problems.*

Not all structures can be solved by molecular replacement. Certain strategies can however be adopted to push its capabilities to the boundaries of what is possible. Careful editing of the input model by removing non conserved, flexible loops can make a big difference. Breaking a flexible model down into multiple rigid domains that can be used in a multicopy search can be a vital ingredient for a successful structure solution. Other suggestions are available from the program documentation.

4. *Interpreting the result of a TLS refinement.*

By default, the ADPs written by phenix.refine are the total ADPs rather than residual ADPs (which can be negative). This convention allows for easy viewing of the results of structure refinement in molecular graphics programs.

5. *Definition of NCS restraints in refinement.*

In order to increase the data to parameter ratio during refinement, multiple copies of the protein within the asymmetric unit can be restrained to have a similar conformation. These NCS restraints can be set up automatically by phenix.refine, or defined manually by the user.

6. *Weight optimization in restrained macromolecular refinement.*

The weight that determines the relative contribution of the x-ray target with respect to the restraint terms is determined automatically. The procedure works well in most cases, but a manual optimization of this weight can be used if necessary. Changing the weight manually can be performed via the following command: `phenix.refine data.mtz model.pdb wxc_scale=5`

Rerunning the refinement job with various values for the weight *wxc_scale* and a careful monitoring of the free R-value will give an indication of a suitable value for the weight. The same manual manipulation of the weighting for isotropic ADP restraints can be achieved with the *wxu_scale* parameter.

Acknowledgments

PHENIX can be downloaded from http://www.phenix.online.org, and is freely available to nonprofit researchers. The open source crystallographic library (the CCTBX) is available from http://cctbx.sourceforge.net/.

The authors gratefully acknowledge the financial support of NIH/NIGMS through grants 5P01GM063210, 5P50GM062412, 5R01GM071939, and the PHENIX industrial consortium. This work was supported in part by the U.S. Department of Energy under Contract No. DE-AC02-05CH11231.

References

1. Page, R., Grzechnik, S. K., Canaves, J. M., Spraggon, G., Kreusch, A., Kuhn, P., Stevens, R. C., and Lesley, S. A. (2003) Shotgun crystallization strategy for structural genomics: an optimized two-tiered crystallization screen against the Thermotoga maritima proteome. *Acta Cryst.* **D59**, 1028–1037.

2. Snell, G., Cork, C., Nordmeyer, R., Cornell, E., Meigs, G., Yegian, D., Jaklevic, J., Jin, J., Stevens, R. C., and Earnest, T. (2004) Automated sample mounting and alignment system for biological crystallography at a synchrotron source. *Structure* **12**, 537–545.

3. Adams, P. D., Grosse-Kunstleve, R. W., and Brunger, A. T. (2003) Computational aspects of high throughput crystallographic macromolecular structure determination. *Methods Biochem. Anal.* **44**, 75–87.

4. Terwilliger, T. C., and Berendzen, J. (1999) Automated MAD and MIR structure solution. *Acta Cryst.* **D55**, 849–861.

5. de la Fortelle, E., and Bricogne, G. (1997) Maximum-likelihood heavy-atom parameter refinement for multiple isomorphous replacement and multiwavelength anomalous diffraction methods. *Meth. Enzymol.* **276**, 472–494.

6. Brunzelle, J. S., Shafaee, P., Yang, X., Weigand, S., Ren, Z., and Anderson, W. F. (2003) Automated crystallographic system for high throughput protein structure determination. *Acta Cryst.* **D59**, 1138–1144.

7. Schneider, T. R., and Sheldrick, G. M. (2002) Substructure solution with SHELXD. *Acta Cryst.* **D58**, 1772–1779.

8. Ness, S. R., de Graaff, R. A., Abrahams, J. P., and Pannu, N. S. (2004) CRANK: new methods for automated macromolecular crystal structure solution. *Structure* **12**, 1753–1761.

9. Holton, J., and Alber, T. (2004) Automated protein crystal structure determination using ELVES. *Proc. Natl. Acad. Sci. USA* **101**, 1537–1542.

10. Panjikar, S., Parthasarathy, V., Lamzin, V. S., Weiss, M. S., and Tucker, P. A. (2005) Auto-Rickshaw: an automated crystal structure determination platform as an efficient tool for the validation of an X-ray diffraction experiment. *Acta Cryst.* **D61**, 449–457.

11. http://www.hwi.buffalo.edu/BnP/

12. Navaza, J. (1994) AMoRe: an automated package for molecular replacement. *Acta Cryst.* **A50**, 157–163.

13. McCoy, A. J., Grosse-Kunstleve, R. W., Storoni, L. C., and Read, R. J. (2005) Likelihood-enhanced fast translation functions. *Acta Cryst.* **D61**, 458–464.

14. Kissinger, C. R., Gehlhaar, D. K., and Fogel, D. B. (1999) Rapid automated molecular replacement by evolutionary search. *Acta Cryst.* **D55**, 484–491.

15. Vagin, A., and Teplyakov, A. (2000) An approach to multi-copy search in molecular replacement. *Acta Cryst.* **A56**, 1622–1624.

16. Perrakis, A., Morris, R., and Lamzin, V. S. (1999) Automated protein model building combined with iterative structure refinement. *Nat. Struct. Biol.* **6**, 458–463.

17. Terwilliger, T. C. (2003) Automated main-chain model building by template matching and iterative fragment extension. *Acta Cryst.* **D59**, 38–44.

18. Terwilliger, T. C. (2003) Automated side-chain model building and sequence assignment by template matching. *Acta Cryst.* **D59**, 45–49.

19. Holton, T., Ioerger, T. R., Christopher, J. A., and Sacchettini, J. C. (2000) Determining protein structure from electron-density maps using pattern matching. *Acta Cryst.* **D56**, 722–734.

20. Levitt, D. G. (2001) A new software routine that automates the fitting of protein X-ray crystallographic electron-density maps. *Acta Cryst.* **D57**, 1013–1019.

21. Jones, T. A., Zou, J. Y., Cowan, S. W., and Kjeldgaard, M. (1991) Improved methods for building protein models in electron density maps and the location of errors in these models. *Acta Cryst.* **A47**, 110–119.

22. McRee, D. E. (1999) XtalView/Xfit—a versatile program for manipulating atomic coordinates and electron density. *J. Struct. Biol.* **125**, 156–165.

23. Emsley, P., and Cowtan, K. (2004) Coot: model-building tools for molecular graphics. *Acta Cryst.* **D60**, 2126–2132.

24. Turk, D. (1992) *Weiterentwicklung eines Programms fuer Molekuelgraphik und Elektrondichte-Manipulation und seine Anwendung auf verschiedene Protein-Strukturaufklaerungen.* Technical University of Munich, Munich.

25. Adams, P. D., Grosse-Kunstleve, R. W., Hung, L.-W., Ioerger, T. R., McCoy, A. J., Moriarty, N. W., Read, R. J., Sacchettini, J. C., Sauter, N. K., and Terwilliger, T. C. (2002) PHENIX: building new software for automated crystallographic structure determination. *Acta Cryst.* **D58**, 1948–1954.

26. Adams, P. D., Gopal, K., Grosse-Kunstleve, R. W., Hung, L. W., Ioerger, T. R., McCoy, A. J., Moriarty, N. W., Pai, R. K., Read, R. J., and Romo, T. D., et al. (2004) Recent developments in the PHENIX software for automated crystallographic structure determination. *J. Synchrotron Radiat.* **11**, 53–55.

27. Grosse-Kunstleve, R. W., Sauter, N. K., Moriarty, N. W., and Adams, P. D. (2002) The Computational Crystallography Toolbox: crystallographic algorithms in a reusable software framework. *J. Appl. Crystallogr.* **35**, 126–136.

28. Grosse-Kunstleve, R. W., and Adams, P. D. (2003) Substructure search procedures for macromolecular structures. *Acta Cryst.* **D59**, 1966–1973.

29. Weeks, C. M., and Miller, R. (1999) Optimizing Shake-and-Bake for proteins. *Acta Cryst.* **D55**, 492–500.

30. Read, R. (2001) Pushing the boundaries of molecular replacement with maximum likelihood. *Acta Cryst.* **D57**, 1373–1382.

31. Schomaker, V., and Trueblood, K. (1968) On rigid-body motion of molecules in crystals. *Acta Cryst.* **B24**, 63.

32. Winn, M. D., Isupov, M. N., and Murshudov, G. N. (2001) Use of TLS parameters to model anisotropic displacements in macromolecular refinement. *Acta Cryst.* **D57**, 122–133.

33. Brunger, A. T., Adams, P. D., and Rice, L. M. (1999) Annealing in crystallography: a powerful optimization tool. *Prog. Biophys. Mol. Biol.* **72**, 135–155.

34. Rice, L. M., and Brunger, A. T. (1994) Torsion angle dynamics: reduced variable conformational sampling enhances crystallographic structure refinement. *Proteins* **19**, 277–290.

35. Berman, H. M., Westbrook, J., Feng, Z., Gilliland, G., Bhat, T. N., Weissig, H., Shindyalov, I. N., and Bourne, P.E. (2000) The protein data bank. *Nucl. Acids Res.* **28**, 235–242.

36. Bernstein, F. C., Koetzle, T. F., Williams, G. J., Meyer, E. F., Jr., Brice, M. D., Rodgers, J. R., Kennard, O., Shimanouchi, T., and Tasumi, M. (1977) The Protein Data Bank: a computer-based archival file for macromolecular structures. *J. Mol. Biol.* **112**, 535–542.

37. Vagin, A. A., Steiner, R. A., Lebedev, A. A., Potterton, L., McNicholas, S., Long, F., and Murshudov, G. N. (2004) REFMAC5 dictionary: organization of prior chemical knowledge and guidelines for its use. *Acta Cryst.* **D57**, 2184–2195.

38. Weininger, D. (1988) SMILES 1. Introduction and encoding rules. *J. Chem. Inf. Comput. Sci.* **28**, 31.

39. Morris, R. J., Zwart, P. H., Cohen, S., Fernandez, F. J., Kakaris, M., Kirillova, O., Vonrhein, C., Perrakis, A., and Lamzin, V. S. (2004) Breaking good resolutions with ARP/wARP. *J. Synchrotr. Radiat.* **11**, 56–59.

40. Fisher, R. G., and Sweet, R. M. (1980) Treatment of diffraction data from crystals twinned by merohedry as intended. *Acta Cryst.* **A36**, 755–760.

41. Yeates, T. O. (1988) Simple statistics for intensity data from twinned specimens. *Acta Cryst.* **A44**, 142–144.

42. Yeates, T. O. (1997) Detecting and overcoming crystal twinning. *Meth. Enzymol.* **276**, 344–358.

43. Lebedev, A. A., Vagin, A. A., and Murshudov, G. N. (2006) Intensity statistics in twinned crystals with examples from the PDB. *Acta Cryst.* **D62**, 83–95.

44. Hyman, J., Chen, H., Di Fiore, P. P., De Camilli, P., and Brunger, A. T. (2000) Epsin 1 undergoes nucleocytosolic shuttling and its eps15 interactor NH(2)-terminal homology (ENTH) domain, structurally similar to Armadillo and HEAT repeats, interacts with the transcription factor promyelocytic leukemia Zn(2)+ finger protein (PLZF). *J. Cell Biol.* **149**, 537–546.

45. Adolph, H. W., Zwart, P., Meijers, R., Hubatsch, I., Kiefer, M., Lamzin, V., and Cedergren-Zeppezauer, E. (2000) Structural basis for substrate specificity differences of horse liver alcohol dehydrogenase isozymes. *Biochemistry* **39**, 12885–12897.

46. Golovin, A., Oldfield, T. J., Tate, J. G., Velankar, S., Barton, G. J., Boutselakis, H., Dimitropoulos, D., Fillon, J., Hussain, A., and Ionides, J. M., et al. (2004) E-MSD: an integrated data resource for bioinformatics. *Nucl. Acids Res.* **32**, D211–216.

47. Sutton, R. B., Ernst, J. A., and Brunger, A. T. (1999) Crystal structure of the cytosolic C2A-C2B domains of synaptotagmin III. Implications for Ca(+2)-independent snare complex interaction. *J. Cell Biol.* **147**, 589–598.

48. Carr, P. D., Gustin, S. E., Church, A. P., Murphy, J. M., Ford, S. C., Mann, D. A., Woltring, D. M., Walker, I., Ollis, D. L., and Young, I. G. (2001) Structure of the complete extracellular domain of the common beta subunit of the human GM-CSF, IL-3, and IL-5 receptors reveals a novel dimer configuration. *Cell* **104**, 291–300.

49. Zwart, P. (2005) Anomalous signal indicators in protein crystallography. *Acta Cryst.* **D61**, 1437–1448.

50. Brunger, A. T., Adams, P. D., Clore, G. M., DeLano, W. L., Gros, P., Grosse-Kunstleve, R. W., Jiang, J. S., Kuszewski, J., Nilges, M., and Pannu, N. S., et al. (1998) Crystallography & NMR system: a new software suite for macromolecular structure determination. *Acta Cryst.* **D54**, 905–921.

51. Potterton, L., McNicholas, S., Krissinel, E., Gruber, J., Cowtan, K., Emsley, P., Murshudov, G. N., Cohen, S., Perrakis, A., and Noble, M. (2004) Developments in the CCP4 molecular-graphics project. *Acta Cryst.* **D60**, 2288–2294.

52. Merritt, E. A. (1999) Comparing anisotropic displacement parameters in protein structures. *Acta Cryst.* **D55**, 1997–2004.

Chapter 29

NMR Screening for Rapid Protein Characterization in Structural Proteomics

Justine M. Hill

In the age of structural proteomics when protein structures are targeted on a genome-wide scale, the identification of proteins that are amenable to analysis using x-ray crystallography or NMR spectroscopy is the key to high throughput structure determination. NMR screening is a beneficial part of a structural proteomics pipeline because of its ability to provide detailed biophysical information about the protein targets under investigation at an early stage of the structure determination process. This chapter describes efficient methods for the production of uniformly ^{15}N-labeled proteins for NMR screening using both conventional IPTG induction and autoinduction approaches in *E. coli*. Details of sample preparation for NMR and the acquisition of 1D ^{1}H NMR and 2D ^{1}H-^{15}N HSQC spectra to assess the structural characteristics and suitability of proteins for further structural studies are also provided.

1. Introduction

The efficient identification of proteins that are amenable to structure determination using x-ray crystallography or NMR spectroscopy is a key issue in structural proteomics and structural biology in general. NMR is increasingly used as a screening tool to identify folded proteins that are promising targets for three-dimensional structure elucidation *(1–9)*. Some structural proteomics projects primarily use 1D ^{1}H NMR spectra of unlabeled proteins that are used for initial crystallization trials *(1)*. However, 2D ^{1}H-^{15}N HSQC spectra of ^{15}N-labeled proteins are more routinely used *(2–9)* due to their improved resolution and higher information content. In the ^{1}H-^{15}N HSQC spectrum, one peak is expected for each backbone and side chain NH group in a protein, providing a diagnostic fingerprint. The dispersion pattern, intensity, and number of observed cross peaks reports directly on the folded state and overall sample quality. As the number of peaks in the HSQC spectrum corresponds approximately to the number of residues in the protein under investigation, conformational heterogeneity can easily be detected by a surplus of peaks.

Samples for NMR screening are most commonly expressed in *E. coli* grown in modified M9 medium containing ^{15}NH$_4$Cl as the sole nitrogen source *(3–6)* or using cell-free expression systems *(7,10,11)*. Recent advances such as the development of autoinduction medium *(12)* have facilitated high throughput protein production in *E. coli*. Autoinduction medium takes advantage of the *lac* operon in *E. coli* that inhibits lactose induction in the presence of glucose. These media are an optimized blend of carbon sources (glucose, glycerol, and lactose) that promote culture growth to high cell densities, followed by lactose-induced protein expression *(12)*. Autoinduction is more convenient than IPTG induction because the expression strain is simply inoculated into autoinducing medium and grown to saturation without the need to monitor culture growth and add inducer at the appropriate time. Furthermore, the cell mass and target protein yield are often increased severalfold compared with conventional protocols using IPTG induction *(12)*. This methodology has been widely adopted in structural proteomics initiatives and is readily modified for the production of isotopically labeled proteins for NMR analysis *(13)*.

This chapter focuses on the application of NMR as a diagnostic tool for the rapid identification of well-folded proteins for structure determination. Methods for the efficient production of uniformly ^{15}N-labeled proteins in *E. coli* by both conventional IPTG induction and autoinduction are described. Details of the preparation of samples for NMR analysis, and their screening using 1D ^{1}H NMR and 2D ^{1}H-^{15}N HSQC spectra are also provided. These simple NMR experiments provide a fast and reliable assessment of the amenability of the protein to further structural analysis.

2. Materials

2.1. M9 Medium for Uniform ^{15}N-Labeling

1. M9 salts (5×): Dissolve 34 g anhydrous Na_2HPO_4, 15 g anhydrous KH_2PO_4 and 2.5 g NaCl in water to a final volume of 1 L (see Note 1). Sterilize by autoclaving and store at room temperature for up to 6 months.
2. Glucose (50% w/v): Dissolve 125 g of D-glucose in water to a final volume of 250 ml. Glucose is slow to dissolve and requires stirring for several hours at room temperature. Sterilize the solution by passing it through a 0.2 μm filter and store at room temperature.
3. $MgSO_4$ (1 M): Dissolve 24.7 g of $MgSO_4 \cdot 7H_2O$ in water to a final volume of 100 ml. Sterilize by autoclaving and store at room temperature.
4. Ammonium chloride (^{15}N, 99%) can be purchased from a number of suppliers, including Cambridge Isotope Labs, Sigma-Aldrich, Silantes, and Spectra Stable Isotopes.
5. Vitamin solution (5 mg/ml): Dissolve 0.5 g thiamine hydrochloride and 0.5 g nicotinic acid in 100 ml water. Filter sterilize and store at 4°C.
6. Trace element solution (1,000×): 40.8 mM $CaCl_2 \cdot 2H_2O$, 21.6 mM $FeSO_4 \cdot 7H_2O$, 6.1 mM $MnCl_2 \cdot 4H_2O$, 3.4 mM $CoCl_2 \cdot 6H_2O$, 2.4 mM $ZnSO_4 \cdot 7H_2O$, 1.8 mM $CuCl_2 \cdot 2H_2O$, 0.3 mM boric acid, and 0.2 mM $(NH_4)_6Mo_7O_{24} \cdot 4H_2O$ *(14)* (see Note 2). To 100 ml of water in a 500-ml Erlenmeyer flask, add the salts in the order listed in Table 29.1 and cover with a perforated parafilm top. Fully dissolve each salt by stirring before adding the next. Filter sterilize and store in the dark at room temperature.
7. Antibiotics for plasmid selection (1,000×): Ampicillin (100 mg/ml in water) or kanamycin (35 mg/ml in water) is selective for most of the T7-based expression vectors that are commonly used. Filter sterilize and store at 4°C.
8. IPTG (isopropyl-β-D-thiogalactopyranoside)

Table 29.1 Recipe for 1,000× trace element solution

	per 100 ml
$CaCl_2 \cdot 2H_2O$	0.60 g
$FeSO_4 \cdot 7H_2O$	0.60 g
$MnCl_2 \cdot 4H_2O$	0.12 g
$CoCl_2 \cdot 6H_2O$	0.08 g
$ZnSO_4 \cdot 7H_2O$	0.07 g
$CuCl_2 \cdot 2H_2O$	0.03 g
Boric acid	2 mg
$(NH_4)_6Mo_7O_{24} \cdot 4H_2O$	25 mg
EDTA	0.50 g

2.2. Autoinduction Medium for Uniform ^{15}N-Labeling

The stock solutions required for autoinduction medium were originally defined by Studier *(12)*. Here the same naming convention is used, where "N" identifies a variant of the media used for ^{15}N-labeling.

1. 20× N: 1 M Na_2HPO_4, 1 M KH_2PO_4 and 0.1 M Na_2SO_4. Dissolve 14.2 g Na_2SO_4, 136 g KH_2PO_4 and 142 g Na_2HPO_4 in water to a final volume of 1 L. Add in sequence to a beaker and stir until all dissolved. Sterilize by autoclaving and store at room temperature.
2. 50× 5052 (5052 = 0.5% glycerol, 0.05% glucose, 0.2% α-lactose): 250 g glycerol (weigh into a beaker), 730 ml water, 25 g D-glucose, and 100 g α-lactose. Add in sequence in a beaker and stir until all dissolved. Lactose is slow to dissolve and may take two hours or more at room temperature. Sterilize by autoclaving and store at room temperature.
3. The other components of autoinduction medium including 1 M $MgSO_4$, $^{15}NH_4Cl$, vitamins, trace element solution (1,000×), and antibiotics (1,000×) are prepared as described in Section 2.1.

3. Methods

3.1. Uniform ^{15}N-Labeling in M9 Medium

This protocol describes the preparation of uniformly ^{15}N-labeled protein in M9 minimal medium on a 1 L scale (see Note 3).

1. Inoculate 5 ml of LB medium containing the appropriate antibiotic with a single colony from a freshly transformed plate of *E. coli* BL21(DE3) cells containing plasmid for overexpression of the target protein.
2. Grow at 37°C overnight with shaking at approximately 200 rpm.
3. In a sterile 250-ml Erlenmeyer flask, prepare 50 ml of M9 minimal medium from sterile concentrated stock solutions as described in Table 29.2 (see Note 4).
4. To adapt the cells to minimal medium, transfer 20 μl of the overnight culture in rich growth medium to 2 ml of M9 minimal medium. Grow at 37°C for 6–8 hours until the culture is visibly turbid.

Table 29.2 Recipe for M9 minimal medium

	per 50 ml	**per 1 L**
Sterile water	40 ml	780 ml
5× M9 salts	10 ml	200 ml
Vitamins	300 μl	6 ml
Glucose (50% w/v)	600 μl	12 ml
Antibiotic (1,000×)	50 μl	1 ml
Trace elements (1,000×)	50 μl	1 ml
1 M MgSO$_4$	150 μl	3 ml
^{15}NH$_4$Cl	50 mg	1 g

5. Transfer the entire 2 ml minimal culture to the remaining 48 ml minimal medium and continue growth at 37°C overnight. This will provide a starter culture for large-scale expression the following day (see Note 5).

6. Prepare 1 L of M9 minimal medium as described in Table 29.2, and divide 500-ml aliquots into sterile 2-L Erlenmeyer flasks.

7. Add 25 ml of the starter culture to each flask and grow at 37°C until the optical density (OD)$_{600}$ is approximately 1.0. This will take 2–3 hours.

8. Induce protein expression by the addition of IPTG. The authors typically use a 1 mM final concentration of IPTG, with overexpression carried out at either 37°C for 5 hours or at 20°C overnight (see Note 6).

9. Harvest the cells by centrifugation at 5,000g, 4°C for 15 minutes.

10. Purify the target protein using the appropriate method.

11. Prepare a sample for NMR analysis as described in Section 3.3.

3.2. Uniform ^{15}N-Labeling in Autoinduction Medium

This protocol describes the preparation of uniformly ^{15}N-labeled protein in autoinduction medium on a 1 L scale (see Notes 7 and 8).

1. Inoculate 5 ml of LB medium containing the appropriate antibiotic with a single colony from a freshly transformed plate of *E. coli* BL21(DE3) cells containing plasmid for overexpression of the target protein (see Note 9).

2. Grow at 37°C overnight with shaking at approximately 200 rpm.

3. In a sterile 250-ml Erlenmeyer flask, prepare 50 ml of NG minimal medium from sterile concentrated stock solutions as described in Table 29.3 (see Note 10).

4. Transfer 20 μl of the overnight culture in rich growth medium to 2 ml of NG minimal medium. Grow at 37°C for 6–8 hours until the culture is visibly turbid.

5. Transfer the entire 2 ml minimal culture to the remaining 48 ml NG medium and continue growth at 37°C overnight. This provides a starter culture for large-scale expression the following day (see Note 5).

6. Prepare 1 L of N-5052 autoinduction medium from sterile concentrated stock solutions as described in Table 29.4, and divide 500-ml aliquots into sterile 2-L Erlenmeyer flasks.

Table 29.3 NG defined medium for growth to saturation

	per 50 ml
Sterile water	47 ml
Vitamins	300 μl
1 M MgSO$_4$	50 μl
Trace elements (1,000×)	50 μl
Glucose (50% w/v)	500 μl
20× N	2.5 ml
Antibiotic (1,000×)	50 μl
^{15}NH$_4$Cl	125 mg

Table 29.4 N-5052 defined medium for autoinduction

	per 1 L
Sterile water	922 ml
Vitamins	6 ml
1 M MgSO$_4$	1 ml
Trace elements (1,000×)	1 ml
50× 5052	20 ml
20× N	50 ml
Antibiotic (1,000×)	1 ml
^{15}NH$_4$Cl	2.5 g

7. Add 25 ml of the starter culture to each flask and grow in a refrigerated incubator at 30°C with shaking at 200 rpm for 24 hours (see Notes 11 and 12).
8. Harvest the cells by centrifugation at 5,000g, 4°C for 15 minutes.
9. Purify target protein using the appropriate method.
10. Prepare a sample for NMR analysis as described in Section 3.3.

3.3. NMR Sample Preparation

1. Dialyze the purified protein against a suitable buffer for NMR. The authors typically use 50 mM sodium phosphate or 10 mM sodium acetate with a pH of 5, 6, or 7 (depending on the pI of the target protein), 150 mM NaCl, 1 mM DTT, and 50 μM NaN$_3$ (see Note 13). If sufficient protein is available, it is advantageous to screen the protein in several buffer conditions.
2. Concentrate the dialyzed protein to 0.2–1.0 mM in a final volume of 500 μl using a centrifugal concentrator such as Amicon Ultra-4 (Millipore, Billerica, MA) with an appropriate molecular weight cutoff.
3. Add 50 μl of D$_2$O for deuterium lock of the NMR spectrometer and transfer the entire sample into a 5 mm NMR tube (535PP; Wilmad, Buena, NJ). If the amount of protein is limited, the sample can be further concentrated and placed in a 300 μl Shigemi tube. Store at 4°C prior to NMR data collection.

3.4. NMR Screening

1. Record 1D ^{1}H NMR and 2D ^{1}H-^{15}N heteronuclear single quantum coherence (HSQC) spectra of the target protein. The authors routinely perform NMR measurements at 25°C on a 500- or 600-MHz spectrometer equipped with a z-shielded gradient triple-resonance probe. Typical 2D data sets are recorded as 1,024×256 complex points with 8–32 scans per increment depending on the sample concentration. Quadrature detection in the indirect dimension is achieved by the States-TPPI method, and a binomial 3–9–19 sequence with water flip-back is used for water suppression. The ^{1}H dimension is referenced to H$_2$O at 25°C, and the carrier frequency of the ^{15}N dimension is set to 118 ppm. 1D ^{1}H NMR spectra can be recorded in a time frame of a few minutes, and measurement time for the HSQC is typically 30–60 minutes.

2. Process the NMR data using TopSpin (Bruker, Billerica, MA), NMRPipe/NMRDraw *(15)* or other suitable software package.

3. Examine the spectra to evaluate the protein's suitability for structure determination. NMR spectral quality is assessed based primarily on spectral dispersion, line widths, and number of resolved peaks observed compared with the number expected from the amino acid sequence.

4. To optimize sample conditions, pH titrations or titrations with cofactors as well as variation of temperature may be performed and monitored using NMR. Following the appearance of the spectra over time can also be used to assess protein stability under the screening conditions.

5. After NMR measurements, the samples can be recovered and used for other biophysical or functional studies.

Representative examples of 1D ^{1}H NMR and 2D ^{1}H-^{15}N HSQC spectra are shown in Fig. 29.1. In 1D ^{1}H NMR spectra, the dispersion of signals in the amide proton (6–10 ppm), α-proton (3.5–6 ppm) and methyl proton (–0.5 to 1.5 ppm) regions, provides the main indicators of folded globular proteins (see Fig. 29.1A,B). In contrast, an unfolded protein shows a smaller dispersion of backbone amide chemical shifts. The appearance of intense peaks near 8.3 ppm is an indicator of disorder in a protein (see Fig. 29.1C), as this region is characteristic of backbone amides in a random-coil configuration.

The positions of peaks in 2D ^{1}H-^{15}N HSQC spectra are indicative of structured or disordered proteins in a similar way to that described for the one-dimensional spectrum. In the spectrum of an unfolded protein, all signals cluster in a characteristic "blob" around 8.3 ppm with little dispersion in either dimension, and side chain NH peaks are generally degenerate. For a well-folded protein, the HSQC spectrum should show well-dispersed peaks of equal intensity, with the number of peaks corresponding to that calculated from the protein sequence (see Fig. 29.1A,B). These spectra indicated that the proteins were amenable to structure determination using NMR methods *(16,17)*. For poorly to moderately folded proteins, HSQC spectra often show a lack of dispersion of peaks with varying intensities and peak numbers inconsistent with the protein sequence (see Fig. 29.1C). Furthermore, line width is highly dependent on the tumbling rate of a molecule in solution and therefore on its molecular weight. Thus, a protein exhibiting a poor HSQC could suggest that it is forming higher-order oligomers or nonspecific aggregates under

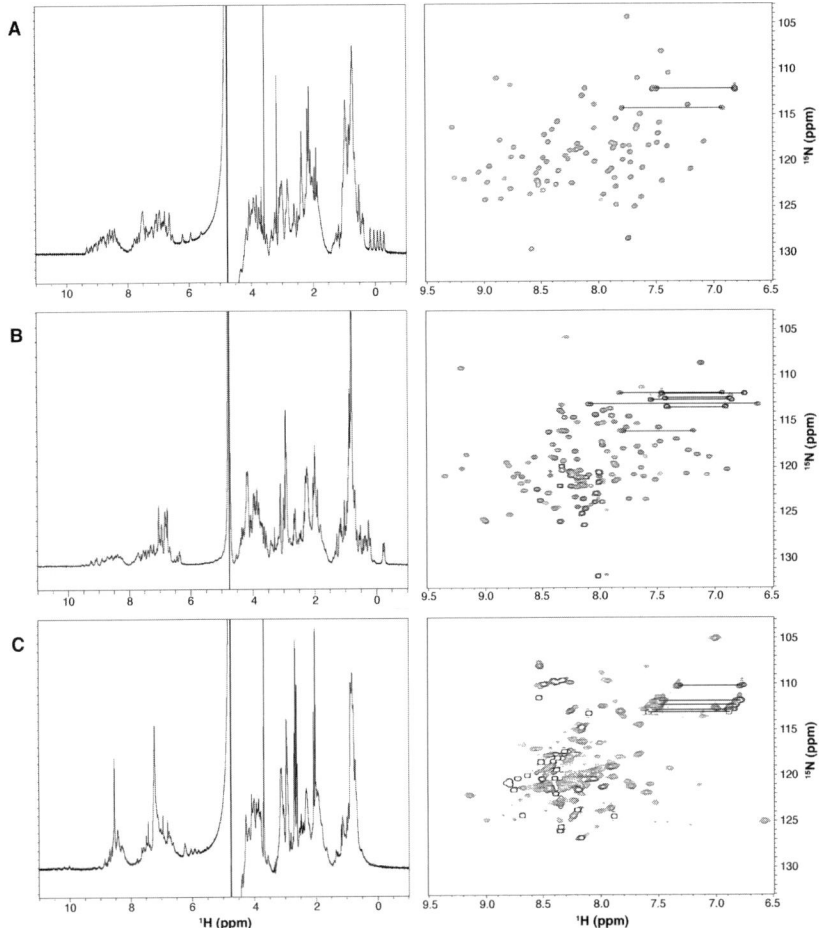

Fig. 29.1 Representative examples of 1D ^{1}H NMR *(left)* and 2D ^{1}H-^{15}N HSQC spectra *(right)* of ^{15}N-labeled proteins. The proteins are: **(A)** ASC2, an 11-kDa pyrin domain-only protein *(17)*, **(B)** PEA-15, 15-kDa *(16)*, and **(C)** a 15-kDa target protein from the macrophage structural proteomics initiative at the University of Queensland (see Meng et al: "Overview of the Pipeline for Structural and Functional Characterization of Macrophage Proteins at the University of Queensland" in this volume). These spectra were recorded at 25°C on a Bruker Avance 600 MHz spectrometer. Peaks corresponding to side chain NH$_2$ groups are connected by horizontal lines.

the NMR screening conditions. This category is not readily amenable to NMR structural analysis without further optimization of the protein construct and/or buffer conditions.

4. Notes

1. Unless otherwise stated, all solutions should be prepared using distilled and deionized water that has a resistivity of 18.2 MΩ·cm. Bacterial growth reagents, antibiotics, and routine laboratory chemicals were obtained from Sigma-Aldrich (St. Louis, MO) or other major distributors.

2. Trace metals are required for maximal growth in fully defined media. Addition of the trace element solution ensures that the large number of metal ion containing enzymes in *E. coli* can function optimally.

3. This protocol can readily be adapted for uniform $^{13}C/^{15}N$-labeling of proteins and for use with D_2O for the production of deuterated proteins *(18)*. To prepare minimal medium for production of $^{13}C/^{15}N$-labeled proteins, substitute the 50% w/v glucose solution (Section 2.1.) with ^{13}C-glucose added at 2 g per L.

4. Media should be used within 24 hours of assembly from sterile stock solutions.

5. The volume of the starter culture can be scaled according to the desired volume of the large-scale growth. In the author's experience, good results are obtained from using a starter culture that is approximately 5% of the final volume.

6. Conditions for optimal growth should be determined in small-scale cultures prior to isotopic labeling. Several different parameters can critically influence the total and soluble protein expression including choice of expression vector and expression strain, temperature, OD at induction, concentration of inducer, and the induction time.

7. This procedure is very efficient for well expressed proteins, requiring only a 250-ml culture to produce sufficient material for ^{1}H-^{15}N HSQC screening in several buffer conditions.

8. $^{13}C/^{15}N$-labeled proteins can also be produced using autoinduction medium with ^{13}C-glycerol as the carbon source *(12,13)*. Glucose cannot be used because it prevents autoinduction. Although the cost of ^{13}C-labeled precursors with this approach is higher than for conventional methods, this is compensated by the ease and reproducibility of the autoinduction medium.

9. If the expression vector uses a T7*lac* promoter, an *E. coli* host strain without a pLysS plasmid is recommended. The combination of T7 lysozyme (expressed by the pLysS plasmid) and the *lac* repressor causes significantly reduced protein expression in autoinduction medium.

10. NG is a defined minimal medium for growth to saturation with little or no induction of target protein expression. This is a variant of PG (also previously known as P-0.5G) *(12)*. Overnight cultures in NG can be used to make freezer stocks that remain viable indefinitely and generate cultures that produce high levels of target protein.

11. Cells should be grown to saturation when using autoinduction. Higher saturation densities mean that cultures may be quite dense after overnight incubation but not yet induced, so care must be taken not to harvest low-temperature cultures before they have saturated *(12)*. Saturation is usually reached in 8–10 hours if the cultures are incubated at 37°C. When lower temperatures are used, saturation may only be reached by incubation for 24 hours or more.

12. Aeration, particularly during the induction phase, is an important parameter for efficient expression in shake flasks. Yields can be improved by the use of baffled flasks, and the volume of the culture should not exceed 25% of the flask volume.

13. The protein's theoretically calculated isoelectric point (pI) should be considered when choosing a buffer for NMR analysis, as the solubility of

a protein tends to decrease in solutions with a pH near its pI. Buffers with a slightly acidic pH are preferred to reduce the chemical exchange rate of the amide protons with water. If the protein contains a large number of cysteine residues, the concentration of DTT can be increased to 10 mM. Sodium azide (NaN_3), added to inhibit bacterial growth, is highly toxic and gloves should always be worn during sample preparation.

Acknowledgments

The author's research was supported by an RD Wright Biomedical Career Development Award (401748) and project grant (351503) from the Australian National Health and Medical Research Council (NHMRC).

References

1. Page, R., Peti, W., Wilson, I. A., Stevens, R. C., and Wüthrich, K. (2005) NMR screening and crystal quality of bacterially expressed prokaryotic and eukaryotic proteins in a structural genomics pipeline. *Proc. Natl. Acad. Sci. USA* **102,** 1901–1905.
2. Rehm, T., Huber, R., and Holak, T. A. (2002) Application of NMR in structural proteomics: screening for proteins amenable to structural analysis. *Structure* **10,** 1613–1618.
3. Yee, A., Chang, X., Pineda-Lucena, A., Wu, B., Semesi, A., Le, B., Ramelot, T., Lee, G. M., Bhattacharyya, S., Gutierrez, P., Denisov, A., Lee, C., Cort, J. R., Kozlov, G., Liao, J., Finak, G., Chen, L., Wishart, D., Lee, W., McIntosh, L. P., Gehring, K., Kennedy, M. A., Edwards, A. M., and Arrowsmith, C. H. (2002) An NMR approach to structural proteomics. *Proc. Natl. Acad. Sci. USA* **99,** 1825–1830.
4. Peti, W., Etezady-Esfarjani, T., Herrmann, T., Klock, H. E., Lesley, S. A., and Wüthrich, K. (2004) NMR for structural proteomics of Thermotoga maritima: screening and structure determination. *J. Struct. Funct. Genomics* **5,** 205–215.
5. Yee, A. A., Savachenko, A., Ignachenko, A., Lukin, J., Xu, X., Skarina, T., Evdokimova, E., Liu, C. S., Semesi, A., Guido, V., Edwards, A. M., and Arrowsmith, C. H. (2005) NMR and x-ray crystallography, complementary tools in structural proteomics of small proteins. *J. Am. Chem. Soc.* **127,** 16512–16517.
6. Snyder, D. A., Chen, Y., Denissova, N. G., Acton, T., Aramini, J. M., Ciano, M., Karlin, R., Liu, J., Manor, P., Rajan, P. A., Rossi, P., Swapna, G. V. T., Xiao, R., Rost, B., Hunt, J., and Montelione, G. T. (2005) Comparisons of NMR spectral quality and success in crystallization demonstrate that NMR and x-ray crystallography are complementary methods for small protein structure determination. *J. Am. Chem. Soc.* **127,** 16505–16511.
7. Tyler, R. C., Aceti, D. J., Bingman, C. A., Cornilescu, C. C., Fox, B. G., Frederick, R. O., Jeon, W. B., Lee, M. S., Newman, C. S., Peterson, F. C., Phillips, G. N., Shahan, M. N., Singh, S., Song, J., Sreenath, H. K., Tyler, E. M., Ulrich, E. L., Vinarov, D. A., Vojtik, F. C., Volkman, B. F., Wrobel, R. L., Zhao, Q., and Markley, J. L. (2005) Comparison of cell-based and cell-free protocols for producing target proteins from the *Arabidopsis thaliana* genome for structural studies. *Proteins* **59,** 633–643.
8. Ab, E., Atkinson, A. R., Banci, L., Bertini, I., Ciofi-Baffoni, S., Brunner, K., Diercks, T., Dötsch, V., Engelke, F., Folkers, G. E., Griesinger, C., Gronwald, W., Günther, U., Habeck, M., de Jong, R. N., Kalbitzer, H. R., Kieffer, B., Leeflang, B. R., Loss, S., Luchinat, C., Marquardsen, T., Moskau, D., Neidig, K.-P., Nilges, M., Piccioli, M., Pierattelli, R., Rieping, W., Schippmann, T., Schwalbe, H., Travé, G., Trenner, J., Wöhnert, J., Zweckstetter, M., and Kaptein, R. (2006) NMR in the SPINE structural proteomics project. *Acta Cryst.* **62,** 1150–1161.

9. Hondoh, T., Kato, A., Yokoyama, S., and Kuroda, Y. (2006) Computer-aided NMR assay for detecting natively folded structural domains. *Protein Sci.* **15,** 871–883.

10. Kigawa, T., Yabuki, T., Matsuda, N., Matsuda, T., Nakajima, R., Tanaka, A., and Yokoyama, S. (2004) Preparation of *Escherichia coli* cell extract for highly productive cell-free protein expression. *J. Struct. Funct. Genomics* **5,** 63–68.

11. Endo, Y., and Sawasaki, T. (2003) High throughput, genome-scale protein production method based on the wheat germ cell-free expression system. *Biotechnol. Adv.* **21,** 695–713.

12. Studier, F. W. (2005) Protein production by autoinduction in high-density shaking cultures. *Protein Expr. Purif.* **41,** 207–234.

13. Tyler, R. C., Sreenath, H. K., Singh, S., Aceti, D. J., Bingman, C. A., Markley, J. L., and Fox, B. G. (2005) Autoinduction medium for the production of [U-^{15}N]- and [U-^{13}C, U-^{15}N]-labeled proteins for NMR screening and structure determination. *Protein Expr. Purif.* **40,** 268–278.

14. Cai, M., Huang, Y., Sakaguchi, K., Clore, G. M., Gronenborn, A. M., and Craigie, R. (1998) An efficient and cost-effective isotope labeling protocol for proteins expressed in *Escherichia coli. J. Biomol. NMR* **11,** 97–102.

15. Delaglio, F., Grzesiek, S., Vuister, G. W., Zhu, G., Pfeifer, J. and Bax, A. (1995) NMRPipe: a multidimensional spectral processing system based on UNIX pipes *J. Biomol. NMR* **6,** 227–293.

16. Hill, J. M., Vaidyanathan, H., Ramos, J. W., Ginsberg, M. H., and Werner, M. H. (2002) Recognition of ERK MAP kinase by PEA-15 reveals a common docking site within the death domain and death effector domain. *EMBO J.* **21,** 6494–6504.

17. Natarajan, A., Ghose, R., and Hill, J. M. (2006) Structure and dynamics of ASC2, a pyrin domain-only protein that regulates inflammatory signalling. *J. Biol. Chem.* **281,** 31863–31875.

18. Tugarinov, V., Kanelis, V., and Kay, L. E. (2006) Isotope labeling strategies for the study of high-molecular-weight proteins by solution NMR spectroscopy. *Nature Protocols* **1,** 749–754.

Chapter 30

Microcoil NMR Spectroscopy: a Novel Tool for Biological High Throughput NMR Spectroscopy

Russell E. Hopson and Wolfgang Peti

Microcoil NMR spectroscopy is based on the increase of coil sensitivity for smaller coil diameters (approximately 1/d). Microcoil NMR probes deliver a remarkable mass-based sensitivity increase (8- to 12-fold) when compared with commonly used 5-mm NMR probes. Although microcoil NMR probes are a well established analytical tool for small molecule liquid-state NMR spectroscopy, after spectroscopy only recently have microcoil NMR probes become available for biomolecular NMR spectroscopy. This chapter highlights differences between commercially available microcoil NMR probes suitable for biomolecular NMR spectroscopy. Furthermore, it provides practical guidance for the use of microcoil probes and shows direct applications for structural biology and structural genomics, such as optimal target screening and structure determination, among others.

1. Introduction

In the past decade, liquid-state nuclear magnetic resonance (NMR) spectroscopy has evolved into a powerful tool for the determination of biomolecular structure and dynamics (1). However, one of the major drawbacks of NMR spectroscopy is its inherent low sensitivity. Sample concentrations of approximately 0.5–1 mM are required for routine structural and dynamical investigations of proteins. Preparation of these large amounts of protein is costly due to the need of ^{15}N and ^{13}C labeling, accomplished by using $^{15}NH_4Cl$ and $[^{13}C_6]$-D-glucose as the sole nitrogen and carbon sources during protein expression, respectively (2). This is especially critical in structural genomics environment, where hundreds to thousands of targets are produced in a high throughput manner (3–6). Moreover, not all proteins can be expressed in yields sufficient for structural studies using NMR spectroscopy (7). Therefore, one of the primary goals for technical development in NMR spectroscopy has always been to increase the sensitivity of the NMR instrumentation, which, in turn, will decrease the amount of protein required for NMR measurements. The five most important parameters that effect the sensitivity of the NMR measurements are the following: (1) the static magnetic field (the signal-to-noise ratio is proportional to $B_0^{3/2}$, highest field commercially available: 22.3 T); (2) the

From: *Methods in Molecular Biology, Vol. 426: Structural Proteomics: High-throughput Methods*
Edited by: B. Kobe, M. Guss and T. Huber © Humana Press, Totowa, NJ

sample concentration (for NMR studies, protein concentrations of approximately 0.5 mM or more are usually needed); (3) the electronic noise of the spectrometer and probe; (4) the coil sensitivity; and (5) new pulse sequence methods that increase the sensitivity by using distinctive magnetization transfer pathways *(8)*.

Recently introduced cryoprobe technology *(9–11)* has reduced this concentration requirement threefold, down to approximately 0.3 mM for complete structure determination for proteins up to 20 kDa, or even less for protein:protein or protein:small molecule interaction studies. However, cryoprobes have multiple drawbacks. First, they are approximately six times more expensive than conventional probes. Second, the sensitivity is highly dependent on the salt concentration of the sample; increasing the salt concentration from 50 to 300 mM decreases the sensitivity of the cryoprobe by a third. Most importantly, they still require hundreds of microliters of sample (300–500 μl), which corresponds to milligrams of protein even for approximately 0.2 mM concentrated samples. This chapter highlights how the increased mass sensitivity of microcoil NMR probes, which is based on improved coil sensitivity, can be used in structural genomics and structural biology applications.

It has been shown that the mass sensitivity (S/N as a function of detected nuclei) is approximately 10 times higher in microcoil NMR probes than in commonly used 5-mm Helmholtz coil NMR probes *(12,13)*. There are two reasons for this improved sensitivity. First, the increase in the signal-to-noise ratio of microcoils with a fixed length-to-diameter ratio improves as the coil diameter decreases, i.e., the sensitivity of the NMR measurement is, to a first approximation, inversely proportional to the coil diameter (approximately 1/d) *(14)*:

$$\frac{B_1}{i} = \frac{\mu_0 n}{d\sqrt{1 + \left(\dfrac{h}{d}\right)^2}} \tag{1}$$

The sensitivity of solenoid coils is given by Equation 1, where B_1 is the applied radio frequency field, i the current unit, μ_0 the permeability in vacuum, n the number of turns, d the diameter of the coil and h the length of the coil. The equation is valid for probes in a regime in which the traditional rules of radiofrequency skin effect apply. The smaller the coil, the greater is the transverse magnetic field strength per unit volume. To achieve a balance between mass sensitivity and sample load, commercially available microcoil NMR probes, usable for biomolecular NMR spectroscopy and therefore in a structural genomics setting, utilize 1-mm inner diameter coils with 3- to 5-mm length. The total flow-cell volume of these microcoils is 5–10 μl, having an active NMR volume of 1.5–5.0 μl. This volume is approximately 180 times less than that for standard 5 mm probes *(15–17)*.

Second, some microcoil NMR probes contain horizontal flow-cells and employ solenoidal transceiver coils *(14)*. Due to stronger coupling with the sample, solenoid coils have a two to three times higher sensitivity when compared with Helmholtz coils. However, a significant drawback of solenoid coils is that they have low RF-field homogeneity, which results in broad resonance signal line widths that are not suitable for high-resolution NMR spectroscopy.

This limitation was recently overcome using an innovative design in which the solenoid coils were embedded in a perflourinated magnetic susceptibility matching fluid *(13)*. Using this setup, these embedded solenoid coils showed the same line widths as Helmholtz coils (approximately 0.7 Hz), and thus these microcoil probes are now suitable for high-resolution NMR measurements. Moreover, these probes show tremendous proton sensitivity and inherent short pulses, which opens up a new avenue for pulse sequence design for the assignment of biomacromolecules. Solenoidal coils can achieve much shorter pulse lengths relative to standard 5 mm RF coils, which is of special importance for measurements with high field NMR spectrometers that are widely used for biomolecular NMR.

In addition to improved mass-based sensitivity, microcoil NMR probes also have a number of significant advantages *(12)*. First, they are three times less expensive than cryoprobes. Second, the sensitivity is almost independent of the salt concentration, experimentally tested up to 500 mM, often important for biological samples. Third, water suppression using microcoil probes is easily achieved. Fourth, microcoil probes are readily installed and require little maintenance. Specifically, a tremendous advantage of the microcoil probe is its ease of shimming. Since the microcoils are perpendicular to the applied B_0 field and are extremely small, their shimming is very different, yet very easily accomplished compared to regular probes. For shimming, one only needs to adjust the low-order shims, i.e., x, y, x^2-y^2, xy, and z. Finally, and most importantly for valuable biological samples, only small amounts of sample, 10–20 µl of 0.5- to 2-mM protein samples, which corresponds to micrograms of protein, are required for NMR measurements.

Although microcoil NMR probes have been available commercially for several years for small molecule work, their use for biomolecular NMR data acquisition has never been exploited until recently *(16–18)*. For biomolecular protein NMR spectroscopy, the most widely used conventional probe design is a triple resonance inverse probe with actively shielded single- or triple-axis gradients. The inner coil is tuned to ^{1}H and ^{2}H frequency with the outer coil tuned to the ^{13}C and ^{15}N frequencies. The microcoil probe designed for biomolecular NMR experiments utilizes a quadruple tuned circuit with an actively shielded single-axis gradient. The microcoil is tuned to ^{1}H, ^{2}H, ^{13}C, and ^{15}N frequencies.

One of the major goals of structural genomics is to dramatically reduce the cost per structure. Using microcoil NMR probes for investigating dynamics, protein:protein interactions and for structure determination of small- or medium-sized proteins enables the achievement of this aim in a much more rapid manner. More importantly, it enables the structure determination of proteins whose small expression yields have prevented them from being determined before, due to low intrinsic expression yields or the inability to optimize expression, as is routine in high throughput protein production pipelines. Previously, large scale expression of isotopically labeled proteins suitable for NMR structure determination was too expensive in these systems. Now, with commercially available growth media (Spectra Stable Isotopes, Columbia, MD; and Cambridge Isotope Laboratory, Andover, MA) and with microcoil NMR probe technology described in this chapter, NMR is a tool that can be exploited to study these proteins.

2. Materials

2.1. NMR Microcoil Probes

Quadruple tuned HCN z-gradient microcoil probes are commercially available from MRM/Protasis (www.micronmr.com) and Bruker BioSpin (www.bruker-biospin.com/nmr/products/1mm_microprobe.html). Although both probes offer approximately the same sensitivity enhancement due to their small RF-coil diameter (both use 1 mm diameter microcoils), Bruker BioSpin's 1-mm microcoil probe does not use a solenoid-coil, but rather a horizontal saddle design coil and therefore neglects the additional sensitivity gain provided by the solenoid coil design. For both solenoid and saddle coil geometries, the sensitivity increases inversely with the coil diameter. However, because the Bruker BioSpin 1-mm microcoil probe (HCN z-gradient) follows the traditional horizontal coil design, small disposable capillary-like tubes can be readily used for the measurements, which can be advantageous, especially in automated setups. These small tubes can either be filled manually via a small v-shaped "funnel-like" top part, which is conveniently attached onto the disposable tubes, or automatically via a modified Gilson 215 liquid handling unit. These Bruker BioSpin 1-mm microcoil probes are currently used in the North East Structural Genomics Consortium for protein structure determination and the Joint Center for Structural Genomics for protein screening. The MRM microcoil probe gains additional sensitivity by using a solenoid coil (two- to threefold S/N increase); however, this setup requires a flow-through design. To achieve comparable B_0 homogeneity with horizontal probes for solenoid coil-based NMR probes, the solenoid coil needs to be embedded in a perfluorinated magnetic susceptibility matching fluid. The CF matching-fluid causes a strong background signal in ^{13}C detected experiments. The setup requires that the protein is injected into the probe by means of a gas tight syringe or a small automatically driven syringe pump and not NMR tubes. Especially for highly concentrated samples this can be a disadvantage, because small air bubbles can form that compromise the shimming and water suppression of the NMR experiments. If air bubbles form, the sample needs to be removed from the probe, collected; and after a few wash injections with sample buffer the sample can then be re-injected. Critically, the authors never saw the formation of air bubbles in the probe during the measurements, clearly indicating that the surface tension of the small capillary tubes protect the sample chamber. This also indicates that diffusion is not a problem in microcoil NMR probes. A dramatic advantage of the solenoidal coils is that they show tremendous proton sensitivity and inherently short pulses *(12,13,19,20)*, much shorter than standard 5-mm RF-coils. These enhanced RF-capabilities enable, e.g., larger broad band opens opportunities for new pulse sequence development.

The first solenoid microcoil NMR probes for biomolecular NMR spectroscopy used a 5-µl full volume and 1.5- to 3.0-µl active volume, which was believed to provide a good balance between mass sensitivity and sample load. However, these probes were originally designed for small-molecule NMR. In recent years, numerous modifications in the design of microcoil NMR probes have been introduced, most notably a 10-µl biomolecular microcoil NMR probes. These probes still take advantage of the small-diameter coil (1 mm) and the solenoid coil design, but have a three to four times larger active volume, critical for increased sensitivity. This is important as previous measurements with microcoil probes clearly indicated that a decreased protein concentration is necessary to make this method suitable for a large number of proteins. Not many proteins can be readily concentrated to the 2- to 5-mM concentration required for measurements on these first-generation microcoil probes. Using the improved probes, approximately 0.5- to 1-mM sample concentrations should be practical for 1D ^{1}H NMR measurements, whereas it will still be necessary to increase to concentration to approximately 2 mM for multidimensional spectroscopy for the measurement to be carried out in a timely manner. In addition, Bruker BioSpin recently announced a 1.7-mm cryo-microcoil NMR probe requiring a 30-μl sample volume for measurements. For biomolecular NMR spectroscopy, these "larger" cryo-microcoil NMR probes might provide the optimum sample volume to sensitivity to sample handling ratio.

2.2. Cleaning Is Critical for the Long-Term Performance of Flow-Through Microcoil NMR Probes

An important aspect in the daily routine use of horizontal flow-through microcoil NMR probes is cleaning the probe on a regular basis. The cleaning intervals depend on the number of samples and nature of the protein samples. Since no NMR tubes are used that can be cleaned outside of the NMR magnet, this needs to be done with the probe either inside the magnet or removed from the magnet and replaced with a regular 5-mm probe during the cleaning procedure. As stated, it is very easy to shim microcoil NMR probes; therefore, only minimal time is needed to change between the two NMR probes. Cleaning a microcoil NMR probe is a two-step process. First, about 10–20 ml of water is flushed automatically through the probe, using a mechanically pressurized syringe. Second, overnight very dilute 1–3% detergent solution (Contrad-70 or Liquinox) is slowly flushed through the microcoil NMR probe and then followed by another extensive water wash. Occasionally, additional 0.1 M sodium hydroxide solution washes or 1 M hydrochloric acid washes are also useful to remove precipitated protein that was not removed by the detergent solutions.

To protect the microcoil probe and dramatically increase the time between cleaning, a filter system can be used prior to the injection port, which was designed by Mark O'Neil-Johnson (Sequoia Sciences, St. Louis, MO) (21). A gas tight Hamilton 1702N (22s needle gauge, 2 needle length, point style 3–blunt tip) is the easiest method for loading a sample. Although automatic loading setups are available that include leap system liquid handling systems among others, the authors feel that it is easier to use syringes for loading, especially for biological samples: 2-μm titanium filters (UpChurch Scientific, Oak Harbor, WA), which are routinely used in most FPLC and HPLC system as inline filters, are ideally used as filters before sample injection. Critically, titanium filters are very inert to biological samples and the authors have not yet detected any interactions with proteins. Finally, these filters can easily be cleaned in 0.1 M sodium hydroxide solutions under steady sonication and are reusable for very long periods.

3. Methods

Several acquisition parameters need to be considered when setting up NMR experiments using microcoil NMR probes, which are different from widely used room temperature 5-mm probes and also 5-mm cryoprobes. Importantly, although small adaptations, such as a slight decrease in the proton power levels, are also needed for cryoprobes, power adaptations for microcoil probes are much more significant. Only extremely low power is needed to achieve 90° magnetization rotations (<5 W for all nuclei; e.g., for cryoprobes power levels are usually in the 20 W range for protons). This is true for all channels independently using either 100 W ^{1}H or 300/500 W broadband linear amplifiers. One important advantage of the dramatically low power requirement is that all subsequent pulse calibrations for the experiments using a microcoil NMR probe can be directly calculated. Amplifiers are highly linear in these low power ranges, even without correction tables, which are used to overcome this problem with non-microcoil NMR probes. However, there are certain jumps in linearity, because of the change of attenuators in the power setup at +20 and +40 dB (Bruker BLAH and BLAX amplifiers), which lead to errors in these calculations and consequently to incomplete excitation. Therefore, for proton pulses usually a low- and a high-power pulse calibration is necessary.

Water suppression can be achieved using presaturation (22), Watergate techniques (23), or coherence order selective pulse sequences (24,25), using low-power adaptation during their respective setups. For highly automated NMR measurement setups, currently used in structural genomic consortia, it is important to automate as many steps as possible. Therefore, it might seem to be a disadvantage to use microcoil NMR probes with horizontal flow-through design, which do not allow for rapid, automated gradient shimming (26). However, because the microcoils are perpendicular to the applied B_0 field and are extremely small, their shimming is very rapidly accomplished compared with regular probes. For shimming, one only needs to adjust the low-order shims, that is, x, y, x2-y2, xy, and z, which can easily be automated using small au-programs and the tune parameters using Bruker TopSpin software. The 1-mm Bruker microcoil, due to its regular vertical setup, allows for gradient shimming. Figure 30-1 shows a comparison of a two-dimensional 2D [^{1}H,^{15}N] HSQC spectrum of the identical protein sample measured in a 5-mm NMR probe and a microcoil NMR probe.

It is apparent that microcoil probes are best suited for mass limited samples rather than concentration limited samples. This represents a potential problem for biomolecular microcoil NMR spectroscopy, since not all proteins are soluble in high enough concentrations. However, if high enough concentrations can be achieved, microcoil NMR probes allow for the measurements of mass limited samples, i.e., obtained from protein expression in a eukaryotic expression system or high throughput bacterial system, in which expressed protein quantities are often limited. High throughput expression methods are used in most structural genomics consortia, especially for expression screening. Nevertheless after rapid low volume expression screening often large-scale expression and purification is necessary to achieve first biophysical characterization of the samples. It has recently been shown that NMR spectroscopy can be elegantly used to screen proteins for their ability to crystallize or to be applicable for structure determination by NMR spectroscopy (27–30).

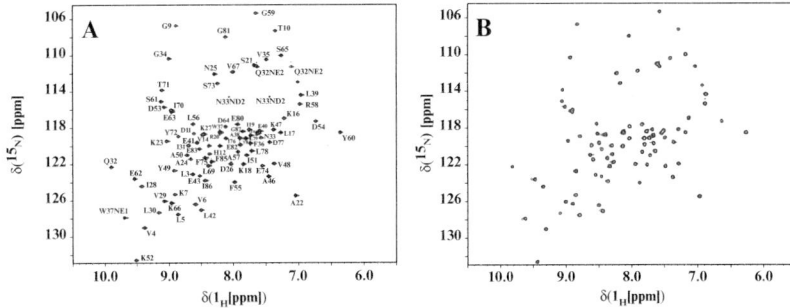

Fig. 30.1 Comparison of the two-dimensional [^{1}H,^{15}N] HSQC spectrum of TM0979 (PDBid: 1RHX, BMRBid: 6010, 87 residues, 9.7 kDa). **A.** Fully annotated [^{1}H,^{15}N] HSQC spectrum recorded with a 5-mm TXI HCN z-gradient probe (Bruker) using a 2-mM sample of TM0979: 8 scans, 1k × 128 complex points, experiment time: 30 minutes. B: HCN z-gradient CapNMR® (MRM Corp., Savoy, IL) microcoil probe using 5 μl of the 2-mM sample used in **(A)** 16 scans, 1k × 128 complex points, experiment time: 60 minutes; measurement temperature for both spectra was 313 K *(16)*.

Microcoil NMR spectroscopy is a unique tool that enables this biophysical characterization, an important step in a typical structural genomics pipeline decision tree, to be made *before* elaborate macroscale protein expression and purification *(31)*.

3.1. Miniaturized NMR Screening Using Microcoil NMR Probes

Only continued miniaturization of expression, purification, NMR screening, and crystallization trials will lead to further reduction of the cost per protein structure and, thus, achieve one of the primary goals of the NIH Protein Structure Initiative (PSI, www.nigms.nih.gov/psi/). Microcoil NMR spectroscopy is perfectly suited to play a critical role as a miniaturization tool, especially for small- and medium-sized proteins *(17)*.

In structural genomics centers, the implementation of robotics and parallelization has enabled researchers to process large numbers of protein targets for structural studies in a short time with good cost efficiency. However, the amount of protein needed for such studies is still substantial, both for NMR screening for folded globular proteins and crystallization trials. This situation has been improved for x-ray crystallography by nanocrystallization, which is now routinely used to screen thousands of crystallization conditions with only micrograms of protein. Further miniaturization of structural genomics pipelines requires replacement of macroexpression and standard NMR methods by microexpression and microcoil NMR technologies.

NMR spectroscopy is one of the few tools that enables biochemists to test the folded state of their protein sample in a rapid manner. Alternative techniques for characterization of the biophysical state of the protein are circular dichroism (CD) spectroscopy, dynamic light scattering (DLS), and deuterium exchange MS *(32,33)*. Although CD spectroscopy and dynamic light scattering can be used, respectively, to characterize the regular secondary structure content and multimeric or aggregated states of proteins, NMR spectroscopy can provide information on the global fold of proteins. In the 1D ^{1}H NMR spectra, the dispersion of the NMR signals in the regions of the methyl protons (–0.5 to

1.5 ppm), α-protons (3.5–6 ppm), and amide protons (6–10 ppm) provides the main indicators of folded globular proteins.

In the Joint Center of Structural Genomics (JCSG), a pilot project was carried out to test if a microcoil NMR probe would be suitable for folded protein screening, directly after microscale test expression, but prior to macroscale expression. Details of the experiments are provided in *(17)*. It was clearly shown that for proteins that could be concentrated to at least 1 mM, spectra can be recorded within a reasonable time frame (less than 4 hours). Due to the small sample volume, high protein concentrations still imply incredibly low amounts of purified protein (30–400 µg). Therefore, it was concluded that microcoil NMR spectroscopy is a valuable tool in this setup. Currently the JCSG is using a Bruker 1-mm microcoil TXI z-gradient probe for their NMR screening efforts.

Lastly, independent of the type of probe used for the NMR screening, it is critical to emphasize that the screening buffer plays a critical role for the interpretation of the screening results and correlation with crystallization success. If buffers are used that are better suited for x-ray crystallography, such as 10 mM Tris pH 8.0, 100 mM NaCl, a good correlation between high-quality NMR spectra and crystallization success can be achieved *(28)*. However, if buffers are used that favor NMR conditions, such as low pH (5.5–6.5) sodium phosphate and 100–300 mM NaCl, recent report show that NMR screening results do not correlate well with the crystallization results *(6,28–30)*.

3.2. Microcoil NMR Probes Enable the Design of Novel Pulse-Sequences: A Novel Aliphatic-Aromatic HCCH-TOCSY Experiment

Taking advantage of the superior radiofrequency capabilities of microcoil NMR probes, which are a result of the unique solenoid coil design, TOCSY transfer between the aliphatic–aromatic regions in a HCCH-TOCSY spectrum has been demonstrated *(16)* (Fig. 30.2). This experiment now allows for the complete side chain assignment of all amino acids, including aromatics, within a single spectrum. The obstacle for aliphatic–aromatic carbon transfer has always been the large carbon chemical shift differences between aliphatic and aromatic residues (average approximately 13,200 Hz at 14.1 T), creating a requirement for high RF-power in order to establish a full TOCSY transfer across these aliphatic–aromatic residues *(34–37)*. This broad TOCSY transfer capability could not be achieved using traditional 5-mm room temperature or cryoprobes. The success of this experiment using the biomolecular microcoil NMR probe has made the aromatic side chain assignment, including connectivities with the aliphatic side chains, possible using a single NMR experiment. This greatly accelerates the aromatic side chain assignment, and can easily be employed for highly concentrated protein samples.

The increased RF capability of microcoil NMR probes could also be employed for biomolecular RNA NMR spectroscopy. In RNA the ^{15}N chemicals shifts of base nitrogen atoms show a large dispersion in the ^{15}N dimension and the RF capabilities of current 5-mm NMR probes are not sufficient to connect these resonances by means of scalar interaction spectroscopy. Again, the dramatically increased RF capabilities of a microcoil probe would enable a direct correlation spectroscopy following the described example for proteins.

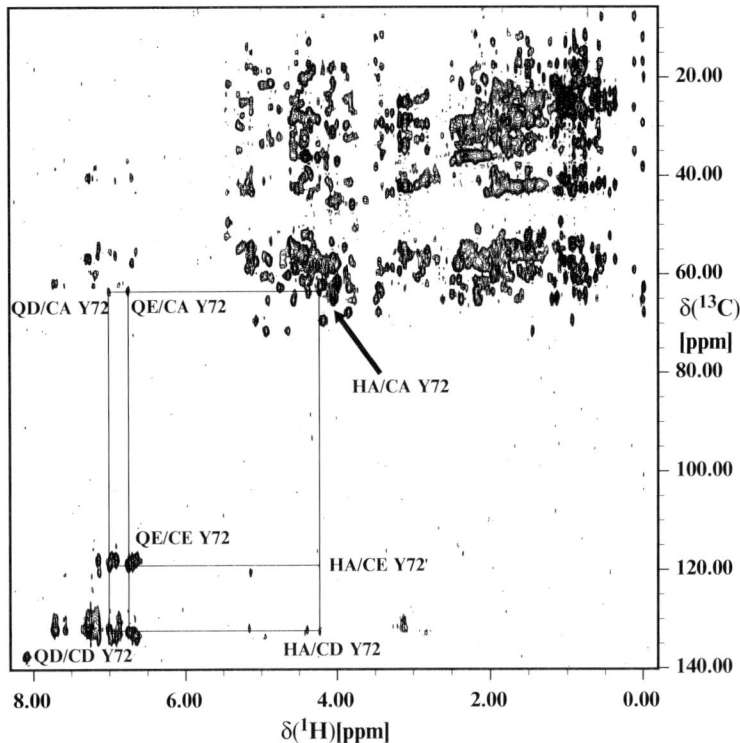

Fig. 30.2 Two-dimensional [^{1}H,^{13}C] aliphatic-aromatic HCCH-TOCSY FLOSPY-16 spectrum recorded with a mixing time of t_m = 11.77 ms. The assignment of tyrosine 72 based on the Cα chemical shift is indicated in the spectrum. (10 mM ^{13}C/^{15}N TM0979 (PDBid: 1RHX, BMRBid: 6010, 87 residues, 9.7 kDa), 313 K, measurement time: 30 hours) *(16).*

3.3. Three-Dimensional Structure Determination Using Microcoil NMR Probes

Recently, the North East Structural Genomics (NESG) consortium has shown that it is possible to determine the three-dimensional (3D) structure of a small protein using a 1-mm Bruker HCN z-gradient microcoil NMR probe *(38)*. Rossi et al. presented the solution structure of a protein coded by gene locus MM1357 of *Methanosarcina mazei* at the 2006 Structural Genomics Keystone Symposium (Keystone, Colorado). This gene encodes a small 5-beta stranded 68-residue protein (TrEMBL ID: Q8PX65_METMA; NESG ID: MaR30, PDB ID: 1yez). The group reported that they used a ^{13}C, ^{15}N-labeled sample (72 μg dissolved in 6 μL, which resulted in a 1.4-mM sample) to achieve nearly complete backbone and side chain assignment, before solving the 3D structure of the protein. The group recently published their results. *(39)* They further report that they had to use about twice the NMR time (19 d vs. 9.5 d) that they usually need for a complete NMR data set, which enabled them to solve the structure of a protein of this small size. However, only 5% of the mass of the protein was used when compared with a regular setup using a conventional 5-mm NMR probe and a Shigemi NMR tube (300 μL). This is the first time a 3D protein structure was solved from less than 100 μg of protein, which shows that for highly soluble small to medium-sized proteins, microcoil NMR spectroscopy is a highly productive and cost efficient method. This opens up new possibilities for the characterization of highly soluble proteins that can only be expressed at low concentration, e.g., when expressed in mammalian expression systems.

4. Conclusion

Microcoil NMR spectroscopy is an exceedingly useful and easy to implement biophysical tool for proteins that can be highly concentrated. Concentrated protein samples (0.5–1 mM and higher) can be rapidly measured due to the excellent mass-based sensitivity of 1 mm microcoil NMR probes, which is 8–12 times increased when compared with regularly used 5-mm NMR probes. This opens up new possibilities for biomolecular NMR spectroscopy, by means of NMR screening, novel experiment design, and low mass 3D structure determination.

Acknowledgments

The authors thank Mark O'Neil-Johnson (Sequoia Sciences) for support and encouragement for the microcoil measurements and Timothy Peck (M. R. M.) for discussions. W. P. thanks his laboratory members, especially Tingting Ju, Matthew Kelker, and Barbara Dancheck, for discussions and critical reading of the manuscript. W. P. is the Manning Assistant Professor for Medical Science at Brown University.

References

1. Wüthrich, K. (2003) NMR studies of structure and function of biological macromolecules (Nobel lecture). *J. Biomol. NMR* **27**, 13–39.
2. Hewitt, L., and McDonnell, J. M. (2004) Screening and optimizing protein production in *E. coli*. *Methods Mol. Biol.* **278**, 1–16.
3. Lesley, S. A., Kuhn, P., Godzik, A., Deacon, A. M., Mathews, I., Kreusch, A., Spraggon, G., Klock, H. E., McMullan, D., Shin, T., Vincent, J., Robb, A., Brinen, L. S., Miller, M. D., McPhillips, T. M., Miller, M. A., Scheibe, D., Canaves, J. M., Guda, C., Jaroszewski, L., Selby, T. L., Elsliger, M. A., Wooley, J., Taylor, S. S., Hodgson, K. O., Wilson, I. A., Schultz, P. G., and Stevens, R. C. (2002) Structural genomics of the *Thermotoga maritima* proteome implemented in a high throughput structure determination pipeline. *Proc. Natl. Acad. Sci. USA* **99**, 11664–11669.
4. Lesley, S. A., and Wilson, I. A. (2005) Protein production and crystallization at the joint center for structural genomics. *J. Struct. Funct. Genom.* **6**, 71–79.
5. Kennedy, M. A., Montelione, G. T., Arrowsmith, C. H., and Markley, J. L. (2002) Role for NMR in structural genomics. *J. Struct. Funct. Genom.* **2**, 155–169.
6. Yee, A., Gutmanas, A., and Arrowsmith, C. H. (2006) Solution NMR in structural genomics. *Curr. Opin. Struct. Biol.* **6**, 611–617.
7. Peti, W., and Page, R. (2007) Strategies to maximize heterologous protein expression in *Escherichia coli* with minimal cost. *Protein Expr. Purif.* **51**, 1–10.
8. Pervushin, K., Riek, R., Wider, G., and Wüthrich, K. (1997) Attenuated T2 relaxation by mutual cancellation of dipole-dipole coupling and chemical shift anisotropy indicates an avenue to NMR structures of very large biological macromolecules in solution. *Proc. Natl. Acad. Sci. USA* **94**, 12366–12371.
9. Styles, P., Soffe, N. F., Scott, C. A., Cragg, D. A., Row, F., White, D. J., and White, P. C. J. (1984) A high-resolution NMR probe in which the coil and preamplifier are cooled with liquid helium. *J. Magn. Reson.* **60**, 397–404.

10. Styles, P., Soffe, N. F., and Scott, C. A. (1989) An improved cryogenically cooled probe for high-resolution NMR. *J. Magn. Reson.* **84**, 376–378.

11. Hajduk, P. J., Gerfin, T., Boehlen, J. M., Haberli, M., Marek, D., and Fesik, S. W. (1999) High throughput nuclear magnetic resonance-based screening. *J. Med. Chem.* **42**, 2315–2317.

12. Olson, D. L., Norcross, J. A., O'Neil-Johnson, M., Molitor, P. F., Detlefsen, D. J., Wilson, A. G., and Peck, T. L. (2004) Microflow NMR: concepts and capabilities. *Anal. Chem.* **76**, 2966–2974.

13. Olson, D. L., Peck, T. L., Webb, A. G., Magin, R. L., and Sweedler, J. V. (1995) High-resolution microcoil 1h-nmr for mass-limited, nanoliter-volume samples. *Science* **270**, 1967–1970.

14. Peck, T. L., Magin, R. L., and Lauterbur, P. C. (1995) Design and analysis of microcoils for NMR microscopy. *J. Magn. Reson. B* **108**, 114–124.

15. Li, Y., Logan, T. M., Edison, A. S., and Webb, A. (2003) Design of small volume HX and triple-resonance probes for improved limits of detection in protein NMR experiments. *J. Magn. Reson.* **164**, 128–135.

16. Peti, W., Norcross, J., Eldridge, G., and O'Neil-Johnson, M. (2004) Biomolecular NMR using a microcoil NMR probe—new technique for the chemical shift assignment of aromatic side chains in proteins. *J. Am. Chem. Soc.* **126**, 5873–5878.

17. Peti, W., Page, R., Moy, K., O'Neil-Johnson, M., Wilson, I. A., Stevens, R. C., and Wüthrich, K. (2005) Towards miniaturization of a structural genomics pipeline using microexpression and microcoil NMR. *J. Struct. Funct. Genom.* **6**, 259–267.

18. Schroeder, F. C., and Gronquist, M. (2006) Extending the scope of NMR spectroscopy with microcoil probes. *Angew Chem. Int. Ed. Engl.* **45**, 7122–7131.

19. Olson, D. L., Lacey, M. E., and Sweedler, J. V. (1998) High-resolution microcoil NMR for analysis of mass-limited, nanoliter samples. *Anal. Chem.* **70**, 645–650.

20. Webb, A. G. (2005) Microcoil nuclear magnetic resonance spectroscopy. *J. Pharm. Biomed. Anal.* **38**, 892–903.

21. Eldridge, G. R., Vervoort, H. C., Lee, C. M., Cremin, P. A., Williams, C. T., Hart, S. M., Goering, M. G., O'Neil-Johnson, M., and Zeng, L. (2002) High throughput method for the production and analysis of large natural product libraries for drug discovery. *Anal. Chem.* **74**, 3963–3971.

22. Wider, G., Hosur, R. V., and Wüthrich, K. (1983) Suppression of the solvent resonance in 2D NMR spectra of proteins in H_2O solution. *J. Magn. Reson.* **52**, 130–135.

23. Piotto, M., Saudek, V., and Sklenar, V. (1992) Gradient-tailored excitation for single-quantum NMR spectroscopy of aqueous solutions. *J. Biomol. NMR* **2**, 661–665.

24. Kay, L. E. (1995) Field gradient techniques in NMR spectroscopy. *Curr. Opin. Struct. Biol.* **5**, 674–681.

25. Schleucher, J., Schwendinger, M., Sattler, M., Schmidt, P., Schedletzky, O., Glaser, S. J., Sorensen, O. W., and Griesinger, C. (1994) A general enhancement scheme in heteronuclear multidimensional NMR employing pulsed field gradients. *J. Biomol. NMR* **4**, 301–306.

26. Weiger, M., Speck, T., and Fey, M. (2006) Gradient shimming with spectrum optimisation. *J. Magn. Reson.* **182**, 38–48.

27. Rehm, T., Huber, R., and Holak, T. A. (2002) Application of NMR in structural proteomics: screening for proteins amenable to structural analysis. *Structure* **10**, 1613–1618.

28. Page, R., Peti, W., Wilson, I. A., Stevens, R. C., and Wüthrich, K. (2005) NMR screening and crystal quality of bacterially expressed prokaryotic and eukaryotic proteins in a structural genomics pipeline. *Proc. Natl. Acad. Sci. USA* **102**, 1901–1905.

29. Snyder, D. A., Chen, Y., Denissova, N. G., Acton, T., Aramini, J. M., Ciano, M., Karlin, R., Liu, J., Manor, P., Rajan, P. A., Rossi, P., Swapna, G. V., Xiao, R., Rost, B.,

Hunt, J., and Montelione, G. T. (2005) Comparisons of NMR spectral quality and success in crystallization demonstrate that NMR and X-ray crystallography are complementary methods for small protein structure determination. *J. Am. Chem. Soc.* **127**, 16505–16511.

30. Yee, A. A., Savchenko, A., Ignachenko, A., Lukin, J., Xu, X., Skarina, T., Evdokimova, E., Liu, C. S., Semesi, A., Guido, V., Edwards, A. M., and Arrowsmith, C. H. (2005) NMR and X-ray crystallography, complementary tools in structural proteomics of small proteins. *J. Am. Chem. Soc.* **127**, 16512–16517.

31. Page, R., Moy, K., Sims, E. C., Velasquez, J., McManus, B., Grittini, C., Clayton, T. L., and Stevens, R. C. (2004) Scalable high throughput microexpression device for recombinant proteins. *Biotechniques* **37**, 364, 366, 368 passim.

32. Pantazatos, D., Kim, J. S., Klock, H. E., Stevens, R. C., Wilson, I. A., Lesley, S. A., and Woods, V. L., Jr. (2004) Rapid refinement of crystallographic protein construct definition employing enhanced hydrogen/deuterium exchange MS. *Proc. Natl. Acad. Sci. USA* **101**, 751–756.

33. Spraggon, G., Pantazatos, D., Klock, H. E., Wilson, I. A., Woods, V. L., Jr., and Lesley, S. A. (2004) On the use of DXMS to produce more crystallizable proteins: structures of the *T. maritima* proteins TM0160 and TM1171. *Protein Sci.* **13**, 3187–3199.

34. Carlomagno, T., Maurer, M., Sattler, M., Schwendinger, M. G., Glaser, S. J., and Griesinger, C. (1996) PLUSH TACSY: Homonuclear planar TACSY with two-band selective shaped pulses applied to C-alpha, C' transfer, and C-beta, C-aromatic correlations. *J. Biomol. NMR* **8**, 161–170.

35. Grzesiek, S., and Bax, A. (1995) Audio-frequency NMR in a nutating frame. Application to the assignment of phenylalanine residues in isotopically enriched proteins. *J. Am. Chem. Soc.* **117**, 6527–6531.

36. Prompers, J. J., Groenewegen, A., Hilbers, C. W., and Pepermans, H. A. M. (1998) Two-dimensional NMR experiments for the assignment of aromatic side chains in 13c-labeled proteins. *J. Magn. Reson.* **130**, 68–75.

37. Yamazaki, T., Forman-Kay, J. D., and Kay, L. E. (1993) Two-dimensional NMR experiments for correlating $^{13}C\beta$ and $^{1}H\delta/\varepsilon$ chemical shifts of aromatic residues in ^{13}C-labeled proteins via scalar couplings. *J. Am. Chem. Soc.* **115**, 11054–11055.

38. Aramini, J. M., Rossi, P., Anklin, C., and Montelione, G. T. (2006) in Experimental NMR Conference Poster Presentation, Asilomar, CA.

39. Aramini, J. M., Rossi, P., Anklin, C., Xiao, R., and Montelione, G. T. (2007) Microgram-scale protein structure determination by NMR. *Nature Methods* **4**, 491–493.

Chapter 31

Protein Structure Determination Using a Combination of Cross-linking, Mass Spectrometry, and Molecular Modeling

Dmitri Mouradov, Gordon King, Ian L. Ross, Jade K. Forwood, David A. Hume, Andrea Sinz, Jennifer L. Martin, Bostjan Kobe, and Thomas Huber

Cross-linking in combination with mass spectrometry can be used as a tool for structural modeling of protein complexes and multidomain proteins. Although cross-links represent only weak structural constraints, the combination of a limited set of experimental cross-links with molecular docking/modeling is often sufficient to determine the structure of a protein complex or multidomain protein at low resolution.

1. Introduction

Most current structural genomics initiatives have focused on high throughput structure determination of individual proteins using x-ray crystallography and NMR spectroscopy. As our understanding of protein folds becomes more comprehensive the next step of proteomics will have to include multiprotein and transient complexes that are essential for many cellular functions. Greater understanding of protein–protein interactions will lead to a deeper understanding of the regulation of cellular processes and significant benefits to biotechnology.

Even though some high-resolution initiatives focus on protein complexes, their success rates are low, judging by deposits in the Protein Data Bank (PDB). These low success rates are associated with the difficulty in crystallizing complexes. An emerging approach is to derive a set of sparse distance constraints using chemical cross-linkers, to map out residues in the protein interaction interface, and then assemble the protein complex structure by computational means. This approach is particularly powerful when partial structural information is available, for example, the structures of individual proteins in a complex or individual domains in a multidomain protein. In such cases, this technique allows high throughput low-resolution structure determination, even with a limited number of approximate distance constraints.

Chemical cross-linkers have been successfully used for many years to derive low-resolution structural information of proteins. Such cross-linking experiments gave insight into interaction modes of large protein complexes as well as identified interacting proteins. However, the demonstration of the

From: *Methods in Molecular Biology, Vol. 426: Structural Proteomics: High-throughput Methods*
Edited by: B. Kobe, M. Guss and T. Huber © Humana Press, Totowa, NJ

presence of an interaction does not include information about how the proteins dock with respect to each other.

Recent advances in mass spectrometry (MS) *(1)* have allowed for analysis of a large number of enzymatically digested peptides *(2)* and hence identification of the exact insertion points of low-abundance cross-links *(3–6)* and have opened up a new perspective on the use of cross-linkers in combination with computational structure prediction *(7,8)*. The approach uses chemical cross-linking information with molecular docking, so that the cross-links are treated as explicit constraints in the calculations. By using a simple rigid body docking algorithm and a small number of constraints it is possible to narrow down the conformation of a protein complex to only a few configurations and hence identify the interface of interaction.

1.1. Cross-linkers

Many types of cross-linkers are commercially available, offering a wide variety of probing tools. Cross-linking reagents can be divided into four general classes: homobifunctional, heterobifunctional, trifunctional, and zero length *(9)*.

Homobifunctional cross-linkers contain a spacer (of various lengths) linking two identical functional groups (Fig. 31.1A). These cross-linkers are designed for simple one-step reactions. Heterobifunctional cross-linkers contain two different functional groups linked by a spacer (see Fig. 31.1B). Having two different reactive groups allows greater control over the cross-linking reaction, enabling the incorporation of each reactive group in separate steps. For example, a NHS ester reactive group can be conjugated with amine groups on one protein, whereas a second photoreactive group will be brought to reaction once the binding partner has been added.

In the case of trifunctional reagents (see Fig. 31.1C), the third functional group can either incorporate an affinity tag *(10)* or another reactive group to cross-link a third site. The presence of an affinity tag, such as biotin, allows a much simpler isolation of inserted cross-linked peptides after enzymatic digestion. Sulfo-SBED is a trifunctional cross-linker that has been used in mapping protein interfaces *(11)*.

Zero-length cross-linkers *(12)* covalently link two proteins without the incorporation of a linker (see Fig. 31.1D).

1.2. Functional Groups

With cross-linking becoming widely used for numerous purposes, hundreds of cross-linking reagents have become commercially available; however, their reactions with proteins are based on a limited number of organic reactions. A large variety of cross-linking reagents are created by mixing and matching these functional groups and varying the spacer lengths between them. The most commonly used functional groups are described here. Although most functional groups are highly specific, it must be noted that nonspecific reactions may also occur at various physiological conditions.

1.2.1. Amine-Reactive Cross-linking Reagents (NHS Esters, Imidoesters)

Amine-reactive cross-linking reagents are among the first side chain–specific cross-linking reagents to be developed. The chemistry of amine-reactive cross-linkers is based on reaction with primary amines. Two main amine-reactive functional groups are used; NHS esters and imidoesters.

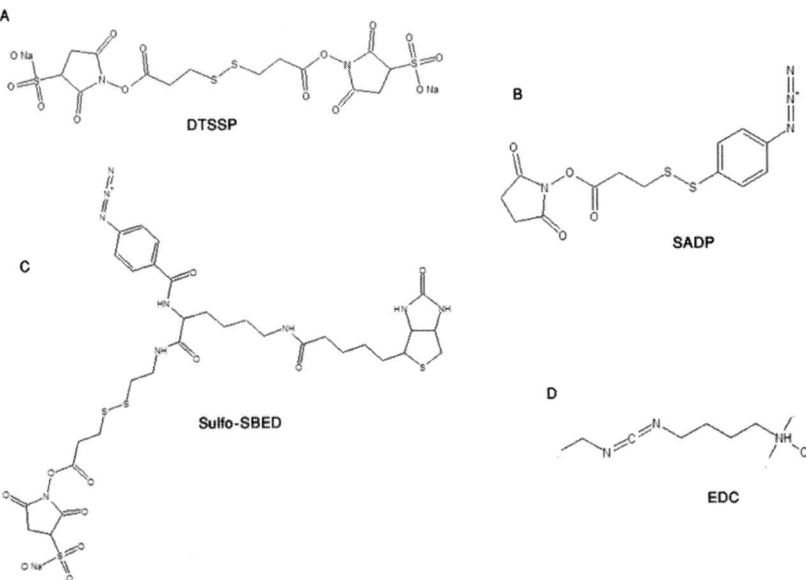

Fig. 31.1 Examples of commercially available cross-linkers. **A.** DTSSP is a homobifunctional amine reactive cross-linker. **B.** SADP is a heterobifunctional cross-linker with amine and sulfhydryl-reactive groups. **C.** Sulfo-SBED is a trifunctional cross-linker containing an amine reactive and photoreactive group as well as a biotin tag. **D.** EDC is a zero length cross-linker, which reacts with amine and carboxylic acid groups.

NHS esters react with ε-amines on lysine residues and free α-amines of N-termini from proteins, forming amide bonds (Fig. 31.2A) *(13,14)*. The reaction can be followed by monitoring NHS release by UV spectroscopy at 260 nm. Both water-soluble and -insoluble reagents are available, with the water-soluble counterpart containing a sulfonate group. Reactions are usually performed at temperatures between 4 °C and room temperature, at a pH between 7 and 9. The reaction is usually quenched with primary amine-containing buffers such as Tris *(7)*. NHS esters might also react with serines and tyrosines *(15)*.

A disadvantage of amine reactive cross-linkers is that a reacted lysine is cleaved with a much lower frequency by trypsin (the most widely used proteolytic enzyme in this field). This results in larger fragments after digestion that are more difficult to identify by mass spectrometry.

Imidoesters also react with ε-amines and free α-amines of N-termini, but form an amidine linkage. Imidoesters react very rapidly with amines at alkaline pH; however, they also have a short half-life for hydrolysis *(16)*. It has been shown that the reaction forming the amidine bond is reversible at high pH *(17)*. Diimidoesters were used to cross-link neighboring proteins in the 30S ribosomal subunit, identifying interactions between S7–S9 and S13–S19 subunits *(18)*.

1.2.2. Maleimides (Sulfhydryl-Reactive Reagents)

Maleimides target thiol groups of cysteines forming thioether bonds (see Fig. 31.2B). Such cross-linkers have been successfully utilized to map interaction sites *(19)*. There are many advantages of using maleimide groups in cross-linking, including high yield of cleavable products and the ability to

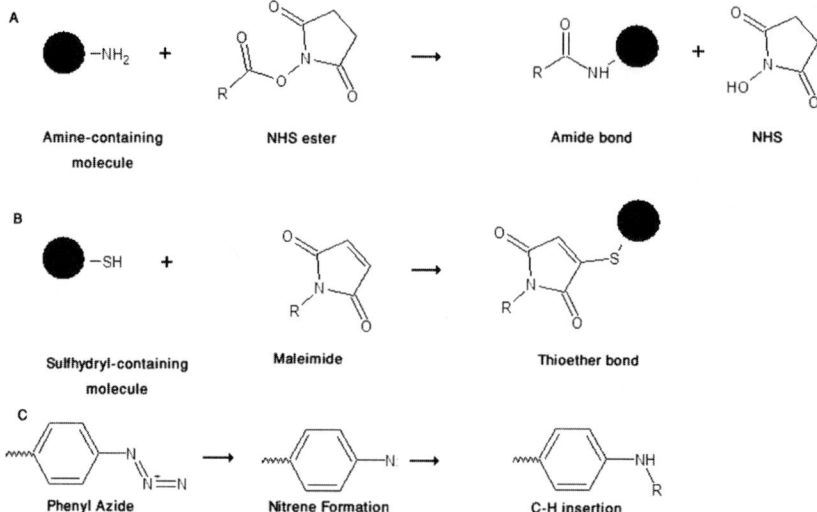

Fig. 31.2 Many chemical reactions are possible between cross-linkers and proteins. **A.** Reaction of NHS-ester cross-linker and an amine forming an amide bond. **B.** Reaction of maleimide and a sulfhydryl forming a thioether bond. **C.** One of many possible radical reactions of phenyl azide.

perform sequential coupling. At pH 6.5–7.5 maleimides react specifically with sulfhydryl groups to form irreversible bonds *(20)*; however, at more basic pH a reaction may occur with amines. An advantage of using maleimides over amine reactive groups is that the reaction leaves lysines accessible for tryptic digestion. Sulfhydryls can be introduced into proteins by reacting primary amines with 2-iminothiolane as the number of free thiols in most proteins is low.

1.2.3. Photoreactive Reagents

Azides and other photoreactive groups become reactive in the presence of UV or visible light. The radical reaction at the presence of light forms a nitrene group that can be inserted into C-H (see Fig. 31.2C) or N-H. Heterobifunctional and trifunctional cross-linkers containing a photoreactive group allow a simple two-step reaction in which one reactive site of the cross-linker is chemically conjugated in the absence of light, whereas the photoreactive site is conjugated to the protein after addition of the binding partner. This minimizes the potential for self-reactivity and increases the probability that any of the cross-linked partners is present as a specific heterodimer. Another advantage is that the nonspecific reactivity of the photoreactive site allows a variety of residues to be good candidates for cross-linking studies, by contrast to cross-linkers that require a specific amino acid for reaction. The nonspecific reactivity becomes a disadvantage for identification of the exact points of cross-linking because the range of possible reaction sites is much larger. Another disadvantage of photoreactive cross-linking reagents is that the radical reaction might cause random insertions into the protein. Hence, all species are present in very low abundance, which greatly complicates the analysis of the resulting fragments.

1.3. Cross-linked Products

The insertion of a chemical cross-linking reagent into proteins can result in a multitude of products. It is important to understand the various types of cross-linked products in order to successfully identify them. Upon proteolysis one of three types of cross-linking products (Fig. 31.3) can be formed; these have been designated type 0, 1, or 2 *(21)*. Type 0 (or a dead end) product results from a cross-linker that has reacted to a peptide at one of its reactive sites but is hydrolyzed at the other end, thus losing its ability to form a cross-link. This type of product provides information on the surface accessibility of the amino acid. A type 1 (or cyclic) product results when both reactive groups cross-link within the same peptide following trypsin digestion. Intrapeptide products provide limited distance information as cross-linked amino acids are generally close in primary structure. A type 2 or interpeptide cross-linked product results from cross-linking of two residues that are either well separated in primary structure, with cleavage sites between them, or on separate protein molecules. These products usually provide the most useful set of distance constraints.

1.4. Identification of Cross-links

The most significant challenge in structural studies using cross-linkers is the actual identification of cross-linked peptides in the mass spectrum. The challenges of analyzing cross-linked products arise from the relatively low abundance of cross-linked material in comparison to underivatized peptides, which leads to difficulties in assignment, especially if no sequence data (MSMS) are available. Assignment by mass alone is problematic in cases when masses can be assigned to native peptides as well as different cross-linked products. Tandem MS (MS/MS) is also challenging as complex spectra are formed when two cross-linked peptides are fragmented. Over the past few years, various approaches have been devised to address this problem. Innovations devised to address this problem include isotope labeling of cross-linkers, introduction of

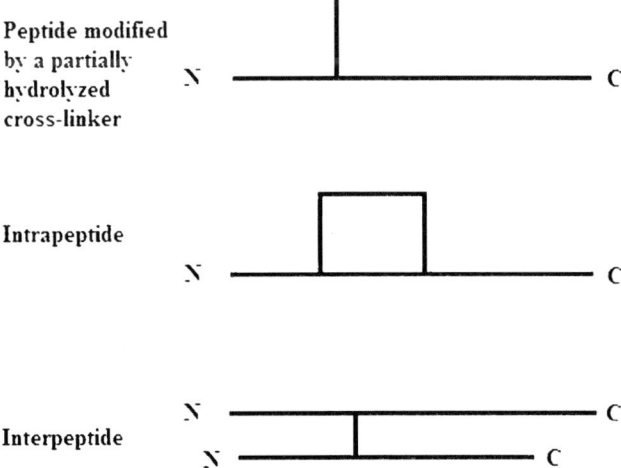

Peptide modified by a partially hydrolyzed cross-linker

Intrapeptide

Interpeptide

Fig. 31.3 The three main types of cross-linking products formed by bifunctional cross-linking reagents.

cleavable cross-linkers and addition of affinity tags to cross-linkers (9). These methods provide different approaches to facilitate correct identification of cross-linked species.

1.4.1. Isotope Labeling

Isotope labeling is a common approach in mass spectrometry. In this case, the use of a mixture of "light" and "heavy" cross-linking reagents allows for easy visual identification of cross-link doublets in a spectrum using standard mass spectrometric ionization techniques, such as matrix-assisted laser desorption/ionization-time of flight (MALDI-TOF) or electrospray ionization (ESI). Isotope labeling involves the use of a 1:1 mixture of identical cross-linking agents differing, for example, only by the number of deuterium atoms of their chemical composition. Cross-linking agents, such as BS^3-d_0 and BS^3-d_4, which differ by the presence of four deuterium atoms, are commercially available (22). The assumption is that the chemical reactivities of the isotope-labeled reagents are identical, thereby yielding doublets separated by the defined mass difference (e.g., 4.025 amu) in the mass spectrum. This usually enables unambiguous discrimination between underivatized peptide peaks and those modified by the specific incorporation of the synthetic cross-linker.

1.4.2. Cleavable Cross-linking Reagents

Since fragmentation spectra provide much more reliable peptide identifications than the parent mass alone, methods to enable cross-linker cleavage (thereby reconstituting linear peptides with interpretable fragmentation spectra) are desirable. This can be achieved by the insertion of a disulfide bond or carboxylic ester into the cross-linker. One such cross-linker is DTSSP (3,3'-Dithiobis[sulfosuc cinimidylpropionate]), which contains a thiol cleavable bond. Comparing the mass spectra of unreduced and reduced peptide allows for quick identification of putative cross-linked fragments (23,24). Masses corresponding to peptides that disappear after reduction are listed as putative cross-linked peptides, and further identification can be carried out by finding the two halves of the cross-linked peptides in the reduced sample.

1.4.3. Affinity-Tagged Cross-linkers

An obvious solution to the problem of low abundance cross-links in a peptide digest is to enrich or purify only those peptides containing the cross-link. Trifunctional cross-linkers containing an affinity tag are powerful tools for cross-linker identification and assignment. With a simple purification step (avidin column in the case of a biotin-tagged cross-linker) non–cross-linked native peptides are eluted, leaving only peptides with bound cross-linker for mass spectrometric analysis. This allows for low abundant cross-linked peptides to be concentrated without native peptides dominating the resulting spectra (11).

1.4.4. MALDI-TOF/TOF-MS of Cleavable Cross-linkers

At the forefront of cross-link identification methods a new promising protocol is being developed combining the DTSSP cleavable cross-linker and MALDI TOF/TOF-MS. This breakthrough uses the characteristic asymmetric breaking of disulfide bonds. A distinct 66 amu-doublet is observed after fragmentation of the disulfide bond from a DTSSP cross-linker (Fig. 31.4). The asymmetric fragmentation of the DTSSP cross-linker gives rise to dehydroalanine and thiocysteine analogues that are 54 and 120 mass units larger than the parent peptides.

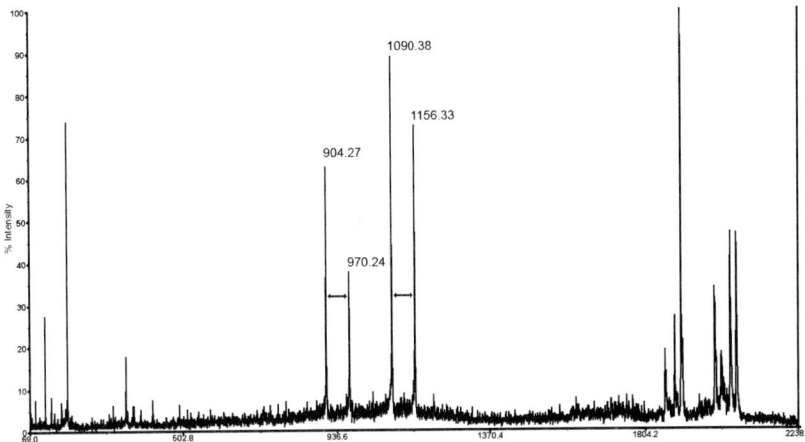

Fig. 31.4 Unique fragmentation of DTSSP cross-linked fragments using MALDI TOF/TOF-MS. The two sets of peaks, which are 66 amu apart from each other, represent the asymmetrical fragmentation of each half of the cross-linked species.

1.5. Mass Spectrometry

Various mass spectrometric methods are available for identification of cross-linked species. Each technique has its own advantages, so various techniques are sometimes required for a thorough analysis. A liquid chromatography (LC) separation step before mass spectrometry has become an invaluable tool as it reduces ion suppression.

1.5.1. MALDI-TOF Mass Spectrometry

One of the MS analyses available to identify possible cross-linked peptides is MALDI-TOF mass spectrometry and has been utilized in numerous studies (3,10,25,26). By comparing modified and non-modified digests one can assign potential cross-linked peptides by mass alone (assuming a well-calibrated instrument). MALDI-TOF mass spectra allow for a quick identification of putative cross-links after overlaying the spectra. Another advantage of this method is the small amount of material needed to carry out the analysis. The biggest disadvantage of using MALDI-MS is the preference for arginine-containing sequences as opposed to lysine-containing sequences. Lysine-containing peptides are often suppressed and their signal is indistinguishable from background noise, whereas arginine has a greater ionization efficiency derived from its more basic side chain (27). This problem of preference can be overcome by the use of an alternative protease to trypsin or by converting lysines to homo-arginines using O-methylisourea (28).

Recently MALDI-LC/MS/MS using a MALDI-TOF/TOF mass spectrometer has been proved to be an invaluable tool in identifying DTSSP cross-linked peptides (or other disulfide-based cleavable cross-linkers). This type of mass spectrometer is valuable for identifying all types of cross-linkers using a preceding LC separation step and MS/MS sequencing. However, it is the distinct fragmentation pattern for DTSSP cross-links that makes this cross-linker such an invaluable tool. The downside is that this method has the potential of generating a large number of spectra that must be examined, and no automated tools are yet available.

1.5.2. Electrospray Ionization

ESI mass spectrometers are also widely used for analysis of cross-linked peptides *(7)*. With a separation technique such as an LC system attached, ESI-MS presents a powerful tool for the identification of cross-linked peptides. Another advantage of ESI is the lack of preference for specific residues, which is prevalent in MALDI-TOF MS.

1.5.3. Fourier Transform Ion Cyclotron Resonance

Fourier transform ion cyclotron resonance (FTICR) mass spectrometers have also been utilized for cross-linking studies *(4)*. FTICR offers both ultra-high resolution and a gas-phase purification step. Its ultra-high accuracy allows for unambiguous identification of cross-linked species by mass alone *(29)*.

1.6. Software

Assigning mass spectrometry mass lists to cross-linked peptides can be done via various internet tools such as Automatic Spectrum Assignment Program (ASAP), MS2Assign *(21)*, and MS2PRO (all three can be found at http://roswell.ca.sandia.gov/~mmyoung/index.html). Such tools allow for analysis of various cross-linker reagents, proteases, and MS equipment. MS2PRO and MS2Assign can also be used to assign MS/MS fragmentation of underivatized and cross-linked peptides. In the authors' lab, custom software was developed to help identify cross-links from mass spectra.

1.7. General Methodology

Two general approaches can be applied to cross-linking studies: top-down and bottom-up *(9)*. The top-down approach analyzes intact cross-linked proteins, in which gas phase purification and fragmentation are carried out within the mass spectrometer. The more widely used bottom-up approach involves separation of cross-linked species by size exclusion, proteolytic digestion, and mass spectrometry (Fig. 31.5).

The introduction of cross-links into protein complexes can be carried out in a one- or two-step reaction depending on the chosen cross-linking reagent. Homobifunctional cross-linkers are mostly employed in a one-step reaction where both functional groups react with the protein complex. Using heterobifunctional reagents can allow for a two-step reaction in which one of the functional groups is reacted with one of the proteins in the complex, the second protein is added and the cross-linking conditions are changed to allow the other functional group to react. This process increases the chance for formation of inter-protein cross-links.

This section provides a protocol for the bottom-up approach using homobifunctional cross-linking reagents. The protein complex of interest is mixed with a cross-linking mixture and allowed to react. After quenching, the resulting cross-linked and non–cross-linked proteins are run on SDS-PAGE for separation. The band of interest is excised and used for *in-gel* digestion. The resulting peptides are submitted to mass spectrometric analysis and the resulting masses are analyzed by using software packages to identify cross-linked products. Computational modeling and/or docking is carried out based on distance constraints from identified cross-links.

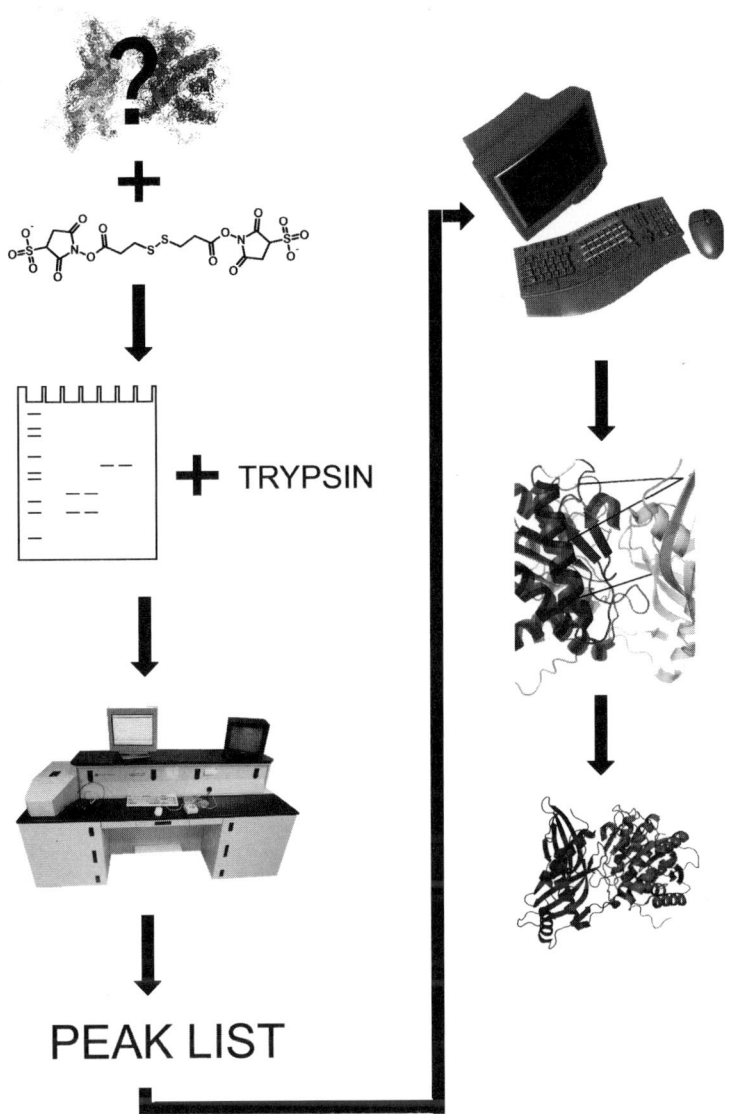

Fig. 31.5 The overall process for protein complex structure determination using the BS^3 or DTSSP cross-linker and mass spectrometry. **A.** The addition of BS^3 or DTSSP (homobifunctional lysine specific cross-linker) to the complex. **B.** The cross-linked complex is separated on an SDS-PAGE gel for in-gel digestion and extraction of peptides. **C.** The extracted peptides were analyzed by MALDI-TOF-MS and LC/ESI-MS. From the mass spectra a list of masses is extracted. **D.** These masses are then assigned using an in-house program that searches for possible cross-links, and these cross-links are used as distance constraints for rigid body docking. **E.** The best structure is determined using a scoring function based on hydrophobic interactions.

2. Materials

2.1. Chemical Cross-linking

1. Bis(sulfosuccinimidyl) suberate cross-linker (S5799; Sigma, St. Louis, MO). Store at −80 °C.

2. DTSSP cross-linker (21578; Pierce, Rockford, IL) Store at −80 °C.
3. 100 mM Tris-HCl pH 8.0.
4. Protein buffer (primary amine free buffer with pH between 7 and 9).

2.2. SDS-PAGE

1. Pre-cast 4–20% SDS-PAGE gel (nonreducing).
2. 5× running buffer: 125 mM Tris, 960 mM glycine, 0.5% (w/v) SDS. Store at room temperature.
3. Prestained molecular weight markers (Bio-Rad, Hercules, CA).
4. Coomassie Blue stain: 0.5 g Coomassie Blue (0.1% w/v in final solution, 50 ml acetic acid (10% v/v in final solution), 100 ml methanol (20% v/v in final solution), 350 ml water.
5. Destain buffer: 400 μl ethanol, 100 μl acetic acid, 500 μl water.

2.3. In-gel Digestion and Extraction

1. Wash buffer: 50% CH_3CN, 50 mM NH_4HCO_3.
2. 0.5 mg/mL trypsin (Sigma). Store at −80 °C.
3. 50 mM NH_4HCO_3.
4. Extraction buffer: 60% CH_3CN/0.1% TFA.

2.4. MALDI-MS

1. CHCA (α-cyano-4-hydroxycinnamic acid) matrix (#G2037A; Agilent Technologies, Santa Clara, CA).
2. ProteoMass peptide and protein MALDI-MS calibration mix (#MSCAL1-1KT; Sigma).
3. 1 M DTT stock.

2.5. LC/MS

1. C18 capillary column (Agilent).
2. 40% CH_3CN/0.1% acetic acid.
3. 100% H_2O/0.1% acetic acid.
4. 100% CH_3CN/0.1% acetic acid.

2.6. Peak Assignment

1. Preferred peak assignment program capable of assigning cross-linked peptides (http://roswell.ca.sandia.gov/mmyoung/index.html).

2.7. Modeling

1. Structure of each protein/domain in Protein Data Bank format.
2. Rigid body docking algorithm capable of accepting distance constraints between atoms as an input.

3. Methods

This section describes the method used to determine the structure of latexin: carboxypeptidase A (CPA) complex (Mouradov et al., 2006) and mouse acyl-CoA thioesterase (using BS^3 and DTSSP cross-linkers, and MALDI-TOF and ESI mass spectrometry).

3.1. Cross-linking

1. Combine $100\,\mu l$ complex (~3 mg/mL at pH 7–9) with 10× molar excess of cross-linking solution BS^3 (S5799; Sigma) or DTSSP cross-linkers (21578; Pierce) in separate Eppendorf cups.
2. Incubate for 2 hours at room temperature (see Note 1).
3. Quench reaction using 100 mM Tris buffer (pH 8).

3.2. SDS-PAGE and In-gel Digestion and Extraction

1. Prepare running buffer by diluting 100 mL of 5× with 400 ml of water.
2. Add running buffer to upper and lower chambers of gel unit. Load markers and $40\,\mu l$ of sample in multiple lanes onto a Gradipore precast SDS-PAGE gel.
3. Separate out cross-linked complex from higher molecular aggregates by running for approximately 1.5 hours at 170 V.
4. Stain with Coomassie Brilliant Blue G250 or R250 and destain to visualize the bands (Fig. 31.6).
5. Excise the bands corresponding to your complex. You may pool two or more excised bands into one Eppendorf cup. Excise at least two bands for best results.
6. Wash the excised bands several times with of $200\,\mu l$ of 50% CH_3CN, 50 mM NH_4HCO_3. Then dry and incubate in $5\,\mu l$ of 0.5 mg/ml trypsin (Sigma) and $200\,\mu l$ of 50 mM NH_4HCO_3 at 37 °C overnight for digestion.
7. Transfer the digestion into clean Eppendorf cup. Perform three peptide extractions by adding $200\,\mu l$ of 60% CH_3CN/0.1% TFA to the gels, shaking at 200 rpm for 30 minutes, centrifuge at 1000 g for 1 minute, and then pool.
8. Concentrate pooled sample using a SpeedVac or Millipore ZipTips C18 (see Note 2) system and resuspend in approximately $60\,\mu l$ 60% CH_3CN/0.1% TFA.

3.3. MALDI Mass Spectrometry

1. Separate the sample cross-linked with DTSSP in two parts. One of the two samples is reduced with 10 mM DTT for 5–10 minutes.
2. Combine your digested sample in a 1:2 ratio with the CHCA matrix (Sigma) (see Note 3).
3. Prepare $1\,\mu l$ of analyte/matrix mixture onto 96-well plate and allow to fully dry.

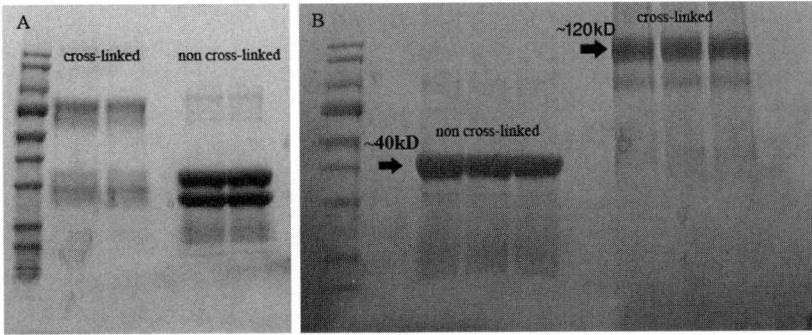

Fig. 31.6 SDS-PAGE of cross-linked and non–cross-linked samples. **A.** Cross-linking carboxypeptidase A and latexin shows the formation of a 1-to-1 complex. **B.** Cross-linking of murine acyl-CoA thioesterase shows the formation of a cross-linked trimer.

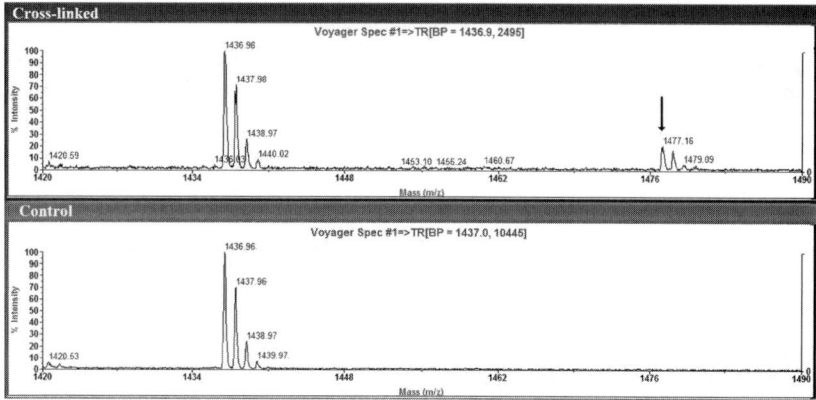

Fig. 31.7 A MALDI-TOF mass spectrum from cross-linked and non–cross-linked tryptically digested murine acyl-CoA thioesterase. The arrow show a mass linked with a potential cross-linked fragment. These assignments are made based on mass alone and must be checked against a non–cross-linked control.

4. Prepare 1 µl of the low molecular weight calibration mixture (Sigma) onto the 96-well plate and allow to dry.
5. Load the sample plate and apply calibration. Each sample is initially analyzed by adding approximately 500 laser shots to one mass spectrum. Export the peak list (Fig. 31.7). For best results de-isotope the series first.

3.4. LC/MS

1. Separated the sample first by reverse phase HPLC (Agilent) using a C18 capillary column. The elution gradient is 0–60% (v/v) acetonitrile in 0.1 acetic acid over 45 minutes with a flow gradient of 0.1 µl/min. The ES spectrum is recorded on an Applied Biosystems QSTAR Pulsar mass spectrometer.
2. Manually go though spectra and extract masses (see Note 4).

3.5. Peptide Assignment

1. Analyze the set of *m/z* peaks obtained from the mass spectra using an in-house program that assigns *m/z* values to possible cross-linked peptide fragments from amino acid sequences (assignment programs are available on the Web) (see Notes 5 and 6).
2. Identify putative cross-linked peptides by comparison of reduced and non-reduced DTSSP spectra.
3. Cross-check putatively assigned cross-linked fragments with the original spectra for validation of "real" peaks by identifying multiple charged states (see Note 7).

3.6. Docking with Distance Constraints

1. The authors' in-house docking algorithm uses a systematic six-dimensional search over all rotations in steps of 5 degrees and all Cartesian translations

of 1.0 Å up to ±66 Å along each coordinate axis. The distance constraint between Cα–Cα distance of cross-linked lysine residues is estimated at 25 Å for the BS^3 and DTSSP cross-linkers. A linear scaling grid cell algorithm with geometric hashing is used to check for any intermolecular residue pairs in close spatial proximity, and thus to exclude those models with steric overlap.

2. Score models by a simple hydrophobic energy score that counts the number of contacts (<8 Å) between hydrophobic amino acids. Consider the top 1,000 structures.

3. Cluster the top 1,000 structures based on the root mean square deviations (RMSD) of the coordinates of Cα atoms. The group with the highest average hydrophobic contacts is considered. The best docked structure is considered to be the highest scoring model based on the scoring function in that group (Figs. 31.8 and 31.9).

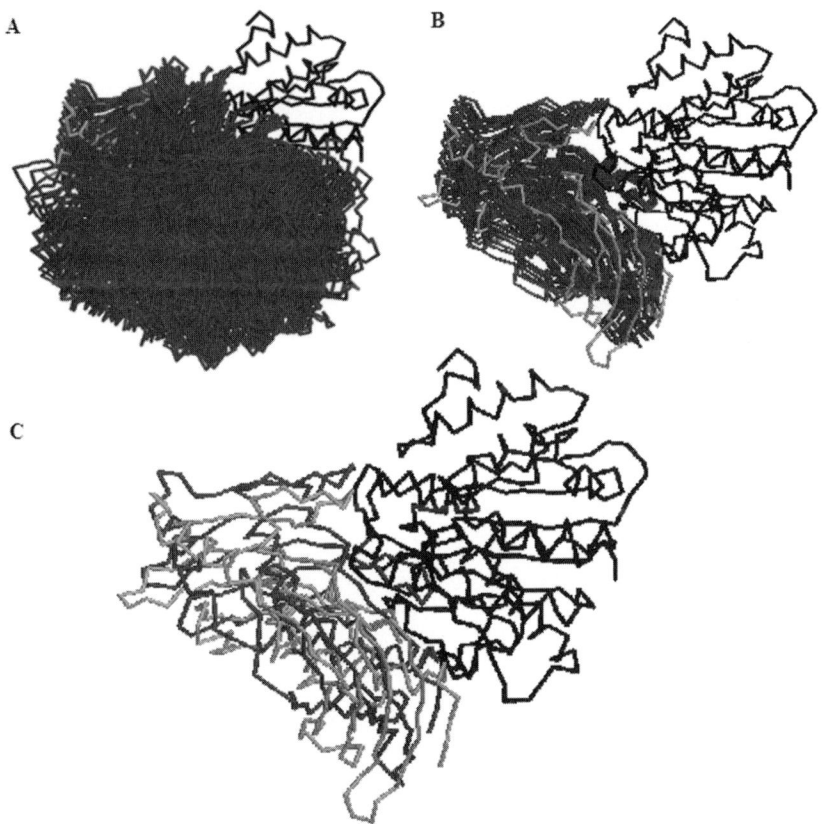

Fig. 31.8 Docking of carboxypeptidase A *(dark gray)* to latexin *(black)* using in-house rigid body docking software. **A.** The top 1,000 structures are shown. **B.** The top scoring cluster based on average hydrophobic interaction scoring. **C.** The top scoring docked structure superimposed onto the crystal structure configuration *(light gray)*. The crystal structure of this complex became available shortly after this model was built *(30)*, which allowed a direct comparison.

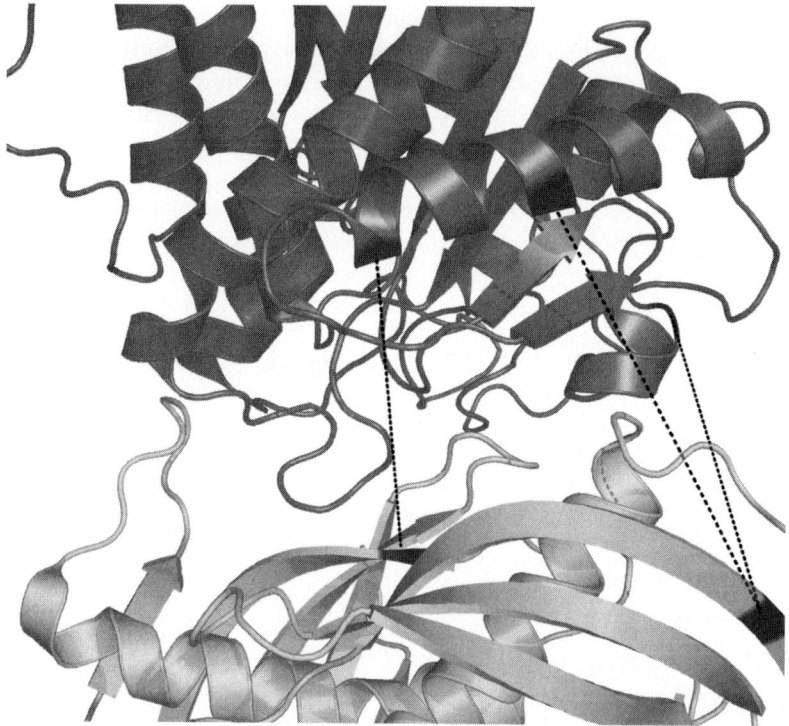

Fig. 31.9 Close-up of the identified interface between latexin *(light)* and carboxypeptidase A *(dark)* as predicted via docking. The dashed lines represent cross-links and the black regions represent location of cross-linked residues.

4. Notes

1. Cross-linking for 2 hours may cause covalently linked higher-order polymers of complexes that tend to naturally form polymers. In order to reduce formation of higher order polymers the reaction must be quenched earlier. A time-course with various quenching times is the best way to find the optimal reaction time.

2. ZipTips (C18 for peptides, C4 for proteins) are preferred for concentrating samples in preparation for mass spectrometry as it de-salts the sample. Salt interferes with both MALDI and ESI mass spectrometry.

3. MALDI-TOF mass spectra can be improved through a process of lysine guanidination *(28)*. MALDI-TOF mass spectra of tryptic digests are dominated by arginine-containing peptides, whereas lysine terminated peptides give signals with lower intensities. By using *O*-methylisourea, one can convert lysine to homoarginine, preventing this problem.

4. Although software is available to extract masses from ESI files, it is strongly recommended to go through the spectra manually. Most automated software extracts many false-positive masses.

5. A maximum allowed error must be defined between the observed masses and expected masses of cross-linked peptides. The error will depend on mass accuracy of the mass spectrometer that is being used. As cross-linked

peptides are in a much lower abundance than native peptides; their peak intensity is considerably lower, and therefore might have a greater error.

6. Assigning mass to peaks of cross-linked proteins can be done via various available internet tools such as ASAP. The authors developed their own tool to meet their requirements.

7. Without sequencing data, cross-link assignment can be a challenging exercise. With mass-only assignments it comes down to how confident one is of the assignment. This involves checking the potential cross-linked mass against native peptide masses as well as trypsin self-cleavage fragments.

Tandem MS (MS/MS) analysis can be added as an extra step. However, fragmentation of two cross-linked peptides creates very complex spectra. New techniques are currently being perfected using cleavable cross-linkers to add another degree of certainty to the assignment as well as being able to create simpler fragmentation patterns using MS/MS.

References

1. Mann, M., and Talbo, G. (1996) Developments in matrix-assisted laser desorption ionization peptide mass spectrometry. *Curr. Opin. Biotechnol.* **7**, 11–19.
2. McLafferty, F. W., Fridriksson, E. K., Horn, D. M., Lewis, M. A., and Zubarev, R. A. (1999) Biochemistry: biomolecule mass spectrometry. *Science* **284**, 1289–1290.
3. Pearson, K. M., Pannell, L. K., and Fales, H. M. (2002) Intramolecular cross-linking experiments on cytochrome c and ribonuclease A using an isotope multiplet method. *Rapid Commun. Mass Spectrom.* **16**, 149–159.
4. Dihazi, G. H., and Sinz, A. (2003) Mapping low-resolution three-dimensional protein structures using chemical cross-linking and Fourier transform ion-cyclotron resonance mass spectrometry. *Rapid Commun. Mass Spectrom.* **17**, 2005–2014.
5. Kruppa, G. H., Schoeniger, J., and Young, M. M. (2003) A top down approach to protein structural studies using chemical cross-linking and Fourier transform mass spectrometry. *Rapid Commun. Mass Spectrom.* **17**, 155–162.
6. Chen, X. H., Chen, Y. H., and Anderson, V. E. (1999) Protein cross-links: Universal isolation and characterization by isotopic derivatization and electrospray ionization mass spectrometry. *Anal. Biochem.* **273**, 192–203.
7. Mouradov, D., Craven, A., Forwood, J. K., Flanagan, J. U., Garcia-Castellanos, R., Gomis-Ruth, F. X., Hume, D. A., Martin, J. L., Kobe, B., and Huber, T. (2006) Modeling the structure of latexin-carboxypeptidase A complex based on chemical cross-linking and molecular docking. *Protein Eng. Des. Sel.* **19**, 9–16.
8. Young, M. M., Tang, N., Hempel, J. C., Oshiro, C. M., Taylor, E. W., Kuntz, I. D., Gibson, B. W., and Dollinger, G. (2000) High throughput protein fold identification by using experimental constraints derived from intramolecular cross-links and mass spectrometry. *Proc. Natl. Acad. Sci. USA* **97**, 5802–5806.
9. Sinz, A. (2006) Chemical cross-linking and mass spectrometry to map three-dimensional protein structures and protein-protein interactions. *Mass Spectrom. Rev.* **25**, 663–682.
10. Trester-Zedlitz, M., Kamada, K., Burley, S. K., Fenyo, D., Chait, B. T., and Muir, T. W. (2003) A modular cross-linking approach for exploring protein interactions. *J. Am. Chem. Soc.* **125**, 2416–2425.
11. Sinz, A., Kalkhof, S., and Ihling, C. (2005) Mapping protein interfaces by a trifunctional cross-linker combined with MALDI-TOF and ESI-FTICR mass spectrometry. *J. Am. Soc. Mass Spectrom.* **16**, 1921–1931.
12. Duan, X., and Sheardown, H. (2006) Dendrimer crosslinked collagen as a corneal tissue engineering scaffold: mechanical properties and corneal epithelial cell interactions. *Biomaterials* **27**, 4608–4617.

13. Bragg, P. D., and Hou, C. (1975) Subunit composition, function, and spatial arrangement in Ca2+-activated and Mg2+-activated adenosine triphosphatases of *Escherichia coli* and *Salmonella typhimurium. Arch. Biochem. Biophys.* **167,** 311–321.

14. Lomant, A. J., and Fairbanks, G. (1976) Chemical probes of extended biological structures: synthesis and properties of cleavable protein cross-linking reagent [dithiobis(succinimidyl-s-35 propionate). *J. Mol. Biol.* **104,** 243–261.

15. Swaim, C. L., Smith, J. B., and Smith, D. L. (2004) Unexpected products from the reaction of the synthetic cross-linker 3,3'-dithiobis(sulfosuccinimidyl propionate), DTSSP with peptides. *J. Am. Soc. Mass Spectrom.* **15,** 736–749.

16. Hunter, M. J., and Ludwig, M. L. (1962) Reaction of imidoesters with proteins and related small molecules. *J. Am. Chem. Soc.* **84,** 3491.

17. Liu, S. C., Fairbanks, G., and Palek, J. (1977) Spontaneous, reversible protein cross-linking in human erythrocyte-membrane-temperature and Ph-dependence. *Biochemistry* **16,** 4066–4074.

18. Lutter, L. C., Bode, U., and Kurland, C. G. (1974) Ribosomal-protein neighborhoods. 3. Cooperativity of assembly. *Mol. Gen. Genet.* **129,** 167–176.

19. Giron-Monzon, L., Manelyte, L., Ahrends, R., Kirsch, D., Spengler, B., and Friedhoff, P. (2004) Mapping protein-protein interactions between MutL and MutH by cross-linking. *J. Biol. Chem.* **279,** 49338–49345.

20. Partis, M. D., Griffiths, D. G., Roberts, G. C., and Beechey, R. B. (1983) Cross-linking of protein by omega-maleimido alkanoyl n-hydroxysuccinimido esters. *J. Protein Chem.* **2,** 263–277.

21. Schilling, B., Row, R. H., Gibson, B. W., Guo, X., and Young, M. M. (2003) MS2Assign, automated assignment and nomenclature of tandem mass spectra of chemically crosslinked peptides. *J. Am. Soc. Mass Spectrom.* **14,** 834–850.

22. Muller, D. R., Schindler, P., Towbin, H., Wirth, U., Voshol, H., Hoving, S., and Steinmetz, M. O. (2001) Isotope tagged cross linking reagents. A new tool in mass spectrometric protein interaction analysis. *Anal. Chem.* **73,** 1927–1934.

23. Davidson, W. S., and Hilliard, G. M. (2003) The spatial organization of apolipoprotein A-I on the edge of discoidal high density lipoprotein particles: a mass spectrometry study. *J. Biol. Chem.* **278,** 27199–27207.

24. Bennett, K. L., Kussmann, M., Bjork, P., Godzwon, M., Mikkelsen, M., Sorensen, P., and Roepstorff, P. (2000) Chemical cross-linking with thiol-cleavable reagents combined with differential mass spectrometric peptide mapping: a novel approach to assess intermolecular protein contacts. *Protein Sci.* **9,** 1503–1518.

25. Sinz, A., and Wang, K. (2001) Mapping protein interfaces with a fluorogenic cross-linker and mass spectrometry: application to nebulin-calmodulin complexes. *Biochemistry* **40,** 7903–7913.

26. Itoh, Y., Cai, K., and Khorana, H. G. (2001) Mapping of contact sites in complex formation between light-activated rhodopsin and transducin by covalent crosslinking: Use of a chemically preactivated reagent. *Proc. Natl. Acad. Sci. USA* **98,** 4883–4887.

27. Krause, E., Wenschuh, H., and Jungblut, P. R. (1999) The dominance of arginine-containing peptides in MALDI-derived tryptic mass fingerprints of proteins. *Anal. Chem.* **71,** 4160–4165.

28. Beardsley, R. L., and Reilly, J. P. (2002) Optimization of guanidination procedures for MALDI mass mapping. *Anal. Chem.* **74,** 1884–1890.

29. Schulz, D. M., Ihling, C., Clore, G. M., and Sinz, A. (2004) Mapping the topology and determination of a low-resolution three-dimensional structure of the calmodulin-melittin complex by chemical cross-linking and high-resolution FTICRMS: direct demonstration of multiple binding modes. *Biochemistry* **43,** 4703–4715.

30. Pallares, L., Bonet, R., Garcia-Castellanos, R., Ventura, S., Aviles, F. X., Vendrell, J., and Gomis-Ruth, F. X. (2005) Structure of human carboxypeptidase A4 with its endogenous protein inhibitor, latexin. *Proc. Natl. Acad. Sci. USA* **102,** 3978–3983.

Section V

Structural Proteomics Initiatives Overviews

Chapter 32

Structural Genomics of Minimal Organisms: Pipeline and Results

Sung-Hou Kim, Dong-Hae Shin, Rosalind Kim, Paul Adams, and John-Marc Chandonia

The initial objective of the Berkeley Structural Genomics Center was to obtain a near complete three-dimensional (3D) structural information of all soluble proteins of two minimal organisms, closely related pathogens *Mycoplasma genitalium* and *M. pneumoniae*. The former has fewer than 500 genes and the latter has fewer than 700 genes. A semiautomated structural genomics pipeline was set up from target selection, cloning, expression, purification, and ultimately structural determination. At the time of this writing, structural information of more than 93% of all soluble proteins of *M. genitalium* is available. This chapter summarizes the approaches taken by the authors' center.

1. Introduction

1.1. Mission

The Protein Structure Initiative (PSI) of US National Institutes of Health (NIH) aims to obtain structural information on all proteins derivable from their DNA sequences (www.nigms.nih.gov/psi/). The objective of the pilot phase (PSI-1) is summarized as follows: (1) to perform pilot studies to develop high throughput methods and protocols to proceed from cloning to structure determination for representatives of diverse protein-sequence families with no sequence similarities to proteins of known structures; (2) identify critical areas and steps for further development to achieve a high throughput operation; and (3) obtain the metrics for assessing the magnitude and scale required for the production phase of PSI (PSI-2) to achieve the overall PSI objective of a comprehensive coverage of the protein structure space.

1.2. Objective

In the pilot phase, the Berkeley Structural Genomics Center (BSGC) set the goal of obtaining structural information of a near complete set of all soluble proteins in two related minimal organisms (the pathogens, *Mycoplasma pneumoniae* [MP] and *Mycoplasma genitalium* [MG], with ~700 and ~500 genes, respectively). This objective is accomplished for MG at a greater than 90% level.

From: *Methods in Molecular Biology, Vol. 426: Structural Proteomics: High-throughput Methods*
Edited by: B. Kobe, M. Guss and T. Huber © Humana Press, Totowa, NJ

1.3. Pipeline

To achieve this, the authors have developed methods and protocols to automate or parallelize many processes from cloning the target genes to structure
determination. Overall pipeline schemes for the single-path approach used in
the initial 2-year period and the multiple-path approach used for the rest of the
PSI-1 period are shown in Fig. 32.1.

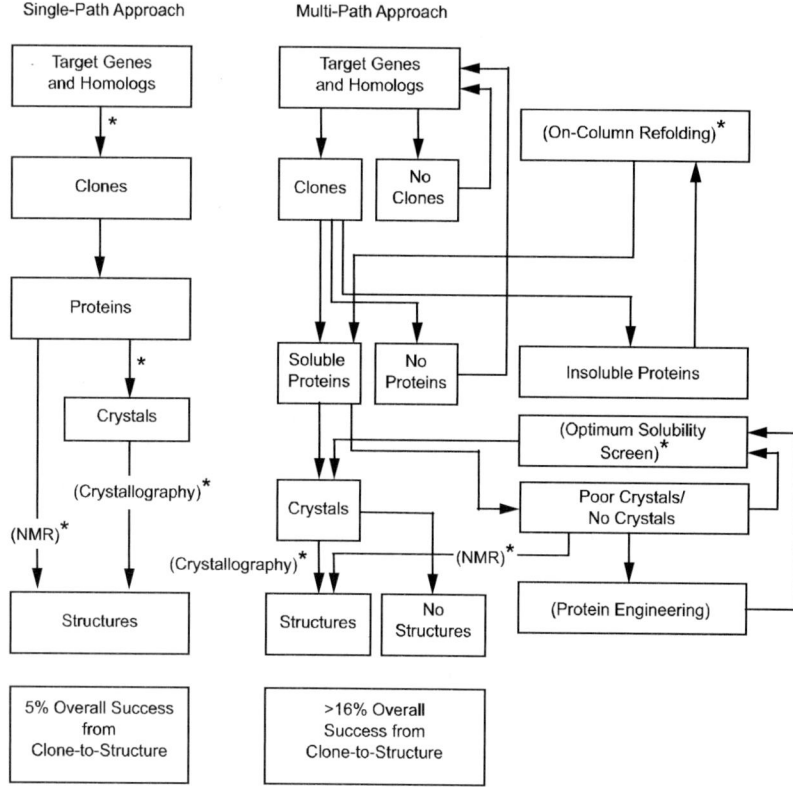

Fig. 32.1 Single-path approach vs. multipath approach for soluble proteins. A large
number of target genes and their homologues (coding for proteins with no sequence
homologies to the proteins of known structures) were selected (see Target Selection),
and the majority of them could be cloned. Of these, some were overexpressed as
proteins in soluble form, protein aggregates, or insoluble inclusion bodies. In the
single-path mode (low-hanging fruits), only the soluble proteins were screened for
crystallization or NMR studies, and of these, only some yielded structures. The
overall success rate for the single-path approach was about 5%. By contrast, in
the multipath mode, those clones not expressing, or underexpressing could be
recloned with different constructs and/or into different vectors to obtain additional
overexpressing clones; proteins that aggregate or could not be concentrated underwent
optimum solubility screening (see the following) to find optimum conditions in which
they were soluble and homogeneous. Those proteins that were insoluble underwent
an on-column refolding process (see the following). Proteins that were chemically and
conformationally homogeneous were used for NMR or crystallization studies. BSGC
experience shows that for this multipath approach the overall success rate increased to
about 16%. (*Processes that are automated or semiautomated.)

2. Metrics and Lessons Learned

Based on BSGC's results during the PSI-1 period, the metrics and lessons required for a structural genomics approach to a large-scale structure determination effort were learned; some were predicted and others were unexpected and surprising. Although some details may be different from those of other PSI-1 centers, the general conclusions of metrics and lessons are expected to be valid. They are summarized in the following:

1. The steps required to proceed from cloning a gene encoding a protein to determining its 3D structure can be divided into two distinct categories: (1) those in which the underlying science and technologies are well understood (thus, *automatable* by instrumentation or programming); and (2) those in which the underlying science is only partially known and the outcome of the processes are unpredictable. The most practical approach for steps of the second category is *multivariable screenings*.

2. The *single-path approach* (see Fig. 32.1), whereby for a large number of diverse genes one single optimized path is taken from cloning to structure determination, has less than a 5% success rate on average in discovering structures of "unique" proteins, the proteins without sequence similarity to those of known structures in the Protein Data Bank (PDB) *(1)*.

3. The *multipath approach* (see Fig. 32.1), in which feedback loops and multifactor screenings are employed for one or more critical steps in the path for challenging proteins that fail by a single-path approach, has a 16% or higher success rate for discovering the structure of unique proteins.

4. Approximately half of the structures of unique proteins revealed new folds, and the remaining half are "remote homologues," structures similar to known structures without sequence similarity (similar structure without sequence similarity) of known folds.

5. Approximately two thirds of the structures of "hypothetical proteins" (proteins that have no sequence homologues among the proteins of known function) infer one or a few possible molecular functions that could be experimentally tested.

6. The protein fold space can be mapped in 3D space based on pairwise structural similarities *(2,3)*; thus providing a platform for representing all protein structures, "the protein structure universe."

3. Selection of Target Proteins for High Throughput Structural Studies

3.1. Method

A structural genomics target is a protein that is selected to determine its 3D structure. BSGC targets during the PSI-1 period include *Mycoplasma* proteins as well as their sequence homologues from other prokaryotes. In general, all rounds of target selection involved three common steps. Since almost all *MG* genes have their homologues in *MP*, each step was started with the set of 677 *MP* open reading frames (ORFs) described in the original annotation of the genome *(4)*. Each ORF was then augmented with a family of homologues from available, fully sequenced prokaryotic genomes

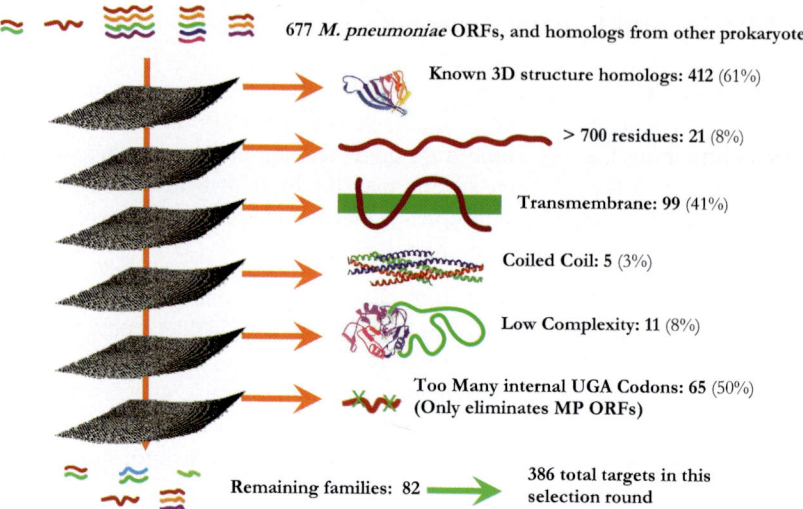

Fig. 32.2 Target selection scheme for *Mycoplasma* genes used at BSGC. Criteria of "filtering" were changed among different rounds of target selection (*See Color Plates*).

to make a target set. First, all target sets recognizably homologous to proteins of known structure in PDB were removed from further consideration. Next, target sets of proteins that were predicted to be unsuitable for high throughput study (e.g., those with predicted transmembrane helices or low-complexity regions) were eliminated. Finally, specific targets were chosen from among proteins in the remaining target sets. The number of targets chosen per family varied among selection rounds. A flow diagram of target selection for a sample round is shown in Fig. 32.2.

3.2. Databases

The following databases were used in selection of targets:

1. MP: Each step was started with the set of 677 *MP* ORFs described in the original annotation of the genome *(4)*.
2. knownstr: a database of sequences from proteins of known structure. This database contained sequences of proteins released by PDB, sequences of proteins deposited in the PDB and made available while the structure is still "on hold," and sequences from TargetDB *(5)*, for which a structure has been solved by another structural genomics center. Sequences of BSGC targets that have progressed to the "Traceable Map" stage were also included, as this usually indicates the structure will soon be completed. The database was updated prior to each target selection round.
3. snr: the nonredundant set of protein sequences from Swiss-Prot *(6)*. All sequences were included in the swissprot, trembl, and trembl_new files (downloaded July 30, 2001 for round 2; November 30, 2001 for round 3; October 21, 2002 for rounds 4–5 and 7–8; and February 26, 2004 for round 6) from Swiss-Prot, which had been filtered with the SEG *(7)* and PFILT *(8)* programs using default options. The filtering was done to reduce the chance of profile corruption *(9)*, which can lead to inaccurate results.

4. Available genomes: NCBI database of proteins from sequenced bacterial and archaeal genomes (ftp://ftp.ncbi.nih.gov/genomes/Bacteria). Targets were only chosen from genomes for which the BSGC had access to purified genomic DNA. These species are listed on the BSGC web site (http://www.strgen.org).

3.3. Identification of Known Structures

At the beginning of each round of target selection, all *MP* proteins and their homologues were considered potential targets. These were then removed from consideration if they were detectably homologous to other proteins of known structure.

1. In each automated target selection round, sequences of all *MP* ORFs were compared with the knownstr database using several sequence comparison tools such as PSI-BLAST *(10)*. PSI-BLAST position-specific scoring matrices (PSSMs) were constructed for each *MP* ORF using 10 rounds of searching the authors' "snr" database with a matrix inclusion threshold E-value of 10^{-2} in most rounds.
2. The PSSMs were used to search the knownstr database, and any hits with an E-value of 10^{-1} or below were eliminated from consideration as targets. This significance threshold was chosen to increase the likelihood of detecting more remote homologues, even though it had some risk of false-positives being removed from the target list.
3. After the second round for target selection, the matrix inclusion threshold was increased to enhance the possibility of identifying remote homologues, at the risk of a higher rate of corrupted PSSMs.
4. Because of the latter possibility, BLAST *(11)* and Pfam *(12)* in target selection rounds 3–6 were also used. All *MP* ORFs with a BLAST hit against knownstr with an E-value of 10^{-1} or below were eliminated from consideration as targets, in addition to those already eliminated by PSI-BLAST.
5. Pfam was also used to detect known structures. The HMMER tool *(13)* was used to compare the Pfam_ls library of hidden Markov models to both the knownstr database and the database of *MP* ORFS, using the family-specific "trusted cutoff" score as a cutoff for assigning significance. All ORFs that had a significant hit to a Pfam family that had also matched at least one known structure were eliminated from consideration.

3.4. Identifying MP Targets Predicted To Be Less Tractable for High Throughput Study

1. As the next step in each target selection round, *MP* proteins and domains that were likely to be predictably less tractable for high throughput study were eliminated. These included proteins with regions of amino acids predicted to be in transmembrane segments, coiled coils, and regions of low complexity. The predictions were made by the SEG program (version dated May 24, 2000) for proteins with low-complexity regions spanning more than 20% of the protein lengths, the CCP program (written by J. Kuzio at NCBI, version dated June 14, 1998), using the algorithm of Lupas *(14)*, for proteins with coiled coil regions, and two programs to identify

transmembrane regions, TMHMM 2.0a *(15)* and PHDhtm *(16)* version 2.1 (October 1998). Any transmembrane region predicted by either program eliminated an *MP* ORF from consideration as a target in rounds 2–5.

2. Potential targets that were long and therefore likely to be challenging also were eliminated; in earlier rounds (round 1–2) of target selection, the length cutoff was 400 amino acids, and in later rounds (round 3–8) it was increased to 700 amino acids.

3. Finally, proteins annotated as ribosomal components were excluded, as these were expected to be unlikely to be stable in the absence of binding partners.

3.5. Identifying Homologues of MP Proteins as Targets

In addition to the *MP* proteins themselves, homologous proteins from other prokaryotes were also chosen as targets. Each *MP* protein (or predicted domain in round 6) that passed through the described filters was used to search the database of available genomes using PSI-BLAST. PSI-BLAST PSSMs were constructed for each *MP* ORF using 10 rounds of searching the nonredundant sequence database "snr" (as described) with default parameters; the PSSMs were then used to search the database of genomes. BLAST version 2.2.4 was also used (with default parameters) in rounds 4–8 to search the genome database. All proteins identified by BLAST or PSI-BLAST with E-values more significant than 10^{-4}, with the region of local similarity covering at least 50 residues, were considered as possible targets.

3.6. Other Factors Considered

Potential targets from *MP* were always selected if they passed an additional screen to ensure they could be expressed in the *Escherichia coli* expression system used at the BSGC. *MP* and other related Mollicutes such as *Ureaplasma urealyticum* can use UGA codons to encode the amino acid tryptophan, whereas UGA is a stop codon in *E. coli*. Thus, cloned *MP* proteins with this codon express truncated proteins in *E. coli*. In cases in which a UGA codon was within about 30 bases of either end of the gene, it could easily be mutated to a UGG codon during cloning, using mutating PCR primers. Other UGA codons, called internal UGA codons, could only be mutated in a more difficult multistep cloning procedure.

When there were too many homologous targets, high priority was given to targets from thermophiles and halophiles, as these were expected to be experimentally more tractable, for example, being partially purified by heating the *E. coli* lysate.

3.7. Target Deselection

The BSGC only seeks to solve structures for protein domains for which the structure cannot be reliably predicted via bioinformatic methods. Therefore, the authors deselect and stop work on targets whose structures of similar proteins have been solved by other groups. Most deselection analysis steps are automated. However, the final decision on whether to stop work on a target is performed manually to decrease work lost due to potential false-positives. This automated analysis and manual review are both performed weekly. More details of the rationale behind this two-step approach are given elsewhere *(17)*.

4. Protein Production

For this purpose, *E. coli* recombinant expression systems are the best option in terms of economy and ease of protein production. Prokaryotic cell-free protein synthesis has also been used occasionally.

4.1. Cloning

For the past 2 years, BSGC has used an in-house version of the ligation independent-cloning (LIC) methodology *(18)*. This LIC system provides both efficient high throughput cloning and flexibility in fusion construction. The LIC method relies on common linker sequences to anneal and join the target segments to the vector. A tobacco etch virus (TEV) protease cleavage site allows cleavage of the fusion tag. The fusions (MBP, GST, TRX, NusA) are utilized primarily for enhancing soluble expression. In addition to the N-terminal His_6 tag, there is a PCR-based method of preparing targets containing a C-terminal His_6 tag that can be used in cases in which the N-terminal His_6 tag is ineffective. The simplicity of the present LIC cloning scheme allows for most of the experimental steps to be performed robotically in groups of 96 targets. The following steps are currently automated on the Biomek 2000 (Beckman Coulter, Fullerton, CA) robot: PCR reaction setup and cleanup, PCR product analysis by E-gel 96 (Invitrogen, Carlsbad, CA), LIC reaction and transformation, mini-expression setup, clone preservation in agar stab, and plasmid preparation.

4.2. Small-Scale Expression

After transformation into the expression host, two colonies are selected and grown in an autoinducing medium *(19)*. Cells are grown in a 96 deep-well plate overnight, spun down, resuspended, and sonicated using the Misonix 3000 sonicator (Misonix, Farmingdale, NY). The lysate is spun, and both soluble and insoluble fractions are run on SDS/PAGE. Presently, all steps, from PCR reaction for 96 targets to analysis of level of expression of the targets are automated and can be achieved in 5 days *(20)*.

4.3. Large-Scale Preparation of Cell Paste

Previously Luria broth with isopropyl-β-D-1-thiogalactopyranoside (IPTG) induction was used. This required the manual addition of IPTG. For the past 2 years an auto-inducing medium (developed by William Studier from Brookhaven National Lab) has been used that induces expression by balancing the levels of glucose and lactose as carbon sources. This formulation spontaneously induces high levels of target protein without the need to monitor growth and increases the soluble expression of target proteins. Two types of autoinducing media are used: (1) ZYP: native medium, and (2) PASM: medium for labeling the target protein with seleno-methionine.

4.4. Protein Purification

Parallel purification is performed on three AKTA Explorer work stations (GE Healthcare, Piscataway, NJ). The authors have recently changed their protocol to the following: Five targets are sequentially purified through three columns in an automated way using the AKTA Explorer with 3D Kit (software necessary for programming these steps). The three columns being used are: HisTrap metal-chelating—desalting column—HiTrap Q/S HP 5 ml column (GE Healthcare). This method takes 10 hours to complete.

4.5. Quality Control Assessments

All purified proteins undergo quality control steps 1–5 as listed in Table 32.1. One-dimensional NMR is performed on proteins smaller than 45 kDa that did not crystallize.

4.6. Protein Production Summary

As mentioned, BSGC is unique in that two minimal organisms (*MP* and *MG*) with the smallest genome size were chosen as the authors' target. The authors' targets have no sequence similarity to those of known structures. Even with a small starting pool of targets, a multipath approach eventually allowed the authors to produce most of their targets. For the minimal organism MP with 677 full-length predicted proteins, after filtering out the proteins that are structural homologues of known structures, those containing transmembrane domains, coiled coils, low-complexity regions, and multiple UGA codons, there remained 82 full-length target genes. Up to 10 homologues for each gene from other organisms were added to make a total of 386 targets. Out of 386 targets, 318 were successfully cloned and 261 clones gave good expression. From those, 191 proteins were purified in good quality and amounts suitable for crystallization screening.

Table 32.1 Protein characterization

Parameters	Method
1. Purity	SDS/PAGE stained with Coomassie Brilliant Blue R
2. Monodispersity	Dynamic light scattering (DynaPro 99; Wyatt Technology, Santa Barbara, CA)
3. Aggregation state	Native gel (Phast system; GE Healthcare, Piscataway, NJ)
	Analytical size exclusion chromatography (G4000SWxl; Tosohaas Corp., Montgomeryville, PA)
4. Molecular weight	Mass spectrometry (MALDI-TOF, Voyager DE; Applied Biosystems, Foster City, CA)
5. Bound elements	ICP-MS (University of Georgia, Athens, GA)
6. Functionality	Panel of enzymatic assays (in collaboration with A. Yakunin, University of Toronto, Toronto, Canada)
7. 1-D NMR	Bruker DRX 500 NMR spectrometer using an 11 (one-one) pulse sequence (D. Wemmer, University of California, Berkeley, CA)

5. Technical Development for Challenging Proteins

5.1. Heat Shock and High Salt Growth

Overexpression of many heterologous proteins results in production of refractive bodies, also known as inclusion bodies (IB). The level of these insoluble proteins can sometimes be reduced by lowering the growth temperature upon induction; changing the media composition; expressing the protein as a fusion with MBP, GST, thioredoxin, or NusA (21,22); and inducing the expression of chaperones. Other approaches for reducing IB production are salt and heat stress, which induce complementing defense mechanisms in bacterial cells, including intracellular accumulation of osmolytes or synthesis of heat-shock proteins, respectively (23,24). Simple heat shock before induction is known to enhance the solubility of some recombinant proteins produced in E. coli (25). Some osmolytes behave as "chemical chaperones" by promoting the correct folding of unfolded proteins in vitro and in the cell (26–29). These two elements, heat shock and high salt media, have been combined to increase the fraction of soluble protein produced from targets.

A protocol has been tested that combines heat shock and high salt growth (30). The cells were grown in the presence of 0.5 M NaCl and incubated at 47 °C at the beginning of induction with IPTG for 20 minutes. The temperature was then decreased to 20 °C for overnight growth. These cells expressed only soluble protein, although the total level of expression was 10-fold lower than when grown under "normal" conditions. This soluble sample was crystallized, and its structure was solved (31).

5.2. On-Column Refolding

Inclusion body formation, as mentioned, can be minimized or avoided by applying complex efforts to enhance production of soluble protein. On the other hand, protein production from inclusion bodies has a number of merits. They are: (1) produced in high yields, even those that are toxic for bacterial cells; (2) generally protected from proteolytic degradation; and (3) easily purified and solubilized. The main challenge is to convert inclusion bodies to properly folded, biologically active proteins. The authors have developed an on-column chemical refolding method (32) for insoluble His-tagged proteins expressed in E. coli partly based on the method described by Rozema and Gellman (33). IBs solubilized in urea are first bound to a metal-chelating affinity column and exposed to a detergent wash to prevent misfolding. This is followed by a β-cyclodextrin wash that removes the detergent and promotes correct folding (34). The target protein is eluted with imidazole, and then goes through further purification steps—IEX and/or SEC—before evaluation by dynamic light scattering (DLS). As an example, 10 of the PSI-1 targets from BSGC that expressed insoluble protein were purified using this method. Three of the 10 targets could not be refolded, but 30–100% refolding was obtained from the other seven. All refolded proteins were subjected to DLS analysis, and five of seven refolded proteins were monodisperse. Six of the seven refolded proteins were able to produce crystals of varying qualities.

5.3. Optimum Solubility Screen

For structural studies, the first step after a protein is purified is to concentrate it in its purification buffer to a concentration suitable for crystallization or NMR studies. This step fails in about 25% of cases because the protein aggregates and precipitates; this adverse phenomenon is totally unpredictable. Inspired by a screen for NMR studies *(35)*, a screening method *(36)* was developed in which a panel of buffers as well as many additives were tested to obtain the most homogeneous and monodisperse solution for each protein that usually aggregates and cannot be concentrated prior to setting up crystallization screens.

A panel of 24 buffers was tested using the hanging-drop method and vapor diffusion equilibrium. After monitoring precipitation, the conditions leading to clear drops were selected for DLS characterization. For this part of the screen, only 24 μl of protein (with concentration ranging from 3 to 10 mg/ml) are required. If the DLS results are not optimal, a series of additives are tested in the presence of the best buffer selected from the initial screen, and again DLS is used to determine the best condition. The OS screen has been performed on 14 samples of cytoplasmic proteins that had aggregated as measured by DLS and had precipitated upon concentration or could not be concentrated. The OS screen indicated that out of the 14 protein samples, the DLS of 11 of them could be improved in different buffers, and in some cases, an additive further improved DLS. Nine of these proteins subsequently could be crystallized.

6. Structure Determination

The overall flow of the process for determining 3D structures of proteins is well established. Although much of the science involved is understood and many of the key techniques are well developed, the underlying science is not well understood in some steps, and so the outcome appears almost stochastic and unpredictable. Thus, from an engineering and automation point of view, the component steps in the process from purified protein to 3D structure can be divided into two categories: (1) the steps that are automatable and can be operated in a high throughput mode; and (2) the steps that can only be processed in a multipath approach by screening a large number of conditions, factors, and paths to increase the probability of success for such steps. The success rate for the single-path approach from purified protein to unique structure is, on average, about 9%. In the PSI-1 pilot stage, despite the limited manual multipath approach, the success rate was increased to approximately 27% (the corresponding success rate from clone to structure is about 5% and 16%, respectively) with additional multipath steps and automation. The overall pipeline at BSGC is schematically shown in Fig. 32.3.

6.1. Crystallization

The science of protein crystallization is not well understood. Currently, the most successful and practical method for finding protein crystallization conditions is to screen a large number of conditions through the sparse matrix crystallization screening method *(37)* and its commercially available variations (e.g., from Hampton Research, Aliso Viejo, CA). During this process, the Hydra Plus One

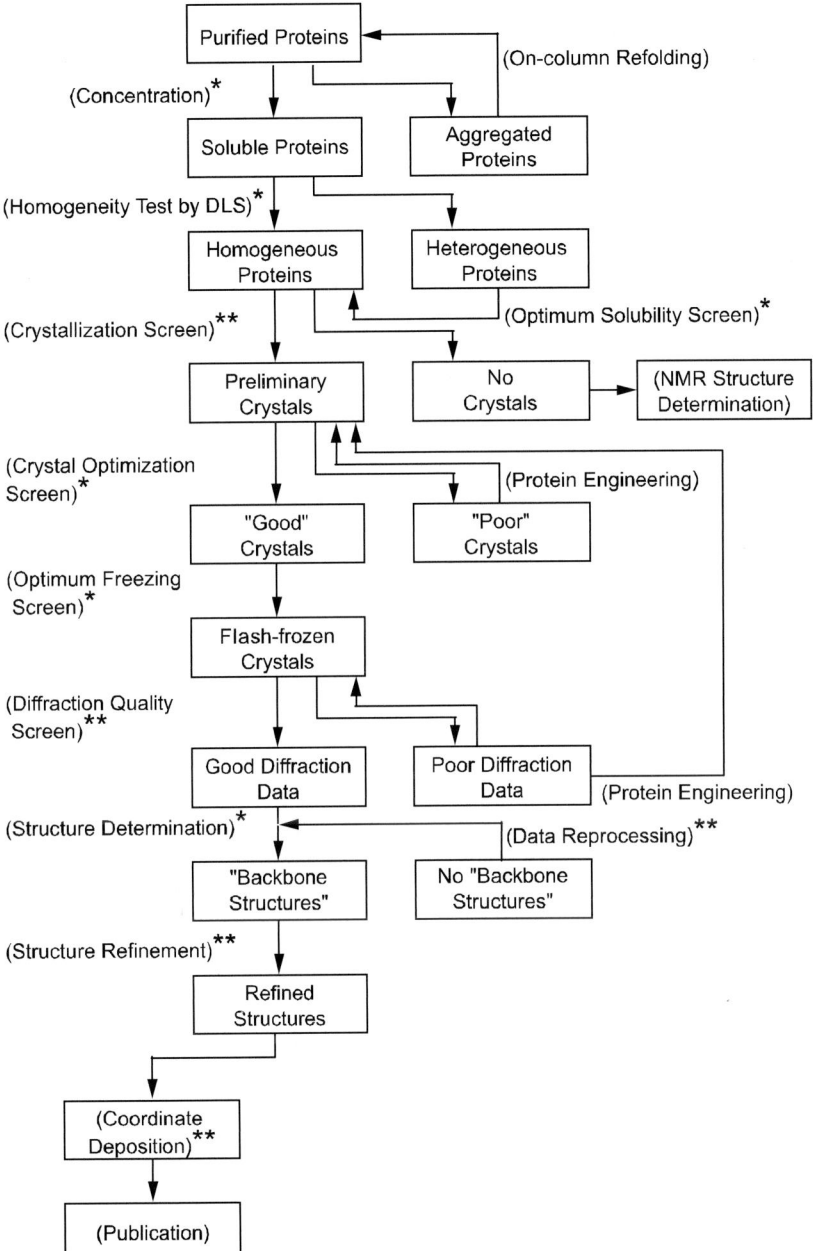

Fig. 32.3 Multipath flow diagram for the process from pure proteins to three-dimensional structures. The process for each step is in parenthesis. The automated steps are marked by **; steps that are partially automated or steps for which screens have been developed but have not been automated are marked by *.

(Matrix Technologies, Hudson, NH) and Phoenix Liquid Handling System (Art Robbins Instruments, Sunnyvale, CA) crystallization robots with 96-well plates are used. The authors routinely screen 4×96 crystallization conditions at two temperatures. Once one or more promising crystal hits are found, the hit conditions are optimized using a protocol developed by the authors to fine-tune the conditions to obtain good diffraction quality crystals. As a result of the on-column refolding step and optimum solubility screen, the success rate for the purified-protein-to-diffraction-quality-crystal process is about 27%.

6.2. Diffraction Data Collection

Many of the steps in diffraction data collection at Advanced Light Source, Lawrence Berkeley National Laboratory, are hardware and software assisted. They include the robotized automatic mounting of frozen crystals, point-and-click crystal centering, and the capability to screen frozen crystals to search for well-diffracting crystals.

6.3. Structure Solution

Once experimental data are collected, high throughput methods are applied to solve and complete a structure. The authors routinely use software developed for structural genomics efforts, such as HySS for substructure determination *(38)*. This software is part of the PHENIX package. The high level of automation that HySS provides makes it possible to determine a substructure at the beamline immediately after data have been collected and processed. Once the anomalous substructure has been located, phase calculation and substructure refinement are performed, using SOLVE *(39)*, MLPHARE *(40)*, or CNS *(41)* as dictated by data quality. In more challenging cases the SHARP *(42)* program is used. The pipeline is shown in Fig. 32.4.

The results of phasing are continued into phase improvement by density modification, using CNS *(41)*, DM *(43)*, and RESOLVE *(44)*. Visual inspection of the electron density map is used to determine whether more experimental data should be collected. For model building the authors use automatic software when possible. If data extend to 2.2 Å, the ARP/warp *(45)* suite is used, and in a favorable case 90% of the model is built. With lower resolution data (between 3 and 2.2 Å) the RESOLVE software is used to build an initial model, typically at least 50% of the main chain is built. The model is then used as a basis for manual model completion. In cases of poor data quality or resolutions below 3.0 Å, a manual model building is used. Structure refinement and model completion makes use of the standard refinement tools: CNS and REFMAC *(46)*, automated water assignment, and manual rebuilding if necessary.

7. Summary of BSGC Throughput during the PSI-1 Pilot Phase

- For the minimal organism *MP* with 677 full-length predicted proteins, after filtering out the proteins that are structural homologues of known structures, those containing transmembrane domains, coiled coils, low complexity regions, and multiple UGA codons, there remained 82 full-length target genes. Up to 10 homologues for each gene from other organisms were added to make a total of 386 targets. Of those, 318 were successfully cloned (not counting clones of domains of full length proteins). The authors' overall success rates are shown in Table 32.2.

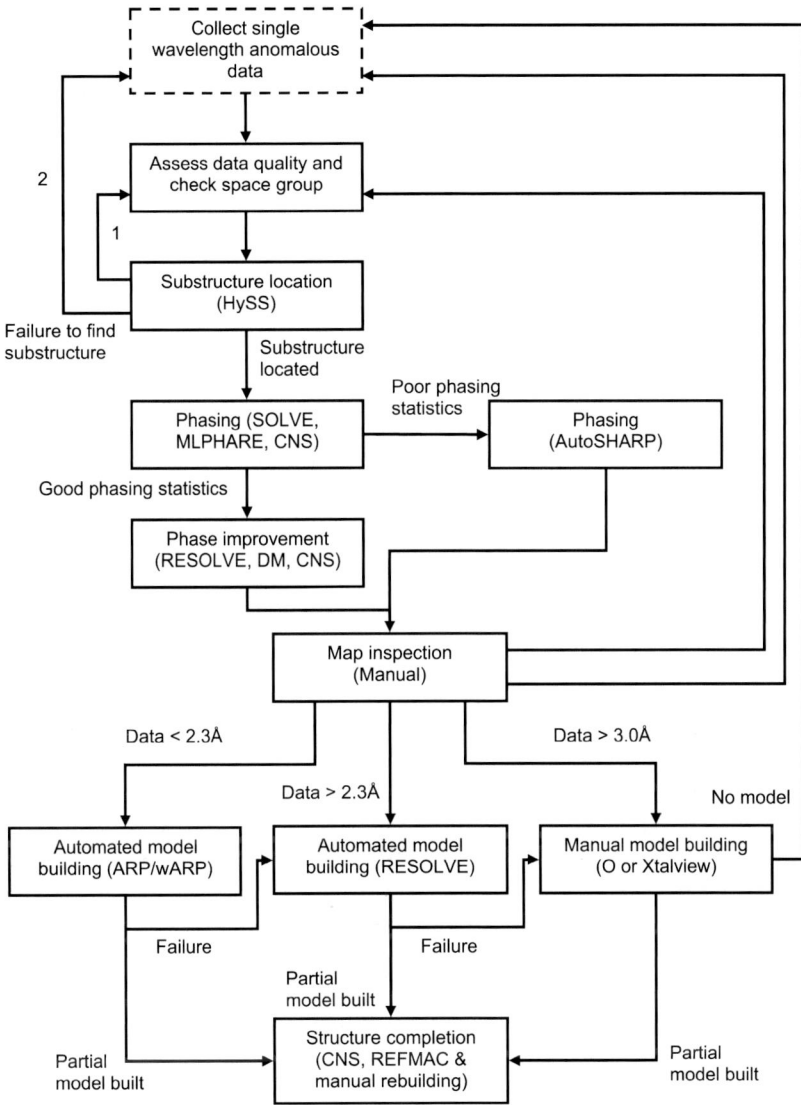

Fig. 32.4 Flowchart for structure solution, model building, and structure completion used by BSGC in the PSI pilot phase.

Table 32.2 BSGC success rate for full-length target proteins during the PSI-1 pilot phase

Full-length genes cloned	Soluble expression	Purified proteins	Crystallized	Structures solved
318	261	191	104	93

• BSGC solved structures: Almost all of BSGC targets are "unique" in that the majority of the targets have no sequence homologues among proteins of known structures. Thus the authors have had a high rate of discovering many new protein folds. A number of these also revealed unexpected bound ligands, suggesting their possible biochemical functions, and others have unusual oligomeric structures not predicted by genetic or biochemical methods. Thus, the majority of BSGC structures belong to one of four categories: (1) hypothetical proteins with novel folds, (2) proteins with novel folds that suggest their molecular functions, (3) proteins with "unique" sequences that reveal novel folds, and (4) hypothetical proteins with known folds ("remote homologues"). The protein structures in categories 2 and 4 can infer possible molecular functions *(47)*.

8. The Protein Structure Space and the Structural Proteome of a Minimal Organism

When the genomic sequence of the first organism was completed, development of computational methods to analyze the sequenced genes became the key for extracting valuable new information, most of which was totally unpredicted and unexpected. The critical importance of the computational methods became even more evident as more genomic sequences became available. As was the case with sequence genomics, the development of computational methods for analysis of the 3D structures of proteins is going to be the key to mining valuable information from the 3D structures of proteins obtained from PSI and other sources *(48)*. Toward this objective, the authors have developed a computational process to represent all unique protein structures in a multidimensional space based on structural similarities and in 3D space for approximate visual representation of the multidimensional structural space.

8.1. The Protein Structure Space Mapping

The PSI objective of near-complete coverage of protein structure space needs a representation method of the space. It has been shown recently *(2,3)* that the protein structure space can be "mapped" in three dimensions as a visual approximation of multidimensional space of the protein structure space, in which all the known and newly determined protein structures are distributed in a highly organized way. Furthermore, the demographic distribution of the protein structures in the map is understandable from the viewpoints of protein architectural features and protein fold evolution. Thus, this representation of the protein structure space provides a *unified platform* on which all the protein structures of the PSI, as well as others, can be mapped to reveal the demography of protein structures, and various structural information, functional information, and evolutionary information can be mapped and mined computationally, once such computational tools are developed.

8.2. Mapping of the Protein Structure Space

One of the major objectives of PSI is to obtain a broad coverage of the protein structure space (Fig. 32.5). To conceptualize the space, and derive new information from the demographic distribution of protein structures in the space, it is useful to define the space in terms of structural similarity. Calculating all pairwise structural similarity for all nonredundant protein structures (~2,000) in PDB, and converting them to structural dissimilarity scores, the authors were able to map the protein structure space in 3D space as a visual approximation of the full-dimensional representation (see Fig. 32.5). To accomplish this, the

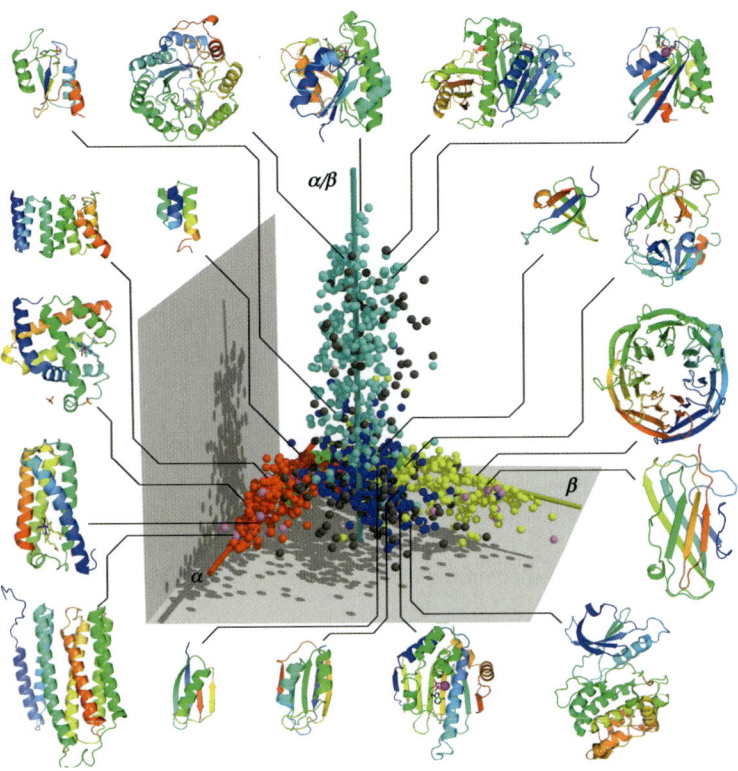

Fig. 32.5 Global representation of protein fold space. All together, 1,898 unique protein structure families are represented by spheres in the three-dimensional space. α, β, and α/β class structures follow three elongated directions denoted by α, β, and α/β feature axes *(3)*. Class designation for each structure family according to the SCOP *(49)* database is indicated by red for α-class, yellow for β-class, blue for α+β class, cyan for α/β class, pink for membrane proteins, and black for multidomain proteins. In most cases, SCOP classification approximately agrees with the demographics of the protein fold space. Some sample structures are shown.

authors used the mathematical method known as multidimensional scaling *(2,3)*. In the structural space, each point represents a unique protein structure family. In this space, each point is located in the space that best fits all pairwise distances between the point and all the rest. Two points are close to each other when their structures are similar. The following observations were made:

1. The protein structure space is sparsely populated, and all protein structures are confined to four elongated regions, each characterized by particular architectural features of proteins. This observation strongly suggests that evolution of proteins may have been strongly restricted by the requirement of architectural stability of proteins.

2. Short and poorly structured proteins are mapped near the "origin," and the size of proteins and the extent of secondary structure, or supersecondary structure, elements generally increase along each feature axis, as indicated in Fig. 32.5. This suggests that these trends may be related to protein-fold evolution.

3. The three feature axes or the three "eigen vector" axes provide a completely general and objective way of classifying protein structures, thus providing a new demographic similar to library cataloguing. Furthermore, these structure features represented by axes are easily computed from protein structure information without solving any structural alignment optimization algorithms, so they may serve as basic feature vectors for a fast, generalized, and automatic protein structure classifier.

4. All new structures from the PSI program and others map roughly within the "envelope" defined by the structural space originally found using approximately 2,000 nonredundant structures of PDB, suggesting that the "protein structure universe" is finite.

This type of representation of protein structure space provides a unified platform on which one can map all the PSI structures and others, to globally visualize the structural relationship among them, identify the regions of different structural population densities for suggesting additional new structures needed, and infer possible protein-fold evolution. Furthermore, all biochemical and biophysical functions can be mapped on the space to obtain a global view of the molecular fold/function relationship on a global level.

8.3. Structural Coverage of a Minimal Organism

At the start of PSI-1, about 2/3 of the MG proteins had no structural information, of which the majority (about 43% of total) were predicted to be soluble proteins. At the end of PSI-1, the authors now have structural information for over 90% of the soluble proteins of this minimal organism (Fig. 32.6) *(50)*.

Further analysis of this and other structural proteomes of small prokaryotes reveals an interesting conservation pattern for protein fold for proteins of particular functional categories, details of which were described recently *(50)*.

8.4. Structural Families Found in the Minimal Organism

The unique structural families represented by all the soluble MG/MP proteins and their homologues are mapped on the protein structure space (Fig. 32.7).

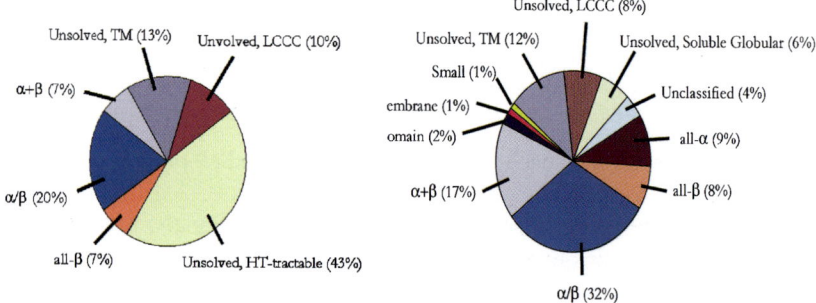

Fig. 32.6 Left. At the start of PSI-1, three-dimensional (3D) fold information was available for 34% of the proteins in *Mycoplasma genitalium*; the rest of the proteins belonged to membrane proteins (13%), low-complexity proteins (10%), and soluble proteins of unknown 3-D folds (43%). **Right.** By the end of PSI-1, 3D fold information was available for over 90% of soluble proteins in this minimal organism.

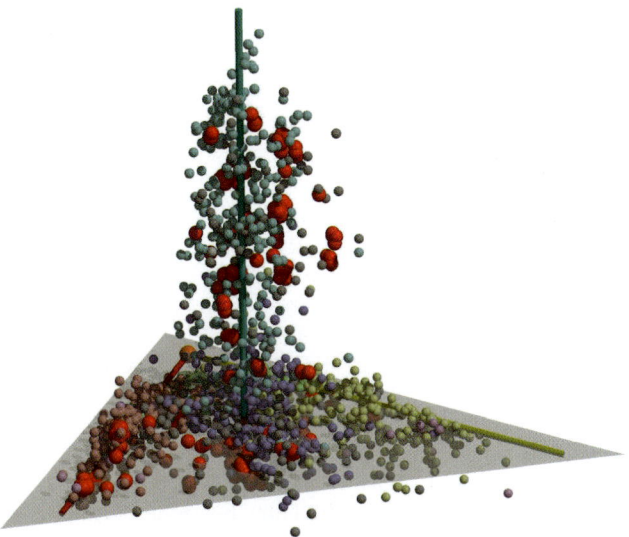

Fig. 32.7 Protein structure families determined at BSGC *(red)* mapped on the protein structure universe. Most of the BSGC structures had no sequence homologues in PDB structure database. About one half of them had new folds and occupy empty spaces in the protein structure universe, and the other half turned out to be "remote homologues" of structures of known folds and occupy the same or very close to preoccupied locations.

As expected, they are located within the envelope defined earlier by the 1,898 nonredundant structures of PDB. There appears to be a paucity in β-class proteins and more abundant usage of α/β-class proteins in these minimal organisms.

Acknowledgments

This work is supported by grants from the NIH (1-P50-GM62412 and 1-R01-GM073109). The authors are grateful to a large number of colleagues who participated in various aspects of BSGC's PSI-1 program, such as high throughput cloning (H. Yokota and B. Gold) and expression (M. Henriquez and B. Martinez), large-scale production and characterization of proteins (C. Huang, Y. Lou, N. Oganesyan, and A. DeGiovanni), crystallization (J. Jancarik, I. Ankoudinova, and H. Hyun) and structure determination (D. Das, J. Liu, V. Oganesyan, and Q. Xian), and structural space mapping (S. Jun, G. Sims, J. Hou, and I.-G. Choi), with the guidance of S. Brenner, D. Wemmer, T. Earnest, D. McKay, and C. Hutchison, Jr.

References

1. Berman, H. M., Westbrook, J., Feng, Z., Gilliland, G., Bhat, T. N., Weissig, H., Shindyalov, I. N., and Bourne, P. E. (2000) The Protein Data Bank. *Nucleic Acids Res.* **28,** 235–242.
2. Hou, J., Sims, G. E., Zhang, C., and Kim, S. -H. (2003) A global representation of the protein fold space. *Proc. Natl. Acad. Sci. USA* **100,** 2386–2390.

3. Hou, J., Jun, S.-R., Zhang, C., and Kim, S.-H. (2005). Global mapping of the protein structure space and application in structure-based inference of protein function. *Proc. Natl. Acad. Sci. U.S.A.* **102,** 3651–3656.

4. Himmelreich, R., Hilbert, H., Plagens, H., Pirkl, E., Li, B. C., and Herrmann, R. (1996) Complete sequence analysis of the genome of the bacterium *Mycoplasma pneumoniae. Nucleic Acids Res.* **24,** 4420–4449.

5. Chen, L., Oughtred, R., Berman, H. M., and Westbrook, J. (2004) TargetDB: a target registration database for structural genomics projects. *Bioinformatics* **20,** 2860–2862.

6. Boeckmann, B., Bairoch, A., Apweiler, R., Blatter, M. C., Estreicher, A., Gasteiger, E., Martin, M. J., Michoud, K., O'Donovan, C., Phan, I., Pilbout, S., and Schneider, M. (2003) The SWISS-PROT protein knowledge base and its supplement TrEMBL in 2003. *Nucleic Acids Res.* **31,** 365–370.

7. Wootton, J. C. (1994) Nonglobular domains in protein sequences: automated segmentation using complexity measures. *Comput. Chem.* **18,** 269–285.

8. Jones, D. T., and Swindells, M. B. (2002) Getting the most from PSI–BLAST. *Trends Biochem. Sci.* **27,** 161–164.

9. Schaffer, A. A., Aravind, L., Madden, T. L., Shavirin, S., Spouge, J. L., Wolf, Y. I., Koonin, E. V., and Altschul, S. F. (2001) Improving the accuracy of PSI-BLAST protein database searches with composition-based statistics and other refinements. *Nucleic Acids Res.* **29,** 2994–3005.

10. Altschul, S. F., Madden, T. L., Schaffer, A. A., Zhang, J., Zhang, Z., Miller, W., and Lipman, D. J. (1997) Gapped BLAST and PSI-BLAST: a new generation of protein database search program. *Nucleic Acids Res.* **25,** 3389–3402.

11. Altschul, S. F., Gish, W., Miller, W., Myers, E. W., and Lipman, D. J. (1990) Basic local alignment search tool. *J. Mol. Biol.* **215,** 403–410.

12. Bateman, A., Coin, L., Durbin, R., Finn, R. D., Hollich, V., Griffiths-Jones, S., Khanna, A., Marshall, M., Moxon, S., Sonnhammer, E. L., Studholme, D. J., Yeats, C., and Eddy, S. R. (2004) The Pfam protein families database. *Nucleic Acids Res.* **32,** D138–141.

13. Eddy, S. R. (1998) Profile hidden Markov models. *Bioinformatics* **14,** 755–763.

14. Lupas, A. (1996) Prediction and analysis of coiled-coil structures. *Methods Enzymol.* **266,** 513–525.

15. Krogh, A., Larsson, B., von Heijne, G., and Sonnhammer, E. L. (2001) Predicting transmembrane protein topology with a hidden Markov model: application to complete genomes. *J. Mol. Biol.* **305,** 567–580.

16. Rost, B., Casadio, R., Fariselli, P., and Sander, C. (1995) Transmembrane helices predicted at 95% accuracy. *Protein Sci.* **4,** 521–533.

17. Chandonia, J. M., Kim, S. H., and Brenner, S. E. (2005) Target selection and deselection at the Berkeley Structural Genomics Center. *Proteins* **62,** 356–370.

18. Aslanidis, C., and De Jong, P. J. (1990). Ligation-independent cloning of PCR products (LIC-PCR). *Nucleic Acids Res.* **20,** 6069–6074.

19. Studier, W. (2005) Protein production by auto-induction in high density shaking cultures. *Protein Expr. Purif.* **41,** 207–234.

20. Nguyen, H., Martinez, B., Oganesyan, N., and Kim, R. (2004) An automated small-scale protein expression and purification screening provides beneficial information for protein production. *J. Struct. Funct. Genom.* **5,** 23–27.

21. Sachdev, D., and Chirgwin, J. M. (1998) Solubility of proteins isolated from inclusion bodies is enhanced by fusion to maltose-binding protein or thioredoxin. *Protein Express. Purif.* **12,** 122–132.

22. Harrison, S. C. (2004) Whither structural biology? *Nat. Struct. Mol. Biol.* **11,** 12–15.

23. Kempf, B., and Bremer, E. (1998) Uptake and synthesis of compatible solutes as microbial stress responses to high-osmolality environments. *Arch. Microbiol.* **170,** 319–330.

24. Bukau, B., and Horwich, A. L. (1998) The Hsp70 and Hsp60 chaperone machines. *Cell* **92,** 351–366.

25. Chen, J., Acton, T. B., Basu, S. K., Montelione, G. T., and Inouye, M. (2002) Enhancement of the solubility of proteins overexpressed in *Escherichia coli* by heat shock. *J. Mol. Microbiol. Biotech.* **4,** 519–524.

26. Samuel D., Kumar, T. K, Ganesh, G., Jayaraman, G., Yang, P. W., Chang, M. M., Trivedi, V. D., Wang, S. L., Hwang, K. C., and Chang, D. K., and Yu, C. (2000) Proline inhibits aggregation during protein refolding. *Protein Sci.* **9,** 344–352.

27. Yang, D. S., Yip, C. M., Huang, T. H, Chakrabartty, A., and Fraser, P. E. (1999) Manipulating the amyloid-beta aggregation pathway with chemical chaperones. *J. Biol. Chem.* **274,** 32970–32974.

28. Voziyan, P. A., and Fisher, M. T. (2000) Chaperonin-assisted folding of glutamine synthetase under nonpermissive conditions: off-pathway aggregation propensity does not determine the co-chaperonin requirement. *Protein Sci.* **9,** 2405–2412.

29. Diamant, S., Eliahu, N., Rosenthal, D., and Goloubinoff, P. (2001) Chemical chaperones regulate molecular chaperones *in vitro* and in cells under combined salt and heat stresses. *J. Biol. Chem.* **276,** 39586–39591.

30. Oganesyan, N., Ankoudinova, I., Kim, S.-H., and Kim, R. (2007) Effect of osmotic stress and heat shock in recombinant protein overexpression and crystallization. *Protein Express. Purif.* **52(2),** 280–285.

31. Das, D., Oganesyan, N., Yokota, H., Pufan, R., Kim, R., and Kim, S.-H. (2004) Crystal structure of the conserved hypothetical protein MPN330 (GI: 1674200) from *Mycoplasma pneumoniae. Proteins Struc. Func. Bioinf.* **58,** 504–508.

32. Oganesyan, N., Kim, S.–H., and Kim, R. (2004) On-column chemical refolding of proteins. *PharmaGenomics* **4,** 22–26.

33. Rozema, D., and Gellman, S.H. (1996) Artificial chaperone-assisted refolding of denatured-renatured lysozyme: modulation of the competition between renaturation and aggregation. *Biochemistry* **35,** 15760–15771.

34. Daugherty, D. L., Rozema, D., Hanson, P. E., and Gellman, S. H. (1998) Artificial chaperone-assisted refolding of citrate synthase. *J. Biol. Chem.* **273,** 33961–33971.

35. Lepre, C. A., and Moore, J. M. (1998) Microdrop screening: A rapid method to optimize solvent conditions for NMR spectroscopy of proteins. *J. Biomol. NMR* **12,** 493–499.

36. Jancarik, J., Pufan, R., Hong, C., Kim, R., Kim, S.–H. (2004) Optimum Solubility (OS) Screening: an efficient method to optimize buffer conditions for homogeneity and crystallization of proteins. *Acta Cryst.* **D60,** 1670–1673.

37. Jancarik, J. and Kim, S. H. (1991) Sparse matrix sampling: a screening method for crystallization of proteins. *J. Appl. Cryst.* **2,** 409–411.

38. Grosse-Kunstleve, R. W., and Adams, P. D. (2003) Substructure search procedures for macromolecular structures. *Acta Cryst.* **D59,** 1966–1973.

39. Terwilliger, T. C., and Berendzen, J. (1999) Automated MAD and MIR structure solution. *Acta Crystallogr. D Biol. Crystallogr.* **55,** 849–861.

40. Collaborative Computational Project, Number 4 (1994) The CCP4 suite: programs for protein crystallography. *Acta Crystallogr. D Biol. Crystallogr.* **50,** 760–763.

41. Brunger, A. T., Adams, P. D., Clore, G. M, DeLano, W. L., Gros, P., Grosse-Kunstleve, R. W., Jiang, J. S., Kuszewski, J., Nilges, M., Pannu, N. S., Read, R. J., Rice, L. M., Simonson, T., and Warren, G. L. (1998) Crystallography & NMR system: a new software suite for macromolecular structure determination. *Acta Cryst.* **D54,** 905–921.

42. de La Fortelle, E., and Bricogne, G. (1997) Maximum-likelihood heavy-atom parameter refinement in the MIR and MAD methods. *Methods Enzymol.* **276,** 472–494.

43. Cowtan, K. (1999) Error estimation and bias correction in phase-improvement calculations. *Acta Cryst.* **D55,** 1555–1567.

44. Terwilliger, T. C. (2000) Maximum likelihood density modification. *Acta Cryst.* **D56,** 965–972.

45. Perrakis, A., Morris, R., and Lamzin, V. S. (1999) Automated protein model building combined with iterative structure refinement. *Nat. Struct. Biol.* **6,** 458–463.

46. Murshudov, G. N., Vagin, A. A., and Dodson, E. J. (1997) Refinement of macromolecular structures by the maximum-likelihood method. *Acta Cryst.* **D53,** 240–255.

47. Kim, S. H., Shin, D. H., Choi, I. G., Schulze-Gahmen, U., Chen, S., and Kim, R. (2003) Structure-based functional inference in structural genomics. *J. Struct. Funct. Genom.* **4,** 129–135.

48. Kim, S.-H., Shin, D. H., Liu, J., Oganesyan, V., Chen, S., Xu, Q. S., Kim, J.-S., Das, D., Schulze-Gahmen, U., Holbrook, S. R., Holbrook, E. L., Martinez, B. A., Oganesyan, N., DeGiovanni, A., Lou, Y., Henriquez, M., Huang, C., Jancarik, J., Pufan, R., Choi, I.-C., Chandonia, J.-M., Hou, J., Gold, B., Yokota, H., Brenner, S. E., Adams, P. A., and Kim, R. (2005) Structural genomics of minimal organisms and protein fold space. *J. Struct. Funct. Genomics.* **6,** 63–70.

49. Murzin, A. G., Brenner, S. E., Hubbard, T., and Chothia, C. (1995) SCOP: a structural classification of proteins database for the investigation of sequences and structures. *J. Mol. Biol.* **247,** 536–540.

50. Chandonia, J. M., and Kim, S. H. (2006) Structural proteomics of minimal organisms: conservation of protein fold usage and evolutionary implications *BMC Struct. Biol.* **6,** 7–22.

Chapter 33

Structural Genomics of Pathogenic Protozoa: an Overview

Erkang Fan, David Baker, Stanley Fields, Michael H. Gelb, Frederick S. Buckner, Wesley C. Van Voorhis, Eric Phizicky, Mark Dumont, Christopher Mehlin, Elizabeth Grayhack, Mark Sullivan, Christophe Verlinde, George DeTitta, Deirdre R. Meldrum, Ethan A. Merritt, Thomas Earnest, Michael Soltis, Frank Zucker, Peter J. Myler, Lori Schoenfeld, David Kim, Liz Worthey, Doug LaCount, Marissa Vignali, Jizhen Li, Somnath Mondal, Archna Massey, Brian Carroll, Stacey Gulde, Joseph Luft, Larry DeSoto, Mark Holl, Jonathan Caruthers, Jürgen Bosch, Mark Robien, Tracy Arakaki, Margaret Holmes, Isolde Le Trong, and Wim G. J. Hol

The Structural Genomics of Pathogenic Protozoa (SGPP) Consortium aimed to determine crystal structures of proteins from trypanosomatid and malaria parasites in a high throughput manner. The pipeline of target selection, protein production, crystallization, and structure determination, is sketched. Special emphasis is given to a number of technology developments including domain prediction, the use of "co-crystallants," and capillary crystallization. "Fragment cocktail crystallography" for medical structural genomics is also described.

1. Introduction

The Structural Genomics of Pathogenic Protozoa (SGPP) Consortium (www.sgpp.org) focused on the determination of crystal structures of proteins from major eukaryotic tropical pathogenic protozoa, specifically:

- *Plasmodium* spp., in particular *Plasmodium falciparum*, which causes the most lethal form of malaria, and also *P. vivax*.
- *Leishmania* spp., causing various forms of leishmaniasis throughout the tropics and subtropics
- *Trypanosoma brucei*, the causative agent of sleeping sickness in Africa
- *Trypanosoma cruzi*, responsible for Chagas disease in Latin America

The importance of the pathogens under investigation becomes clear from the alarming statistics from the WHO web site (www.who.int/topics/en/):

- Malaria: Each year, approximately 300 to 500 million malaria infections lead to over 1 million deaths, and about 90% of these occur in Africa, especially among young children. The rapid spread of resistance to

From: *Methods in Molecular Biology, Vol. 426: Structural Proteomics: High-throughput Methods*
Edited by: B. Kobe, M. Guss and T. Huber © Humana Press, Totowa, NJ

antimalarial drugs, coupled with widespread poverty and weak health infrastructure means that mortality from malaria continues to rise in developing countries.

• Leishmaniasis is endemic in 88 countries on five continents with approximately 12 million persons infected. Visceral leishmaniasis, or "kala azar," results in a mortality rate of nearly 100% if left untreated. Since 1993 there has been a significant increase in both the geographical area in which the diseases occur as well as the number of people infected. Particularly worrisome is the tendency of coinfection with HIV and *Leishmania donovani*, the causative agent of visceral leishmaniasis.

• Sleeping sickness, or African trypanosomiasis, is caused by the *T. brucei* parasite and is a daily threat to more than 60 million people. Occurring only in sub-Saharan Africa, it is estimated that 300,000 to 500,000 people have the disease. The disease is fatal if not treated.

• Chagas disease, another form of trypanosomiasis, is caused by the *T. cruzi* parasite and has a wide distribution in Central and South America. Early diagnosis and treatment can be difficult, but are again essential. It is endemic in 21 countries, with 16–18 million people infected and 100 million people at risk.

The available treatments for the three "trypanosomatid" infections are often ineffective, toxic, and/or difficult to administer, and resistance to antimalarial drugs is widespread globally. Research to define new therapeutic targets will aid the discovery of better treatments for these neglected diseases. A structural genomics effort on these protozoan organisms is likely to provide critical information on the precise architecture of potential drug targets. With this in mind, the targets of SGPP are chosen in part on the basis of medical relevance, and partly on the basis of expectations for discovering novel sequence-to-fold relationships as outlined in the Protein Structure Initiative strategy (www.nigms.nih.gov/Initiatives/PSI).

The other major mission of SGPP was the development of new methodologies for high throuput structural genomics projects. All of the organisms targeted by SGPP are eukaryotes, which are well known to be difficult targets for structural biology. Additionally, the genomes of these organisms have only recently been sequenced and are still in the process of annotation. Because of these complicating factors, a substantial effort has been devoted to the development and testing of technologies and procedures for optimizing the various steps in the structural genomics effort.

2. Pipeline Overview

The main pipeline of SGPP consists of target selection, protein production, protein crystallization, x-ray data collection, and structure determination. These pipeline units are supported by an informatics unit that performs data archiving, sharing, and mining (Fig. 33.1). During a period of 4 years, a high-capacity and high throughput structural genomics pipeline was established for obtaining crystal structures of the targeted protozoan parasites, as evident by a significant increase in production seen throughout the SGPP pipeline during the time period from September 2004 through August 2005 (Table 33.1).

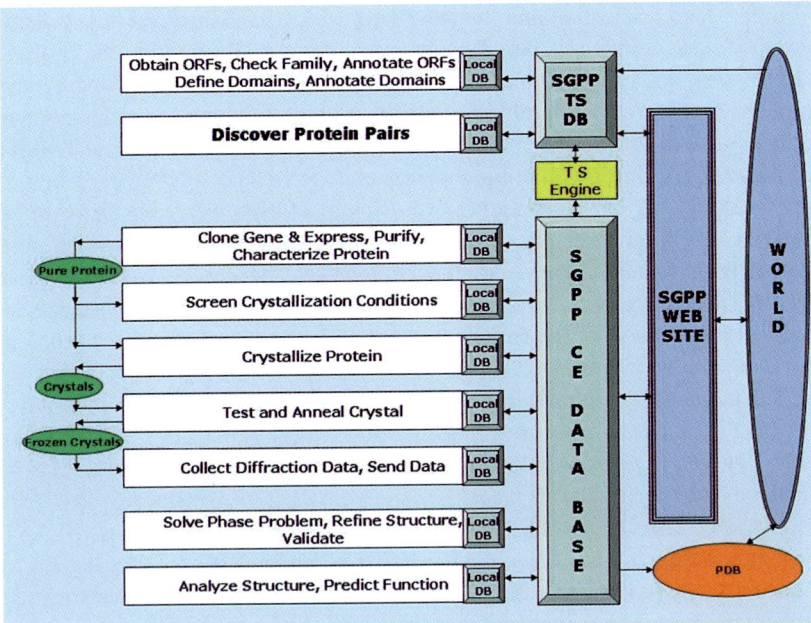

Fig. 33.1 Schematic diagram of the SGPP pipeline and data flow. DB, database; ORF, open reading frame; PDB, protein data bank; TS, target selection (CE, Central).

Table 33.1 Progress of all targets in SGPP (membrane proteins not included)

	As of August 2004: total successful targets	As of August 2005: total successful targets	Increase over 1 year (percent)
Cloned	5,471	10,767	97
Purified	404	839	108
Hits in crystal screen	100	194	94
Targets sent to crystallographers	57	123	116
Diffraction quality crystals	40	85	113
Total solved structures	15	40	167

2.1. Target Selection

One challenge for target selection in SGPP was that the genome sequences of the targeted organisms were not complete at the start of SGPP in 2001, with the complete genome sequence of *P. falciparum* 3D7 published in 2002 *(1)*, and the *L. major* Friedlin, *T. brucei* 927, and *T. cruzi* CL-Brener (jointly called the "Tritryp") genomes were published only toward the end of the SGPP project in 2005 *(2–5)*. It was very fortunate that several members of the SGPP consortium were actively involved in the genome sequencing and gene prediction and annotation projects for the trypanosomatids. This allowed the authors to select targets from extensively manually curated gene annotations of these organisms.

In the Target Selection unit, target criteria were established and targets were grouped into various sets based upon several criteria. Early on in the project, targets were segregated into predicted soluble proteins and integral membrane proteins (IMPs), using algorithms capable of identifying transmembrane spanning regions (TMSRs). The "solubles" are composed of targets with no TMSRs or those containing an N-terminal signal peptide (which was removed before primer design). Targets containing two or more TMSRs were considered to be IMPs. Criteria for selecting soluble targets included length, pI, disorder, and size of Pfam family. Criteria varied per "target set." Target sets were usually at least 96 in size and were assigned "set identifiers" in the SGPP status database so that the progress of the entire set could be tracked throughout the pipeline for comparative purposes.

The target selection process for soluble proteins addresses three underlying themes: (1) the discovery of new sequence-to-fold relationships; (2) the identification and expression of protein pairs; and (3) the generation of targets of medical and functional relevance.

To concentrate on proteins that are likely to represent novel folds, BLAST searches were carried out against the NCBI nonredundant protein database to create a score profile from the resulting multiple-sequence alignment for each potential target. These profiles were then used to search the PDB to quantify identity and similarity of the target sequence to proteins of known structure. In general, proteins were excluded from further analysis if the sequence identity was greater than 30%.

Information for protein-pair targets came from two sources. Experimentally, the SGPP consortium collaborated with industrial partners using yeast two-hybrid techniques to analyze numerous interactions in *P. falciparum (6)*. Computationally, from a database of interacting proteins in yeast, the authors' industrial partner provided a list of homologues in Tritryps that are likely to be engaged in pairwise interactions. These results from both sources were filtered for characteristics that would make it difficult to obtain soluble and crystallized protein pairs, such as size, significant stretches of low-complexity regions, and so on. However, the rejection criteria are less stringent than for the single solubles since, for example, low-complexity regions might become ordered in the course of complex formation.

For targets with medical and functional relevance, several approaches were implemented. One consisted of enzyme homologue discovery. Proteins annotated with an EC number in the genome database, those belonging to a cluster of orthologous genes (COG) that also contained a protein with an EC number, or those with PSI-BLAST e-values of 10^{-6} in the BRENDA enzyme database (www.brenda.uni-koeln.de), were selected as a set of "enzyme-like" targets. Second, homologues of known drug targets were identified in the genome database. Third, important protein targets were also solicited from the research community. The SGPP home page (www.sgpp.org) has an area where members of the research community can nominate proteins that they feel should be prioritized for expression, crystallization, and structure determination. Fourth, functional genomics information from publicly available sources, such as the RNAi Database at trypanofan.path.cam.ac.uk/ were consulted to select essential genes that lead to abnormal phenotypes.

For all full-length targets, the SGPP consortium applied computational domain prediction technologies and fed back chunks (defined as a single or linear combination of predicted soluble domains) into the central target selection database (discussed further in Section 3.1.). Using different combinations of target selection criteria *(7)*, target sets were created for use by the SGPP pipeline. Overall, a total of approximately 19,000 soluble and 949 IMP-targets have been selected for entry into the amplification, cloning, expression, protein purification, crystallization, diffraction, and structure determination pipeline.

2.2. Protein Production

The SGPP consortium took the advantage of its two separate protein production centers for a controlled examination of protein production methodologies, which were critical issues in achieving successful production and crystallization of the generally difficult eukaryotic targets studied in SGPP. The University of Rochester (UR) center developed and used extensively a cleavable N-terminal His$_6$-tag vector, whereas the University of Washington (UW) center employed a vector encoding a noncleavable N-terminal His$_6$-tag. The UW protein expression unit also developed new vectors for expression in mammalian and insect cells. The Rochester protein production unit also constructed a system of vectors for easy coexpression of two proteins *(8)*. The results and procedures for cloning and expressing 1,000 malaria genes for protein production has been described by Mehlin et al. *(9)*. It appeared that production of soluble protein is favored by small size, large Pfam family and a pI below 8. In the later years of SGPP, both its protein production centers moved toward attacking the more difficult, medically relevant targets, as it was felt that structural genomics programs in general were neglecting these targets in favor of ones that were more immediately tractable. Technical innovations developed in the SGPP for high throughput protein production included cloning grills *(10)*, ligase-independent cloning vectors (Grayhack et al., unpublished) and flash-freezing of proteins in PCR plates *(11)*.

The organization of SGPP allowed a side-to-side comparison of cleavable vs. noncleavable expression systems. Initially, the two protein centers in SGPP were working on different targets, but lately an effort has been made to clone the same targets into the two vector systems. Although it was initially believed that the gains from this would be marginal, it was of interest to observe that these small sequence differences can have a profound effect on its solubility for some proteins, for example phosphodiesterases (PDEs).

Due to their demonstrated importance to the parasites and the ability for phosphodiesterase inhibitors (e.g., Viagra, Cialis) to be developed clinically, PDEs are high-value, potential drug targets in these organisms. SGPP took a set of 450 PDE variants as targets, which included multiple truncation variants of several of these enzymes, and put them into both the cleavable and uncleavable-tag vector systems, generating a total of 900 protein variants. A total of 13 of them were stable with only three of these soluble *within both* vector systems. (For the sake of this experiment, soluble proteins were only counted if they expressed to high enough levels and were isolated and shipped off for crystal screening trials.) The cleavable tag vector performed marginally better than the uncleavable-tag vector: generating ten as opposed to six soluble proteins. What is truly intriguing about this work, however, is that a total of ten targets were soluble in one vector system but not in the other.

In addition, factors that affect soluble expression of these large numbers of eukaryotic proteins in *E. coli* were also revealed to some extent. For example, for plasmodium proteins, higher molecular weight, greater protein disorder, more basic isoelectric point, and a lack of homology to *E. coli* proteins all seem to correlate independently with difficulties in expression. In contrast, codon usage, and the percentage of adenosines and thymidines (which is high in *P. falciparum*) did not appear to affect soluble expression significantly *(9)*.

Overall, the two expression centers of SGPP successfully cloned over 10,700 targets out of approximately 11,150 for which primers were designed and ordered during the course of four years, and reached a throughput of producing more than 1,000 soluble eukaryotic protein samples (including repeated production for certain targets) in its final year in sufficient quantities for crystallization experiments.

2.3. Protein Crystallization

The protein crystallization process in SGPP consisted of two stages. The first stage was the high throughout screening (HTS) of initial lead conditions that may produce protein crystals. This was performed by the SGPP unit within the HTS crystallization screening laboratory at the Hauptman-Woodward Medical Research Institute (HWI) (www.hwi.buffalo.edu) *(12)*. The second stage was the follow-up crystal growth, which was performed by the crystallization unit of SGPP at the UW.

The HWI HTS lab uses robotic liquid handling systems extensively. Using a total of 400 μl of sample solution, the liquid handling systems prepare 1,536 well plates that are prepared with oil and 1,536 unique crystallization cocktail solutions. Each plate screens a single protein sample combined with 1,536 unique crystallization cocktail solutions mixed with protein under oil. Results from these micro batch under oil experiments are recorded (saved as images) over a 4-week time course. For SGPP, all images were screened manually for potential leads of crystallization. Cumulatively, the HWI HTS laboratory has been able to identify lead conditions for 194 out of more than 800 unique targets for SGPP, with very often multiple leads per targets.

The Crystal Growth Unit of SGPP followed two parallel paths. One was developing an optimal follow-up for the HTS hits found at HWI. The second was developing a new method for robotic protein crystallization in capillaries, which is briefly described in Section 3. Protein samples received from the SGPP Protein Production Units were immediately characterized systematically by UV spectroscopy, SDS PAGE, native gels, dynamic light scattering, and, for a limited number of cases, limited proteolysis. All the characterization data was archived in a database for future reference and data mining. Cumulatively, the SGPP Crystallization Unit received more than 1,100 protein samples for over 800 different targets. Diffraction quality crystals were obtained for approximately 123 of these targets and delivered to down stream units for data collection and structure determination. Interestingly, the degree of fragmentation in limited proteolysis experiments appeared to be a good predictor of the crystallization success of protein targets: the more resistant to proteolysis, the greater the probability of crystal growth (O. Kalyuzhniy and W. Hol, unpublished observations).

2.4. Data Collection, Processing, and Structure Determination

SGPP data collection was done at two synchrotron sources: the Advanced Light Source (ALS) at the Lawrence Berkeley National Laboratory and the Stanford Synchrotron Radiation Laboratory (SSRL). Both synchrotron labs developed robotic equipment for crystal handling and loading including automated crystal annealing, streamlining procedures for users including remote data collection, and software for control of data collection, processing, and analysis *(13–15)*. Strategies for improving crystal quality included crystal annealing techniques *(16,17)* and cocrystallization with ligands. SGPP developed a special database (XRAYDB; Bosch et al., unpublished), which was used to track mounted crystals, cryoprotection procedures, results of screens at synchrotrons, and other early data in the structure determination process.

For high throughput data processing HKL2000 *(18)* and the ELVES automated scripts by Holton and Alber *(19)* were used. For structure determination, a variety of modern structural biology techniques were employed including SeMet-based phasing, bromide soak and other heavy atom phasing procedures, molecular replacement phasing and autotracing. Overall, the SGPP consortium collected 549 sets of diffraction data for 123 unique targets, and solved approximately 40 protein structures from the targeted organisms. In addition, a significant effort was made in developing "fragment cocktail crystallography" as described in Section 4.

SGPP explored the use of structural predictions from the ROSETTA program (boinc.bakerlab.org/rosetta) in high throughput structure determination. A specific example is SGPP target Tcru010945AAA from *T. cruzi*. This target has 27% sequence identity to an arginase with known structure (PDB entry 1cev) *(20)*. Attempts to use "1cev" for molecular replacement failed however. SAD phasing based on 9 Se sites yielded a relatively poor initial map, which nevertheless eventually led to a fully refined structure using SOLVE/RESOLVE *(21)* with the help from ROSETTA for prediction of loops and other connecting segments consistent with the partial model. ROSETTA was able to generate candidate loop conformations for the missing portions that agree quite well with the electron density in the map from SAD phasing, and it seems clear that automation of this approach could save much manual effort.

The authors also used ROSETTA to generate a full prediction for this same target Tcru010945AAA starting only from the sequence. This was not fully an *ab initio* prediction, as ROSETTA was able to identify and use suspected homologues, including 1cev, in the PDB. The CCP4 program MOLREP was able to find a molecular replacement solution from this model, and the resulting map is traceable. This is in distinct contrast to the failure to generate a usable map using the 1cev structure directly as a molecular replacement probe. This example does not address the more ambitious goal of solving a crystal structure using a purely *ab initio* prediction, but it does illustrate the value of ROSETTA in reducing the amount of manual work required on the part of a crystallographer.

2.5. Informatics

In addition to developing and maintaining several local databases in separate SGPP units, the central informatics unit of SGPP also maintained three major central databases:

1. Target prioritization: Applies criteria selected by investigators to score targets identified by target selection, domain prediction, and yeast two-hybrid. Available criteria included the number of transmembrane helices predicted, Pfam family size, percent identity of nearest PDB and human orthologues, nearest orthologue with diffraction quality crystals in TargetDB (targetdb.pdb.org), sequence length, number and extent of low complexity and disordered regions, amino acid composition, subcellular localization, and curated values such as medical and functional relevance.
2. Target status: Reports overall progress for each ORF and each chunk of an ORF: on the web in a table and in XML to the public; and in shared files for internal data mining by SGPP researchers. This database acquired data directly from Sesame and local databases in each unit from target selection to structure determination.
3. Sesame central tracking database: Tracks each action taken on each target, including repeated steps. Sesame is a three-tiered database system developed at the University of Wisconsin, Madison, by Zsolt Zolnai in the Center for Eukaryotic Structural Genomics (CESG) (www.uwstructuralgenomics.org). A copy of Sesame was maintained in Seattle.

Once sufficient data were collected in the central databases, the informatics unit also performed data mining to answer some critical and interesting questions related to achieving higher success rate for the authors' structural genomics pipeline. For example, data mining on results of HWI's HTS crystallization screening included (F. Zucker et al., unpublished): (1) Polyethylene glycols (PEGs) are the most effective component for protein crystallization in SGPP, both alone and in combinations with salt or other organic chemicals. (2) Proteins with high disorder or hydropathy do not crystallize well in the full screen. No individual condition or class of conditions was found that improved their crystallization. (3) Other predicted properties, such as isoelectric point or charge in a given condition, did not significantly affect crystallization rates. (4) Dynamic light scattering (DLS) was a reasonably good predictor of crystallizability, as expected. Proteins with high polydispersity or multiple DLS peaks did not crystallize well. However, sensitivity to proteolysis was a better indicator (O. Kalyuzhniy and W. Hol, unpublished observations). (5) Protein concentration had a minor and inconsistent effect. (6) Proteins annotated as enzyme-like and proteins found in Pfam crystallized better than other proteins, especially than proteins annotated as hypothetical.

Overall, the central informatics units of SGPP archived and searched information of over 21,000 potential protein targets under consideration, approximately 13,000 active targets, more than 33,000 recorded actions, as well as detailed protocols in various SGPP units.

3. Technology Development

As stated, methodology development was also a major task within SGPP in hope to help achieve high throughputs and high success rates for structural genomics projects. These developmental projects were implemented at various places throughout the entire SGPP pipeline. They included (in order of pipeline progression from target selection to structure determination and annotation):

- Computational domain parsing and chunk selection
- Experimental yeast two-hybrid selection of protein pairs in *Plasmodium falciparum (6)*
- A "multispecies approach" for target cloning and expression of plasmodium and trypanosomatids, taking in particular advantage of the large number of closely related *Leishmania* species for finding a target-variant that yields soluble protein
- Whole gene synthesis
- Protein pair expression *(8)*
- Integral membrane protein cloning, expression, and purification in *E. coli*
- Single-chain antibody selection, production, and complex formation against target proteins
- "Co-crystallant" design, synthesis, and testing, as well as ligand screening; specific ligands for improving protein cocrystallization
- Capillary crystal-growth robotics
- Crystal annealing, healing, and screening robotics
- Fragment cocktails for crystallography *(33)*

A few of these "special SGPP projects" are selectively described in the following.

3.1. Computational Domain Prediction and Chunk Selection

"Chunk" has a unique meaning in SGPP in that it refers to fragments of a protein that may be solubly expressed. A chunk could be a single domain or a linear combination of several domains of a target protein. The selection of chunks as targets for the SGPP pipeline played a significant role in the target selection process of SGPP, especially due to the fact that the authors were often dealing with hard to express eukaryotic protein targets. In particular, proteins from the targeted organisms frequently contain genes coding for major insertions of amino acids that are absent in other organisms. Two fully automated domain prediction methods, Ginzu and RosettaDOM *(7)*, were implemented in SGPP for chunk selection. Ginzu has been used to predict domains in nearly all SGPP targets using a hierarchical procedure that assigns domains based on homology to known structures and protein families using successively less confident methods. RosettaDOM relies on information only in the query sequence by using the Rosetta *de novo* structure prediction method to build three-dimensional models, and then applying a structure based domain assignment algorithm to parse each model into domains. Domain boundaries that are consistently assigned in the models are predicted to be the actual domain boundaries. Both methods were top performers among automated methods in CASP6 (Critical Assessment of Techniques for Protein Structure Prediction), and performed well even when compared with human predictors *(22)*.

Experimentally, computational chunking has been performed on 253 targets within the SGPP pipeline giving a total of 1,495 chunks. These chunks were subjected to overexpression. Among them, 91 targets with no full-length expression resulted in one or more expressible chunks. For 44 targets that we were not able to obtain soluble samples of the full-length protein, one or more soluble chunks for each target were purified in sufficient quantities for downstream pipeline use. In summary, SGPP rescued a significant number of targets for further study using chunking. Although the overall expression success rate per chunk is roughly the same with or without chunking for a random target in the genomes of SGPP organisms, the fact that targets might be rescued that did not result in full-length proteins using computational chunking is significant for general biological studies of the parasite proteins. This may be especially important for potential drug development targets of these parasites. For example, within the 253 targets that have been chunked, 57 are potential drug targets, which consist mainly of homologues of known enzymes. Through chunking, 28% (16 targets) gave soluble chunks but not the full-length sample; therefore, for really important protein targets, chunking will significantly enhance the chance of producing soluble parts of those proteins for future studies.

3.2. Design, Synthesis, and Systematic Testing of Co-Crystallants

Protein crystallization has a long history of discovery that very specific additives are indispensable for obtaining diffraction quality crystals for particular proteins. Additives can be inorganic compounds such as zinc chloride or small organic molecules such as phenol. Their beneficial effect can be due to: (1) a change in dielectric, leading to changes in solubility of the protein; (2) the formation of a specific complex with the protein, which can either reduce conformational heterogeneity of the protein or lead to a packing that involves the additive; or (3) the formation of very weak complexes on one face of the crystal, thereby slowing down the growth in that direction and allowing other crystal faces to grow (23). The Biological Macromolecule Crystallization Database (BMCD) (24) lists 383 different additives reported in the literature, many of which are already present in crystallization screens. Most of them are ions, alcohols, carbohydrates, and surfactants.

However, there are in principle combinatorial chemistry opportunities to increase the repertoire of additives. Additives that operate through mechanism (2) above, which the authors like to refer to as "co-crystallants", are of great interest because incorporation of elements such as Br or Se in these compounds should, in addition to possibly promoting crystal growth, also allow for MAD/SAD phasing. As a special project in SGPP, the design, synthesis, and testing of a pilot set of such potential co-crystallants were carried out.

The authors incorporated two classes of bromine-containing synthetic co-crystallants in their pilot collection of compounds (Fig. 33.2). The design principle included the consideration of: (1) a water soluble moiety (the guanidine moiety, or one water-soluble function group attached to the imidazolidinone moiety); (2) a bromine-containing moiety for potential MAD/SAD phasing; and (3) a diversification point for incorporation of a variety of chemical substitutions to provide potentially beneficial interactions with protein targets. The synthesis of these two classes compounds was published (25,26), and 44 compounds were prepared for the authors' initial test.

An initial test of co-crystallants with 10 proteins in solutions containing 10% DMSO (v/v) at 10 mM each compound led to heavy precipitation of protein.

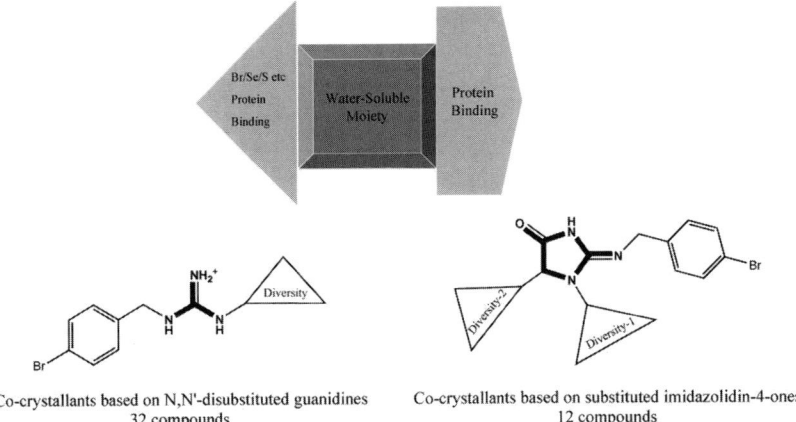

Co-crystallants based on N,N'-disubstituted guanidines
32 compounds

Co-crystallants based on substituted imidazolidin-4-ones
12 compounds

Fig. 33.2 Two classes of synthetic co-crystallants.

A second study investigated the effectiveness of these co-crystallants on crystallization using the 96 crystallization cocktails in the Hampton Research Index screen using a co-crystallant concentration that was lowered from the 10 mM in the initial study to reduce the macromolecules' precipitation. Seven co-crystallants were used at 2 and 0.4 mM concentrations. The extent of macro-molecular precipitation was slightly less at 2 mM than at 0.4 mM co-crystallant. The average precipitation for all nine of the proteins studied against the seven co-crystallants decreased from 65% in the initial study to 34% in this study. The extent of precipitation was independent of co-crystallant and dependent upon the macromolecule. After removing the crystallization hits that were identified in the control experiments, there were 36 unique crystallization conditions that appear to require the presence of the co-crystallant. Two other SGPP proteins produced crystalline hits in the presence of co-crystallants.

The crystalline outcomes produced by these two macromolecules and 42% of the lysozyme crystallization conditions all contained the same co-crystallant. It appears that these co-crystallants do have the potential to positively affect the crystallization of protein samples that could otherwise not easily be achieved. However, more systematic studies are needed to fully reveal the pros and cons of using synthetic co-crystallants in a structural genomics setting.

3.3. Capillary Crystal-Growth Robotics

Development of a robotics system for protein crystal growth in capillaries was another special project in SGPP. The rationale for this approach was three-fold:

1. In capillaries, crystallization space can be traversed in a unique manner for optimum crystal growth. One way this can be achieved is by placing air gaps between fluids to permit first free liquid interface diffusion followed by vapor diffusion in glass capillaries. Alternatively, water evaporation through plastic capillary walls provides a second opportunity for optimizing protein crystal growth experiments by removing water

from the crystallization volume, thereby increasing both protein and precipitation concentration simultaneously.

2. The Meldrum group designed and developed a special instrument, the ACAPELLA *(27,28)*, able to deliver into capillaries low (50 nl and possibly down to the 5 nl range) protein and precipitant volumes to yield many experiments per volume of reagent. This is due to the fact that the instrument can precisely control the number of individual 100 pl liquid droplets delivered by piezoelectric dispensing;

3. By growing crystals in plastic capillaries, the potential exists to automate crystal growth and crystal mounting completely. This would involve:

- Taking images for evaluating the progress of crystal growth
- Analyzing and ranking the images by computational procedures
- Designing optimization strategies
- Filling capillaries to obtain diffraction-quality crystals
- Making images of data collection size crystals
- Freezing the crystals in the capillary *in situ*
- Using the images at the synchrotron beam lines to fully automatically center crystals in the x-ray beam

In this ideal approach the crystal would never have to be touched manually and the entire process from protein solution up to and including mounted crystals, crystal centering, and data collection can be fully automated.

The "ACAPELLA-5K" capillary-filling robot (Fig. 33.3) is able to fill 5,000 capillaries in 8 hours with volumes of around 0.5 to 1 µl *(27,28)*. Fully operational capabilities include aspiration of submicroliter volumes in capillaries, followed by delivering even smaller additional volumes of other solutions from several different dispensing piezo electric units. The filled capillaries can be photographed at several time points during this process, the volumes can be mixed in the capillary, and the filled capillaries are stored in holders, which allow easy photography to follow crystal growth. Encouragingly, a few initial tests showed that three entirely different proteins grew as beautifully shaped crystals in capillaries filled by this piece of equipment, which was designed for a different purpose *(27,28)*.

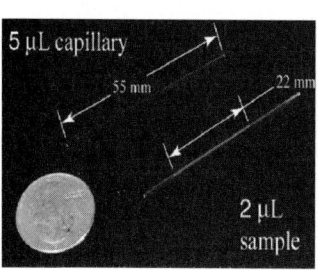 

Fig. 33.3 The ACAPELLA capillary crystallization robot. **Left.** 5 µl capillary format (0.2–2 µl sample volumes possible) for high throughput operations in the ACAPELLA instrument. **Middle.** 100 pl droplets being dispensed from piezoelectric reagent dispensers into glass capillaries. **Right.** ACAPELLA-5K general-purpose submicroliter automated fluid handling system *(27)* showing the core processor in the front.

A series of hardware upgrades were implemented on the ACAPELLA capillary processor instrument platform to assess overall system robustness for high throughput reliability and effectiveness in generating optimal crystallization conditions within plastic capillaries. Although these design improvements worked as intended, there are fundamental limitations in the ability of the piezo dispenser technology to dispense an important set of crystallization reagents. Based on the experimental results from loading capillaries, the hardware architecture is being be reevaluated to address shortcomings discovered in the original architecture.

Encouraging preliminary studies showed that crystals could be flash-frozen *inside* the capillary, completely surrounded by liquid. Recording of excellent data up to 1.8 Å resolution appeared possible for hen egg white lysozyme crystals flash frozen in capillaries. For the *T. brucei* RNA ligase 1 (REL1) catalytic domain *(29)* anomalous differences to 2.5 Å could be recorded of SeMet REL1 crystals (Bosch et al., unpublished results). This shows the great potential of capillary crystal growth combined with "in capillary" flash freezing for *ab initio* structure determination.

4. Medical Structural Genomics

Although the majority of the SGPP studies were geared toward coverage of protein fold space, part of the targets in SGPP were selected for potential applications in drug development. Moreover, we explored an additional direction, called "medical structural genomics," that can have a major impact on drug development for neglected parasitic diseases since it includes discovery of binding modes of small molecule compounds to potential drug target proteins.

Therefore, in the later stages of SGPP, the authors created a library of compounds for carrying out fragment based ligand discovery by x-ray crystallography. This effort is based on an approach initiated with earlier crystallographic studies with small molecule cocktails *(30,31)*. The studies described by Verlinde et al. *(30)* were carried out prior to 1992 at the University of Groningen, The Netherlands. This approach is related to later studies entitled "SAR by NMR" *(32)*, which described how NMR screening can identify small ligands binding at adjacent locations on the protein surface, and how these small ligands, or "fragments," could be linked to obtain high-affinity ligands *(32)*. In brief, assuming one has crystals that allow for access to the binding site of interest, one can soak the crystal with a high concentration cocktail of small but shapewise diverse fragments prior to x-ray diffraction measurements. Quite often, one of the fragments will bind at or near the target site. Because of the shape diversity, it should then be possible to identify which fragment in the cocktail has bound by inspection of the resulting electron density. Obtaining crystal structures of potential drug targets from these organisms with small organic molecules bound is of even greater importance to drug development efforts than structures of unliganded proteins.

So far, the SGPP consortium has selected about 700 small molecules for creating fragment cocktails using selection criteria as described in the literature *(33)*. These 700 molecules were grouped into about 70 cocktails based on differences in molecular mass and shape. The authors have prepared the first 30 cocktails of approximately 10 compounds each for "fragment cocktail crystallography." Initial attempts of soaking these cocktails in buffers containing 10% DMSO proved to be quite successful. Soaking three inital SGPP proteins: Lmaj004144AAA, Pfal005984AAA, and Tbru015777AAA in cocktail solutions gave high-resolution datasets that allowed identification of unique ligands in particular cocktails for all three proteins (see www.sgpp.org and www.msgpp.org for additional examples). In the case of Tbru015777AAA, a nucleoside 2-deoxyribosyltransferase from *T. brucei*, several ligands bound to the active site were identified *(33)*.

The success in applying fragment cocktail crystallography in a structural genomics setting will open up new avenues for future drug development efforts against the parasites targeted by the SGPP consortium. This intersection of protein structure space and chemical space is currently pursued in the Medical Structural Genomics of Pathogenic Protozoa (MSGPP) program project (see www.msgpp.org).

5. Summary

In the 4-year period, the SGPP consortium established a high throughput structural genomics pipeline for obtaining protein crystal structures from four major disease-causing protozoa: *Plasmodium falciparum* and three trypanosomatid parasites. A large number of new protein crystal structures were solved for these organisms. Thousands of plasmids (the *P. falciparum* expression constructs have been deposited with the MR4 repository; www.malaria.mr4.org). Hundreds of protein samples, and experimental protocols are also important results assisting in the battle against several neglected parasitic diseases.

Acknowledgments

The authors acknowledge the numerous contributions to SGPP by Ken Stuart, Martin Berg, Keith Hodgson, past SGPP members, advisory committee members, collaborators, industrial partners, and the Protein Structure Initiative (PSI) of NIH (grant GM064655 from NIGMS and NIAID).

References

1. Gardner, M. J., Hall, N., Fung, E., White, O., Berriman, M., Hyman, R. W., Carlton, J. M., Pain, A., Nelson, K. E., Bowman, S., Paulsen, I. T., James, K., Eisen, J. A., Rutherford, K., Salzberg, S. L., Craig, A., Kyes, S., Chan, M. S., Nene, V., Shallom, S. J., Suh, B., Peterson, J., Angiuoli, S., Pertea, M., Allen, J., Selengut, J., Haft, D., Mather, M. W., Vaidya, A. B., Martin, D. M., Fairlamb, A. H., Fraunholz, M. J., Roos, D. S., Ralph, S. A., McFadden, G. I., Cummings, L. M., Subramanian, G. M., Mungall, C., Venter, J. C., Carucci, D. J., Hoffman, S. L., Newbold, C., Davis, R. W., Fraser, C. M., and Barrell, B. (2002) Genome sequence of the human malaria parasite *Plasmodium falciparum. Nature* **419,** 498–511.
2. Berriman, M., Ghedin, E., Hertz-Fowler, C., Blandin, G., Renauld, H., Bartholomeu, D. C., Lennard, N. J., Caler, E., Hamlin, N. E., Haas, B., Bohme, U., Hannick, L., Aslett, M. A., Shallom, J., Marcello, L., Hou, L., Wickstead, B., Alsmark, U. C., Arrowsmith, C., Atkin, R. J., Barron, A. J., Bringaud, F., Brooks, K., Carrington, M., Cherevach, I., Chillingworth, T. J., Churcher, C., Clark, L. N., Corton, C. H.,

Cronin, A., Davies, R. M., Doggett, J., Djikeng, A., Feldblyum, T., Field, M. C., Fraser, A., Goodhead, I., Hance, Z., Harper, D., Harris, B. R., Hauser, H., Hostetler, J., Ivens, A., Jagels, K., Johnson, D., Johnson, J., Jones, K., Kerhornou, A. X., Koo, H., Larke, N., Landfear, S., Larkin, C., Leech, V., Line, A., Lord, A., Macleod, A., Mooney, P. J., Moule, S., Martin, D. M., Morgan, G. W., Mungall, K., Norbertczak, H., Ormond, D., Pai, G., Peacock, C. S., Peterson, J., Quail, M. A., Rabbinowitsch, E., Rajandream, M. A., Reitter, C., Salzberg, S. L., Sanders, M., Schobel, S., Sharp, S., Simmonds, M., Simpson, A. J., Tallon, L., Turner, C. M., Tait, A., Tivey, A. R., Van Aken, S., Walker, D., Wanless, D., Wang, S., White, B., White, O., Whitehead, S., Woodward, J., Wortman, J., Adams, M. D., Embley, T. M., Gull, K., Ullu, E., Barry, J. D., Fairlamb, A. H., Opperdoes, F., Barrell, B. G., Donelson, J. E., Hall, N., Fraser, C. M., Melville, S. E., and El-Sayed, N. M. (2005) The genome of the African trypanosome *Trypanosoma brucei. Science* **309,** 416–422.

3. El-Sayed, N. M., Myler, P. J., Bartholomeu, D. C., Nilsson, D., Aggarwal, G., Tran, A. N., Ghedin, E., Worthey, E. A., Delcher, A. L., Blandin, G., Westenberger, S. J., Caler, E., Cerqueira, G. C., Branche, C., Haas, B., Anupama, A., Arner, E., Aslund, L., Attipoe, P., Bontempi, E., Bringaud, F., Burton, P., Cadag, E., Campbell, D. A., Carrington, M., Crabtree, J., Darban, H., da Silveira, J. F., de Jong, P., Edwards, K., Englund, P. T., Fazelina, G., Feldblyum, T., Ferella, M., Frasch, A. C., Gull, K., Horn, D., Hou, L., Huang, Y., Kindlund, E., Klingbeil, M., Kluge, S., Koo, H., Lacerda, D., Levin, M. J., Lorenzi, H., Louie, T., Machado, C. R., McCulloch, R., McKenna, A., Mizuno, Y., Mottram, J. C., Nelson, S., Ochaya, S., Osoegawa, K., Pai, G., Parsons, M., Pentony, M., Pettersson, U., Pop, M., Ramirez, J. L., Rinta, J., Robertson, L., Salzberg, S. L., Sanchez, D. O., Seyler, A., Sharma, R., Shetty, J., Simpson, A. J., Sisk, E., Tammi, M. T., Tarleton, R., Teixeira, S., Van Aken, S., Vogt, C., Ward, P. N., Wickstead, B., Wortman, J., White, O., Fraser, C. M., Stuart, K. D., and Andersson, B. (2005) The genome sequence of *Trypanosoma cruzi*, etiologic agent of Chagas disease. *Science* **309,** 409–415.

4. El-Sayed, N. M., Myler, P. J., Blandin, G., Berriman, M., Crabtree, J., Aggarwal, G., Caler, E., Renauld, H., Worthey, E. A., Hertz-Fowler, C., Ghedin, E., Peacock, C., Bartholomeu, D. C., Haas, B. J., Tran, A. N., Wortman, J. R., Alsmark, U. C., Angiuoli, S., Anupama, A., Badger, J., Bringaud, F., Cadag, E., Carlton, J. M., Cerqueira, G. C., Creasy, T., Delcher, A. L., Djikeng, A., Embley, T. M., Hauser, C., Ivens, A. C., Kummerfeld, S. K., Pereira-Leal, J. B., Nilsson, D., Peterson, J., Salzberg, S. L., Shallom, J., Silva, J. C., Sundaram, J., Westenberger, S., White, O., Melville, S. E., Donelson, J. E., Andersson, B., Stuart, K. D., and Hall, N. (2005) Comparative genomics of trypanosomatid parasitic protozoa. *Science* **309,** 404–409.

5. Ivens, A. C., Peacock, C. S., Worthey, E. A., Murphy, L., Aggarwal, G., Berriman, M., Sisk, E., Rajandream, M. A., Adlem, E., Aert, R., Anupama, A., Apostolou, Z., Attipoe, P., Bason, N., Bauser, C., Beck, A., Beverley, S. M., Bianchettin, G., Borzym, K., Bothe, G., Bruschi, C. V., Collins, M., Cadag, E., Ciarloni, L., Clayton, C., Coulson, R. M., Cronin, A., Cruz, A. K., Davies, R. M., De Gaudenzi, J., Dobson, D. E., Duesterhoeft, A., Fazelina, G., Fosker, N., Frasch, A. C., Fraser, A., Fuchs, M., Gabel, C., Goble, A., Goffeau, A., Harris, D., Hertz-Fowler, C., Hilbert, H., Horn, D., Huang, Y., Klages, S., Knights, A., Kube, M., Larke, N., Litvin, L., Lord, A., Louie, T., Marra, M., Masuy, D., Matthews, K., Michaeli, S., Mottram, J. C., Muller-Auer, S., Munden, H., Nelson, S., Norbertczak, H., Oliver, K., O'Neil, S., Pentony, M., Pohl, T. M., Price, C., Purnelle, B., Quail, M. A., Rabbinowitsch, E., Reinhardt, R., Rieger, M., Rinta, J., Robben, J., Robertson, L., Ruiz, J. C., Rutter, S., Saunders, D., Schafer, M., Schein, J., Schwartz, D. C., Seeger, K., Seyler, A., Sharp, S., Shin, H., Sivam, D., Squares, R., Squares, S., Tosato, V., Vogt, C., Volckaert, G., Wambutt, R., Warren, T., Wedler, H., Woodward, J., Zhou, S., Zimmermann, W., Smith, D. F., Blackwell, J. M., Stuart, K. D., Barrell, B., and Myler, P. J. (2005) The genome of the kinetoplastic parasite, *Leishmania* major. *Science* **309,** 436–442.

6. LaCount, D. J., Vignali, M., Chettier, R., Phansalkar, A., Bell, R., Hesselberth, J. R., Schoenfeld, L. W., Ota, I., Sahasrabudhe, S., Kurschner, C., Fields, S., and Hughes, R. E. (2005) A protein interaction network of the malaria parasite *Plasmodium falciparum. Nature* **438,** 103–107.

7. Kim, D. E., Chivian, D., Malmstrom, L., and Baker, D. (2005) Automated prediction of domain boundaries in CASP6 targets using Ginzu and RosettaDOM. *Proteins* **61,** 193–200.

8. Alexandrov, A., Vignali, M., LaCount, D. J., Quartley, E., de Vries, C., De Rosa, D., Babulski, J., Mitchell, S. F., Schoenfeld, L. W., Fields, S., Hol, W. G., Dumont, M. E., Phizicky, E. M., and Grayhack, E. J. (2004) A facile method for high throughput co-expression of protein pairs. *Mol. Cell. Proteomics* **3,** 934–938.

9. Mehlin, C., Boni, E. E., Buckner, F. S., Engel, L., Feist, T., Gelb, M. H., Haji, L., Kim, D., Liu, C., Mueller, N., Myler, P., Reddy, J. T., Sampson, J., Subramanian, E., Van Voorhis, W. C., Worthey, E. A., Zucker, F., and Hol, W. G. J. (2006) Heterologous Expression of Proteins from *Plasmodium falciparum:* results from 1000 genes. *Mol. Biochem. Parasitol.* **148,** 144–160.

10. Mehlin, C., Boni, E. E., Andreyka, J., and Terry, R. W. (2004) Cloning grills: high throughput cloning for structural genomics. *J. Struct. Funct. Genomics* **5,** 59–61.

11. Deng, J., Davies, D. R., Wisedchaisri, G., Wu, M., Hol, W. G. J., and Mehlin, C. (2004) An improved protocol for rapid freezing of protein samples for long-term storage. *Acta Crystallogr. D* **60,** 203–204.

12. Luft, J. R., Collins, R. J., Fehrman, N. A., Lauricella, A. M., Veatch, C. K., and DeTitta, G. T. (2003) A deliberate approach to screening for initial crystallization conditions of biological macromolecules. *J Struct Biol* **142,** 170–179.

13. Rupp, B., Segelke, B. W., Krupka, H. I., Lekin, T., Schafer, J., Zemla, A., Toppani, D., Snell, G., and Earnest, T. (2002) The TB structural genomics consortium crystallization facility: towards automation from protein to electron density. *Acta Crystallogr. D* **58,** 1514–1518.

14. Cork, C., O'Neill, J., Taylor, J., and Earnest, T. (2006) Advanced beamline automation for biological crystallography experiments. *Acta Crystallogr. D* **62,** 852–858.

15. Ellis, P. J., Cohen, A. E., and Soltis, S. M. (2003) Beamstop with integrated X-ray sensor. *J. Synchrotron Radiat.* **10,** 287–288.

16. Yeh, J. I., and Hol, W. G. J. (1998) A flash annealing technique to improve diffraction limits and lower mosaicity in crystals of glycerol kinase. *Acta Crystallogr. D* **54,** 479–480.

17. Harp, J. M., Timm, D. E., and Bunick, G. J. (1998) Macromolecular crystal annealing: overcoming increased mosaicity associated with cryocrystallography. *Acta Crystallogr. D* **54,** 622–628.

18. Otwinowski, Z., and Minor, W. (1997) Processing of x-ray diffraction data collected in oscillation mode. *Methods Enzymol.* **276,** 307–326.

19. Holton, J., and Alber, T. (2004) Automated protein crystal structure determination using ELVES. *Proc. Natl. Acad. Sci. USA* **101,** 1537–1542.

20. Bewley, M. C., Jeffrey, P. D., Patchett, M. L., Kanyo, Z. F., and Baker, E. N. (1999) Crystal structures of Bacillus caldovelox arginase in complex with substrate and inhibitors reveal new insights into activation, inhibition and catalysis in the arginase superfamily. *Structure* **7,** 435–448.

21. Terwilliger, T. C. (2003) SOLVE and RESOLVE: automated structure solution and density modification. *Methods Enzymol.* **374,** 22–37.

22. Kryshtafovych, A., Venclovas, C., Fidelis, K., and Moult, J. (2005) Progress over the first decade of CASP experiments. *Proteins* **61,** 225–236.

23. Mikol, V., and Giege, R. (1992) in (Ducruix, A., and Giege, R., eds.), *Crystallization of Nucleic Acids and Proteins.* Oxford University Press, New York, 219–239.

24. Gilliland, G. L., Tung, M., Blakeslee, D. M., and Ladner, J. E. (1994) Biological Macromolecule Crystallization Database, Version 3.0: new features, data and the NASA archive for protein crystal growth data. *Acta Crystallogr. D* **50,** 408–414.

25. Li, J., Zhang, Z., and Fan, E. (2004) Solid-phase synthesis of 1,5-substituted 2-N-Alkylamino)=imidazolidin-4-ones. *Tetrahedron Lett.* **45**, 1267–1269.

26. Li, J., Zhang, G., Zhang, Z., and Fan, E. (2003) TFA sensitive arylsulfonylthiourea-assisted synthesis of N, N'-substituted guanidines. *J. Org. Chem.* **68**, 1611–1614.

27. Meldrum, D. R., Evensen, H. T., Pence, W. H., Moody, S. E., Cunningham, D. L., and Wiktor, P. J. (2000) ACAPELLA-1K, a capillary-based submicroliter automated fluid handling system for genome analysis. *Genome Res.* **10**, 95–104.

28. Arutunian, E. B., Meldrum, D. R., Friedman, N. A., and Moody, S. E. (1998) Flexible software architecture for user-interface and machine control in laboratory automation. *Biotechniques* **25**, 698–702, 704–695.

29. Deng, J., Schnaufer, A., Salavati, R., Stuart, K., and Hol, W. G. J. (2004) High Resolution crystal structure of an editosome enzyme from *Trypanosoma brucei*: RNA editing ligase I. *J. Mol. Biol.* **343**, 601–613.

30. Verlinde, C. L. M. J., Kim, H., Bernstein, B. E., Mande, S. C., and Hol, W. G. J. (1997) Antitrypanosomiasis drug development based on structures of glycolytic enzymes. In *Structure-Based Drug Design* (Veerapandian, P., ed.), pp. 365–394. Marcel Dekker, New York.

31. Nienaber, V. L., Richardson, P. L., Klighofer, V., Bouska, J. J., Giranda, V. L., and Greer, J. (2000) Discovering novel ligands for macromolecules using X-ray crystallographic screening. *Nat. Biotechnol.* **18**, 1105–1108.

32. Shuker, S. B., Hajduk, P. J., Meadows, R. P., and Fesik, S. W. (1996) Discovering high-affinity ligands for proteins: SAR by NMR. *Science* **274**, 1531–1534.

33. Bosch, J., Robien, M. A., Mehlin, C., Boni, E., Riechers, A., Buckner, F. S., Van Voorhis, W. C., Myler, P. J., Worthey, E. A., DeTitta, G., Luft, J. R., Lauricella, A., Gulde, S., Anderson, L. A., Kalyuzhniy, O., Neely, H. M., Ross, J., Earnest, T. N., Soltis, M., Schoenfeld, L., Zucker, F., Merritt, E. A., Fan, E., Verlinde, C. L., and Hol, W. G. (2006) Using fragment cocktail crystallography to assist inhibitor design of *Trypanosoma brucei* nucleoside 2-deoxyribosyltransferase. *J. Med. Chem.* **49**, 5939–5946.

Chapter 34

High Throughput Crystallography at SGC Toronto: an Overview

Alexey Bochkarev and Wolfram Tempel

The completion of the human genome allows the analysis, for the first time, of biological systems in the context of entire gene families. For enzymes, this approach permits the exploration of complex substrate specificity networks that often exhibit considerable overlap within and between protein families. The case for a family-based approach to protein studies is compelling, given the prospect of exploiting these specificities for various purposes, such as the development of therapeutic reagents. The Structural Genomics Consortium (SGC) was created to determine the structures of proteins with relevance to human health and place the structures into the public domain without restriction on use. The SGC operates out of the Universities of Toronto and Oxford, and Karolinska Institutet, each working on nonoverlapping protein target lists. The SGC focus on human protein families requires a repertoire of crystallography methods that differ from those adopted by structural genomics projects that are focused on filling out protein fold space. The key differences are heavier reliance on in house x-ray sources for diffraction data collection and predominant use of molecular replacement for phase determination. As projects such as the US Protein Structure Initiative and others fill the PDB with representatives of most major fold families, the SGC approach will become an increasingly useful model for many structural biology laboratories in the future. Technical details of the flow of samples and data within the high throughput (HTP) environment at SGC Toronto are presented, and provide a useful paradigm for the organization of collaborative or shared x-ray instrumentation facilities.

1. Introduction

The Structural Genomics Consortium Toronto (SGC Toronto) and SGC centers at the University of Oxford and Karolinska Institutet constitute a large-scale structural genomics operation. The selection of structure targets

From: *Methods in Molecular Biology, Vol. 426: Structural Proteomics: High-throughput Methods*
Edited by: B. Kobe, M. Guss and T. Huber © Humana Press, Totowa, NJ

and their pursuit distinguishes SGC from traditional structural genomics. With an aim of comprehensive and detailed coverage of selected human protein families, the SGC produces high-resolution protein models primarily by means of single crystal x-ray diffraction. Although reliance on protein crystallography is common among many structural genomics projects, closer analysis reveals significant differences.

Like other structural genomics projects, SGC follows a specialist approach to task distribution within its structure pipeline. Although this approach is shared across all SGC sites, details of its implementation vary among the centers. At the SGC sites, cloning is performed by teams that provide expression constructs for all targets of their specific centers. At SGC Toronto, all recombinant proteins are then expressed, purified, and crystallized within groups that specialize in a selected subset of target proteins, such as families of proteins related by sequence. Family-specific expertise ensures that the length of gene constructs is optimized to enhance the proteins' propensity for crystallization and also that relevant cofactors and inhibitors are used in the crystallization process. At all SGC sites, these specialized target groups benefit from the local biophysics groups, which help identify small molecule ligands or specific conditions such as pH or salt concentration that may improve the suitability of protein samples for structure determination (1). Members of the specialized target groups also mount and flash-freeze (2) promising crystals for diffraction screening and data collection. Frozen crystal samples are then provided to the Protein Crystallography (PX) group for further analysis.

A complex communication network complements the sample flow into the PX group. Within SGC Toronto, for example, each member of the PX group has developed a close working relationship with one of the specialized target groups and has thereby become its primary contact for any crystallographic issues. An array of mechanisms for the exchange of information and data is now in use, and tailored to the needs of each group. Thus, solutions based on paper records, spreadsheets, and a relational database system are used to alert crystallographers that crystals are available for diffraction screening and provide feedback on the results to the specialized target groups. However, these tools merely facilitate communication. Frequent discussions, both casual and formal, are essential to help educate each other about capabilities and limitations of available techniques at different check points between target selection and structure solution. This one-on-one strategy helps establish an efficient route to a target's structure solution. It has proved to be essential for troubleshooting any failures to obtain crystals of a particular target or their lack of diffraction quality.

At SGC Toronto, crystallographic computing relies on commodity hardware, gigabit Ethernet networking, the most widely used crystallographic software packages and a refined set of system management tools. Crystallographic data, such as diffraction images and model coordinates, are stored on a central file server. They are thus available to interested and authorized parties at computer workstations in the labs.

In the 2 years from the start of operations on July 1, 2004, SGC Toronto has deposited 120 structures in the Protein Data Bank (PDB) (3). Eighty-five percent of the structures were solved by molecular replacement, and 60% of the structures were solved and refined using data collected at the in-house rotating anode source. Across the other SGC sites, the proportion of structures determined by the method of molecular replacement is similar, but there is a greater preference to use synchrotron radiation for data collection. At SGC Toronto, nearly 1,000 crystals are screened and approximately 30 full diffraction data sets collected per month, whereas typically six to eight crystallographic models are deposited to the PDB during that period. This level of throughput is based on a combination of powerful x-ray sources, automation, and efficient resource management.

2. Overview of Methods

2.1. Crystal Mounting and Storage

Protein crystals are mounted into nylon loops (CryoLoops, Hampton Research, Aliso Viejo, CA) held on hollow pins, at a distance of 18 mm to the bottom of a ferromagnetic base that will fit on the goniometer (see Note 1). Once frozen, they are held as a set of 12 in labeled cassettes designed for the ACTOR robot (see Section 2.2.). Cassettes, in turn, are stored in racks of five cassettes in a cryogenic storage vessel (35VHC, Taylor-Wharton). The storage vessels can accommodate six such racks and are located in a room adjacent to the x-ray source. Cassettes are manually deposited into and retrieved from the storage vessels. The location of specific crystals that require screening is communicated, depending on the groups and individuals involved, through either a paper or electronic version of a scoring sheet. On the form, the experimenter mounting and freezing the crystals provides details such as cassette label and location, protein sample description, crystallization tray label, well coordinates, and cryogenic treatment.

2.2. In-House Diffraction Equipment

All equipment described in this paragraph was supplied by Rigaku/Molecular Structure Corp. (The Woodlands, TX), unless otherwise noted. X-rays are generated using a state-of-the-art rotating copper anode (FR-E). Its left-hand port (robot side) is equipped for robot-controlled crystal diffraction screening: confocal optics (VariMax High Flux) with an (exchangeable) 0.2-mm collimator, a square 300-mm detector that houses three image plates and dual readers (R-axis HTS), a single axis goniometer and a crystal mounting robot (ACTOR). The right-hand port (HighRes side) is primarily used for data collection: confocal optics (VariMax High Resolution) with a 0.1-mm collimator, a dual image plate detector (R-axis IV++), and a partial χ goniometer. Liquid nitrogen-based crystal cooling equipment for each side was procured from Oxford Cryosystems Ltd.

2.3. Diffraction Screening

For automated diffraction screening on the robot side, two 0.5° oscillation images at $\phi = 0°$ and $\phi = 90°$ are exposed for 70 seconds (see Note 2) each at a crystal-to-detector distance of 150 mm (see Note 3). Sixty crystals can be screened in a single program run. Including pin mounting, loop centering, x-ray exposure, and detector read-out, each crystal sample requires approximately 8 minutes. Therefore, a full run takes approximately 8 hours, often during overnight hours. A form (described in Section 2.1.) holds fields for comments on crystal diffraction, to be filled by the crystallographer after screening (see Note 4). The screening results are then discussed with the responsible scientist(s) in the specialized target groups. Crystallographers complement the feedback on the form with more detailed recommendations and/or assistance in crystal optimization and freezing.

2.4. Diffraction Data Collection

For data collection on the HiRes side of the rotating anode generator (see Section 2.2.), typically a 180° scan in 0.5° steps is performed around ω with an exposure time of 2 minutes (see Note 5). The crystal to detector distance is adjusted according to the expected high resolution limit (see Note 6). This mode of data collection accounts for the majority of structures deposited in the PDB by SGC Toronto.

In addition, synchrotron data collection is also scheduled at a rate of 1 to 2 days per month (see Note 7). It is usually considered in situations when crystals do not diffract to sufficiently high resolution on the in-house source or when phasing by anomalous scattering requires an x-ray wavelength that is only available at a tunable source (see Note 8). A preliminary processing and analysis of the data is performed (see the following) and again feedback is provided to the crystallizers on the suitability of the data/crystal and a recommendation for further experiments, if needed.

2.5. Crystallographic Computing

Servers and workstations (see Note 9) rely on Gentoo Linux (www.gentoo. org) system software and are connected by gigabit Ethernet networking. The server provides centralized LDAP (www.openldap.org) user authentication and several terabytes of network-attached disk storage (NAS) (see Note 10) for data and computer programs (see Note 11).

Data reduction almost exclusively relies on the HKL2000 software suite *(4)*. PHASER *(5)* and MOLREP *(6)* are commonly used for molecular replacement (see Note 12). *De novo* phasing is usually accomplished with SOLVE/RESOLVE *(7)*, BnP *(8)*, or SHELXC/D/E *(9,10)* in application of the single wavelength anomalous scattering *(11)* method. Data quality permitting, a preliminary model is next obtained automatically by RESOLVE or ARP/WARP *(12)*. The model is refined in REFMAC *(13)* or CNS *(14)*. O *(15)*, XFIT *(16)*, and COOT *(17)* are used for graphics-assisted modeling (see Note 13). Model geometry is assessed with programs such as PROCHECK *(18)* and MOLPROBITY *(19)*.

3. Notes

1. CryoLoops must be prepared in a consistent manner in order to avoid loop-to-loop variations and consequent failure of the ACTOR robot to accurately position the crystal in the beam. The authors found it useful to have them assembled and cleaned for re-use by a single person for the entire operation.
2. Shorter exposure times do not result in time savings due to the detector's read-out time for single-head reading. For faster, but less accurate dual-head reading, this time can be reduced.

3. Under this configuration, the data resolution at the detector's edges is approximately 2.00 Å.

4. Diffraction images are written to networked storage and can therefore be inspected on workstations across SGC Toronto.

5. Here, too, detector read-out time limits the time savings of shorter exposures.

6. The minimum distance of 120 mm allows for the capture of 1.78 Å resolution data at the detector's edges.

7. Diffraction experiments are usually performed by SGC Toronto personnel but "mail-in" programs offered by several synchrotron sources have been used successfully and continue to be actively explored.

8. A rotating anode source with a chromium target *(20)* is available at the University of Toronto. Members of the PX group have successfully exploited the increased anomalous signal of sulfur at longer wavelength (2.29 Å) for *de novo* protein phasing. The authors are also aware of the utility of this alternative target material for selenomethionine phasing *(21)*.

9. One dual-processor × 86 workstation per crystallographer proved generally sufficient. Remote login and usage of idle computing cycles of other crystallographers' workstations exists in exceptional cases as an option.

10. Storage is provided by multiple disks in several RAID5 arrays. Precise implementation is subject to frequent changes due to increased space demand and technological advances. Diffraction images kept on-line for convenience make up the bulk of storage requirements.

11. This server-client architecture greatly simplifies updates of computer programs and the user database.

12. Molecular replacement is attempted given the existence of suitable models, based on BLAST *(22)* alignment of the target's peptide sequence against the closest match in the PDB. Crystallography and nuclear magnetic resonance–based models have been used successfully. Failure of molecular replacement in any given case will be followed by consultations between the involved members of the PX and specialized target groups with the aim of an alternative phasing strategy, such as the preparation of a selenomethionine derivative of the target.

13. It is noteworthy that no hardware-assisted three-dimensional viewing is involved in this process.

Acknowledgments

The Structural Genomics Consortium is a public private charitable partnership that receives funds from the Wellcome Trust, Genome Canada, GlaxoSmithKline, the Swedish Agency for Innovation Systems, the Canadian Institutes for Health Research, the Ontario Research and Development Challenge Fund, the Knut and Alice Wallenberg Foundation, the Ontario Innovation Trust, the Swedish Foundation for Strategic Research, the Petrus and Augusta Hedlund's Foundation, and the Canadian Foundation for Innovation.

References

1. Vedadi, M., Niesen, F. H., Allali-Hassani, A., Fedorov, O. Y., Finerty, P. J.,Jr, Wasney, G. A., Yeung, R., Arrowsmith, C., Ball, L. J., Berglund, H., Hui, R.,

Marsden, B. D., Nordlund, P., Sundstrom, M., Weigelt, J., and Edwards, A. M. (2006) Chemical screening methods to identify ligands that promote protein stability, protein crystallization, and structure determination. *Proc. Natl. Acad. Sci. USA* **103**, 15835–15840.

2. Hope, H. (1988) Cryocrystallography of biological macromolecules: a generally applicable method. *Acta Crystallogr. B* **44**, 22–26.

3. Berman, H. M., Westbrook, J., Feng, Z., Gilliland, G., Bhat, T. N., Weissig, H., Shindyalov, I. N., and Bourne, P. E. (2000) The Protein Data Bank. *Nucleic Acids Res.* **28**, 235–242.

4. Otwinowski, Z., and Minor, W. (1997) Processing of X-ray diffraction data collected in oscillation mode. *Methods Enzymol.* **276**, 307–326.

5. McCoy, A. J., Grosse-Kunstleve, R. W., Storoni, L. C., and Read, R. J. (2005) Likelihood-enhanced fast translation functions. *Acta Crystallogr.* **61**, 458–464.

6. Vagin, A., and Teplyakov, A. (1997) MOLREP: an automated program for molecular replacement. *J. Appl. Crystallogr.* **30**, 1022–1025.

7. Terwilliger, T. C. (2003) SOLVE and RESOLVE: automated structure solution and density modification. *Methods Enzymol.* **374**, 22–37.

8. Weeks, C. M., Blessing, R. H., Miller, R., Mungee, R., Potter, S. A., Rappleye, J., Smith, G. D., Xu, H., and Furey, W. (2002) Towards automated protein structure determination: BnP, the SnB-PHASES interface. *Z. Kristallogr.* **217**, 686–693.

9. Schneider, T. R., and Sheldrick, G. M. (2002) Substructure solution with SHELXD. *Acta Crystallogr.* **58**, 1772–1779.

10. Sheldrick, G. (2002) Macromolecular phasing with SHELXE. *Z. Kristallogr.* **217**, 644–650.

11. Wang, B. C. (1985) Resolution of phase ambiguity in macromolecular crystallography. *Methods Enzymol.* **115**, 90–112.

12. Perrakis, A., Morris, R., and Lamzin, V. S. (1999) Automated protein model building combined with iterative structure refinement. *Nat. Struct. Biol.* **6**, 458–463.

13. Murshudov, G. N., Vagin, A. A. and Dodson, E. J. (1997) Refinement of macromolecular structures by the maximum-likelihood method. *Acta Crystallogr.* **53**, 240–255.

14. Brunger, A. T., Adams, P. D., Clore, G. M., DeLano, W. L., Gros, P., Grosse-Kunstleve, R. W., Jiang, J. S., Kuszewski, J., Nilges, M., Pannu, N. S., Read, R. J., Rice, L. M., Simonson, T., and Warren, G. L. (1998) Crystallography & NMR system: a new software suite for macromolecular structure determination. *Acta Crystallogr.* **54**, 905–921.

15. Jones, T. A., Zou, J. Y., Cowan, S. W., and Kjeldgaard, M. (1991) Improved methods for building protein models in electron-density maps and the location of errors in these models. *Acta Crystallogr. A* **47**, 110–119.

16. McRee, D. E. (1999) XtalView Xfit: a versatile program for manipulating atomic coordinates and electron density. *J. Struct. Biol.* **125**, 156–165.

17. Emsley, P., and Cowtan, K. (2004) Coot: model-building tools for molecular graphics. *Acta Crystallogr.* **60**, 2126–2132.

18. Laskowski, R. A., Macarthur, M. W., Moss, D. S., And Thornton, J. M. (1993) Procheck: a program to check the stereochemical quality of protein structures. *J. Appl. Crystallogr.* **26**, 283–291.

19. Davis, I. W., Murray, L. W., Richardson, J. S., and Richardson, D. C. (2004) MolProbity: structure validation and all-atom contact analysis for nucleic acids and their complexes. *Nucleic Acids Res.* **32**, W615–W619.

20. Yang, C., Pflugrath, J., Courville, D., Stence, C., and Ferrara, J. D. (2003) Away from the edge: SAD phasing from the sulfur anomalous signal measured in-house with chromium radiation. *Acta Crystallogr.* **59**, 1943–1957.

21. Xu, H., Yang, C., Chen, L., Kataeva, I. A., Tempel, W., Lee, D., Habel, J. E., Nguyen, D., Pflugrath, J. W., Ferrara, J. D., Arendall, W. B.,3rd, Richardson, J. S., Richardson, D. C., Liu, Z. J., Newton, M. G., Rose, J. P., and Wang, B. C. (2005) Away from the edge II: in-house Se-SAS phasing with chromium radiation. *Acta Crystallogr.* **61,** 960–966.

22. Altschul, S. F., Gish, W., Miller, W., Myers, E. W. and Lipman, D. J. (1990) Basic local alignment search tool. *J. Mol. Biol.* **215,** 403–410.

Chapter 35

The Structural Biology and Genomics Platform in Strasbourg: an Overview

Didier Busso, Jean-Claude Thierry, and Dino Moras

This chapter describes the modules and facilities of the Structural Biology and Genomics Platform (SBGP), Strasbourg, France. The platform consists of three modules (cloning, mini-expression screening; optimization-large scale protein production; characterization, crystallization) with dedicated scientists, and other facilities for purifying recombinant proteins and solving three-dimensional (3D) structures. Strong collaborations have been established with the Integrative Bioinformatics and Genomics group, located in the same institition, for target selection and domains definition. To handle large numbers of samples, classical and new protocols were adapted to automation, increasing reproducibility and reducing error risks as well. Using the platform and its facilities, over 2,000 expression vectors have been constructed and more than 40 novel structures, of mostly human proteins, have been solved.

1. Introduction

Structural genomics (SG) programs have been initiated worldwide with the aim of solving 3D structure of proteins of living organisms (www.isgo.org). The different goals that SG initiatives have are: (1) having a complete view of protein folds of a minimal organism that, in turn, may help in assigning function for hypothetical proteins that represent a large fraction of encoded proteins (1,2); (2) having a deeper understanding of protein function that will facilitate pharmaceutical drug development (3–5), and (3) elucidating molecular mechanism(s) implicating functional complexes in higher eukaryotes to understand fundamental aspects of living organisms that may serve, in turn, for biomedical application and biotechnologies (6). Whatever the strategy adopted, the analysis of large number of parameters generated to identify suitable conditions to produce samples for structure determination required

From: *Methods in Molecular Biology, Vol. 426: Structural Proteomics: High-throughput Methods*
Edited by: B. Kobe, M. Guss and T. Huber © Humana Press, Totowa, NJ

new developments at each step of the process from gene cloning through 3D structure determination using parallelization and automation as much as possible *(7–11)*.

The Structural Biology and Genomics Department (IGBMC, Illkirch, France) is implicated in recent SG programs through the Structure Proteomics IN Europe (SPINE) and SPINE2-Complexes networks (www.spineurope.org/ and www.spine2.eu/) with the goal of solving crystal structures of protein family members related to human health, alone or in complex with partner(s), including ligand, protein, or nucleic acid. To achieve this goal, the authors have implemented a medium-throughput platform (SBGP) (lbgs.u-strasbg.fr/sbgp/) to assist challenging research projects. The SBGP integrates standard bench protocols to automation especially for cloning, recombinant protein expression screening, and protein crystallization, and offers facilities and assistance for production, purification, and characterization of proteins. The following sections describe strategies from target selection to crystallization and 3D structure determination, and detail the cloning procedures for which a robust pipeline has been established. An overview of the platform organization is given in Fig. 35.1.

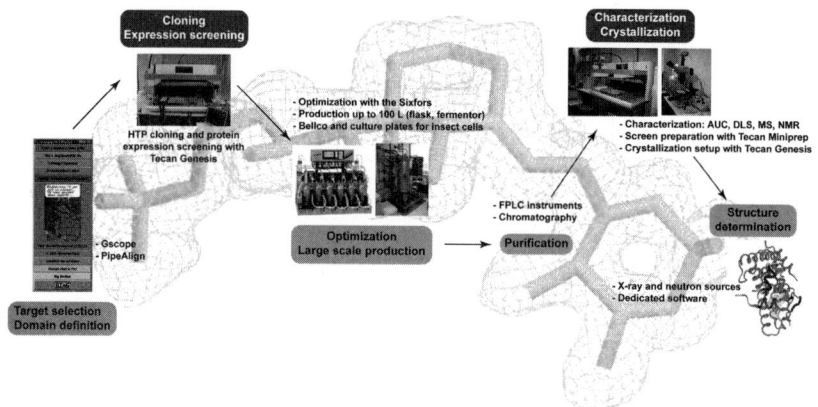

Fig. 35.1 Flowchart of the Structural Biology and Genomics Platform (SBGP) and facilities. SBGP consists of modules that are largely automated with dedicated staff *(dark gray boxes with white text)* and facilities *(light gray boxes with black text)*. The "Optimization—Large scale production" module is operated by a dedicated engineer whereas the "Purification" module consists of equipment shared by the department members. For protein characterization, a dedicated department staff is in charge of AUC (analytical ultracentrifugation), DLS (dynamic light scattering), and NMR (nuclear magnetic resonance) equipment. MS (mass spectrometry) measurements are carried out with BioOrganic Mass Spectrometry Laboratory members (Strasbourg, France). The target selection and domains definition is conducted by the BioInformatics Platform (bips.u-strasbg.fr/en/) and the Integrative Bioinformatics and Genomics laboratory (O. Poch, IGBMC, France) in collaboration with research managers. Structure determination is performed by scientists using available x-ray sources and software as described in Section 3.4.5.

2. Materials

2.1. Target Selection

1. DNA sources used for PCR amplification varied from genomic DNA (microorganisms), to cDNA already subcloned in a plasmid.
2. Protein domains are defined by combining results obtained with different software such as DisEMBL1.4 (dis.embl.de/), GlobPlot2 (globplot.embl. de/), FoldIndex (bip.weizmann.ac.il/fldbin/findex), RONN (www.strubi. ox.ac.uk/RONN) for disorder prediction, and Jpred (www.compbio.dundee. ac.uk/~www-jpred/submit.html) for secondary structure prediction. Selection of homologous sequences in various organisms is performed by using a local library of software (PipeAlign) (bips.u-strasbg.fr/PipeAlign/). This software is free and is available on the Web.

2.2. Cloning

2.2.1. Primer Design

Primer design is automated using computational GScope software developed in-house (R. Ripp, in preparation).

2.2.2. PCR: Reaction, Analysis, and Purification

1. Phusion DNA polymerase and buffer (F530L; Finnzymes, Espoo, Finland).
2. Taq DNA polymerase and buffer (M0267; New England Biolabs, Beverly, MA).
3. Primers were from Sigma-Aldrich (St Louis, MO).
4. Agarose gels are run in 1× TAE buffer: TAE/L of 50×: 242 g Tris base, 57.1 mL glacial acetic acid, 100 ml 0.5 M ethylenediaminetetraacetic acid (EDTA), pH 8.0. Store at room temperature. Prepare the 1× TAE solution by diluting the 50× stock with double distilled water. When making 0.5 M stock solution of EDTA, add sodium hydroxide pellets to adjust pH to 8.0.
5. MicroAmp optical 96-well PCR plate (N801-0560; Applied Biosystems, Foster City, CA).
6. UV-star 96-well plate (655801; Greiner BioOne, Kremsmünster, Austria).
7. NucleoFast PCR purification kit (743100.10; Macherey Nagel, Düren, Germany).
8. Impact2 eight-channel pipettor (2130, Matrix, Hudson, NH).
9. I-cycler PCR machine (Bio-Rad, Hercules, CA).
10. Tecan Genesis Workstation robot (Tecan, Maennedorf, Switzerland) equipped with four low-volume needles and four standard-volume needles, a rotary shaker, a vacuum manifold, and a cooling block.
11. DNA quantification was calculated by measuring $OD_{260\,nm}$ with either the GENios microplate reader (Tecan) or the NanoDrop spectrophotometer (ND-1000; NanoDrop, Wilmington, DE).

2.2.3. Gateway Cloning

1. Luria Bertani (LB) medium (12780-052; Invitrogen, Carlsbad, CA): 20 g/L. Sterilize by autoclaving. Store at room temperature.

2. LB-agar medium (22700-025, Invitrogen): 32 g/L. Sterilize by autoclaving. Store at room temperature.
3. Antibiotic stock solutions: ampicillin at 100 mg/ml and gentamicin at 7 mg/ml. Antibiotic stock solutions are prepared by dissolving appropriate amount of powder in double distilled water. Dissolved antibiotics are aseptically filtered through an 0.22 μm filter, aliquoted, and stored at −20 °C.
4. BP clonase mix (11789-021, Invitrogen).
5. LR clonase mix (11791-043, Invitrogen).
6. pDONR207 plasmid (12213013, Invitrogen).
7. *Escherichia coli* DH5α chemically competent cells (see Note 1).
8. AirPore sealing tape (19571; Qiagen, Hilden, Germany).
9. Disposables: six-well culture plates (CLS3506; Sigma-Aldrich), 96-deep well culture plate (969799901; TreffLab, Degersheim, Switzerland), and 24-deep well culture plate (7701-5102; Whatman, Middlesex, UK).
10. Tecan Genesis Workstation robot (Tecan) equipped with four low-volume needles and four standard-volume needles, a rotary shaker, a vacuum manifold, and a cooling block.
11. NucleoSpin robot-96 plasmid kit (740708.4; Macherey Nagel).
12. Impact2 eight-channel pipettor (Matrix).

2.3. Protein Crystallization

1. Crystallization plates: CrystalQuick 96-well plate (609101; Greiner BioOne), Innovaplate SD-2 plate (11978-1; Innovadyne, Santa Rosa, CA), and PZero 3:1 plate (Corning CLS3555; Sigma-Aldrich).
2. Optical sealing tape (CLS6575; Sigma-Aldrich).
3. 96-deep well culture plate (969799901; TreffLab).
4. Sealing mat for 96-deep well plate (4421; Matrix).
5. Tecan Genesis Workstation robot (Tecan) equipped with 8 low-volume needles and a TeMo 96-channel dispensers.
6. Cartesian Honeybee 81 nanoliter crystallization robot (Genomic Solutions, Cambridgeshire, United Kingdom).
7. Tecan Miniprep robot (Tecan).

3. Methods

3.1. Target Selection

The Structural Biology and Genomics Department selects targets corresponding to human protein family members related to health such as transcription factors, nuclear receptors, and corepressors/coactivators. The authors' approach consists of using bioinformatic tools developed locally as well as free access Web-based tools to define protein domains for a given target and select homologous sequences in a wide range of sequenced organisms, including eukaryotes *(Mus musculus, Danio rerio, Drosophila melanogaster, Alvinella pompejana, Caenorhabditis elegans*, and *Saccharomyces cerevisiae)* as well as microorganisms such as *Pyrococcus furiosus, Thermus thermophilus, Encephalitozoon cuniculi*, and others.

3.2. Cloning

Because proteins are so different in their behavior, it is impossible to predict whether a protein of interest will be well expressed, soluble, easy to purify, and amenable to crystallization. SBGP combines different strategies well documented to favor soluble expression of wild-type recombinant proteins when expressed heterogeneously, such as using different expression systems *(12,13)* or a wide range of fusion partners *(14)*, or changing host strains as well as growth conditions *(15–18)*. Although the last described strategy does not require further subcloning to be carried out, the comparison of fusion partners' impact on protein solubility or changing the expression system led to subcloning the open reading frame (ORF) of interest in a library of expression vectors, which becomes laborious when handling a large number of genes. In recent years, new cloning technologies have been developed to expedite parallel cloning often required in SG approaches *(19)*. SBGP has chosen to use the Gateway cloning method because it is site specific, directional, and based on conservative recombination, eliminating the requirement to work with restriction enzymes and ligase *(20)*. Moreover, after DNA sequencing of the Entry vector, the Gateway technology allows a rapid subcloning of the ORF into virtually any expression vectors without further verification, as is the case for other ligation independent cloning methods *(21,22)*. During the past years, SBGP has developed sets of Gateway-based vectors to: (1) screen fusion partner impact on recombinant protein behavior when expressed in *E. coli*; (2) express protein in insect cells infected by recombinant baculovirus; and (3) be able to express up to six proteins at a time in *E. coli (23)*.

Basically, the Gateway technology can be divided into three steps. First, the DNA sequence corresponding to the ORF of interest is flanked by specific *att*B1 and *att*B2 recombination sites during PCR. Second, the resulting PCR product, after purification, is cloned into a Donor vector during the BP reaction where *att*B sites recombine with *att*P sites of the Donor vector (pDONR207). This reaction generates an Entry clone with the ORF flanked by *att*L1 and *att*L2 sites. Third, after validation by DNA sequencing, the ORF can be transferred easily into any appropriate Destination vectors during the LR reaction where *att*L sites recombine with *att*R sites of the Destination vector(s). The LR reaction generates expression vectors and restores *att*B sites flanking the ORF.

The following sections describe the cloning protocols the authors have developed and adapted for automation (Fig. 35.2).

3.2.1. Primer Design

Specific primers to amplify ORFs of interest are computationally designed using the GScope software. Forward primers contained the following sequence: four guanines (G) followed by the 25-nucleotide *att*B1 sequence (ACAAGTTTGTACAAAAAAGCAGGCT), two extra nucleotides to maintain the reading frame, optional specific signals such as a sequence encoding protease recognition site and/or restriction site, and at least an 18 to 25 nucleotides gene-specific sequence with a Tm = 60°C (see Note 2). Reverse primers contained the following sequence: four guanines (G) followed by the 25-nucleotide *att*B2 sequence (ACCACTTTGTACAAGAAAGCTGGGT), one extra nucleotide to maintain the reading frame once the LR recombination reaction occurred (important when using vectors encoding a C-terminal fusion), optional specific signal such as a restriction site and a Stop codon (when using

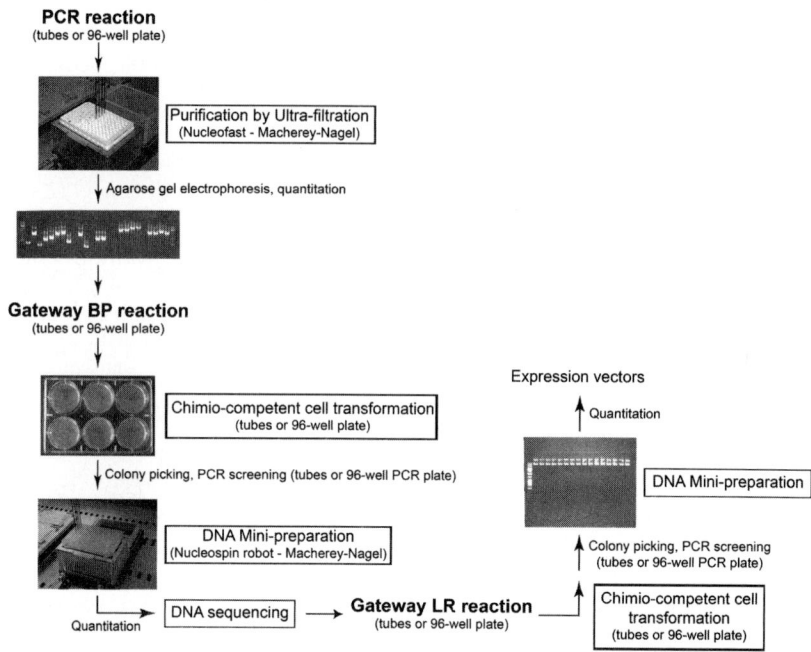

Fig. 35.2 Schematic representation of the cloning strategy. Each step of the cloning strategy is indicated. Steps that are fully automated are boxed and other steps require manual input. Roughly, 1 week of one full-time equivalent is needed to complete the process for 96 constructs besides DNA sequencing. DNA sequencing is performed by an in-house facility.

vectors that do not encode a C-terminal fusion), and at least an 18 to 25 nucleotides gene-specific sequence with a Tm = 60 °C.

3.2.2. PCR: Reaction, Analysis, and Purification

PCR reactions are performed using a multi-channel pipettor either in 96-well plate or tubes.

1. Primers are suspended in pure water to have a final concentration of 100 μM, and stored at −20 °C.
2. PCR is performed in a 20-μl reaction with 0.4 U of Phusion DNA polymerase according to manufacturer's instructions (see Note 3).
3. PCR reactions are analyzed on agarose gel run in 1× TAE buffer.
4. PCR products purification is performed with the NucleoFast kit following manufacturer's instruction on the Tecan Genesis Workstation robot. Cleaned PCR products are suspended in 50 μl of pure water.
6. After analysis on agarose gel, PCR products are quantified using either the GENios microplate reader with UV-compatible microplate or the NanoDrop spectrophotometer (see Note 4).

3.2.3. Gateway Cloning

BP and LR reactions are performed using a multichannel pipettor either in 96-well plate or tubes.

1. BP is performed for 16 hours at room temperature in a 5-μl reaction containing 1 μl of BP Clonase in 1× buffer plus 100 ng of gentamicin

resistant pDONR207 plasmid and 30 fmol of purified PCR product (see Step 3.2.2.).

2. After a 10-minute incubation at 37°C in the presence of 1.3 µg of proteinase K, 50 µl of DH5α chemically competent cells are transformed with the reaction and cells are plated on LB + antibiotic plates (see Note 5).

3. After overnight incubation at 37°C, two individual colonies are inoculated into a starter culture consisting of 100 µl of 2× LB medium + antibiotic dispensed into a 96-deep well culture plate. Plates are sealed with AirPore tape and cells were grown for 90 minutes at 37°C in a microplate rotary shaker (Infors, Bottmingen, Switzerland) at 350 rpm.

4. Characterization of positives clones is done by PCR screening in a 20-µl reaction with 2.5 U of Taq DNA polymerase according to manufacturer's instruction using 1.5 µl of the starter culture (see step 3) as DNA template (see Note 6).

5. After analysis on an agarose gel, 10 µl of the starter culture corresponding to the positive clones are used to inoculate a 2.5-ml culture in 2× LB medium + antibiotic dispensed into 24-deep well culture plate. Plates are sealed with AirPore tape and incubated at 37°C for 16 hours in a microplate rotary shaker (Infors) at 350 rpm.

6. DNA minipreparations are performed with the NucleoSpin robot-96 plasmid kit following manufacturer's instruction on the Tecan Genesis Workstation robot. DNA is eluted in 50 µl of elution buffer and quantified as described in Section 3.2.2.

7. Entry clones are sequenced using the authors' DNA sequencing facility and sequence analysis is done using the GScope software.

8. For positive Entry clones, LR reactions are performed for 1 hour at room temperature in a final volume of 5 µl containing 1 µl of LR Clonase in 1× buffer plus 100 ng of Destination vector and 100 ng of Entry clone.

9. Analysis and preparation of expression vectors are performed by repeating steps 2 to 6.

10. Obtained expression vectors are used for recombinant protein expression screening as described in this volume *(24)*.

3.3. Protein Production

Besides classical scaled-up expression using Erlenmeyer flasks, SBGP places emphasis on culture optimization to obtain enough material for structural studies. The authors have implemented a multireactors system (Sixfors; Infors) consisting of six 500-ml fermenters individually controlled for temperature, stirring, air flow, and reagent delivery. By growing the cells in classical media with the Sixfors, the authors were able to reach an $OD_{600nm} = 12$ corresponding, in cell mass, to a 6-L culture in classical flasks. Thus, the Sixfors allowed parallel testing of culture parameters as well as performing reproducible culture batches, producing enough cell paste to carry out further protein purification. After validation, culture parameters defined using the Sixfors could be scaled for the large-scale fermenter (100 L; Infors), to produce poorly expressed proteins at levels necessary for structural studies.

For purification, cells are suspended in lysis buffer and disrupted either by sonication or a microfluidizer processor (M-110EH; Microfluidics, Newton, MA). Microfluidization offers the advantages of continuous operation, short processing times to minimize product degradation, and reproducibility with a high yield of undamaged proteins *(25)*. After clarification, proteins are purified by chromatography, including affinity, ion exchange, and size exclusion resins, on FPLC instruments (Biologic; BioRad). To increase the throughput and optimize common purification steps such as affinity and size exclusion chromatography, an Aktaxpress system (GE Healthcare, Piscataway, NJ) has been implemented for parallel purification. After purification, proteins are characterized by electrophoresis on native or SDS gels and by biophysical methods, including dynamic light scattering (DLS) and analytical ultracentifugation. The laboratory is also developing mass spectroscopy approaches for high throughput sample characterization, in collaboration with Alain van Dorsselaer's group *(26)*.

3.4. Protein Crystallization and Structure Determination

Before setting-up crystallization trials, purified proteins are concentrated on Amicon Ultra, Centricon or Centriprep concentrators (Millipore, Billerica, MA) and tested by DLS (Wyatt Technology, Santa Barbara, CA) as a predictor for favorable crystallization conditions *(27,28)*. Thus, suitable protein concentration for crystallization is determined using the precrystallization test (Hampton Research, Aliso Viejo, CA).

Finding the appropriate crystallization condition is still empirical and requires screening of numerous conditions *(29)*. The authors have a library of 1,104 different conditions stored into 96-deep well blocks that combines commercially available kits (from Hampton Research: Index, Salt RX, Natrix, and Membfac; from deCODE Genetics, Reykjavik, Iceland: Wizard I and II ; and from Qiagen, Valencia, CA: The Classics, The Anions, The Cations, The AmSO4, and The JCSG+), and in-house designed and prepared kits using the authors' library of chemicals with a Tecan Miniprep robot (NR-LBD 96, the PEG-Ion-pH, and the PEG-Ion) *(30)*.

3.4.1. Crystallization Screen Preparation Using the Tecan Miniprep Robot

Different tests (i.e., evaporation over time, speed of pipetting, execution time) have been made to define that the optimal volume to use in the 96-deep well plate containing the crystallization screen is 0.8 ml.

1. 50 ml conical tubes containing appropriate stock solutions, prepared and stored by SBGP, are placed on specific holders on the Miniprep's deck.
2. Destination 96-deep well culture plate for preparing crystallization solutions is placed on the robot.
3. An Excel spreadsheet with final concentration of each solution for each well of the 96-deep well plate is prepared.
4. The robot dispenses first water then the appropriate volume of stock solutions accordingly to the Excel table.
5. After the liquid handling, the 96-deep well plate containing the crystallization screen is covered with a specific mat and stored at 4 °C.

3.4.2. Crystallization Set-up Using the Tecan Genesis Workstation Robot

Crystallization plates from different commercial sources were tested, and higher rates of crystallization were achieved using the CrystalQuick crystallization plate from Greiner BioOne (D. Busso, in preparation). The CrystalQuick crystallization plate presents three drop support per well accommodating the possibility of crystallizing up to three proteins, or three ratios of reservoir/protein for a given protein, at a time per crystallization condition.

To avoid evaporation during the liquid transfer process, a cover has been implemented to present only one row of the crystallization plate at a time and fully cover the plate at the end of the process.

1. Place the crystallization plate, cover, and 96-deep well plate containing crystallization solutions on the robot's deck.
2. Select the different variables (i.e., volume of crystallization condition, volume of reservoir to be mixed with the protein, volume of protein, number of proteins).
3. The robot aspirates the crystallization solutions from the 96-deep well plate and dispenses them into the crystallization plate (typically, 100 µl).
4. Place protein dispensed into eight Eppendorf tubes on the robot's deck.
5. The robot aspirates the appropriate volume of protein (0.5–2 µl), then the appropriate volume of reservoir solution (0.5–2 µl), and dispenses both solutions into drop support (see Note 7).
6. Plates were bar-coded and manually sealed with optical transparent film before storage at 17 and/or 4 °C.

3.4.3. Crystallization Set-up Using the Cartesian Robot

1. Place the crystallization plate (Innovaplate SD-2 or the PZero 3:1) and the 96-deep well plate containing crystallization solutions on the Tecan's deck.
2. The TeMo 96-channel dispensers aspirates the 96 crystallization solutions at once and dispenses them into the crystallization plate (typically, 70 µl).
3. Place the crystallization plate in the humidity chamber of the Cartesian robot and place the protein in a PCR tube.
4. The robot aspirates, with 8 needles, crystallization solution and dispenses them into drop support (typically, 0.1–0.2 µl) in 20 minutes.
5. The ninth needle aspirates the protein and dispenses it on fly in the drop support in less than 2 minutes.
6. The plates are stored as described in Section 3.4.2. step 6.

3.4.4. Crystallization Plate Visualization

If one protein was screened per day against the 1,104 crystallization conditions, the authors would generate 345 plates per month. With a viewing schedule of six times through a 1-month lifetime of a plate, about 20,000 drops per protein (i.e., about 600,000 image per month), on average, need to be viewed. Thus, a prototyped system was developed in the authors' laboratory that consists of a CCD camera imaging system utilizing a microscope (Leica, Solms, Germany), and an x,y controlled tray adapted to the microplates with bar-code recognition. The system is capable of imaging a 96-well plate in 3 minutes and displays an image gallery for visual crystal tracking on a desktop computer screen *(31)*. With the increase of the throughput, the next step will be a temperature-controlled storage system for automated imaging with integrated scheduling and crystal detection software *(32,33)*. The benefit of such a system will be to keep track of drop history (i.e., protein/reservoir ratio, drop volume, reservoir volume, well position, crystallization condition) associated with imaging linked to the SQL database developed from protein information to protein purification.

3.4.5. X-Ray Sources and Crystal Structure Determination

The screening for crystal quality, to either refine crystallization conditions or perform data collection, involves in-house equipment and synchrotron facilities. The authors are equipped with MicroMax007 high flux generator, two confocal optics, CCD camera and image plate, cryocooling devices, and Free Mounting System™, which allows a humidity control to optimize crystals with respect to their diffraction properties. Short- and long-term projects strongly benefit from the synchrotron facilities in Europe. Access to the beam lines is essentially provided through the procedures of Block Allocation Groups, allowing flexibility in terms of time availability and beam line specificities. Crystal structure determination, from data processing, phase determination using multiple anomalous dispersion or molecular replacement methods, to refinement, structure validation, and analysis, make use of a number of integrated packages and programs running on Unix, Linux, and/or Silicon Graphics platforms.

4. Conclusion

This chapter provides an overview of the authors' platform facilities and describes specific protocols developed and/or adapted to automation. With the pipeline described in this chapter, the authors are able to handle large number of samples from cloning to crystallization. The implementation of such a platform has allowed SBGP to support numerous research projects. Figure 35.3 shows that about 2,500 expression vectors (800 using the Gateway technology), corresponding to different domains and fusion partners for given targets, have been constructed. More than 40 novel crystal structures have been deposited corresponding to mammalian and mainly human proteins. It is hoped that the recombinant protein expression screening method recently validated described in chapter 10 in the present book *(24)* will help fill the gap between the number of expression vectors constructed and those tested for expression. Since a large fraction of the proteins of interest are from eukaryotic sources, efforts are being placed on improving the throughput of recombinant proteins that may be expressed in eukaryotic systems such as the insect cell system, adapting described procedures *(34,35)*. Moreover, future targets such as protein–protein complexes that are produced in low yields led the authors to implement crystallization trials in the nanoliter range as reported elsewhere *(36)*, and the adaptation of other crystallization methods to robotics, such as crystallization within cubic phases, under gel, or using microfluidic technology *(37)*, which may help macromolecular complex crystallization.

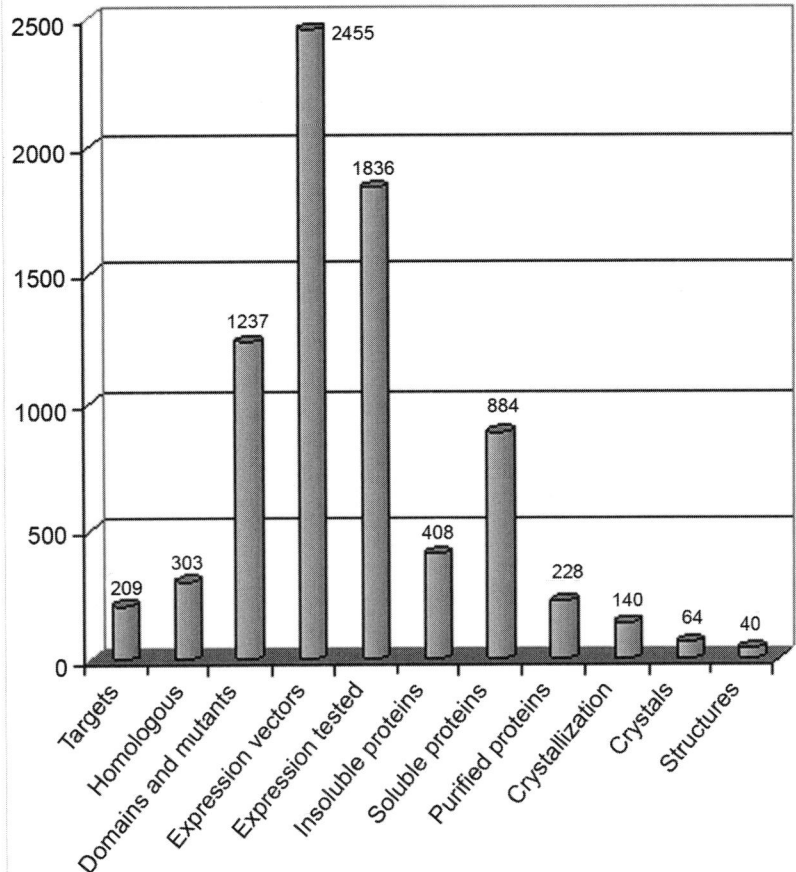

Fig. 35.3 Step by step summary of the results. The different steps of the pipeline are mentioned along the horizontal axis following chronological order. The vertical axis gives the absolute number for a given step and the exact number for each step is written on top of each bar. Number given for "Purified proteins," "Crystallization," "Crystals," and "Structures" are directly related to the number of selected targets. Number of insoluble proteins is given for information since they do not undergo further purification. The number of structures corresponds to lead molecules and, for example, does not include the isomorphous ligand complexes.

5. Notes

1. *E. coli* DH5α cells are prepared accordingly to Inoue et al. *(38)*.
2. To reduce primer expenses, the authors designed generic forward primers having the following sequence: four guanines (G) followed by the 25-nucleotide *att*B1 sequence (ACAAGTTTGTACAAAAAAGCAGGCT), two extra nucleotides to maintain the reading frame, and the sequence encoding appropriate protease recognition site (thrombin, TEV, protease 3C). Having the protease recognition site located downstream of the att sequence after recombination circumvents a caveat of the Gateway technology, which is to incorporate the translated *att* sequence (TSLYKKAG) into the protein of

interest. Besides generic primers, a specific forward primer is designed for each ORF to be amplified with the following sequence: sequence encoding protease recognition site (to allow annealing with generic primer during PCR), and at least an 18 to 25 nucleotides gene-specific sequence with a Tm = 60 °C.

3. When PCR is performed with the generic forward primer, perform the first PCR as described in Section 3.2.2. using the specific primer and the reverse primer. After analysis, use 1 µl of the first PCR as a template to perform a second PCR as described in Section 3.2.2. using the generic primer and the reverse primer. PCR cycles were performed as described by the manufacturer.

4. When using the GENios microplate reader, dilute PCR product five times in double distilled water in a final volume of 100 µl. After measuring the OD_{260nm}, calculate the DNA concentration using the following formula:

$$[DNA](M) = ((OD_{260mm} \times 4 \times 5 \times 50).10^{-3})/(660 \times L)$$

with the factor 4 to correct the path length, the factor 5 to correct the dilution, the factor 50 to calculate the amount of DNA, the factor 660 corresponding to the molecular mass of a base pair, and L corresponding to the length of the PCR product in base pairs.

5. Transformation reactions are performed on the Tecan Genesis Workstation robot following the protocol described by Busso et al. in this volume (24) using 50 µl of DH5α chemical competent cells, adding 150 µl of LB buffer and spreading 50 µl of the transformation mix onto LB-Agar + antibiotic (gentamicin at 7 µg/ml for BP reactions and ampicillin at 100 µg/ml for LR reactions) dispensed into six-well culture plates.

6. For PCR screening, primers have been designed to anneal at the recombination sites present both on pDONR207 and recombinant Destination plasmids.

7. Aspiration first of the protein and then the reservoir solution avoids any protein solution contamination and dispensing both solutions at a time ensure homogeneous mixing.

Acknowledgments

The authors are grateful to Jean-Pierre Samama for useful discussions especially concerning the x-ray sources and the structure determination section of the manuscript. The authors are sincerely grateful to Structural Biology and Genomics Department members participating in platform elaboration, in particular to Pierre Poussin-Courmontagne, Loubna Salim, Matthieu Stierlé, and Edouard Troesch (SBGP staff). Special thanks go to Rosalind Kim for critical reading of the manuscript and useful discussions. This work was supported by funds from RNG through the Genopole program, and by the European Commission as SPINE EEC QLG2-CT-2002-00988 and SPINE2-complexes LSHG-CT-2006-031220.

References

1. Kim, S. H., Shin, D. H., Liu, J., Oganesyan, V., Chen, S., Xu, Q. S., Kim, J. S., Das, D., Schulze-Gahmen, U., Holbrook, S. R., Holbrook, E. L., Martinez, B. A., Oganesyan, N., DeGiovanni, A., Lou, Y., Henriquez, M., Huang, C., Jancarik, J.,

Pufan, R., Choi, I. G., Chandonia, J. M., Hou, J., Gold, B., Yokota, H., Brenner, S. E., Adams, P. D., and Kim, R. (2005) Structural genomics of minimal organisms and protein fold space. *J. Struct. Funct. Genomics* **6,** 63–70.

2. Chandonia, J. M., and Kim, S. H. (2006) Structural proteomics of minimal organisms: conservation of protein fold usage and evolutionary implications. *BMC Struct. Biol.* **6,** 7.

3. Hajduk, P. J., Huth, J. R., and Tse, C. (2005) Predicting protein druggability. *Drug. Discov. Today* **10,** 1675–1682.

4. Arcus, V. L., Lott, J. S., Johnston, J. M., and Baker, E. N. (2006) The potential impact of structural genomics on tuberculosis drug discovery. *Drug. Discov. Today* **11,** 28–34.

5. Scapin, G. (2006) Structural biology and drug discovery. *Curr. Pharm. Des.* **12,** 2087–2097.

6. Baumeister, W., and Steven, A. C. (2000) Macromolecular electron microscopy in the era of structural genomics. *Trends Biochem. Sci.* **25,** 624–631.

7. Acton, T. B., Gunsalus, K. C., Xiao, R., Ma, L. C., Aramini, J., Baran, M. C., Chiang, Y. W., Climent, T., Cooper, B., Denissova, N. G., Douglas, S. M., Everett, J. K., Ho, C. K., Macapagal, D., Rajan, P. K., Shastry, R., Shih, L. Y., Swapna, G. V., Wilson, M., Wu, M., Gerstein, M., Inouye, M., Hunt, J. F., and Montelione, G. T. (2005) Robotic cloning and protein production platform of the northeast structural genomics consortium. *Methods Enzymol.* **394,** 210–243.

8. Vincentelli, R., Canaan, S., Offant, J., Cambillau, C., and Bignon, C. (2005) Automated expression and solubility screening of His-tagged proteins in 96-well format. *Anal. Biochem.* **346,** 77–84.

9. Cork, C., O'Neill, J., Taylor, J., and T., E. (2006) Advanced beamline automation for biological crystallography experiments. *Acta Cryst.* **D62,** 852–858.

10. Su, X. D., Liang, Y., Li, L., Nan, J., Brostromer, E., Liu, P., Dong, Y., and Xian, D. (2006) A large-scale, high-efficiency and low-cost platform for structural genomics studies. *Acta Cryst.* **D62,** 843–851.

11. Ueno, G., Kanda, H., Hirose, R., Ida, K., Kumasaka, T., and Yamamoto, M. (2006) RIKEN structural genomics beamlines at the SPring-8; high throughput protein crystallography with automated beamline operation. *J. Struct. Funct. Genomics* **7,** 15–22.

12. Yokoyama, S. (2003) Protein expression systems for structural genomics and proteomics. *Curr. Opin. Chem. Biol.* **7,** 39–43.

13. Falzon, L., Suzuki, M., and Inouye, M. (2006) Finding one of a kind: advances in single-protein production. *Curr. Opin. Biotechnol.* **17,** 347–352.

14. Esposito, D., and Chatterjee, D. K. (2006) Enhancement of soluble protein expression through the use of fusion tags. *Curr. Opin. Biotechnol.* **17,** 353–358.

15. Braun, P., Hu, Y., Shen, B., Halleck, A., Koundinya, M., Harlow, E., and LaBaer, J. (2002) Proteome-scale purification of human proteins from bacteria. *Proc. Natl. Acad. Sci. USA* **99,** 2654–2659.

16. Hammarstrom, M., Hellgren, N., van Den Berg, S., Berglund, H., and Hard, T. (2002) Rapid screening for improved solubility of small human proteins produced as fusion proteins in *Escherichia coli*. *Protein Sci.* **11,** 313–321.

17. Sorensen, H. P., and Mortensen, K. K. (2005) Advanced genetic strategies for recombinant protein expression in *Escherichia coli*. *J. Biotechnol.* **115,** 113–128.

18. Berrow, N. S., Büssow, K., Coutard, B., Diprose, J., Ekberg, M., Folkers, G. E., Levy, N., Lieu, V., Owens, R. J., Peleg, Y., Pinaglia, C., Quevillon-Cheruel, S., Salim, L., Scheich, C., R., V., and Busso, D. (2006) Recombinant protein expression and solubility screening in *Escherichia coli*: a comparative study. *Acta Cryst.* **D62,** 1218–1226.

19. Hartley, J. L. (2006) Cloning technologies for protein expression and purification. *Curr. Opin. Biotechnol.* **17,** 359–366.

20. Hartley, J. L., Temple, G. F., and Brasch, M. A. (2000) DNA cloning using in vitro site-specific recombination. *Genome Res.* **10,** 1788–1795.

21. Aslanidis, C., and de Jong, P. J. (1990) Ligation-independent cloning of PCR products (LIC-PCR). *Nucleic Acids Res.* **18,** 6069–6074.

22. van den Ent, F., and Lowe, J. (2006) RF cloning: a restriction-free method for inserting target genes into plasmids. *J. Biochem. Biophys. Methods* **67,** 67–74.

23. Busso, D., Delagoutte-Busso, B., and Moras, D. (2005) Construction of a set Gateway-based destination vectors for high-throughput cloning and expression screening in *Escherichia coli. Anal. Biochem.* **343,** 313–321.

24. Busso, D., Stierlé, M., Thierry, J. C., and Moras, D. Automated recombinant protein expression screening in *Escherichia coli. Methods Mol. Biol.* in press.

25. Middelberg, A. P. (1995) Process-scale disruption of microorganisms. *Biotechnol. Adv.* **13,** 491–551.

26. Sanglier, S., Bourguet, W., Germain, P., Chavant, V., Moras, D., Gronemeyer, H., Potier, N., and Van Dorsselaer, A. (2004) Monitoring ligand-mediated nuclear receptor-coregulator interactions by noncovalent mass spectrometry. *Eur. J. Biochem.* **271,** 4958–4967.

27. Wilson, W. W. (2003) Light scattering as a diagnostic for protein crystal growth: a practical approach. *J. Struct. Biol.* **142,** 56–65.

28. Jancarik, J., Pufan, R., Hong, C., Kim, S. H., and Kim, R. (2004) Optimum solubility (OS) screening: an efficient method to optimize buffer conditions for homogeneity and crystallization of proteins. *Acta Cryst.* **D60,** 1670–1673.

29. Jancarik, J., and Kim, S. H. (1991) Sparse matrix sampling: a screening method for crystallization of proteins. *J. Appl. Cryst.* **24,** 409–411.

30. Newman, J. (2004) Novel buffer systems for macromolecular crystallization. *Acta Cryst.* **D60,** 610–612.

31. Busso, D., Poussin-Courmontagne, P., Rosé, D., Ripp, R., Litt, A., Thierry, J. C., and Moras, D. (2005) Structural genomics of eukaryotic targets at a laboratory scale. *J. Struct. Funct. Genomics* **6,** 81–88.

32. Hosfield, D., Palan, J., Hilgers, M., Scheibe, D., McRee, D. E., and Stevens, R. C. (2003) A fully integrated protein crystallization platform for small-molecule drug discovery. *J. Struct. Biol.* **142,** 207–217.

33. Mayo, C. J., Diprose, J. M., Walter, T. S., Berry, I. M., Wilson, J., Owens, R. J., Jones, E. Y., Harlos, K., Stuart, D. I., and Esnouf, R. M. (2005) Benefits of automated crystallization plate tracking, imaging, and analysis. *Structure* **13,** 175–182.

34. Bahia, D., Cheung, R., Buchs, M., Geisse, S., and Hunt, I. (2005) Optimisation of insect cell growth in deep-well blocks: development of a high-throughput insect cell expression screen. *Protein Expr. Purif.* **39,** 61–70.

35. McCall, E. J., Danielsson, A., Hardern, I. M., Dartsch, C., Hicks, R., Wahlberg, J. M., and Abbott, W. M. (2005) Improvements to the throughput of recombinant protein expression in the baculovirus/insect cell system. *Protein Expr. Purif.* **42,** 29–36.

36. Hui, R., and Edwards, A. (2003) High-throughput protein crystallization. *J. Struct. Biol.* **142,** 154–161.

37. Hansen, C. L., and Quake, S. R. (2003) Microfluidics in structural biology: smaller, faster … better. *Curr. Opin. Struct. Biol.* **13,** 538–544.

38. Inoue, H., Nojima, H., and Okayama, H. (1990) High efficiency transformation of *Escherichia coli* with plasmids. *Gene* **96,** 23–28.

Chapter 36

Bacterial Structural Genomics Initiative: Overview of Methods and Technologies Applied to the Process of Structure Determination

Miroslaw Cygler, Ming-ni Hung, John Wagner, and Allan Matte

The focus over the last several years on increasing the number of three-dimensional structures of macromolecules by implementation of high throughput methodology has led to the establishment of dedicated structural genomics programs around the world. These worldwide efforts have in turn led to development of novel, parallelized approaches to cloning, expression, purification, and crystallization of proteins. This chapter describes in some detail the approaches and protocols that have been implemented in the Bacterial Structural Genomics Initiative.

1. Introduction

Over the last several years substantial funding has been dedicated worldwide toward establishing methodologies for speeding up the process of determining the three-dimensional (3D) structure of proteins starting from the DNA sequence of their corresponding genes. Although the scientific aims of various groups differ, achieving their objectives depends on speeding up the rate-limiting steps in the overall process, either the generation of well diffracting protein crystals, or the preparation of well behaved, concentrated protein samples for NMR. The sequence-to-structure path can be divided into several discrete steps and much effort has been and is still dedicated to continued development and improvement of the procedures used to parallelize and automate each of these steps. The overall approaches of different groups engaged in structural genomics efforts are similar, concentrating on using robotics to accomplish higher throughput through parallelization, and are well described in the literature *(1–8)*. However, the detailed implementation of the strategies employed differs among the different groups. For example, three general methods have been used for cloning genes into expression plasmids: restriction enzyme–based methods, ligation-independent cloning, and the recombination based method (Gateway system). Similarly, different groups have developed their own in-house crystallization screens based on statistical analysis of their own large-scale crystallization trials *(9–11)*.

From: *Methods in Molecular Biology, Vol. 426: Structural Proteomics: High-throughput Methods*
Edited by: B. Kobe, M. Guss and T. Huber © Humana Press, Totowa, NJ

The authors' efforts have focused on bacterial proteins with the goals of improving their knowledge of enzymatic mechanisms and characterizing proteins of unknown function both structurally and functionally *(12)*. The targets were selected from the three genomes of *Escherichia coli* that were first sequenced, K-12 *(13)*, O157:H7 EDL933 *(14)*, and CFT073 *(15)*.

2. Materials

2.1. Cloning Procedures

1. Forward and reverse primers (Integrated DNA Technologies, Coralville, IA).
2. Vectors: derived from pGEX-4T1 (GE Healthcare, Baie d'Urfé, QC) and pET15b (EMD Biosciences; Novagen, Madison, WI) parent vectors.
3. Biomek FX workstation (Beckman-Coulter, Mississauga, ON).
4. PCR master mix: *E. coli* genomic DNA template (American Type Culture Collection, Manassas, VA), 10× Pfu reaction buffer, dNTPs, and Pfu DNA polymerase (Stratagene, La Jolla, CA).
5. GeneAmp 2700 thermocycler (Applied Biosystems, Foster City, CA).
6. Gel Doc™ XR Gel documentation system (Bio-Rad Laboratories, Mississauga, ON).
7. E-gel 96 buffer-less 2% agarose gels (Invitrogen Canada, Burlington, ON).
8. Montage™ PCR$_{96}$ filter plate, MultiScreen vacuum manifold (Millipore Corp., Mississauga, ON).
9. T4 DNA ligase (New England Biolabs, Ipswich, MA).
10. Chemically competent *E. coli* DH5α cells.
11. Montage™ DNA plasmid mini-prep$_{96}$ kit (Millipore Corp.).
12. Chemically competent *E. coli* BL21(DE3) and BL21.
13. TrakMates 2D barcoded storage tubes (Matrix Technologies, Hudson, NH).

2.2. Small-Scale Protein Expression and Purification Testing

1. Ni-NTA (Qiagen, Mississauga, ON).
2. Glutathione Sepharose (GE Healthcare).
3. SDS-PAGE: Criterion Dodeca Cell electrophoresis unit and pre-cast Criterion gradient gels (Bio-Rad).
4. 96-well deep blocks (Whatman, Florham Park, NJ).
5. Airpore tape (Qiagen).
6. Lysis buffer: 50 mM Tris-HCl pH 7.5, 400 mM NaCl, 10 mM β-mercaptoethanol, 5% (v/v) glycerol, 20 mM imidazole, 1× BugBuster protein extraction reagent (Novagen), 50 μg/ml lysozyme, 6.25 U/ml Benzonase nuclease (Novagen), and Complete™ protease inhibitor cocktail (Roche Diagnostics, Mississauga, ON).
7. Filter plate (Whatman).
8. 96-well Unifilter plate (Whatman) or AcroPrep 96 filter plate (Pall Life Sciences, East Hills, NY).
9. Equilibration buffer: 50 mM Tris-HCl, pH 7.5, 400 mM NaCl, 10 mM imidazole.
10. Wash buffer (50 mM Tris-HCl pH 7.5, 400 mM NaCl, 40 mM imidazole.
11. Corning-Costar round bottom plate (Corning Life Sciences, Acton, MA).
12. Elution buffer (50 mM Tris-HCl pH 7.5, 400 mM NaCl, 250 mM imidazole, 10 mM β-mercaptoethanol.

13. Mutiscreen HTS 96-well immobilon-P membrane plate installed on the Biomek vacuum manifold.
14. HRP-conjugated tetra-His antibody (Qiagen).
15. Chemiluminescence reagents (Perkin-Elmer, Wellesley, MA).
16. 96-well plate reader (MicroBeta workstation; Perkin-Elmer).

2.3. Protein Production and Purification

2.3.1. Ni-NTA Affinity Purification

1. Media: Terrific Broth (TB), 2YT, Circle Grow (Q-Biogene, Irvine CA), auto-inducing media; *E. coli metA-* auxotroph strain DL41 or DL41(DE3).
2. LeMaster medium.
3. Ni-Sepharose (GE Healthcare).
4. Talon cobalt resin (Clontech, Mountain View, CA).
5. Lysis buffer: 50 mM Tris-HCl pH 7.5, 400 mM NaCl, 5% [w/v] glycerol, 0.5% [w/v] Triton X-100, 20 mM imidazole, 10 mM β-mercaptoethanol (His-tag fusions) or the same buffer containing 20 mM DTT and no imidazole (GST fusions) supplemented with the protease inhibitors Pefabloc (AEBSF) at 0.1 mM (100 mM stock, 4 °C), benzamidine at 0.5 mM (100 mM stock, −20 °C), and 10 μM leupeptin (10 mM stock, 20 °C).
6. Econo column (Bio-Rad).
7. Wash buffer 1: lysis buffer without detergent and protease inhibitors.
8. Wash buffer 2: 50 mM Tris-HCl pH 7.5, 1.0 M NaCl, 5% [w/v] glycerol, 20 mM imidazole, 10 mM BME.
9. Wash buffer 3: 50 mM Tris-HCl pH 7.5, 0.4 M NaCl, 5% [w/v] glycerol, 40 mM imidazole, 10 mM BME.
10. Elution buffer: 50 mM Tris-HCl pH 7.5, 0.2 M NaCl, 5% [w/v] glycerol, 250 mM imidazole, 10 mM BME.
11. Protease inhibitor cocktail (#P-8849; Sigma-Aldrich).

2.3.2. GST-Tag Purification
1. GST-beads (Glutathione-Sepharose 4B; GE Healthcare).
2. Lysis buffer: 50 mM Tris-HCl pH 7.5, 0.4 M NaCl, 0.5% (v/v) Triton X-100, 5% (w/v) glycerol, 10 mM DTT + protease inhibitors.
3. Bio-Rad Econo column.
4. Wash buffer 1: 50 mM Tris-HCl pH 7.5, 1.0 M NaCl, 0.5% (v/v) Triton X-100, 5% (w/v) glycerol, 10 mM DTT.
5. Wash buffer 2: 50 mM Tris-HCl pH 7.5, 0.4 M NaCl, 5% (w/v) glycerol, 10 mM DTT.
6. Wash buffer 3 (TEV cleavage buffer): 50 mM Tris-HCl pH 8.0, 150 mM NaCl, 0.5 mM EDTA, 1 mM DTT.

2.3.3. TEV Protease Cleavage on Beads
1. Protease Inhibitor Cocktail (#P-8849; Sigma).

2.3.4. GST Fusion Elution and TEV Cleavage in Solution
1. Exchange buffer: 20 mM Tris-HCl pH 7.5, 0.2 M NaCl, 5% [w/v] glycerol, 5 mM DTT.
2. Final protein buffer: 20 mM Tris-HCl pH 7.5, 0.2 M NaCl, 5% [w/v] glycerol, 5 mM DTT.

2.4. Quality Control of Purified Proteins

1. Agilent 1100 Series LC/MSD (Agilent Technologies, Palo Alto, CA).
2. Wyatt DLS plate reader (Wyatt Technology, Advance, NC).

2.5. Limited Proteolysis

1. Proteases: trypsin, chymotrypsin, endo Glu-C, elastase, and Lys-C (sequencing grade, Roche Molecular Biologicals) prepared as stock solutions at 5 mg/ml in 50 mM Tris-Cl buffer pH 7.5 (8.5 for elastase) and stored at −20 °C.
2. Protease inhibitor cocktail (#P-8849; Sigma).

2.6. Protein Crystallization Screening

1. Hydra-Plus-One liquid handling robot (Apogent Discoveries, Hudson, NH).
2. Intelliplate 96-well sitting-drop crystallization plate (Art Robbins Instruments, Sunnyvale, CA).
3. Screens Hampton I, II and Index (Hampton Research, Aliso Viejo, CA), or PEG, Anions, MBclass, and Ammonium Sulfate Suites (Qiagen, Mississauga, ON), the PACT suite, or the authors' own in-house developed screen of 392 conditions.
4. Beckman FX robot (Beckman Coulter Canada., Mississauga, ON).
5. 96-well Intelliplate (Art Robbins Instruments).
6. Viewseal pressure sensitive adhesive (Greiner Bio-One, Monroe, NC).
7. CrystalFarm™ 400 imaging system (Bruker AXS, Madison, WI).
8. 24-well VDX hanging drop plates (Hampton Research).
9. MPAX crystallization robot (Douglas Instruments, Berkshire, UK).

2.7. Structure Determination

1. HTC imaging plate detector (Rigaku-MSC, The Woodlands, TX) mounted on a microfocus 007 rotating anode generator (Rigaku-MSC).
2. Automounter pucks (Boyd Technologies, Manchester, CA).
3. CrystalCap HT™ pins and caps (Hampton Research).

3. Methods

3.1. Target Selection

The criteria used for selection of targets for cloning and expression define the objectives of a structural genomics effort. At the onset of the authors' project emphasis was placed on determining structures of *E. coli* enzymes that are involved in Small Molecule Metabolic Pathways (SMMPs). A cursory look through the databases dedicated to *E. coli* (EcoCyc, KEGG, and EchoBase) *(16–18)* showed that many pathways had substantial gaps in structural knowledge of participating enzymes. The information compiled in *(19)* was of much help in selection of proteins from SMMPs. The second group targeted for structural studies included proteins for which there was no functional assignment. Here the compilation of data on the *E. coli* proteome by Riley and coworkers *(20,21)* was extremely helpful. The proteins from this category were ranked according to the size of the PFAM sequence family (www.sanger.ac.uk/Software/Pfam) *(22)* to which they belonged, with those corresponding to large families having higher priority. In some cases where a larger protein obviously consisted of one or more domains, as annotated within ExPasy *(23,24)*, these were cloned and expressed separately for crystallography or NMR.

3.2. SPEX Db Database

Working in parallel with a large number of targets through a complex set of experimental procedures requires a dedicated Laboratory Information Management System (LIMS) to preserve and coordinate the experimental information pertaining to every step in the process. To this end a Structural Proteomics Experimental Database (SPEX Db) was developed *(25)*. This database is accessible through any Web browser (sgen.bri.nrc.ca/brimsg/bsgi.html). All participants in the project, independent of their physical location, input experimental details into the database and have immediate access to all the information about the target protein with which they are working. The database allows a rapid overview of the entire project and is indispensable in the decision-making process. In addition to containing all experimental details, the system also regularly scans entries in the Protein Data Bank and in the depository of Structural Genomics projects Target DB (RCSB) and compares them with the entries in SPEX Db. This function provides constantly updated information on the status of proteins homologous to the authors' targets. Indeed, SPEX Db is now used for all ongoing projects in the authors' laboratory.

Another key element of SPEX Db is its use in tracking materials related to the genomics project. A large quantity of materials are generated from the different steps in the pipeline, from DNA primers and samples of plasmid DNA, to glycerol stocks of expression clones, to samples of purified proteins and crystals stored for x-ray data collection. Pages corresponding to these various materials are generated and can be searched, so that for example, a particular glycerol stock can be readily found and used to prepare a culture for protein production. The authors have adopted the use of TrakMates 2D bar-coded storage tubes (Matrix Technologies, Hudson, NH) for storing glycerol stocks, with the corresponding barcode number recorded on the glycerol stock entry within the database.

3.3. Cloning Procedures

Protein sequences are analyzed to identify those having signal sequences using the program SignalP v3.0 *(26)* or transmembrane segments, as predicted using TMHMM *(27)* to identify and remove these regions prior to cloning. Since all the proteins the authors are interested in are of bacterial origin, the expression system of choice is *E. coli*. Therefore, the simplest cloning strategy involving restriction enzymes was adopted.

Software scripts are used to execute the following steps. First, the DNA sequence corresponding to the ORF of interest is identified from the most recent *E. coli* genome sequence maintained at NCBI. These sequences are then analyzed for the presence of internal restriction sites, namely, *Bam*HI, *Bgl*II, *Eco*RI, and *Mfe*I. As well, the ATG start codon is removed from the ORF sequence to avoid the possibility of a secondary transcription initiation site. Another script generates the forward and reverse primer sequences and adds the appropriate restriction enzyme site to each primer as well as a short 2-6 nucleotide "tail" before each site. A stop codon is added to the reverse primer as necessary. The script selects primers of appropriate length, with consideration given to maintaining a T_m of 70–72 °C. Forward and reverse primers are synthesized in 2× 96-well plates, with individual wells corresponding to the forward and reverse primers for each gene. The amount of DNA in each well of the plates is normalized to 4 nmol, so the primers can be resuspended in the same volume of buffer prior to PCR.

To be able to use the same PCR product for insertion into several expression vectors that encode different tags and protease cleavage sites, the vectors were engineered to contain the same in-frame cloning sites, namely *BamHI* and *EcoRI*. The vectors were derived from pGEX-4T1 and pET15b parent vectors. Some vectors were further modified to include a tobacco etch virus (TEV) protease cleavage sequence *(28)*. A variety of tags were tested to evaluate their effectiveness in affinity purification and for improving the solubility of expressed proteins. The authors' routine procedure consists of cloning each target into three expression vectors: pFO4—MGSSHis$_6$GS-ORF, pJW234—MGSSHis$_8$**DYDIPTT***ENLYFQ*↓*GS*-ORF, and pRL652—GST-*ENLYFQ*↓*GS*-ORF. The pJW234 vector contains a seven-residue spacer (bold) between the His$_8$ tag and the TEV recognition site (underlined, arrow indicates the cleavage site) to improve cleavage efficiency *(29,30)*. For generating fusion proteins with C-terminal His$_6$ tags the vector pRL574 – ORF-His$_8$ is utilized. All these vectors are based on an IPTG-inducible T7 (pFO4, pJW234) or *tac* promoter (pRL652). Corresponding vectors containing a thrombin cleavage site are used when cleavage with TEV protease is problematic. Cloning of targets is performed in 96-well plate format and the consecutive steps are executed with an automated liquid handling dual-bridge Biomek FX workstation. This liquid handling robot is equipped with 96-channel (96-tip) and eight channel (8-tip) pipetting heads, variable-speed orbital shaker, tip washing station, vacuum manifold, gripper for moving plates, tip loader, and deck that can accommodate at least 14 pieces of labware.

Lyophilized primers are resuspended in 80 μl of 10 mM Tris-HCl, pH 8.0. The following components are combined in a PCR plate–PCR master mix, and are dispensed using the eight-channel head, whereas the forward and reverse primers are added using the 96-channel head. The PCR plate is then transferred to a GeneAmp 2700 thermocycler. The following PCR conditions are used: after an initial denaturation at 95 °C for 2 minutes, 30 cycles of 95 °C for 1 minute, 55 °C for 1 minute, and 72 °C for 6 minutes were followed by a final incubation at 72 °C for 10 minutes. PCR products are verified following amplification using E-gel 96 bufferless 2% agarose gels (see Note 1).

Once a digital image of the gel has been made using the gel documentation system, the expected and observed sizes of the PCR products are compared using in-house software to evaluate the individual amplicons (Fig. 36.1). PCR products are purified on the robot using the 96-channel head to dispense the PCR products to a Montage™ PCR$_{96}$ filter plate, which is placed on the vacuum manifold. Following application of a vacuum of 10˝ Hg for 30 minutes to filter the PCR reaction solution, 50 μl of 10 mM Tris buffer is dispensed to the filter plate to resuspend the PCR products. The resuspended PCR products are then transferred to a Costar Cone round bottom plate and are ready for restriction enzyme digestion.

Purified products are then digested with restriction enzymes (*BamHI/EcoRI*, *BamHI/MfeI*, *BglII/EcoRI*, or *BglII/MfeI*) at 37 °C. Digested products are then

Fig. 36.1 PCR product analysis by agarose gel electrophoresis showing the sizes of amplicons. The green (correct size) or magenta (incorrect size or missing) lines mark the expected positions of the fragments on the gel. Markers are shown in the mid-section of each panel.

again purified using a Montage™ PCR$_{96}$ filter plate as described and then ligated into *BamHI/EcoRI* digested, dephosphorylated vector DNA using T4 DNA ligase at room temperature for 4 hours and the ligation products transformed into chemically competent *E. coli* DH5α cells (see Note 2). Following transformation and growth on LB amp agar plates, three individual colonies are inoculated into corresponding wells of three, 96-well deep blocks containing 1 ml LB amp media. After overnight growth, a 2-µl portion of each culture is PCR-screened using vector-specific primers, PCR products are detected by agarose gel electrophoresis (E-gel 96) and positive clones from each of three screening plates are combined into a master 96-well deep block containing 1 ml LB amp media for DNA plasmid miniprep and glycerol stocks. Glycerol stocks are then prepared from overnight cultures and stored in TrakMates 2D-barcoded tubes. Plasmid DNA minipreps are performed using the Montage™ DNA plasmid miniprep$_{96}$ kit (see Note 3), and are then used to transform chemically competent cells, either *E. coli* BL21(DE3) for pFO4 and pJW234 vectors, or *E. coli* BL21 for the pRL652 vector. Following transformation, plates are incubated overnight at 37 °C. A colony is picked from each plate and transferred to 1 ml LB amp media in a 96-well deep block and grown overnight at 37 °C. Glycerol stocks are then prepared as described. A portion of the same culture is used as an inoculum for expression testing as described in the following.

3.4. Small-Scale Protein Expression and Purification Testing

Once the expression clones have been generated, it is necessary to determine the expected level of protein expression for each clone, as well as the solubility of the expressed protein. The authors' initial procedures involved culturing the expression clones in 24-well deep blocks, performing manual small-scale purifications using Ni-NTA or glutathione Sepharose resins with a vacuum manifold and Qiagen filter plates, and analyzing the total protein, soluble protein, fusion protein bound to the affinity matrix, and the protein following elution by SDS-PAGE with a Criterion Dodeca Cell electrophoresis unit and precast Criterion gradient gels. Although very workable and informative, this approach suffers from the fact that it is not amenable to automation and is quite laborious. To perform small-scale expression analysis in a more automated manner, the protocol was adapted to run on the Biomek FX platform. To avoid the manual manipulations associated with gel electrophoresis, two methods were devised to detect proteins eluted from the affinity resins, using either an anti-His$_4$ antibody, or through determination of the quantity of eluted protein using a colorimetric assay *(31)*. In this procedure, cultures are grown using 1 ml LB amp media in 96-well deep blocks sealed with Airpore tape to facilitate aeration. A 20-μl culture inoculum is added to each well and the blocks are agitated at 250 rpm at 37 °C for 2 hours. After 2 hours, IPTG is added to a final concentration of 100 μM and the cultures continued at room temperature (~22 °C) for an additional 16 hours, prior to harvesting the cells using a plate rotor (15 minutes, 2,500 rpm, 4 °C). To facilitate cell lysis, the pellets are first frozen at −20 °C overnight.

Following thawing of the pellets, each pellet is resuspended in 100 μl lysis buffer. The deep well block is shaken at 600 rpm for 20 minutes to effect cell lysis. The lysate is then transferred from the deep well block to a filter plate to obtain the soluble protein fraction that will be used for purification (see Note 4). To each well of a 96-well Unifilter plate or AcroPrep 96 filter plate is added 50 μl of Ni-NTA slurry for His-tag fusion purifications or 50 μl Glutathione Sepharose 4B slurry for GST-fusion purification. Excess ethanol is removed by applying a vacuum of 8″ Hg for 20 seconds. The resin is equilibrated by adding 2 × 100 μl of equilibration buffer, followed by application of a vacuum of 8″ Hg for 30 seconds. Ni-NTA resin is then incubated with the lysate for 30 minutes at room temperature. Each well is washed using 3 × 200 μl of wash buffer, applying a vacuum of 10″ Hg for 1 minute after each wash. Following washing, the filter plate is placed over a round bottom plate in the vacuum manifold collar and proteins are eluted using 3 × 50 μl volumes of elution buffer, incubating for 10 minutes before applying vacuum for each elution.

There are two ways in which the level of protein expression is evaluated. The authors use both a dot-blot immunodetection method, as well as protein determination using the Bradford protein assay *(31)* on fractions following elution. These assays can be performed either under native (i.e., soluble protein) or denaturing conditions (total protein). For the robotic dot-blot assay, purified protein samples are dispensed using the multichannel head onto a Mutiscreen HTS 96-well immobilon-P membrane plate installed on the Biomek vacuum manifold, followed by application of vacuum and three washing steps. Detection of His-tagged protein is performed using the HRP-conjugated tetra-His antibody following the recommended blocking and washing procedure, and developed using chemiluminescence reagents (see Note 5). Luminescence is detected and quantified on a 96-well plate reader. An example of a dot-blot result following small-scale purification of His-tagged protein is shown in Table 36.1. Alternatively, the protein concentration of the eluates is quantified by a Bradford assay *(31)* using a 10-μl aliquot of the eluate and BioRad protein assay reagent in a 200-μl volume. Protein quantitation is determined

Table 36.1 Results of a small-scale purification experiment of his-tagged proteins followed by dot-blot analysis to determine protein expression.

A	1	2	3	4	5	6	7	8	9	10	11	12
A	APAG_ pMN193	YAHO_ pMN194	BOLA_ pMN195	C0888_ pMN196	YCAR_ pMN197	C1215_ pMN198	SFAB_ pMN199	C1268_ pMN200	YCGW_ pMN201	YBCQ_ pMN202	C1619_ pMN203	YCGL_ pMN204
B	YCIN_ pMN205	C1914_ pMN206	C1919_ pMN207	C2944_ pMN208	C3176_ pMN209	C3646_ pMN210	HYBF_ pMN211	TDCR_ pMN212	INSB_ pMN213	PPDC_ pMN214	C0271_ pMN215	C0311_ pMN216
C	C0357_ pMN217	C0946_ pMN218	C0962_ pMN219	C0964_ pMN220	C1109_ pMN221	C1217_ pMN222	MCHC_ pMN223	C1281_ pMN224	C1501_ pMN225	C1503_ pMN226	C1528_ pMN227	vector
D	C1889_ pMN229	FLIT_ pMN230	C2396_ pMN231	C2397_ pMN232	C2410_ pMN233	C2415_ pMN234	EUTC_ pMN235	C3242_ pMN236	C3271_ pMN237	C3385_ pMN238	C3519_ pMN239	IUCB_ pMN240
E	C4226_ pMN241	C4511_ pMN242	C4522_ pMN243	C4540_ pMN244	C4542_ pMN245	C4577_ pMN246	DGT_ pMN247	YAJK_ pMN248	GIPA_ pMN249	TORS_ pMN250	FLGD_ pMN251	YNHC_ pMN252
F	SDAA_ pMN253	FLIH_ pMN254	FLIK_ pMN255	C2426_ pMN256	WZZB_ pMN257	C2562_ pMN258	WCAL_ pMN259	CPSB_ pMN260	GATZ_ pMN261	SDAB_ pMN262	YGGG_ pMN263	IUCD_ pMN264
G	YDCM_ pMN265	C3870_ pMN266	RFAP_ pMN267	KDTA_ pMN268	WECD_ pMN269	YHDG_ pMN270	YBBF_ pMN271	YBBT_ pMN272	YGCA_ pMN273	YHCJ_ pMN274	YNHE_ pMN275	YLEB_ pMN276
H	CITA_ pMN277	HTRB_ pMN278	FLGJ_ pMN279	SOHB_ pMN280	SPPA_ pMN281	MSBB_ pMN282	C2394_ pMN283	MREC_ pMN284	YIGR_ pMN285	YGJO_ pMN286	YHHU_ pMN287	BGLX_ pMN288
B												
A	79970	<0	108448	2850	15281	<0	31655	14242	<0	28011	446787	<0
B	95355	<0	2299	225	4296	<0	7568	<0	9766	180353	7253	84240
C	17402	<0	20501	128187	18090	<0	<0	<0	<0	50289	49877	(–) control
D	298653	1703	6103	<0	<0	165309	49711	176861	17327	<0	1715	<0
E	<0	198335	<0	<0	<0	<0	<0	<0	<0	<0	<0	32919
F	147651	109216	<0	<0	<0	<0	<0	<0	<0	3039	<0	26122
G	2053	29620	<0	<0	236322	58441	<0	106796	15877	265955	4518	<0
H	15653	698	65720	<0	<0	38845	201210	<0	<0	305603	43176	(+) control

[a]The His-tagged expression clones under study.
[b]The detected levels of fluorescence following detection with an anti-His antibody. Yellow indicates high expression, cyan low expression, and gray no expression under the conditions tested.

by absorbance at 595 nm using a 96-well plate reader and related to a standard curve of bovine serum albumin (BSA). If one desires to obtain the molecular weights of the expressed proteins as determined by the colorimetric or immunodetection assays, the eluted proteins from the automated 1-ml scale protein purification method can be analyzed by SDS-PAGE.

3.5. Protein Production and Purification

Proteins that are sufficiently well expressed and soluble, as determined by the small scale dot-blot or Bradford assays, are expressed in 0.5–1 l cultures, the volume dependent on the estimated expression level. Protein productions are carried out in flasks using rich buffered media, such as Terrific Broth (TB), 2YT, or Circle Grow or the autoinducing media developed by F.W. Studier *(32)*. The postinduction culture is maintained at room temperature (23 °C). In general, 2–20 mg of purified protein per liter culture are obtained. For SeMet protein production, plasmid DNA is transformed in the *E. coli metA-* auxotroph strain DL41 or DL41(DE3) and grown in LeMaster medium containing 25 mg/L of L-SeMet *(33)*. At times, the metabolic suppression of methionine synthesis is used by expressing the protein in *E. coli* BL21 or M15 backgrounds using M9 minimal medium in the presence of 50 mg/L L-SeMet *(34)*.

3.5.1. Typical Protein Production Protocol in TB or 2YT Media
3.5.1.1. Inoculum
1. Cells from glycerol stock or colony from plate into 50-ml medium
2. Grow at 37 °C with shaking overnight (~16 hours)

3.5.1.2. Culture
1. 1 L for medium/high expression, 2 L for low expression
2. Add 50 ml o/n culture per liter
3. Preinduce at 37 °C until OD_{600} 0.5–0.8
4. Induce with 100 μM IPTG
5. Postinduction at 20 °C for 5 hours (high/medium expression) or o/n (low expression)
6. Pellet cells at 4,000 *g*/20 minutes/4 °C
7. Resuspend cell pellet with 40 ml/L culture of lysis buffer
8. Store cells o/n at −20 °C in 50-ml tubes

3.5.2. Typical Affinity Purification Protocols
3.5.2.1. Ni-NTA Affinity Purification, No Cleavage (see Note 6)
1. Ni-NTA beads (Qiagen): 0.5 ml of slurry (low expressing protein) to 2.0 ml slurry (high expressing clone), corresponding to ~0.25–1.0 ml bed volume of resin.
2. Wash beads in deionized water and then in lysis buffer (above).
3. Add washed beads to protein sample in Econo column and incubate with occasional mixing for 30 minutes at 4 °C.
4. Setup columns on column support at RT and collect flow-through.
5. Add 50 ml wash buffer 1, incubate for 5 minutes, and collect wash fraction. Save 10 μl of eluant for SDS-PAGE.
6. Add 25 ml wash buffer 2, 5-minute incubation. Save 10 μl of eluant for SDS-PAGE.
7. Add 25 ml wash buffer 3. Incubate for 5 minutes and collect flow-through. Save 10 μl of eluant and 50 μl of beads for SDS-PAGE.

8. Elute bound proteins using elution buffer. Incubate for 10 minutes and collect 3 × 3 ml fractions.

9. Add 3 µl of protease inhibitor cocktail to each elution fraction.

10. Keep beads at 4 °C and check on SDS-PAGE (Beads-after-elution).

3.5.2.2. GST-tag Purification, Cleavage on Beads

1. GST-beads: 4 ml of slurry (75%; ~3 ml bed volume of resin).

2. Wash beads in deionized water and then in lysis buffer.

3. Add washed beads to protein sample in an Econo column and incubate with occasional mixing for 30 minutes at 4 °C.

4. Setup columns on column support at RT and collect flow-through fraction.

5. Add 50 ml of wash buffer 1.

6. Add 25 ml of wash buffer 2. Save 10 µL of eluant for SDS-PAGE.

7. Add 25 ml of wash buffer 3, TEV cleavage buffer. Save 10 µL of eluant for SDS-PAGE.

8. Add 5 ml of TEV cleavage buffer (above). Save 20 µl beads for Bradford assay and SDS-PAGE (Beads-before cleavage).

3.5.2.3. TEV protease cleavage on beads (see Note 7)

1. Incubate at TEV protease:protein ratio of 1:100 (w/w, based on the Bradford protein assay) at RT for 3 hours with occasional mixing. Evaluate cleavage efficiency by Bradford protein assay of protein eluate. If yield of protein is less than expected, incubate TEV protease:protein 1:100 (w/w) at 4 °C/overnight to complete cleavage.

2. Add 10 µl of protease inhibitor cocktail/5 ml elution fraction to stop the cleavage reaction.

3. Incubate each 5-ml fraction with 50 µl washed Ni-NTA beads to remove His-tagged TEV protease.

3.5.2.4. GST Fusion Elution and TEV Cleavage in Solution (Alternative to Section 3.5.2.3)

1. Incubate beads in TEV protease cleavage buffer containing 10 mM glutathione for 15 minutes at room temperature and collect the eluate.

2. Proceed with TEV cleavage in solution under conditions as in Section 3.5.2.3.: Monitor cleavage by SDS-PAGE.

3. Dialyze o/n against final protein buffer to remove glutathione.

4. Incubate with washed GST beads to bind cleaved GST and residual GST fusion protein.

5. Incubate each 5-ml fraction with 50 µl washed Ni-NTA beads to remove His-tagged TEV protease.

3.6. Quality Control of Purified Proteins

The purity of the final protein fraction is characterized by SDS-PAGE and its aggregation state evaluated by native-PAGE. The mass of the purified protein is verified by ESI-MS using an Agilent 1100 Series LC/MSD. The status of the protein in solution is subsequently tested by dynamic light scattering (DLS) using a Wyatt plate reader at a protein concentration 1–2 mg/ml in the elution buffer. If the protein shows monomodal, monodisperse behavior, proceed to exchange the protein into the final protein buffer with further concentration to approximately 8–10 mg/ml. This sample is tested again by DLS to determine if further aggregation takes place as the protein concentration increases. Monodisperse protein samples are forwarded to crystallization trials. If the initial sample is polydisperse investigate the effect of changing conditions such as pH, buffer system, salt concentration, presence of ions and, when known, substrates, cofactors, and inhibitors, on the mono/polydispersity of the protein sample. This has become a crucial step in preparation of a new protein for crystallization trials as it helps to stabilize and improve behavior of some of the authors' proteins (Fig. 36.2). In cases in which an additive is identified that is important to the behavior of the protein sample, it is added as a component to the lysis buffer during the next purification of that particular protein.

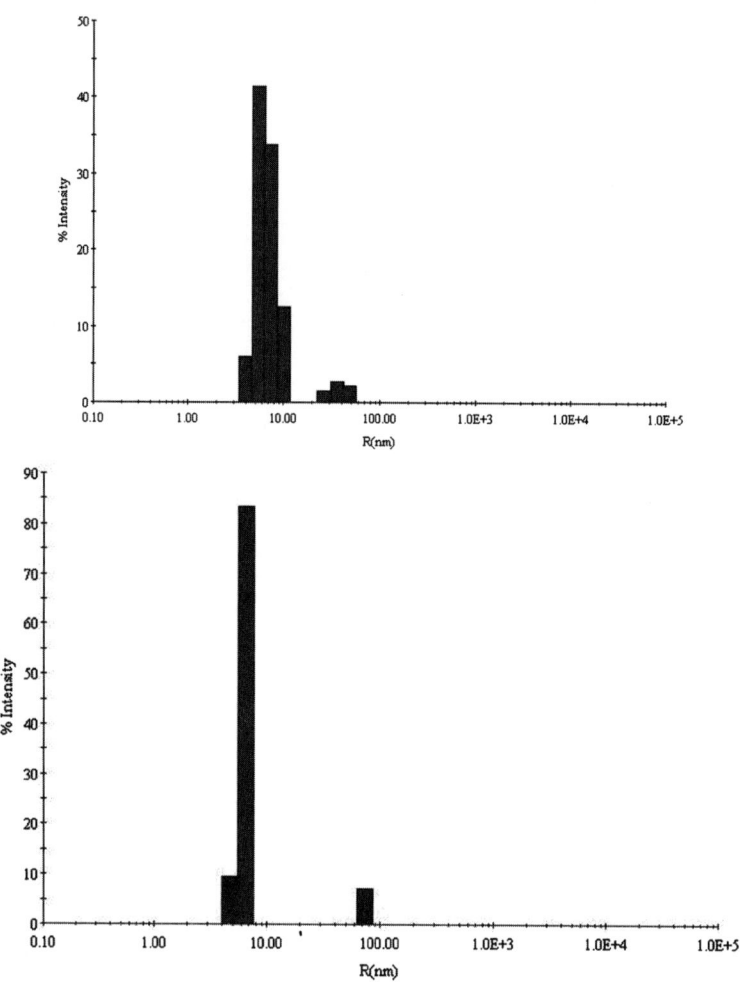

Fig. 36.2 Dynamic light scattering of *E. coli* YgjO, a putative 16S rRNA methyl-transferase. **A.** The initial sample was concentrated to 6 mg/ml in a buffer containing 20 mM Tris-HCl pH 7.5, 0.2 M NaCl, 4% (w/v) glycerol, 5 mM DTT, with a measured polydispersity of 26%. **B.** Following addition of the substrate *S*-adenosylmethionine (SAM) at a concentration of 1 mM, the polydispersity dropped to 9%, and yielded a single crystallization hit.

3.7. Limited Proteolysis Identifies Core Domains

Many proteins possess a domain organization that in some cases is prohibitive to crystallizing the intact protein *(35)*. This domain organization can sometimes be inferred by sequence comparisons or detected by observation of one or more lower molecular weight bands by SDS-PAGE following purification and storage of the sample. Alternatively, flexible parts of a protein molecule can sometimes be "clipped" from the N- or C-terminal ends, resulting in improved crystals. As part of the authors' standard quality control procedures on each purified protein, the sensitivity or resistance of each protein is tested to a battery of proteases. Five proteases are commonly used: chymoptrypsin, trypsin, lysC, endo-GluC, and elastase. Aliquots of protein are incubated with each of these proteases at a ratio of 100:1 (w/w) and incubated at room temperature for 30 minutes. Following digestion, samples are boiled for 5 minutes and subjected to SDS-PAGE. One looks for a major band having a smaller molecular weight compared with the undigested control sample, and can be especially interesting if it is common to two or more of the proteases used to perform the digestion (Fig. 36.3). This band is subsequently transferred to a PVDF membrane and subjected to N-terminal sequencing. In some cases the same digested sample is also analyzed by ESI-MS. The corresponding DNA fragment can then be cloned, and the protein expressed, purified, and tested for crystallization.

3.7.1. Protocol for Limited Proteolysis

Five proteases are used: trypsin, chymotrypsin, endo Glu-C, elastase, and Lys-C.
1. Dilute proteases to a working concentration of 0.02 mg/ml prior to use.
2. Adjust protein concentration to 2 mg/ml with buffer.
3. For simple SDS-PAGE analysis, mix 5 μl of protein sample with 5 μl of protease yielding a ratio of 100:1 (w/w) in a 0.5-ml centrifuge tube. For N-terminal sequencing a protein sample of 10 μg should suffice. For LC-MS analysis, a protein sample of 100 μg should be used.
4. Incubate protein-protease mixture at RT for 30 minutes.
5. Stop proteolysis by either adding 1 μl of Sigma protease inhibitor cocktail, mixing and store at −20 °C, or adding 2 μl 5× SDS-PAGE loading buffer and boiling for 5 minutes.

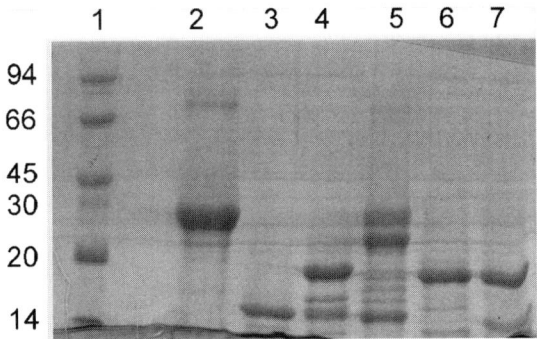

Fig. 36.3 Limited proteolysis and SDS-PAGE of HypB from *E. coli* O157:H7, using the protocol as described in the text. Lane 1, LMW markers; lane 2, undigested control; lane 3, trypsin; lane 4, chymotrypsin; lane 5, endo-gluC; lane 5, elastase; lane 6, endo-lysC.

6. Centrifuge tubes (21,000 g, 10 minutes, RT) in a table-top centrifuge and load digests on a 12.5% SDS-PAGE gel.

3.8. Protein Crystallization Screening

Crystallization screening is performed using a Hydra-Plus-One liquid handling robot equipped with a nano-dispense needle capable of delivering protein volumes as low as 100 nl. Crystallization solutions are either from commercial suppliers, including screens Hampton I, II, and Index, or PEG, Anions, MBclass, and Ammonium Sulfate suites, the PACT suite *(11)* or the authors' own in-house developed screen of 392 conditions. In each case, stock crystallization solutions in 10-ml screw-capped tubes are first dispensed to 96-deep-well blocks using the Beckman FX robot, sealed with cap-mats and stored at 20 °C.

Typically, a minimum of approximately 400 conditions are screened initially with a new protein sample, and as many as 600 conditions can be used, depending on the quantity of protein available. For crystallization screening the authors use 96-well Intelliplates each with two sample wells for sitting drop vapor diffusion. First, 70–80 µl of reservoir solution is dispensed from the deep well block to 96 wells of the Intelliplate. Next, a 400-nl volume of reservoir solution is dispensed from the deep well block to the smaller sample depression. Finally, the nanodispense head dispenses 300 nl of protein to the sample depression. If a specific ligand or cofactor is known for the protein being crystallized, the second depression is used with the protein and ligand, typically at a ligand concentration of 2–5 mM. After dispensing, the drops are carefully checked and the plate sealed using Viewseal pressure sensitive adhesive. The crystallization plates are placed in a CrystalFarm™ 400 imaging system, and imaged on a schedule on days 1, 2, 4, 8, 22, and 60. A composite image is generated of each drop from five to six individual images taken through a depth of approximately 0.5 mm (Fig. 36.4). These images are written in jpg format, and can be accessed via a Web interface by individual users through their accounts. Users score their images, according to whether the drops are clear, or show precipitate, microcrystals, or macrocrystals. For more promising hits, the trays are removed and the crystal harvested for in-house x-ray diffraction testing. The CrystalFarm™ software allows for analysis of crystallization conditions versus events in the drops (e.g., crystals, precipitate, clear) and assembling a pseudophase diagram. This information can be used to further optimize crystallization conditions.

The authors' current optimization strategies employ either 96-well plates or 24-well VDX hanging drop plates. An effort is first made to reproduce the best crystallization hit(s) following screening. This is made more amenable by having in-house prepared reagent stock solutions, as it has been found that crystallization hits arising from commercial screens are sometimes difficult to reproduce without having exactly the same reagents in hand. This can be especially true with regard to PEG solutions, which are inherently less stable, and in which it is not unusual to see differences in crystallization behavior using the same molecular weight PEG from different suppliers. All of the usual optimization strategies are employed: streak-, micro- and macro-seeding, fine grid screening (pH, salt concentration, and precipitant concentration), additive screening, varying the drop ratios and crystallization methods (e.g., hanging drop, microbatch under oil). Such grid screening can be readily set up using the IMPAX crystallization robot.

Fig. 36.4 Crystal images of (**A**) *Neisseria gonorrhoeae* PoaD and (**B**) *Escherichia coli* WzzB obtained using the CrystalFarm™ visualization robot.

As one would expect, a number of proteins do not yield promising hits, or in some cases, no hits at all following the initial screening. At the level of crystallization screening, alternative strategies are employed. Plates are set at 4 °C rather than 20 °C to determine the effect of temperature on precipitation behavior and crystallization. As well, microbatch crystallization is employed using 96-well Impact plates covered with paraffin oil as an alternative to sitting drop vapor diffusion. These plates are set and imaged using the same equipment as for the Intelliplates. In other cases in which no promising precipitation pattern is observed, it is necessary to go back and further optimize the protein buffer or genetically modify the protein by making deletions or point mutants (see Note 8).

3.9. Structure Determination

Crystals are harvested directly from 96-well sitting drop or 24-well hanging drop trays, and tested for diffraction properties on a HTC imaging plate detector mounted on a microfocus 007 rotating anode generator. Initially, crystals are tested directly from the drop without cryoprotection to get an idea of the raw crystal quality. The reservoir solution is then tested to see whether or not it will be a suitable cryo solvent as is. There are two primary strategies for cryoprotection of crystals. One is to add glycerol to a final concentration of 20–25% to the reservoir solution; the other is to increase the concentration of precipitant (polyethylene glycol, MPD, or salt) to obtain a satisfactory freeze. It has been found *(36)* that in a number of cases crystal annealing, either by reintroducing the cooled crystal for 10 seconds to 1 minute in cryosolution or by blocking the cold stream for 2–3 seconds, can result in a dramatic improvement of spot shape and sometimes in diffraction resolution.

Once a crystal has been identified that appears suitable for structural analysis, it is stored in a $N_2(l)$ Dewar for synchrotron x-ray data collection. Stored crystals are logged into the SPEX Db database so that an electronic record is available. Recently, the ALS-design of automounter pucks was adopted, in which up to 16 crystals can be stored on each puck. Automounter-compatible pins, CrystalCap HT™, with a base-to-loop distance of 22 mm are used to mount the crystals. These pins are compatible with the X12B and X29 automounters at the National Synchrotron Light Source, BNL.

In most cases, structure determination from SeMet-labeled protein crystals *(33)* is fairly routine. In many cases Se-SAD is sufficient to obtain an interpretable electron density map. Heavy atom anomalous scatterers are identified using SOLVE *(37)*, SHELX-D *(38)*, or BnP *(39)* and phases calculated using SOLVE, SHARP *(40)*, or SHELX-E *(41)*. Density modification is usually performed using RESOLVE *(42)*, which in some cases is also utilized to generate the initial model. Where higher resolution (<2.5 Å) is available, ArpWarp *(43)* is usually employed for auto-building.

Another method successfully applied in the authors' program is sodium bromide or sodium iodide soaks *(44)*. To this end the protein crystal is soaked for 0.5–1 minute in a cryoprotecting solution containing 0.5–1 M NaBr or NaI, flash frozen, and data collected in the SAD regime. This method has been used to solve, for example, the structure of *E. coli* HisB (NaBr) *(45)* and *C. jejuni* PglD (NaI, unpublished).

If the crystal contains other heavier atoms, for example, Zn, Cd, Fe, they can also be used as a source for the anomalous diffraction. Examples from the authors' program include *E. coli* WecD *(46)* and *Methanobacterium thermoautotrophicum* HisI *(47)*, both solved using the anomalous signal of Zn.

3.10. Examples of Protein Structures and Their Complexes with Ligands

This section presents three examples of structures determined in the authors' program that display different structural and biochemical insights gained from these investigations.

E. coli L-histidinol dehydrogenase (HisD) *(48)* catalyzes the last two steps in the biosynthesis of L-histidine: sequential NAD-dependent oxidations of L-histidinol to L-histidinal and then to L-histidine. This ancient biosynthetic pathway found in bacteria, archaebacteria, fungi, and plants provided a paradigm for the operon, transcriptional regulation of gene expression and feedback inhibition of a pathway. HisD functions as a homodimer and requires the presence of one Zn^{2+} cation per monomer. The 3D structure of *E. coli* HisD in the apo state as well as complexes with histidinol, histidine, Zn^{2+}, and NAD^+ (Fig. 36.5) have been determined. Each monomer is made of four domains,

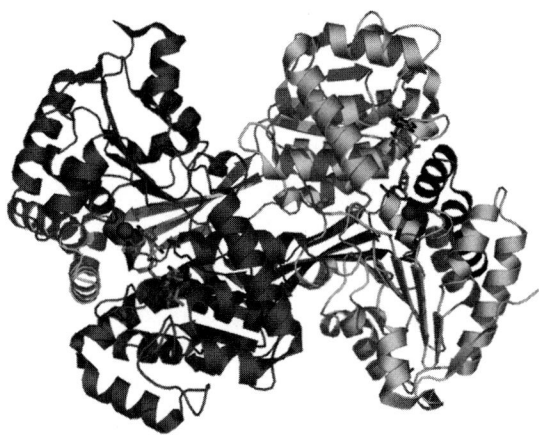

Fig. 36.5 Structure of the HisD dimer (PDB 1KAE) depicting the domain swapping. One subunit is colored slate and the second is colored wheat. NADP and L-histidinol ligands are shown in stick representation and colored by CPK, with the Zn^{2+} ions shown as pink spheres. Figure was prepared using PyMol (www.pymol.org).

two of which display a very similar, incomplete Rossmann fold, suggesting an ancient event of gene duplication. The dimer is intertwined, possibly indicating domain swapping. Residues from both monomers form the active site. Zn^{2+} plays a crucial role in substrate binding but is not directly involved in catalysis. The active site residue His327 participates in acid-base catalysis, whereas Glu326 activates a water molecule. NAD^+ binds weakly to one of the Rossmann fold domains in a manner different from that previously observed for other proteins having a Rossmann fold.

  E. coli N-succinylarginine dihydrolase (AstB) *(49)* is an enzyme from the ammonia-producing arginine succinyl transferase (AST) pathway, the major pathway in *E. coli* and related bacteria for arginine catabolism as a sole nitrogen source. This pathway consists of five steps, each catalyzed by a distinct enzyme. AstB catalyzes the second reaction step. The enzyme exhibits a pseudo fivefold symmetric α/β propeller fold of circularly arranged ββαβ-modules enclosing the active site. The crystal structure indicates clearly that this enzyme belongs to the amidinotransferase (AT) superfamily and that the active site contains a Cys—His-Glu triad characteristic of the AT superfamily. Structures of the complexes of AstB with the reaction product and a Cys365Ser mutant with bound *N*-succinylarginine substrate (Fig. 36.6) suggest a catalytic mechanism that consists of two cycles of hydrolysis and ammonia release, with each cycle utilizing a mechanism similar to that proposed for arginine deiminases. Like other members of the AT superfamily of enzymes, AstB possesses a flexible loop that is disordered in the absence of substrate and assumes an ordered conformation upon substrate binding, shielding the ligand from the bulk solvent, thereby controlling substrate access and product release.

  An *E. coli* 23S rRNA 2'O-methyltransferase (RlmB) *(50)* catalyzes the methylation of guanosine 2251, a modification conserved in the peptidyl-transferase domain of 23S rRNA. RlmB consists of an N-terminal domain connected by a flexible extended linker to a catalytic C-terminal domain and forms a dimer in solution. The C-terminal domain displays a divergent methyltransferase fold, with a unique knotted region (Fig. 36.7), and lacks the

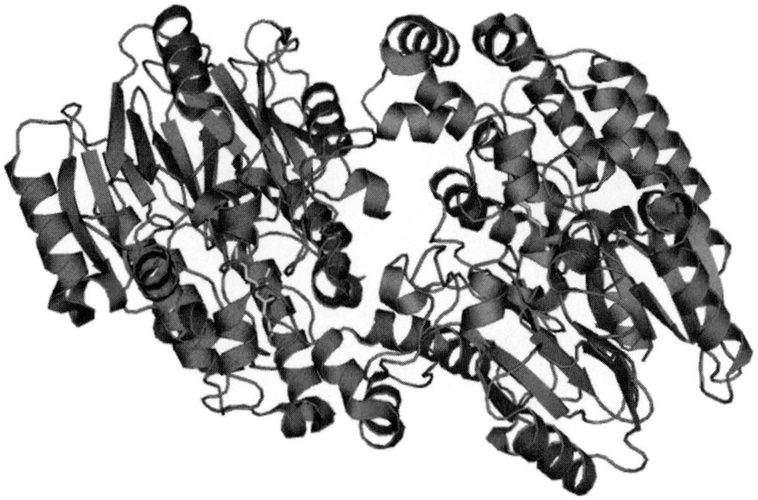

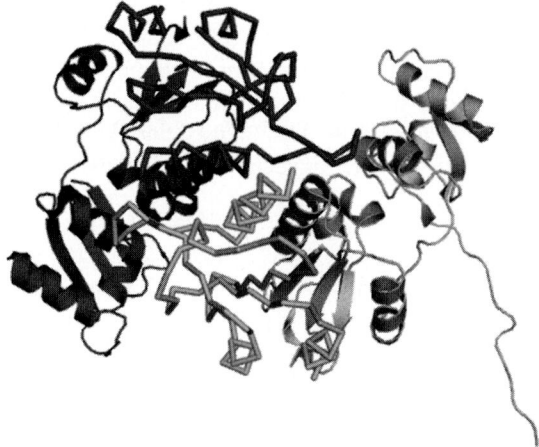

Fig. 36.6 Structure of the AstB dimer (PDB 1YNH), with the *N*-succinyl ornithine product shown in stick representation and colored yellow.

Fig. 36.7 Structure of the RlmB dimer (PDB 1GZ0), with the deep knot region (residues 159–243) shown in ribbon representation.

classical AdoMet binding site features. The N-terminal domain is similar to ribosomal proteins L7 and L30, suggesting a role in 23S rRNA recognition. The conserved residues in this novel family of 2'O-methyltransferases cluster in the knotted region, suggesting the location of the catalytic and AdoMet binding sites.

4. Notes

1. Use of E-gel-96 offer a number of advantages for high throughput analysis over conventional agarose gels, in terms of gel loading, the number of samples that can be analyzed, the speed of electrophoresis, and the fact that ethidium bromide is included in the gel for direct visualization following

electrophoresis. A disadvantage of this system, however, is their relatively poor resolving power due to the small size of the gel.

2. One of the most laborious aspects of high throughput cloning is the large amount of plate spreading that needs to be done following transformation. The authors have tried to manually spread the plates using glass beads, or use the Biomek FX eight-channel head to dispense cells on smaller-surface area six-well tissue culture plates. One difficulty with glass beads is that one often does not obtain a uniform distribution of colonies on the plates; cells tend to clump together, resulting in clusters of colonies. In using the robot for plating, it is critical to properly dilute the cells to have well separated colonies in each well when using a six-well tissue culture plate.

3. To run the Montage™ DNA plasmid mini-prep$_{96}$ kit on the Biomek FX, program the plasmid prep protocol on the software, assemble the Millipore manifold on the robot by placing the plasmid plate inside the vacuum manifold (the lysate clearing plate on top of the manifold), and place the deep well block containing the cell pellet on the orbital shaker. The protocol starts by dispensing solution I into the deep well block and resuspending the cells by automatic shaking. Solutions II and III are dispensed into the well, followed with repeated pipetting for thorough resuspension. The cell lysate is transferred to a vacuum manifold and filtered by vacuum with cell debris remaining on the clearing plate, and collection of the supernatant in the plasmid plate. The robot gripper then removes the clearing plate, moves the plasmid plate to the top of the manifold, and applies the vacuum to filter any remaining liquid in the DNA plate. Following one step of washing with water, the DNA is resuspended in 10 mM Tris buffer, and transferred to a Costar Cone round bottom plate.

4. The authors' early experiments with the Qiagen filter plates included a centrifugation step prior to application of the lysate to the filter plate, as the combination of unlysed cells and cell debris often resulted in clogging of the plates. Even with the newer filter plates, it is important to realize that all wells do not filter at the same rate; it depends to some extent on the culture density within individual wells. A suitable vacuum must be obtained to avoid clogging the wells of the filter plate, but at the same time allowing for filtering of the lysate. A vacuum of 10″ Hg for 30 minutes is optimal.

5. Choice of the specific anti-His antibody is important to the success of this experiment. One wishes to minimize the amount of background in the assay, which requires as specific an antibody as can be readily obtained. Other anti-His antibodies, such as anti-His$_3$ have been tried, but the anti-His$_4$ antibody from Qiagen has been found to be suitable.

6. There are several important points to keep in mind regarding purification of His-tagged proteins. The bed volume of the resin should be minimized, such that it binds the available fusion protein, but not in excess. This results in a cleaner product following elution. Often one can exceed the manufacturer's cited binding capacity, meaning that less resin is required than one might expect. The authors have tested Ni-NTA (used routinely), Ni-Sepharose (GE Healthcare), and Talon cobalt-based (Clonetech) resins. In some cases Ni-Sepharose can result in a cleaner protein preparation than N-NTA, although this is protein dependent. In some cases it also necessary to adjust the imidazole concentration in the wash buffer, either increasing it (in the case of too many impurities), or decreasing it (in the case of poor

556 Cygler et al.

binding to the resin). For proteins that tend to precipitate due to lack of solubility, it is sometimes necessary to double the volume of the elution buffer such that the protein is eluted in a more dilute form.

7. Cleavage of fusion proteins using TEV protease can produce variable results, depending on the nature of the fusion protein (His-tag vs GST-tag), cleavage on the beads or following elution, the composition of the cleavage buffer, temperature, and ratio of TEV protease to fusion protein used. The most important parameter in enhancing cleavage appears to be the ratio of TEV protease:fusion protein, with ratios as high as 1:50 or even 1:20 sometimes being needed to obtain effective cleavage. Although this can consume large amounts of protease, the protease is readily removed using its His-tag, and due to its high specificity, does not result in secondary cleavage of the fusion protein. If the fusion protein is stable, cleavage at 30 °C for 2 hours can be tried as an alternative to cleavage at room temperature.

8. Truncations can be based on the results of limited proteolysis, or designed based on a multiple sequence alignment. Small truncations of residues at the N- or C-terminal ends of a protein can have a positive influence on crystallization behavior. Normally one wishes to design truncations such that they do not include elements of secondary structure, so the use of secondary structure predictions based on a multiple alignment, or computing a homology model (if possible) are recommended to aid this process. Similar ideas apply to the introduction of point mutations to aid in protein crystallization, with the added caveat that one would often like these residues to be on the exposed surface of the protein, so a prediction of surface accessibility based on a multiple alignment is in order.

Acknowledgments

The authors acknowledge the important contributions of a number of lab members, both past and present, to the procedures and protocols described in this chapter, including Yunge Li, Véronique Sauvé, Christine Munger, Eunice Ajamian, Erumbi S. Rangarajan, Pietro Iannuzzi, Guy Nadeau, Robert Larocque, Patrice Bouchard, Stephane Raymond, Jayaraman Sivaraman, Nicholas O'Toole, Joseph D. Schrag, Vladimir V. Lunin, Gurvan Michel, Joao Barbosa, Ante Tocilj, and Frederic Ouelette.

References

1. Rupp, B. (2003). High throughput crystallography at an affordable cost: the TB Structural Genomics Consortium Crystallization Facility. *Acc. Chem. Res.* **36,** 173–181.

2. Busso, D., Kim, R., and Kim, S. H. (2003). Expression of soluble recombinant proteins in a cell-free system using a 96-well format. *J. Biochem. Biophys. Meth.* **55,** 233–240.

3. Dieckman, L. J., Hanly, W. C., and Collart, E. R. (2006). Strategies for high throughput gene cloning and expression. *Genet. Eng. (NY)* **27,** 179–190.

4. Lesley, S. A. (2001). High throughput proteomics: protein expression and purification in the postgenomic world. *Protein Expr. Purif.* **22,** 159–164.

5. Peti, W., Page, R., Moy, K., O'Neil-Johnson, M., Wilson, I. A., Stevens, R. C., and Wuthrich, K. (2005). Towards miniaturization of a structural genomics pipeline using micro-expression and microcoil NMR. *J. Struct. Funct. Genom.* **6,** 259–267.

6. Bhikhabhai, R., Sjoberg, A., Hedkvist, L., Galin, M., Liljedahl, P., Frigard, T., Pettersson, N., Nilsson, M., Sigrell-Simon, J. A., and Markeland-Johansson, C. (2005). Production of milligram quantities of affinity tagged-proteins using automated multistep chromatographic purification. *J. Chromatogr. A* **1080,** 83–92.

7. Scheich, C., Sievert, V., and Bussow, K. (2003). An automated method for high throughput protein purification applied to a comparison of His-tag and GST-tag affinity chromatography. *BMC Biotechnol.* **3,** 12.

8. Kim, Y., Dementieva, I., Zhou, M., Wu, R., Lezondra, L., Quartey, P., Joachimiak, G., Korolev, O., Li, H., and Joachimiak, A. (2004). Automation of protein purification for structural genomics. *J. Struct. Funct. Genomics* **5,** 111–118.

9. Page, R., and Stevens, R. C. (2004). Crystallization data mining in structural genomics: using positive and negative results to optimize protein crystallization screens. *Methods* **34,** 373–389.

10. Kimber, M. S., Vallee, F., Houston, S., Necakov, A., Skarina, T., Evdokimova, E., Beasley, S., Christendat, D., Savchenko, A., Arrowsmith, C. H., Vedadi, M., Gerstein, M., and Edwards, A. M. (2003). Data mining crystallization databases: knowledge-based approaches to optimize protein crystal screens. *Proteins* **51,** 562–568.

11. Newman, J., Egan, D., Walter, T. S., Meged, R., Berry, I., Ben, J. M., Sussman, J. L., Stuart, D. I., and Perrakis, A. (2005). Towards rationalization of crystallization screening for small- to medium-sized academic laboratories: the PACT/JCSG+ strategy. *Acta Crystallogr. D. Biol. Crystallogr.* **61,** 1426–1431.

12. Matte, A., Sivaraman, J., Ekiel, I., Gehring, K., Jia, Z., and Cygler, M. (2003). Contribution of structural genomics to understanding the biology of *Escherichia coli. J. Bacteriol.* **185,** 3994–4002.

13. Blattner, F. R., Plunkett, G., Bloch, C. A., Perna, N. T., Burland, V., Riley, M., Collado-Vides, J., Glasner, J. D., Rode, C. K., Mayhew, G. F., Gregor, J., Davis, N. W., Kirkpatrick, H. A., Goeden, M. A., Rose, D. J., Mau, B., and Shao, Y. (1997). The complete genome sequence of *Escherichia coli* K-12. *Science* **277,** 1453–1474.

14. Perna, N. T., Plunkett, G. I., Blattner, F. R., Mau, B., and Blattner, F. R. (2001). Genome sequence of enterohaemorrhagic *Escherichia coli* O157:H7. *Nature* **409,** 529–533.

15. Welch, R. A., Burland, V., Plunkett, G., III, Redford, P., Roesch, P., Rasko, D., Buckles, E. L., Liou, S. R., Boutin, A., Hackett, J., Stroud, D., Mayhew, G. F., Rose, D. J., Zhou, S., Schwartz, D. C., Perna, N. T., Mobley, H. L., Donnenberg, M. S., and Blattner, F. R. (2002). Extensive mosaic structure revealed by the complete genome sequence of uropathogenic *Escherichia coli. Proc. Natl. Acad. Sci. USA* **99,** 17020–17024.

16. Keseler, I. M., Collado-Vides, J., Gama-Castro, S., Ingraham, J., Paley, S., Paulsen, I. T., Peralta-Gil, M., and Karp, P. D. (2005). EcoCyc: a comprehensive database resource for *Escherichia coli. Nucleic Acids Res.* **33,** D334–D337.

17. Kanehisa, M., Goto, S., Hattori, M., Oki-Kinoshita, K. F., Itoh, M., Kawashima, S., Katayama, T., Araki, M., and Hirakawa, M. (2006). From genomics to chemical genomics: new developments in KEGG. *Nucleic Acids Res.* **34,** D354–D357.

18. Misra, R. V., Horler, R. S., Reindl, W., Goryanin, I. I., and Thomas, G. H. (2005). EchoBASE: an integrated post-genomic database for *Escherichia coli. Nucleic Acids Res.* **33,** D329–D333.

19. Teichmann, S. A., Rison, S. C., Thornton, J. M., Riley, M., Gough, J., and Chothia, C. (2001). The evolution and structural anatomy of the small molecule metabolic pathways in *Escherichia coli. J. Mol. Biol.* **311,** 693–708.

20. Riley, M., Abe, T., Arnaud, M. B., Berlyn, M. K., Blattner, F. R., Chaudhuri, R. R., Glasner, J. D., Horiuchi, T., Keseler, I. M., Kosuge, T., Mori, H., Perna, N. T., Plunkett, G., III, Rudd, K. E., Serres, M. H., Thomas, G. H., Thomson, N. R., Wishart, D., and Wanner, B. L. (2006). *Escherichia coli* K-12: a cooperatively developed annotation snapshot—2005. *Nucleic Acids Res.* **34,** 1–9.

21. Riley, M., and Serres, M. H. (2000). Interim report on genomics of *Escherichia coli. Annu. Rev. Microbiol.* **54**, 341–411.

22. Finn, R. D., Mistry, J., Schuster-Bockler, B., Griffiths-Jones, S., Hollich, V., Lassmann, T., Moxon, S., Marshall, M., Khanna, A., Durbin, R., Eddy, S. R., Sonnhammer, E. L., and Bateman, A. (2006). Pfam: clans, web tools and services. *Nucleic Acids Res.* **34,** D247–D251.

23. Bairoch, A., Apweiler, R., Wu, C. H., Barker, W. C., Boeckmann, B., Ferro, S., Gasteiger, E., Huang, H., Lopez, R., Magrane, M., Martin, M. J., Natale, D. A., O'Donovan, C., Redaschi, N., and Yeh, L. S. (2005). The Universal Protein Resource (UniProt). *Nucleic Acids Res.* **33**, D154–D159.

24. Gasteiger, E., Gattiker, A., Hoogland, C., Ivanyi, I., Appel, R. D., and Bairoch, A. (2003). ExPASy: the proteomics server for in-depth protein knowledge and analysis. *Nucleic Acids Res.* **31,** 3784–3788.

25. Raymond, S., O'Toole, N., and Cygler, M. (2004). A data management system for structural genomics. *Proteome. Sci.* **2,** 4.

26. Bendtsen, J. D., Nielsen, H., Von, H. G., and Brunak, S. (2004). Improved prediction of signal peptides: SignalP 3.0. *J. Mol. Biol.* **340,** 783–795.

27. Kall, L., Krogh, A., and Sonnhammer, E. L. (2004). A combined transmembrane topology and signal peptide prediction method. *J. Mol. Biol.* **338,** 1027–1036.

28. Kapust, R. B., Tozser, J., Fox, J. D., Anderson, D. E., Cherry, S., Copeland, T. D., and Waugh, D. S. (2001). Tobacco etch virus protease: mechanism of autolysis and rational design of stable mutants with wild-type catalytic proficiency. *Protein Eng.* **14,** 993–1000.

29. Dougherty, W. G., Carrington, J. C., Cary, S. M., and Parks, T. D. (1988). Biochemical and mutational analysis of a plant virus polyprotein cleavage site. *EMBO J.* **7,** 1281–1287.

30. Carrington, J. C., and Dougherty, W. G. (1988). A viral cleavage cassette: Identification of amino acid sequences required for tobacco etch virus polyprotein processing. *Proc. Natl. Acad. Sci. USA* **85,** 3391–3395.

31. Bradford, M. M. (1976). A rapid and sensitive method for the quantitation of microgram quantities of protein-dye binding. *Anal. Biochem.* **72,** 248–254.

32. Studier, F. W. (2005). Protein production by auto-induction in high density shaking cultures. *Protein Expr. Purif.* **41,** 207–234.

33. Hendrickson, W. A., Horton, J. R., and LeMaster, D. M. (1990). Selenomethionyl proteins produced for analysis by multiwavelength anomalous diffraction (MAD): a vehicle for direct determination of three-dimensional structure. *EMBO J.* **9,** 1665–1672.

34. Doublie, S. (1997). Preparation of selenomethionyl proteins for phase determination. *Meth. Enzymol.* **276,** 523–530.

35. Koth, C. M., Orlicky, S. M., Larson, S. M., and Edwards, A. M. (2003). Use of limited proteolysis to identify protein domains suitable for structural analysis. *Meth. Enzymol.* **368,** 77–84.

36. Heras, B., and Martin, J. L. (2005). Post-crystallization treatments for improving diffraction quality of protein crystals. *Acta Crystallogr. D. Biol. Crystallogr.* **61,** 1173–1180.

37. Terwilliger, T. C. (2002). Automated structure solution, density modification and model building. *Acta Crystallogr. D Biol. Crystallogr.* **58,** 1937–1940.

38. Schneider, T. R., and Sheldrick, G. M. (2002). Substructure solution with SHELXD. *Acta Crystallogr. D. Biol. Crystallogr.* **58,** 1772–1779.

39. Weeks, C. M., and Miller, R. (1999). Optimizing Shake-and-Bake for proteins. *Acta Crystallogr.* **D55,** 492–500.

40. Bricogne, G., Vonrhein, C., Flensburg, C., Schiltz, M., and Paciorek, W. (2003). Generation, representation and flow of phase information in structure determination: recent developments in and around SHARP 2.0. *Acta Crystallogr. D. Biol. Crystallogr.* **59,** 2023–2030.

41. Sheldrick, G. M. (2002). Macromolecular phasing with SHELXE. *Z. Kristallogr.* **217,** 644–650.

42. Terwilliger, T. C. (2000). Maximum-likelihood density modification. *Acta Crystallogr.* **D56,** 965–972.

43. Perrakis, A., Morris, R., and Lamzin, V. S. (1999). Automated protein model building combined with iterative structure refinement. *Nat. Struct. Biol.* **6,** 458–463.

44. Dauter, Z., Li, M., and Wlodawer, A. (2000). Practical experience with the use of halides for phasing macromolecular structures: a powerful tool for structural genomics. *Acta Crystallogr.* **D57,** 239–249.

45. Rangarajan, E. S., Proteau, A., Wagner, J., Hung, M. N., Matte, A., and Cygler, M. (2006). *E. coli* histidinol phosphate phosphatase: Structural snapshots along the reaction pathway. *J. Biol. Chem.* **281,** 37930–37941.

46. Hung, M. N., Rangarajan, E., Munger, C., Nadeau, G., Sulea, T., and Matte, A. (2006). Crystal structure of TDP-fucosamine acetyltransferase (WecD) from *Escherichia coli*, an enzyme required for enterobacterial common antigen synthesis. *J. Bacteriol.* **188,** 5606–5617.

47. Sivaraman, J., Myers, R. S., Boju, L., Sulea, T., Cygler, M., Jo Davisson, V., and Schrag, J. D. (2005). Crystal structure of *Methanobacterium thermoautotrophicum* phosphoribosyl-AMP cyclohydrolase HisI. *Biochemistry* **44,** 10071–10080.

48. Barbosa, J. A., Sivaraman, J., Li, Y., Larocque, R., Matte, A., Schrag, J. D., and Cygler, M. (2002). Mechanism of action and NAD+-binding mode revealed by the crystal structure of L-histidinol dehydrogenase. *Proc. Natl. Acad. Sci. USA.* **99,** 1859–1864.

49. Tocilj, A., Schrag, J. D., Li, Y., Schneider, B. L., Reitzer, L., Matte, A., and Cygler, M. (2005). Crystal structure of N-succinylarginine dihydrolase AstB, bound to substrate and product, an enzyme from the arginine catabolic pathway of *Escherichia coli. J. Biol. Chem.* **280,** 15800–15808.

50. Michel, G., Sauvé, V., Larocque, R., Li, Y., Matte, A., and Cygler, M. (2002). The structure of the RlmB 23S rRNA methyltransferase reveals a new methyltransferase fold with a unique knot. *Structure* **10,** 1303–1315.

Chapter 37

High Throughput Protein Production and Crystallization at NYSGXRC

Michael J. Sauder, Marc E. Rutter, Kevin Bain, Isabelle Rooney, Tarun Gheyi, Shane Atwell, Devon A. Thompson, Spencer Emtage, and Stephen K. Burley

Phase II of the Protein Structure Initiative, funded by the NIH NIGMS (National Institute of General Medical Sciences), is a 5-year effort to determine thousands of protein structures. The New York SGX Research Center for Structural Genomics (NYSGXRC) is one of the four large-scale production centers tasked with determining 100–200 structures annually. Almost all protein production is carried out using the high throughput structural biology platform at SGX Pharmaceuticals (SGX), which supplies 120 or more ultrapure proteins per month for NYSGXRC crystallization and structure determination activities. Protocols for PCR, cloning, expression/solubility testing, fermentation, purification, and crystallization are described. General protocols and detailed experimental results for each target are updated weekly at the public PepcDB website (pepcdb.pdb.org/), and all NYSGXRC clones should be available in 2008 through the PlasmID resource operated by the Harvard Institute of Proteomics.

1. Introduction

The NYSGXRC is one of four large-scale production centers funded by the NIH NIGMS for phase II of the Protein Structure Initiative (PSI-2). The goal of the PSI-2 is the experimental determination of 3,000 protein structures in 5 years, beginning July 2005. This challenge is being addressed by leveraging the high throughput protein production and structure determination platforms developed during a 5-year pilot phase (PSI-1). For the NYSGXRC, the protein production platform was created by SGX Pharmaceuticals in 1999–2001 (then called Structural GenomiX) and was utilized for supplying PSI-1 proteins when SGX became a member of the New York Consortium in 2002 (1).

In addition to other internal and external collaborations, the SGX platform routinely supplies more than 120 PSI-2 proteins per month for crystallization, which supports a structure determination rate of over 150 structures per year. This chapter describes the methods used by SGX for cloning, expressing, purifying, and crystallizing proteins for the NYSGXRC.

From: *Methods in Molecular Biology, Vol. 426: Structural Proteomics: High-throughput Methods*
Edited by: B. Kobe, M. Guss and T. Huber © Humana Press, Totowa, NJ

2. Materials

Numerous techniques are described in the following sections, and specific buffers and reagents are defined in each relevant section.

All PSI-2 Centers, including the NYSGXRC, are required to publicize information about target status and experimental protocols. PepcDB (Protein Expression Purification and Crystallization Database) is maintained by the Protein Data Bank (PDB; www.pdb.org), which also serves as the world-wide repository of macromolecular structure information. On a weekly basis, NYSGXRC provides an update to PepcDB of the authors' general protocols and the status of each target, the clones that have been made (and their DNA sequences), results from expression/solubility testing, fermentation and purification details, molecular weight as determined by electrospray ionization mass spectrometry (ESI-MS), crystallization conditions of crystals that have been screened at the SGX-CAT company beamline at the Advanced Photon Source or the National Synchrotron Light Source (NSLS), and which targets have had their structures deposited in the PDB.

An NIH-funded reagent repository is being created at the Harvard Institute of Proteomics (HIP) to store glycerol stocks of clones that have been generated as part of the PSI. During 2007–2008, NYSGXRC plans to ship the approximately 8,000 clones that it has generated to the repository and will ship all new clones thereafter for the duration of PSI-2. These clones represent human and rodent protein phosphatases, phosphatases from various pathogenic organisms, community-nominated targets from the enolase *(2)* and amidohydrolase *(3)* superfamilies, and thousands of bacterial/archaeal genes from more than 100 different organisms.

The NIGMS-funded KnowledgeBase will shortly be serving as a structural genomics portal, providing search and browsing capabilities linking the data in PepcDB, HIP, PDB, UniProt, and other data sources. The goal will be to provide easy access to structural information and facilitate the interpretation and use of these structures by biologists.

Each target may have multiple construct boundaries cloned into multiple vectors, and may be expressed as either SeMet-labeled or native protein. All clones have a unique identifier in the authors' Laboratory Information Management System (LIMS), and each new fermentation attempt has a unique protein order ID (PID). Split pools in purification can also be numbered and specific steps tracked for each pool. Numeric data such as molecular weight, estimated purity, yield, and final concentration, and images of gels, gel filtration chromatograms, and mass spectra are stored in LIMS and accessible to all consortium members. Much of these data are also accessible from the public PepcDB web site.

3. Methods

3.1. Expression of Recombinant Proteins in *E. coli*

3.1.1. Targets and Primers

Roughly 70% of the work from each PSI Center is focused on maximizing structural coverage of protein sequence space, 15% is focused on targets of biomedical relevance defined by each Center, and the final 15% of effort is on targets nominated by the broader research community. Following target selection, NYSGXRC targets are uploaded into the SGX LIMS in batch mode; target data include the full-length DNA sequence, Genbank or UniProt accession number, species name, and Pfam target domain or other family identifier. Expression construct boundaries are manually chosen based on inspection of automated sequence analysis data, which includes PSI-BLAST, PDB-BLAST, secondary structure and disorder prediction, Pfam domain boundaries, signal peptide and transmembrane helix prediction, and multiple sequence alignment/conservation analysis. Short, single domain genes are typically expressed as full-length proteins, whereas larger, multidomain proteins receive more detailed analysis and multiple construct choices. Forward and reverse PCR primer pairs for each expression construct are automatically designed, with an option to individually check each sequence and make minor adjustments in length to achieve a more reasonable GC content and a GC clamp at the 3′end. Primers are typically 18–26 bases. They are ordered in 96-well plates with a desalting-step for purity and delivered at a concentration of 150 μM. Subsequent working dilution primer plates are made at 10 μM and arrayed in the same wells to match the upcoming PCR.

3.1.2. PCR

The primary PCR reaction is carried out in 96-well plates. In preparation for the primary PCR reaction, the authors have established a series of LIMS queries that create a spreadsheet containing primer names, primer location coordinates in the primer plates, primer melting temperatures, primer pair PCR product lengths, 96-well coordinates for the PCR products, DNA template to use, and so on. In addition, a printout is generated for the PCR reaction recipe, PCR protocol, and a legend of base-pair–sized products corresponding with the PCR wells.

For the primary PCR, a standard master mix (based on a 50-μl reaction) is prepared on ice that contains water, 10× PCR buffer (AccuPrime buffer; Invitrogen, Carlsbad CA), and enzyme (AccuPrime, Invitrogen). It is distributed to each of the 96 PCR plate wells, also maintained on ice. The corresponding forward and reverse primers are in matching 96-well plates at 10 μM concentration. A 12-channel pipettor and eight pipettes are utilized to transfer 2 μl of each primer into the corresponding PCR reaction. Subsequently, the DNA template is added to the PCR reaction, often aliquoted one well at a time (depending on the commonality of template source across the 96-well primary PCR plate). After all the components have been added to the wells, the 96-well primary PCR plates are sealed and subjected to standard PCR thermocycling conditions.

Following completion of the PCR reaction, all 50 μl of the primary PCR reaction are electrophoresed on agarose gels (1% TAE, ethidium bromide), visualized in ultraviolet light, and extracted from the gel (plugged) by hand, to isolate a single fragment of the correct size. The bands are extracted from the gel by cutting a 1-ml pipette tip on its taper to approximate the same diameter as the gel band width, plugged with that, and transferred to a "Gel Extraction" 96-deep well block. This Gel Extraction block is processed with a standard Qiagen (Valencia, CA) Gel Extraction procedure (#28704) that is performed under vacuum on the benchtop. The purified primary PCR products are eluted in a final volume of approximately 60 μl in 10 mM Tris HCl, pH 8.5.

NYSGXRC is supplementing in-house PCR efforts with total gene synthesis from a commercial vendor. The cost of synthesis has been steadily dropping and the authors find synthesis to be cost-effective compared with working with eukaryotic cDNAs, and it should be competitive with using bacterial genomic DNAs shortly. More information on our experience with total gene synthesis is available in Note 1.

3.1.3. Topogation and Transformation

For PSI-2 work, SGX utilizes a custom (Invitrogen) topoisomerase adapted vector that the authors designate BC. The BC vector incorporates a T7 promoter (4,5), a gene-conferring kanamycin (kan) resistance, and a C-terminal 6× histidine tag. Codons for the oligopeptides MSL and EGHHHHHH are added N- and C-terminal to the inserted sequence, respectively. Ligation with the equivalent of 1 µl of Topo-adapted BC vector per reaction is carried out by adding water and a salt solution (200 mM NaCl and 1 mM $MgCl_2$; final concentration), then 4.5 µl of the mix is distributed to each well of a new 96-well plate. A 12-channel pipette is used to transfer 1.5 µl of the gel-extracted PCR products into this Topogation mix. The cloning reaction takes 5 minutes at room temperature. Following the Topogation reaction, 5 µl of the topogation is transformed into 50 µl of Top10 competent cells (Invitrogen) by heat shock (45 seconds at 42 °C). After heat shock the cells are immediately put back on ice for two minutes, then 250 µl SOC is added and the cells are placed in a 37 °C shaking incubator for one hour. The transformation cells are plated by hand from each individual well onto 96 different LB-agar/kan plates and the cells are spread using glass beads. The plates are incubated overnight at 37 °C for colony growth.

3.1.4. Picking and Screening Cloned Colonies

Since the cloning vector and our PCR products are both blunt-ended, the fragment of interest can ligate in either the correct orientation or backward. Consequently, clones are screened for fragment orientation using diagnostic PCR (dPCR). To screen for the colonies that contain the gene fragment in the correct orientation, 12 colonies are picked from each transformation plate into 100 µl of LB media and grown (shaking at 400 rpm) for approximately 4 hours. Diagnostic PCR is carried out using 2 µl of this *E. coli* growth, in a total reaction volume of 20 µl. A forward primer (T7) that corresponds with the vector and a gene-specific reverse primer that corresponds with the 3′ end of the target insert are utilized. Clones that contain the insert in the correct orientation give rise to a PCR product of the correct size (plus the T7/vector added length), whereas clones that have incorporated the insert in the backward orientation yield no PCR product at all. The diagnostic PCR set-up consists of 12 reactions (one row in a 96-well plate) for each of the individual 96 clonings. This task amounts to 12 96-well PCR plates to screen by dPCR. Subsequent to subjecting the reactions to standard PCR thermocycling conditions (32 cycles), 20 µl of the dPCR products are loaded onto 12 1% agarose TAE 96-well EGels (Invitrogen), electrophoresed, and photographed. The photos are analyzed using custom software written by CalBay, which allows one to click the gel photo on the computer screen to select three positives for each row of 12 picks per gene. This software, when finished, compiles the positive picks into one list for a Qiagen 3000 robot to cherry pick the 12 96-well plates of diagnostic picks into three 96-deep well blocks of consolidated positives.

The 3 × 96 positive colonies are inoculated into 1 ml of LB/kan media in 96-well plates and grown with 400 rpm shaking 37 °C for 20–24 hours on a HiGro (Genomic Solutions GeneMachines Ann Arbor, MI). The plates are pelleted (3,000 rpm, 10 minutes) and the supernatant decanted. The standard Promega miniprep kit (#A7510) on a Qiagen 8000 robot is used and the purified plasmids are eluted in 100 μl of 10 mM Tris HCl pH 8.5.

3.1.5. Transforming Minipreps into Expression Cells

Three 96-well plates (one per plate of minipreps) containing 20 μl of competent BL21(DE3) RIL CodonPlus cells per each well are thawed. These expression cells contain a plasmid that encodes for the rare arginine, isoleucine, and leucine codons, which has chloramphenicol 30 mg/ml (chlor) resistance. Two microliters of miniprep DNA is added to the thawed cells with a 12-channel multipipettor. The cells are transformed using standard heat-shock protocol (ice for 20 minutes, heat shock at 42 °C for 45 seconds, ice for 2 minutes), and 100 μl SOC is added. Transformations are grown in 96-well format shaking at 400 rpm at 37 °C for 1 hour and then plated on LB/kan agar plates. They are plated by dripping 10 μl onto a tilted plate and letting it run down in a streak. The LB-agar plates are incubated at 37 °C overnight for colony growth. The next morning reveals a line of individual colonies.

3.1.6. Recombinant Protein Expression

Inoculations for expression are made from an overnight culture. As such, one colony per miniprep transformation is first picked into 1 ml of LB/kan/chlor media and grown in 96-well blocks, shaking at 400 rpm, 37 °C overnight. The next morning, glycerol stock plates are made from the overnight growths, and also 5 μl of the overnight growth is used to inoculate into a fresh 1-ml aliquot of LB/Kan/Chlor media in a 96-well block format. This is grown shaking 400 rpm at 37 °C, until the cultures reach an OD between 0.6 and 1.0, at which time IPTG is added to a final concentration of 1 mM. At this time the deep-well blocks are incubated, shaking at 400 rpm at the reduced temperature of 20 °C for 22 hours of expression, then the cultures are pelleted (3,000 rpm, 10 minutes) and decanted.

At the same time the LB/kan/chlor media is inoculated, a duplicate plate is made containing an autoinduction ZYP media (6). The ZYP/kan/chlor cultures are shaken side by side with the LB cultures. However, when the LB cultures are taken out to add IPTG, the ZYP cultures are merely transferred to a 20 °C 400 rpm shaker, and they continue shaking for another 22 hours. Then, like the LB cultures, they are pelleted and decanted.

3.1.7. Expression and Solubility Testing

The cell pellets from expression are each resuspended in 225 μl of Resuspension Buffer (50 mM Tris pH 7.6, 0.1% Tween, 100 mM NaCl, 20 mM imidazole, 0.2% IGEPAL). The resuspension is sonicated (5 bursts for 1 second each) using a Matrical SonicMan. Ten microliters of the resuspension is removed postsonication (Total Sample), added to 10 μl SDS loading dye, and saved to run on BioRad 24-well SDS gels. The remaining lysate is pelleted for 5,000 rpm at 10 minutes. Subsequently, a Qiagen 3000 robot is utilized to perform one-step affinity purification in 96-well format. The expression cell supernatants are added to 20 μl of Qiagen Nickel-NTA Magnetic Agarose beads, bound to the His-tagged proteins in the lysate by shaking for 30 minutes, then magnetized to remove the Resuspension Buffer. Next, the Nickel-NTA/protein complex is washed twice with 200 μl of wash buffer (same as resuspension buffer but without IGEPAL), shaken again, and magnetized to remove the wash buffer. Finally, the 96-well plate is removed from the Qiagen 3000 and the His-tag protein is eluted from the beads by hand by adding 20 μl of Elution Buffer (same as wash plus 200 mM imidazole), magnetizing, and then pipetting off the 20 μl (soluble sample) to a 96-well plate containing 10 μl SDS loading dye.

The total and soluble samples are electrophoresed on 24-well PAGE-SDS gels (BioRad). The LB samples are loaded side by side with their respective ZYP samples. Totals are loaded together on one gel and solubles are loaded together on another gel. After running, these gels are rinsed in 10% acetic acid for 10 minutes, then fresh 10% acetic acid is added along with 10 μl SYPRO orange protein stain and the gels are gently shaken for 1 hour. Gels are imaged on a transilluminator with a UV light, photographed, analyzed and annotated by hand. Results are entered into the spreadsheets that were generated for the primary PCR. Data for each of the 96 targets/constructs is uploaded into LIMS from the spreadsheet, including: expression score (high, medium, low, none), solubility score (high, medium, low, marginal, none), observed molecular weight, best media (LB or ZYP), plate coordinates of the final clone selected, and so on. Individual tube glycerol stocks are generated and used for all scale-up fermentations. Following scale-up and purification, if mass spectrometer analysis indicates a mass discrepancy compared to the theoretical sequence, or when crystal hits are observed, the clone is sequenced and the theoretical clone sequence is superseded by the actual sequence in LIMS.

Many of the authors' community-nominated targets are from the enolase and amidohydrolase superfamilies, so they chose to evaluate the benefit of additional divalent metal cations during protein expression. This analysis is described in Note 2.

3.1.8. NYSGXRC Molecular Biology Metrics

Normal throughput is one 96-well plate of targets (primary PCR plate) per person in 3–4 weeks. In general there are 3–4 FTEs working on high throughput cloning and *E. coli* expression/solubility testing. This work plan ensures testing of 3–4 96-well plates per month. The success rate for obtaining soluble protein is highly variable and depends on the class/family of targets being attempted. However, on average, 40–50% of the clones attempted show sufficient expression and solubility to be passed to fermentation.

3.2. NYSGXRC High Throughput *E. coli* Fermentation

3.2.1. Expression Vector

E. coli constructs for the NYSGXRC project are cloned into a custom Topoisomerase-adapted expression vector that utilizes a T7 promoter *(4,5)* and carries kanamycin resistance. Plasmids containing a gene encoding the protein of interest are transformed into BL21(DE3) Codon+/RIL competent cells that carry chloramphenicol resistance. Kanamycin/chloramphenicol antibiotics are used for selection in all subsequent large-scale fermentation steps.

3.2.2. Large-Scale Fermentation of Selenomethionine Labeled Proteins

Results from small-scale expression/solubility testing are used to determine the optimal media type and volume required to generate 10 mg or more of purified protein for each construct. Two different media types, LB and Studier autoinducing (hereafter referred to as ZYP), are tested to access the expression and solubility of protein being expressed. Successful clones are then passed to fermentation and the media designation for large-scale fermentation is determined based on the media type that exhibited the highest solubility score under small-scale conditions.

In year 2 of PSI-2, the NYSGXRC consortium modified the protein production process and now produces all proteins as SeMet-labeled rather than native. A 2× high yield SeMet media (hereafter referred to as HY) from Medicilon (Chicago, IL; see Note 3) is used to generate SeMet labeled protein. In this process, prepackaged packets of growth media, inhibitory amino acids/seleno-methionine, mineral/trace metal mix, vitamins, and IPTG are used to carry out large-scale fermentations.

Small-scale expression/solubility testing has been carried out in HY SeMet media to accurately access relative expression levels and amount of soluble protein present under a SeMet labeling process. Clones that are passed to fermentation are scaled-up in fermentation according to the small-scale expression/solubility score. High expression/high solubility clones are grown at large-scale in 1 L of HY SeMet media, whereas clones passed with expression/solubility scores of High/Med, through Low/Low are grown in 2 L of HY SeMet media. Depending on the yield, these targets may be reprocessed through other salvage pathways such as new construct boundaries or expression as a Smt3 (SUMO) fusion protein *(7)* (see Note 4).

Glycerol stocks are used to generate 37 °C overnight cultures in HY media. 2-L baffled shake flasks containing 1 L of HY plus vitamins are then inoculated with 10 ml of the overnight culture and grown to an $OD_{600} = 1.0–1.5$, at which time inhibitory amino acids and SeMet are added to the culture. After 20 minutes, 0.4 mM IPTG is added to each flask and the culture is grown for 16 hours at 22 °C. Cultures are then harvested the next morning by centrifugation at 6,000 rpm ($10,000\,g$) for 10 minutes, the supernatant discarded, and the cell pellet weighed. The pellets are then transferred to 50-ml polypropylene conical tubes, and stored at −80 °C until required for purification.

The amount of SeMet added prior to induction depends on the methionine content of the protein. All purified proteins are evaluated by ESI-MS, so the authors have a semiquantitative measure of SeMet incorporation to indicate when there is a heterogeneous mixture of incompletely labeled species. After evaluating 60, 90, and 120 mg/L SeMet (see Note 3), the authors have standardized the use of 90 mg/L SeMet (HY). If the protein has a high methionine content, or the standard protocol results in a high population of partially labeled species, it has been found that 120 mg/L SeMet (HY+) usually leads to complete labeling. For the majority of proteins, the level of SeMet incorporation is close to 100%.

Clones that do not yield a sufficient amount of properly labeled protein in HY SeMet media, or simply do not express soluble protein, are repeated in either LB or ZYP autoinducing media to produce native protein (which requires heavy atom or sulfur SAD phasing techniques for structure determination).

3.2.3. Large-Scale Fermentation of Native Proteins

When LB media is used for large scale growth, 500 ml of LB (in a 2-L baffled flask) is inoculated with 15 ml of an 37 °C overnight culture, cells are grown at 37 °C to $OD_{600} = 0.8$ at and then induced with 0.4 mM IPTG for 16 hours at 22 °C. When ZYP media is used for large scale growth, 250 ml of ZYP (in a 1-L baffled flask) is inoculated with 5 ml of an 37 °C overnight culture, cells are grown at 37 °C for 5–6 hours, and then the temperature is shifted to 22 °C for 16 hours. Regardless of media type, cells are then harvested the next morning by centrifugation for 10 minutes at 6,500 rpm ($10,000\,g$). Supernatant is discarded and the cells pellets are weighed, transferred to 50-ml polypropylene conical tubes, and stored at −80 °C until required for purification.

3.2.4. Fermentation Throughput and Yields

Weekly fermentation capacity for NYSGXRC averages 60 unique clones per week (or roughly 240 fermentations per month). Final optical density at harvest ranges from $OD_{600} \sim 4$ for LB media, $OD_{600} \sim 12$ for ZYP media and $OD_{600} \sim 5$ for HY media. Pellet weights average 7 g/L for LB media, 17 g/L for ZYP media, and 8 g/L for HY media.

3.3. NYSGXRC Protein Purification

All proteins are first processed by the following standard method. Results are then evaluated, and additional purification techniques employed only if necessary.

3.3.1. Cell Lysis

Cells are lysed by sonication, using an automated, custom built machine in 50 mM Tris-HCl, pH 7.5, 500 mM NaCl, 25 mM imidazole, 0.1% Tween 20, in the presence of protease inhibitor cocktail (Sigma, St. Louis, MO) and DNAase; 5 ml buffer is used per gram of pelleted cells.

3.3.2. Nickel Affinity

Crude lysate is centrifuged for 30 minutes at 17,000 g at 4 °C to pellet debris. The decanted supernatant is then incubated with Ni^{2+} agarose (Qiagen) for 30 minutes; 5 ml packed gel is used for each purification. The agarose is poured into 50-ml glass columns (Kontes Glass, from VWR International, West Chester, PA) and washed with 10 column volumes wash buffer (50 mM Tris-HCl, pH 7.8, 500 mM NaCl, 10 mM imidazole, 10% glycerol) and His tagged protein eluted with wash buffer containing 500 mM imidazole. Fractions containing the desired protein are identified by SDS PAGE and pooled. Proteins expressed in the BC vector, which have a noncleavable His tag, are then concentrated for gel filtration. Protein purification of Smt3 fusion proteins is described in Note 4, and experience with pressure-induced disaggregation of insoluble proteins is described in Note 5.

3.3.3. Gel Filtration

Protein from nickel affinity is concentrated to 5 ml and gel filtered on an S75 or S200 column, with 10 mM HEPES, pH 7.5, 150 mM NaCl, 10% glycerol, and 5 mM DTT as the mobile phase. Six sequential gel filtrations are run overnight using an automated program and an AKTA FPLC (Pharmacia). Three FPLCs give a purification capacity of 18 proteins per day. Fractions containing the target protein are identified by SDS PAGE, pooled, and concentrated to 5 mg/ml for storage at −80 °C. A representative gel filtration is shown in Fig. 37.1. In 2007, this procedure was replaced with a 4 module AKTAxpress system (GE Healthcare Life Sciences aktaxpress.com Piscataway, NJ)

3.3.4. Protein Quality Control

Protein purity is assessed by SDS PAGE, Matrix Assisted Laser Desorption/Ionization (MALDI), Electrospray ionization (ESI) mass spectrometry, and analytical gel filtration following concentration. All data and spectra are stored in LIMS. On average, about 6–7% of the PSI-2 proteins fail following mass spectrometric analysis.

Protein samples that are less than 90% pure undergo further purification, typically with anion or cation exchange chromatography. If the measured protein mass does not match the predicted molecular weight (>100 ppm

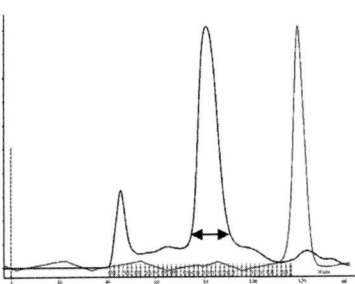 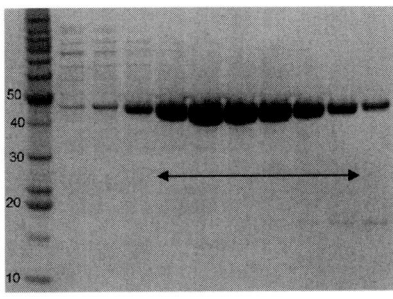

Fig. 37.1 Representative gel filtration on S75 following Smt3 cleavage and Ni^{2+} affinity. Arrows indicate fractions containing the protein of interest.

mass accuracy), the protein sample is digested using a specific protease and identified using tandem mass spectrometry. The protein is confirmed to be the intended protein, and the clone is sequenced. In the unlikely event that the mass discrepant protein is the result of a mix-up at an earlier stage of the process, the LIMS records are corrected and the work on the intended target is initiated.

When multiple proteins of a specific class (e.g., phosphatases) are being processed, the authors review results to determine the stage in purification at which proteins are lost, and then develop salvage pathways to address this issue. One such salvage pathway is screening to optimize purification buffers. The authors' method for buffer screening is detailed in Note 6. Another class of targets that presented some unique challenges were the enolases and amidohydrolases, part of the community-nominated target program. Since these are metal-dependent enzymes, the benefit to small-scale protein expression and solubility of using a cocktail of divalent metal cations (see Note 2) was evaluated. Large-scale fermentation was conducted in the presence of extra metal cations if it was determined to be beneficial.

3.3.5. Protein Purification Metrics
NYSGXRC averaged approximately 120 successful purifications of almost exclusively SeMet-labeled proteins each month during August through November 2006, and this purification rate was maintained during 2007 and beyond. Purification of more than approximately 1,400 proteins per year should easily guarantee suitable diffracting crystals to average more than 150 structures per year. Purification success rates are strongly dependent on protein family and range from 62% for protein phosphatases to 67% for general targets to 80% for enolases/amidohydrolases (the majority of the community-nominated targets).

3.4. SGX Crystallization

Crystallization of NYSGXRC proteins at SGX involves three steps: initial screening, optimization, and phasing. During initial screening proteins are screened for crystallization in commercially available broad screens. During optimization, initial hits are optimized for crystal growth, cryoprotection, and diffraction. Some proteins require further modification of crystallization to obtain phasing information for structure determination.

3.4.1. Screening

Proteins are screened in three commercially available 96-well screens at room temperature. A review of crystallization data from the authors' laboratory indicates that most of the screens used in the past (Hampton Crystal Screen I and II, Hampton Index, Jena JBS screens, and Emerald Wizard I and II) produce the same number of hits, with only Hampton's Index screen being clearly distinguishable from the rest in terms of the number of hits leading to crystal structures. This screen might give better crystals since it is the only screen that incorporates varied pH in its sparse matrix design. The authors' current choices for screens are (in order) Qiagen Nextal Classics II, Classics I, and ComPAS screens (the first two being essentially identical to Hampton's Index and Crystal Screen I/II, respectively).

Proteins are screened at approximately 10 mg/ml in 1 μl + 1 μl sitting drop vapor diffusion trials using Robbins' Intelliplate. The optimal protein concentration is first evaluated by using a PEG-based precipitation test: 1 μl of protein and 20% PEG 6K are mixed and observed under a microscope; if it is heavily precipitated, the protein is diluted and the test repeated until a light precipitate is observed. If the drop is clear, the protein is concentrated and the test repeated. All trials are set up on a Tecan Freedom liquid handler and after plate sealing stored in a custom Rigaku Minstrel gantry/imager for automated imaging of the crystallization trials at days 1, 3, 7, 14, 30, and 90.

Review of crystallization images and trays directly under a microscope are used for both scoring and identification of crystalline hits. Automated scoring of images is not used to identify hits or eliminate clear drops. All information regarding protein contents, screen information, well contents, scores, and comments are tracked in LIMS for easy review and data mining.

If crystals of sufficient size are identified (>50 μm on the shortest side), these are harvested, transferred to a solution containing 70–80% well solution and 20–30% cryoprotectant (typically glycerol and ethylene glycol) and flash frozen in liquid nitrogen for data collection. The data for more than half of the authors' PSI-2 structures are obtained from crystals harvested directly from initial screens.

3.4.2. Crystal Optimization

Crystalline leads that are either not large enough or not of sufficient quality to yield useable x-ray diffraction data proceed to optimization. Most optimization is done in 24-well plates using 1 μl + 1 μl sitting drop vapor diffusion. Various components of the lead conditions are optimized one at a time by varying precipitant concentration and (in approximate order): pH, salt concentration, precipitant (e.g., the molecular weight of the PEG or by substitution of ionic precipitants), protein concentration, additives, and buffer type. Additional techniques include seeding (usually streak seeding), detergent screening, and varying drop size or type of diffusion (sitting, hanging or under oil). Optimization experiments are visualized using a microscope.

If crystals are of sufficient size and do not diffract well, an effort is made to optimize the cryoprotectant solution.

In the absence of an in-house setup to mount loops or examine the freezing behavior in loops, the authors freeze in small, thin-wall Eppendorf tubes as a substitute. First, 100 μl of a cryoprotectant solution is prepared containing either a mixture of well solution and cryoprotectant reagent or a freshly prepared solution that mimics the well solution but also has cryoprotectant included.

Typically 15–30% solutions are tested in 5% increments: 5 µl of the solution is transferred to a 650 µl thin-walled Eppendorf (PCR tubes). This tube is plunged into liquid nitrogen and either examined under a microscope while still in liquid nitrogen or removed and examined under a microscope. The technique is used to determine the minimum percentage of each cryoprotectant needed for the crystallization conditions and has been very reliable.

Optimization of cryoprotectant solutions starts with testing two concentrations of the following cryoprotectants: glycerol, ethylene glycol, PEG 400, and MPD. Additional optimization procedures include incorporating the cryoprotectant into the crystallization conditions, sequential transfer through increasing cryoprotectant concentrations, and crystal dehydration by transfer into solutions (with or without multi-hour incubations) containing increased precipitant concentrations.

3.4.3. Phasing

Since the standard fermentation media produces selenomethionine incorporated protein, experiments specific for obtaining phases are not usually needed. However, for proteins that have no methionines or the few methionines are predicted to be in flexible regions, additional manipulations are sometimes necessary to assist structure determination. The techniques currently in use are growing larger crystals (>200 µm per side) for sulfur anomalous phasing or heavy atom soaking. As a last resort, additional methionines may be added by substituting Leu, Val, or Ile residues with Met. This may be done in-house using QuikChange or by total gene synthesis (usually performed as a preventative measure).

3.4.4. Crystallization and LIMS

One aspect of the authors' process that differs significantly from typical crystallization labs is that their LIMS system tracks all aspects of crystallization, including every protein prep used, every tray set up, every crystallization condition (and well score), every crystal harvested and screened (including screening diffraction images), and every dataset collected. This coverage allows very powerful database queries to help decide where effort should be applied to maximize productivity and prevents well-behaving targets from falling through the cracks, a serious problem for a large project such as the Protein Structure Initiative. It also gives a much more complete view of the crystallization process since instead of only having information about the one protein preparation, crystallization condition, cryoprotectant, and dataset that lead to the final structure, the authors have a general view of the types of proteins, conditions, and cryoprotectants that work. In some cases these are very narrow parameters, but more often there is a broad range of conditions and reagents that work. It is difficult to represent this detailed information in public databases such as the NIH-funded PepcDB, but the authors plan to make much of our detailed crystallization results publicly available in the near future.

NYSGXRC also takes advantage of SGX-CAT at the Advanced Photon Source (APS) in Chicago, a dedicated mail-in protein crystallography beamline owned and operated by SGX Pharmaceuticals. Twenty-five percent of beamtime is reserved for general users, which include any academic groups doing structural genomics. Crystals are sent by FedEx, users can view screening images and define collection parameters, and datasets are returned on a hard drive.

4. Notes

1. Gene synthesis. Over the approximately 15 months ending December 1, 2006, expressing clones for over 1,500 distinct targets were generated, and 40% of these were from synthetic genes codon-optimized for expression in *E. coli*. Over five dozen structures have been determined that were derived from synthetic clones. Extensive use of gene synthesis allows the NYSGXRC to include targets for which genomic DNA and/or cDNA are not available. In some cases, the organism is unknown because the target sequence was obtained from environmental (or metagenomic) sequencing projects. Total gene synthesis also provides the ability to do protein engineering entirely at the bioinformatics stage and delete lengthy disordered regions, make inactivating (or activating) mutations, and add extra methionines to facilitate crystallographic phasing using selenomethionine-labeled protein. If PCR fails for multiple members of a target family, the authors often have the gene synthesized instead. Synthetic genes were supplied by a commercial vendor, Codon Devices (codondevices.com, Cambridge, MA).

2. Metal cocktails for expressing metal-dependent enzymes. Small-scale expression/solubility testing of amidohydrolase/enolase (AH/EN) family members was conducted with a cocktail of divalent cations (100 mM each of $ZnSO_4$, $MnCl_2$, $CaCl_2$, and $NiCl_2$); 1.2-ml LB cultures were inoculated with a single colony and cultures grown for approximately 4 hours at 37 °C. The cultures were then used to inoculate 200 µl of ZYP autoinduction media. The LB cultures were induced with IPTG and the temperature reduced to 22 °C for protein expression. After 4 hours, the temperature of the ZYP cultures was reduced to 22 °C for autoinduction and expression. After 22 hours, cultures were pelleted by centrifugation, resuspended in lysis buffer, and lysed by sonication. Soluble His tagged protein was extracted on nickel beads and analyzed by SDS-PAGE. For 27/45 (60%) of cases, the cation cocktail improved the yield of soluble AH/EN protein, whereas in eight cases (17%) the yield was reduced. For expressing clones with extremely low (or no) detectable soluble protein at small scale, performing the large-scale fermentation with the metal cocktail did not produce enough protein to allow purification of quantities suitable for characterization or crystallization (in 18/18 cases).

3. Optimizing SeMet labeling. High Yield SeMet media kit (product # MD0450004) was obtained from Medicilon (Chicago, IL). The authors' initial protocol using 60 mg/L SeMet with the 2× high-yield media (HY) resulted in a number of proteins with incomplete SeMet incorporation. Adding 50% more SeMet was sufficient for most proteins, although the authors found that proteins with many methionines required even more SeMet. An example is given in Fig. 37.2, in which a 280-residue protein (NP_891225; Q7WEE2_BORBR) with 12 methionines was expressed in the presence of 60, 90, and 120 mg/L selenomethionine. The left panel shows that less than half the total protein had full SeMet incorporation at all 12 methionine sites and there was a significant population of species with only two, one, or zero labeled methionines. A 50% increase in SeMet prior to induction resulted in most of the protein being fully incorporated, although there was still a small population of species with very little labeling. Doubling the amount of SeMet (to 120 mg/L) resulted in almost complete incorporation, with

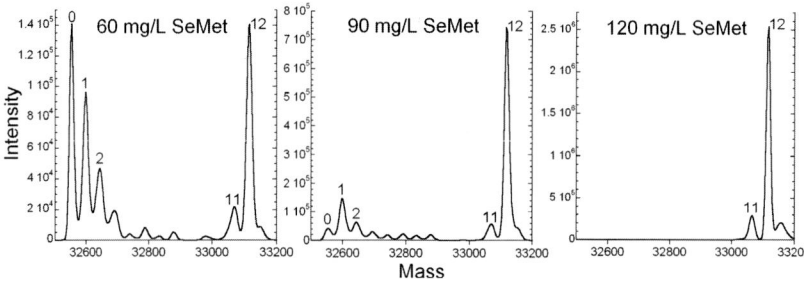

Fig. 37.2 Electrospray ionization mass spectra of purified protein containing 4.3% methionines grown in the presence of 60, 90, or 120 mg/L selenomethionine (SeMet).

only a small population of molecules lacking SeMet at one of the sites. The SGX LIMS has been modified to automatically calculate the number of methionines in each clone. If there are no methionines in the construct, a fermentation order will default to a non-SeMet media and prevent users from ordering a fermentation using HY media. If there are up to nine methionines, it defaults to using 90 mg/L SeMet, and if there are 10 or more methionines in the protein, it autoselects 120 mg/L SeMet. This minimizes manual intervention, reduces errors, and simplifies the work of the protein characterization and crystallography groups because it minimizes incomplete SeMet incorporation. The most unusual protein the authors have purified, in terms of methionine content, was a 172-residue fragment predicted to be a 4-helix bundle (UniProt: Q5YMI4_NOCFA; Genbank: YP_121971) that contained 19 methionines (11%). There was only a small population of protein molecules with full selenomethionine incorporated when growth was performed with 60 mg/L SeMet; increasing the amount of SeMet to 90 mg/L shifted the population so that the majority of the protein demonstrated full incorporation.

4. SUMO fusion. Proteins expressed in the His_6-Smt3-fusion vector must be cleaved using the Ulp1 enzyme (7). After identification by SDS PAGE, fractions containing the Smt3 fusion protein are pooled. The Smt3 tag is cleaved by overnight incubation with Ulp1 enzyme in dialysis against the wash buffer at 4 °C. Dialysis is necessary to remove imidazole, which would inhibit subsequent removal of Smt3 from the preparation. Cleavage is confirmed by SDS PAGE. Cleaved Smt3 and uncleaved fusion protein are then removed by a second incubation with Ni^{2+} agarose, which is then poured into a glass column and washed; protein which flows through the column includes the cleaved target protein, free of Smt3 (Fig. 37.3).

5. Pressure-induced disaggregation. SGX evaluated a commercially available device that uses high-pressure to disaggregate or refold proteins. Although the technique may work for a fraction of proteins after extensive optimization of the buffer and other experimental conditions, the authors found that the technique was not amenable to high throughput processing of a large number of insoluble or marginally soluble proteins from inclusion bodies. As a practical matter, use of high-pressure appeared to generate smaller aggregates, but aggregates nonetheless that are not considered optimal candidates for subsequent crystallization.

6. Buffer optimization. Solubility at lysis was identified at the stage in purification where most phosphatases failed. Literature analysis and in-house experience were used to construct a 96-condition fractional multifactorial screen (Table 37.1) to examine the effects of the listed conditions and additives on solubility. A 200-ml culture of the chosen phosphatase was distributed in 2-ml aliquots in a 96-well plate. Cells were pelleted by centrifugation, resuspended in the test buffers, and lysed by sonication. Debris was removed by centrifugation, and the supernatants decanted to a clean plate. His tagged protein was isolated on Nickel beads (Qiagen). After washing, protein eluted with 250 mM imidazole was analyzed by SDS PAGE (see Fig. 37.3). An optimum buffer was identified and confirmed by secondary screening around these conditions.

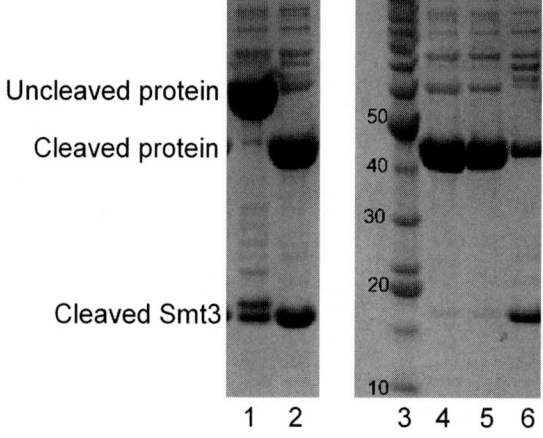

Fig. 37.3 Affinity purification of Smt3 tagged protein. Crude lysate was incubated with Ni^{2+} agarose (Qiagen) for 30 minutes. Agarose was washed with wash buffer (50 mM Tris-HCl, pH 7.8, 500 mM NaCl, 10 mM imidazole, 10% glycerol, 10 mM methionine) and His tagged protein eluted with wash buffer containing 500 mM imidazole (lane 1). The Smt3 tag was then removed by overnight incubation with Ulp1 enzyme in dialysis against CB2A (lane 2). Cleaved Smt3 and uncleaved protein were removed by a second incubation with Ni^{2+} agarose (lane 3: markers; lane 4: flow-through from second Ni^{2+} affinity, containing cleaved protein; lane 5: elution from second affinity, containing cleaved Smt3).

Table 37.1 Components of buffer screen for protein phosphatases

Buffer	Salt	Detergent	Additives	Reductant	pH
50 mM Tris,	150 mM NaCl	0.2% BOG*	10 mM MgCl$_2$	βMe	6.5
50 mM NaH$_2$PO$_4$	250 mM NaCl	0.1% CHAPS	200 mM arginine		7.5
	500 mM NaCl	0.1% NP40	1 mM EDTA		8.5
	250 mM KCl		200 µM vanadate		
			50 mM NaF		
			10 mM methionine		
			10% ethylene glycol		
			1 M NDSB 256		

*βMe, beta mercaptoethanol; BOG, βoctyl-glucoside; NDSB, non-detergent sulfobetaine; NP40, Nonidet P40.

Acknowledgments

This work is supported by the Protein Structure Initiative of the National Institute for General Medical Sciences under grant U54-GM74945. The authors thank the members of NYSGXRC for their continued efforts: the SGX informatics, protein production and crystallization teams, the SGX-CAT beamline staff at the APS, the laboratories of Steve Almo at Albert Einstein College of Medicine, S. Swaminathan at Brookhaven National Laboratory, Andrej Sali at University of California San Francisco, and Mark Chance at Case Western Reserve University.

References

1. Bonanno, J. B., Almo, S. C., Bresnick, A., Chance, M. R., Fiser, A., Swaminathan, S., Jiang, J., Studier, F. W., Shapiro, L., Lima, C. D., Gaasterland, T. M., Sali, A., Bain, K., Feil, I., Gao, X., Lorimer, D., Ramos, A., Sauder, J. M., Wasserman, S. R., Emtage, S., D'Amico, K. L., and Burley, S. K. (2005) New York-Structural GenomiX Research Consortium (NYSGXRC): a large scale center for the protein structure initiative. *J. Struct. Funct. Genom.* **6,** 225–232.

2. Gerlt, J. A., Babbitt, P. C., and Rayment, I. (2005) Divergent evolution in the enolase superfamily: the interplay of mechanism and specificity. *Arch. Biochem. Biophys.* **433,** 59–70.

3. Seibert, C. M., and Raushel, F. M. (2005) Structural and catalytic diversity within the amidohydrolase superfamily. *Biochemistry* **44,** 6383–6391.

4. Studier, F. W., and Moffatt, B. A. (1986) Use of bacteriophage T7 RNA polymerase to direct selective high-level expression of cloned genes. *J. Mol. Biol.* **189,** 113–130.

5. Studier, F. W., Rosenberg, A. H., Dunn, J. J., and Dubendorff, J. W. (1990) Use of T7 RNA polymerase to direct expression of cloned genes. *Methods Enzymol.* **185,** 60–89.

6. Studier, F. W. (2005) Protein production by auto-induction in high density shaking cultures. *Protein Expr. Purif.* **41,** 207–234.

7. Mossessova, E., and Lima, C. D. (2000) Ulp1-SUMO crystal structure and genetic analysis reveal conserved interactions and a regulatory element essential for cell growth in yeast. *Mol. Cell* **5,** 865–876.

Chapter 38

Overview of the Pipeline for Structural and Functional Characterization of Macrophage Proteins at the University of Queensland

Weining Meng, Jade K. Forwood, Gregor Guncar, Gautier Robin, Nathan P. Cowieson, Pawel Listwan, Dmitri Mouradov, Gordon King, Ian L. Ross, Jodie Robinson, Munish Puri, Justine M. Hill, Stuart Kellie, Thomas Huber, David A. Hume, Jennifer L. Martin, and Bostjan Kobe

This chapter describes the methodology adopted in a project aimed at structural and functional characterization of proteins that potentially play an important role in mammalian macrophages. The methodology that underpins this project is applicable to both small research groups and larger structural genomics consortia. Gene products with putative roles in macrophage function are identified using gene expression information obtained via DNA microarray technology. Specific targets for structural and functional characterization are then selected based on a set of criteria aimed at maximizing insight into function. The target proteins are cloned using a modification of Gateway® cloning technology, expressed with hexa-histidine tags in *E. coli*, and purified to homogeneity using a combination of affinity and size exclusion chromatography. Purified proteins are finally subjected to crystallization trials and/or NMR-based screening to identify candidates for structure determination. Where crystallography and NMR approaches are unsuccessful, chemical cross-linking is employed to obtain structural information. This resulting structural information is used to guide cell biology experiments to further investigate the cellular and molecular function of the targets in macrophage biology. Jointly, the data sheds light on the molecular and cellular functions of macrophage proteins.

1. Introduction

Structural genomics initiatives aim to provide a comprehensive view of the protein structure universe by determining structures of representative proteins from every protein family *(1)*. Achieving such a goal requires large,

From: *Methods in Molecular Biology, Vol. 426: Structural Proteomics: High-throughput Methods*
Edited by: B. Kobe, M. Guss and T. Huber © Humana Press, Totowa, NJ

coordinated research teams and substantial funding *(2)*. However, the parallel processing and high throughput approaches adopted by structural genomics initiatives can also be applied to projects of smaller scale, and promise faster and more cost-effective results. Furthermore, a smaller team can identify a niche in the worldwide structural genomics initiative through careful protein target selection. The authors have applied these ideas to a project aimed at the structural characterization of proteins that play important roles in macrophage function.

Macrophages are cells representing the first line of defense against pathogens and play crucial roles in both innate and acquired immunity. They comprise 15–20% of cells in most organs, and are particularly abundant at the routes of pathogen entry such as lung, skin, gut, and genitourinary tract *(3)*. Macrophages detect pathogens by receptors that recognize generic nonmammalian structures, including cell wall components (e.g., lipopolysaccharide, LPS; peptidoglycans; lipoteichoic acids) and microbial DNA (e.g., unmethylated CpG motifs) *(4)*. Upon recognition, the macrophage engulfs and destroys the foreign organism, at the same time activating a spectrum of genes, creating a hostile extracellular environment in the host. Additional cells are recruited to the site of invasion and an appropriate acquired immune response is primed dependent on the class of pathogen. However, some pathogens have been able to evade these defenses, and in some cases, such as *Mycobacterium tuberculosis*, take advantage of the macrophage as a portal of infection and replicate or survive within this cell type. This can lead to life-threatening conditions such as disseminated intravascular coagulation, hypotension, and pathological fever *(5)*. In chronic local infections, or response to inflammation caused by noninfectious agents that activate macrophages but cannot be cleared, macrophage products cause local tissue destruction; the wasting disease known as cachexia *(6)*.

A detailed knowledge of the regulation of macrophage function forms the basis for the development of two classes of therapeutics. On one hand, it may be desirable to amplify the toxic function of macrophages to destroy microorganisms or tumor cells more effectively. Alternatively, selective suppression of components of the macrophage activation response offers approaches to treatment of acute conditions such as septicemia and toxic shock, and chronic conditions such as arthritis, atherosclerosis, and obstructive lung disease *(7)*.

To understand better the process of macrophage activation, a program was undertaken to structurally and functionally characterize novel proteins involved in macrophage activation *(8)*. This chapter presents the methodology of the authors' pipeline (Fig. 38.1), focusing on target selection, cloning, expression, purification, and structural characterization of proteins involved in macrophage activation. The pipeline is applicable to both small research groups within academia and larger consortia.

2. Methods

2.1. Target Selection

Proteins with likely roles in macrophage function are identified based on expression profiling using DNA microarrays. The selected proteins are: (1) expressed selectively in mouse macrophages, and/or (2) transcriptionally

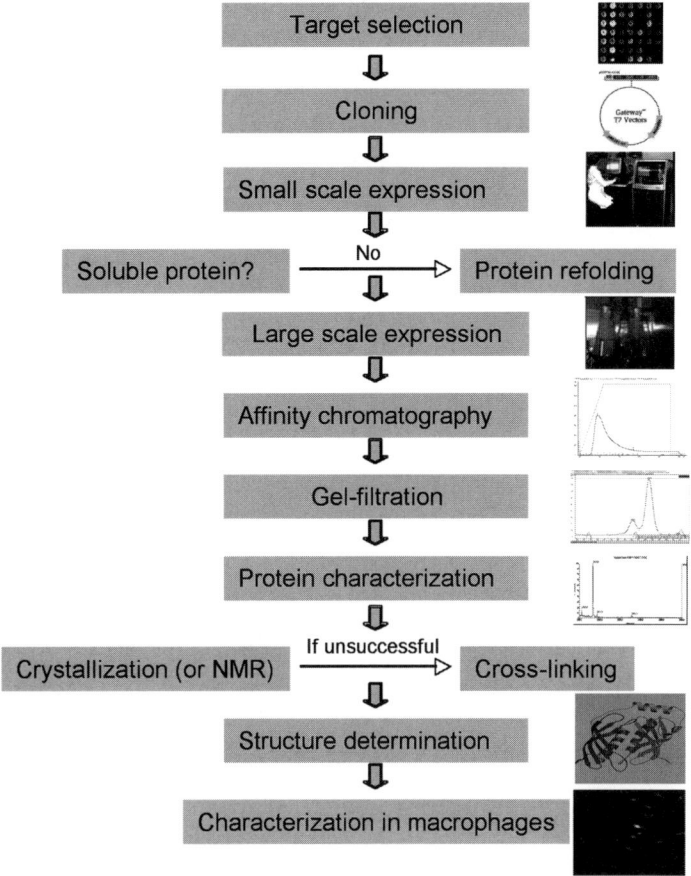

Fig. 38.1 A schematic diagram describing the methodologies applied in the pipeline.

regulated following stimulation of mouse macrophages *in vitro*, and/or (*3*) transcriptionally regulated in mouse models of arthritis or chronic obstructive pulmonary disease. Only proteins with human homologues are chosen for further study to ensure medical relevance. To maximize the value of the determined structures, only the proteins with less than 30% amino acid sequence identity to proteins with known three-dimensional structures (Protein Data Bank; PDB) (*9*) are selected. Protein fold recognition servers such as PHYRE (*10*) and FUGUE (*11*) are used to further examine the predicted structures. Target proteins are prioritized based on their expected suitability for structural studies, using a number of properties inferred from the sequence and functional annotation (e.g., the presence of putative transmembrane regions predicted by TMHMM, sequence length, isoelectric point, percentage of charged residues, and hydropathy index) (*12*). An examination of the relevant literature is carried out on top-ranking proteins to produce the final target list. The details of the target selection procedure, the associated Web tools, and customization options are presented elsewhere in this volume.

2.2. Cloning

For all high throughput steps involving liquid handling, the authors use the Biomek 2000 automated liquid handling workstation and 96-well plates. The Triple Master PCR system (Eppendorf) is employed for all PCR amplifications. One shot TOP10 chemically competent *E. coli* cells (Invitrogen, Carlsbad, CA) are used for all transformation reactions. The primer sequences are generated automatically using an in-house Perl script that takes target nucleotide sequences as an input, generating primer sequences that start at the termini, contain additional bases to reach an annealing temperature of 55 °C, and end in a cytosine or guanine.

The target genes are amplified by PCR using a macrophage cDNA pool or cDNAs from FANTOM2 clonesets *(13)*. The PCR products are purified from primers and other buffer components using the Montage 96-well PCR purification kit (Millipore, Billerica, MA) set up on the Biomek 2000 workstation.

The PCR products are cloned into expression vectors using a modification of the Gateway recombinatorial cloning methodology that allows expression of the recombinant proteins with a short hexa-histidine tag *(14,15)*. The purified PCR products are cloned into the Gateway entry vector pDONR-221 (Invitrogen) using a two-step PCR followed by a recombinatorial LR reaction *(15)*. A gene-specific primer containing a linker of 12 nucleotides is used in the first PCR step, and a BP-reaction "universal adapter primer" *(15)* containing hexa-histidine tag is employed in the second PCR step. The reaction enzyme BP Clonase facilitates the recombination between a specific sequence (attB) included in the product of second PCR and the attP sequence of the donor vector, pDONR-221. Following the BP reaction, One Shot TOP10 chemically competent *E. coli* cells (Invitrogen) are used to transform the reaction mixture. The colonies are grown overnight in LB media containing kanamycin and the plasmids are then purified using the Plasmid Miniprep96 Montage kit (Millipore). The constructs are analyzed by restriction enzyme digestion with BsrG1. The genes contained within the pDONR vector are transferred into the pDEST14 expression vector using the LR reaction. The expression vectors are assayed for correct insertion of the gene by digestion with the restriction enzyme BsrG1, followed by electrophoresis. The positive clones are additionally confirmed by DNA sequencing (Australian Genome Research Facility, Brisbane, QLD, Australia).

2.3. Expression and Purification

As reported by the worldwide SG Centers TargetDB Statistics Report (October 2, 2006), only about 25% (30% for prokaryotes and viruses, 15% for eukaryotes) of clones can be successfully expressed and purified. The main hurdle is protein solubility in the non-native host. The authors use conventional expression and purification procedures for the soluble targets, whereas protein refolding (see Section 2.4.) is used as a salvage pathway for the insoluble proteins. An estimation of the protein expression and solubility in *E. coli* is achieved rapidly in 1-ml cultures followed by purification in 96-well format. The analysis with an automated electrophoresis instrument (Caliper 96 Bioanalyzer) enables prompt quantification of the yields and the subsequent choice of targets suitable for large-scale purification or refolding. The expression and purification protocol is outlined below followed by the protein refolding procedure in Section 2.4.

2.3.1. Small-Scale Expression and Purification

Expression vectors are transformed into chemically competent *E. coli*. BL21(DE3)pLysS cells. The proteins are expressed in 1-ml autoinduction media *(16)* in 96-deep well plates at 30 °C and 18 °C. Growth is monitored at 600 nm to determine the optimum incubation time (~20 hours at 30 °C, ~70 hours at 18 °C). The purification is automated on a Biomek 2000 using the Automated MagneHis™ Protein Purification System Protocol #EP011 (Promega, Madison, WI). The cell pellets are resuspended, then lysed using detergent lysis buffer (FastBreak™ Cell Lysis Reagent, Promega), and MagneHis™ Ni-Particles (Promega) are employed to initiate binding of the His-tagged recombinant target protein. A MagneBot 96 Magnetic Device (Promega) is used to allow the Ni-particles to be captured by the magnet. Finally, the proteins are eluted from the resin in 100 μl of 25 mM Hepes (pH 7.4), 150 mM NaCl, 250 mM imidazole. The protein samples are analyzed on Caliper 96 Bioanalyzer to assess the size, purity, and yield. Targets with accurate size are ranked according to the yield and selected for large-scale expression and purification.

2.3.2. Large-Scale Expression and Purification

Target proteins selected from the small-scale expression trials are produced in large scale (2–4 liters of autoinduction media). The cultures are grown in the conditions that yielded the highest amount of purified protein in small-scale expression. Following affinity chromatography an additional step of size exclusion chromatography (SEC) is used to purify the proteins to homogeneity. All purification steps are performed at 4 °C. A TALON cobalt affinity resin (Scientifix, Clayton, VIC, Australia) or the nickel-based affinity column, HisTrap™FF (GE Healthcare, Piscataway, NJ) is employed in the affinity step (the wash and elution buffers are 100 mM Hepes (pH 7.4) and 150 mM NaCl, containing 20 mM or 300 mM imidazole, respectively), and the proteins are then loaded onto the SEC column S200 HiLoad™ 16/60 Superdex™ (GE Healthcare). The protein collected from SEC in 100 mM Hepes (pH 7.4), 150 mM NaCl is concentrated using Amicon Ultra Centrifugal Filter Devices (Millipore) typically to approximately 20 mg/ml for protein characterization and crystallization.

2.4. Protein Refolding

Insoluble protein expression may be the biggest bottleneck limiting structure genomics initiatives *(17)*. Although 20–60% of proteins expressed in *E. coli* result in insoluble inclusion bodies *(18–20)*, many of these proteins may be amenable to refolding. The authors have developed a matrix-assisted refolding approach, in which correctly folded proteins are distinguished from misfolded proteins by their elution from affinity resin *(21)*. Proteins that are subjected to refolding while bound to metal affinity resin are often resistant to elution by imidazole. The authors hypothesized that misfolded proteins formed hydrophobic interactions with the surface of the resin. This difference in binding properties between folded and misfolded proteins is the basis for separating the two in this assay. Briefly, a chaotrope is used to solubilize inclusion bodies from bacterial fermentation and His-tagged protein is bound to metal affinity chromatography resin. The chaotrope is removed by washing the resin in a renaturing buffer and correctly folded protein is subsequently eluted using imidazole. SDS-PAGE is used to compare the quantity of protein in the soluble fraction with that remaining on the resin. This represents the measure of refolding efficiency. The assay is amenable to automation on a liquid handling workstation. The details of this procedure are presented elsewhere in this volume.

2.5. Protein Characterization

SDS-PAGE, size exclusion chromatography, mass spectrometry, and circular dichroism (CD) spectroscopy are used to characterize the proteins after purification. Samples of each step of large-scale expression and purification are analyzed by SDS-PAGE to provide a qualitative estimate of purity and reveal any proteolytic degradation or large disparities in protein size. Calibrated size exclusion chromatography gives an estimate of the oligomerization state and the presence of aggregation. Analysis by mass spectrometry on a Voyager DE STR MALDI-TOF or Applied Biosystems QSTAR Pulsar mass spectrometers is used to examine the exact molecular weights of the purified proteins. Finally, proteins are analyzed by CD spectroscopy. The authors have used both a conventional CD instrument (J-810, Jasco, Easton, MD) and synchrotron radiation CD (SRCD; Daresbury Synchrotron Laboratory, Daresbury, UK). SRCD is particularly suitable as a high throughput method, because there are fewer limitations on buffer components and data collection is faster (22). CD spectra yield information on protein secondary structure, and are therefore particularly useful for identifying proteins with large proportions of unstructured regions that are unlikely to yield useful high-resolution structural information.

2.6. NMR Spectroscopy

For a subset of proteins with a molecular weight less than 20 kDa, NMR spectroscopy is also used to assess their suitability for structure determination. Uniformly ^{15}N-labeled proteins are overexpressed in *E. coli* BL21(DE3) cells using modified autoinduction media containing ^{15}NH$_4$Cl (2.5 g/L) as the sole nitrogen source. Proteins are then purified using the strategy described in the preceding. Samples for NMR screening contain approximately 0.3 mM protein in 50 mM sodium phosphate buffer (pH 7 or pH 4), 150 mM NaCl, and 1 mM DTT in H$_2$O/D$_2$O (9:1). 1D ^{1}H and 2D ^{1}H-^{15}N HSQC NMR spectra are acquired at 25 °C on a Bruker Avance 600 MHz spectrometer equipped with a z-shielded gradient triple resonance probe, and analyzed using NMRPipe/NMRDraw (23). NMR spectral quality and feasibility of three-dimensional structure determination is assessed based on spectral dispersion, line widths, and number of resolved peaks observed compared to the number expected from the amino acid sequence.

2.7. Crystallography

2.7.1. Protein Concentration Optimization

The optimal protein concentration for crystallization screens is determined by setting up a hanging drop vapor diffusion experiment under two conditions: 2.0 M ammonium sulfate, 0.1 M Tris-HCl (pH 8.5) (Hampton Crystal Screen condition 4) and 30% PEG 4000, 0.1 M Tris-HCl (pH 8.5), 0.2 M magnesium chloride (Hampton Crystal Screen condition 6) at different protein concentrations.

The most suitable protein concentration is evaluated after 12 hours incubation at 18 °C as the one yielding light precipitation in at least one of these conditions.

2.7.2. Crystallization

The vapor diffusion technique (hanging and sitting drop experiments) is employed to screen the proteins using commercial (Hampton Research, Emerald Biostructures, Jena Bioscience, Molecular Dimensions) and in-house screens (24,25). The Biomek 2000 robot (Beckman Coulter, Mississauga, ON) is used to prepare the reservoir solutions in the trays, which are stored at 4 °C. Protein crystallization droplets are set up in 96-well plates with a Mosquito nanolitre-dispensing robot (TTP Labtech) and placed at two temperatures (18 and 4 °C). The experiments are monitored using a DeCode Genetics Crystal Monitor. Protein crystals identified are then optimized by setting up focused grid screens (26).

2.7.3. Structure Determination

The x-ray diffraction quality of protein crystals is assessed using an in-house Rigaku FR-E rotating anode generator (Rigaku/MSC, The Woodlands, TX) with a RaxisIV++ image plate detector. Crystals of native protein are flash-cooled in a nitrogen gas stream at approximately 100 K after soaking in a suitable cryoprotectant.

When diffraction quality crystals are obtained and a suitable molecular replacement model is not available, the protein is expressed in minimal media in the presence of SeMet to produce SeMet-labeled protein crystals for use in MAD phasing (27) at a synchrotron. Standard crystallographic packages such as HKL2000 (28), CrystalClear (Rigaku), SOLVE (29), Arp/Warp (30), the CCP4 package (31), and Coot (32) are used to process the data, obtain phase information, build the model, and refine and visualize the structures.

2.8. Cross-linking

Chemical cross-linking is employed to obtain structural information on proteins for which structure determination by x-ray diffraction and NMR are not successful. The chemical cross-linkers BS^3 (bis(sulfosuccinimidyl) suberate; Sigma, S5799) and DTSSP (dithiobis(sulfosuccinimidylpropionate), 21578; Pierce, Rockford, IL) are used to obtain distance constraints. This technique allows high throughput low-resolution structure determination, particularly when some prior structure information is available (e.g., multidomain proteins with known domain structures). The protein of interest is allowed to react with the cross-linker, and after quenching the resulting cross-linked and non-crosslinked controls are separated by SDS-PAGE. The band of interest is excised and used for in-gel digestion with trypsin. The masses of the resulting peptides are analyzed by mass spectrometry with the help of in-house software to identify cross-linked products. Based on the identified distance constraints, models are built using docking and modeling techniques. Details of this technique are given elsewhere in this volume.

2.9. Functional Characterization

The authors' structural studies are complemented by cell biology experiments to further characterize protein function. The DNA encoding the protein is subcloned into a mammalian expression vector such as pDEST21 and transfected into the

RAW264.7 murine macrophage cell line. Two types of experiments are typically carried out in the transfected cells in the first instance: subcellular localization and effect of overexpression on macrophage function. Localization experiments are carried out by a combination of confocal immunofluorescence microscopy and subcellular fractionation, followed by Western blotting. The transfected protein is detected using a V5 tag expressed at the C-terminus. The functional consequences of overexpression of the protein in macrophages are assessed by monitoring the proliferative and cytokine responses to a range of stimuli such as LPS and CSF-1 (colony stimulating factor-1), using real-time PCR and ELISA assays to detect inflammatory genes and proteins, respectively. These experiments provide information on the cellular functions of the proteins, complementing structural data that usually shed light on the molecular function, as well as providing functional data for proteins that fail to yield structural data.

3. Progress of the Pipeline

3.1. Statistics

So far, the authors have processed 318 macrophage proteins, of which 220 have been successfully cloned and 52 expressed in a soluble form in *E. coli*, and entered crystallization trials. Examples of structures resulting from this pipeline include latexin (1.8 Å resolution) *(33)* and long chain acyl-CoA thioesterase (2.4 Å resolution) (Fig. 38.2) *(34)*.

3.2. Functions Revealed

Latexin is the only known mammalian carboxypeptidase inhibitor. The authors have shown that latexin is expressed constitutively at high basal levels in mouse macrophages and can be further upregulated by stimulation of the cells with growth factor or proinflammatory stimuli *(33)*. The crystal structure of latexin (see Fig. 38.2) unexpectedly revealed structural similarities with the

(A)   (B)

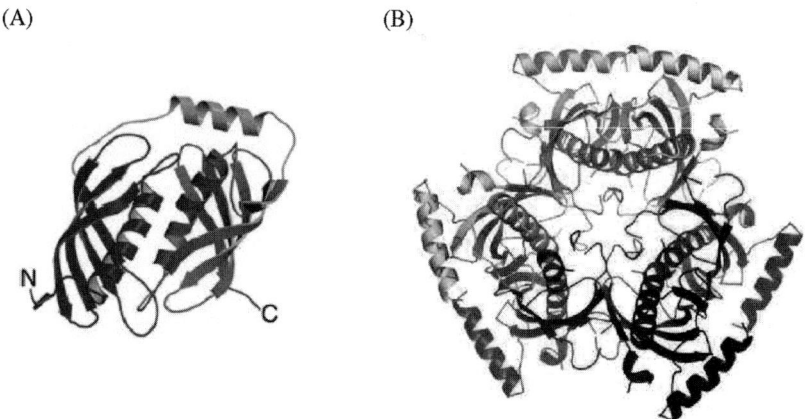

Fig. 38.2 Structures of (**A**) latexin *(33)* and (**B**) ACOT7.

cysteine protease inhibitor cystatin *(33)*. Together, the data suggest a role for latexin in the regulation of proteolysis during inflammation.

Acyl-CoA thioesterases (ACOTs) are a family of enzymes that are conserved through evolution from bacteria to mammals. These proteins catalyze the hydrolysis of acyl-CoA moieties to the respective fatty acid constituents and coenzyme A *(35)*. Long-chain acyl-CoAs are intermediates in lipid metabolism and regulators of cellular processes including ion transport, vesicle trafficking, protein phosphorylation, and gene expression *(35–37)*. Mouse ACOT7 contains two thioesterase domains in tandem. The authors have determined the crystal structures of each domain separately, and modeled the full-length protein using distance constraints based on chemical cross-linking (see Fig. 38.2). The structure explains the requirement of the two domains for enzymatic activity and the structural basis for long-chain acyl-CoA specificity *(34)*.

Acknowledgments

This work was supported by Australian Research Council (ARC; to JLM and BK). BK is an ARC Federation Fellow and a National Health and Medical Research Council (NHMRC) Honorary Research Fellow. JMH is the recipient of an RD Wright Biomedical Career Development Award from the NHMRC. MP thanks DEST for the Australia-Asia Fellowship.

References

1. Burley, S. K. (2000) An overview of structural genomics. *Nat. Struct. Biol.* **7,** 932–934.
2. Chandonia, J. M., and Brenner, S. E. (2006) The impact of structural genomics: expectations and outcomes. *Science* **311,** 347–351.
3. Gordon, S., Crocker, P. R., Morris, L., Lee, S. H., Perry, V. H., and Hume, D. A. (1986) Localization and function of tissue macrophages. *Ciba Found. Symp.* **118,** 54–67.
4. Hoffmann, J. A., Kafatos, F. C., Janeway, C. A., and Ezekowitz, R. A. (1999) Phylogenetic perspectives in innate immunity. *Science* **284,** 1313–1318.
5. Mammen, E. F. (2000) Disseminated intravascular coagulation (DIC). *Clin. Lab. Sci.* **13,** 239–245.
6. Morley, J. E., Thomas, D. R., and Wilson, M. M. (2006) Cachexia: pathophysiology and clinical relevance. *Am. J. Clin. Nutr.* **83,** 735–743.
7. Duffield, J. S. (2003) The inflammatory macrophage: a story of Jekyll and Hyde. *Clin. Sci. (Lond.)* **104,** 27–38.
8. Puri, M., Robin, G., Cowieson, N., Forwood, J. K., Listwan, P., Hu, S. H., Guncar, G., Huber, T., Kellie, S., Hume, D. A., Kobe, B., and Martin, J. L. (2006) Focusing in on structural genomics: the University of Queensland structural biology pipeline. *Biomol. Eng.* **23,** 281–289.
9. Berman, H. M., Westbrook, J., Feng, Z., Gilliland, G., Bhat, T. N., Weissig, H., Shindyalov, I. N., and Bourne, P. E. (2000) The Protein Data Bank. *Nucleic Acids Res.* **28,** 235–242.
10. Fleming, K., Kelley, L. A., Islam, S. A., MacCallum, R. M., Muller, A., Pazos, F., and Sternberg, M. J. (2006) The proteome: structure, function and evolution. *Philos. Trans. R. Soc. Lond. B Biol. Sci.* **361,** 441–451.
11. Shi J, B. T., and Mizuguchi K. (2001) FUGUE: sequence-structure homology recognition using environment-specific substitution tables and structure- dependent gap penalties. *J. Mol. Biol.* **310,** 243–257.

12. Chen, Y., Yu, P., Luo, J., and Jiang, Y. (2003) Secreted protein prediction system combining CJ-SPHMM, TMHMM, and PSORT. *Mamm. Genome* **14,** 859–865.

13. Bono, H., Kasukawa, T., Furuno, M., Hayashizaki, Y., and Okazaki, Y. (2002) FANTOM DB: database of Functional Annotation of RIKEN Mouse cDNA Clones. *Nucleic Acids Res.* **30,** 116–118.

14. Yokoyama, S. (2003) Protein expression systems for structural genomics and proteomics. *Curr. Opin. Chem. Biol.* **7,** 39–43.

15. Listwan, P., Cowieson, N., Kurz, M., Hume, D. A., Martin, J. L., and Kobe, B. (2005) Modification of recombinatorial cloning for small affinity tag fusion protein construct generation. *Anal. Biochem* **346,** 327–329.

16. Studier, F. W. (2005) Protein production by auto-induction in high density shaking cultures. *Protein Expr. Purif.* **41,** 207–234.

17. Gilbert, M., and Albala, J. S. (2002) Accelerating code to function: sizing up the protein production line. *Curr. Opin. Chem. Biol.* **6,** 102–105.

18. Moy, S., Dieckman, L., Schiffer, M., Maltsev, N., Yu, G. X., and Collart, F. R. (2004) Genome-scale expression of proteins from *Bacillus subtilis. J. Struct. Funct. Genomics* **5,** 103–109.

19. Huang, R. Y., Boulton, S. J., Vidal, M., Almo, S. C., Bresnick, A. R., and Chance, M. R. (2003) High throughput expression, purification, and characterization of recombinant *Caenorhabditis elegans* proteins. *Biochem. Biophys. Res. Commun.* **307,** 928–934.

20. Christendat, D., Yee, A., Dharamsi, A., Kluger, Y., Gerstein, M., Arrowsmith, C. H., and Edwards, A. M. (2000) Structural proteomics: prospects for high throughput sample preparation. *Prog. Biophys. Mol. Biol.* **73,** 339–345.

21. Cowieson, N. P., Wensley, B., Listwan, P., Hume, D. A., Kobe, B., and Martin, J. L. (2006) An automatable screen for the rapid identification of proteins amenable to refolding. *Proteomics* **6,** 1750–1757.

22. Miles, A. J., and Wallace, B. A. (2006) Synchrotron radiation circular dichroism spectroscopy of proteins and applications in structural and functional genomics. *Chem. Soc. Rev.* **35,** 39–51.

23. Delaglio, F., Grzesiek, S., Vuister, G. W., Zhu, G., Pfeifer, J., and Bax, A. (1995) NMRPipe: a multidimensional spectral processing system based on UNIX pipes. *J. Biomol. NMR* **6,** 227–293.

24. Page, R., Grzechnik, S. K., Canaves, J. M., Spraggon, G., Kreusch, A., Kuhn, P., Stevens, R. C., and Lesley, S. A. (2003) Shotgun crystallization strategy for structural genomics: an optimized two-tiered crystallization screen against the *Thermotoga maritima* proteome. *Acta Crystallogr. D Biol. Crystallogr.* **59,** 1028–1037.

25. Majeed, S., Ofek, G., Belachew, A., Huang, C. C., Zhou, T., and Kwong, P. D. (2003) Enhancing protein crystallization through precipitant synergy. *Structure* **11,** 1061–1070.

26. Senger, A. B., and Mueser, T. C. (2005) A method for the rapid preparation of custom grid-screen crystallization trays using standardized pipetting maps is presented. *J. Appl. Cryst.* **38,** 847–850.

27. Hendrickson, W. (1999) Maturation of MAD phasing for the determination of macromolecular structures. *J. Synchrotron. Radiat.* **6,** 845–851.

28. Otwinowski, Z., and Minor, W. (1997) Processing of X-ray diffraction data collected in oscillation mode. *Methods Enzymol.* **276,** 307–326.

29. Terwilliger, T. C., and Berendzen, J. (1999) Automated MAD and MIR structure solution. *Acta Crystallogr. D Biol. Crystallogr.* **55,** 849–861.

30. Morris, R. J., Perrakis, A., and Lamzin, V. S. (2003) ARP/wARP and automatic interpretation of protein electron density maps. *Methods Enzymol.* **374,** 229–244.

31. CCP4 (1994) The CCP4 suite: programs for protein crystallography. *Acta Crystallogr. D Biol. Crystallogr.* **50,** 760–763.

32. Emsley, P., and Cowtan, K. (2004) Coot: model-building tools for molecular graphics. *Acta Crystallogr. D Biol. Crystallogr.* **60,** 2126–2132.

33. Aagaard, A., Listwan, P., Cowieson, N., Huber, T., Ravasi, T., Wells, C. A., Flanagan, J. U., Kellie, S., Hume, D. A., Kobe, B., and Martin, J. L. (2005) An inflammatory role for the mammalian carboxypeptidase inhibitor latexin: relationship to cystatins and the tumor suppressor TIG1. *Structure* **13,** 309–317.

34. Forwood, J. K., Thakur, A. S., Guncar, G., Marfori, M., Mouradov, D., Meng, W., Robinson, J., Huber, T., Kellie, S., Martin, J. L., Hume, D. A., and Kobe, B. (2007) Structural basis for recruitment of tandem hotdog domains in acyl-CoA thioesterase 7 and its role in inflammation. *Proc. Natl. Acad. Sci. U.S.A.* **104,** 10382–10387.

35. Hunt, M. C., and Alexson, S. E. (2002) The role Acyl-CoA thioesterases play in mediating intracellular lipid metabolism. *Prog. Lipid Res.* **41,** 99–130.

36. Faergeman, N. J., and Knudsen, J. (1997) Role of long-chain fatty acyl-CoA esters in the regulation of metabolism and in cell signalling. *Biochem. J.* **323,** 1–12.

37. Yamada, J. (2005) Long-chain acyl-CoA hydrolase in the brain. *Amino Acids* **28,** 273–278.

Chapter 39

Structural Genomics of the Bacterial Mobile Metagenome: an Overview

Andrew Robinson, Amy P. Guilfoyle, Visaahini Sureshan, Michael Howell, Stephen J. Harrop, Yan Boucher, Hatch W. Stokes, Paul M. G. Curmi, and Bridget C. Mabbutt

Mobile gene cassettes collectively carry a highly diverse pool of novel genes, ostensibly for purposes of microbial adaptation. At the sequence level, putative functions can only be assigned to a minority of carried ORFs due to their inherent novelty. Having established these mobilized genes code for folded and functional proteins, the authors have recently adopted the procedures of structural genomics to efficiently sample their structures, thereby scoping their functional range. This chapter outlines protocols used to produce cassette-associated genes as recombinant proteins in *Escherichia coli* and crystallization procedures based on the dual screen/pH optimization approach of the SECSG (SouthEast Collaboratory for Structural Genomics). Crystal structures solved to date have defined unique members of enzyme fold classes associated with transport and nucleotide metabolism.

1. Introduction

The acknowledged extent of lateral gene transfer (LGT) presents challenges to genomics studies of microorganisms: no one cell line can be considered to be representative of the gene content of a species as a whole. For many species, the proportion of genes acquired by LGT approaches 20%, and it is this LGT fraction that often contains a high degree of sequence novelty *(1,2)*. Thus, in contrast to a genome overall, up to 90% of mobile genes within a microorganism can be classified as unique or matching only to conserved hypothetical families. From an evolutionary perspective, these laterally transferred genes are an important component of the genome, since they are likely to encode proteins with functions adaptive to the cell. As a result, there is growing interest in analyzing this specific pool of genes, the bacterial mobile metagenome, in a more systematic manner.

One important contributor to the mobility of genes that lends itself to concerted analysis is the integron/gene cassette system *(3)*. The two components of this gene capture system are the integron, and its units of capture, mobile gene cassettes. The integron encodes a DNA integrase to carry out site-specific recombination between a specific site (*attI*, also part of the integron) and a second site associated with the mobile gene cassettes. Gene cassettes are the

From: *Methods in Molecular Biology, Vol. 426: Structural Proteomics: High-throughput Methods*
Edited by: B. Kobe, M. Guss and T. Huber © Humana Press, Totowa, NJ

smallest known mobilizable units of DNA, usually comprising only a single ORF and the recombination site (a 59-base element, or *attC*).

Integrons are a common feature of bacterial genomes, independently capturing many cassettes to generate tandem arrays containing hundreds of cassettes. By targeting the recombination sites within these tandem arrays, the recent technique of cassette PCR has allowed the authors to selectively recover gene cassettes from metagenomic DNA, mixed cultured populations, and defined bacterial cell lines *(4,5)*. Recovery of cassette-associated genes by this method has the added advantage of allowing isolation of complete genes in the absence of prior sequence information.

With the extensive application and development of cassette PCR, it has become clear that the mobile gene cassette pool is a particularly rich source of protein diversity and novelty *(5,6)*. The minority of recovered sequences for which a putative function can be assigned include DNA modification proteins and enzymes of secondary metabolism, implying that the mobile gene pool acts as a reservoir for adaptive functions. When prepared in recombinant form, the gene products of these novel sequence variants form fully folded proteins with demonstrable catalytic activity *(7)*. Consequently there is now great interest in applying structural genomics approaches to the discovery of both new protein structures and proteins with new substrate specificities from this genetic source. It is only through elucidation of three-dimensional structures and the subsequent identification of structural homologues that the functional diversity resident within the bacterial metagenome can be fully revealed.

2. Materials

Environmental ORFs were isolated by cassette PCR from soil, sediment, biomass, or water samples collected from a variety of locations in Australia and Antarctica *(4)*. The second genetic source comprises fosmid libraries created from integron-associated genes located within DNA of *Vibrio* strains isolated from mudcrab larvae in aquaculture *(5)*. The encoded protein targets range in size from 8 to 44 kDa and include membrane-associated domains.

3. Methods

3.1. Target Selection

1. Target protein sequences are screened for potential signal peptides using the program SignalP *(8)*. Where a signal peptide is indicated (~14% of metagenome sequences), this segment is removed from the target sequence prior to further analysis.
2. Target sequences are partitioned into priority groups according to the number of transmembrane regions predicted: Group 1 (none), Group 2 (single region), or Group 3 (multiple). The prediction of a transmembrane sequence is considered significant if recognized by at least two of the programs DAS *(9)*, TMHMM *(10)*, or HMMTOP *(11)*. Group 2 targets may be pursued in truncated form; Group 3 targets are removed from the pipeline.

3. Group 1 targets are further partitioned according to extent of disordered sequence segments. Disordered regions are identified where more than 10 residues show disorder propensity, as indicated by the program GlobPlot2 *(12)*. Targets with no disordered regions are allocated highest priority (Group 1a), and those for which a globular domain is predicted are also retained (Group 1b).

4. Sequence homologues for Group 1 targets are identified using BLAST searches *(13)* of the nonredundant, environmental nonredundant, Swissprot and PDB databases. The targets of this pipeline are generally of relatively novel sequence (to date, ~15% match proteins with a determined function), and all proceed downstream.

3.2. Cloning

E. coli expression is most appropriate for the production of these microbially derived genes. ORFs are cloned into a conventional *E. coli* expression vector for production as recombinant proteins, usually chosen so as to engineer a cleavable N-terminal His_6-tag for affinity purification. A direct fusion PCR method (In-Fusion product range; Clontech, Mountain View, CA) amenable to high-throughput cloning has recently been adopted, in which single-stranded regions of homology are created between the PCR product and the linearized vector.

1. PCR is conducted on each gene of interest using a high-fidelity DNA polymerase and primers that introduce 15 bp extensions homologous to the ends of the chosen vector. Meanwhile, the expression vector is linearized by restriction enzyme digestion (two enzymes) and gel-purified.

2. Purified PCR product and digested vector are combined (2:1 molar ratio) in the presence of In-Fusion reagent (Clontech). The recombinant vector is then cloned into Fusion-Blue competent cells (Clontech).

3. Successful clones are identified within plated transformants using PCR screening to confirm DNA insertion (eight colonies screened/target), the sequence of which is subsequently verified.

3.3. Protein Expression and Solubility Studies

Protein expression for each target is optimized following transformation of the expression vector into a panel of host cells. Target ORFs of the mobile metagenomics pipeline do not often contain rare codons.

1. Isolated plasmid is transformed into all of the following cells: BL21(DE3)Star (Invitrogen, Mt Waverley, VIC), Rosetta2(DE3), and BL21(DE3) pLysS (Merck, Kilsyth, VIC). Within the metagenomics pipeline, the BL21(DE3) pLysS has proved the most successful expression host for yield and solubility.

2. Small-scale expression (2 mL) is carried out at 25 °C, employing both autoinduction and IPTG-induced (0.5 mM added at OD_{600} 0.5) expression.

3. Yield of soluble protein is assessed for each target/expression host combination from pelleted aliquots taken pre- and post-induction. Samples are repeatedly subjected to freeze-thawing (in the presence of lysozyme and DNase I) and soluble and insoluble protein fractions analyzed by SDS-PAGE.

3.4. Protein Purification and Buffer Trials

The expression host found to be most favorable for each target is used to produce recombinant product for rapid batch affinity purification. Buffer trials are conducted early on to optimize downstream procedures.

1. Cell pellets from large-scale (1–4 L) growths are lysed (French press), and purified protein isolated from the soluble fraction on a nickel-based affinity column (HisTrap, GE Healthcare, Rydalmere, NSW). Wash and elution buffers contain 40 and 500 mM imidazole, respectively.

2. Aliquots of each protein target are placed in dialysis buttons (50 μL), covered with dialysis membrane and exchanged into six buffers in a 24-well plate. Buffers sampled (at 50 mM) are: sodium acetate (pH 4.0 and 5.0), MES (pH 6.0), HEPES (pH 7.0), and Tris-HCl (pH 8.0 and 9.0), each in the presence of 0, 250, 500, and 750 mM NaCl. Following further exchange (second plate of identical buffer wells), the optimal buffer condition is noted as that containing no protein precipitate at the lowest NaCl concentration.

3. For subsequent preparations, each target protein is dialyzed into its optimized buffer following affinity chromatography. An additional step of size-exclusion chromatography is used to purify some targets, particularly to assess viability for NMR.

3.5. Screening for Crystallization

Although commercial sparse-matrix screens are sometimes used, crystallization screens incorporate the efficient two-step screening/optimization strategy developed by the SouthEast Collaboratory for Structural Genomics (SECSG) *(14)*. The procedure utilizes pre-screening with a precipitant screen to streamline subsequent optimization of pH.

1. Protein solutions (1.5 μl, 5–20 mg/ml) are mixed 1:1 with SECSG Precipitant Screen reagents (i.e. 60 conditions) placed above reservoirs (80 μl) of each screen reagent in round wells of a 96-format microplate.

2. Extent of phase separation or precipitate observed within protein drops is scored after overnight equilibration (20 °C). A precipitant is selected where it provides greatest coverage of phase space for the protein of interest, that is, it produces clear drops at low concentrations and shows precipitate or crystalline material at high concentrations.

3. Having identified the appropriate precipitant, the relevant screen (SECSGI-XII) combining precipitant range (15–29%) with buffer (pH 3.2–9.4) is prepared *(14)*. Formats of crystallisation screens are as for pre-screening (step 1).

4. Fine screening and seeding are pursued to optimize crystals for diffraction.

3.6. Protein Crystallography

1. Diffraction data are usually collected in a synchrotron beam line in conjunction with in-house systems. Structures solved have involved molecular replacement with sequence homologues and heavy atom phasing, but the authors anticipate a future need for seleno-methionine derivatization.

Table 39.1 Target progress within the metagenomics pipeline.

	Total	PCR	Cloned	Expressed	Soluble	Buffer trial	Crystallized	Diffraction	Structure
Number attempted	325	94	74	49	38	38	20	6	2
Successful targets	—	96%	88%	78%	58%	55%	45%	66%	—
Number environmental	108	20	14	14	7	7	4	2	1
Number *Vibrio*	217	70	46	24	15	14	5	2	1

Target numbers are shown as of August 2006.

Calculation of electron density maps, model building, and refinement uses standard processing software suites.
2. Identification of structural homologues (e.g. using DALI program) *(15)* is used to obtain first understanding of biological functions of the cassette gene derived targets.

4. Progress of the Pipeline

4.1 Statistics

Progress to date on total of 325 ORFs is depicted in Table 39.1. The mobile gene targets are proving highly amenable to the conventional cloning and screening procedures utilized in structural genomics pipelines. Analysis of a selection of Group 1 soluble targets by size-exclusion chromatography reveals the majority of these recombinant proteins to exist as oligomers in solution. These have yielded relatively poor NMR spectra *(16)* and the majority of targets have consequently been pursued by x-ray crystallography. Two structures have been solved to date, Bal32a at 1.85 Å (PDB file: 1TUH) *(16)*, and *i*MazG at 1.8 Å *(17)*, both depicted in Fig. 39.1.

4.2. Functions Revealed

The structure of Bal32a reveals a dimer of a common α+β barrel fold used for transport or metabolism of hydrophobic substrates in a range of biochemical functions. In this protein, derived from contaminated soil, the substrate site is uniquely organized and buried more deeply than previously seen for this cone-shaped protein fold *(16)*.

The structure of *i*MazG is a tetramer, and is the first representative of a new (11th) subfamily of MazG NTP-Pases. Although there is less than 32% sequence homology to other MazG enzymes, the backbone structure and Mg^{2+}-binding site are seen to be tightly conserved in *i*MazG. However, a unique substrate-binding site is identified for the cassette-derived enzyme *(17)*. Biochemical assay of *i*MazG confirms that appropriate nucleotide dNTP substrates (dCTP and dATP) are broken down *in vitro* by the recombinant enzyme.

A   B

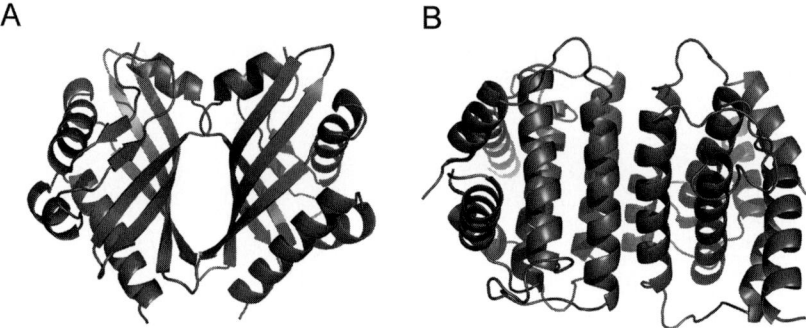

Fig. 39.1 Ribbon diagrams depicting structures of mobile metagenome targets determined by x-ray crystallography. **A.** Bal32a (1.85 Å), a dimeric protein discovered within industrially-contaminated soil. The α+β fold consists of a highly bent β-sheet flanked by α-helices and supports a sizeable cavity, believed to be an enzymatic active site. **B.** *i*MazG (1.8 Å), a tetrameric protein encoded within the integron of *Vibrio* DAT 722. Although the all α-helical structure of *i*MazG is similar to that of other MazG proteins, its substrate-binding site appears to be unique.

Acknowledgments

The authors are grateful to Moreland Gibbs, Robert Willows, Anwar Sunna, and Meghna Sobti for useful discussions and provision of materials. This work was funded by the Australian Research Council (Discovery scheme), the National Health and Medical Research Council, and Macquarie University. AR acknowledges receipt of a RAACE (Macquarie University) postgraduate scholarship.

References

1. Ochman, H., Lawrence, J. G., and Groisman, E. A. (2000) Lateral gene transfer and the nature of bacterial innovation. *Nature* **405,** 299–304.
2. Philippe, H., and Douady, C. J. (2003) Horizontal gene transfer and phylogenetics. *Curr. Opin. Microbiol.* **6,** 498–505.
3. Hall, R. M., and Stokes, H. W. (1993) Integrons: novel DNA elements which capture genes by site-specific recombination. *Genetica* **90,** 115–132.
4. Stokes, H. W., Holmes, A. J, Nield, B. S., Holley, M. P., Nevalainen, K. M. H., Mabbutt, B. C., and Gillings, M. R. (2001) Gene cassette PCR: sequence-independent recovery of entire genes from environmental DNA. *Appl. Env. Microbiol.* **67,** 5240–5246.
5. Boucher, Y., Nesbø, C. L., Joss, M. J., Robinson, A., Mabbutt, B. C., Gillings, M. R., Doolittle, W. F., and Stokes, H. W. (2006) Recovery and evolutionary analysis of complete integron gene cassette arrays from *Vibrio. BMC Evol. Biol.* **6,** 3.
6. Holmes, A. J., Gillings, M. R., Nield, B. S., Mabbutt, B. C., Nevalainen, K. M. H., and Stokes, H. W. (2003) The gene cassette metagenome is a basic resource for bacterial genome evolution. *Environ. Microbiol.* **5,** 383–394.
7. Nield, B. S., Willows, R. D., Torda, A. E., Gillings, M. R., Holmes, A. J., Nevalainen, K. M. H., Stokes, H. W., and Mabbutt, B. C. (2004) New enzymes from environmental cassette arrays: functional attributes of a phosphotransferase and a RNA-methyltransferase. *Protein Sci.* **13,** 1651–1659.

8. Bendtsen, J. D., Nielsen, H., von Heijne, G., and Brunak, S. (2004) Improved prediction of signal peptides: SignalP 3.0. *J. Mol. Biol.* **340,** 783–795.

9. Cserzo, M., Wallin, E., Simon, I., von Heijne, G., and Elofsson, A. (1997) Prediction of transmembrane alpha-helices in prokaryotic membrane proteins: the dense alignment surface method. *Protein Eng.* **10,** 673–676.

10. Krogh, A., Larsson, B., von Heijne, G., and Sonnhammer, E. L. (2001) Predicting transmembrane protein topology with a hidden Markov model: application to complete genomes. *J. Mol. Biol.* **305,** 567–580.

11. Tusnady, G. E., and Simon, I. (1998) Principles governing amino acid composition of integral membrane proteins: application to topology prediction. *J. Mol. Biol.* **283,** 489–506.

12. Linding, R., Russell, R. B., Neduva, V., and Gibson, T. J. (2003) GlobPlot: Exploring protein sequences for globularity and disorder. *Nucleic Acids Res.* **31,** 3701–3708.

13. Altschul, S. F., Gish, W., Miller, W., Myers, E. W., and Lipman, D. J. (1990) Gapped BLAST and PSI-BLAST: a new generation of protein database search programs. *J. Mol. Biol.* **215,** 403–410.

14. Shah, A. K., Liu, Z. J., Stewart, P. D., Schubot, F. D., Rose, J. P., Newton, M. G., and Wang, B. C. (2005) On increasing protein-crystallization throughput for X-ray diffraction studies. *Acta Crystallogr. D.* **61,** 123–129.

15. Holm, L., and Sander, C. (1993) Protein structure comparison by alignment of distance matrices. *J. Mol. Biol.* **233,** 123–138.

16. Robinson, A., Wu, P. S.-C., Harrop, S. J., Schaeffer, P. M., Dosztányi, Z., Gillings, M. R., Holmes, A. J., Nevalainen, K. M. H., Stokes, H. W., Otting, G., Dixon, N. E., Curmi, P. M. G., and Mabbutt, B. C. (2005) Integron-associated mobile gene cassettes code for folded proteins: the structure of Bal32a, a new member of the adaptable α+β barrel family. *J. Mol. Biol.* **346,** 1229–1241.

17. Robinson, A., Guilfoyle, A. P., Harrop, S. J., Boucher, Y., Stokes, H. W., Curmi, P. M. G., and Mabbutt, B. C. (2007) A putative house-cleaning enzyme encoded within an integron array: 1.8 Å crystal structure defines a new MazG subtype. Molecular Microbiol., in press.

Index

Printed in the United States of America